Recombinant DNA

SECOND EDITION

James D. Watson
COLD SPRING HARBOR LABORATORY

Michael Gilman
ARIAD PHARMACEUTICALS, INC.

Jan Witkowski
BANBURY CENTER, COLD SPRING HARBOR LABORATORY

Mark Zoller
ARIAD PHARMACEUTICALS, INC.

SCIENTIFIC
AMERICAN
BOOKS

Distributed by

W. H. Freeman and Company

New York

The cover illustration, by Marvin Mattelson, symbolizes some of the elements of this book. The DNA double helix is, of course, central to the book, as it is to the cover illustration. The blocks are double-stranded DNA fragments synthesized by the polymerase chain reaction, a technique that has revolutionized the way molecular genetics experiments are done. The number of fragments doubles repeatedly, going off into the distance (see Chapter 6). The coat colors of the mice running down the helix (in the same direction but with opposite polarity!), are changing from albino to chimeric, then chimeric to agouti. These coat color changes show mice in which genetic engineering has been used to knock out a specific gene. The experiment is shown more realistically in Figure 14-9.

Library of Congress Cataloging-in-Publication Data.

Recombinant DNA/James D. Watson ... [et al.]. — 2nd ed.
    p.   cm.
  Includes bibliographical references and index.
  ISBN 0-7167-1994-0. — ISBN 0-7167-2282-8 (pbk.)
  1. Recombinant DNA.   I. Watson, James D., 1928–
QH442.R37   1992
574.87′3282—dc20   91-38483
                                                                      CIP

Copyright © 1983 by James D. Watson, John Tooze, and David T. Kurtz

Copyright © 1992 by James D. Watson, Michael Gilman, Jan Witkowski, and Mark Zoller

Printed in the United States of America

Scientific American Books is a subsidiary of Scientific American, Inc.

Distributed by W. H. Freeman and Company, 41 Madison Avenue, New York, New York 10010

Seventh printing, 1998

Contents

5 METHODS OF CREATING RECOMBINANT DNA MOLECULES *63*

Analysis of Cloned Genes

6 THE POLYMERASE CHAIN REACTION *79*

7 THE ISOLATION OF CLONED GENES *99*

New Tools for Studying Gene Function

15 GENETIC ENGINEERING OF PLANTS 273

Analysis of Important Biological Processes by Using Recombinant DNA

16 MOLECULES OF IMMUNE RECOGNITION 293

17 MOVING SIGNALS ACROSS MEMBRANES 313

Application of Recombinant DNA to Biotechnology

Impact of Recombinant DNA on Human Genetics

Preface

Application of recombinant DNA techniques to biology is bringing about a revolution in our understanding of living organisms. There is no field of experimental biology that is untouched by the power we now have to isolate, analyze, and manipulate genes. When the first edition of *Recombinant DNA* was published in 1983, recombinant DNA techniques were already being used extensively for the analysis of viral and bacterial genetics, but dissection of eukaryotic genes was only just beginning. There were hints of what was to come. The concept of the gene as a continuous stretch of DNA had been shattered with the discovery of introns, but alternative splicing and genes-within-genes were yet to be revealed. Identification of cellular oncogenes seemed to promise an understanding of cancer, but the mechanisms of their action—and the existence of tumor suppressor genes—were still subjects for speculation. A handful of genetic diseases were being analyzed at the molecular level, but the isolation of the disease genes and the development of gene therapy were yet to come.

Our aim in writing the second edition of *Recombinant DNA* is to show how recombinant DNA techniques have led to the explosion in our knowledge of fundamental biological processes. As in the first edition, which was subtitled *A Short Course*, we provide a concise presentation of the methods, underlying concepts, and far-reaching applications of recombinant DNA technology. The field has grown since the publication of the first edition, and so has our book. But even though our previous subtitle may be inappropriate for this enlarged edition, our approach to the material has remained true to the spirit of the "short course": as before, the uninitiated will find access to the field of recombinant DNA here.

The book is now divided into six major sections. The first five chapters, which are largely unchanged from the first edition, provide a historical introduction to the early development of recombinant DNA technology, up to the point when studies of eukaryotic organisms began in earnest. In the next section we describe in detail the methods currently used to clone and analyze genes, and devote an entire chapter to the polymerase chain reaction, which has had an extraordinary impact on research. The great power of recombinant DNA techniques comes from the ability to explore gene functions by manipulating genes and then introducing them back into cells. The third section of the book discusses how this is done in mammalian cells, yeast, mice, and plants. The fourth section describes the progress these manipulations have allowed in key areas of biology. Here the range of recombinant DNA applications is demonstrated, from the analysis of cell cycle control and embryonic development, to the isolation of genes involved with brain function. Indeed, these techniques have spawned a whole industry—biotechnology. In the fifth section, we describe some of its accomplishments, including the development of genetically engineered pharmaceutical and agricultural products, and the studies of the human immunodeficiency virus that are leading the attack on AIDS. The differences between the first and second editions are perhaps most evident in the final section, where we describe the revolution in human molecular genetics and the ways in which recombinant DNA techniques are providing new methods for diagnosis and treatment of human inherited diseases.

The topics that are covered and the approach we take to describing them make this book suitable for undergraduate and graduate students in molecular biology, cell biology, biochemistry, genetics, or biotechnology courses; for medical students and physicians; and for others who have an interest in recombinant DNA techniques—for example, forensic scientists, patent attorneys, and science journalists.

Textbooks dealing with biochemistry, molecular genetics, and molecular biology usually present information without describing the experiments done to obtain it. We think that this is a pity, because designing and doing experiments is exciting and fun. As in the first edition, we have used real experiments to illustrate important biological phenomena, and we have plundered our colleagues' papers for interesting examples. Figures are used profusely to try to make complex real-life experiments intelligible, but inevitably we have not been able to present all the subtle details. Those who want to explore these details will find the experiments in the research papers listed at the end of each chapter, and the review papers we cite will provide an entry point to each topic.

This book is atypical in another regard. Because we do not consider it primarily a textbook for conveying undisputed facts about molecular biology, we have been able to include exciting research at the cutting edge of biology. The interpretation of experimental data often changes with time, so the reader should bear in mind that future research might require modification of some of the ideas we present. This is all part and parcel of doing research, because a science that does not change is a dead science. Modern experimental research in biology is an ever-changing dynamic enterprise, and we hope that *Recombinant DNA* conveys the excitement of the continuing process of discovering how organisms work.

Acknowledgments

Those who most deserve our thanks are our families, from whom we were taken for many days by this book, and our colleagues at Cold Spring Harbor Laboratory and Genentech, who sometimes had a difficult time communicating with us when we were preoccupied with writing.

We are grateful to the many friends and colleagues who read our manuscript, criticized, corrected, and provided information. They include Sue Alpert, French Anderson, Avi Ashkenazi, David Beach, Martin Bobrow, Tom Caskey, Jeff Chamberlain, Irvin Chen, Francis Collins, Alan Coulson, David Cox, Ken Culver, Kay Davies, Jim Eberwine, Stan Fields, Uta Francke, Ted Friedman, Bruce Futcher, Peter Gergen, Richard Gibbs, Paul Godowski, Andre Goffeau, Takashi Gojobori, Morris Goodman, Alan Handyside, Kenshi Hayashi, Dan Hartl, Andrew Hiatt, Tom Hynes, Paula Jardieu, Karen Johnson, Dan Klessig, Jeff Kuret, Mike Laspia, Philip Leder, Fred Ledley, Pal Maliga, Vincent Marchesi, Rob Martienssen, Dusty Miller, Rick Myers, David Nelson, Karoly Nikolics, Luis Parada, Scott Putney, Don Rio, Liz Robertson, David Schlessinger, Matt Scott, John Sulston, Shirley Tilghman, Barbara Trask, Rebecca Ward, Robin Weiss, Jim Wells, Tom White, and Bob Williamson. Any errors, nevertheless, are ours, and not theirs.

Elizabeth Zayatz, the development editor, had the unenviable job of trying to make us concentrate on the work at hand when we wanted to be doing other things. She did wonderfully well keeping us at it and became a trusted friend. Bill O'Neal and Jodi Simpson, the manuscript editors, smoothed awkward passages and made us think more carefully about the information we were trying to convey. Janet Tannenbaum, the project editor, combed the manuscript and graphics with a thoroughness that must have required a magnifying glass. She guided us through a forest of edited manuscript, galleys, and page proofs. At times it seemed that never a day went by without an overnight package from Janet arriving on our desks. Alison Lew gave the text, illustrations, and cover their design and Bill Page looked after the art program—tasks that rapidly assumed epic proportions as we all worked to produce the figures that are an essential complement to the text. The figures were rendered by Network Graphics and Tomo Narashima. The beautiful and remarkable cover art is by Marvin Mattelson. Listening to him expound on surrealism and watching him sketch were highlights of the production. Julia De Rosa masterfully coordinated the production process to allow rapid completion of the project. Linda Chaput, President of Scientific American Books, was always patient and understanding. Her guidance and advice were invaluable when the going got tough, and our discussions with her were rewarding, and fun.

James D. Watson   Michael Gilman
Jan Witkowski   Mark Zoller
December, 1991

1

Establishing the Role of Genes Within Cells

There is no substance so important as DNA. Because it carries within its structure the hereditary information that determines the structures of proteins, it is the prime molecule of life. The instructions that direct cells to grow and divide are encoded by it; so are the messages that bring about the differentiation of fertilized eggs into the multitude of specialized cells that are necessary for the successful functioning of higher plants and animals. And because it has been present in virtually an infinite number of interchangeable chemical species, DNA has provided the basis for the evolutionary process that has generated the many millions of different life-forms that have occupied the earth since the first living organisms came into existence some 3 to 4 billion years ago.

This extraordinary capacity of altered DNA molecules to give rise to new life-forms that are better adapted for survival than were their immediate progenitors has made possible the emergence of our own species, with its ability to perceive the nature of its environment and to utilize this information to build the civilizations of modern man. As a result of our ability for rapid conceptual thought, we have for several centuries been asking ever deeper questions about the nature of inanimate objects like water, rocks, and air, as well as about the stars of surrounding space. And biology, the science of living objects, which only

40 years ago was generally perceived to be a much inferior science, has swiftly come of age. By now there exists an almost total consensus of informed minds that the essence of life can be explained by the same laws of physics and chemistry that have helped us understand, for example, why apples fall to the ground and why the moon does not, or why water is transformed into gaseous vapor when its boiling point is exceeded.

The key to our optimism that all secrets of life are within the grasp of future generations of perceptive biologists is the ever accelerating speed at which we have been able to probe the secrets of DNA. Now we know so much that it is difficult to remember the intellectual chaos that still existed in 1944—when DNA was first reported to carry genetic information—and that was to disappear effectively only in 1953, when the structure of DNA was revealed to be a complementary double helix. Since then it has been clear to all that the "brain," so to speak, of all cells is DNA: From DNA issue the commands that regulate the nature and number of virtually every type of cellular molecule. So, increasingly, our attention has been devoted to unlocking the information within DNA. We know that if we can reveal the exact form of our genetic blueprints, we shall have taken a giant step toward eventually understanding the many complex sets of interconnected chemical reactions that cause fertilized eggs to develop into highly complex multicellular organisms.

Before we examine DNA itself in more detail, we shall first look briefly at the cells in which it resides, to determine the nature of the commands that DNA must generate.

The Building Blocks of All Life Are Cells

The smallest irreducible units of life are cells. They were first seen over 300 years ago, soon after the construction of the first microscopes. By the middle of the nineteenth century, it had become clear that all living organisms are built from cells.

Generally, cells are very small, with diameters much less than 1 mm, so they are invisible to the naked eye. In the simplest cells, bacteria, a cell wall surrounds a very thin fatty acid–containing outer (*plasma*) membrane which in turn surrounds a superficially unstructured inner region. Within this inner region is located the bacterial DNA that carries the genetic information. The plasma membrane is effectively impermeable, except to selected food molecules and ions, so the inner cell contents are contained and are not lost to the outside. The integrity of this outer membrane is thus essential for the life of the cell, and it is constructed in such a way that minor accidental tears or openings are sealed automatically, like punctures in a self-sealing automobile tire.

In virtually all cells other than bacteria, the inner cellular mass is partitioned into a membrane-bounded, spherical body called the *nucleus* and an outer surrounding *cytoplasm*. In the nucleus is located the cellular DNA in the form of coiled rods known as *chromosomes*. Cells that contain a nucleus are referred to as *eukaryotic cells,* whereas the nuclei-free bacteria and their close relatives, the blue-green algae, are known as *prokaryotic cells.*

Cells Are Tiny Expandable Factories That Simultaneously Synthesize Several Thousand Different Molecules

The essence of a cell is its ability to grow and divide to produce progeny cells, which are likewise capable of generating new cellular molecules and replicating themselves. To perform these functions, cells must be chemically very sophisticated; indeed, even the very simplest cells contain more than 2500 different molecules. Thus, cells are, in effect, tiny factories that grow by taking in simple molecular building blocks, like glucose and carbon dioxide, and somehow converting them into the many diverse carbon-containing molecules that are required for cellular functioning. In growing and dividing, cells also require an external source of energy to ensure that the cellular chemical reactions proceed in the desired direction of biosynthesis. Cells are therefore governed by the same laws of thermodynamics that describe the energies of atoms to the physicist and the energies of molecules to the chemist. For most cells, the energy input necessary for their maintenance and growth is gained from breaking down food molecules, although cells that are capable of photosynthesis use the energy of sunlight directly.

A Cell's Molecules Can Be Divided into Small Molecules and Macromolecules

The molecules of a cell fall into two very different size classes. One class is the so-called *small molecules:* the various sugars, amino acids, and fatty acids. At least 750 different types of small molecules are found in virtually every cell. The second class of cellular molecules is the *macromolecules,* among which the proteins and nucleic acids are the most important. These are invariably polymeric molecules that are formed by joining together specific types of small molecules (such as amino acids) into long chains (such as proteins). *Polymeric* molecules contain a large number of these *monomeric* subunits, and are frequently hundreds to thousands of times larger than the small molecules, which typically contain somewhere between 10 to 50 atoms precisely linked together. The number of different macromolecules in most bacterial small cells is even larger than the number of different small molecules. Our best guess now is that there are in excess of 2000 distinct macromolecular species.

Thus, just the number and sizes of their molecules make even the simplest cells extraordinarily complex from a chemical viewpoint. The chemical uniqueness of cells lies less, however, in the inherent complexity of their individual molecules than in the nature of the chemical reactions that transform one cellular molecule into another. The most important of these reactions are those that (1) lead to the breakdown of food molecules into molecular units that can be reassembled into vital cellular components, (2) harness much of the energy released by the breakdown of food (or absorption of light) into energy-rich molecules that later pass on their excess energy to ensure that the chemical reactions of the cell proceed in the desired direction, (3) synthesize the small-molecule building blocks of the cellular macromolecules, and (4) assemble these monomeric building blocks into highly ordered macromolecules.

Special Cellular Catalysts Called Enzymes Effectively Determine the Chemical Reactions That Occur in Cells

Each of a cell's several thousand different molecules is potentially capable of a very large number of chemical reactions with other cellular molecules. Yet in any cell, only a very tiny fraction of these potential reactions takes place at measurable rates, because the possibility of most chemical reactions occurring unaided at the temperatures at which life exists is very low. The cellular chemical reactions that do occur at rates compatible with life are greatly speeded up by *enzymes,* special catalytic molecules that are unique to cells. Like all other catalysts, enzymes are not used up in the course of chemical reactions, and a given enzyme molecule may function many thousands of times in a single second. Most enzymes are highly specific and catalyze only a single type of chemical reaction; conversely, most chemical reactions in cells are catalyzed only by one specific enzyme.

The chemical identity of enzymes was initially obscure, with many scientists believing that enzymes might represent a still undiscovered class of biological molecules. By 1935, however, it was clear that all enzymes were proteins, and today it is known that in fact the vast majority of proteins in all cells have enzymatic roles. Enzymes act by binding together on their surfaces the potential partners of a chemical reaction, thus greatly speeding up the rates at which these reactants can collide and undergo a chemical reaction. In enzyme-catalyzed reactions, certain key atoms of the enzyme often directly participate by temporarily forming chemical bonds with the participating reagents (called enzyme *substrates*), thereby forming metastable chemical intermediates that have a heightened potential for chemical reactivity.

A Given Protein Possesses a Unique Sequence of Amino Acids Along Its Polypeptide Chain

The realization that all enzymes are proteins greatly heightened the interest of chemists in establishing the precise details of protein structure. Already by 1905,

FIGURE **1-1**
The amino acid building blocks of a protein. Chain lengths vary from 5 to more than 4000 amino acids.

through the work of Emil Fischer in Germany, proteins were known to be polymeric molecules built up from amino acids linked to each other by *peptide bonds* to form linear polypeptide chains (Figure 1-1). The exact number of different amino acids remained in question until about 1940, when it was established that the vast majority of proteins were built up from a mixture of the same 20 amino acids, with the percentage of a given amino acid varying from one protein to another (Table 1-1). These findings made it likely that each polypeptide chain was characterized by the unique sequence of its amino acids, a conjecture that was first shown to be correct in 1951, when Frederick Sanger in Cambridge, England, reported the order of the amino acids along one of the two polypeptide

chains that constitute the hormone insulin (Figure 1-2).

Most proteins contain only one polypeptide chain. However, there are many other proteins that are formed through the aggregation of separately synthesized chains that have different sequences. The oxygen-carrying protein hemoglobin, for example, is formed by the aggregation of four polypeptide chains, two with a specific α sequence and two containing the β sequence. The sizes of different proteins vary greatly: Their component polypeptide chains contain between as few as 5 amino acids and as many as 4000.

Though the vast majority of proteins are enzymes, many proteins have structural roles as well. They help, for example, to form the essential fabric of the outer plasma membrane and that of the nuclear membrane. Proteins such as the fibrous collagen aid in building the connective tissue between cells. Other proteins— for example, actin and tropomyosin—are constituents of muscle fibers. Still others, among them calmodulin, function to bind ions, like Ca^{2+}, that regulate the activities of other enzymes. Finally, there are proteins that function outside the cell such as insulin, a hormone, and antibodies that are responsible for maintaining the body's immune response.

The Functioning of an Enzyme Demands a Precise Folding of Its Polypeptide Chain

Once a polypeptide chain is put together, weak chemical interactions between the specific side groups of its amino acids cause it to fold up into a unique three-dimensional form whose exact shape is a function of its amino acid sequence. Whereas many proteins fold up into long, rodlike shapes, most enzymes are globular, with cavities into which their substrates can fit in the way that keys fit into locks. Such cavities, bounded by the appropriate amino acid side groups, enable enzymes to bring their substrates into the close proximity that will permit them to react chemically with one another. The specific amino acid sequence of a given enzyme is thus very important. If inappropriate amino acids are present, then the polypeptide chain cannot fold up to form the properly shaped catalytic cavity. By having available 20 of these amino

TABLE **1-1**
The 20 Amino Acids in Proteins

AMINO ACID	THREE-LETTER ABBREVIATION	SINGLE-LETTER CODE
Glycine	Gly	G
Alanine	Ala	A
Valine	Val	V
Isoleucine	Ile	I
Leucine	Leu	L
Serine	Ser	S
Threonine	Thr	T
Proline	Pro	P
Aspartic acid	Asp	D
Glutamic acid	Glu	E
Lysine	Lys	K
Arginine	Arg	R
Asparagine	Asn	N
Glutamine	Gln	Q
Cysteine	Cys	C
Methionine	Met	M
Tryptophan	Trp	W
Phenylalanine	Phe	F
Tyrosine	Tyr	Y
Histidine	His	H

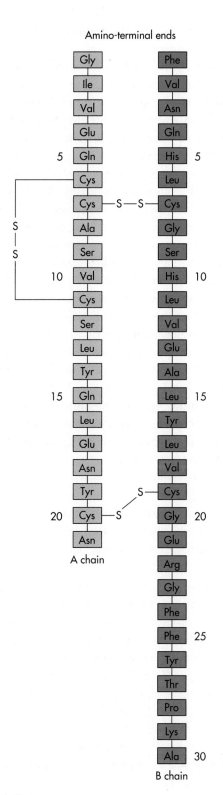

FIGURE **1-2**

The amino acid composition of bovine insulin. The insulin molecule has two amino acid chains linked by two chemical bonds between sulfur atoms. Human insulin differs at three positions, two in the A chain and one in the B chain.

acid building blocks, each with a unique shape and with its own chemical properties, and by being able to form polypeptide chains of variable lengths, a cell has the potential to evolve new proteins possessing catalytic cavities with shapes complementary to virtually any potential substrate.

Activation of Molecules to High-Energy Forms Promotes Their Chemical Reactivity

Enzymes catalyze both the forward and backward directions of a given chemical reaction. Their presence alone does not cause the chemical reactions that they catalyze to move in directions that provide orderly cell growth and division. The favored direction of a chemical reaction is actually determined by the thermodynamic law that a reaction moves in the direction that maximizes the generation of heat and molecular disorder. At first glance, many cellular chemical reactions appear to flaunt this rule. Within growing cells, for example, the polymerization of amino acids into polypeptides is greatly favored over the energetically more likely breakdown of polypeptide chains into their component amino acid monomers. In reality, however, the laws of thermodynamics are never even momentarily violated in cells. Instead cells have evolved specific enzymes that push energetically unfavorable reactions to completion by coupling them to reactions that are much more energetically favorable. The net result of both reactions is the release of energy that usually becomes dissipated as heat. Coupling is brought about when an enzyme causes a substrate that is to be consumed (for example, one of the nucleotide monomers for DNA synthesis) to react with one of the cell's small-molecule energy donors (say, ATP) to give rise to an energized activated substrate molecule. The subsequent decomposition of this activated substrate in the synthetic reaction will release significant energy, which will result in the favoring of the synthetic reaction (in our example, DNA synthesis) over the reverse degradative reaction (DNA breakdown).

The need for cells to carry on simultaneously many different coupled reactions requires cells to devote a significant part of their total metabolic activity to the constant replenishment of their ATP supply. In animal cells and in most bacteria, the formation of ATP is coupled with the breakdown of food molecules that are even richer in energy. In cells that carry out photosynthesis, the energy of sunlight is used directly to make ATP.

Cellular Metabolism Can Be Visualized Through Metabolic Maps

Certain key small molecules of the cell serve as substrates for many different enzymes and thus can be incorporated into many other molecules. However, there are a number of metabolites that function only as intermediates in the biosynthesis of a given complex amino acid, such as tryptophan, or as intermediary metabolites in a quite different metabolic pathway, such as the one through which tryptophan is broken down as a food source. Thus, when we diagram all the various known enzymatic pathways that link together the small molecules of a cell, the pattern (*metabolic map*) we perceive is not one of hopeless complexity. Instead we find a most revealing picture of how cells use food molecules to provide the active building blocks, as well as the energy, for the biosynthetic reactions that build up the small-molecule monomeric precursors of the complex polysaccharides like glycogen, or of the even more complicated proteins and nucleic acids.

We also see that for every small molecule a cell possesses, there must exist at least one enzyme that directly catalyzes its synthesis, and quite often another enzyme that can degrade it when it is not needed. The number of different enzymes in a cell must thus greatly exceed the number of different small molecules. All the enzymes, like their respective small molecules, are not present in equal amounts; the enzymes whose substrates are present only in small amounts (for example, vitamins) are likewise present only in small amounts.

Enzymes Cannot Determine the Order of the Amino Acids in Polypeptide Chains

Although enzymes completely determine the specificity of the chemical reactions between small molecules, there is no way that they can be used to determine the order of amino acids in the thousands of different proteins that every cell possesses. An average-sized polypeptide chain contains several hundred amino acids arranged in a unique irregular sequence. If the ordering of the amino acids in such a chain were carried out by enzymes, there would have to be an enormous number of these enzymes, each capable of recognizing a large number of contiguous amino acids. In turn, each of these hypothetical "amino-acid-sequence-recognizing enzymes" would have to be put together by its own set of different "sequence-recognizing" enzymes, and so forth. This type of scheme obviously cannot work, and we are led to the inescapable conclusion that cells must contain specific "information-bearing" molecules, analogous perhaps to the molds of the sculptor or to the master plates of the lithographer. Such molecules must encode the ordering information so that it can be used to select the correct amino acids in the course of polypeptide synthesis. These information-bearing molecules, moreover, must somehow also be able to synthesize new copies of themselves, so that when a growing cell splits into two daughter cells, each of the progeny cells possesses copies of the master molds (or templates).

As thus described, these putative information-bearing molecules are clearly very similar to the chromosomes that carry the genes that control our heredity. We shall in fact soon show that chromosomes contain the information that is used to determine the order of amino acids in cellular proteins. Here we simply want to emphasize that the cells' possession of a discrete hereditary system is not an accident of evolution. Rather, the existence of discrete genetic molecules that determine the specificity of proteins was an essential ingredient for the emergence of life itself.

Mendel's Breeding Experiments with the Pea Plant First Revealed the Discreteness of Genetic Determinants (Genes)

The knowledge that many human traits like the color of our eyes and the shape of our faces have been passed on to us from our parents must go back to the days of early humans. Such inheritable characteristics are of course not limited to humans, and long before the current science of the study of heredity or genetics was born, plant and animal breeders sought to improve the quality of their domestic plants and animals. Hereditary traits were not effectively analyzed, however, until controlled matings were made between individuals with well-defined traits. This was first successfully accomplished by the Austrian monk Gregor Mendel, who in the early 1860s in Brno, Czechoslovakia, crossed pea plants of differing morphologies. Most importantly, he not only noted the appearance of specific traits in the first generation of progeny, but he went on to make and study crosses between these progeny as well as among the progeny and the original parent pea plants.

Mendel's experiments led him to conclude that many traits of the pea were under the control of two distinct factors (later named *genes*), one coming from the male parent and the other from the female parent. Mendel further noted that the various traits he studied were not *linked* together and must thus have been borne on separate hereditary units. He realized that some genes (*dominant genes*) express themselves when they are present in only one copy, whereas other genes (*recessive genes*) require two copies for expression.

An important further step in the early development of genetics was the distinction between *phenotype* and *genotype*. First made explicit by Johannsen, the genotype is the complete genetic composition of an organism, while its phenotype is the physical expression of that genotype.

Chromosomes Are the Cellular Bearers of Heredity

About the same time that Mendel made his observations, it became known that heredity is transmitted through the egg and sperm. And since a sperm contains relatively little cytoplasmic material in comparison with the amount of material in its nucleus, the obvious conjecture was that the function of the nucleus was to carry the hereditary determinants of a cell. Soon afterward, the chromosomes within the nucleus were made visible through the means of special dyes (*chromo-* comes from the Greek for "color"), and the cells of members of a given species were found to contain a constant number of chromosomes. This number was seen to exactly double prior to the cell division

process (*mitosis*), which in turn reduced the chromosomes back to their exact original count. Within a given cell type, a variety of chromosomes of different sizes were observed, with each distinct type usually being present in two copies (pairs of *homologous chromosomes*). A few years later, the number of chromosomes in the sex cells of sperm and eggs was shown to be exactly N (the *haploid* number), half the 2N (the *diploid* number) found in *somatic* (nonsex) cells. The partitioning of chromosomes during mitosis was found to be exact: Each daughter cell receives one copy of each chromosome present in the parental cell. In contrast, during the formation of sex cells (*meiosis*), the chromosome number is reduced to N (Figure 1-3). The fertilization process between sperm and egg thus restores the 2N chromosome number characteristic of somatic cells, with one chromosome in each pair coming from the male parent and the other from the female parent. Chromosomes thus behaved exactly as they should if they were the bearers of Mendel's genes. Surprisingly, however, the hypothesis that parents each contribute half of an offspring's genes was not proposed in this form until 1902.

The first trait to be assigned to a chromosome was sex itself. The work was done at Columbia University, where in 1905 Nettie Stevens and Edmund Wilson discovered the existence of the so-called *sex chromosomes*. Their research showed that one chromosome, the X, is present in two copies in females but in only one copy in males, who also carry a morphologically distinct Y chromosome. During the halving of chromosome numbers as the sex cells are formed, all eggs necessarily receive single X chromosomes, while sperm receive either an X or a Y chromosome. Fertilization by a sperm containing an X chromosome generates female (XX) progeny, whereas fertilization by a sperm containing a Y chromosome yields male (XY) progeny. The proposal that sex traits are located on a single pair of chromosomes neatly explained the 1:1 ratio of males and females.

It was natural to speculate that all traits, not just those determining sex, might be chromosomally located. The initial confirmation came very soon, between 1910 and 1915, from genetic crosses with the red-eyed fruitfly *Drosophila* in T. H. Morgan's laboratory at Columbia University (Figure 1-4). In the first such experiments, a mutated gene leading to white eyes was located, or *mapped,* on the X chromosome.

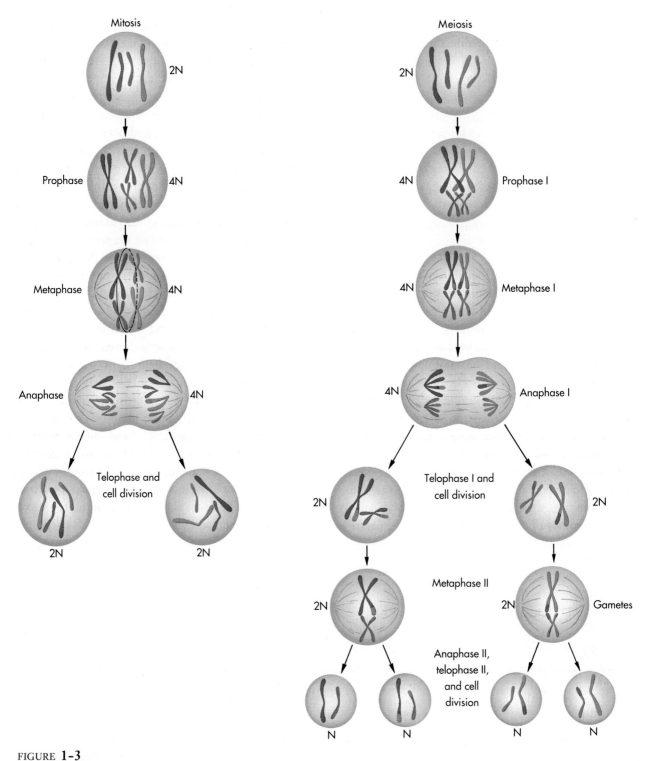

FIGURE 1-3

A comparison of cell division by mitosis and meiosis. Mitosis, somatic cell division, leads to two identical daughter cells that each have the same number of chromosomes as the parent cell. Meiosis, sex cell division, produces four daughter cells (gametes) that each have half the number of chromosomes as the parent cell. Crossing over, shown in prophase I, occurs during meiosis.

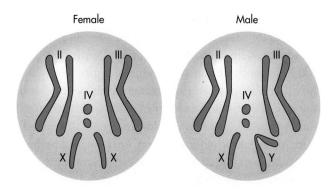

FIGURE **1-4**
The chromosomes of male and female *Drosophila*.

Over the next several years, many more mutants were mapped to either the sex chromosome or to one of the nonsex chromosomes (*autosomes*). The coexistence of two mutants on the same chromosome was shown by their tendency to be linked together in genetic crosses. But this linkage was never complete; it was sometimes broken as a consequence of a physical exchange of chromosome parts (called *crossing over*) that occurred when homologous chromosomes came together in pairs (*synapsis*) during meiosis before forming haploid sex cells (Figure 1-3). Most important, crossing over provided a way of determining the order of genes along chromosomes. The farther apart two genes are along a chromosome, the greater the chance that a crossover will occur between them and thus situate them on opposite chromosomes (Figure 1-5). The frequency with which this separation (reassortment) occurs is proportional to the distance between the two genes and, thus, can be used to determine their positions relative to each other on the chromosome. The arrangements of genes along chromosomes were always found to be linear, and by 1912 the hypothesis that chromosomes are the bearers of heredity was virtually unassailable.

Early breeding experiments utilized genetically stable variants (mutants) that appeared spontaneously at low frequency. We now know that many spontaneous mutants result from changes in single genes. Such genes are mutant genes, as opposed to normal, or *wild-type*, genes. In 1926 Hermann Muller and Lewis Stadler independently discovered that x-rays induce mutations, and thereby provided geneticists with a much larger number of mutant genes than were available when the so-called *spontaneous mutations* were the only source of variability.

As genetic experiments grew in momentum, it became apparent that a very large number of different genes were located on each chromosome; estimates for the number of genes present on all the chromosomes in a given sperm or egg soon exceeded 500. By the late 1930s, it seemed probable that at least 1500 genes were contained within the four chromosomes of *Drosophila*.

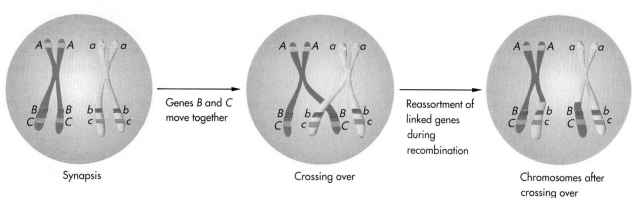

FIGURE **1-5**
Reassortment of genes by crossing over. *B* and *C* (and *b* and *c*) are linked because they move together during recombination. Gene *A* at the other end of the chromosome is seen to be only loosely linked to (*B* + *C*) because a crossover has occurred somewhere between *A* and (*B* + *C*).

At first it was taken for granted that crossing over occurred only between genes, because chromosomes were conceived as being constructed like children's beads; that is, genes were thought to be held together by connectors that easily broke and rejoined. By the late 1930s, though, when many thousands of progeny from single *Drosophila* crosses were examined, scientists discovered rare examples of apparent crossing over *within* a gene that determines eye color. At that time there was no way of knowing that the two mutants being examined contained mutations in the same gene. The mutations might have been in adjacent genes that were functionally related and that therefore produced seemingly identical changes (such as white eyes). Could crossing over cause rearrangements within genes themselves? The answer had to await the development of systems that employed many, many mutations in the same gene, and that allowed millions—not merely thousands—of progeny from the same genetic cross to be examined.

The One Gene, One Protein Hypothesis Is Developed

Geneticists first focused on finding new hereditary traits and mapping their respective genes to precise chromosomal locations. The traits they studied were obvious morphological ones like eye color and wing shape, markers whose chemical basis was (and often still is) totally obscure. How most mutations lead to altered phenotypes necessarily remained unknown. As early as 1909, however, the English physician Archibald Garrod noted that several human hereditary traits were metabolic diseases characterized by the failure of known chemical reactions to take place. (Today one of the best understood of such genetic diseases is *phenylketonuria*, in which the amino acid phenylalanine cannot be converted to the related amino acid tyrosine. This "error" leads to a buildup in the blood of the toxic intermediary metabolite phenylpyruvate.) Garrod hypothesized that such metabolic diseases, which he called "inborn errors of metabolism," were due to the absence of specific enzymes that were synthesized under the direction of the wild-type genes. If Garrod's hunch was correct, then for every enzyme, and perhaps for each protein that a cell possesses, there must exist a corresponding gene.

Because so few details of cellular metabolism were then understood, some 30 years had to pass before definitive experiments could be done to prove the one gene, one enzyme relationship. These studies were done in the early 1940s at Stanford University by the geneticist George Beadle and the biochemist Edward Tatum on the mold *Neurospora.* This microorganism normally grows on a simple diet of glucose and inorganic ions. However, exposure to x-rays and ultraviolet light, agents that were by then known to increase vastly the rate at which mutant genes arise, led to production of mutant *Neurospora* strains that multiplied only when their normal diets were supplemented with additional food molecules (*growth factors*) that were specific to each mutant cell. Some of these mutant *Neurospora* cells required specific amino acids like arginine or cysteine, whereas others required a particular vitamin or one of the purine or pyrimidine building blocks of the nucleic acids. In each case, the specific metabolic requirements were afterwards shown to be due to the absence of one of the enzymes involved in the specific metabolic pathway that led to the synthesis of the growth factors. A few years later, the one gene, one enzyme concept was extended in general when the molecular defect that causes the human disease *sickle-cell anemia* was shown by Linus Pauling at the California Institute of Technology to be due to chemically altered (mutant) hemoglobin molecules.

Reading List

General

Watson, J. D., N. Hopkins, J. Roberts, J. A. Steitz, and A. Weiner. *Molecular Biology of the Gene,* 4th ed. Benjamin-Cummings, Menlo Park, Calif., 1987.

Stryer, L. *Biochemistry,* 3rd ed. Freeman, New York, 1988.

Zubay, G. *Biochemistry,* 2nd ed. Macmillan, New York, 1988.

Alberts, B., D. Bray, J. Lewis, M. Raff, K. Roberts, and J. D. Watson. *Molecular Biology of the Cell,* 2nd ed. Garland, New York, 1989.

Darnell, J., H. Lodish, D. Baltimore. *Molecular Cell Biology,* 2nd ed. Freeman, New York, 1990.

Lehninger, A. *Principles of Biochemistry.* Worth, New York, 1982.

Original Research Papers

PROTEIN STRUCTURE

Sanger, F., and H. Tuppy. "The amino acid sequence in the phenylalanyl chain of insulin." *Biochem. J.,* 49: 463–490 (1951).

Sanger, F., and E. O. P. Thompson. "The amino acid sequence in the glycyl chain of insulin." *Biochem. J.,* 53: 353–374 (1953).

GENES AND THE CHROMOSOMAL THEORY OF HEREDITY

Mendel, G. English translation of Mendel's experiments in plant hybridization, reprinted in: Peters, J. A., ed. *Classic Papers in Genetics.* Prentice-Hall, Englewood Cliffs, N.J., 1959; and Stern, C., and E. R. Sherwood, eds. *The Origin of Genetics.* Freeman, San Francisco, 1966.

Stevens, N. M. "Studies in spermatogenesis with especial reference to the 'accessory' [sex] chromosome." *Carn. Inst. Wash.,* publ. 36, pp. 1–32 (1905).

Morgan, T. H. "Sex-limited inheritance in *Drosophila.*" *Science,* 32: 120–122 (1910).

Morgan, T. H., A. H. Sturtevant, H. J. Muller, and C. Bridges. *The Mechanism of Mendelian Heredity.* Henry Holt & Co., New York, 1915.

Muller, H. J. "Artificial transmutation of the gene." *Science,* 46: 84–87 (1927).

Wilson, E. B. *The Cell in Development and Heredity,* 3rd ed. Macmillan, New York, 1928.

Stadler, L. J. "Mutations in barley induced by x-rays and radium." *Science,* 68: 186–187 (1928).

Allen, G. E. *Thomas Hunt Morgan: The Man and His Science.* Princeton University Press, Princeton, N. J. 1978.

THE ONE GENE, ONE PROTEIN HYPOTHESIS

Garrod, A. E. "Inborn errors of metabolism." *Lancet,* 2: 1–7, 73–79, 142–148, 214–220 (1908). (Also published as a book by Oxford University Press, London, 1909.)

Beadle, G. W., and E. L. Tatum. "Genetic control of biochemical reactions in *Neurospora.*" *Proc. Natl. Acad. Sci. USA,* 27: 499–506 (1941).

Pauling, L., H. A. Itano, S. J. Singer, and I. C. Wells. "Sickle cell anemia: a molecular disease." *Science,* 110: 543–548 (1949).

CHAPTER

2

DNA Is the Primary Genetic Material

The realization that genes determine the structure of proteins was a very important milestone in the development of genetics, but it did not have any immediate consequences. So long as the molecular structure of the gene was unknown, there was no way to think constructively about gene-protein relations. In fact, as recently as 1950 there was no general agreement on which class of molecules genes belonged to. Nevertheless, the best guess was that the gene consisted of *deoxyribonucleic acid,* a still poorly understood polymeric macromolecule that was just starting to be called by its abbreviation, *DNA.*

DNA Is Sited on Chromosomes

For many years it was hoped that as microscopes improved, it might eventually be possible to see genes sitting side by side along chromosomes. But even with the advent in the early 1940s of the first electron microscopes, which had a potential resolution over 100 times greater than that of light microscopes, there were disappointments. The first electron microscope pictures of chromosomes showed no repeating pattern at the molecular level; this suggested a highly irregular gene structure that would not be simple to interpret. Attempts to purify

13

chromosomes away from other cellular constituents were much more informative, although it was impossible to obtain really pure chromosomes.

Two main chromosomal components were almost invariably found: (1) deoxyribonucleic acid (DNA) and (2) a class of small, positively charged proteins known as the *histones;* these, being basic, neutralized the acidity of DNA. DNA had been known to be a major constituent of the nucleus (hence the name *nucleic* acid) ever since its discovery in 1869 by the Swiss scientist Frederick Miescher. In the 1920s, with the DNA-specific purple dye developed by the German chemist Robert Feulgen, DNA was found to be sited on the chromosomes. DNA therefore had the location expected for a genetic material. In contrast, the histones could apparently be ruled out as genetic components because they were absent from many sperm, which contained instead even smaller basic proteins, the *protamines.* But most biochemists were not inclined to focus attention on DNA. They thought it would not be nearly as specific as the proteins, of which they knew an unlimited number could be constructed by chaining together the 20 amino acids in different orders. So it was widely believed that some minor and not yet well characterized protein component of the chromosomes might be found to be the true genetic material.

Cells Contain RNA as well as DNA

Already late in the nineteenth century it had been discovered that cells have a second kind of nucleic acid—what we now call *ribonucleic acid (RNA).* Unlike DNA, which is located primarily in the nucleus, RNA is found in abundance in the cytoplasm as well as in the nucleus. Within the nucleus, RNA is concentrated in a few dense granules (*nucleoli*) that are attached to chromosomes.

Both DNA and RNA resemble proteins in that they are constructed from many smaller building blocks linked end to end. However, *nucleotides,* the building blocks of nucleic acid, are more complex than any amino acid. Each nucleotide contains a phosphate group, a sugar moiety, and either a *purine* or a *pyrimidine* base (flat, ring-shaped molecules containing carbon and nitrogen) (Figure 2-1). When nucleotides are linked together in large numbers, they are called *polynucleotides.*

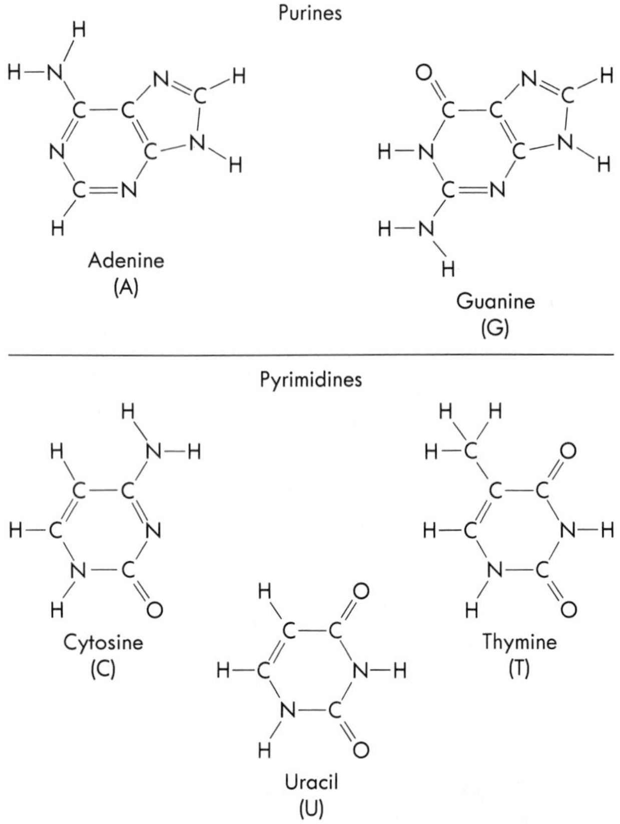

FIGURE **2-1**
Bases of nucleic acids.

Early on, the sugar component of RNA was known to be different from that of DNA. Yet it was not until the 1920s that the work of Phoebus Levine of the Rockefeller Institute revealed that the sugar of DNA is *deoxyribose* (hence the name *deoxyribonucleic acid*) while the sugar in RNA is ribose. Two purines and two pyrimidines are found in both DNA and RNA. The two purines, *adenine* and *guanine,* are used in both DNA and RNA; the pyrimidine *cytosine* is likewise found in DNA and RNA. However, the pyrimidine *thymine* is found only in DNA, while the structurally similar pyrimidine *uracil* appears in RNA. Figure 2-2 shows the three constituents of a nucleotide.

In both DNA and RNA, the nucleotides are linked together to form very long polynucleotide chains. The linkage consists of chemical bonds running from the phosphate group of one nucleotide to the deoxyribose (or ribose) group of the adjacent nucleotide. Each deoxyribose (ribose) residue contains several atoms to which phosphate groups might attach, and there was initially much difficulty in identifying the exact atoms that are bridged by the phosphate groups.

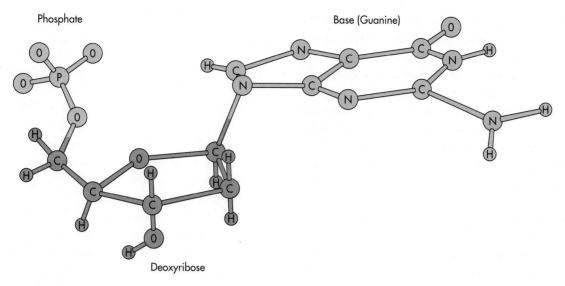

Phosphate

Base (Guanine)

Deoxyribose

FIGURE **2-2**
A DNA nucleotide. The base is attached to a deoxyribose ring that is in turn bonded to a phosphate group. In DNA molecules, nucleotides are linked together to form long chains by bonds running from the phosphate group of one nucleotide to the deoxyribose group of the adjacent nucleotide. The base shown here, guanine, can be replaced by any one of the other three DNA bases—adenine, cytosine, or thymine.

Another question that went unanswered for a long time was how the four different nucleotides are ordered along a DNA (or RNA) molecule. No methods existed for even estimating the exact amounts of the four nucleotides within DNA or RNA; and as late as the 1940s, the possibility could not be dismissed that DNA and RNA had regular repeating structures in which each base was repeated every four nucleotides along the polynucleotide chain. More interesting, however, was the alternative possibility, that there were a very large number of different DNA and RNA molecules, each with its own specific irregular sequence of bases. If this were the way these molecules were constructed, DNA and RNA could have encoded in their varying base sequences the massive amount of information needed to specify the order of the very large number of amino acid sequences found in the proteins of the living world.

A Biological Assay for Genetic Molecules Is Discovered

DNA did not begin to get the serious attention that, with hindsight, we now realize it should have had until it was shown through a biological assay to be able to alter the heredity of certain bacteria. The development of this assay was not at all premeditated, but arose in 1928 out of the studies of the English microbiologist Fred Griffith on the pathogenicity of the bacterium *Diplococcus pneumomiae*, which causes pneumonia. Griffith made the unexpected observation that heat-killed pathogenic cells, when mixed with living nonpathogenic cells and injected into animals, were able to *transform* a small percentage of the nonpathogenic cells into pathogenic cells. In becoming pathogenic, the nonvirulent cells acquired the thick, outer, polysaccharide-rich cell wall (the capsule) that somehow confers pathogenicity to cells that possess it. Griffith thus discovered the existence of an active (genetic?) substance that remained undamaged by the lethal exposure of pathogenic cells to heat, and that could later move into nonpathogenic cells and direct them to make capsules.

Griffith himself did not try seriously to identify the active principle. This task was taken up by Oswald Avery, whose scientific career at the Rockfeller Institute in New York had been principally spent working out the chemistry of bacterial outer capsules. When he began working on the "transforming factor," Avery thought it likely that the active substance would be a complex polysaccharide that in some way primed the synthesis of more polysaccharides of the same kind. As his first step, he showed that the active factor could be extracted from heat-killed cells that were broken—a necessary precondition for the later isolation of the factor from other molecules. Then followed a decade of intensive studies that finally ended with Avery and his younger colleagues, Maclyn McCarty and Colin MacLeod, concluding that the transforming factor was a DNA molecule. Not only was DNA the predominant molecule in their most purified preparation of the transforming factor, but the transforming activity was specifically destroyed by a highly purified preparation of DNase, a then just discovered enzyme that specifically breaks down DNA. In contrast, the transforming activity was unaffected by exposure to enzymes that degraded proteins, or to enzymes that degraded RNA.

Avery's experiments, first announced in 1944, had been so carefully done that the conclusion that DNA was the transforming factor was considered indisputable by a majority of scientists. However, some skeptics preferred to believe that Avery and his colleagues had somehow missed seeing the "genetic protein," and that DNA was required for activity in their assay only because it functioned as an unspecific scaffold to which the real protein genes were fixed. Upon reflection, though, it seems that the pinpointing of DNA should not have been unexpected. By the time of Avery's experiments, DNA was known to be very large molecule containing hundreds of nucleotides. If the sequences of the four main nucleotides were found to be irregular, then the number of potentially different DNA sequences would be the astronomically large 4^n(n = the number of nucleotides in a chain).

The only point in question was the generality of Avery's observation. Were all genes made of DNA, or were there other genetic molecules that functioned in other situations? Clearly the matter could be quickly resolved if it proved possible to change the heredity of other life-forms through the addition of specific DNA molecules. At that time, though, there was no way to isolate undamaged DNA molecules from most plant and animal sources, and it was not possible to extend transformation to other organisms. A group in France did claim that it had been able to use DNA to change the plumage of ducks that had grown up from eggs into which DNA had been injected. The eggs, however, had been obtained from a local country market and were of unclear ancestry; as a result, no one took the claims of "transduction" seriously.

Viruses Are Packaged Genetic Elements That Move from Cell to Cell

Interest in DNA had also risen as a result of its discovery in several highly purified viruses. The nature of these tiny disease-causing particles, which multiply only in living cells, was long disputed. Some scientists considered them a sort of naked gene; others preferred to think of them as the smallest form of life. Only when it became possible to purify them away from cellular debris and look at them in the electron microscope did their nature begin to be revealed. They were clearly not minute cells; rather, they lost their identity as discrete particles when they multiplied within cells. The best guess, therefore, was that they were parasites at the genetic level, and that by studying how they multiplied, researchers might develop definitive systems for analyzing gene structure and replication.

At this point a collection of physicists, chemists, and biologists (the "phage group") turned their attention to the growth cycle of the viruses that multiply in bacteria—the *bacteriophages* (-*phage* is from the Greek for "eating"). Most favored for study was a group of phages named *T1*, *T2*, *T3*, and λ, which multiply within the common intestinal bacterium *Escherichia coli* (*E. coli*). A single parental phage particle can multiply to several hundred progeny particles within roughly 20 minutes. Analysis of the genetic properties of these phages started when mutants arose during the multiplication cycle. When several independently arising mutant phages infected a single bacterium, some of the progeny phages that were produced appeared normal. It was reasoned that these normal phages could arise only by mixing of the genes from different mutant phages present in a single cell. Subsequent experiments employing many different mutants suggested that each virus particle contained several different genes linearly arranged along the viral chromosome.

By purely genetic experiments, however, scientist could not decide whether it was the DNA or one of the protein components that carried the genetic specificity. This point was not settled until 1952, when at Cold Spring Harbor, New York, Alfred Hershey and Martha Chase showed that only the DNA of phages entered the host bacteria. Their surrounding protein

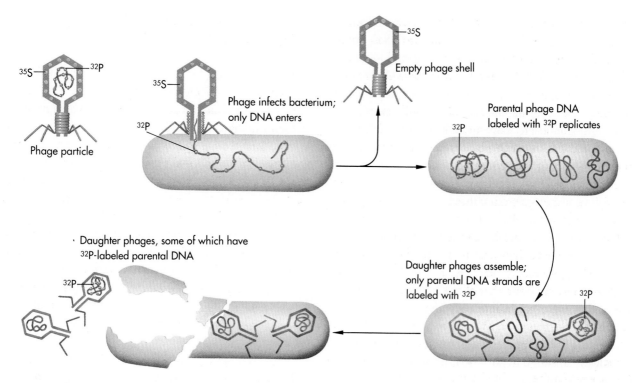

FIGURE **2-3**
Hershey and Chase used radioactively labeled bacteriophage T2 to demonstrate that its genetic material is DNA, not protein.

coats remained outside and thus could be ruled out as potential genetic material (Figure 2-3).

Molecules with Complementary Sizes and Shapes Attract One Another

Even before the exact structure of DNA became known, there was speculation about the nature of the attractive forces that might bring together the appropriate monomeric precursors of proteins and nucleic acids on the surfaces of their respective master molds, or templates. By far the most likely candidates were the so-called weak bonds, as opposed to the covalent bonds that link atoms together in molecules. Enzymes are not needed to either make or break weak bonds; these bonds are spontaneously made and broken at the temperatures of living cells and are the forces that hold molecules together when they aggregate into liquids like water or into the rigidly defined crystals of the solid state. Weak bonds also provide the attractive forces that hold together antibodies and their respective antigens. The polypeptide chains of individual antibodies are folded in such a way that they produce cavities that are complementary in shape to the molecular groups on the surfaces of the antigens to which they bind. Such complementarity permits a large number of different weak bonds to form and produces a correspondingly strong affinity between the antigens and their respective antibodies.

Weak bonds fall into two major classes. One class (*ionic bonds* and *hydrogen bonds*) depends in essence on the attractive forces between atoms of opposite charges. The second class, the *van der Waals attractive forces,* develops when molecules with complementary shapes come into close proximity. Both these categories of attractive forces are thus quite specific, and, particularly when several weak bonds are present simultaneously, their collective energies should be sufficient to ensure that only the correct nucleotide is inserted in the appropriate slot on the template surface of the genetic molecule. Given the nature of weak bonds, the product of a template-controlled polymerization process should not be a molecule identical in shape with its template, but rather one whose shape is exactly complementary in outline. The question of how a polynucleotide chain could be folded so as to be complementary in shape to another polynucleotide chain could not, however, be asked in the absence of knowledge about the covalent bonds that hold the nucleotides of DNA together.

It was thus not surprising that geneticists, impatient with the slow pace at which the fundamental chemistry of DNA was then being established, and curious about the nature of the forces leading to the pairing of homologous chromosomes prior to crossing over, asked whether there might exist still undiscovered chemical forces that were applicable only to macromolecules and that generated attractive forces between identical molecules. This hypothesis was, however, regarded as chemically unsound by serious structural chemists like Linus Pauling, who in 1940 wrote a brief statement on that subject with the theoretical physicist Max Delbrück. They dismissed as silliness the concept of the attraction of like molecules, and stated that gene duplication would necessarily involve the synthesis of molecules with complementary outlines.

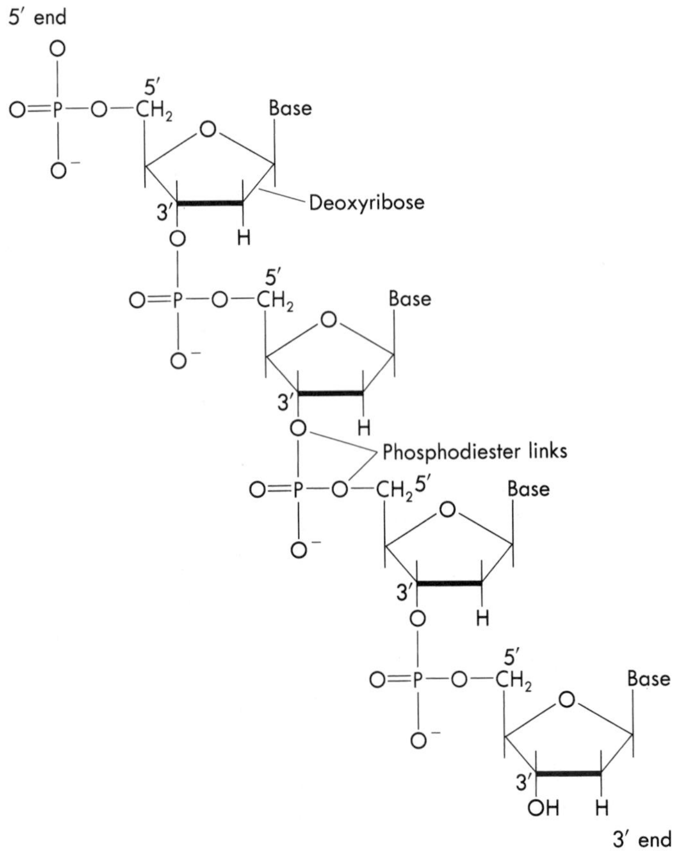

FIGURE **2-4**
Regular phosphodiester bonds between sugar and phosphate groups form the backbone of DNA.

The Diameter of DNA Is Established

Although in the late 1930s Swedish physical chemists had already obtained evidence that DNA was asymmetrical from its behavior in solution, direct measurements of its size became possible only when the electron microscope came into general use in the years following the end of World War II. All carefully prepared samples showed extremely elongated molecules many thousands of angstroms ($Å = 10^{-10}$ meter) in length and approximately 20 Å thick. All the molecules were unbranched, which confirmed the highly regular backbone structure proposed by organic chemists (see below). From the length and the fact that each nucleotide base was just over 3 Å, it was clear that most DNA molecules were composed of many thousands of nucleotides and were very possibly much larger than any other natural polymeric molecules.

The Nucleotides of DNA and RNA Are Linked Together by Regular 5'-3' Phosphodiester Bonds

The knowledge that the transforming factor was DNA became general just as World War II was ending, when scientists, then dissociating themselves from military research, began looking around for new problems to take on. At that time Alexander Todd, already one of England's most effective chemists, decided to focus on the chemistry of complex nucleotides related to those found in DNA. By the early 1950s, his large research group at the Chemical Laboratories of Cambridge University established the precise phosphate-ester linkages that bound the nucleotides together. Their results were appealingly simple. These linkages were always the same, with the phosphate group connecting the 5' carbon atom of one deoxyribose residue to the

3′ carbon atom of the deoxyribose in the adjacent nucleotide (Figure 2-4). No traces of any unusual bonds were found, and Todd's group concluded that the polynucleotide chains of DNA, like the polypeptide chains of proteins, are strictly linear molecules.

It took longer to settle the question of the nature of the linkages within RNA; its structure was not discovered until two years later, in 1955. Like DNA, RNA was found to have a highly regular backbone that employs only 5′-3′ phosphodiester links to hold together its component nucleotides.

The Composition of Bases of DNA from Different Organisms Varies Greatly

Chromatographic separation methods that allowed exact measurement of the proportion of the various amino acids in proteins were first applied successfully in England in the early 1940s. Soon afterward, these methods were extended to determining the amounts of the purine and pyrimidine bases in nucleic acids. The bases of DNA were first quantitatively analyzed by Edwin Chargaff in his laboratory at the College of Physicians and Surgeons of Columbia University. By 1951 it was not only clear that the four bases were not present in equal numbers but also apparent that their relative amounts could vary greatly among distantly related organisms. Chargaff also noted that the amounts of the four bases did not vary independently but that for all the species he looked at, the amounts of the purine adenosine (A) were very close to if not identical with the amounts of the pyrimidine thymine (T). Similarly, the amounts of the second purine, guanine (G), were always very similar to if not identical with those of the second pyrimidine, cytosine (C). The number of purine groups in DNA was thus approximately equal to the number of pyrimidines.

It was not immediately known whether there was any deep significance to the equivalence of the purines and pyrimidines; much less whether the A = T and G = C relationships *(Chargaff's rules)* were important. To start with, the separation methods were imperfect, and it was impossible to state with certainty that the ratios of these components of DNA were exactly equal. There was, moreover, the report that the bacterial virus (phage) T2 lacked any cytosine. If this was true,

then the equivalence of purines and pyrimidines that Chargaff had noticed might not be general enough to be meaningful. Soon afterward, in 1952, the novel nature of the phage T2 DNA was explained. Although cytosine was indeed absent, there was present instead a modified cytosinelike base, and most importantly, its amount was equal to that of guanine. The significance of the A = T and G = C observations, however, remained as unclear as before; and Chargaff was initially known primarily for his demonstration that DNA molecules from different species could have dramatically different base compositions. Because all DNA molecules were not the same, most likely there existed, even within the same cell, a large number of different DNA molecules, each having its own unique nucleotide sequence.

DNA Has a Highly Regular Shape

In one sense DNA chains are very regular: They contain repeating sugar (deoxyribose)-phosphate residues that are always linked together by exactly the same chemical bonds. These identical repeating groups form the "backbone" of DNA. On the other hand, the four different bases of DNA can be attached in any order along the backbone, and this variability gives DNA molecules a high degree of individuality (or specificity). So, depending on which part of a DNA molecule we focus on, we may view it as regular or as irregular. More important, however, is how DNA "views" itself. Does its chain fold up into a regular configuration dominated by its regular backbone? If so, the configuration would most likely be a helical one in which all the sugar-phosphate groups would have identical chemical environments. On the other hand, if the chemistry of the bases dominates the DNA structure, we would fear that no two chains would have identical three-dimensional configurations. In this case, the task of figuring out how each of these differently shaped molecules could serve as a template for the formation of another DNA chain would have been beyond our reach.

The only direct way to examine the three-dimensional structure of DNA was to see how it diffracts (bends) x-rays. Dry DNA has the appearance of irregular white fluffs of cotton; but it becomes highly tacky when it takes on water, and it can then be drawn out into thin fibers. Within these fibers, the individual long, thin DNA molecules line up parallel to one another. In structural analysis, such fibers are placed in the path of an x-ray beam, and the pattern of the diffracting rays is recorded on photographic film. DNA was first examined this way in 1938 by the Englishman William Astbury, using material prepared in Sweden by E. Hammerstein, who had developed procedures that allowed DNA of very high molecular weight to be isolated from thymus glands. Astbury found that DNA did indeed yield a distinctive diffraction pattern, and so the individual DNA molecules must have had some preferred orientation. However, the individual diffraction spots were not sharp like those produced by crystalline material, and the possibility remained that DNA chains never assumed a precise configuration common to all chains. After World War II, Astbury resumed taking x-ray pictures of DNA and obtained some patterns that were considerably better defined. From these he proposed that the individual purine and pyrimidine bases were stacked perpendicular to the long axis of their molecules as if they were a pile of pennies. His better diffraction patterns still remained far from crystalline, though, and the precise structure of the DNA remained in question.

Then in 1950 the physicist Maurice Wilkins, who was working at Kings College, London, with DNA that had been carefully prepared in Bern by the Swiss chemist R. Signer, obtained a truly crystalline diffraction pattern. The individual DNA molecules that came together to form the crystalline fibers must have been very similar in form, or they would not have been able to pack together so regularly. It thus became certain that DNA does have a precise structure, the determination of which might begin to reveal the manner in which DNA functions as a template.

The Fundamental Unit of DNA Consists of Two Intertwined Polynucleotide Chains (the Double Helix)

The data obtained from the fiber diffraction pattern of a molecule as complicated as DNA cannot by itself provide sufficient information to reveal the molecular structure. Inspection of such patterns, however, often provides key parameters that strongly demarcate the outlines of the molecule under investigation. This proved to be the case with DNA, where the key x-ray patterns turned out to be not those obtained from the crystalline DNA fibers, but ones obtained from the less ordered aggregates that form when DNA fibers are exposed to a higher relative humidity and take up more water. These paracrystalline patterns were first seen by Rosalind Franklin, a colleague of Wilkins, working also in London. Her pictures revealed a dominant crosslike pattern, the telltale mark of a helix. Thus, despite the presence of an irregular sequence of bases, the sugar-phosphate backbone of DNA nevertheless assumed a helical configuration. A separate nucleotide was found every 3.4 Å along its fiber axis, with 10 nucleotides, or 34 Å, being required for every turn of the helix.

A very important inference came out of the routine measurement of the diameter of the helix. Given the

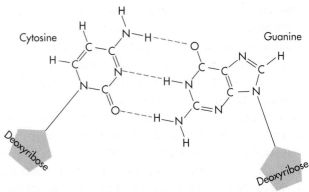

FIGURE 2-5
Hydrogen bonding between the adenine-thymine and guanine-cytosine base pairs.

measured density of DNA, the estimated 20-Å diameter of DNA was far too large for a DNA molecule containing only one chain. The fundamental unit of DNA must consist instead of two intertwined chains; these could be further shown from the diffraction patterns to run in opposite directions. This was indeed a surprising result, for prior to the use of x-ray diffraction methods, no chemist had ever suspected that DNA was a multichained molecule.

The Double Helix Is Held Together by Hydrogen Bonds Between Base Pairs

How the two chains are held together in the DNA molecule could not be ascertained from the x-ray data alone. The final clues came in the spring of 1953, when James Watson and Francis Crick, then working at the Cavendish Laboratory of Cambridge University, built three-dimensional models of DNA to look for the energetically most favorable configurations compatible with the helical parameters provided by the x-ray data. This approach quickly led them to the conclusion that the sugar-phosphate backbones are on the outside of the DNA molecule and that the purine and pyrimidine bases are on the inside, oriented in such a way that they can form hydrogen bonds to bases on opposing chains. The exact hydrogen-bonding pattern used in DNA suddenly emerged with the realization from model building that if a purine on one chain is always hydrogen-bonded to a pyrimidine on the other chain, the area occupied by the paired bases would always be the same throughout the length of the DNA molecule. Seen from the outside, DNA is a very regular-appearing structure, despite the irregular sequence of bases on any one chain. Equally important, the two purines adenine and guanine do not unselectively bond to the two pyrimidines thymine and cytosine. Adenine (A) can pair only with thymine (T), while guanine (G) can bond only with cytosine (C) (Figure 2-5). Each of these base pairs possesses a symmetry that permits it to be inserted into the double helix in two ways (A=T and T=A; G≡C and C≡G); thus along any given DNA chain, all four bases can exist in all possible permutations of sequence (Figure 2-6).

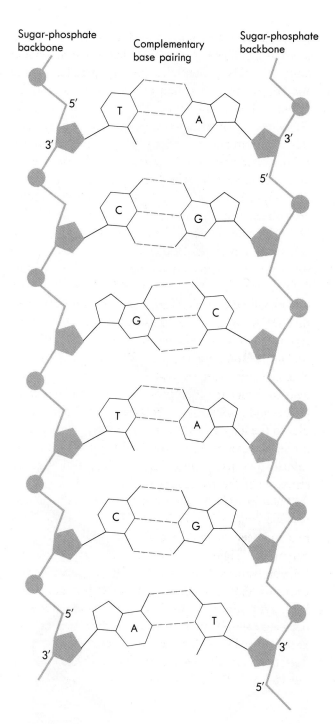

FIGURE **2-6**
The base pairing of two DNA chains.

Because of this specific base pairing, if the sequence of one chain (for example, TCGCAT) is known, that of its partner (AGCGTA) is also known. The opposing sequences are referred to as *complementary*, and the

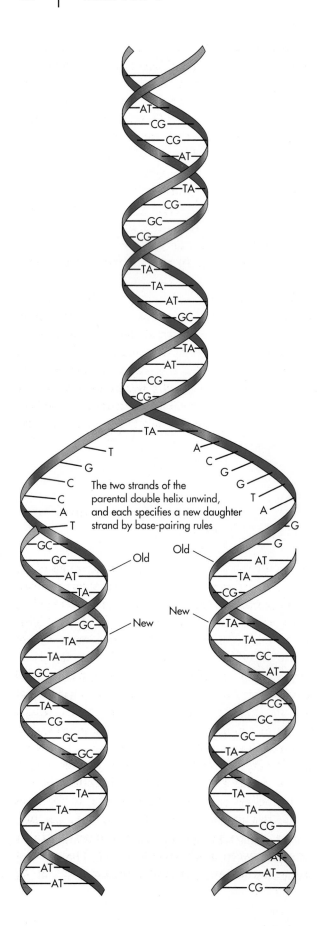

The two strands of the parental double helix unwind, and each specifies a new daughter strand by base-pairing rules

Old

Old

New

New

corresponding polynucleotide partners as *complementary chains*. Despite the relative weakness of the hydrogen bonds holding the base pairs together, each DNA molecule contains so many base pairs that the complementary chains never spontaneously separate under physiological conditions. If, however, DNA is exposed to near-boiling temperatures, so many base pairs fall apart that the double helix separates into its two complementary chains (this process is called *denaturation*). The existence of the double helix provides a structural chemical explanation for Chargaff's rules: A = T and G = C. Only with complementary base pairing could all the backbone sugar-phosphate groups have identical orientations and permit DNA to have the same structure with any sequence of bases.

The Complementary Nature of DNA Is at the Heart of Its Capacity for Self-Replication

Before we knew what genes looked like, it was almost impossible to speculate wisely about how they could be exactly duplicated prior to cell division. Any proposal had to be general. As we mentioned earlier, Linus Pauling and Max Delbrück suggested in 1940 that the surface of the gene somehow acts as a positive mold, or template, for the formation of a molecule of complementary (negative) shape, in much the same way that material can be molded around a piece of sculpture for the purpose of making a cast. The complementarily shaped negative could then serve as the template for the formation of its own complement, thereby producing an identical copy of the original mold.

Thus the realization that the two chains of DNA had complementary shapes caused great excitement. It was promptly proposed that the two strands of the double helix should be regarded as a pair of positive and negative templates, each specifying its comple-

FIGURE **2-7**
Identical daughter double helices are generated through the semiconservative replication of DNA.

ment and thereby capable of generating two daughter DNA molecules with sequences identical with those of the parental double helix (Figure 2-7). If this was indeed the way DNA duplicates, the process could be confirmed by proof that the parental strands separate before duplicating themselves and that each daughter molecule contains one of the parental chains.

The elucidation of the structure of DNA thus accomplished in one decisive act far more than we could ever have prudently predicted. Before the double helix was found, we hoped that determination of the correct structure of DNA would provide a firm foundation upon which a model of DNA replication could be erected. Instead, discovery of the double helix brought with it as a gigantic bonus a powerful clue, if not the total answer, to the long-desired goal of understanding gene duplication.

Of course, the possibility still existed that the self-complementary structure of the double helix had nothing to do with how it was synthesized. Virtually no one, however, really believed this alternative. The finding of the first highly plausible example of a template had to be an occasion for joy, not for scepticism. The time had also passed for any further questioning about what genes were. The name of the game was DNA.

DNA Replication Is Found to Be Semiconservative

It took some five years before there was firm evidence of strand separation. Proof came from the experiments of Matthew Meselson and Franklin Stahl at the California Institute of Technology. They had the clever idea of using density differences to separate parental DNA molecules from daughter molecules. They first grew cultures of the bacterium *E. coli* in a medium highly enriched in the heavy isotopes ^{13}C and ^{15}N. By virtue of its isotopic content, the DNA in these bacteria was much heavier than the normal light DNA coming from cells grown in the presence of the far more abundant natural isotopes ^{12}C and ^{14}N. Because of its greater density, the heavier DNA could be clearly separated from the light DNA by high-speed centrifugation in cesium chloride.

When heavy-DNA-containing cells were transferred to a normal "light" medium and allowed to multiply for one generation, all the heavy DNA was replaced by DNA of a density that was halfway between heavy and light. The disappearance of the heavy DNA indicated that DNA replication is not a conservative process in which the complementary strands of the double helix stay together. Instead, its replacement by the hybrid-density DNA implied a semiconservative replication process, in which the two heavy parental strands separate to serve as templates for complementary light strands, and each daughter molecule has one heavy (parent) strand and one light strand.

Whether the complementary strands completely separate before replication starts was not immediately known. Later, abundant electron microscope evidence of Y-shaped replication forks indicated that strand separation and replication go hand in hand. As soon as a section of double helix begins to separate for replication, the resulting single-stranded regions are quickly used as templates and become new double-helical regions.

DNA Molecules Can Be Renatured as well as Denatured

Under physiological conditions, the two strands of the double helix almost never come apart spontaneously. If, however, double helices are exposed to near-boiling temperatures or to extremes of pH (pH < 3 or pH > 10), they quickly fall apart (are denatured) into their component single strands. At first, denaturation was regarded as essentially irreversible, but by 1960 Julius Marmur, Paul Doty, and their coworkers at Harvard showed that the complementary single strands recombine to form native double helices when they are kept for several hours at subdenaturing conditions (approximately 65°C). Such annealing *renaturation* events are very specific and produce perfect double helices only when the sequences of the bases of the two combining strands are exactly complementary. Imperfect double helices, however, can be formed between nearly complementary molecules at less stringent (lower) annealing temperatures. So by observing the extent of such imperfect renaturation events, researchers can determine the genetic relationships among DNA from different species.

Renaturation also can be induced to occur between DNA and RNA chains with complementary sequences. It was through the preparation of such DNA–RNA hybrid double helices that in 1961 Ben Hall and Sol Spiegelman obtained definitive proof that DNA functions in protein synthesis by serving as the template for the formation of complementary RNA chains (Chapter 3).

G·C Base Pairs Fall Apart Less Easily Than Their A·T Equivalents

Three hydrogen bonds link guanine to cytosine, whereas two hydrogen bonds hold together the adenine-thymine base pairs. Double helices containing a prevalence of G·C base pairs are thus more stable (that is, they denature at higher temperatures) than helices in which A·T base pairs predominate. In fact, the proportion of A + T to G + C within DNA specimens can be directly ascertained by measuring the temperature at which half the DNA molecules fall apart into their component single strands.

Palindromes Promote Intrastrand Hydrogen Bonding

Not only do base pairs form between bases on opposing strands, but they can also form between bases of single chains that have nearby inverted repetitious sequences (*palindromes*) that allow the formation of hydrogen-bonded hairpin loops (Figure 2-8). The possibility thus exists that the momentary denaturation of palindromic regions often leads to the formation of semistable cruciform loops (stem loops) able to interact with specific DNA-binding proteins.

5-Methylcytosine Can Replace Cytosine in DNA

In many higher plant and animal DNAs a significant fraction of the cytosine residues exist in a modified form in which a methyl group is attached to the 5 carbon atom of the pyrimidine ring (5-methylcytosine)(Figure 2-9). Such methyl groups do not affect the way their respective molecules can hydrogen-bond, and the base pairs formed by 5-methylcytosine with guanine are equivalent in strength to those formed by cytosine. In eukaryotic DNA, cytosine res-

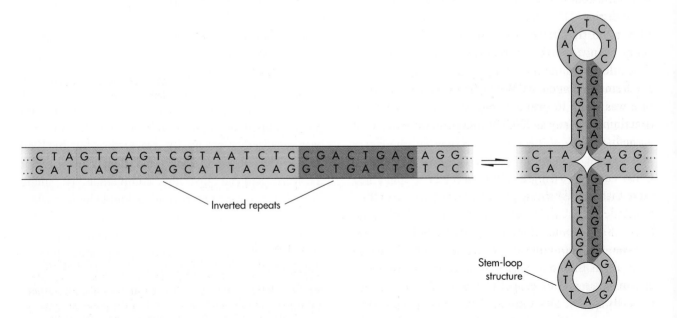

FIGURE **2-8**
A segment of DNA containing an inverted repeat can locally denature and then renature to form stem-loop structures.

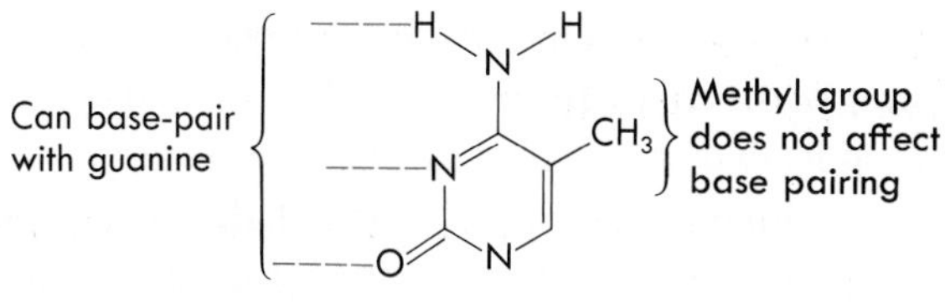

FIGURE **2-9**
5-Methylcytosine.

idues that contain methyl groups are always located next to guanine residues on the same chain [(5′)CpG(3′), where p is the 3′-to-5′ phosphodiester bond between adjacent nucleotides], and DNA molecules that have relatively many (5′)CpG(3′) segments are more methylated than those in which the (5′)CpG(3′) dinucleotide is relatively rare. At first it was uncertain how 5-methyl-C arose, but now it is clear that methyl groups are added after their respective DNA chains are synthesized. 5-Methyl-C is particularly common in higher plants, where some 20 percent of the total cytosine residues become methylated. Highly methylated DNA is usually associated with genetically silent regions of the chromosome.

Chromosomes Contain Single DNA Molecules

When the double helix was found, it was believed that many distinct DNA molecules were used to construct all but the smallest chromosome. This picture suddenly changed with the realization that long DNA molecules are inherently fragile and easily break into much smaller fragments. When, by the late 1950s, great care was taken to prevent shearing, DNA molecules containing as many as 200,000 base pairs were detected immediately.

Now the best guess is that the chromosome of *E. coli* contains a single DNA molecule made up of more than 4 million base pairs. Likewise, a single DNA molecule is thought to exist within each of the even larger chromosomes of higher plants and animals; these chromosomes contain on the average some 20 times more DNA than occurs in the *E. coli* chromosome. A chromosome is thus properly defined as a single, genetically specific DNA molecule to which are attached large numbers of proteins that are involved in maintaining chromosome structure and regulating gene expression.

Viruses Are Sources of Homogeneous DNA Molecules

Until recently, the only DNA molecules that had been seriously studied were isolated from DNA viruses, each of which contains a single DNA molecule. Such DNA may be either linear or circular in shape, with replication generally starting at a unique internal site and moving away bidirectionally until the duplication

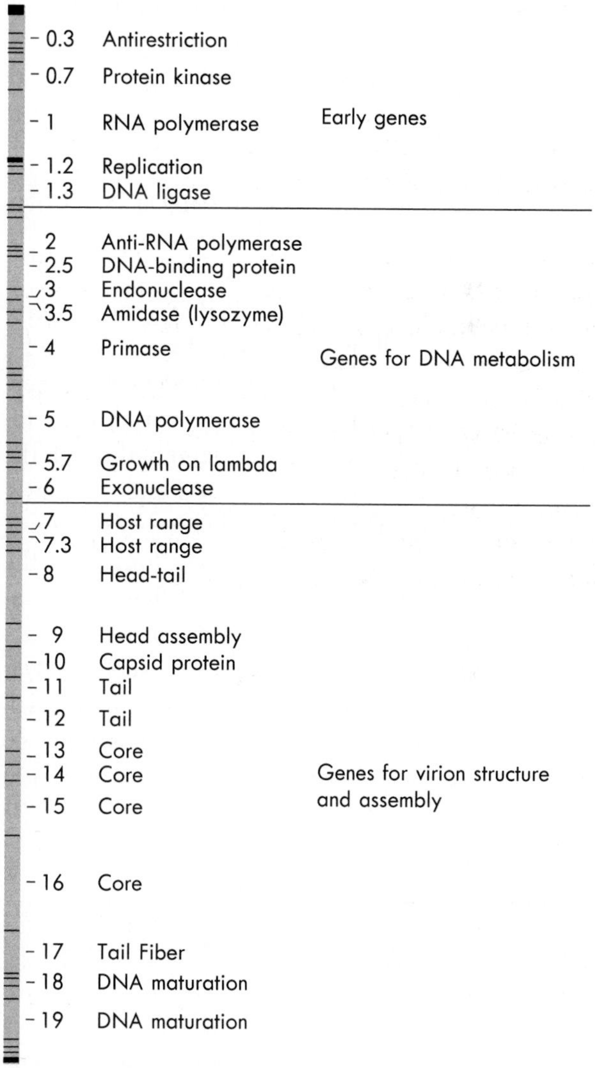

FIGURE **2-10**
A genetic and physical map of the DNA of bacteriophage T7. The positions of the 50 genes (the thin black lines) are drawn to scale according to their positions in the nucleotide sequence, and the functions of some of the genes are given. The "early genes" are those necessary during the early stages of host infection. (Decimal designations reflect genes that were mapped after the original 19 positions had been assigned.) (Dunn and Studier, 1983).

process is completed. The various DNA phages, particularly those that multiply in *E. coli*, were the favored source of DNA in early studies, both because they are readily grown in large amounts and because many have relatively small DNA molecules that do not easily break in solution. The well-studied linear DNA of phage T7, for example, consists of 39,936 base pairs, along which approximately 50 genes have been mapped (Figure 2-10). From DNA sequence analysis the number of amino acids in each of the polypeptide products of these genes has been determined. Over 92 percent of the base pairs in this DNA are used to specify these products, so the individual genes are very close to one another.

Phage λ DNA Can Insert into a Specific Site Along the *E. coli* Chromosome

At first it was generally believed that after its entry into a bacterial cell, phage DNA always initiated a lytic multiplication cycle that resulted in the generation of hundreds of progeny phage within each infected cell. Later it became clear, largely through the work of André Lwoff at the Institute Pasteur in Paris, that certain phage DNAs have the alternative possibility of being inserted in a *prophage* form into the chromosome of their bacterial host, where they become effectively indistinguishable from normal bacterial genes. By now the best known of such *lysogenic* phages is phage λ, the linear DNA molecule of which is 48,513 base pairs in length and contains some 60 genes (Figure 2-11). Before λ DNA inserts itself into

the *E. coli* chromosome, it circularizes by the base-pairing of the complementary single-stranded tails that exist at its two ends, called the *cohesive ends*, or *cos sites*. The resulting circular λ DNA molecule then recombines into the *E. coli* chromosome by a crossover event with a specific group of bases along the *E. coli* chromosome (Figure 2-12). The resulting linear prophage remains virtually inert genetically until it is provoked by a signal (usually some form of damage to the host chromosome such as exposure to UV light) that leads to a reverse crossover event that releases λ DNA from the *E. coli* chromosome, allowing it to initiate a lytic cycle of multiplication.

Abnormal Transducing Phages Provide Unique Segments of Bacterial Chromosomes

The chromosomes of *E. coli*—and perhaps of all bacteria—are circular, with bidirectional DNA replication always initiated at a specific site. They are much too long to visualize in their entirety in the electron microscope, and without genetic tricks there would be no method to select any specific section to study. Even the most careful isolation procedures necessarily shear bacterial chromosomes into tens of pieces with no two fragments having the same ends. Luckily, by 1965 some high-powered bacterial genetics changed this bleak picture. Careful genetic examination of certain phages revealed that a small percentage genetically recombined their DNA with that of their host

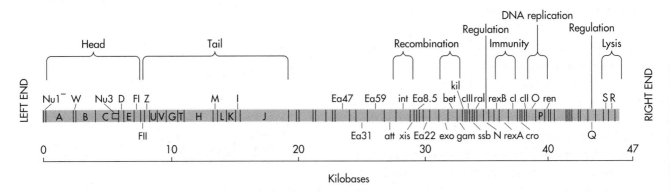

FIGURE **2-11**
The organization of the bacteriophage λ genome. The positions of the λ genes are shown along the central bar (colored), and some of their functions are indicated above (Daniels, Sanger, and Coulson, 1983).

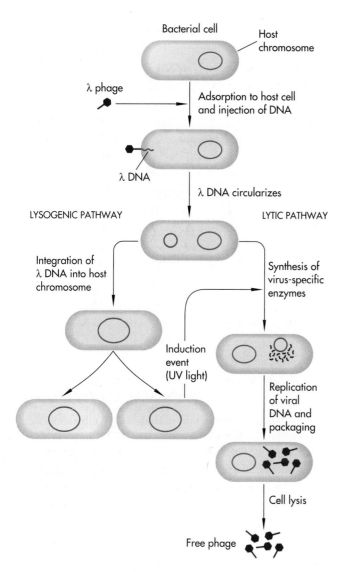

FIGURE **2-12**
Insertion of λ phage DNA into the *E. coli* chromosome. The λ DNA circularizes by the base pairing of the complementary single-stranded tails at its ends. The circular molecule then recombines into the *E. coli* chromosome by a crossover event with specific bases in the chromosome (lysogenic pathway). Subsequent damage to the host chromosome (such as exposure to UV light) frees the phage DNA and permits resumption of the lytic cycle.

mally cannot make. With genetic tricks, specific transducing phages carrying from one to several desired bacterial genes can be isolated.

Plasmids Are Autonomously Replicating Minichromosomes

In addition to their main chromosomes (with 4 million base pairs), many bacteria possess large numbers of tiny circular DNA molecules that may contain only several thousand base pairs. These minichromosomes, called *plasmids,* were first noticed as genetic elements that were not linked to the main chromosome and that carried genes that conveyed resistance to antibiotics such as the tetracyclines or kanamycin (Figure 2-13). That these genes were found on plasmids as opposed to main-chromosomal DNA was not a matter of chance. Antibiotic resistance requires relatively

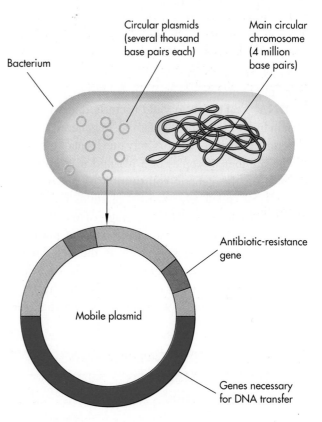

FIGURE **2-13**
Plasmids are small, autonomously replicating bacterial DNA molecules containing genes conveying resistance to specific antibiotics. Most plasmids are mobile, but nonmobilizable plasmids also exist.

bacterial cells, to yield abnormal phages in which fragments of bacterial DNA were inserted in the phage chromosomes. The phages carrying these hybrid chromosomes are called *transducing phages.* If these phages then infect another strain of bacteria, the bacterial DNA they bring from the previous host can program this strain of bacteria to manufacture proteins it nor-

large amounts of the enzymes that chemically neutralize the antibiotics. By being located on many plasmids, their respective genes are present in much higher numbers of copies (their *copy number*) than they would be if they were located on the main chromosome.

Certain plasmids, called *episomes,* have the ability to move on and off the main chromosomal elements. How episomes jump on or off chromosomes was long a mystery. Now it is clear that this capacity often reflects the possession of mobile genetic elements whose movements are accomplished through the fusion of two independently replicating DNA units (*replicons*) (Chapter 10). Plasmids that can integrate into the bacterial chromosome can be transferred from one bacterium to another when the cells mate and a copy of the "male" chromosome is transferred to the "female" cell. Some plasmids, however, are unable to integrate into the bacterial chromosome, so they cannot be transferred from cell to cell during mating. They are called *nonmobilizable,* and once a gene is on such a plasmid it cannot easily move.

Because plasmid DNA is so much smaller than even highly fragmented chromosomal DNA, it is easily separable, and highly purified plasmid DNA is readily obtained. In the laboratory, when plasmid DNA is added to plasmid-free bacteria in the presence of Ca^{2+}, the DNA is taken up to yield bacteria that will soon contain many copies of the plasmid. In general, a given bacterial cell usually harbors only one form of plasmid.

The number of copies of a plasmid in a host cell depends on the genetic constitution of the plasmid and cell. So-called *relaxed-control* plasmids may multiply until each cell has on the average 10 to 200 copies of the plasmid. In contrast, *stringent-control* plasmids replicate at about the same rate as the cell's main chromosome and are present in only one or a few copies per cell. The relaxed plasmids are the ones used for recombinant DNA research, as we shall see later.

Circular DNA Molecules May be Supercoiled

So long as a DNA molecule has a linear form, its conformation is not closely controlled by the exact rotation of successive nucleotides around the helical

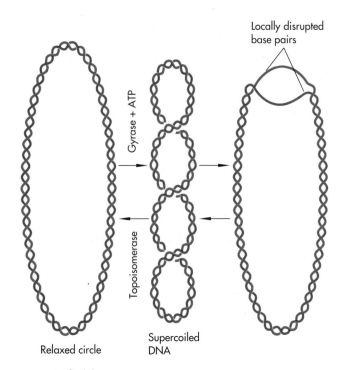

FIGURE **2-14**
A relaxed circular DNA molecule can be twisted into a negatively supercoiled molecule by the action of DNA gyrase. The reverse reaction is catalyzed by topoisomerase ("nicking-closing" enzyme). The strain in the negatively supercoiled form can be relieved by local disruption of the double helix to produce single-stranded regions, thereby returning the molecule to its relaxed form. The degree of supercoiling determines how compact the circular DNA molecule will become; molecules with increasing degrees of supercoiling are increasingly compact and therefore migrate more rapidly during electrophoresis through a gel (Figure 2-15).

axis. The original x-ray photographs of DNA suggested an approximate 36° rotation between successive nucleotides within the *B form* of DNA—the form that exists in highly hydrated fibers and that is thus presumed to be the form that DNA takes up in solution. At that time (1953) it was considered largely irrelevant whether the rotation angle of DNA was, say, 34.5° as opposed to 36°. But it was later discovered that the two ends of a linear DNA may become covalently bound to each other, and the rotation angle became a much more crucial parameter. With circular DNA, any appreciable change in angle of twist between successive nucleotides leads the molecules to adopt *su-*

percoiled configurations. In negative supercoiling, the number of negative supercoils exactly equals the number of added twists of the double helix brought about by an increased rotation angle. To relieve the strain of negative supercoiling, small AT-rich sections of double-helical DNA may separate into single strands. They thus become susceptible to cleavage by specific DNA-cutting enzymes that attack only single-stranded, as opposed to double-helical, DNA (Figure 2-14).

Supercoiled DNA (whether negatively or positively twisted) adopts a more compact configuration than its "relaxed" equivalents. It thus moves faster than "relaxed" DNA when sedimented in an ultracentrifuge or subjected to gel electrophoresis (page 66) by an electrical field (Figure 2-15). At first, supercoiling was

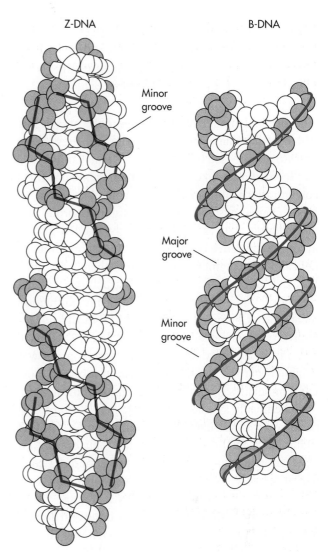

FIGURE **2-16**
Left-handed (Z-) DNA forms a zigzag helix.

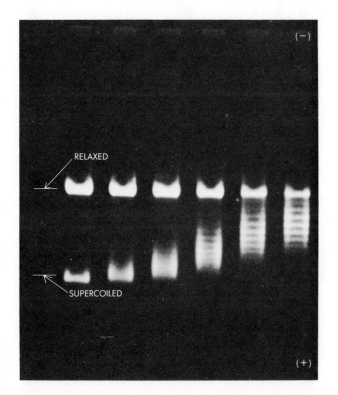

FIGURE **2-15**
Agarose gel electrophoresis can separate DNA molecules with different amounts of supercoiling. Completely supercoiled DNA has the greatest mobility in agarose gels; molecules with progressively fewer superhelical turns migrate progressively more slowly. (Courtesy of James C. Wang.)

regarded as an almost accidental feature of the conditions used to study DNA in the laboratory. With time, however, it has become clear that in cells the degree of supercoiling is strongly controlled by specific enzymes (*topoisomerases* and *gyrases*) that are capable of either adding or subtracting supercoiled twists in DNA. Now it is suspected that most circular DNA molecules (such as plasmid DNA) within cells are negatively supercoiled, with the resulting sections of separation between strands conceivably allowing DNA regulatory proteins access to the DNA strands.

Most Double Helices Are Right-Handed, but Under Special Conditions Certain DNA Nucleotide Sequences Lead to Left-Handed Helices

The original model of the double helix was right-handed (the chains turn to the right as they move upward) as opposed to left-handed. Right-handed helices seemed more likely because their stereo chemical configurations appeared more stable than those of DNA chains twisted to the left. Rigorous proof that most double helices are in fact right-handed had to await the existence of truly crystalline, short, double-helical DNA segments made from chemically synthesized oligonucleotides. Such structures can be rigorously solved (determined) by x-ray diffraction techniques. Chemically synthesized DNA did not effectively become available until the late 1970s (page 69), and it was only in 1979 that rigorous proof was obtained that the solution form (the B form) of most DNA molecules is right-handed. Almost ironically, the first chemically synthesized double helix structure to be solved was left-handed. Its component GCGCGC chains formed left-handed double helices (*Z-form DNA*) in which successive G-C units adopted a configuration very different from that adopted by successive nucleotides in B-form DNA (Figure 2-16). Soon afterwards, the crystal structure of several double helices formed by oligonucleotides containing all four bases revealed B-form right-handed DNA.

Double helices composed of GCGC sequences twist in the left-handed direction only in high concentrations of salt; in low salt concentrations, they exist in the right-handed B form. Interestingly, the methylation of C residues in GCGC chains stabilizes the left-handed form, permitting its existence under salt conditions similar to those in vivo. Whether significant amounts of biologically relevant left-handed DNA will be found in cells remains to be seen. One possible biological role for left-handed DNA might be based on the fact that small regions of Z-DNA relieve the superhelical strain in a negatively supercoiled DNA molecule, in the same way that small regions of single-stranded DNA can. These local regions of Z-DNA may bind to regulatory proteins that are specific for Z-DNA (Figure 2-17).

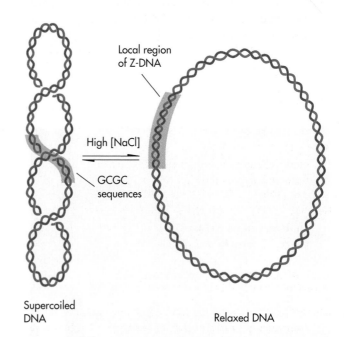

FIGURE **2-17**
Local regions of left-handed (Z-) DNA can relieve the strain of a supertwisted DNA molecule in the same way that local single-stranded regions can.

Reading List

General

Cairns, J., G. S. Stent, and J. D. Watson, eds. *Phage and the Origins of Molecular Biology.* Cold Spring Harbor Laboratory, Cold Spring Harbor, N. Y., 1966.

Watson, J. D. *The Double Helix.* Atheneum, New York, 1968. (Text and trade paperback editions.) New American Library, New York, 1969. (Paperback.)

Olby, R. *The Path to Double Helix.* University of Washington Press, Seattle, 1974.

Portugal, F. H., and J. S. Cohen. *A Century of DNA: A History of the Discovery of the Structure and Function of the Genetic Substance.* MIT Press, Cambridge, Mass., 1977. (Paperback edition, 1980.)

Stent, G. S., and R. Calendar. *Molecular Genetics: An Introductory Narrative,* 2nd ed. Freeman, San Francisco, 1978.

Judson, H. F. *The Eighth Day of Creation.* Simon and Schuster, New York, 1980.

Watson, J. D. *The Double Helix: A Norton Critical Edition.* G. S. Stent, ed. Norton, New York, 1980.

Hendrix, R., J. Roberts, F. W. Stahl, and R. Weisberg, eds. *Lambda* II. Cold Spring Harbor Laboratory. Cold Spring Harbor, N. Y., 1983.

McCarty, M. *The Transforming Principle: Discovering that Genes Are Made of DNA.* Norton, New York, 1985.

Crick, F. C. *What Mad Pursuit: A Personal View of Scientific Discovery.* Basic Books, New York, 1988.

Witkowski, J. A. "Fifty years on: molecular biology's hall of fame." *Trends Biotechnol.,* 6: 234–243 (1988).

Brock, T. D. *The Emergence of Bacterial Genetics.* Cold Spring Harbor Laboratory Press, Cold Spring Harbor, N. Y. 1990.

Kornberg, A., and T. A. Baker. *DNA Replication,* 2nd ed. Freeman, San Francisco, 1992.

Original Research Papers

DNA AS THE PRIMARY GENETIC MATERIAL

Avery, O. T., C. M. MacLeod, and M. MacCarty. "Studies on the chemical nature of the substance inducing transformation of pneumococcal types." *J. Exp. Med.,* 79: 137–158 (1944).

Hershey, A. D., and M. Chase. "Independent functions of viral protein and nucleic acid in growth of bacteriophage." *J. Gen. Physiol.,* 36: 39–56 (1952).

POLYNUCLEOTIDE CHEMISTRY AND ELECTRON MICROSCOPY

Chargaff, E. "Structure and function of nucleic acids as cell constituents." *Fed. Proc.,* 10: 654–659 (1951). [Review]

Brown, D. M., and A. R. Todd. "Nucleotides, part X. Some observations on structure and chemical behaviour of the nucleic acids." *J. Chem. Soc.,* pt. 1: 52–58 (1952).

Williams, R. C. "Electronmicroscopy of sodium desoxyribonucleate by use of a new freeze-drying method." *Biochim. Biophys. Acta,* 9: 237–239 (1952).

Dekker, C. A., A. M. Michaelson, and A. R. Todd. "Nucleotides, part XIX. Pyrimidine deoxyribonucleoside diphosphates." *J. Chem. Soc.,* pt. 1: 947–951 (1953).

Wyatt, G. R., and S. S. Cohen. "The bases of the nucleic acids of some bacterial and animal viruses: the occurrence of 5-hydroxymethylcytosine." *Biochem. J.,* 55: 774–782 (1953).

THE DOUBLE HELIX

Franklin, R. E., and R. G. Gosling. "Molecular configuration in sodium thymonucleate." *Nature,* 171: 740–741 (1953).

Watson, J. D., and F. H. C. Crick. "Molecular structure of nucleic acids: a structure for deoxyribose nucleic acid." *Nature,* 171: 737–738 (1953).

Watson, J. D., and F. H. C. Crick. "Genetical implications of the structure of deoxyribonucleic acid." *Nature,* 171: 964–967 (1953).

Watson, J. D., and F. H. C. Crick. "The structure of DNA." *Cold Spring Harbor Symp. Quant. Biol.,* 18: 123–131 (1953).

Wilkins, M. H. F., A. R. Stokes, and H. R. Wilson. "Molecular structure of deoxypentose nucleic acids." *Nature,* 171: 738–740 (1953).

Crick, F. H. C., and J. D. Watson. "The complementary structure of deoxyribonucleic acid." *Proc. Roy. Soc., A,* 223: 80–96 (1954).

REPLICATION OF DNA

Pauling, L., and M. Delbrück. "The nature of the intermolecular forces operative in biological processes." *Science,* 92: 77–79 (1940).

Meselson, M., and F. W. Stahl. "The replication of DNA in a *Escherichia coli.*"*Proc. Natl. Acad. Sci. USA,* 44: 671–682 (1958).

DNA DENATURATION AND RENATURATION

Doty, P., J. Marmur, J. Eigner, and C. Schildkraut. "Strand separation and specific recombination in deoxyribonucleic acids: physical chemical studies." *Proc. Natl. Acad. Sci. USA,* 46: 461–476 (1960).

Marmur, J., and L. Lane. "Strand separation and specific recombination in deoxyribonucleic acids: biological studies." *Proc. Natl. Acad. Sci. USA,* 46: 453–461 (1960).

Hall, B. D., and S. Spiegelman. "Sequence complementarity of T2-DNA and T2-specific RNA." *Proc. Natl. Acad. Sci. USA,* 47: 137–146 (1961).

Schildkraut, C. L., J. Marmur, and P. Doty. "The formation of hybrid DNA molecules, and their use in studies of DNA homologies." *J. Mol. Biol.,* 3: 595–617 (1961).

5-METHYLCYTOSINE AND GENE REGULATION

Wyatt, G. R. "The purine and pyrimidine composition of deoxypentose nucleic acids." *Biochem. J.,* 48: 584–590 (1951).

Doskočil, J., and F. Šorm. "Distribution of 5-methylcytosine in pyrimidine sequences of deoxyribonucleic acids." *Biochim. Biophys. Acta,* 55: 953–959 (1962).

Holliday, R., and J. E. Pugh. "DNA modification mechanisms and gene activity during development." *Science,* 187: 226–232 (1975).

Riggs, A. D. "X inactivation, differentiation, and DNA methylation." *Cytogenet. Cell. Genet.,* 14: 9–25 (1975).

Sager, R., and R. Kitchin. "Selective silencing of eukaryotic DNA." *Science,* 189: 426–433 (1975).

Gruenbaum, Y., H. Cedar, and A. Razin. "Substrate and sequence specificity of eukaryotic DNA methylase." *Nature,* 295: 620–622 (1982).

CHROMOSOMAL AND PLASMID DNA

Marmur, J., R. Rownd, S. Falkow, L. S. Baron, C. Schildkraut, and P. Doty. "The nature of intergeneric episomal infection." *Proc. Natl. Acad. Sci. USA,* 47: 972–979 (1961).

Cairns, J. "The bacterial chromosome and its manner of replication as seen by autoradiography." *J. Mol. Biol.,* 4: 407–409 (1963).

Watanabe, T. "Infectious heredity of multiple drug resistance in bacteria." *Bact. Rev.,* 27: 87–115 (1963).

Cohen, S. N., and C. A. Miller. "Multiple molecular species of circular R-factor DNA isolated from *Escherichia coli.*" *Nature,* 224: 1273–1277 (1969).

LYTIC, LYSOGENIC, AND
TRANSDUCING BACTERIOPHAGES

Lwoff, A. "Lysogeny." *Bact. Rev.,* 17: 269–337 (1953).

Matsushiro, A. "Specialized transduction of tryptophan markers in *Escherichia coli* K12 by bacteriophage φ 80." *Virology,* 19: 475–482 (1963).

Daniels, D., F. Sanger, and A. R. Coulson. "Features of bacteriophage lambda: analysis of the complete nucleotide sequence." *Cold Spring Harbor Symp. Quant. Biol.,* 47: 1009–024 (1983).

Dunn, J. J., and F. W. Studier. "The complete nucleotide sequence of bacteriophage T7 DNA, and the locations of T7 genetic elements." *J. Mol. Biol,* 166: 477–535 (1983).

SUPERCOILING

Vinograd, J., J. Lebowitz, R. Radloff, R. Watson, and P. Laipis. "The twisted circular form of polyoma DNA." *Proc. Natl. Acad. Sci. USA,* 53: 1104–1111 (1965).

Wang, J. C. "Interaction between DNA and an *Escherichia coli* protein W." *J. Mol. Biol.,* 55: 523–533 (1971).

Champoux, J. J., and R. Dulbecco. "An activity from mammalian cells that untwists superhelical DNA—a possible swivel for DNA replication." *Proc. Natl. Acad. Sci. USA,* 69: 143–146 (1972).

Gellert, M., K. Mizuuchi, M. H. O'Dea, and H. A. Nash. "DNA gyrase: an enzyme that introduces superhelical turns into DNA." *Proc. Natl. Acad. Sci. USA,* 73: 3872–3876 (1976).

Wang, J. C., L. J. Peck, and K. Becherer. "DNA supercoiling and its effects on DNA structure and function." *Cold Spring Harbor Symp. Quant. Biol.,* 47: 251–257 (1983).

LEFT-HANDED DNA

Pohl, F. M., and T. M. Jovin. "Salt-induced co-operative conformation change of a synthetic DNA: equilibrium and kinetic studies with poly (dG-dC)." *J. Mol. Biol.,* 67: 375–396 (1972).

Wang, A. H. -J., G. J. Quibley, F. J. Kolpak, J. L. Crawford, J. H. van Boom, G. van der Marel, and A. Rich. "Molecular structure of a left-handed DNA fragment at atomic resolution." *Nature,* 282: 680–686 (1979).

Wing, R. M., H. R. Drew, T. Takano, C. Broka, S. Tanaka, K. Itakura, and R. E. Dickerson. "Crystal structure analysis of a complete turn of B-DNA." *Nature,* 287: 755–758 (1980).

Behe, M., and G. Felsenfeld. "Effects of methylation on a synthetic polynucleotide: the B-Z transition in poly (dG-m⁵dC)·poly (dG-m⁵dC)." *Proc. Natl. Acad. Sci. USA,* 78: 1619–1623 (1981).

Dickerson, R. E., H. R. Drew, B. N. Conner, R. M. Wing, A. V. Fratini, and M. L. Kopka. "The anatomy of A-, B-, and Z-DNA." *Science,* 216: 475–485 (1982).

Peck, L. J., A. Nordheim, A. Rich, and J. C. Wang. "Flipping of cloned d(pCpG)n·d(pCpG)n DNA sequences from a right- to a left-handed helical structure by salt, Co(III), or negative supercoiling." *Proc. Natl. Acad. Sci. USA,* 79: 4560–4564 (1982).

3

Elucidation of the Genetic Code

Once the double helix was identified, it became possible to speculate more precisely on the one gene, one protein relationship. First of all, the genetic information in DNA had to be conveyed solely by the linear sequences of the four bases (A, T, G, and C) in the DNA alphabet. Therefore gene mutations had to represent changes in the sequence of bases, either through the substitution of one base pair for another, or through addition or deletion of one or many base pairs (Figure 3-1). Mutant proteins in turn clearly represented changes in the amino acid sequence, the simplest mutants being proteins in which one amino acid was replaced by another.

A Mutation in the Hemoglobin Molecule Is Traced to a Single Amino Acid Replacement

The first experiments showing mutant proteins bearing single amino acid replacements focused on the sickle hemoglobin molecules produced in humans suffering from the genetic disease sickle-cell anemia. Working in Cambridge, England, in 1957, Vernon Ingram analyzed both the α and the β chains that make up the $\alpha_2\beta_2$ form of adult hemoglobin molecules. No changes were found

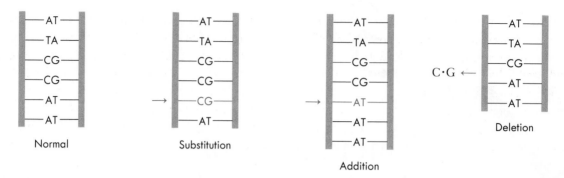

FIGURE **3-1**

The three mechanisms of mutation: substitution, addition, and deletion of a base pair in a DNA strand.

in the sickle α chains, but each sickle β chain differed from the normal wild-type hemoglobin β chain through a specific amino acid substitution (glutamic acid → valine) that had occurred at a unique site (position 6) on the β chain. This discovery hinted that many mutations might represent single base-pair changes, as opposed to more drastic alterations in the base sequence. Probing deeper into the gene-protein relationship in sickle-cell anemia was not, however, possible at that time, because there was no conceivable way to isolate the DNA coding for the respective hemoglobin chain.

The Use of *E. Coli* Leads to the Rapid Development of Fine-Structure Genetics

The great speed at which the basic facts of molecular genetics emerged following the discovery of the double helix was possible only because of a prior decision taken collectively in the mid-1940s by a small group of younger scientists interested in the nature of the gene. This decision was to focus, whenever possible, on genetic experiments with the bacterium *Escherichia coli* (*E. coli*) and its various phages. Though some *E. coli* strains cause disease, other totally nonpathogenic strains (for example, B and K12) had become well-adapted for laboratory use. They grew in culture extremely well, dividing as frequently as once every 20 minutes. They thus provided almost perfect systems for studying the organization of genes in the simplest known cells and also served as ideal vehicles for observing more closely the nature of viruses.

The first rigorous genetic experiments with *E. coli* and its phages were undertaken by Max Delbrück and Salvador Luria. In the early 1940s they observed that some *E. coli* cells mutated to become resistant to specific phages, and they made accurate measurements of the spontaneous mutation rates. Soon afterwards the first phage mutants were isolated by Alfred Hershey; he showed that they recombined genetically. Yet it was not until the structure of DNA was known that the genetic structure of a single phage was thoroughly explored. The decisive experiments were done at Purdue University by Seymour Benzer, who isolated many hundreds of mutations within the *r*II gene of T4 to test whether genetic recombination (crossing over) occurs within the genes themselves. He soon found this to be the case; in fact, he discovered that *most* crossing over takes place within, rather than between, genes.

Shortly afterwards, the same conclusion was reached from genetic studies of *E. coli* itself. Successful crosses between mutant bacteria were first done at Yale University in 1946 by Joshua Lederberg and Edward Tatum. After they mixed together pairs of different *E. coli* mutants, each bearing several different mutations causing nutritional deficiencies, they obtained progeny bacteria that lacked the nutritional requirements of either of their respective parents. Lederberg and Tatum soon established a tentative genetic map, and the stage was set for a growing number of other geneticists to join Lederberg in exploiting the fact that very large numbers of progeny bacteria could be obtained quickly from a single genetic cross. Genetics could thus be studied much more easily with bacteria than with any higher organism. Though most of the early research went to determining the order of the various genes along the single *E. coli* chromosome and establishing the existence of two different sexes, attention later turned to the structure of single bacterial genes. The situation proved to be similar to that in the *r*II gene of bacteriophage T4, with the mutations in each such bacterial gene mapping in a strictly linear order. This was to be expected if the mutable sites were the successive base pairs of the DNA molecules.

The Gene and Its Polypeptide Products Are Colinear

The gene could now be precisely defined as the collection of adjacent nucleotides that specify the amino acid sequences of the cellular polypeptide chains. Simplicity argued that the corresponding nucleotide and amino acid sequences would be colinear, and this hypothesis was soon confirmed by correlation of the relative locations of mutations in a gene with the locations of changes in its polypeptide products. The

best early data were obtained at Stanford University by Charles Yanofsky, who studied mutations in the *E. coli* gene coding for tryptophan synthetase, an enzyme needed to make the amino acid tryptophan. He demonstrated very convincingly that the relative position of each amino acid replacement matched the relative position of its respective mutation along the genetic map (Figure 3-2). The molecular processes underlying colinearity, however, were not at all obvious, because the 20 different amino acids far exceeded the number of different nucleotides in DNA. A one-to-one correspondence between nucleotides and amino acids could not exist. Instead, *groups* of nucleotides must somehow specify (code for) each amino acid.

RNA Carries Information from DNA to the Cytoplasmic Sites of Protein Synthesis

A direct template role for DNA in the ordering of amino acids in proteins was known to be impossible, because almost all DNA is located on the chromosomes in the nucleus, whereas most, if not all, cell

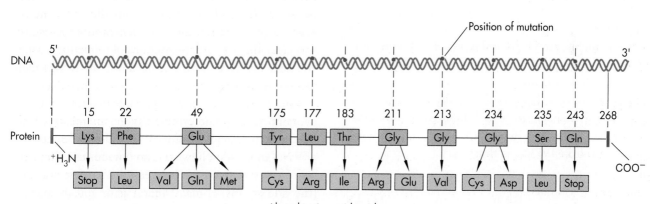

Altered amino acid residues

FIGURE **3-2**

The order of mutations in a gene coding for tryptophan synthetase is the same as the order of amino acid changes in the gene's polypeptide product. The dots on the DNA strand indicate the positions of mutations; the numbers below indicate the positions of the changes in the amino acids. The amino acids that appear in these positions in normal chains are shown in the boxes below the numbers, and the amino acids resulting from the mutations appear in the bottom row of boxes. Some mutations caused synthesis of the protein chain to be terminated (Stop). (Mutations occurring in the same position on the DNA strand in different *E. coli* cells can produce different amino acid changes, as happened here for the 49th, 211th, and 234th amino acids.) Evidence of this kind showed that a gene and its polypeptide product are colinear.

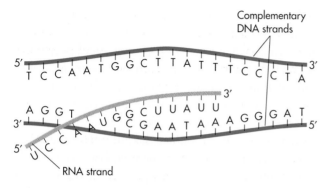

FIGURE **3-3**
An RNA strand is synthesized (transcribed) upon a locally single-stranded region of DNA.

protein synthesis occurs in the cytoplasm. The genetic information of DNA (the nucleotide sequence) thus had to be transferred to an intermediate molecule, which would then move into the cytoplasm, where it would order the amino acids. Speculation that this intermediate molecule was RNA became serious as soon as the double helix was discovered. For one thing, the cytoplasm of cells that made large numbers of proteins always contained large amounts of RNA. Even more importantly, the sugar-phosphate backbones of DNA and RNA were known to be quite similar, and it was easy to imagine the synthesis of single RNA chains upon single-stranded DNA templates to yield unstable hybrid molecules in which one strand was DNA and the other strand was RNA (Figure 3-3). Here it is important to note that the unique base of RNA, uracil (U), is chemically very similar to thymine (T) in that it specifically base-pairs to adenine. The relationship between DNA, RNA, and protein, as conceived in 1953, was thus:

$$DNA \xrightarrow{TRANSCRIPTION} RNA \xrightarrow{TRANSLATION} PROTEIN$$

REPLICATION

in which single DNA chains serve as templates for either complementary DNA molecules (in the process of DNA replication) or complementary RNA molecules (in the process of *transcription*). In turn, the RNA molecules serve as the templates that order the amino acids within the polypeptide chains of proteins during the process of *translation,* so named because the nu-

cleotide language of nucleic acids is translated into the amino acid language of proteins. This scheme—that information flows from DNA to RNA to protein—become known as the central dogma of molecular biology.

How Do Amino Acids Line Up on RNA Templates?

The groups of nucleotides that code for an amino acid are called *codons*. From the beginning of codon research it seemed likely that most, if not all, codons were composed of three adjacent nucleotides. Groups of two nucleotides could be arranged in only 16 different permutations ($4 \times 4 = 16$), 4 too few to code for the 20 different amino acids. Groups of three (AAA, AAC, AAU, etc.), however, could result in 64 independent permutations ($4 \times 4 \times 4 = 64$), many more than are logically needed to specify all the amino acids. So there was speculation about whether the codons might overlap in such a way that given bases would help to specify more than one amino acid. If that were true, then there would be restrictions governing which amino acids could be linked together. The first known amino acid sequences were eagerly scanned by the physicist George Gamow, to see whether some amino acids never occurred next to each other. By 1957 it was clear that no such restrictions of sequence existed and that the successive codons along an RNA chain did not overlap.

Initially it was considered possible that the sequence of amino acids was determined by the manner in which specific amino acids fit into cavities along the surfaces of RNA molecules; however, no obvious complementarity was found to exist between the specific portions (the side groups) of many amino acids and the purine and pyrimidine bases of RNA. The side groups of the amino acids leucine and valine, for example, cannot form any hydrogen bonds, and unless they were somehow modified they would not be expected to be attracted to any RNA template. This dilemma led Francis Crick to propose by early 1955 that many, if not all, amino acids first had to be attached to some form of adapter molecules before they could chemically bind to an RNA template. Testing the adapter hypothesis, however, was impossible until techniques were developed for dissecting biochemically the process by which amino acids become incorporated into growing polypeptides.

Roles of Enzymes and Templates in the Synthesis of Nucleic Acids and Proteins

The initial reaction of geneticists to the double helix was pure delight at seeing how it could function as a template. In a real sense their 50-year quest was over. In contrast, the biochemists, who were then working out how enzymes participated in the synthesis of the nucleotides and amino acids, saw their role as really just beginning. They realized that making phosphodiester and peptide bonds would also require specific enzymes. The discovery of such enzymes would demand finding conditions in which, say, DNA, RNA, or protein is made in extracts of disrupted cells. The first such experiments were difficult, because although most cells contain the enzymes to make proteins and nucleic acids, cells frequently also possess active enzymes that can break these molecules down. In the first successful experiments, only very small amounts of DNA, RNA, and protein were made. For example, test tube–made DNA could be detected only by using radioactively labeled precursors (such as [^{14}C]thymine) to distinguish it from the much larger amounts of unlabeled DNA that preexisted in the cells used to make the "cell-free extracts." However, the possession of an active extract in which, say, DNA was made allowed further experimentation with extracts that had been fractionated into various components, to discover both the exact precursors and the nature of the enzyme or enzymes needed for the assembly process. Such experiments would also, it was hoped, reveal the chemical identities of the templates required in the synthetic process.

By 1960, both DNA and RNA had been successfully synthesized in highly purified cell-free extracts, and the nature of their immediate precursors was firmly established. In both cases, the precursors were nucleoside triphosphates, nucleotides containing three adjacent phosphate groups. Two of the phosphate groups are split off when adjacent nucleotides are linked together, and the energy present in the broken bonds is used to make the phosphodiester links of the sugar-phosphate backbone. The enzymes that directly make the phosphodiester bonds are called *polymerases* (because of the polymeric nature of the nucleic acids). The enzymes that make DNA are called *DNA polymerases,* and those that make RNA are known as *RNA polymerases.*

DNA polymerases catalyze the formation of DNA only in the presence of preexisting DNA templates, and the newly made DNA chains contain sequences complementary to their templates. For RNA polymerase to make RNA, DNA must be present; its template role was likewise shown by finding complementarity between DNA templates and RNA products.

Proteins Are Synthesized from the Amino Terminus to the Carboxyl Terminus

One end of a polypeptide has an amino acid with a free amino group, while the other end bears an amino acid with a free carboxyl group. In 1961 Howard Dintzis at Johns Hopkins University showed that polypeptide chains grow by stepwise addition of single amino acids, starting with the amino-terminal amino acid and finishing with the carboxyl-terminal amino acid. Though all proteins are synthesized beginning with the amino acid methionine, one or several amino-terminal amino acids are frequently cleaved away by *proteolytic* (protein-degrading) enzymes to produce functional polypeptide products that have different amino-terminal amino acids from those of the primary translation products.

Three Forms of RNA Are Involved in Protein Synthesis

Fractionation of active cell-free extracts that had incorporated radioactively labeled amino acids into polypeptides revealed the temporary attachment of the newly made protein to ribosomes, the semispherical 200-Å-diameter cytoplasmic particles that contain

TABLE 3-1
RNA Molecules in *E. coli*

TYPE	RELATIVE AMOUNT (%)	MASS (KILODALTONS)*
Ribosomal RNA (rRNA)	80	1.2×10^3
		5.5×10^2
		3.6×10^1
Transfer RNA (tRNA)	15	2.5×10^1
Messenger RNA (mRNA)	5	Heterogeneous

* The largest and smallest rRNAs are part of the large (50s) subunit; the medium-sized rRNA belongs to the small (30s) subunit.

RNA (Table 3-1). All ribosomes were found to be built from two subunits of unequal size (their molecular weights, or MWs, are 1 and 2 million). The smaller (30S) subunit contains a 16S RNA molecule and 21 proteins, and the larger (50S) subunit two RNA molecules (5S and 23S) and 34 proteins. Until 1960 it was generally believed that *ribosomal RNA (rRNA)* molecules were the templates that ordered the amino acids. But, though the involvement of ribosomes in protein synthesis was indisputable, no plausible hypothesis could be proposed as to how ribosomal RNA chains of fixed sizes could specify polypeptides that showed so much variation in size; some chains contained as few as 50 amino acids, whereas others contained up to 2000.

Then, to everyone's great surprise, it was discovered that none of the ribosomal RNA components had a template role. Instead, the true templates represented a minor fraction—2 to 5 percent—of all cellular RNA. Because they carried the specificity of the genes to the cytoplasm, these templates were named *messenger RNA (mRNA)*. In the cytoplasm, the ribosome moves along the mRNA, so that successive codons are brought into position for ordering their

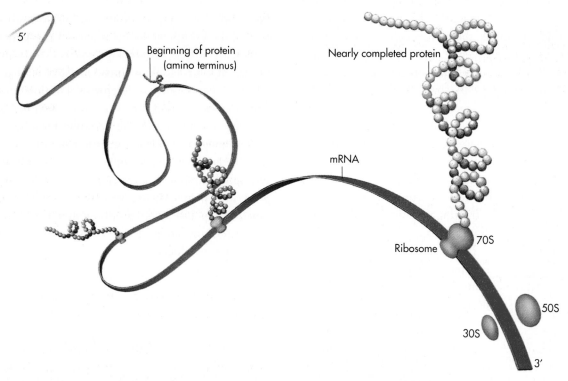

FIGURE 3-4
Messenger RNA carries genetic information from the DNA to the ribosomes, where it is translated into protein. The polypeptide chains are elongated as ribosomes move along the mRNA molecules, with the 5′ ends of the mRNA being translated first.

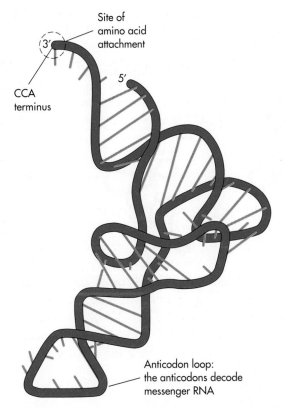

Site of amino acid attachment

3′

CCA terminus

5′

Anticodon loop: the anticodons decode messenger RNA

FIGURE **3-5**
The structure of a transfer RNA molecule. Base pairing within the single-stranded molecule gives it its distinctive shape. The anticodon loop is the portion that decodes messenger RNA. An amino acid attaches to the CCA bases at the 3′ end of the chain.

respective amino acids (Figure 3-4). Ribosomes are thus factories which by themselves have no specificity and which can attach to any mRNA molecule.

Equally important was the discovery that prior to their incorporation into protein, the amino acids are chemically linked to small RNA molecules called *transfer RNA (tRNA)*. The amino acid–tRNA complexes line up next to the codons of mRNA, with the actual recognition and binding being mediated by tRNA components. No contacts exist between the individual amino acids and the mRNA codons. Molecules of tRNA are thus the adapters whose existence had been predicted several years before by Francis Crick.

For each of the 20 different amino acids, a specific enzyme catalyzes its linkage to the 3′ end of its specific tRNA molecule. The binding of tRNA to mRNA is mediated by sets of internal nucleotides on the tRNA that have sequences complementary to their respective codons on the mRNA. These tRNA sequences are called *anticodons* (Figures 3-5 and 3-6).

Genetic Evidence Reveals That Codons Contain Three Bases

An exhaustive genetic study of a large number of phage T4 mutants that contained additions or deletions of single base pairs led Sydney Brenner and Francis Crick in 1961 to the major statement that each codon contains three bases. In Cambridge, England, they made genetic crosses to show that the addition or subtraction of either one or two base pairs invariably led to highly abnormal nonfunctional proteins. In contrast, if three base pairs were either added or subtracted, the resulting proteins frequently were totally active. They concluded, as we now know correctly, that the genetic code is read in stepwise groups of three base pairs. If one or two bases are added or deleted, the *reading frame* is upset, leading to the use of a completely new collection of codons that invariably code for amino acid sequences that make no functional sense. In contrast, when groups of three base pairs are inserted or deleted, the resulting protein, now containing one more or one less amino acid, remains otherwise unchanged and often retains full biological activity.

RNA Chains Are Both Synthesized and Translated in a 5′-to-3′ Direction

Every nucleic acid chain has a direction defined by the orientation of its sugar-phosphate backbone. The end terminating with the 5′ carbon atom is called the *5′ end*, while the end terminating with the 3′ carbon atom is called the *3′ end*. All RNA chains, as well as DNA chains, grow in the *5′-to-3′* direction. Translation

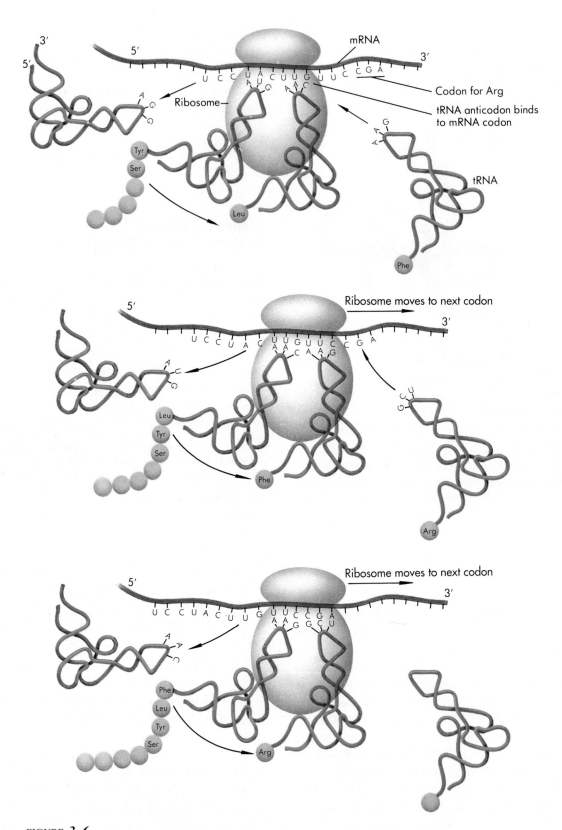

FIGURE **3-6**
At the ribosome, the codons of a messenger RNA molecule base-pair with the anticodons of transfer RNAs, which are charged with amino acids.

occurs in the same direction, and as we will see later (Chapter 4), in bacteria RNA chains can begin to be translated long before their synthesis is complete. The end of a gene at which transcription begins is called the 5′ end, reflecting the 5′-to-3′ order of its mRNA product. Correspondingly, the end of the gene at which transcription ceases is called the 3′ end.

Synthetic mRNA Is Used to Make the Codon Assignments

The realization that ribosomes are by themselves unspecific and become programmed to make specific proteins only by binding mRNA molecules led Marshall Nirenberg and Heinrich Matthaei in 1961 to do a historic experiment. They used as mRNA an enzymatically made regular polynucleotide, poly (U), which consisted of a long, repeated sequence of nucleotides containing the base uracil. When poly (U) was added to cell extracts containing ribosomes depleted of normal mRNA, they observed that only polyphenylalanine was synthesized. UUU thus coded for the amino acid phenylalanine. Soon poly (A) was found to code for strings of lysine residues, while poly (C) yielded polypeptides containing only proline. Over the next several years synthetic polynucleotides containing random mixtures of two or more nucleotides were used to tentatively decode many other codons.

The Genetic Code Is Fully Deciphered by June 1966

Most of the remaining unidentified codons were established when H. Gobind Khorana found ways to make codons by using repeating copolymers (for example, GUGUGU..., AAGAAG..., and GUUGUU...). By 1966 the search for the genetic code was over, and it had been unambiguously shown that (1) all codons contain three successive nucleotides, (2) many amino acids are specified by more than one codon (the so-called *degeneracy of the code*), and (3) 61

TABLE 3-2					
The Genetic Code					

FIRST POSITION (5′ END)	SECOND POSITION				THIRD POSITION (3′ END)
	U	C	A	G	
U	Phe	Ser	Tyr	Cys	U
	Phe	Ser	Tyr	Cys	C
	Leu	Ser	Stop	Stop	A
	Leu	Ser	Stop	Trp	G
C	Leu	Pro	His	Arg	U
	Leu	Pro	His	Arg	C
	Leu	Pro	Gln	Arg	A
	Leu	Pro	Gln	Arg	G
A	Ile	Thr	Asn	Ser	U
	Ile	Thr	Asn	Ser	C
	Ile	Thr	Lys	Arg	A
	Met	Thr	Lys	Arg	G
G	Val	Ala	Asp	Gly	U
	Val	Ala	Asp	Gly	C
	Val	Ala	Glu	Gly	A
	Val	Ala	Glu	Gly	G

Note: Given the position of the bases in a codon, it is possible to find the corresponding amino acid. For example, the codon (5′) AUG (3′) on mRNA specifies methionine, whereas CAU specifies histidine. UAA, UAG, and UGA are termination signals. AUG is part of the initiation signal, and it codes for internal methionines as well. (From L. Stryer, *Biochemistry*, 3d ed., W. H. Freeman, 1988.)

of the 64 possible combinations of the three bases are used to code for specific amino acids (Table 3-2). The three combinations that do not specify any amino acid (UAA, UAG, UGA) were all found to code for stop signals that indicate chain termination.

The finding of stop codons at first created the expectation that specific start codons might also exist, especially since it was becoming more and more certain that all proteins begin with the amino acid methionine. But there is only one methionine codon (AUG), and it codes for internally located methionine as well as initiator methionine. AUGs that are used to start polypeptides are all closely preceded by a purine-rich sequence (for example, AGGA) that may help to position the starting AUG opposite the ribosomal cavity containing the initiating amino acid–tRNA complex.

"Wobble" Frequently Permits Single tRNA Species to Recognize Multiple Codons

Initially it seemed highly probable that anticodons bind codons by means of three A—U and/or G—C hydrogen bonds that are identical to the A—T and G—C bonds that bind the two strands of the double helix together. But soon experimental results began to show that single tRNA species can bind to, say, both UUU and UUC codons. This suggested to Francis Crick that although the first two bases in a codon always pair in a DNA-like fashion, the pairing in the third position is less restrictive, so that nonstandard base pairing (wobble) is allowed for the third bases. Additional types of pairing became possible, opening up the possibility that there need not always be a distinct tRNA species for each of the 61 codons corresponding to amino acids. To date, the most complete data come from yeasts that have been shown, from a combination of genetic and DNA sequence data, to contain approximately 45 tRNA species. Thus, many yeast tRNA molecules have to recognize more than one codon.

Great variation exists in the relative amounts of particular tRNA species that are present in a given cell. In part, this variation reflects differences in the abundance of the amino acids the tRNAs specify. For example, the amino acids methionine and tryptophan occur relatively rarely in most proteins, and comparatively small amounts of their respective tRNAs are present. Moreover, when more than one tRNA form exists for a given amino acid, these different tRNA forms tend not to be present in equal amounts. This suggested that the more numerous tRNAs recognize the more commonly used codons for a given amino acid. This supposition was proved correct when methods for determining the precise nucleotide sequences of genes became available (Chapter 5). The rate at which an mRNA message is translated into its corresponding polypeptide chain thus may be controlled in part by whether it contains codons that are recognized by the more commonly available tRNA forms.

How Universal Is the Genetic Code?

Virtually all the experiments used to decipher the genetic code employed ribosomes and tRNA molecules from *E. coli*. It could thus be asked whether mRNA molecules are always translated into the same amino acid sequences, independent of the source of the translation machinery. At the start, the answer was thought to be yes, for it was hard to imagine how the code could change during the course of evolution. By now the initial expectations that the genetic code for chromosomal DNA would prove to be universal have been rigorously confirmed in a large variety of organisms, ranging from the simplest prokaryotes to the most complex eukaryotes. An interesting exception, however, occurs in the genetic code used by the DNA from mitochondria. Although for many years it was believed that DNA is located only in the nucleus, by the early 1960s it had become clear that the cytoplasmic organelles, the *mitochondria* and the *chloroplasts,* both possess their own unique DNA molecules. Now it is generally believed that mitochondria and chloroplasts represent the descendents of primitive bacterial cells that became symbiotically engulfed by primitive ancestors of the present eukaryotic organisms and increased the host's ATP-generating capacity (Chapter 22). For the most part, the genetic code used by mitochondria is identical with that used by nuclear DNA. However, UGA, a stop codon for nuclear DNA, is read as tryptophan in mitochondria. In addition, mitochondrial AUA is read as methionine, whereas AUA is read as isoleucine in nuclear DNA.

These differences are due to the relatively small number of different tRNAs coded by mitochondrial DNA. Only 22 different tRNA species are present in mitochondria, in contrast to the more than 40 tRNAs that are available for ordinary translation. In many cases, just the first two bases in a codon are actually read, with the base in the third position playing no role in the tRNA selection process.

Average-Sized Genes Contain at Least 1200 Base Pairs

Because all codons were found to contain three base pairs, it was obvious that the number of base pairs in a gene must be at least three times the number of amino acids in its respective polypeptide. An average-sized protein of 400 amino acids was thus thought to require a section of DNA consisting of some 1200 nucleotide pairs. Because this number was found to be much smaller than the number of base pairs in even the smallest DNA molecule, it was concluded that most DNA molecules contain many genes.

Mutations Change the Base Sequence of DNA

Mutations arise as a change in the coding sequence of a gene. *Substitutions* occur when one base is replaced by another. In *transitions,* a purine is substituted for a purine, or a pyrimidine for a pyrimidine. In *transversions,* a purine is substituted for a pyrimidine or vice versa. *Insertions* and *deletions* are the addition and removal of one or more bases, respectively. Different mutations will have differing consequences for the function of the protein. A *nonsense mutation,* resulting from a point mutation that converts a codon to a stop codon, produces premature termination of the polypeptide chain and usually a nonfunctional protein. Because of the redundancy of the genetic code, substitutions may not lead to the incorporation of an incorrect amino acid in the protein, and even when an incorrect amino acid is used (a *missense mutation*), it may have little effect on the function of the protein unless it is in a critical portion of the protein. Deletions and insertions usually have very drastic effects on proteins because they alter the triplet groupings in which the bases are read. Mutations are important for two reasons. They are responsible for inherited disorders and other diseases such as cancer that involve alterations in genes. At the same time, mutations are the source of phenotypic variation on which natural selection acts. The process in which mutation produces variability and natural selection favors any resulting advantageous variants is the driving force of evolution.

Suppressor tRNAs Cause Misreading of the Genetic Code

Before the discovery of the double helix, *suppressor genes* had already been identified. These genes somehow have the potential to nullify the effects of specific mutations in a large variety of different genes. Now it is understood that many suppressor genes act by causing occasional misreadings of the genetic code.

The nonsense mutations can be suppressed by mutant tRNA suppressor genes. The tRNAs coded for by these genes can be charged normally with an amino acid, but because of substitutions in their bases, the anticodons of these tRNAs are now complementary to one of the stop codons. For example, there are mutant genes for tyrosine tRNA that code for the anticodon (3′)AUC(5′) instead of the correct anticodon for tyrosine, (3′)AUG(5′). The mutant anticodon recognizes the stop codon (5′)UAG(3′) instead of the appropriate tyrosine codon (5′)UAC(3′). When a mutant tyrosine tRNA comes across an internal, mutant UAG stop codon, it recognizes the stop codon and inserts a tyrosine into the polypeptide chain, allowing translation to continue. The end result is a full-length, functionally normal protein (Figure 3-7). Cells containing such a mutant tRNA suppressor gene can sur-

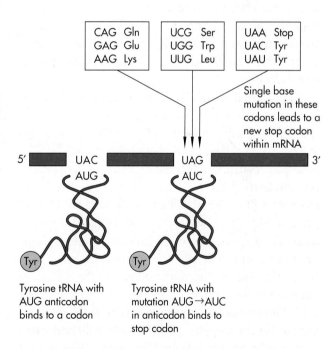

FIGURE **3-7**
Suppressor tRNAs. Mutations to U, A, or G in the first, second, or third bases, respectively, of the codons listed in the boxes will lead to a UAG stop codon (nonsense mutation) within the coding region of an mRNA. One of the codons for tyrosine is (5′)UAC(3′), and the corresponding anticodon is (3′)AUG(5′). If there is a mutation in the anticodon of the tyrosine tRNA so that it becomes (3′)AUC(5′), this mutant tRNA will bind to the UAG stop codon and insert a tyrosine into the growing peptide chain. Thus synthesis of the protein will continue through the stop codon. If a tyrosine at this position restores protein function, then the nonsense mutation is suppressed.

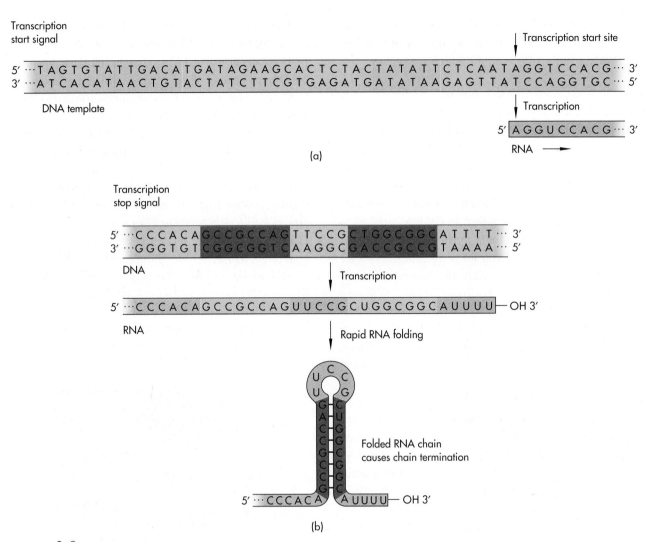

Transcription start signal

Transcription start site

5′ ···TAGTGTATTGACATGATAGAAGCACTCTACTATATTCTCAATAGGTCCACG··· 3′
3′ ···ATCACATAACTGTACTATCTTCGTGAGATGATATAAGAGTTATCCAGGTGC··· 5′

DNA template

Transcription

5′ AGGUCCACG··· 3′

RNA ⟶

(a)

Transcription stop signal

5′ ···CCCACAGCCGCCAGTTCCGCTGGCGGCATTTT··· 3′
3′ ···GGGTGTCGGCGGTCAAGGCGACCGCCGTAAAA··· 5′

DNA

Transcription

5′ ···CCCACAGCCGCCAGUUCCGCUGGCGGCAUUUU— OH 3′

RNA

Rapid RNA folding

Folded RNA chain causes chain termination

5′ ···CCCACA   AUUUU— OH 3′

(b)

FIGURE **3-8**
(a) *E. coli* RNA polymerase recognizes two blocks of sequences, one at approximately −35 and another at about −10 from the start of transcription. (b) The RNA polymerase stops after transcribing a run of U residues that follows a palindrome. The newly transcribed RNA forms a stem-loop structure that probably constitutes a termination signal for RNA polymerase.

vive only if there are multiple copies of the gene for that tRNA. In this way, there remain sufficient numbers of normal tRNA molecules to insert the correct amino acid at the appropriate codons. In the case of tyrosine, there are multiple copies of the tyrosine tRNA gene coding for the (3′)AUG(5′) anticodon in *E. coli* cells, so that there are many functional tyrosine tRNA molecules in addition to the suppressor tyrosine tRNA molecules.

UGA-suppressor tRNAs occasionally insert amino acids at the normal sites of chain termination, leading to many oversized polypeptide chains. We can assume

that such events are deleterious to their respective cells, which may explain why nonsense-suppressor strains generally grow much less well than their normal equivalents.

The Signals for Starting and Stopping the Synthesis of Specific RNA Molecules Are Encoded Within DNA Sequences

The genetic information within a DNA molecule usually serves as the template for the synthesis of a large number of shorter RNA molecules, most of which in turn serve as templates for the synthesis of specific polypeptide chains. In bacteria, specific nucleotide segments (often called *promoters*) are recognized by RNA polymerase molecules that start RNA synthesis (Figure 3-8a). After transcription of a functional RNA chain is finished, a second class of signals leads to the termination of RNA synthesis and the detachment of RNA polymerase molecules from their respective DNA templates. How termination is brought about is still being worked out. RNA chains tend to end with several U residues; just before the site of termination, a nucleotide sequence capable of forming a hairpin loop is usually found (Figure 3-8b). At many termination sites, specific RNA-chain-terminating proteins play a role.

In the transcription-initiation step, the two chains of the double helix come apart, with only one of the two strands at any start site being copied into its RNA complement. The hybrid DNA-RNA double-helical sections generated as transcription proceeds have only a fleeting existence, with the newly made RNA segments quickly peeling off from the transcription complexes, allowing the just-transcribed sections to reassume their native double-helical states. In bacteria, which have no nucleus, the 5' ends of nascent mRNA chains attach to ribosomes long before their respective chains have been completely synthesized, and as will be seen (page 57), the process of transcription and translation can be closely coupled. Regulatory elements for eukaryotic genes are found both close to and distant from the start site.

Increasingly Accurate Systems Are Developed for the *in Vitro* Translation of Exogenously Added mRNAs

It is now more than 20 years since the first exogenously added mRNA molecules were successfully used to

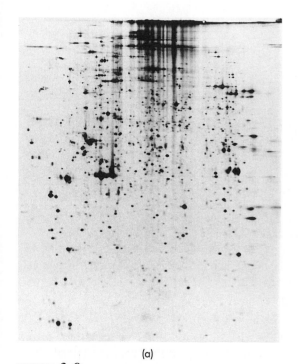
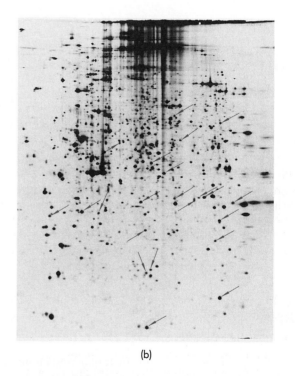

(a)   (b)

FIGURE **3-9**

Two-dimensional gel electrophoresis of labeled proteins produced by growing tissue culture cells in medium containing the amino acid methionine labeled with ^{35}S. A polyacrylamide gel is used to separate the proteins, first by their electrical charge (left to right), and then by their mass (top to bottom). Each spot in the figures represents at least one radioactively labeled protein. (a) Control cells. (b) Cells stimulated to divide by addition of fresh medium. Cells enter mitosis about 20 hours after stimulation. They begin synthesizing increased amounts of protein, or new proteins, in preparation for mitosis; and these proteins can be detected by adding the radioactive amino acid 15 hours after stimulation. The arrows mark some of the proteins that differ between control and stimulated cells. (Courtesy of J. Garrels.)

program the translation of functional protein molecules. In that time, increasingly reliable systems have been developed for studying the in vitro translation of mRNA molecules. At first only polypeptide chains of modest size (from 150 to 300 amino acids) could be made because the early cell-free systems all contained significant amounts of contaminating ribonucleases that degraded the added mRNA templates. Since then, particularly with the development of virtually nuclease-free extracts from reticulocytes (the hemoglobin-synthesizing precursors of red blood cells), it has become possible to use mRNA preparations from many cells to routinely program the synthesis of hundreds of different proteins, including the largest of the cellular polypeptides, which can contain thousands of amino acids. Proteins can now be analyzed by two-dimensional gel electrophoresis (Figure 3-9). Today the presence of a given specific functional mRNA molecule can be routinely tested for by determining whether its polypeptide product is made in a cell-free protein synthesizing extract.

Our ability to translate mRNA preparations so easily into their respective polypeptides does not mean in any sense that we understand in detail at the molecular level how proteins are made. The exact functions of the more than 60 different proteins used in constructing ribosomes remain almost totally unknown. Likewise, the functions of the two ribosomal RNA (rRNA) components of each ribosome remain as unclear as they were in 1960, when these components were found to be structural molecules as opposed to genetic information–carrying molecules.

Reading List

General

The Genetic Code. Cold Spring Harbor Symp. Quant. Biol., vol. 31. Cold Spring Harbor Laboratory, Cold Spring Harbor, N. Y., 1967.

Nomura, M., A. Tissières, and P. Lengyel, eds. *Ribosomes.* Cold Spring Harbor Laboratory, Cold Spring Harbor, N. Y., 1974.

Losick, R., and M. Chamberlin, eds. *RNA Polymerase.* Cold Spring Harbor Laboratory, Cold Spring Harbor, N. Y., 1976.

Original Research Papers

FINE-STRUCTURE GENETICS

Luria, S. E., and M. Delbrück. "Mutations of bacteria from virus sensitivity to virus resistance." *Genetics,* 28: 491–511 (1943).

Hershey, A. D. "Spontaneous mutations in bacteria viruses." *Cold Spring Harbor Symp. Quant. Biol.,* 11: 67–77 (1947).

Lederberg, J., and E. L. Tatum. "Novel genotypes in mixed cultures of biochemical mutants of bacteria." *Cold Spring Harbor Symp. Quant. Biol.,* 11: 113–114 (1947).

Benzer, S. "Fine structure of a genetic region in bacteriophage." *Proc. Natl. Acad. Sci. USA,* 41: 344–354 (1955).

Ingram, V. M. "Gene mutations in human hemoglobin: the chemical difference between normal and sickle cell hemoglobin." *Nature,* 180: 326–328 (1957).

Benzer, S. "On the topology of the genetic fine structure." *Proc. Nat. Acad. Sci. USA,* 45: 1607–1620 (1959).

Yanofsky, C., B. C. Carlton, J. R. Guest, D. R. Helinski, and U. Henning. "On the colinearity of gene structure and protein structure." *Proc. Natl. Acad. Sci. USA,* 51: 266–272 (1964).

DISCOVERY OF DNA-DEPENDENT RNA SYNTHESIS

Hurwitz, J., A. Bresler, and R. Diringer. "The enzymatic incorporation of ribonucleotides into polyribonucleotides and the effect of DNA." *Biochem. Biophys. Res. Comm.,* 3: 15–19 (1960).

Stevens, A. "Incorporation of the adenine ribonucleotide into RNA by cell fractions from *E. coli* B." *Biochem. Biophys. Res. Comm.,* 3: 92–96 (1960).

Weiss, S. B. "Enzymatic incorporation of ribonucleotide triphosphates into the interpolynucleotide linkages of ribonucleic acid." *Proc. Natl. Acad. Sci. USA,* 46: 1020–1030 (1960).

MECHANISMS OF PROTEIN SYNTHESIS

Zamecnik, P. C., and E. B. Keller. "Relationship between phosphate energy donors and incorporation of labeled amino acids into proteins." *J. Biol. Chem.,* 209: 337–354 (1954).

Hoagland, M. B., E. B. Keller, and P. C. Zamecnik. "Enzymatic carboxyl activation of amino acids." *J. Biol. Chem.,* 218: 345–358 (1956).

Crick, F. H. C. "On protein synthesis. Biological replication of macromolecules." *Symp. Soc. Exp. Biol.,* 12: 138–163 (1958).

Hoagland, M. B., M. L. Stephenson, J. F. Scott, L. I. Hecht, and P. C. Zamecnik. "A soluble ribonucleic acid intermediate in protein synthesis." *J. Biol. Chem.,* 231: 241–257 (1958).

Brenner, S., F. Jacob, and M. Meselson. "An unstable intermediate carrying information from genes to ribosomes for protein synthesis." *Nature,* 190: 576–581 (1961).

Dintzis, H. M. "Assembly of the peptide chain of hemoglobin." *Proc. Natl. Acad. Sci. USA,* 47: 247–261 (1961).

Gros, F., H. Hiatt, W. Gilbert, C. G. Kurland, R. W. Risebrough, and J. D. Watson. "Unstable ribonucleic acid revealed by pulse labelling of *Escherichia coli.*" *Nature,* 190: 581–585 (1961).

Jacob, F., and J. Monod. "Genetic regulatory mechanisms in the synthesis of proteins." *J. Mol. Biol.,* 3: 318–356 (1961).

Kim, S. H., F. L. Suddath, F. L. Quigley, A. McPherson, L. Sussman, A. H. J. Wang, N. C. Seeman, and A. Rich. "Three-dimensional tertiary structure of yeast phenylalanine transfer RNA." *Science,* 185: 435–440 (1974).

Robertus, J. D., J. E. Ladner, J. T. Finch, D. Rhodes, R. S. Brown, B. F. C. Clark, and A. Klug. "Structure of yeast phenylalanine tRNA at 3Å resolution." *Nature,* 250: 546–551 (1974).

THE GENETIC CODE

Crick, F. H. C., L. Barnett, S. Brenner, and R. J. Watts-Tobin. "General nature of the genetic code for proteins." *Nature,* 192: 1227–1232 (1961).

Nirenberg, M. W., and J. H. Matthaei. "The dependence of cell-free protein synthesis in *E. coli* upon naturally occurring or synthetic polyribonucleotides." *Proc. Natl. Acad. Sci. USA,* 47: 1588–1602 (1961).

Leder, P., and M. W. Nirenberg. "RNA code words and protein synthesis II: nucleotide sequence of a valine RNA code word." *Proc. Natl. Acad. Sci. USA,* 52: 420–427 (1964).

Nishimura, S., D. S. Jones, and H. G. Khorana. "The in vitro synthesis of a copolypeptide containing two amino acids in alternating sequence dependent upon a DNA-like polymer containing two nucleotides in alternating sequence." *J. Mol. Biol.,* 13: 302–324 (1965).

Crick, F. H. C. "Codon-anticodon pairing: the wobble hypothesis." *J. Mol. Biol.,* 19: 548–555 (1966).

Ikemura, T. "Correlation between the abundance of *Escherichia coli* transfer RNAs and the occurrence of the respective codons in its protein genes." *J. Mol. Biol.,* 146: 1–21 (1981).

MITOCHONDRIAL GENES

Barrell, B. G., S. Anderson, A. T. Bankier, M. H. L. deBruijn, E. Chen, A. R. Coulson, J. Drouin, I. C. Eperon, D. P. Nierlich, B. A. Roe, F. Sanger, P. H. Schreier, A. J. H. Smith, R. Staden, and I. G. Young. "Different pattern of codon recognition by mammalian mitochondrial tRNAs." *Proc. Natl. Acad. Sci. USA,* 77: 3164–3166 (1980).

Bonitz, S. G., R. Berlani, G. Coruzzi, M. Li, G. Macino, F. G. Nobrega, M. P. Nobrega, B. E. Thalenfeld, and A. Tzagoloff. "Codon recognition rules in yeast mitochondria." *Proc. Natl. Acad. Sci. USA,* 77: 3167–3170 (1980).

Anderson, S., A. T. Bankier, B. G. Barrell, M. H. L. deBruijn, A. R. Coulson, J. Drouin, I. C. Eperon, D. P. Nierlich, B. A. Roe, F. Sanger, P. H. Schreier, A. J. H. Smith, R. Staden, and I. G. Young. "Sequence and organization of the human mitochondrial genome." *Nature,* 290: 457–465 (1981).

SUPPRESSOR GENES AND THEIR
SUPPRESSION BY MUTANT tRNAS

Capecchi, M. R., and G. Gussin. "Suppression in vitro: identification of serine-sRNA as a 'nonsense' suppressor." *Science,* 149: 417–422 (1965).

Engelhardt, D. L., R. Webster, R. Wilhelm, and N. Zinder. "In vitro studies on the mechanism of suppression of a nonsense mutation." *Proc. Natl. Acad. Sci. USA,* 54: 1791–1797 (1965).

RNA CHAIN INITIATION AND TERMINATION

Melnikova, A. F., R. Beabealashvilli, and A. D. Mirzabekov. "A study of unwinding of DNA and shielding of the DNA grooves by RNA polymerase by using methylation with dimethylsulphate." *Eur. J. Biochem.,* 83: 301–309 (1978).

Farnham, P. J., and T. Platt. "A model for transcription termination suggested by studies on the *trp* attenuator in vitro using base analogs." *Cell*, 20: 739–48 (1980).

Farnham, P. J., and T. Platt. "Rho-independent termination: dyad symmetry in DNA causes RNA polymerase to pause during transcription in vitro." *Nuc. Acids Res.*, 9: 563–577 (1981).

Birchmeier, C., R. Grosschedl, and M. L. Birnstiel. "Generation of authentic 3' termini of an H2A mRNA in vivo is dependent on a short inverted DNA repeat and on spacer sequences." *Cell*, 28: 739–745 (1982).

Gamper, H. B., and J. E. Hearst. "Size of the unwound region of DNA in *Escherichia coli* RNA polymerase and calf thymus RNA polymerase II ternary complexes." *Cold Spring Harbor Symp. Quant. Biol.*, 47: 447–453 (1983).

IN VITRO SYSTEMS FOR PROTEIN SYNTHESIS

Roberts, B. E., and B. M. Patterson. "Efficient translation of tobacco mosaic virus RNA and rabbit globin 9S RNA in a cell-free system from commercial wheat germ." *Proc. Natl. Acad. Sci. USA*, 70: 2330–2334 (1973).

Pelham, H. R. B., and R. J. Jackson. "An efficient mRNA-dependent translation system from reticulocytes lysates." *Eur. J. Biochem.*, 67: 247–256 (1976).

4

The Genetic Elements That Control Gene Expression

E ven before the basic outline of the genetic code became established, it was obvious that intricate molecular mechanisms must exist in cells to control the numbers of their respective proteins. Within *E. coli,* for example, the relative amounts of the different proteins vary enormously (from less than .01 percent to about 2 percent of the total) depending on their function, even though each protein product is coded by a single gene along the *E. coli* chromosome. A priori we can imagine two ways that the cell might achieve this differential synthesis. The first way would be by the evolution of molecular signals that control the rates at which specific mRNA molecules are transcribed off their DNA templates (*transcriptional control*). The second way would involve molecular devices for controlling the rate at which mRNA molecules, once synthesized, are translated into their polypeptide products (*translational control*). It makes sense for a cell not to make more mRNA molecules than it needs, and most initial attention by molecular biologists focused on whether transcriptional control existed. Here genetic analysis of the so-called induced enzymes of bacteria presented the first key insight, from which has emerged definitive understanding at the molecular level about how mRNA synthesis is regulated.

Repressors Control Inducible Enzyme Synthesis

Bacteria generally exist in environments that change rapidly, and some of the enzymes that they may need at one moment may be useless or even counterproductive soon afterwards. Similarly, they may suddenly require an enzyme whose presence was previously unnecessary. One way for bacteria to meet these challenges would be for them to carry on simultaneously the synthesis of all such enzymes, whether or not the substrates for those enzymes were present. This would be wasteful, however; and in fact bacteria do not operate this way. Instead, many bacterial genes are constructed to function at highly variable rates, so that they make mRNA molecules at appreciable levels only when their genes receive signals from outside to go into action.

The compounds that transmit these signals are called *inducers*. For example, *E. coli* cells normally make the enzyme β-galactosidase (β-gal) at high rates only when its inducer, lactose, is present. (For lactose to function as a food source, it must be cleaved into the simpler sugars glucose and galactose; β-galactosidase is the enzyme that catalyzes this splitting.) The presence of lactose greatly increases the rate at which RNA polymerase can bind to the beginning of the gene that codes for β-galactosidase and initiate the synthesis of the respective mRNA. In turn, the higher amounts of β-galactosidase mRNA lead to correspondingly more β-galactosidase.

Genetic analysis proved crucial to working out the molecular details of this adaptive phenomenon, with the essential clues emerging from the study of mutants that were unable to vary the amount of β-galactosidase. The key findings were that whether lactose was present or absent, some mutants made maximum amounts of enzyme while other mutants produced only traces. Such results led Jacques Monod and François Jacob of the Institut Pasteur in Paris to postulate (1) that a

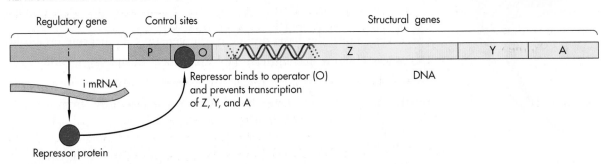

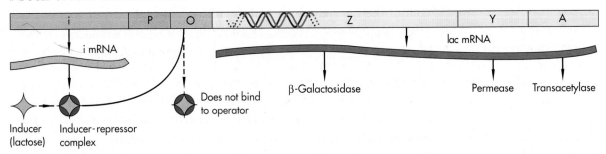

FIGURE **4-1**

Repressors and inducers control the functioning of the genes belonging to the lactose (lac) operon. The regulatory gene (i) codes for the lactose repressor. The P segment of the DNA chain is the "promoter" and is discussed in the next figure.

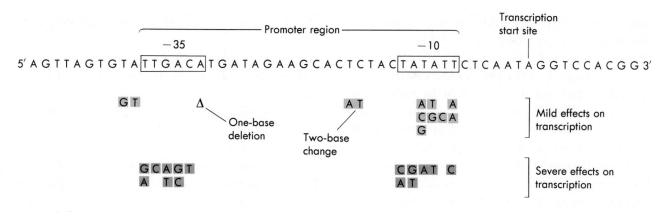

FIGURE **4-2**
Specific DNA sequences are important for efficient transcription of *E. coli* genes by RNA polymerase. The boxed sequences approximately 35 and 10 nucleotides before the transcription start site are highly conserved in all *E. coli* promoters. Mutations in these regions have mild (green) and severe (orange) effects on transcription. The mutations may be changes of single nucleotides or pairs of nucleotides, or a deletion (Δ) may occur.

specific repressor molecule exists that binds near the beginning of the β-galactosidase gene at a specific site called the *operator* and that, by binding to the operator site on the DNA, sterically prevents RNA polymerase from commencing synthesis of β-gal mRNA and (2) that lactose acts as an inducer which, by binding to the repressor, prevents the repressor from binding to the operator. In the presence of lactose, the repressor is inactivated and the mRNA is made. Upon removal of lactose, the repressor regains its ability to bind to the operator DNA and switch off the lactose gene (Figure 4-1).

Bacterial Genes with Related Functions Are Organized into Operons

The *E. coli* gene for β-galactosidase is located adjacent to two additional genes involved in lactose metabolism. One gene codes for lactose permease, a protein that facilitates the specific entry of lactose into bacteria, while the second codes for thiogalactosidase trans-acetylase, an enzyme that may help to remove lactoselike compounds that β-galactosidase cannot split into useful metabolites. The same mRNA that codes for β-galactosidase codes for the permease and the acetylase; thus when lactose is added to *E. coli* cells, the relative amounts of all three proteins rise coor-

dinately. The collections of adjacent genes that are transcribed into single mRNA molecules, together with their control region, are called *operons*. Some operons are large; for example, the 11 proteins involved in the synthesis of the amino acid histidine are all translated off one extremely large mRNA molecule containing over 10,000 nucleotides.

Promoters Are the Start Signals for RNA Synthesis

The site where RNA polymerase binds to the beginning of an operon is called the *promoter*. The maximum rate of transcription from a particular segment of DNA depends on the sequence of bases in its promoter, and the frequency of the initiation of transcription can vary by a factor of at least 1,000. Two highly conserved separate sets of nucleotide blocks make up the *E. coli* promoter (Figure 4-2). One block is located about 10 nucleotides upstream of the mRNA start site, and the other is about 25 nucleotides further upstream (the "stream" flows in the direction of transcription). These blocks are designated −10 and −35 relative to the mRNA start site at +1. The initial step in *E. coli* transcription is believed to be the recognition and binding of an RNA polymerase molecule to the −35 region. Subsequently, the −10 region is thought to *melt* (open up) into its component single strands, allowing transcription to begin at the $+1$ position. Both blocks were initially identified by the existence of point mutations that blocked RNA synthesis. These mutations affect only the synthesis of the mRNA molecule immediately downstream, so promoters are examples of *cis-acting* control elements. In contrast, repressors are not limited in their binding to the DNA molecule that carries their genetic information, so they have been called *trans-acting* control elements.

The operator sequence to which a repressor binds is always close to and may partially overlap the promoter for the operon being controlled (Figure 4-1). The binding of a repressor physically blocks the binding of RNA polymerase to the promoter. The control of a promoter by a repressor is thus an example of negative control. Promoters can also be under the control of positive *effectors* that increase the rate at which mRNA chains are made. Positive control elements most likely act by helping RNA polymerase bind to the promoter and separate the DNA strands in the promoter, facilitating initiation of transcription.

Repressor Molecules Are Normally Made at Constant Rates

Each repressor is coded by a specific gene. The gene for the lactose repressor lies immediately in front of the operon. In other cases, however, the repressor gene is widely separated from the operon genes on which it acts. The rate at which repressors are made is normally unchanging; such invariant synthesis is known as *constitutive synthesis*. The exact rate of this constitutive synthesis is a function of the structure of the promoter of the repressor gene. Normally the promoters of repressor genes function at very low rates, leading to the presence of only a few repressor mRNA molecules in the average cell. However, there exist promoter mutants that allow much higher rates of repressor mRNA synthesis and, correspondingly, much higher numbers of repressor molecules per cell. Even in the presence of high levels of inducer, such mutant cells make smaller-than-usual amounts of the induced proteins.

Repressors Are Isolated and Identified

Because the genetic studies leading to the postulation of repressors were so complete, the role of repressors as key bacterial control elements seemed almost inescapable. Final proof, though, had to await the development of biochemical procedures by which individual repressors could be isolated, chemically identified, and shown to bind specifically to their respective operators. These steps depended on the development of genetic techniques for increasing the number of repressor molecules per cell beyond the few copies normally present. So long as repressors were available only in the amounts that occurred in nature, there was no effective way to isolate them.

Two tricks were used in conquering the lactose repressor. The first was the genetic manipulation of the *E. coli* genome. The beginning of the lac operon and the gene coding for the lac repressor were attached to the phage λ chromosome. When such phages multiply in *E. coli*, several hundred copies of the lac repressor gene are produced, as well as correspondingly large numbers of lac-repressor mRNA molecules. In this way the amounts of repressor per cell were amplified about 10-fold. Still further enrichment came from constructing the β-lac strains with mutant promoters, which overproduced lac-repressor mRNA by another factor of 10. The resulting amounts of repressor became sufficient to allow Walter Gilbert and Benno Müller-Hill to demonstrate at Harvard University in 1966 that the lactose repressor is a protein of MW 38,600 and that it has two specific binding sites—one for lactoselike compounds, the other for DNA containing the lactose operator. Virtually simultaneously, Mark Ptashne, also at Harvard, isolated from phage λ the repressor that controls the rates at which several classes of λ-specific mRNA are made. This repressor is a 26,000-MW polypeptide chain, and it likewise binds only to its specific operator.

Synthesis of the classical λ repressor is itself under the control of a second λ-specific repressorlike protein called *cro*. Cro is a relatively small molecule composed of two identical polypeptide chains with 66 amino acids each. It acts by binding to the promoter of the gene coding for the λ repressor and thus turning off the synthesis of the repressor RNA.

Initial x-ray crystallographic analyses of cro, lambda repressor, *E. coli* CAP (*catabolite activator protein;* discussed in the next section), and the repressor of phage 434 led to models for their binding to DNA. The proteins bind to DNA as dimers. The region of the protein that interacts with the DNA has been localized to a particular α helix that binds in the major groove of the DNA helix. This model has been confirmed by x-ray analysis of the complexes formed between these repressors and their operator sequences carried by synthetic oligonucleotides.

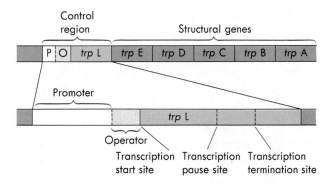

FIGURE **4-3**
The *E. coli* tryptophan (*trp*) operon. Control elements at the beginning of the operon are indicated.

Positive Regulation of Gene Transcription Also Occurs

More recently, several positive regulators of RNA polymerase binding, and hence gene expression, have been isolated from *E. coli* and shown to be proteins. The best-understood positive regulator protein signals to the appropriate genes that glucose is not available as a food source. When glucose is absent, the amounts of the intracellular regulator *cyclic AMP* (*cAMP*) build up. This cAMP then binds to the DNA-binding protein CAP. The resulting cAMP-CAP complexes, by binding to the respective promoters, help to activate operons whose enzymes can break down alternative sugars such as lactose and galactose.

In bacteria, the regulation of DNA functioning is thus controlled in part by the binding of specific regulatory proteins to control sequences situated at the beginnings of their various genes (operons).

Attenuation Is Another Form of Regulation

The expression of many bacterial operons involved in amino acid biosynthesis is also influenced by a process called *attenuation*. This phenomenon was discovered through the elegant experiments by Charles Yanofsky and his coworkers on the tryptophan (*trp*) operon of *E. coli*. This by now exhaustively studied operon consists of a transcriptional control region and five structural genes that encode the enzymes involved

in the last steps of tryptophan biosynthesis (Figure 4-3). The rate at which transcription of *trp* mRNA begins is controlled by a tryptophan-activated repressor molecule that can block the access of RNA polymerase to the *trp* promoter. Once started, however, transcription does not necessarily extend to the end of the operon to produce the very long full-length *trp* mRNA. Instead, incomplete transcription (attenuation) frequently occurs to produce a relatively short 162-base mRNA molecule that codes for a correspondingly small *trp* L (leader) protein. Although the function (if any) of the leader protein has not been discovered, it has the very interesting property of being rich in the amino acid tryptophan, whose sole codon is UGG.

Whether attenuation takes place depends on the exact folding pattern of the nascent mRNA between the transcribing RNA polymerase and the ribosome translating the *trp* L region. The *trp* L mRNA contains stretches of sequence that can base-pair with each other, forming either of two possible hairpin-like structures (Figure 4-4). When tryptophan is plentiful, the ribosome translates freely through the leader, and the nascent mRNA folds into a structure that signals RNA polymerase to terminate transcription. When tryptophan is scarce, there are few activated Trp-tRNAs, and the ribosome stalls at the UGG codons in the leader sequence. Now the mRNA folds into an alternative structure (Figure 4-4). This fold is not recognized by RNA polymerase, which proceeds to

HIGH TRYPTOPHAN LEVEL   LOW TRYPTOPHAN LEVEL

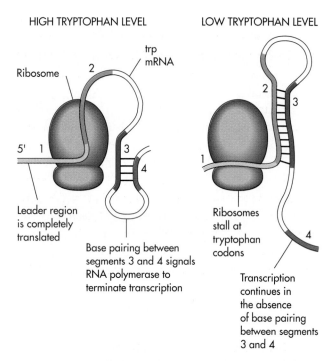

FIGURE **4-4**

A model for attenuation at the *trp* operon. The 5' end of the *trp* operon mRNA, the leader region (region 1), is rich in tryptophan codons. When tryptophan is available, normal translation of this leader sequence occurs. As this happens, the *trp* mRNA forms a stem-loop structure (regions 3 and 4) that apparently does not allow RNA polymerase to continue the transcription of the remainder of the *trp* operon. Note that this "attenuated" state is the *normal* situation; it is only when the tryptophan level drops that the attenuation is relieved. This is believed to occur when a ribosome is stalled trying to translate the leader sequence; the resulting formation of a different stem-loop structure in the next segment of *trp* mRNA (regions 2 and 3) allows the RNA polymerase to continue with the transcription of the remainder of the *trp* operon.

the end of the operon. Thus, under normal nutritional conditions, most *trp* operon transcripts are "attenuated," or terminated early. It is only when the tryptophan level drops that full-length transcripts of the operon are made efficiently. Tryptophan starvation therefore increases expression of the *trp* operon in two ways: by removing the *trp*-activated repressor (thus increasing transcription by a factor of 70) and by relieving the normal attenuation (increasing transcription another 8- to 10-fold). Overall, then, the expression of the operon can be increased approximately 600-fold.

Translational Control Is the Second Means of Controlling Protein Synthesis

The expression of proteins in *E. coli* is also controlled at the level of translation. Efficient initiation of translation depends on the existence of a purine-rich group of six to eight nucleotides just upstream from the AUG initiation codon. The existence of this *ribosome-binding site* was first pointed out in 1974 by John Shine and Larry Dalgarno in Canberra, Australia. They noted its close complementarity in sequence to the 3' end of the 16S rRNA molecules found in bacterial ribosomes. Apparently the initial positioning of mRNA upon the smaller ribosomal subunit requires base pairing between the 16S rRNA chain and the ribosome-binding (*Shine-Dalgarno*) sequences. In general, the most efficiently translated mRNAs have their ribosome-binding sequences centered eight nucleotides upstream from the initiation codon. Mutations in this region, including mutations that place this sequence closer to or farther away from the AUG start codon, can greatly lower the translational efficiency of their respective mRNAs. It is important to note that possession of an appropriately sited Shine-Dalgarno sequence does not guarantee initiation of protein synthesis. Many such sequences are effectively buried in the loop-structures and have no way of interacting with 16S rRNA.

Recent evidence shows that the efficiency with which given Shine-Dalgarno sequences work can be modulated by proteins that bind to them and block their availability. The best-understood example involves the ribosomal proteins (*r-proteins*) of *E. coli* (Figure 4-5). When the rate of r-protein synthesis exceeds the rate at which rRNA is made, free r-proteins accumulate and certain "key" ones bind to the Shine-Dalgarno sequences on the r-protein mRNA molecules. In this way ribosomal proteins are not synthesized faster than they can be used in making ribosomes.

The genes for different sets of r-proteins are contained on a number of different operons in the *E. coli* genome. Each operon encodes its own "key" protein, which inhibits the expression of the entire operon. Key r-proteins also bind to rRNA early during ribosome assembly. The sequences on rRNA that bind these proteins and the sequences on the r-protein mRNA that interact with key proteins are quite similar (Figure 4-6). Translational control thus represents

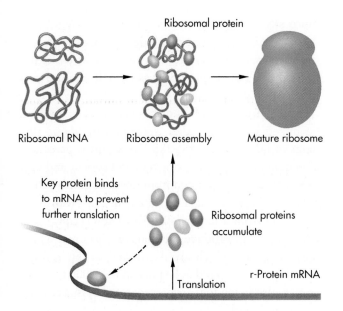

FIGURE **4-5**

Ribosomal protein levels control the translation of ribosomal protein mRNAs. When the rate of r-protein synthesis exceeds the rate of rRNA synthesis, free r-proteins accumulate. Some of them bind to the Shine-Dalgarno sequences on the r-protein mRNAs and prevent further translation. This mechanism ensures that r-proteins are not synthesized faster than they can be used in making ribosomes.

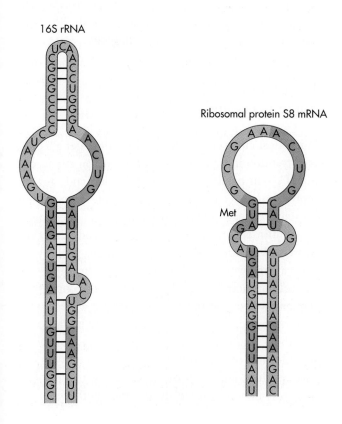

competition between rRNA and r-protein mRNA for the binding of these key proteins. When r-protein mRNAs are rendered untranslatable by the binding of key proteins, they are degraded more rapidly than usual.

Once translation has started, its rate is determined by the availability of the various tRNA species corresponding to the specific codons employed in the mRNA molecules. As discussed in Chapter 3, different tRNA molecules are present in quite different amounts, with those present in larger quantities generally corresponding to more commonly used codons. Messages having a high proportion of codons recognized by rare tRNA molecules are thus translated more slowly than those containing codons recognized by abundant tRNAs.

Probing Gene Regulation in Higher Plants and Animals Presented Early Difficulties

By the late 1960s the question was asked whether the genes of higher cells are also regulated by specific DNA-binding proteins. In particular, would such proteins be the key to understanding the mysteries of embryology, through which fertilized eggs divide and differentiate, eventually giving rise to the highly specific cell types that make up tissues and organs? The only way to proceed was to study single animal or plant cells growing in culture, as opposed to studying whole organisms. Virtually all serious embryologists were hoping that model cell culture systems in which the functions of single genes could be followed during differentiation would soon be developed. Merely watching proteins come and go during differentiation, however, would not by itself be a major step forward. We wanted to discover the signals that would turn a given gene on or off and to determine how so many proteins can have their expression so exquisitely coordinated.

FIGURE **4-6**

Hairpin structures of binding sites on 16S rRNA and r-protein mRNA for the ribosomal protein S8. Regions of homology are indicated in orange.

The only signals already identified in higher organisms were certain steroid hormones that had recently been shown to bind to specific receptor molecules in the cytoplasm. These hormone-receptor complexes then moved to the nucleus, where they bound to chromosomes and somehow turned the synthesis of many specific proteins on or off. All attempts to pursue this phenomenon at the DNA level, however, came to naught. There was no way of finding the particular sections of DNA to which the hormone-receptor complexes were binding and of distinguishing specific binding from nonspecific binding. Moreover, although mutant cells were found in which hormones did not function normally, there was no simple way to map the relevant genes. We could not—and still cannot—make conventional genetic crosses between higher cells growing in culture. As a result, even if operators and repressors existed in higher cells, the possibility of genetically demonstrating them was slim, if not nonexistent. And so long as it remained impossible to isolate from the large chromosomes of higher organisms the specific DNA segments that coded for particular proteins, direct tests for the presence of specific DNA-binding proteins were also out of the question. As we shall see later, the great importance of recombinant DNA techniques has stemmed from the fact that they allow specific fragments of any DNA to be isolated.

Purified *Xenopus* Ribosomal RNA Genes Are Isolated

Even before the advent of recombinant DNA, several perceptive observations by Don Brown in Baltimore and by Max Birnsteil in Edinburgh led to the isolation, in virtually pure form, of the genes coding for the ribosomal RNAs (18S, 28S, and 5S RNAs) of the toad *Xenopus*. This was possible, first of all, because these genes undergo enormous amplification in the very large *Xenopus* oocytes, to the point where they can represent almost 70 percent of the total nuclear DNA, and, secondly, because their base composition is significantly different from the rest of the nuclear DNA, allowing further purification on density (CsCl) gradients. The 5S DNA was purified by exploitation of its high A·T content, which allows it to bind many

more silver ions (Ag^{2+}) than most other DNA can; Ag–DNA is, of course, much denser than DNA, and the 5S genes that migrate with the bulk of DNA in a normal CsCl gradient appear as a small satellite peak in an Ag–CsCl gradient.

The individual 18S and 28S genes were found to be linked closely together, separated by about 1000 bases of DNA. This 18S–28S DNA unit is present in a tandem array of approximately 450 copies, and each 18S–28S unit is separated from the next unit by about 5 kb of *spacer* DNA. The 18S and 28S rRNAs are produced from a larger (40S) RNA transcript, which is processed to produce the mature ribosomal RNAs. The 5S genes, which are not linked to the 18S and 28S genes, were found to be present in 10,000 to 20,000 copies, also in tandem arrays separated by spacers that varied in length.

The purified 5S genes were faithfully transcribed when microinjected into *Xenopus* oocyte nuclei, indicating that the information necessary for accurate initiation and termination of transcription is contained on the purified genes (in other words, the genes have cis-acting controlling elements). These studies on purified ribosomal RNA genes were the prototype of similar experiments on a large number of other eukaryotic genes that became available only after the advent of recombinant DNA technology.

Eukaryotic mRNAs Have Caps and Tails

With most eukaryotic DNA so difficult to get a handle on, attention initially focused primarily on eukaryotic mRNA—in particular, on mRNAs that code for the more abundant cellular molecules like hemoglobin, the chicken egg protein ovalbumin, and the immunoglobulin chains made by antibody-producing cells. Most eukaryotic mRNAs, unlike their prokaryotic equivalents, were found to have long runs of A [*poly (A) tails*] at their 3′ ends. These tails do not come from sequences encoded in the DNA but are added to the ends of transcribed RNAs. Subsequently it was found that the 5′ ends of eukaryotic mRNAs are blocked by the addition of 7-methyl-G$_{ppp}$ caps (7-methylguanosine residues joined to the mRNAs by triphosphate linkages) that are added during the synthesis of the

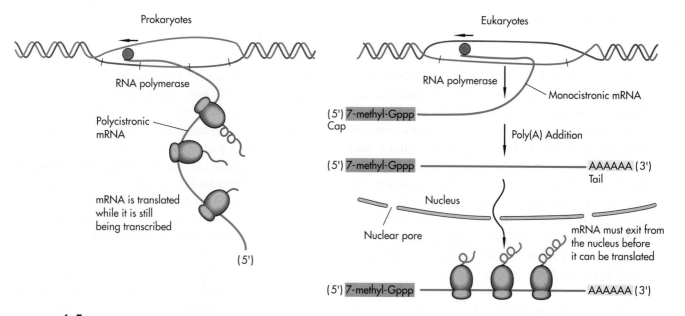

Synthesis and translation of prokaryotic vs. eukaryotic mRNAs. Prokaryotic mRNAs are often polycistronic; that is, they contain information for more than one protein. The 5′ end of the message is being translated (and probably degraded) while the 3′ end is still being transcribed. Eukaryotic mRNAs are monocistronic; one mRNA codes for only one protein. During transcription, methyl caps are added at the 5′ end; and after transcription, a poly(A) tail is added at the 3′ end. The mRNA must then be transported through pores in the nuclear membrane into the cytoplasm, where it can be translated.

primary transcript (Figure 4-7). The 5′ caps may guide ribosomes to begin translation at the correct AUG codon, but the function of the 3′ poly (A) tails is unknown. They may protect eukaryotic mRNA against attack by RNA-degrading enzymes.

Eukaryotes Are Found to Have Three Kinds of RNA Polymerases

The successful synthesis of eukaryotic RNA in the test tube led to the realization that all eukaryotic cells contain three different RNA polymerases, each with a distinct functional role. Ribosomal RNA (rRNA) is transcribed off rDNA genes by the enzyme *RNA polymerase I*. The synthesis of messenger RNA is catalyzed by *RNA polymerase II*, with the subsequent addition of the poly(A) tails carried out by the enzyme *poly(A) polymerase*. Transfer RNA and a variety of smaller nuclear and cytoplasmic RNAs are made by *RNA polymerase III*. This multiplicity of RNA polymerases in eukaryotes contrasts with the single form of RNA polymerase that is responsible for all RNA

synthesis in prokaryotes. The existence of three different types of eukaryotic RNA polymerases suggests the existence of three different types of promoters that can be regulated independently.

Eukaryotic DNA Is Organized into Nucleosomes

Also distinctive of eukaryotic cells is the way the DNA molecules are tightly complexed with proteins to form *chromatin*, nucleoprotein fibers with a beaded appearance. The key to the beadlike organization of chromatin is its histone proteins, of which there are five main classes: *H1, H2A, H2B, H3,* and *H4.* They come together to form *nucleosome* units containing an octomeric core composed of two molecules each of the H2A, H2B, H3, and H4 components, around which some 200 base pairs of DNA are wound. Connecting the nucleosome units are short stretches of linker DNA, where histone H1 is found. Also found within chromatin are a large number of different DNA-binding proteins. These proteins are likely to play a variety of specific roles in regulating gene expression.

The basic nucleosome structure must clearly be modified as transcription passes through individual units. As soon as chromatin began to be examined seriously with the electron microscope, observers claimed that especially "active" chromatin (for example, that making ribosomal RNA) lacked the normal beaded appearance. Until more individual genes could be examined, however, the molecular nature of active chromatin remained unknown.

Animal Viruses Are Model Systems for Gene Expression in Higher Cells

Although for a long time the chromosomal DNA of higher cells seemed virtually impossible to study, strong arguments were made that some real answers might come from investigating how DNA viruses multiply in animal cells. Many of these viruses grew well in cultured cells and could easily be labeled with radioactive isotopes. Moreover, some—like the monkey virus SV40 and the mouse virus polyoma—had remarkably small circular DNA molecules containing maybe as few as 5000 base pairs, coding for fewer than 10 proteins. They thus resembled in genetic simplicity the simple bacterial viruses from which so much had been learned. These viruses were also attractive because several were found to transform certain normal cells into cancerous cells capable of forming tumors in appropriate animals. Each of these viruses thus might contain one or more genes coding for proteins capable of transforming a normal cell into its cancerous equivalent. By studying such viruses, biochemists hoped to learn some of the principles of gene expression in higher cells, as well as a few basic facts about the origin and nature of cancer cells.

Therefore, in the late 1960s, deeper understanding of the DNA tumor viruses became the goal of an increasing number of scientists, many of whom had decided to switch their research from bacteria and bacteriophages to cells of higher animals. By 1970, the SV40 and polyoma viruses were both shown to have a life cycle that is neatly divisible into early and late phases. During the early phase of the SV40 life cycle, the cell makes mRNA coding for a viral protein that accumulates in the cell nucleus. There this viral protein plays a necessary role in SV40 DNA replication. When it was discovered that this protein was the major SV40-coded product that was consistently present in cells made cancerous by the virus, the protein was considered the key to the cancer-causing ability of SV40. Since its discovery, therefore, this protein has been known as the *T* (tumor) *protein* or *T antigen*. The late stage of the SV40 life cycle is marked by the synthesis of SV40 DNA as well as the synthesis of mRNA that codes for the structural proteins of the virus particle's outer shell.

How to confirm these facts at the molecular level, however, was not obvious, because there was as yet no way to relate the SV40 or polyoma virus mRNAs and the corresponding proteins to specific sections of the SV40 and polyoma DNAs. The circular forms of the viral DNAs at first seemed to preclude finding any specific reference point (such as an end) from which, with the electron microscope, the locations of the early and late mRNA could be mapped.

The RNA Tumor Viruses Replicate by Means of a Double-Stranded DNA Intermediate

Equally inaccessible at the molecular level were the genomes of the various RNA tumor viruses (viruses whose genome consists of dimers of identical *RNA* chains, instead of DNA chains). They had suddenly become much more interesting, through the independent discoveries in 1970 by Howard Temin and S. Mizutani and by David Baltimore that their replication involves the *reverse transcription* of the infecting RNA genomes into complementary DNA strands. The enzyme involved, *reverse transcriptase,* is coded for by the viral genomes, which incorporate it into the infectious viral particles so that it is able to act the moment the infecting RNA chromosome enters an appropriate host cell. Soon after its synthesis, the complementary DNA strand acts as a template to make the double-helical DNA representation of the infecting RNA strand. In turn, this DNA becomes integrated as a provirus into host chromosomal DNA, where the viral life cycle can be completed through transcription of the proviral DNA into an RNA strand identical with that found in the infectious virus.

In 1970, tasks such as these—learning how the genes of animal and plant cells are controlled and how cancers arise in these cells—seemed necessarily to be objectives for the distant future. Considerable progress had been made using viruses and bacteria as model systems to study gene regulation and cancer, but it seemed impossible that we would ever attain a really precise understanding at the molecular level of the crucial workings of the cells of higher organisms. The fact that these cells would never be as simple to work with as bacteria or their phages was something we were realizing we might have to live with. There was no economically feasible way to grow animal cells on the same large scale as bacteria, and the technical means to analyze the molecular genetics of these cells were simply not available. Furthermore, the idea that it would be possible to isolate and manipulate specific segments of cellular DNA was unthinkable. All this was to change with the development of recombinant DNA techniques.

Reading List

General

Tooze, J., ed. *Molecular Biology of Tumour Viruses.* Cold Spring Harbor Laboratory, Cold Spring Harbor, N.Y., 1973.

Chromatin. Cold Spring Harbor Symp. Quant. Biol., vol. 42. Cold Spring Harbor Laboratory, Cold Spring Harbor, N.Y., 1978.

Miller, J. H., and W. S. Reznikoff, eds. *The Operon,* 2nd ed. Cold Spring Harbor Laboratory, Cold Spring Harbor, N.Y., 1980.

Beckwith, J., J. Davies, and J. A. Gallant. *Gene Function in Prokaryotes.* Cold Spring Harbor Laboratory, Cold Spring Harbor, N.Y., 1983.

Structures of DNA. Cold Spring Harbor Symp. Quant. Biol., vol. 47. Cold Spring Harbor Laboratory, Cold Spring Harbor, N.Y., 1983.

Brock, T. D. *The Emergence of Bacterial Genetics.* Cold Spring Harbor Laboratory Press, Cold Spring Harbor, N.Y., 1990.

Original Research Papers

REPRESSORS AND OPERATORS

Jacob, F., and J. Monod. "Genetic regulatory mechanisms in the synthesis of proteins." *J. Mol. Biol.,* 3: 318–356 (1961).

Beckwith, J. R., and W. R. Signer. "Transposition of the *lac* region of *E. coli.* I. Inversion of the *lac* operon and transduction of *lac* by ϕ80." *J. Mol. Biol.,* 19: 254–265 (1966).

Gilbert, W., and B. Müller-Hill. "Isolation of the lac repressor." *Proc. Natl. Acad. Sci. USA,* 56: 1891–1898 (1966).

Ptashne, M. "Isolation of the λ phage repressor." *Proc. Natl. Acad. Sci. USA,* 57: 306–313 (1967).

Ptashne, M., and N. Hopkins. "The operators controlled by the λ phage repressor." *Proc. Natl. Acad. Sci. USA,* 60: 1282–1287 (1968).

Shapiro, J., L. MacHattie, L. Eron, G. Ihler, K. Ippen, and J. Beckwith. "Isolation of pure *lac* operon DNA." *Nature,* 224: 768–774 (1969).

Wang, J., M. D. Barkley, and S. Bourgeois. "Measurements of unwinding of *lac* operator by repressor." *Nature,* 251: 247–249 (1974).

Anderson, W. F., D. H. Ohlendorf, Y. Takeda, and B. W. Matthews. "Structure of the cro repressor from bacteriophage λ and its interaction with DNA." *Nature,* 290: 754–758 (1981).

Pabo, C. O., and M. Lewis. "The operator-binding domain of λ repressor: structure and DNA recognition." *Nature,* 298: 443–447 (1982).

Steitz, T. A., D. H. Ohlendorf, D. B. McKay, W. F. Anderson, and B. W. Matthews. "Structural similarity in the DNA-binding domains of catabolite gene activator and *cro* repressor proteins." *Proc. Natl. Acad. Sci. USA,* 79: 3097–3100 (1982).

Kim, R., and S.-H. Kim. "Direct measurement of DNA unwinding angle in specific interaction between *lac* operator and repressor." *Cold Spring Harbor Symp. Quant. Biol.,* 47: 481–484 (1983).

Matthews, B. W., D. H. Ohlendorf, W. F. Anderson, R. G. Fisher, and Y. Takeda. "Cro repressor protein and its interaction with DNA." *Cold Spring Harbor Symp. Quant. Biol.,* 47: 427–433 (1983).

PROMOTERS

Pribnow, D. "Bacteriophage T7 early promoters: nucleotide sequences of two RNA polymerase binding sites." *J. Mol. Biol.,* 99: 419–443 (1975).

Wang, J. C., J. H. Jacobsen, and J.-M. Saucier. "Physiochemical studies on interactions between DNA and RNA polymerase. Unwinding of the DNA helix by *E. coli* RNA polymerase." *Nuc. Acids Res.,* 4: 1225–1241 (1977).

Chamberlin, M. J., W. C. Nierman, J. Wiggs, and N. Neff. "A quantitative assay for bacterial RNA polymerases." *J. Biol. Chem.,* 254: 10061–10069 (1979).

McClure, W. R. "Rate-limiting steps in RNA chain initiation." *Proc. Natl. Acad. Sci. USA,* 77: 5634–5638 (1980).

Youderian, P., S. Bouvier, and M. Susskind. "Sequence determinants of promoter activity." *Cell,* 30: 843–853 (1982).

Ackerson, J. W., and J. D. Gralla. "In vivo expression of *lac* promoter variants with altered −10, −35, and spacer sequences." *Cold Spring Harbor Symp. Quant. Biol.,* 47: 473–476 (1983).

ATTENUATORS

Bertrand, K., I. Korn, F. Lee, T. Platt, C. L. Squires, C. Squires, and C. Yanofsky. "New features of the regulation of the tryptophan operon." *Science,* 189: 22–26 (1975).

Oxender, D. L., G. Zurawski, and C. Yanofsky. "Attenuation in the *Escherichia coli* tryptophan operon: the role of RNA secondary structure involving the Trp codon region." *Proc. Natl. Acad. Sci. USA,* 76: 5524–5528 (1979).

Yanofsky, C. "Attenuation in the control of expression of bacterial operons." *Nature,* 289: 751–758 (1981).

POSITIVE CONTROL

Zubay, G., D. Schwartz, and J. Beckwith. "Mechanism of activation of catabolite-sensitive genes: a positive control system." *Proc. Natl. Acad. Sci. USA,* 66: 104–110 (1970).

Epstein, W., L. B. Rothman, and J. Hesse. "Adenosine 3':5'-cyclic monophosphate as mediator of catabolite repression in *Escherichia coli.*" *Proc. Natl. Acad. Sci. USA,* 72: 2300–2304 (1975).

Simpson, R. B. "Interaction of the cAMP receptor protein with the lac promoter." *Nuc. Acids Res.,* 8: 759–766 (1980).

McKay, D. B., I. T. Weber, and T. A. Steitz. "Structure of catabolite gene activator protein at 2.9Å resolution: incorporation of amino acid sequence and interactions with cyclic-Amp." *J. Biol. Chem.,* 257: 9518–9524 (1982).

TRANSLATIONAL CONTROL

Shine, J., and L. Dalgarno. "The 3'-terminal sequence of *Escherichia coli* 16S ribosomal RNA: complementarity to nonsense triplets and ribosome binding sites." *Proc. Natl. Acad. Sci. USA,* 71: 1342–1346 (1974).

Nomura, M., J. L. Yates, D. Dean, and L. E. Post. "Feedback regulation of ribosomal protein gene expression in *Escherichia coli:* structural homology of ribosomal RNA and ribosomal protein mRNA." *Proc. Natl. Acad. Sci. USA,* 77: 7084–7088 (1980).

ISOLATION OF RIBOSOMAL RNA GENES

Birnstiel, M., J. Speirs, I. Purdom, K. Jones, and U. E. Loening. "Properties and composition of the isolated ribosomal DNA satellite of *Xenopus laevis.*" *Nature,* 219: 454 (1968).

Weinberg, R. A., and S. Penman. "Processing of 45S nucleolar RNA." *J. Mol. Biol.,* 47: 169–178 (1970).

Brown, D. D., and K. Sugimoto. "The structure and evolution of ribosomal and 5S DNAs in *Xenopus laevis* and *Xenopus mulleri.*" *Cold Spring Harbor Symp. Quant. Biol.,* 38: 501–505 (1974).

Brown, D. D., and J. B. Gurdon. "High-fidelity transcription of 5S DNA injected into *Xenopus* oocytes." *Proc. Natl. Acad. Sci. USA,* 74: 2064–2068 (1977).

CAPS AND POLY(A) TAILS

Lim, L., and E. S. Canellakis. "Adenine-rich polymer associated with rabbit reticulocyte messenger RNA." *Nature,* 227: 710–712 (1970).

Darnell, J. E., R. Wall, and R. J. Tushinski. "An adenylic acid-rich sequence in messenger RNA of HeLa cells and its possible relationship to reiterated sites in DNA." *Proc. Natl. Acad. Sci. USA,* 68: 1321–1325 (1971).

Edmonds, M., M. H. Vaughan, Jr., and H. Nakazato. "Polyadenylic acid sequences in the heterogeneous nuclear RNA and rapidly-labeled polyribosomal RNA of HeLa cells: possible evidence for a precursor relationship." *Proc. Natl. Acad. Sci. USA,* 68: 1336–1340 (1971).

Lee, S. Y., J. Mendecki, and G. Brawerman. "A polynucleotide segment rich in adenylic acid in the rapidly-labeled polyribosomal RNA component of mouse sarcoma 180 ascites cells." *Proc. Natl. Acad. Sci. USA,* 68: 1331–1335 (1971).

Aviv, H., and P. Leder. "Purification of biologically active globin messenger RNA by chromatography on oligo-thymidylic acid-cellulose." *Proc. Natl. Acad. Sci. USA,* 69: 1408–1412 (1972).

Shatkin, A. J. "Capping of eucaryotic mRNAs." *Cell,* 9: 645–653 (1976).

NUCLEOSOMES

Kornberg, R. "Chromatin structure: a repeating unit of histones and DNA." *Science,* 184: 868–871 (1974).

Olins, A. L., and D. E. Olins. "Spheroid chromatin units (v bodies)." *Science,* 183: 330–332 (1974).

TUMOR VIRUSES

Black, P. W., W. P. Rowe, H. C. Turner, and R. J. Huebner. "A specific complement-fixing antigen present in SV40 tumor and transformed cells." *Proc. Natl. Acad. Sci. USA,* 50: 1148–1156 (1963).

Benjamin, T. L. "Virus-specific RNA in cells productively infected or transformed by polyoma virus." *J. Mol. Biol.,* 16: 359–373 (1966).

Sambrook, J., H. Westphal, P. R. Srinivasan, and R. Dulbecco. "The integrated state of viral DNA in SV40-transformed cells." *Proc. Natl. Acad. Sci. USA,* 60: 1288–1295 (1968).

Baltimore, D. "Viral RNA-dependent DNA polymerase." *Nature,* 226: 1209–1211 (1970).

Temin, H. M., and S. Mizutani. "Viral RNA-dependent DNA polymerase." *Nature,* 226: 1211–1213 (1970).

5

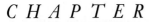

Methods of Creating Recombinant DNA Molecules

Given the length of even the smallest DNA molecules, the isolation of the first repressors signified to many biologists more the end of an era rather than the beginning of a new cycle of important conceptual advances. Unless some radically new tricks emerged to manipulate DNA, there could be no immediate bright future for eukaryotic molecular biologists. So, to avoid possibly marking time, several distinguished contributors to our primary knowledge on the storage of genetic information in DNA left molecular genetics to start up new careers in neurobiology. What they could not have foreseen was the very rapid development over the next several years of the enzymological and chemical techniques that gave rise to recombinant DNA, and the consequent period of scientific excitement and achievement that has seen few if any parallels in the history of biological research.

Nucleic Acid Sequencing Methods Are Developed

For deeper insights about the organization of DNA, methods had to be developed to reveal exact nucleotide sequences—first, of selected regions of a gene, then of an entire gene, and finally of an entire chromosome. The first nucleic acid sequences to be established were not of DNA, but were of the relatively small tRNA molecules that contain 75 to 80 nucleotides. By 1964, the sequence of the yeast alanine tRNA molecule was worked out. To do this, Robert Holley and his colleagues at Cornell University had to find specific enzymes that broke the tRNA chains reproducibly into smaller and smaller discrete fragments, until they could be sequenced directly by simple stepwise degradation procedures.

With each passing year these methodologies greatly improved, and by 1975 the complete sequence of the RNA chromosome[1] of the single-stranded RNA phage MS2 was worked out in Walter Fiers' laboratory in Ghent. For the first time the precise way in which a simple chromosome was put together could be visualized. And for the first time the exact codons that specify the amino acids of the three proteins coded by the three genes of phage MS2 became known, as well as the stop codons that signal chain termination. Few nucleotides separated the three genes, but unexpectedly long untranslated regions (of 129 and 174 bases, respectively) existed at the two ends. Here, as on all other messenger RNA–like molecules, the two physical ends never acted as start or stop signals.

Direct sequencing of any DNA molecule was not then possible, because there was no way to cut DNA at specific points to produce discrete reproducible fragments having unique sequences. The available deoxyribonucleases (*DNases*) all cut DNA into hopelessly heterogeneous collections of small fragments whose order within the original DNA could never be deciphered.

Restriction Enzymes Make Sequence-Specific Cuts in DNA

All the nucleases, the enzymes that were first found to break the phosphodiester bonds of nucleic acids, showed very little sequence dependency; the most specific was the T1 RNase, which was found to cut only next to guanine residues. Highly preferred sites of cleavage on certain RNAs were discovered, but these reflected the way single-stranded RNA molecules fold into complex three-dimensional arrangements rather than any tendency of the enzymes to cut within specific base sequences. The prevailing opinion

was that highly specific nucleases would never be found and that therefore the isolation of discrete DNA fragments, even from viral DNA, would not be possible. The only grounds for thinking differently were observations, beginning as early as 1953, that when DNA molecules from one strain of *E. coli* were introduced into a different *E. coli* strain (for example, *E. coli* strain B versus *E. coli* strain C), they rarely functioned genetically. Instead, the foreign DNAs were almost always quickly fragmented into smaller pieces. Quite infrequently, the infecting DNA molecule would not be broken down, because it had somehow become modified so that it and all its descendants could now multiply on the new bacterial strain. In 1966 chemical analysis of a small viral DNA modified in such a way that it could survive in a different strain of *E. coli* revealed the presence of one to several methylated bases not present in the unmodified DNA. Methylated bases are not inserted as such into growing DNA chains; they arise through the enzymatically catalyzed addition of methyl groups to newly synthesized DNA chains.

The stage was thus set in the late 1960s for Stewart Linn and Werner Arber, working in Geneva, to find in extracts of cells of *E. coli* strain B both a specific *modification enzyme* that methylated unmethylated DNA and a *restriction nuclease* that broke down unmethylated DNA. Over the next several years, the discovery of restriction nucleases and their companion modification methylases in two other *E. coli* strains opened up the possibility that many site-specific nucleases might exist. None of these early *E. coli* restriction enzymes lived up to their finders' first hopes, however, because although the enzymes recognized specific unmethylated sites, they cleaved the DNA at random locations far removed from these sites.

Soon specific restriction nucleases that did cleave at specific sites in DNA were identified. The first was discovered in 1970 by Hamilton Smith of Johns Hopkins University, who followed up his accidental finding

---

[1] RNA replaces DNA as the genetic material in many viruses (for example, tobacco mosaic, influenza, polio, certain RNA phages). Replication follows the same pattern used for DNA, with single RNA chains serving as templates to make chains with complementary sequences. The specific replication enzymes are coded on the viral RNA chromosomes and called RNA *replicases*. Equally important, one of the complementary partners can also function as mRNA by combining with ribosomes and coding directly for the amino acids of the viral proteins.

that the bacterium *Haemophilus influenzae* rapidly broke down foreign phage DNA. This degradative activity was subsequently observed in cell-free extracts and shown to be due to a true restriction nuclease, because the enzyme broke down *E. coli* DNA, whereas it failed to cut up the DNA of the *Haemophilus* cells from which it had been extracted. Highly purified *Hin*dII, as this enzyme is called, was found to bind to the following set of sequences, in which the arrows indicate the exact cleavage sites, and "Py" and "Pu" represent any pyrimidine or purine residue:

$$(5')GTPy{\downarrow}PuAC(3')$$
$$(3')CAPu{\uparrow}PyTG(5')$$

Since then, restriction enzymes that cut specific sequences have been isolated from several hundred bacterial strains, and over 150 different specific cleavage sites have been found (Table 5-1). These enzymes recognize specific sequences of four to eight bases. A given 4-bp site occurs on average every 256, and a 6-bp site every 4096, base pairs. But the distribution of sites is irregular, so that a specific region of DNA may be cut more or less frequently than the statistical average. One factor that affects the frequency with which DNA is cut is its base composition. For example, *Not*I has an eight-base recognition sequence that includes CpG dinucleotides that occur very infrequently in mammalian DNA. It produces fragments between

---

TABLE **5-1**

Some Restriction Enzymes and Their Cleavage Sequences

MICROORGANISM	ENZYME ABBREVIATION	SEQUENCE	NOTES*
Haemophilus aegytius	*Hae*III	5'...G G|C C...3' 3'...C C|G G...5'	1
Thermus aquaticus	*Taq*I	5'...T|C G A...3' 3'...A G C|T...5'	2
Haemophilus haemolyticus	*Hha*I	5'...G C G|C...3' 3'...C|G C G...5'	3
Desulfovibrio desulfuricans	*Dde*I	5'...C|T N A G...3' 3'...G A N T|C...5'	4
Moraxella bovis	*Mbo*II	5'...G A A G A (N)$_8$|...3' 3'...C T T C T (N)$_7$|...5'	5
Escherichia coli	*Eco*RV	5'...G A T|A T C...3' 3'...C T A|T A G...5'	1
	*Eco*RI	5'...G|A A T T C...3' 3'...C T T A A|G...5'	2
Providencia stuarti	*Pst*I	5'...C T G C A|G...3' 3'...G A|C G T C...5'	3
Microcoleus	*Mst*II	5'...C C|T N A G G...3' 3'...G G A N T|C C...5'	4
Nocardia otitidis-caviarum	*Not*I	5'...G C|G G C C G C...3' 3'...C G C C G G|C G...5'	6

* Notes:
1. Enzyme produces blunt ends.
2. The single strand is the 5' strand.
3. The single strand is the 3' strand.
4. The base pair *N* can be any purine or pyrimidine pair.
5. The enzyme does not cut within the recognition sequence, but at whatever sequence lies eight nucleotides 3' to the recognition site.
6. *Not*I has an eight-base recognition sequence and cuts mammalian DNA very infrequently.

1 million and 1.5 million base pairs in size. Such enzymes that produce very large DNA fragments have proved invaluable for long-range physical mapping of mammalian DNA.

Restriction Maps Are Highly Specific

The various fragments generated when a specific viral DNA is cut by a restriction enzyme can be easily separated by using *agarose gel electrophoresis* (Figure 5-1). The rate at which the fragments migrate through

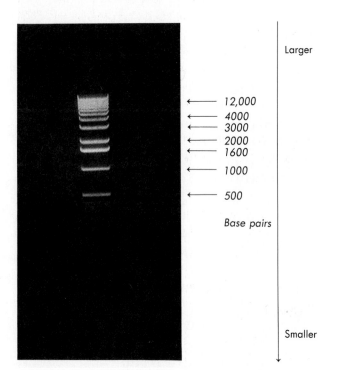

FIGURE **5-1**
DNA molecules of different sizes may be separated by electrophoresis, a process in which an electrical field is used to move the negatively charged DNA molecules through porous agarose gels. Molecules of the same size move at the same speed, and because smaller DNA molecules move faster than larger molecules, they become separated into bands. The bands on this gel contain molecules ranging in size from 500 bp to 12,000 bp, although the bands larger than 4000 bp are not well enough separated from each other to be distinguished. The DNA is stained with ethidium bromide, a molecule that fluoresces when illuminated with UV light.

the gel is a function of their lengths, with small fragments moving much faster than large fragments. Depending on the concentration of agarose, the larger fragments may hardly be able to move into the gel. Restriction fragments move unharmed through such agarose gels and can be eluted as biologically intact double helices. Staining of these gels with dyes that bind to DNA generates a series of bands, each corresponding to a restriction fragment whose molecular weight can be established by calibration with DNA molecules of known weights. Different restriction enzymes necessarily give different restriction fragments for the same viral DNA molecule. In general, the most useful enzymes are the ones that have rare recognition sequences and that therefore produce small numbers of fragments that can easily be separated from one another on the agarose gels.

The first restriction map was obtained in 1971 by Daniel Nathans, a colleague of Hamilton Smith at Johns Hopkins. Nathans used the *Hin*dII enzyme to cut the circular DNA of SV40 into 11 specific fragments. The order in which these 11 fragments occurred in the SV40 DNA could be deduced by studying the patterns of fragments produced as the digestion proceeded to completion. The first cut broke the circular molecule into a linear structure that was then cut into progressively smaller fragments. By following the pattern of production of, first, the overlapping intermediate-sized fragments and, from them, the fragments of the complete digest, Nathans produced a restriction map that located the sites on the circular viral DNA that are attacked by the restriction enzyme (Figure 5-2). Repeating the experiment with other enzymes produced a more detailed map with many different restriction sites.

With this information it became possible to determine the positions of regions of biological importance on the circular viral DNA. For example, by briefly radioactively labeling replicating viral DNA and then digesting it with *Hin*dII, Nathans proved that the replication of SV40 DNA always begins in one specific *Hin*dII fragment and proceeds bidirectionally around the circular DNA molecule. Subsequently, by using other enzymes (including *Eco*RI, which cuts SV40 DNA only once), experimenters precisely located the site of the initiation of DNA replication at some 1700 base pairs away from the *Eco*RI site.

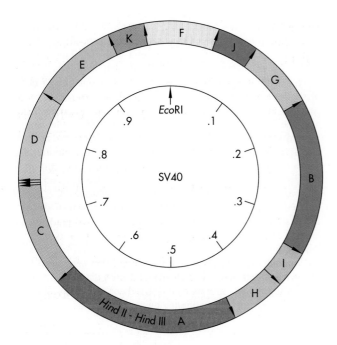

FIGURE **5-2**

A restriction map of SV40 produced circa 1972 with the enzymes *Hin*dII and *Hin*dIII. The inner circle is marked with map units, an early convention for measuring the viral genome. The genome is now known to be 5243 base pairs long, so that each map unit is approximately 524 base pairs. By convention, the single *Eco*RI site is the start of the genome. Each arrow in the outer circle marks a site where *Hin*dII or *Hin*dIII cuts the viral DNA.

The restriction maps and restriction fragments were used to identify on the viral DNA the regions that specify the mRNAs of the viral proteins at different stages during viral replication. To do this, radioactively labeled mRNA from infected cells was isolated at early and late times after infection. Pure restriction fragments of the viral DNA were prepared and denatured so that the two DNA chains of the double helix separated. The mRNA was then mixed with the separated DNA strands in conditions that allowed the RNA to form RNA-DNA double helices with DNA strands that had a complementary base sequence. Such RNA-DNA hybridization experiments revealed that both early and late viral RNA are coded by continuous DNA regions, each spanning about half the total SV40 DNA. The promoters of both the early and the late mRNAs are near the origin of DNA replication, but

synthesis of early mRNA proceeds in one direction, and that of late mRNA proceeds in the other. These techniques, which allowed identification of the genetically significant regions of DNA, were important in themselves. But they also paved the way to the development of extremely useful new methods of DNA sequencing and recombinant DNA techniques.

Restriction Fragments Lead to Powerful New Methods for Sequencing DNA

When the first restriction fragments became available, there was no good method of sequencing them directly. The only realistic way to proceed was to use RNA polymerase to synthesize their complementary RNA chains, on which the elegant new RNA-sequencing procedures of Fred Sanger could be employed. In the mid-1960s Sanger had stopped sequencing proteins and turned his attention to working out fast, simple procedures for sequencing long stretches of RNA. By employing Sanger's procedures, Sherman Weissman at Yale University and Walter Fiers in Ghent established, by the end of 1976, the sequence of more than half the over 5200 base pairs of the DNA of SV40 virus.

A breakthrough came with the advent of methods that allowed sequencing of fragments of DNA of from 100 to 500 base pairs. Sanger devised the first of these direct DNA-sequencing methods, the *plus-minus method*, in 1975. It is based on the elongation of DNA chains with DNA polymerase. With this technique the 5386-bp sequence of the small DNA phage ϕX174 was quickly determined. An equally powerful method based on the chemical degradation of DNA chains was developed at Harvard University by Allan Maxam and Walter Gilbert in 1977 (Figure 5-3). All the 5243 base pairs of SV40 DNA became quickly known, as did those of the small recombinant plasmid pBR322, whose 4362 bases were determined in less than a year by Greg Sutcliffe in Gilbert's laboratory.

Sanger later devised a second method for sequencing DNA, and again, he used enzymatic rather than chemical, techniques. Specific terminators of DNA chain elongation—2′,3′-dideoxynucleoside triphosphates—were synthesized. These (ddNTPs) mole-

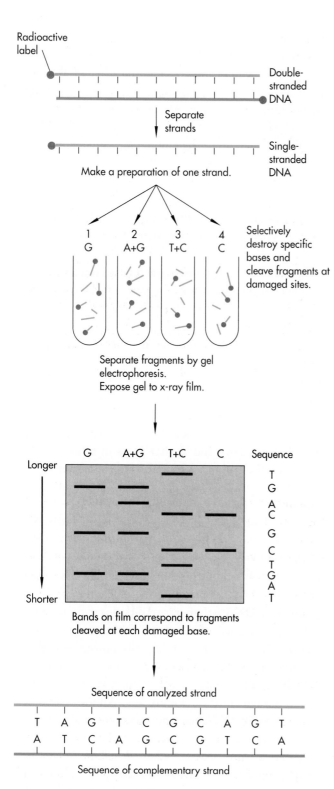

Radioactive label

Double-stranded DNA

Separate strands

Single-stranded DNA

Make a preparation of one strand.

| 1 | 2 | 3 | 4 |
| G | A+G | T+C | C |

Selectively destroy specific bases and cleave fragments at damaged sites.

Separate fragments by gel electrophoresis.
Expose gel to x-ray film.

| G | A+G | T+C | C | Sequence |

Longer

T
G
A
C
G
C
T
G
A
T

Shorter

Bands on film correspond to fragments cleaved at each damaged base.

Sequence of analyzed strand

| T | A | G | T | C | G | C | A | G | T |
| A | T | C | A | G | C | G | T | C | A |

Sequence of complementary strand

FIGURE **5-3**
The Maxam and Gilbert DNA-sequencing procedure. A segment of DNA is labeled at one end with ^{32}P. The labeled DNA is divided into four samples and each sample is treated with a chemical that specifically destroys one or two of the four bases in the DNA. The conditions of the reaction are controlled so that only a few sites are nicked in any one DNA molecule. When these nicked molecules are treated with piperidine, the DNA backbone is broken at the site at which the base had been destroyed. This generates a series of labeled fragments, the lengths of which depend on the distance of the destroyed base from the labeled end of the molecule. For instance, if there are G residues 3, 6, and 9 bases away from the labeled end, then treatment of the DNA strand with chemicals that cleave at G will generate labeled fragments 2, 5, and 8 base in length. The sets of labeled fragments obtained from each of the four reactions are run side by side on an acrylamide gel that separates DNA fragments according to size, and the gel is autoradiographed. The pattern of bands on the x-ray film is read to determine the sequence of the DNA.

cules can be incorporated normally into a growing DNA chain through their 5′ triphosphate groups. However, they cannot form phosphodiester bonds with the next incoming deoxynucleotide triphosphates (dNTPs). When a small amount of a specific dideoxy NTP (say, ddATP) is included along with the four deoxy NTPs normally required in the reaction mixture for DNA synthesis by DNA polymerase, the products are a series of chains that are specifically terminated at the dideoxy residue (Figure 5-4a). Thus four separate reactions, each containing a different dideoxy NTP, can be run, and their products displayed on a high-resolution acrylamide gel (Figure 5-4b). With this technique the 5577-bp sequence of the phage G4, a relative of ϕX174, was determined quickly.

Determining the exact sequence of any segment of DNA of reasonable size is now a feasible project for any molecular biology laboratory. The most extensive sequence so far determined is that for the Epstein-Barr virus, 172, 282 base pairs in length. Continuing developments, including semiautomated machines for running and analyzing sequencing gels, have led to the astounding prospect of sequencing the entire genomes of higher organisms, including humans.

From the sequence of a gene it is a simple matter

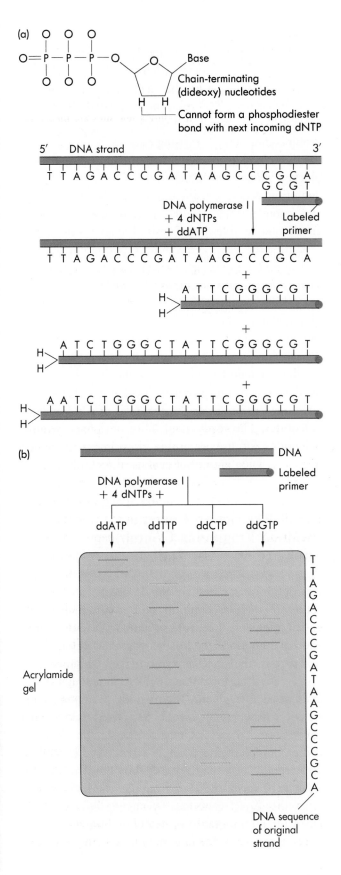

FIGURE **5-4**
The Sanger DNA-sequencing procedure. (a) 2′,3′-Dideoxynucleotides of each of the four bases are prepared. These molecules can be incorporated into DNA by *E. coli* DNA polymerase because they have a normal 5′ triphosphate; however, once incorporated into a growing DNA strand, the dideoxynucleotide (ddNTP) cannot form a phosphodiester bond with the next incoming dNTP. Growth of that particular DNA chain stops. A Sanger sequencing reaction consists of a DNA strand to be sequenced, a short labeled piece of DNA (the primer) that is complementary to the end of that strand, a carefully controlled ratio of one particular dideoxynucleotide with its normal deoxynucleotide, and the other three dNTPs. When DNA polymerase is added, normal polymerization will begin from the primer; when a ddNTP is incorporated, the growth of that chain will stop. If the correct ratio of ddNTP:dNTP is chosen, a series of labeled strands will result, the lengths of which are dependent on the location of a particular base relative to the end of the DNA. (b) A DNA strand to be sequenced, along with labeled primer, is split into four DNA polymerase reactions, each containing one of the four ddNTPs. The resultant labeled fragments are separated by size on an acrylamide gel, and autoradiography is performed; the pattern of the fragments gives the DNA sequence.

to deduce the amino acid sequence of the protein it specifies; in fact, nowadays it is often faster to determine the sequence of a protein by this indirect route rather than by directly sequencing the protein. Whereas protein sequencing can take months and even years, DNA sequencing can often be accomplished in a matter of days.

Oligonucleotides Can Be Synthesized Chemically

The emergence of quick, convenient methods for the synthesis of moderately long oligonucleotides with defined sequences has followed close upon the development of rapid sequencing methods. Chemical synthesis is based on the ability to protect specifically (that is, to prevent having a chemical reaction occur at) either the 5′ or the 3′ end of a mono- or oligonucleotide. This is done by hanging a large blocking

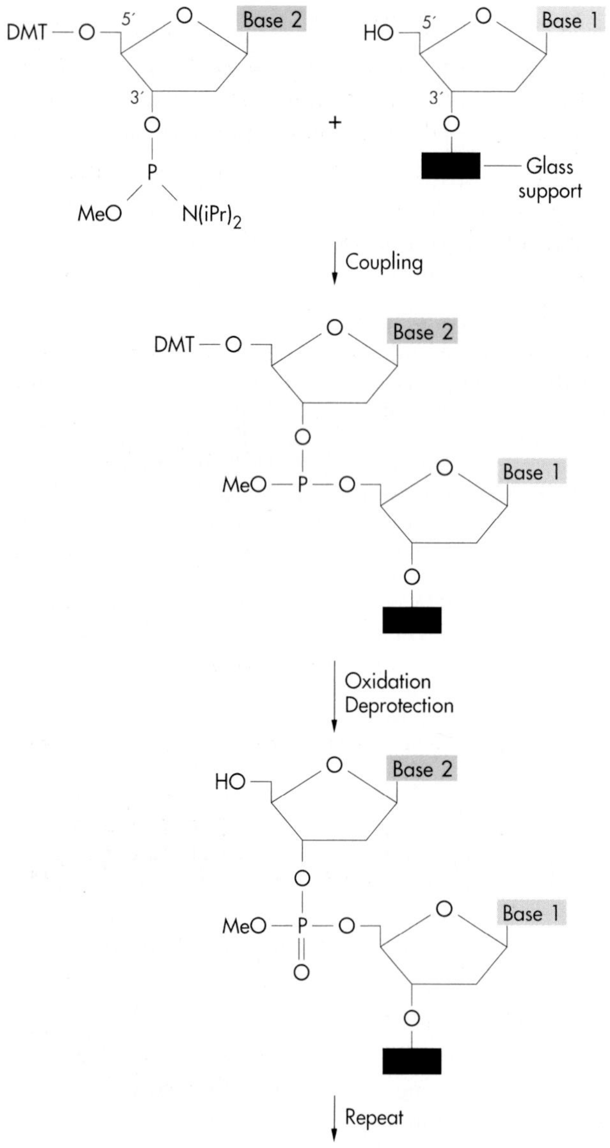

FIGURE **5-5**
Synthesis of an oligonucleotide by solid-phase phosphor-amidite chemistry. The 3' nucleotide is attached to an inert support of small glass beads, which are placed in a reaction vessel. The oligonucleotide is built up one nucleotide at a time from 3' to 5' by a three-step cycle. In the first step, the nucleotide precursor containing base 2 is added to the reaction vessel. The 5' hydroxyl of base 1 couples to the 3' phosphorous of base 2. In the second step, the unstable trivalent phosphite is oxidized to the stable phosphate. In the third step, the dimethoxytrityl (DMT) group that protects the 5' hydroxyl of the newly added nucleotide is removed, thereby completing one cycle. The process is repeated by addition of the next nucleotide precursor. Finally, the completed oligonucleotide is cleaved from the glass support, and groups protecting the phosphates and the bases are removed. iPr, isopropyl.

group onto either the 5' or the 3' hydroxyl (Figure 5-5). Different blocking groups are used: some can be removed with acid, some with base. Thus a 5' blocked mononucleotide can be chemically condensed with a 3'-blocked molecule, resulting in a dinucleotide that is blocked at both ends. Either the 5' or the 3' blocking group is then removed (using either acid or base), and the dinucleotide is reacted with an appropriately unblocked mono- or dinucleotide. This cycle of condensation, removal of one or the other blocking group, and recondensation can be repeated many times until an oligonucleotide of the desired length is obtained.

Until a few years ago, synthesis of oligonucleotides was a time-consuming process that was limited to linking together fewer than 20 nucleotides. Oligonucleotide synthesis is now performed using programmable machines that are capable of synthesizing oligonucleotides as long as 100 bases in about 10 hours. The limiting factors are the progressively lower yields of oligonucleotides with increasing length and the need to purify the products. Techniques have been developed recently for the synthesis of RNA oligonucleotides. The ready availability of defined oligonucleotides for use as probes or primers has made possible a wide variety of analytical techniques.

Many Restriction Enzymes Produce Fragments Containing Sticky (Cohesive) Ends

Restriction enzymes like *Hin*dII break DNA at the center of their recognition sites to produce blunt-ended fragments that are base-paired out to their ends and have no tendency to stick together (Figure 5-6). In contrast, the *Eco*RI enzyme makes staggered cuts that create short four-base single-stranded tails on the ends of each fragment (Figure 5-7). Many other restriction enzymes also make staggered cuts, leaving single-stranded tail sequences specific for each enzyme. Complementary single-stranded tails tend to associate by base pairing and thus are often called *cohesive*, or *sticky*, *ends*. For example, the linear molecules that *Eco*RI generates by cutting circular SV40 DNA often temporarily recyclize by base pairing between their tails. Fragments held together by such

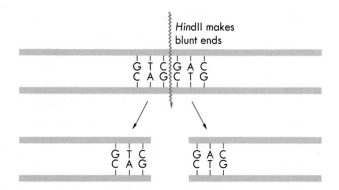

FIGURE **5-6**

The *Hin*dII restriction enzyme cuts DNA at the center of its recognition site, leaving blunt ends.

covered, of which *DNA polymerase III* is now considered the enzyme involved in most DNA chain elongation. Much of the inherent complexity of DNA replication arises because the two chains of the double helix run in opposite directions ($5' \rightarrow 3'$ and $3' \rightarrow 5'$), and their daughter strands must likewise run in opposite directions. Yet the elongation of *all* individual daughter chains occurs in the $5' \rightarrow 3'$ direction. This apparent paradox was resolved by the realization that one daughter strand grows continuously in one direction while the other daughter strand is made discontinuously from smaller pieces that are individually elongated in the opposite direction. This feature ne-

base pairing can be permanently rejoined by adding the enzyme *DNA ligase* to catalyze the formation of new phosphodiester bonds.

Base pairing occurs only between complementary base sequences, so the cohesive AATT ends produced by *Eco*RI will not, for example, pair will the AGCT ends produced by *Hin*dIII. But any two fragments (regardless of their origin) produced by the same enzyme can stick together and later be joined together permanently by the action of DNA ligase (Figure 5-7). Such experiments were first done at Stanford in 1972 by Janet Mertz and Ron Davis, who realized that *Eco*RI in conjunction with DNA ligase would provide a general way to achieve in vitro, site-specific genetic recombination.

Many Enzymes Are Involved in DNA Replication

The enzyme DNA ligase, which may be used to seal together restriction fragments, is but one of many enzymes that are now known to be involved in DNA replication. The first enzyme to make DNA chains in vitro, *DNA polymerase I,* was for almost 10 years believed to be the main—if not the only—polymerase needed to link up deoxynucleotides into DNA chains. In 1967, however, an *E. coli* mutant was found that had almost no DNA polymerase I yet made DNA at normal rates. Within a year, two new DNA polymerases were dis-

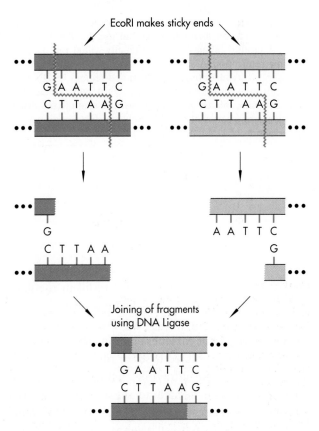

FIGURE **5-7**

The *Eco*RI restriction enzyme makes staggered, symmetrical cuts in DNA away from the center of its recognition site, leaving cohesive, or "sticky," ends. A sticky end produced by *Eco*RI digestion can anneal to any other sticky end produced by *Eco*RI cleavage.

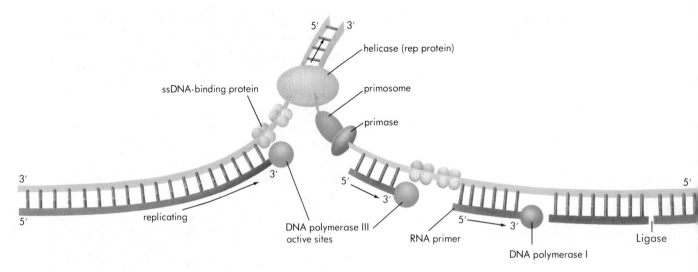

FIGURE **5-8**
The enzymes involved in DNA replication in *E. coli.* Several enzymes have been found to be necessary for DNA replication in *E. coli.* DNA polymerase cannot initiate chains *de novo*, but requires a primer. This is provided by an RNA polymerase called a *primase*, which in association with the *primosome* complex of proteins synthesizes a short stretch of RNA. DNA polymerase III can then take over and use this RNA as a primer to continue the synthesis of DNA. A protein called *helicase* (originally called rep) is necessary to unwind the DNA helix to allow replication. A single-stranded DNA–binding protein (ssDNA-binding protein) is also necessary to stabilize the single-stranded regions of DNA that are transiently formed during the replication process. Finally, since DNA polymerase can synthesize DNA in only the 5′-to-3′ direction, one of the strands must be synthesized discontinuously (the *lagging strand*). This leads to a series of short stretches of DNA with gaps in between. These gaps are filled by the action of DNA polymerase I and sealed with DNA ligase.

cessitates both DNA polymerase I to fill in the gaps and the joining enzyme DNA ligase to seal the completed pieces together. (Figure 5-8).

In addition, other enzymes edit the DNA to remove erroneously incorporated bases or repair chains damaged by agents like ultraviolet light or x-rays. Still more proteins are needed to separate the parental strands at the replication fork, as well as to bind temporarily to single-stranded regions before their conversion to double helices. Also needed are several specific proteins involved in initiating DNA chains.

Most of these proteins were found over the last 20 years in the laboratory of Arthur Kornberg at Stanford University or by his former collaborators working elsewhere.

Sticky Ends May Be Enzymatically Added to Blunt-Ended DNA Molecules

The calf thymus enzyme *terminal transferase*, which adds nucleotides to the 3′ ends of DNA chains, provides a general method for creating cohesive ends on blunt-ended DNA fragments. For example, if polydeoxy A [poly(dA)] is added to the two 3′ ends of one double-stranded fragment, and polydeoxy T [poly(dT)] is added to the 3′ ends of another fragment, the two fragments, when mixed together, can form base pairs between their complementary tails. Appropriate enzymes can be added to fill in any single-stranded gaps, and finally DNA ligase can be used to permanently join the two fragments (Figure 5-9). This procedure, developed in 1971–1972 by Peter Lobban and Dale Kaiser, and by David Jackson and Paul Berg, provides a second general method for creating recombinant DNA molecules. But it introduces regions of poly(dA)·poly(dT) base pairs at the junctions between the fused fragments. Such additional base sequences could affect the function of the joined molecules; and whenever possible, cohesive ends generated by restriction enzyme cuts are used to create recombinant DNA molecules.

Later developments have made it possible to "mix and match" almost any combination of DNA ends by first blunt-ending fragments generated by restriction endonucleases and then ligating small oligonucleotide "linkers" onto the blunt ends. These linkers can be synthesized containing any restriction site for subsequent ligation.

Small Plasmids Are Vectors for the Cloning of Foreign Genes

The realization that *Eco*RI generates specific cohesive ends that can later be sealed up by DNA ligase was followed within a year by the development of the first

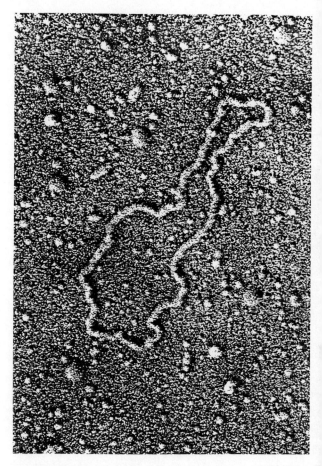

FIGURE **5-10**
Electron micrograph of plasmid pSC101. (Courtesy of Stanley N. Cohen, Stanford University.)

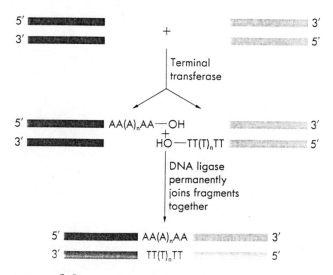

FIGURE **5-9**
Two blunt-ended DNA fragments can be joined together by adding poly(dA) and poly(dT) tails to the ends of the fragments. The complementary tails will form base pairs, and enzymes can be used to fill in any single-stranded gaps and join the fragments permanently.

practical method for systematically cloning specific DNA fragments, regardless of their origin. The essential trick was the random insertion of the *Eco*RI-generated fragments of a DNA molecule into circular plasmid DNA that also had been cut with *Eco*RI. This procedure led to hybrid plasmids, which could be used to infect bacteria. Each bacterial cell acquired a recombinant plasmid carrying a specific foreign DNA fragment. In the first such experiments, carried out in early 1973 by Herbert Boyer and Stanley Cohen and their collaborators at Stanford and the University of California, San Francisco, the small *E. coli* plasmid pSC101 (Figure 5-10) was used because it contained only a single *Eco*RI recognition site and was converted by *Eco*RI into linear molecules. When such DNA was mixed with foreign DNA fragments also possessing cohesive ends generated by *Eco*RI, and DNA ligase

was added, new hybrid plasmids were created (Figure 5.11). Each contained one or more pieces of foreign DNA inserted into the *Eco*RI site of the plasmid.

In the original Boyer-Cohen experiment, the foreign DNA was that of another plasmid, and the recombinant was a new plasmid containing two origins of replication. The possibility then existed of doing further experiments in which all sorts of foreign DNA, both from microbes and from higher plants and animals, would be inserted into these plasmids. For example, the *E. coli* chromosome with its 4 million base pairs contains about 500 different recognition sites for *Eco*RI. By random insertion of the *Eco*RI fragments into pSC101, it would be possible to clone all the *E. coli* genes in the form of fragments that could be easily isolated for subsequent genetic as well as biochemical manipulation.

DNA of Higher Organisms Becomes Open to Molecular Analysis

Most important were the possibilities that recombinant DNA opened up for the analysis of DNA from plants and animals. Although transducing phages carrying specific parts of bacterial chromosomes had long been available, no one had succeeded in constructing transducing animal viruses carrying specific chromosomal genes such as the ones that code for hemoglobin or the muscle proteins actin and myosin. Now by randomly inserting large numbers of different restriction fragments of human DNA into the proper bacterial plasmids (this procedure is called *shotgun cloning*), we had a high likelihood of subsequently generating one or more bacterial clones containing the recombinant plasmid carrying the specific human gene that was wanted. The moment an appropriate selection technique for a specific mRNA species (such as human hemoglobin mRNA) could be devised, finding the right clone would be no problem. The detailed structural and functional characterization of genes should be virtually commonplace.

With the advent of new recombinant DNA procedures, we now had the perfect method of isolating desired DNA restriction fragments, if not the intact genomes of DNA tumor viruses. Previously, when growing small batches of tumor viruses by conventional cell culture methods, we had never been absolutely sure that we were not subjecting ourselves to some risk of cancer. If such a risk existed, it was assumed to be small because the viruses used (like SV40 or polyoma) were quite similar to common human viruses that apparently did not cause tumors. Nonetheless, it would be a relief to researchers to be able to stop growing intact tumor viruses in large batches, and it was clear that recombinant DNA procedures should be taken up by tumor virologists at maximum possible speed.

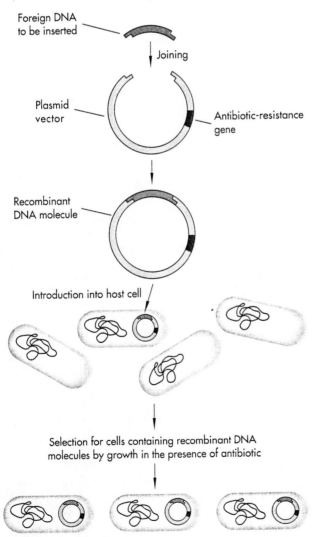

Foreign DNA to be inserted

Joining

Plasmid vector

Antibiotic-resistance gene

Recombinant DNA molecule

Introduction into host cell

Selection for cells containing recombinant DNA molecules by growth in the presence of antibiotic

FIGURE **5-11**
The cloning of DNA in a plasmid.

Scientists Voice Concerns About the Dangers of Unrestricted Gene Cloning

However, in 1971, even as the first plans were being made to carry out gene cloning experiments, concerns began to be raised about the safety of such experiments. These first concerns centered on proposals to clone the genomes of DNA tumor viruses and replicate the recombinant molecules in bacterial cells like *E. coli*. Could *E. coli*, present in the intestinal tract of all human beings, act as a vector to transmit DNA that might cause human cancer? In 1972, it was pointed out that the DNA of mice, and possibly that of human beings, harbored the DNA genomes of latent RNA tumor viruses. Many of the recombinant plasmids that scientists would isolate in their search for specific genes might contain as well these DNA proviruses that might be able to cause human cancer. If so, should all recombinant DNA experiments using vertebrate DNA be regarded as possibly dangerous? The answers were mixed. This was hardly surprising given that the necessary data were not yet available for rational assessment of the risks. There was virtually unanimous agreement, however, that researchers should not have unrestricted freedom to do experiments that might have military consequences.

Thus the first discussions of the conjectured hazards of recombinant DNA techiques took place before the techniques were available to carry out the work. Early discussions aroused no really deep emotions, but the announcement of the first *Eco*RI-pSC101 cloning experiments brought matters to a head. It became clear that the future of studying the molecular basis of genetics lay in using recombinant DNA techniques, and that it would be impossible to go on doing molecular genetics as though recombinant DNA did not exist. The question was whether to move ahead as fast as possible, or to try to devise methods that would allay worries about the risk of these experiments without straitjacketing most future explorations of recombinant DNA.

Guidelines for Recombinant DNA Research Are Proposed at the Asilomar Conference

In July of 1974, a letter appeared in *Science* urging that scientists considering recombinant DNA experiments should pause until there had been further evaluation of the risks involved. This evaluation took place in February of 1975, when a group of more than 100 internationally respected molecular biologists gathered at the Asilomar Conference Center located near Monterey, California. In the absence of knowledge about whether any danger might exist, a nearly unanimous consensus emerged that some restrictions on DNA cloning were appropriate. Amongst other proposals, the recommendation was made that DNA cloning should make use only of bacteria that had been genetically disabled so that they would not grow well outside the test tube.

Afterward, the "Asilomar recommendations" were considered by a special committee appointed by the National Institutes of Health (NIH). In its deliberations, the committee recommended guidelines that effectively precluded the use of recombinant DNA techniques for studing the genes of cancer viruses. These recommendations became codified in official government regulatory guidelines that took force in July 1976 and were administered by the Recombinant DNA Advisory Committee (RAC). A remarkable feature of the discussions about the safety of recombinant DNA was the unprecedented participation of nonscientists in communities like Cambridge, Massachusetts, in the debates. This set a pattern that led to the appointment of lay members to the RAC and has ensured a continuing involvement of nonscientists in discussions about applications of recombinant DNA. These regulations in the United States were paralleled by the establishment of similar bodies in Europe, for example the Genetic Manipulation Advisory Group (GMAG) in the United Kingdom.

Recombinant DNA Comes of Age

Many thought that the NIH regulations were too restrictive, and in some cases scientifically unsound. As more experiments were performed, and more data accumulated, it became increasingly apparent that the actual hazards of recombinant DNA experiments were extremely small. Discussions throughout 1978 led to new, less restrictive NIH regulations that took effect in January 1979 and permitted cloning of viral cancer genes.

Since that time, recombinant DNA technology (which by now has come to include any technique for manipulating DNA or RNA) has fulfilled its promise many times over. As will be apparent throughout this book, the tools of recombinant DNA have been seized eagerly by biologists and have produced a wealth of knowledge in all areas of biology, from control of gene expression to studies of evolution. A new industry, biotechnology, has been created using these same tools to produce drugs and vaccines, and we have learned so much in the past 10 years about the genetics of human beings that comparison with our knowledge prior to 1980 is impossible. The next sections of this book bring us right up to date on new technical advances in cloning genes, and in our understanding of gene structure and gene expression. We will begin with a description of the polymerase chain reaction, a new technique that since the mid-1980s has transformed the ways we design and perform recombinant DNA experiments.

Reading List

General

Wu, R., ed. *Recombinant DNA, Methods in Enzymology,* vol. 68. Academic, New York, 1980.

Hall, S. S. *Invisible Frontiers: The Race to Synthesize a Human Gene.* Tempus, Washington, D.C., 1989.

Kornberg, A., and T. A. Baker. *DNA Replication,* 2nd ed. Freeman, New York, 1992.

Original Research Papers

RNA SEQUENCING

Holley, R. W. "The nucleotide sequence of a nucleic acid." *Sci Am.,* 214(2): 30–39 (1966).

Fiers, W., R. Contreras, F. Duerinck, G. Haegeman, D. Iserentant, J. Merregaert, W. Min Jou, F. Molemans, A. Raeymaekers, V. Berghe, G. Volckaert, and M. Ysebaert. "Complete nucleotide sequence of bacteriophage MS2 RNA: primary and secondary structure of replicase gene." *Nature,* 260: 500–507 (1976). Correction in *Nature,* 260: 810.

RESTRICTION ENZYMES AND MAPS

Linn, S., and W. Arber. "Host specificity of DNA produced by *Escherichia coli,* X. *In vitro* restriction of phage fd replicative form." *Proc. Natl. Acad. Sci. USA,* 59: 1300–1306 (1968).

Meselson, M., and R. Yuan. "DNA restriction enzyme from *E. coli.*" *Nature,* 217: 1110–1114 (1968).

Kelly, T. J., Jr., and H. O. Smith. "A restriction enzyme from *Hemophilus influenzae,* II. Base sequence of the recognition site." *J. Mol. Biol.,* 51: 393–409 (1970).

Smith, H. O., and K. W. Wilcox. "A restriction enzyme from *Hemophilus influenzae,* I. Purification and general properties." *J. Mol. Biol.,* 51: 379–391 (1970).

Danna, K., and D. Nathans. "Specific cleavage of simian virus 40 DNA by restriction endonuclease of *Hemophilus influenzae.*" *Proc. Natl. Acad. Sci. USA,* 68: 2913–2917 (1971).

Sharp, P. A., B. Sugden, and J. Sambrook. "Detection of two restriction endonuclease activities in *Hemophilus parainfluenza* using analytical agarose-ethidium bromide electrophoresis." *Biochemistry,* 12: 3055–3062 (1973).

Roberts, R. J. "Restriction and modification enzymes and their recognition sequences." *Nuc. Acids Res.,* 11: r135–r167 (1983).

SEQUENCING DNA

Sanger, F., G. G. Brownlee, and B. G. Barrel. "A two-dimensional fractionation procedure for radioactive nucleotides." *J. Mol. Biol.,* 13: 373–398 (1965).

Sanger, F., and A. R. Coulson. "A rapid method for determining sequences in DNA by primed synthesis with DNA polymerase." *J. Mol. Biol.,* 94: 444–448 (1975).

Maxam, A. M., and W. Gilbert. "A new method of sequencing DNA." *Proc. Natl. Acad. Sci. USA,* 74: 560–564 (1977).

Sanger, F., G. M. Air, B. G. Barrel, N. L. Brown, A. R. Coulson, J. C. Fiddes, C. A. Hutchison, III, P. M. Slocombe, and M. Smith. "Nucleotide sequence of bacteriophage ϕX174." *Nature,* 265: 678–695 (1977).

Sanger, F., S. Nicklen, and A. R. Coulson. "DNA sequencing with chain-terminating inhibitors." *Proc. Natl. Acad. Sci. USA,* 74: 5463–5467 (1977).

Fiers, W., F. Contreras, G. Haegeman, R. Rogers, A. Vande Voorde, H. Van Heuverswyn, J. Van Herreweghe, G. Volckaert, and M. Ysebaert. "Complete nucleotide sequence of SV40 DNA." *Nature,* 273: 113–120 (1978).

Reddy, V. B., B. Thimmappaya, R. Dhar, K. N. Subramanian, B. S. Zain, J. Pan, P. K. Ghosh, M. L. Celma, and S. M. Weissman. "The genome of simian virus 40." *Science,* 200: 494–502 (1978).

Sutcliffe, G. "Complete nucleotide sequence of the *E. coli* plasmid pBR322." *Cold Spring Harbor Symp. Quant. Biol.,* 43: 77–90 (1979).

OLIGONUCLEOTIDE SYNTHESIS

Heyneker, H. L., J. Shine, H. M. Goodman, H. Boyer, J. Rosenberg, R. E. Dickerson, S. A. Narang, K. Itakura, S. Linn, and A. D. Riggs. "Synthetic *lac* operator is functional *in vivo.*" *Nature,* 263: 748–752 (1976).

Gait, M. J., and R. C. Sheppard. "Rapid synthesis of oligodeoxyribonucleotides: a new solid-phase method." *Nuc. Acids Res.,* 4: 1135–1158 (1977).

Khorana, H. G. "Total synthesis of a gene." *Science,* 203: 614–625 (1979).

Itakura, K., and A. D. Riggs. "Chemical DNA synthesis and recombinant DNA studies." *Science,* 209: 1401–1405 (1980).

ENZYMOLOGY OF DNA REPLICATION

Kornberg, A. "Biologic synthesis of deoxyribonucleic acid." 1959 Nobel prize lecture reprinted in *Science,* 131: 1503–1508 (1960).

Olivera, B. M., and I. R. Lehman. "Linkage of polynucleotides through phosphodiester bonds by an enzyme from *Escherichia coli.*" *Proc. Natl. Acad. Sci. USA,* 57: 1426–1433 (1967).

Weiss, B., and C. C. Richardson. "Enzymatic breakage and joining of deoxyribonucleic acid, I. Repair of single-strand breaks in DNA by an enzyme system from *Escherichia coli* infected with T4 bacteriophage." *Proc. Natl. Acad. Sci. USA,* 57: 1021–1028 (1967).

Zimmerman, S. B., J. W. Little, C. K. Oshinsky, and M. Gellert. "Enzymatic joining of DNA strands: a novel rection of diphosphopyridine nucleotide." *Proc. Natl. Acad. Sci. USA,* 57: 1841–1848 (1967).

THE FIRST RECOMBINANT DNA MOLECULES

Jackson, D., R. Symons, and P. Berg. "Biochemical method for inserting new genetic information into DNA of simian virus 40: Circular SV40 DNA molecules containing lambda phage genes and the galactose operon of *Escherichia coli.*" *Proc. Natl. Acad. Sci. USA,* 69: 2904–2909 (1972).

Mertz, J. E., and R. W. Davis. "Cleavage of DNA by RI restriction endonuclease generates cohesive end." *Proc. Natl. Acad. Sci. USA,* 69: 3370–3374 (1972).

Cohen, S., A. Chang, H. Boyer, and R. Helling. "Construction of biologically functional bacterial plasmids *in vitro.*" *Proc. Natl. Acad. Sci. USA,* 70: 3240–3244 (1973).

Lobban, P., and A. D. Kaiser. "Enzymatic end-to-end joining of DNA molecules." *J. Mol. Biol.,* 79: 453–471 (1973).

THE RECOMBINANT DNA DEBATE

Singer, M., and D. Soll. "Guidelines for DNA hybrid molecules." *Science,* 181: 1174 (1973).

Berg, P., D. Baltimore, H. W. Boyer, S. N. Cohen, R. W. Davis, D. S. Hogness, D. Nathans, R. Roblin, J. D. Watson, S. Weissman, and N. D. Zinder. "Potential biohazards of recombinant DNA molecules." *Science,* 185: 303 (1974).

Berg, P., D. Baltimore, S. Brenner, R. O. Roblin, and M. F. Singer. "Asilomar conference on recombinant DNA molecules." *Science,* 188: 991–994 (1975).

Norman, C. "Genetic manipulation: guidelines issued." *Nature,* 262: 2–4 (1976).

Rogers, M. *Biohazard.* Knopf, New York, 1976.

Department of Health, Education, and Welfare. "Guidelines for research involving recombinant DNA molecules." *Federal Register,* Tuesday, January 29, 1980.

Watson. J. D., and J. Tooze. *The DNA Story.* Freeman, San Francisco, 1981. (A sourcebook for documents on the recombinant DNA debate)

Krimsky, S. *Genetic Alchemy.* MIT Press, Cambridge, Mass., 1983.

Zilinskas, R. A., and B. K. Zimmerman. *The Gene-Splicing Wars: Reflections on the Recombinant DNA Controversy.* Macmillan, New York, 1986.

6

The Polymerase Chain Reaction

The polymerase chain reaction technique (PCR) was devised by Kary Mullis in the mid-1980s and, like DNA sequencing, has revolutionized molecular genetics by making possible a whole new approach to the study and analysis of genes. A major problem in analyzing genes is that they are rare targets in a complex genome that in mammals may contain as many as 100,000 genes. Many of the techniques in molecular genetics are concerned with overcoming this problem. These techniques are very time-consuming, involving cloning and methods for detecting specific DNA sequences (Chapter 7). The polymerase chain reaction has changed all this by enabling us to produce enormous numbers of copies of a *specified* DNA sequence without resorting to cloning. This chapter describes the PCR technique and some of its novel applications. Other applications that are in routine use will be found in many of the following chapters.

The Polymerase Chain Reaction Amplifies Specific Regions of DNA

The PCR exploits certain features of DNA replication. DNA polymerase uses single-stranded DNA as a template for the synthesis of a complementary new

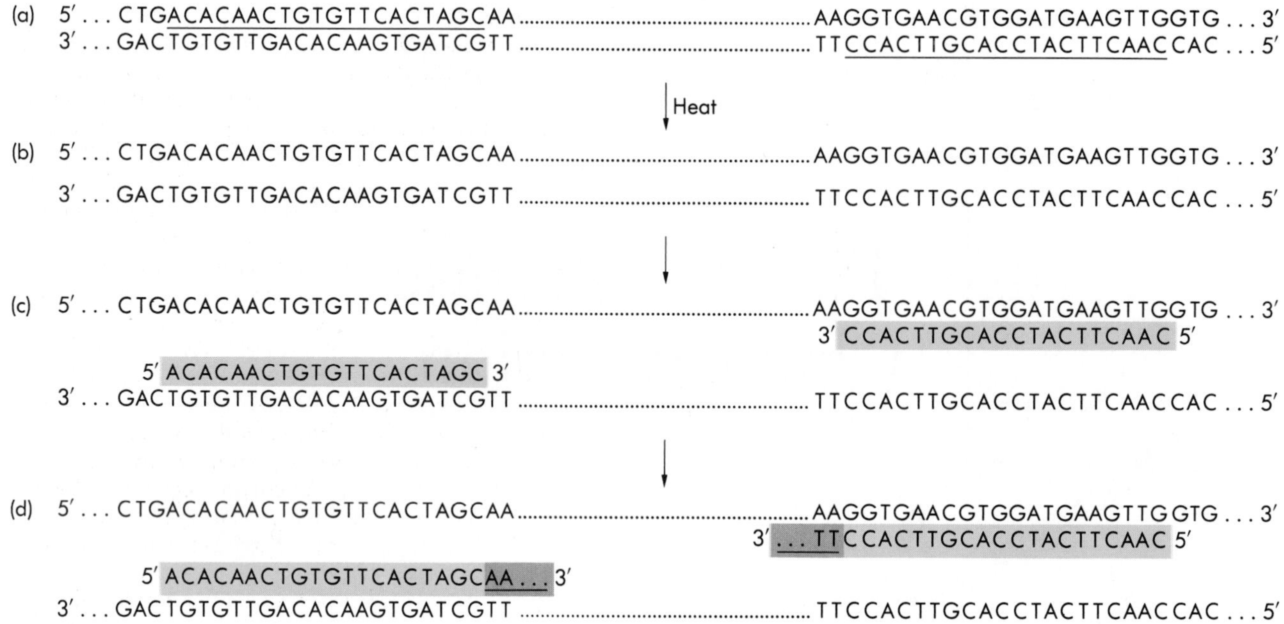

(a) 5'... CTGA<u>CACAACTGTGTTCACTAGC</u>AA ....AAGGTGAACGTGGATGAAGTTGGTG ... 3'
3'... GACTGTGTTGACACAAGTGATCGTT ....TT<u>CCACTTGCACCTACTTCAAC</u>CAC ... 5'

Heat

(b) 5'... CTGACACAACTGTGTTCACTAGCAA ....AAGGTGAACGTGGATGAAGTTGGTG ... 3'

3'... GACTGTGTTGACACAAGTGATCGTT ....TTCCACTTGCACCTACTTCAACCAC ... 5'

(c) 5'... CTGACACAACTGTGTTCACTAGCAA..AAGGTGAACGTGGATGAAGTTGGTG ... 3'
3' CCACTTGCACCTACTTCAAC 5'
5' ACACAACTGTGTTCACTAGC 3'
3'... GACTGTGTTGACACAAGTGATCGTT ....TTCCACTTGCACCTACTTCAACCAC ... 5'

(d) 5'... CTGACACAACTGTGTTCACTAGCAA..AAGGTGAACGTGGATGAAGTTGGTG ... 3'
3'... TTCCACTTGCACCTACTTCAAC 5'
5' ACACAACTGTGTTCACTAGC<u>AA</u>...3'
3'... GACTGTGTTGACACAAGTGATCGTT ....TTCCACTTGCACCTACTTCAACCAC ... 5'

FIGURE **6-1**
Primers for DNA polymerase. (a) The target for amplification, a small section covering 110 bp of the β-globin gene, is shown. Two sequences separated by 60 nucleotides are detailed, and the 20 nucleotides underlined are used as oligonucleotide primers for the PCR. (b) When the DNA is heated, the strands separate. (c) The oligonucleotide primers (shown in green) hybridize specifically to their complementary sequences at the 3' ends of each strand of the target sequence. (d) DNA polymerase uses these primers to begin synthesis of new strands (shown in orange and underlined) complementary to the target DNA sequences in the 5'-to-3' directions.

strand. These single-stranded DNA templates can be produced by simply heating double-stranded DNA to temperatures near boiling. DNA polymerase also requires a small section of double-stranded DNA to initiate ("prime") synthesis (Figure 6-1). Therefore the starting point for DNA synthesis can be specified by supplying an oligonucleotide primer that anneals to the template at that point. This is the first important feature of the PCR—that DNA polymerase can be directed to synthesize a specific region of DNA.

Both DNA strands can serve as templates for synthesis provided an oligonucleotide primer is supplied for each strand. For a PCR, the primers are chosen to flank the region of DNA that is to be amplified so that the newly synthesized strands of DNA, starting at each primer, extend beyond the position of the primer on the opposite strand (Figure 6-2). Therefore new primer binding sites are generated on each newly

synthesized DNA strand. The reaction mixture is again heated to separate the original and newly synthesized strands, which are then available for further cycles of primer hybridization, DNA synthesis, and strand separation. The net result of a PCR is that by the end of n cycles, the reaction contains a theoretical maximum of 2^n double-stranded DNA molecules that are copies of the DNA sequence between the primers (Table 6-1). This is the second important feature of PCR—it results in the "amplification" of the specified region.

Performing a Polymerase Chain Reaction

The PCR is a relatively straightforward laboratory technique, although because the technique is so versatile and the range of applications so wide, it is difficult

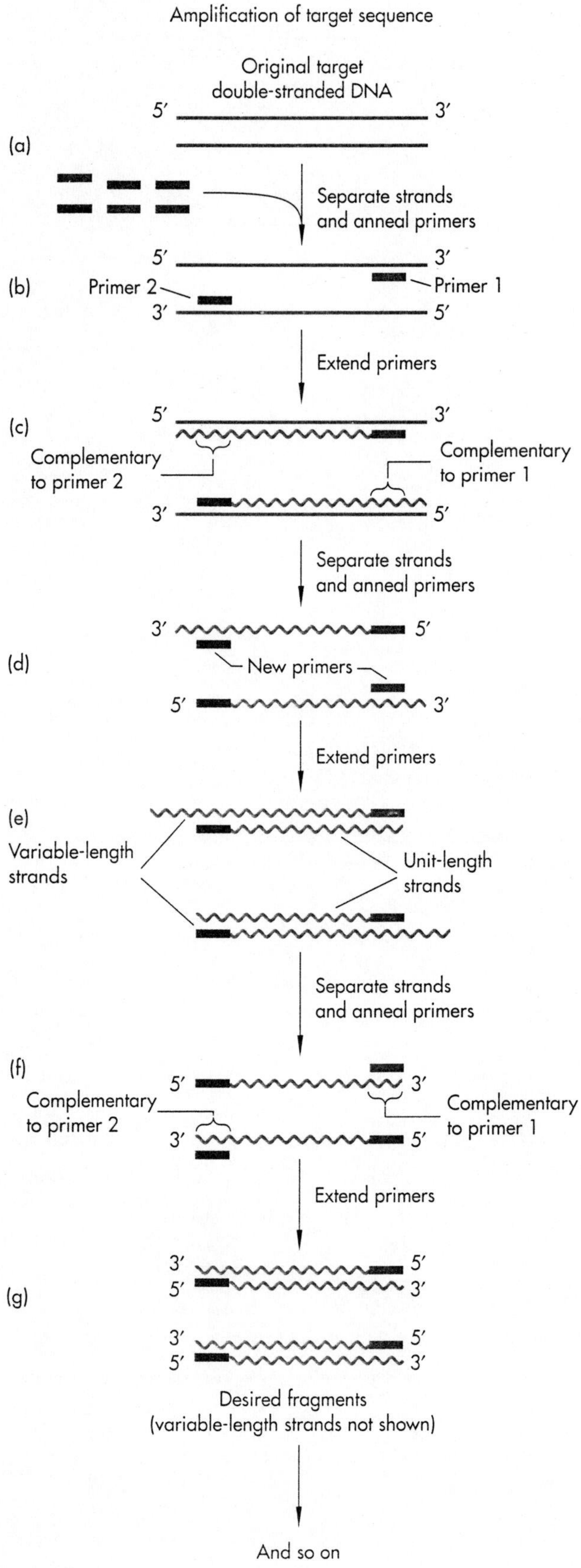

Amplification of target sequence

FIGURE **6-2**

The polymerase chain reaction. (a) The starting material is a double-stranded DNA molecule. (b) The strands are separated by heating the reaction mixture and then cooled so that the primers anneal to the two primer binding sites that flank the target region, one on each strand. (c) *Taq* polymerase synthesizes new strands of DNA, complementary to the template, that extend a variable distance beyond the position of the primer binding site on the other template. (d) The reaction mixture is heated again; the original and newly synthesized DNA strands separate. Four binding sites are now available to the primers, one on each of the two original strands and the two new DNA strands. (To simplify the diagram, subsequent events involving the original strands are omitted.) (e) *Taq* polymerase synthesizes new complementary strands, but the extension of these chains is limited precisely to the target sequence. The two newly synthesized chains thus span exactly the region specified by the primers. (f) The process is repeated, and primers anneal to the newly synthesized strands (and also to the variable length strands, but these are omitted from the figure). (g) *Taq* polymerase synthesizes complementary strands, producing two double-stranded DNA fragments that are identical to the target sequence. The process is repeated. (see figure 6-4).

to give a "typical" example. The starting material for a PCR is DNA that contains the sequence to be amplified. It is not necessary to isolate the sequence to be amplified, because it is defined by the primers used in the reaction. The amount of DNA needed for a PCR is very small. In normal laboratory experiments, less than a microgram of total genomic DNA is sufficient; but as we shall see later, the PCR can be used to amplify sequences from a single DNA molecule. The two oligonucleotide primers directing the starting points for DNA synthesis, DNA polymerase, and a mixture of all four deoxynucleotide precursors are added to a tube containing the DNA. The total volume is usually 100 μL.

The next step in the process is to heat the reaction mixture at about 94°C for 5 minutes (Figure 6-3). At this temperature, the double-stranded DNA molecules separate completely, forming single strands that become the templates for the primers and DNA poly-

TABLE **6-1**

PCR Amplification of DNA Fragment

CYCLE NUMBER	NUMBER OF DOUBLE-STRANDED TARGET MOLECULES
1	0
2	0
3	2
4	4
5	8
6	16
7	32
8	64
9	128
10	256
11	512
12	1024
13	2048
14	4096
15	8192
16	16,384
17	32,768
18	65,536
19	131,072
20	262,144
21	524,288
22	1,048,576
23	2,097,152
24	4,194,304
25	8,388,608
26	16,777,216
27	33,544,432
28	67,108,864
29	134,217,728
30	268,435,456
31	536,870,912
32	1,073,741,824

merase. The temperature is then lowered to allow the oligonucleotide primers to anneal to the complementary sequences in the DNA molecules. This annealing temperature is a key variable in determining the specificity of a PCR, so temperatures and times used vary depending on the sequences to be amplified. This generates the primed templates for DNA polymerase. For the next step, the temperature is raised to 72°C, the optimal temperature for the heat-stable *Taq* DNA polymerase described in the following section. The

temperature is held at 72°C for up to 5 minutes for DNA synthesis to proceed.

At the end of this period, the temperature is raised once more to 94°C, but now for only 20 seconds, so that the short stretches of double-stranded DNA (the original strand and the newly synthesized complementary strand) separate. These single strands become templates for another round of DNA synthesis, and the cycle of heating to separate strands, annealing of primers, and synthesis by DNA polymerase is repeated for as many as 30 to 60 cycles (Figure 6-4).

Taq Polymerase Simplifies and Improves the PCR

Originally, *E. coli* DNA polymerase was used in the PCR, but this enzyme is heat-sensitive and is destroyed at the temperatures needed to separate double-stranded DNA. Therefore fresh enzyme had to be added manually for each cycle, a tedious process. An

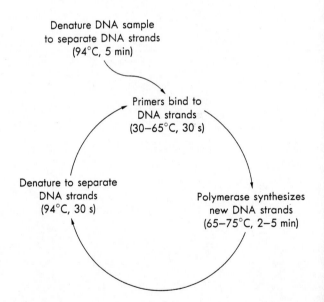

FIGURE **6-3**
The PCR cycle. The DNA sample is heated to separate the DNA strands (initial denaturation), and then the reaction mixture goes through repeated cycles of primer annealing, DNA synthesis, and denaturation. The target sequence doubles in concentration for each cycle.

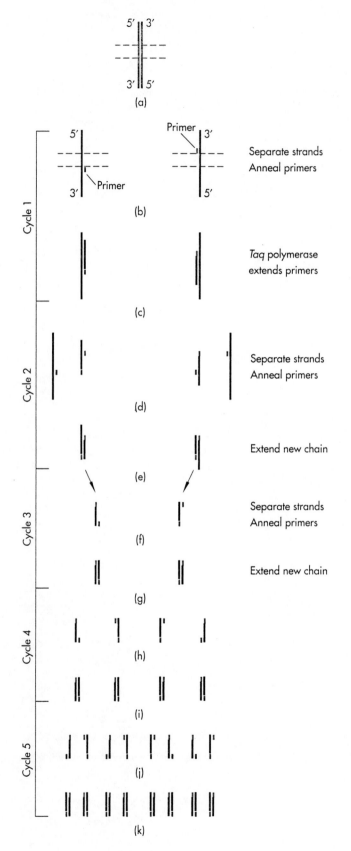

FIGURE 6-4
The polymerase chain reaction leads to amplification of the target sequences (marked with dashed lines in [a] and [b]). The first of the PCR-synthesized target fragments is shown in (g) (corresponding to [g] in figure 6-2). This figure shows how the number of these fragments subsequently doubles for each cycle of the reaction.

important technical advance came with the discovery that bacteria living in hot springs have DNA polymerases that work best at high temperatures. The bacterium *Thermus aquaticus* lives in water at a temperature of 75°C. Its DNA polymerase (*Taq* polymerase) has a temperature optimum of 72°C and is reasonably stable even at 94°C! *Taq* polymerase can be added just once at the start of a reaction and will remain active through a complete set of amplification cycles. This development has allowed the automation of the PCR through the use of thermal cyclers, which are heating blocks that can be programmed to carry out the time and temperature cycles for a PCR. Now the ingredients for a PCR can be placed in a thermal cycler and the reaction carried out without any manual intervention.

The specificity and sensitivity of the PCR is also improved by *Taq* polymerase. At the lower temperature required for *E. coli* DNA polymerase, the primers can anneal at sites where sequences differ slightly from the target sequences (Figure 6-5). Amplification can occur when mismatching primers are close together on opposite strands of DNA. Because correct complementary sequences for the primers are incorporated into the synthesized fragments, an unwanted sequence is produced with ends that precisely match the primers. Such an "incorrect" fragment synthesized in the early cycles of a PCR will thus be efficiently amplified on subsequent cycles. In contrast, annealing of oligonucleotide primers to sites other than the desired ones is significantly reduced at the temperatures used with *Taq* polymerase. Consequently there is no amplification of sequences other than the targeted sequence. The specificity is improved further in the "hot-start" method, where all the reagents save one are heated to 72°C before adding the final ingredient, for example,

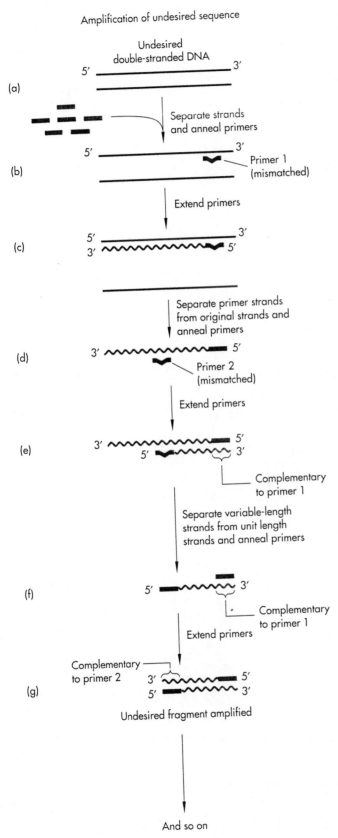

Amplification of undesired sequence

Undesired
double-stranded DNA

(a)

Separate strands
and anneal primers

(b) Primer 1 (mismatched)

Extend primers

(c)

Separate primer strands
from original strands and
anneal primers

(d) Primer 2 (mismatched)

Extend primers

(e) Complementary to primer 1

Separate variable-length
strands from unit length
strands and anneal primers

(f) Complementary to primer 1

Extend primers

Complementary to primer 2

(g)

Undesired fragment amplified

And so on

the *Taq* polymerase. This increase in specificity simplifies the analysis of PCR products. The amplified target fragment can be seen easily on an ethidium bromide–stained gel because the background staining of non-target sequences is eliminated.

Fidelity of DNA Synthesis by *Taq* Polymerase Determines the Accuracy of PCR Amplification

Like all other biochemical processes, DNA replication is not a perfect process, and occasionally DNA polymerase will add an incorrect nucleotide to the growing DNA chain. The rate of misincorporation measured in a naturally replicated DNA molecule is approximately 1 in 10^9 nucleotides. Cells achieve such extraordinary accuracy because the DNA replication machinery removes mismatched nucleotides added to the DNA chain. In vitro *Taq* polymerase does not have this "proofreading" capability, and using the temperatures and salt concentrations typical of a PCR, the enzyme incorporates one incorrect nucleotide for about every 2×10^4 nucleotides incorporated. This is not a serious matter for bulk analysis of PCR products because molecules with the same misincorporated nucleotide will form a very small portion of the total

FIGURE **6-5**
Mismatching primers can lead to the efficient amplification of undesired sequences. Following separation of the original double-stranded DNA (a), oligonucleotide primers can anneal to sequences that differ slightly from the target sequences (b). DNA polymerase will use these mismatched primers and synthesize a complementary strand to the undesired sequences 3′ to the primers (c). The first mismatch will produce a DNA strand of indefinite length that has the first primer incorporated at its 5′ end (d). A mismatch of the second primer to that undesired strand produces a double-stranded DNA molecule, one strand of which has the second primer at its 5′ end and sequences complementary of the first primer at its 3′ end (e). This strand (f) is now a perfect template for subsequent rounds of amplification, and the unwanted DNA increases in concentration just as the target sequence does.

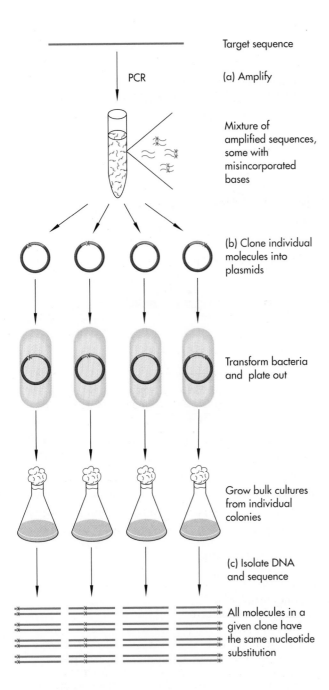

Target sequence

(a) Amplify

Mixture of amplified sequences, some with misincorporated bases

(b) Clone individual molecules into plasmids

Transform bacteria and plate out

Grow bulk cultures from individual colonies

(c) Isolate DNA and sequence

All molecules in a given clone have the same nucleotide substitution

FIGURE **6-6**
Cloning DNA using the PCR and the consequences of polymerase mistakes. (a) *Taq* polymerase misincorporates nucleotides (indicated by an "x") into amplified sequences. (b) Cloning into a vector selects just one DNA molecule from the whole population of PCR molecules. (c) When DNA is prepared from these bacteria and sequenced, all the cloned DNA molecules are the same in each clone. In contrast, the vast majority of the molecules in the original PCR mixture have the correct nucleotide at any particular position, so that sequencing the mixture averages out the substitutions in individual molecules.

number of molecules synthesized. But misincorporation is important if PCR fragments are to be used for cloning (Figure 6-6). Each clone is derived from a single amplified molecule. If this molecule contains one or more misincorporated nucleotides, then all the cloned DNA in that clone will carry the identical "mutation." This problem can be reduced by beginning the PCR with a large, rather than a small, number of template molecules. Fewer cycles of amplification are needed, and less total DNA synthesis takes place.

DNA for the PCR Comes from a Variety of Sources

DNA for a PCR is often total genomic DNA extracted from cells. However, the PCR does not require highly purified DNA, and DNA released by the boiling of cells can be used directly without any purification. The PCR can also be used to study the pattern of gene expression: mRNA is converted to cDNA using reverse transcriptase, and the cDNA then serves as the template for the PCR. DNA sequences do not have to be isolated before amplification by a PCR, because the specificity of the reaction is determined by the oligonucleotide primers. For example, a PCR is a very convenient way of preparing DNA from cloned inserts in plasmid or bacteriophage vectors (Figure 6-7). All that is needed are oligonucleotide primers complementary to the vector sequences on either side of the cloning site. Any insert is easily amplified in this way regardless of its sequence. This PCR method is now being used routinely to screen for clones containing inserts.

DNA is a very stable molecule, provided it is not exposed to nucleases, and DNA from the most extraordinary sources has been used for PCRs. For example, human papilloma virus DNA has been detected in cervical carcinoma biopsies embedded in paraffin for over 40 years. Human archival materials that can provide DNA for PCR are not restricted just to those embedded in paraffin. Blood samples for the neonatal detection of phenylketonuria have routinely been taken by heel-prick of newborns and stored as dried

(a)

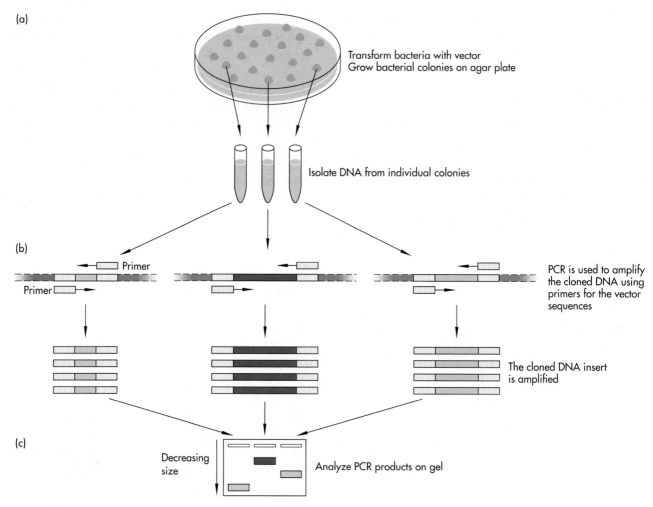

(b)

(c)

FIGURE **6-7**
Amplifying cloned DNA from vectors. (a) Single plaques or colonies are picked for PCR amplification. (b) The sequences flanking the cloning sites of the vector are known, and these sequences are used to design primers that serve as the starting points for *Taq* polymerase. DNA strands are synthesized complementary to the cloned DNA that lies between the primers. (c) The amplified sequences can be characterized by electrophoresis and Southern blotting.

spots on cards for many years. As much as 1 μg of DNA can be extracted from a blood spot; this is more than enough to be used for PCR analyses of human disease genes such as the genes for phenylketonuria and cystic fibrosis. As we shall see later, even DNA isolated from Egyptian mummies several thousand years old has been amplified by the PCR.

PCR is used to Amplify Human-Specific DNA Sequences

When human cells are fused with rodent cells, the human chromosomes are preferentially lost from the hybrid cells, so that hybrid cell lines containing only one human chromosome, or even a part of a human chromosome, can be produced. These cell lines are a convenient source of DNA for cloning if the fragment of chromosome carried by the cells contains the desired gene. So-called *Alu-PCR* provides a very simple means of characterizing and amplifying just the human DNA in these hybrid lines. The method uses primers for repetitive sequences that are inserted at many places in genomes (chapter 8). One of these elements, the *Alu* repeat, is present in as many as 900,000 copies in the human genome. This 300-bp *Alu* repeat is very variable, but it contains a sequence that is human-specific. Two oligonucleotide primers were made, one

in each direction, using the most conserved section of this sequence (Figure 6-8). The primers cannot be used together because they hybridize to each other. However, *Alu* sequences can be present in either direction in DNA, so the primers can be used individually, and human sequences will be amplified if they lie between adjacent *Alu* repeats that are pointing in opposite directions (Figure 6-8). Varying numbers of human DNA fragments are produced depending on the size of the human DNA in the cell line. This technique is widely used for "rescuing" human DNA sequences from other DNAs.

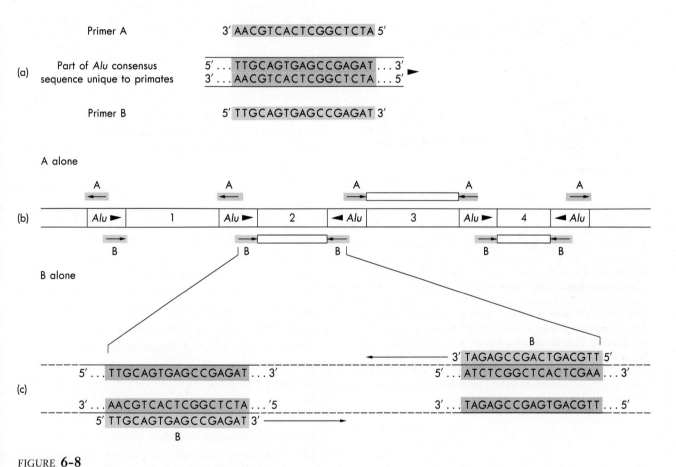

FIGURE **6-8**

Alu-PCR is used to isolate human-specific DNA sequences . (a) The seventeen base pair *Alu* sequence that is used to design the primers is shown; and the two primers, A and B, are shown above and below. The arrowhead indicates the orientation of the *Alu* repeat. (b) A stretch of human DNA about 18 kbp long will contain about five *Alu* sequences, separating four sections of DNA (1–4). The orientation of these *Alu* sequences is shown by the arrowheads. Only section 3 is amplified when primer A alone is used in a PCR, because the *Alu* sequences flanking it are the only ones pointing in opposite directions, allowing the A oligonucleotides to anneal in the correct orientation for PCR amplification. (c) This process is shown in detail for primer B and section 2 of the DNA.

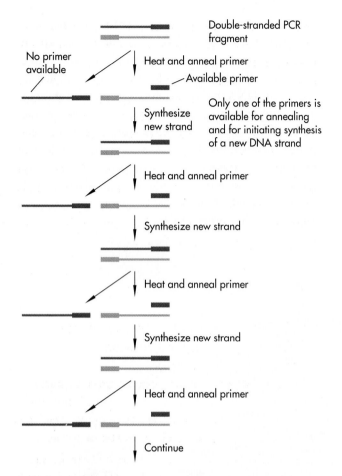

Double-stranded PCR fragment

No primer available

Heat and anneal primer

Available primer

Only one of the primers is available for annealing and for initiating synthesis of a new DNA strand

Synthesize new strand

Heat and anneal primer

Synthesize new strand

Heat and anneal primer

Synthesize new strand

Heat and anneal primer

Continue

FIGURE **6-9**
Asymmetric PCR for producing single-stranded DNA for sequencing. One of the two primers for the PCR is present in limiting concentration. A small number of cycles take place producing double-stranded DNA copies of the target sequence (PCR fragment), until the limiting primer (orange) is depleted. The other primer (purple) continues to initiate DNA synthesis using as templates the strands previously synthesized with the orange primer. These template strands are continually recycled, so that by the end of the reaction, strands initiated with the purple primer are present in great excess over the orange strands. The excess of purple strands remain single-stranded.

PCR Products Can Be Sequenced Directly

Like any DNA, the double-stranded products of a PCR can be sequenced. However, because single-stranded DNA is the best template for Sanger's dideoxy chain termination method (see Chapters 5 and 7), a technique called *asymmetric PCR* has been devised

to produce single-stranded DNA (Figure 6-9). The standard PCR is set up except that the concentration of the two primers differs by a factor of 100. Double-stranded DNA fragments are produced until the limiting primer is used up. The remaining primer continues to anneal and prime DNA synthesis, producing only one of the two strands. Although this strand accumulates at a linear rather than an exponential rate, sufficient single-stranded DNA is produced for sequencing.

Detecting Mutations Using PCR Amplification

Mutations occur in cancers and in inherited disorders. Knowing the nature of the mutation in a patient is important for diagnosis and therapy. PCR amplification is proving invaluable as a tool for screening particular genes for mutations. As we shall see later in this chapter, the PCR can be used to follow the fate of cancer cells after therapy; and in Chapter 27 we shall discuss examples of the increasing number of human inherited disorders being diagnosed in this way.

In 1989, Harold Varmus and J. Michael Bishop were awarded the Nobel prize for their discovery that the oncogenes carried by RNA tumor viruses are part of the normal genetic makeup of the cell; cancers arise when these normal genes mutate (Chapter 18). This major advance in cancer research has been taken further with the recognition that certain cancers are caused by specific and reproducible mutations. For example, mutations in the *ras* oncogene have been recognized in many human cancers, and the PCR has been used to analyze the pattern and frequency of mutations in the human *RAS* genes. (By convention, human oncogenes are written in capital letters.) Samples from large numbers of patients were screened rapidly by the PCR, and the studies revealed that different forms of lymphoid malignancies had different *RAS* mutations.

Genetic analysis has important practical implications for retinoblastoma, a childhood cancer of the eye caused by mutations in a gene at q14 on chromosome 13 ("q" refers to the long "arm" of the chromosome; "14" refers to a particular location on this arm). There is a heritable form of retinoblastoma in which there is a germ line mutation at one of the two retinoblastoma alleles. All the cells of these patients contain one normal and one mutant allele, and a single mutation at the normal allele is all that is required to develop the cancer. Retinoblastoma can also arise spontaneously in patients who have not inherited a mutant allele if a retinal cell suffers two mutations, one in each allele. Thus in families with just one affected member, it has been difficult to determine if the hereditary form of retinoblastoma is involved. This can now be done by using the PCR and sequencing to analyze mutations in tumor tissue and normal tissue from affected individuals. Mutations in some patients are found only in tumor tissue, suggesting that these are spontaneous mutations and that the children of these patients will not be predisposed to developing retinoblastoma. In other cases, the mutation is found in normal tissue as well as in the tumor cells, showing that these patients are born with one mutation in all their cells. The children of these patients will be at a greatly increased risk of developing retinoblastoma because they may inherit a mutated copy of the retinoblastoma gene. These findings are important for genetic counseling.

PCR Amplification Is Used for Monitoring Cancer Therapy

The ability to detect genetic lesions characteristic of tumor cells is a valuable tool for the oncologist trying to determine if a patient being treated for a leukemia is free of malignant cells. The physician would like to stop treatment with cytotoxic drugs or radiation as soon as the cancer is destroyed, and conversely would want to resume treatment as soon as a relapse begins. As we shall learn in Chapter 18, some cancers arise as a result of chromosomal translocations that involve known genes. For example, there is a translocation between chromosomes 14 and 18 in follicular lymphomas. With such mutations as markers, these cancer cells can be detected by conventional Southern blotting if they are present at concentrations of 1 in 100 normal cells, so that a patient judged cancer-free by Southern blotting may still harbor significant numbers of cancer cells. In contrast, PCR techniques are capable of detecting as few as 1 cancer cell in 10^6 normal cells, providing a much more sensitive indicator for the oncologist. The two PCR primers are chosen from the sequences adjacent to the breakpoints on each chromosome. It is only in cells with the translocation that the primers are brought together so that the sequences between them can be amplified (Figure 6-10). A similar strategy has been devised for detecting leukemias using mRNA as the starting material. This has the advantage that mRNA already represents an amplification over genomic DNA in that a cell contains many copies of the mRNA for a gene.

PCR Amplification Is Used to Detect Bacterial and Viral Infections

Similarly, PCR techniques can be used to monitor bacterial or viral infections. Conventional diagnostic procedures are based on the ability to grow organisms in culture or to detect their presence in patients by using antibodies. One difficulty with these types of tests is that the former may require several weeks before diagnosis is possible and the latter is relatively insensitive. These are particularly important considerations for diagnosis of AIDS or for studies of the epidemiology of human immunodeficiency virus (HIV) infections. Here, as in cancer therapy, the aim is to detect rare infected cells among a large population of uninfected cells. To detect HIV, PCR primers are made for sequences in the virus, and a PCR is carried out using DNA extracted from peripheral blood cells. This approach detects integrated HIV DNA in infected cells. The presence of viral RNA is thought to indicate an active infection, and this can be diagnosed by performing PCR using cDNA templates produced by reverse transcription of RNA from infected cells.

Tuberculosis is caused by *Mycobacterium tuberculosis*; diagnosis is difficult because there may be too few organisms in pathological samples for histological diagnosis. Instead, the pathogen has to be identified following growth in culture and antibiotic sensitivity testing, a procedure that takes several weeks. This is clearly a situation in which PCR techniques should

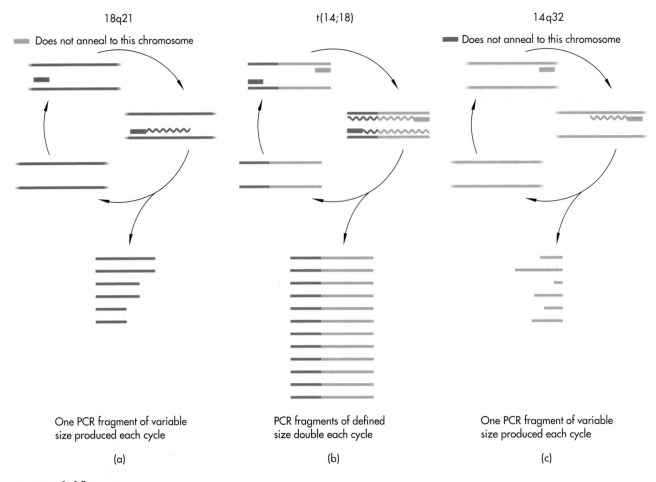

18q21

t(14;18)

14q32

Does not anneal to this chromosome

Does not anneal to this chromosome

One PCR fragment of variable size produced each cycle

PCR fragments of defined size double each cycle

One PCR fragment of variable size produced each cycle

(a)

(b)

(c)

FIGURE **6-10**

Detecting cancer cells using the PCR. In follicular lymphomas, there is a chromosomal rearrangement involving chromosomes 14 and 18. Part of the immunoglobulin heavy chain locus (J_H) on chromosome 14 at q32 is translocated to the locus for the *BCL-2* oncogene on chromosome 18 at q21. The breakpoints on the chromosomes are consistent from one patient to another, so that it is possible to design primers that anneal to chromosomes 18 (a) and 14 (c), 5′ and 3′, respectively, to the breakpoints on the chromosomes. The t(14;18)(q32;q21) translocation ("t" indicates translocation; the "14" and "18" refer to chromosomes 14 and 18) brings the two primers close together on the same double-stranded DNA molecule (b). When DNA from patients having this translocation is amplified, the primers direct the synthesis of a PCR fragment approximately 300 bp long that crosses the boundary of the translocation. This defined fragment doubles in quantity with each cycle. In contrast, although these primers will initiate synthesis of new DNA strands on the normal chromosome 14 (c) and chromosome 18 (a), respectively, these strands accumulate at a linear rate because they can be synthesized using only the starting DNA as a template. They are of variable length because their 3′ ends are not defined.

help. PCR amplification has been performed using primers for a sequence within a gene that is highly conserved in all mycobacterial species. The amplified DNA fragment was hybridized with species-specific probes to identify the specific strain involved. The PCR-based test proved far more rapid than the conventional test and far more sensitive, detecting as few as 10 bacilli in 10^6 eukaryotic cells. It is clear that PCR techniques will soon be adapted to routine diagnostic use in clinical microbiology laboratories.

PCR Amplification Is Used for Sex Determination of Prenatal Cells

One area in which genetic analysis of a small number of cells is very important is in prenatal diagnosis. For inherited X-linked disorders that affect only males, sex determination is the first step in prenatal diagnosis. Male sex determination using DNA is possible because males carry unique sequences on the Y chromosome. Some of these are repeat sequences, already "amplified" relative to normal genes. For example, the 3.5-kb DYZ1 sequence is present on the Y chromosome in as many as 5000 copies. PCR techniques can be used to amplify a 149-bp fragment from the DYZ1 sequence, specific to males. In one experiment to perform prenatal sex determination after in vitro fertilization, researchers removed a single cell from each 10-cell human preimplantation embryo using a micromanipulator, and the DYZ1 sequence was amplified from each cell using 60 cycles of the PCR (Figure 6-11). An amplified fragment was obtained only from cells from male embryos.

This procedure has now been used clinically for families at risk for X-linked inherited disorders, with implantation of biopsied embryos into mothers. The speed of PCR allows biopsy, sex determination, and transfer of the embryos to the mothers in the same day. Usually there are sufficient numbers of female embryos for two embryos to be transferred, and their subsequent progress is monitored by assay of chorionic

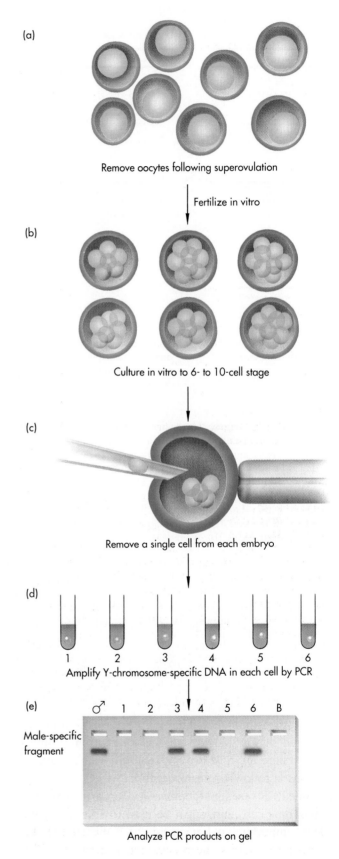

FIGURE **6-11**
Determining sex of fetuses at risk for X-linked inherited disorders. (a) Oocytes are removed from the mother following superovulation and fertilized in vitro. (b) The oocytes that are fertilized successfully are cultured in vitro until there are 6 to 10 cells in each embryo. (c) A hole is made in the zona pellucida and a single cell removed from each embryo. (d) Amplification of the DYZ1 sequence is attempted. (e) Only in DNA from males is the male-specific DYZ1 sequence amplified by PCR, giving rise to a 149-bp, male-specific fragment. The lane marked with the male symbol is a positive control showing the expected fragment; the lane marked B (for "Blank") is from a PCR that included all the reagents but no DNA and is used to detect any contamination. Female embryos are negative (lanes 1, 2, and 5) and are implanted into the mothers.

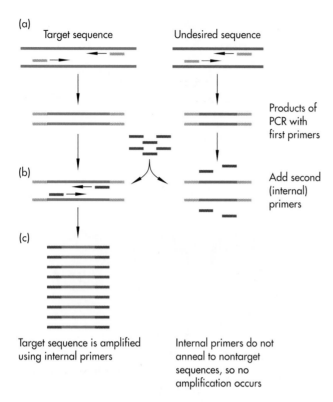

(a) Target sequence   Undesired sequence

Products of PCR with first primers

(b)   Add second (internal) primers

(c)

Target sequence is amplified using internal primers

Internal primers do not anneal to nontarget sequences, so no amplification occurs

FIGURE **6-12**

Nested primers increase the specificity of a PCR. The chance that undesired sequences will be amplified increases with increasing numbers of amplification cycles. This can be reduced by performing two rounds of PCR consecutively, using two different pairs of primers. The known sequence is used to design two pairs of primers, one pair (internal primers) within the region amplified by the first pair. (a) The products of the first PCR include both the target and the undesired sequences. (b) A sample of this reaction mixture is then amplified using the second, internal pair of primers. (c) Only fragments that have sequences complementary to the second set of primers will be amplified. Unwanted regions of DNA amplified by the first primers are very unlikely to contain sequences that will anneal the second pair of primers. In this way, the second round of the PCR selectively amplifies fragments containing the target region.

gonadotrophin levels. The sex of these fetuses is checked by karyological analysis of chorionic villus cells. There are now five apparently normal, healthy girls born following this procedure.

The PCR can amplify *repeated* sequences from single embryonic cells, but can it be used to amplify *single* genes important in human inherited diseases? This question has not yet been answered directly. Instead,

unfertilized human oocytes were used as equivalents for single embryonic cells. Sequences from two genes, those for Duchenne muscular dystrophy and cystic fibrosis, were amplified from these single cells, and the appropriate fragments were detected. In order to detect an amplified fragment from the cystic fibrosis gene, amplification had to be carried out for 80 cycles.

It is possible to determine that a fetus is male by cytological detection of male cells in the peripheral blood of a mother carrying a male fetus. However, the reliability of this procedure is poor and precludes its use for prenatal sex determination. It has been estimated that the number of fetal male cells present in the maternal circulation is less than 1 per 70,000 maternal cells, but as we have seen, PCR techniques are capable of detecting such rare cells. In one study, blood samples were taken from pregnant women as early as after 9 weeks of gestation, DNA extracted, and part of the DYZ1 sequence amplified. After 40 cycles of amplification, an aliquot of the reaction mixture was used for a further 15 to 20 cycles of amplification. The primers used for this second round of amplification were internal to the first pair of primers. So-called *nested primers* reduce significantly the chance of amplifying unwanted sequences when a large number of amplification cycles are used (Figure 6-12). Sexes of all fetuses in this study were determined correctly. Fetal sampling, whether by chorionic villus biopsy or by amniocentesis, carries a small but significant risk of miscarriage, so the advantage of using maternal blood for fetal sexing is clear. Whether this technique can be developed for routine use is another matter. Quite extraordinary lengths had to be taken to prevent contamination of samples by male DNA (see below).

PCR Methods Permit Linkage Analysis Using Single Sperm Cells

Determining recombination between genes at meiosis has been the classic way to map genes ever since its application to *Drosophila* by A. H. Sturtevant and T. H. Morgan. This is difficult for mapping human genes because linkage studies have to be performed in families, where matings cannot be controlled and the number of offspring is low (see Chapters 26 and 30 for a further discussion of human gene mapping).

FIGURE **6-13**

Gene linkage analysis using single sperm. (a) The positions of the parathyroid hormone gene locus (PTH) and $^G\gamma$-globin gene locus (HBG2) are shown on the short arm of human chromosome 11. (b) Single sperm are transferred to tubes, and amplification of the PTH and HBG2 loci is carried out simultaneously using two different sets of primers. (c) The products of the PCR are spotted onto a nitrocellulose filter and tested with probes specific for each of the four possible alleles, *A* and *a* for PTH and *B* and *b* for HBG2. The hybridization pattern is revealed after exposing the filter to x-ray film. (d) The haplotype of each sperm for these alleles can be read from the autoradiograph, and the recombination fraction between the loci can be calculated; hence the genetic distance between the two loci can be determined.

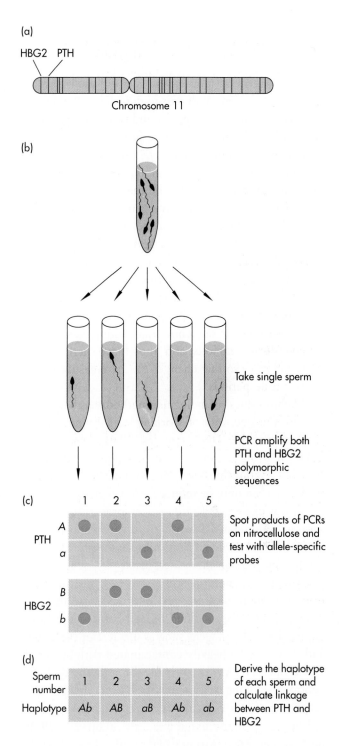

PCR analysis of alleles in sperm offers a way of performing linkage analysis with single cells. A chromosome contained in a sperm is a single meiotic product, and by examining two loci at a time and determining the frequency of recombination between them, it should be possible to derive the genetic distance between the loci. For the geneticist, examining 1000 sperm is like studying a family with 1000 children. These estimates of genetic distances are for *male* chromosomes, an important point, given that recombination frequencies for some genes differ for male and female meioses.

A preliminary experiment using loci with a known recombination rate was performed to determine the feasibility of this approach. The loci chosen, the parathyroid hormone gene locus (PTH) and the $^G\gamma$-globin locus (HBG2) on the short arm of human chromosome 11, have a recombination frequency of about 0.15 based on family analysis. For this PCR experiment a total of 708 sperm from two donors heterozygous at both loci were analyzed. The genes were amplified and then tested with allele-specific oligonucleotide probes to detect the four possible meiotic products (Figure 6-13). To calculate a recombination frequency, technical complications have to be taken into account; for example, two sperm may be in the same tube. As a consequence, a complex statistical treatment of the data is necessary. Nevertheless, the frequency of recombination as calculated in this experiment was 0.16, in very good agreement with the estimate from family studies. The combination of PCR and allele-specific oligonucleotide probes offers a promising method for fine-resolution mapping of human genes. One current limitation is isolating sufficient numbers of single sperm cells for analysis, but this may be overcome by using fluorescence-activated cell sorters.

PCR Techniques Are Used in Studies of Molecular Evolution

Molecular genetic information is being used increasingly in evolutionary studies to determine the degree of relatedness between species and in so doing to build up family trees similar to those produced by classical comparative methods (Chapter 22). It is assumed that as species diverge from a common ancestor, their nucleotide sequences diverge and that measurements of the degree of nucleotide divergence (or, conversely, of nucleotide homology) can be used to determine relationships. Homology can be measured by determining the degree of hybridization between total DNA from individuals, by comparing nucleotide changes in the same gene between species, or by comparing mitochondrial genes. Mitochondrial genes have the advantage that the mitochondrial genome is not rearranged during meiosis and they have high point-mutation rates so that changes can be measured over shorter time periods.

The usefulness of molecular methods has been limited by the necessity of working with living species that can provide DNA samples. Thus, relationships between living organisms can be examined directly, but the relationships of living organisms to extinct organisms can only be inferred. However, tissue samples from extinct and rare species preserved in museums throughout the world are a vast resource, and DNA has been isolated from sources as varied as museum skins, human mummies, dried plants, and even soft tissue specimens in preservatives. Unfortunately, the DNA molecules are very short because of degradation; they are modified because of damage from environmental mutagens such as ultraviolet radiation; and they may be heavily contaminated with bacterial DNA. Such DNA is not suitable for studying by conventional cloning techniques. For example, the average size of DNA fragments cloned from 4000-year-old human mummy tissue was only 90 bp. The PCR has changed the situation dramatically. The PCR efficiently amplifies small fragments of DNA, such as are more likely to remain intact in ancient samples of DNA, and PCR methods can amplify these intact molecules even if very few are present. Remarkably, it seems that *Taq* polymerase can synthesize full-length target sequences, even if copies of the target sequence are fragmented, provided these fragments overlap. As the amplified fragments increase in concentration, two different fragments can anneal to each other by their overlapping sequences, and function as primers. Each fragment is therefore extended using the other fragment as a template, so that even in this situation a complete target sequence is synthesized.

For example, analysis of mitochondrial DNA has been used similarly to elucidate the phylogenetic position of *Thylacinus cynocephalus*, the extinct marsupial wolf. Sequences from the mitochondrial cytochrome *b* and 12S ribosomal RNA genes were amplified using DNA isolated from the hides of museum specimens and compared with the same sequences from living marsupials and the South American opossum. Sequence analysis of the PCR products showed clearly that *Thylacinus cynocephalus* is related more closely to marsupials and justifies its common name of the marsupial wolf.

The PCR has even been used to amplify DNA sequences from fossils. These fossils may be formed under conditions that promote a remarkable degree of preservation. Leaves in shale fossil beds dating from the Miocene Period (18 million years ago) are so well preserved that the leaves are still green. DNA has been extracted from fossilized leaves scraped off the rocks and used for a PCR with primers for part of the ribulose 1,5-biphosphate carboxylase gene (*rbc*L). Astonishingly, the correct size of fragment, 820 bp long, was produced—from DNA molecules 18 million years old. Furthermore, asymmetric PCR was used to produce single-stranded molecules for sequencing. Comparisons of the *rbc*L sequence of the fossil plant with that of existing species identified this extinct Miocene plant as a member of the Magnolia family.

PCR has also been used in studies of the evolution of extant human populations. For example, mitochondrial sequences in DNA from freshly plucked hair were amplified using the PCR and analyzed to determine the genealogical relationships of the !Kung people. These are aboriginal southern Africans who speak click language and who have remained genetically isolated. This study supported the notion of the African origin of human mitochondrial DNA.

Contamination Can Be a Problem in PCR Studies

An unforeseen and unwelcome corollary of the amplification power of the PCR is that minor contamination of the starting material can have serious consequences. For example, in the study of prenatal sex determination using maternal blood samples described previously, there was one positive sample from a non-pregnant negative control. This was attributed to contamination by minute amounts of tissue shed from the skin of a male worker. It proved necessary to have only women preparing and analyzing the DNA samples. In another case, DNA from one of the museum specimens of the marsupial wolf yielded human mitochondrial sequences when amplified. In this case, the contaminating PCR product was readily identified as a contaminant by sequence analysis, but the contamination of ancient human DNA by contemporary human DNA might be much more difficult to detect. Contamination may prove to be a significant problem in forensic applications of the PCR. Biological samples associated with crime scenes are rarely in the pristine state of the materials used to prepare DNA in the laboratory (see Chapter 27 for a discussion of forensic applications of recombinant DNA techniques).

One common source of contamination is the products of previous amplification reactions. A completed PCR reaction mixture may contain as many as 10^{13} amplified fragments, so that even minute volumes such as droplets in an aerosol from a pipette tip contain very large numbers of amplifiable molecules. One solution to contamination problems is to physically separate pre- and postamplification steps. A clever development uses uridine triphosphate (UTP) instead of deoxythmidine and includes the enzyme uracil-N-glycosylase (UNG) in the PCR reaction mixture. UNG will degrade any DNA contamination from a former PCR because the previously amplified DNA contains UTP. UNG is then inactivated by the high temperature of the new PCR, so the products of the new reaction can accumulate. It is always essential to run negative controls that will reveal the presence of contaminating DNA in the reaction reagents in every PCR experiment.

The Polymerase Chain Reaction— A Technical Revolution in Molecular Genetics

The versatility of the polymerase chain reaction is enormous, and the combination of the PCR and sequencing is an extraordinarily powerful tool for the analysis of genes. The most spectacular demonstration of this is the molecular identification of the Magnolia plant from the Miocene Period. Who could have imagined that it would ever be possible to analyze in detail genes from DNA 18 million years old? Other applications of the PCR described later in this book are no less remarkable. For example, it is used for cloning genes, for in vitro mutagenesis, in mapping and sequencing large genomes, and in diagnosing human inherited disorders. In 1985 there were but three research reports on the PCR; 5 years later it is being used in thousands of laboratories. The PCR has indeed revolutionized the practice of molecular genetics.

Reading List

General

Vosberg, H-P. "The polymerase chain reaction: an improved method for the analysis of nucleic acids." *Hum. Gen.,* 83: 1–15 (1989).

Arnheim, N., T. White, and W. E. Rainey. "The application of PCR: organismal and population biology." *BioScience,* 40: 174–182 (1990).

Eisenstein, B. I. "The polymerase chain reaction: a new method of using molecular genetics for medical diagnosis." *N. E. J. Med.,* 322: 178–183 (1990).

Innis, M. A., D. H. Gelfand, J. J. Sninsky, and T. J. White, eds. *PCR Protocols: A Guide to Methods and Applications.* Academic, N.Y., 1990.

Erlich, H. A., D. Gelfand, and J. J. Sninsky. "Recent advances in the polymerase chain reaction." *Science,* 252: 1643–1651 (1991).

Original Research Papers

HISTORICAL

Mullis, K. B. "The unusual origin of the polymerase chain reaction." *Sci. Am.,* 262: 56–65 (1990).

THE POLYMERASE CHAIN REACTION

Saiki, R. K., S. J. Scharf, F. Faloona, K. B. Mullis, G. T. Horn, H. A. Erlich, and N. Arnheim. "Enzymatic amplification of beta-globin sequences and restriction site analysis for diagnosis of sickle cell anemia." *Science,* 230: 1350–1354 (1985).

Mullis, K., and F. Faloona. "Specific synthesis of DNA in vitro via a polymerase catalyzed chain reaction." *Meth. Enzymol.,* 55: 335–350 (1987).

Taq POLYMERASE AND ACCURACY OF THE PCR

Brutlag, D., and A. Kornberg. "Enzymatic synthesis of deoxyribonucleic acid: XXXVI. A proofreading function for the 3′ to 5′ exonuclease activity in deoxyribonucleic acid polymerases." *J. Biol. Chem.,* 247: 241–248 (1972).

Chien, A., D. B. Edgar, and J. M. Trela. "Deoxyribonucleic acid polymerase from the extreme thermophile *Thermus aquaticus.*" *J. Bacteriol.,* 127: 1550–1557 (1976).

Saiki, R. K., D. H. Gelfand, S. Stoffel, S. J. Scharf, R. Higuchi, G. T. Horn, K. B. Mullis, and H. A. Erlich. "Primer-directed enzymatic amplification of DNA with a thermostable DNA polymerase." *Science,* 239: 487–491 (1988).

Keohavong, P., and W. G. Thilly. "Fidelity of DNA polymerases in DNA amplification." *Proc. Natl. Acad. Sci. USA,* 86: 9253–9257 (1989).

Eckert, K. A., and T. A. Kunkel. "High fidelity DNA synthesis by the *Thermus aquaticus* DNA polymerase." *Nuc. Acids Res.,* 18: 3739–3744 (1990).

DESIGNING PRIMERS FOR PCR

Lathe, J. "Synthetic oligonucleotide probes deduced from amino acid sequence data: theoretical and practical considerations". *J. Mol. Biol.,* 183: 1–12 (1985).

Lee, C. C, X. Wu, R. A. Gibbs, R. G. Cook, D. M. Muzny, and C. T. Caskey. "Generation of cDNA probes directed by amino acid sequence: cloning of urate oxidase." *Science,* 239: 1288–1291 (1988).

Wilks, A. F. "Two putative protein-tyrosine kinases identified by application of the polymerase chain reaction." *Proc. Natl. Acad. Sci. USA,* 86: 1603–1607 (1989).

SOURCES OF DNA FOR PCR

McCabe, E. R. B., S-Z. Huang, W. K. Seltzer, and M. L. Law. "DNA microextraction from dried blood spots on filter paper blotters: potential applications to newborn screening." *Hum. Gen.,* 75: 213–216 (1987).

Higuchi, R., C. H. von Beroldingen, G. F. Sensabaugh, and H. A. Erlich. "DNA typing from single hairs." *Nature,* 332: 543–546 (1988).

Lench, N., P. Stanier, and R. Williamson. "Simple non-invasive method to obtain DNA for gene analysis." *Lancet,* 1: 1356–1358 (1988).

Liang, W., and J. P. Johnson. "Rapid plasmid insert amplification with polymerase chain reaction." *Nuc. Acid Res.,* 16: 3579 (1988).

Lyonnet, S., C. Caillaud, F. Rey, M. Berthelon, J. Frenzal, J. Rey, and A. Munnich. "Guthrie cards for detection of point mutations in phenylketonuria." *Lancet,* ii: 507 (1988).

Shibata, D., W. J. Martin, and N. Arnheim. "Analysis of DNA sequences in forty-year-old paraffin-embedded thin-tissue sections: a bridge between molecular biology and classical histology." *Cancer Res.,* 48: 4564–4566 (1988).

Williams, C., L. Weber, R. Williamson, and M. Hjelm. "Guthrie spots for DNA-based carrier testing in cystic fibrosis." *Lancet,* ii: 693 (1988).

SPECIFIC AMPLIFICATION OF HUMAN SEQUENCES

Korenberg, J. R. and M. C. Rykowski. "Human genome organization: Alu, LINES, and the molecular structure of metaphase chromosome bands." *Cell* 53: 391–400 (1988).

Nelson, D. L., S. A. Ledbetter, L. Corbo, M. F. Victoria, R. Ramirez-Solis, T. D. Webster, D. H. Ledbetter, and C. T. Caskey. "*Alu* polymerase chain reaction: a method for rapid isolation of human-specific sequences from complex DNA sources." *Proc. Natl. Acad. Sci. USA,* 86: 6686–6690 (1989).

Ledbetter, S. A., D. L. Nelson, S. T. Warren, and D. H. Ledbetter. "Rapid isolation of DNA probes within specific chromosome regions by interspersed repetitive sequence polymerase chain reaction." *Genomics,* 6: 475–481 (1990).

PCR AND DNA SEQUENCING

Wong, C., C. E. Dowling, R. K. Saiki, R. G. Higuchi, H. A. Erlich, and H. H. Kazazian. "Characterization of β-thalassemia mutations using direct genomic sequencing of amplified single copy DNA." *Nature,* 330: 384–386 (1987).

Gyllenstein, U. B., and H. A. Erlich. "Generation of single-stranded DNA by the polymerase chain reaction and its application to direct sequencing of the *HLA-DQA* locus." *Proc. Natl. Acad. Sci. USA*, 85: 7652–7656 (1988).

Innis, M. A., K. B. Myambo, D. H. Gelfand, and M. A. D. Brow. "DNA sequencing with *Thermus aquaticus* DNA polymerase and direct sequencing of polymerase chain reaction-amplified DNA." *Proc. Natl. Acad. Sci. USA*, 85: 9436–9440 (1988).

Scharf, S. J., G. T. Horn, and H. A. Erlich. "Direct cloning and sequence analysis of enzymatically amplified genomic sequences." *Science*, 233: 1076–1078 (1988).

ANALYZING MUTATIONS IN CANCERS

Burmer, G. C., and L. A. Loeb. "Mutations in the KRAS2 oncogene during progressive stages of human colon carcinoma." *Proc. Natl. Acad. Sci. USA*, 86: 2403–2407 (1989).

Neri, A., D. M. Knowles, A. Greco, F. McCormick, and R. Dalla-Favera. "Analysis of *RAS* oncogene mutations in human lymphoid malignancies." *Proc. Natl. Acad. Sci. USA*, 85: 9268–9272 (1989).

Yandell, D. W., T. A. Campbell, S. H. Dayton, R. Petersen, D. Walton, J. B. Little, A. McConkie-Rosell, E. G. Buckley, and T. P. Dryja. "Oncogenic point mutations in the human retinoblastoma gene: their application to genetic counseling." *N. E. J. Med.*, 321: 1689–1695 (1989).

Neubauer, A., B. Neubauer, and E. Liu. "Polymerase chain reaction based assay to detect allelic loss in human DNA: loss of β-interferon gene in chronic myelogeneous leukemia." *Nuc. Acid Res.*, 18: 993–998 (1990).

MONITORING CANCER THERAPY

Lee, M-S., K-S. Chang, F. Cabanillas, E. J. Freireich, J. M. Trujillo, and S. A. Stass. "Detection of minimal residual cells carrying the t(14;18) by DNA sequence amplification." *Science*, 237: 175–178 (1987).

Crescenzi, M., M. Seto, G. P. Herzig, P. D. Weiss, R. C. Griffith, and S. J. Korsmeyer. "Thermostable DNA polymerase chain amplification of t(14;18) chromosome breakpoints and detection of minimal residual disease." *Proc. Natl. Acad. Sci. USA*, 85: 4869–4873 (1988).

Kawasaki, E. S., S. S. Clark, M. Y. Coyne, S. D. Smith, R. Champlin, O. N. Witte, and F. P. McCormick. "Diagnosis of chronic myeloid and acute leukemias by detection of leukemia-specific mRNA sequences amplified *in vitro*." *Proc. Natl. Acad. Sci. USA*, 85: 5698–5702 (1988).

Lee, M-S., A. LeMaistre, H. M. Kantajian, M. Talpaz, E. J. Freireich, J. M. Trujillo, and S. A. Stass. "Detection of two alternative *bcr/abl* mRNA junctions and minimal residual disease in Philadelphia chromosome positive chronic myelogenous leukemia by polymerase chain reaction." *Blood*, 73: 2165–2170 (1989).

Roth, M. S., J. H. Antin, E. L. Bingham, and D. Ginsburg. "Detection of Philadelphia chromosome-positive cells by polymerase chain reaction following bone marrow transplant for chronic myelogenous leukemia." *Blood*, 74: 882–885 (1989).

USING PCR FOR DIAGNOSIS OF BACTERIAL AND VIRAL INFECTIONS

Laure, F., C. Rouzioux, F. Veber, C. Jacomet, V. Courgnard, S. Blanche, M. Burgard, C. Griscelli, and C. Brechot. "Detection of HIV1 in infants, and children by means of the polymerase chain reaction." *Lancet*, ii: 538–541 (1988).

Ou, C-Y., S. Kwok, S. W. Mitchell, D. H. Mack, J. J. Sninsky, J. W. Krebs, P. Feorino, D. Warfield, and G. Schochetman. "DNA amplification for the direct detection of HIV-1 in DNA of peripheral blood mononuclear cells." *Science*, 239: 295–297 (1988).

Brisson-Noel, A., D. Lecossier, X. Nassif, B. Gicquel, V. Levy-Frebault, and A. J. Hance. "Rapid diagnosis of tuberculosis by amplification of mycobacterial DNA in clinical samples." *Lancet*, ii: 1069–1071 (1989).

Horsburgh, C. R., J. Jason, I. M. Longini, K. H. Mayer, G. Schochetman, G. W. Rutherford, G. R. Seage, III, C-Y. Ou, S. D. Holmberg, C. Schable, A. R. Lifson, J. W. Ward, B. L. Evatt, and H. W. Jaffe. "Duration of human immunodeficiency virus infection before detection of antibody." *Lancet*, ii: 637–640 (1989).

Pang, S., Y. Koyanagi, S. Mikles, C. Wiley, H. V. Vinters, and I. S. Y Chen. "High levels of unintegrated HIV-1 DNA in brain tissue of AIDS dementia patients." *Nature*, 343: 85–89 (1990).

PRENATAL SEX DETERMINATION

Coutelle, C., C. Williams, A. Handyside, K. Hardy, R. Winston, and R. Williamson. "Genetic analysis of DNA from single human oocytes: a model for preimplantation diagnosis of cystic fibrosis." *Brit. Med. J.*, 299: 22–24 (1989).

Handyside, A. H., J. K. Pattinson, R. J. A. Penketh, J. D. A. Delhanty, R. M. L. Winston, and E. G. D. Tuddenham. "Biopsy of human preimplantation embryos and sexing by DNA amplification." *Lancet*, i: 347–349 (1989).

Holding, C., and M. Monk. "Diagnosis of beta-thalassaemia by DNA amplification in single blastomeres from mouse preimplantation embryos." *Lancet*, ii: 532–535 (1989).

Lo, Y-M. D., P. Patel, J. S. Wainscoat, M. Sampietro, M. D. G. Gillmer, and K. A. Fleming. "Prenatal sex determination by DNA amplification from maternal peripheral blood." *Lancet*, ii: 1363–1365 (1989).

Handyside, A., E. H. Kontogianni, K. Hardy, and R. M. L. Winston. "Pregnancies from biopsied human preimplantation embryos sexed by Y-specific DNA amplification." *Nature,* 344: 768–770 (1990).

LINKAGE ANALYSIS USING SINGLE SPERM

Arnheim, N., H. Li, and X. Cui. "PCR analysis of DNA sequences in single cells: single sperm gene mapping and genetic disease diagnosis". *Genomics* 8: 415–419 (1990). [Review]

Li, H., U. B. Gyllensten, X. Cui, R. K. Saiki, H. A. Erlich, and N. Arnheim. "Amplification and analysis of DNA sequences in single human sperm and diploid cells." *Nature,* 335: 414–417 (1988).

Cui, X., H. Li, T. M. Goradia, K. Lange, H. H. Kazazian, D. Galas, and N. Arnheim. "Single-sperm typing: determination of genetic distance between the $^G\gamma$-globin and parathyroid hormone loci by using the polymerase chain reaction and allele-specific oligomers." *Proc. Natl. Acad. Sci. USA,* 86: 9389–9393 (1989).

STUDIES OF MOLECULAR EVOLUTION

Paabo, S., R. G. Higuchi, and A. C. Wilson. "Ancient DNA and the polymerase chain reaction." *J. Biol. Chem.,* 264: 9709–9712 (1989). [Review]

Higuchi, R., B. Bowman, M. Freiberger, O. A. Ryder, and A. C. Wilson. "DNA sequences from the quagga, an extinct member of the horse family." *Nature,* 312: 282–284 (1984).

Paabo, S. "Ancient DNA: extraction, characterization, molecular cloning, and enzymatic amplification." *Proc. Natl. Acad. Sci. USA,* 86: 1939–1943 (1989).

Thomas, R. H., W. Schafner, A. C. Wilson, and S. Paabo. "DNA phylogeny of the extinct marsupial wolf." *Nature,* 340: 465–467 (1989).

Vigilant, L., R. Pennington, H. Harpending, T. D. Kocher, and A. C. Wilson. "Mitochondrial DNA sequences in single hairs from a southern African population." *Proc. Natl. Acad. Sci. USA,* 86: 9350–9354 (1989).

Golenberg, E. M., D. E. Giannasi, M. T. Clegg, C. J. Smiley, M. Durbin, D. Henderson, and G. Zurawski. "Chloroplast DNA sequence from a Miocene *Magnolia* species." *Nature,* 344: 656–658 (1990).

PCR CONTAMINATION

Kwok, S., and R. Higuchi. "Avoiding false positives with PCR." *Nature,* 339: 237–238 (1989). [Review]

Kitchin, P. A., Z. Szotyori, C. Fromholc, and N. Almond. "Avoidance of false positives." *Nature,* 344: 201 (1990).

Longo, M. C., M. S. Berninger, and J. L. Hartley. "Use of uracil DNA glycosylase to control carry-over contamination in polymerase chain reactions." *Gene,* 93: 125–128 (1990).

7

The Isolation of Cloned Genes

In the mid-1970s, our ability to exploit recombinant DNA methods to full potential faced several obstacles. One was the need for development of disabled hosts and vectors that would have no significant probability of surviving outside the laboratory. Guidelines for working with recombinant DNA were established in 1976 by the National Institutes of Health (Chapter 5). The first safe *E. coli* K12 strain was χ_{1776}, named in honor of the United States Bicentennial. Soon after came the first approved cloning vectors, pMB9 and pBR322, derived from naturally occurring bacterial plasmids. These were the primitive tools that molecular biologists used to clone the first genes. As improved bacterial strains and cloning vectors were developed, the effort shifted to cloning cDNAs first and genes second. Through the years, the methodologies for cloning and sequencing DNA have become simpler yet more powerful, so that today, almost any cDNA desired can be cloned. The ability to do this has been aided by improvements in the areas of oligonucleotide synthesis, mammalian cell culture and transformation, and enzymes that operate on RNA and DNA. These methods made it easier to obtain full-length cDNAs and made it possible to clone cDNAs from low-abundance mRNAs. The recent development of PCR (Chapter 6) has expanded our abilities further. In this chapter, we will focus on the experimental details of gene cloning and analysis that will serve as a foundation for the chapters to come.

Improved Bacteria and Vectors Are Developed

In Chapter 5 we learned about the basic techniques for cloning a DNA fragment into a plasmid vector and for introduction of the vector into bacteria. The early bacterial strains, such as χ_{1776}, did not grow very well and were inefficient in their ability to take up plasmid DNA. Subsequently, bacteria were developed with desirable properties for cloning and were approved under the NIH guidelines. As researchers gained more experience with recombinant DNA, the restrictions were greatly relaxed and now a variety of bacterial strains have been constructed, each for a particular use. For example, a strain that takes up DNA with high efficiency is used to propagate cDNA and genomic libraries. A strain that lacks a functional β-galactosidase gene is used to detect the presence of a foreign insert in a cloning vector. Strains containing *pili,* extracellular structures through which filamentous phage infect the cell, are used to produce single-stranded phage DNA for sequencing. A strain carrying a mutation in a protease gene is used to detect a cDNA-encoded protein in expression cloning.

As we will learn, there are a number of ways to introduce plasmid DNA into bacterial cells. The standard procedure is to mix the cells and plasmid DNA in a solution of calcium chloride and then treat the cells to a brief *heat shock.* The use of salts other than calcium chloride and metal ions for the growth and heat shock steps has improved the *efficiency of transformation.* As we will discuss in Chapter 12, DNA can be efficiently introduced into bacteria and other types of cells by electroporation, which transfers DNA through holes that are induced to appear transiently in the cell membrane. In contrast to plasmids, λ phage–derived cloning vectors are packaged in vitro into infectious particles and introduced into bacteria by infection.

Just as new bacterial strains have been introduced over the past 15 years, new plasmid- and phage-derived vectors have been constructed that facilitate identification, sequencing, and functional analysis of a cloned gene or cDNA. The development of these new cloning vectors has been aided by DNA synthesis technology and the availability of restriction enzymes with a wide range of specificities. The structure of a typical cloning vector consists of a set of unique restriction sites (sometimes positioned within a short nucleotide sequence called a *polylinker*) for insertion of DNA and a replication origin so that the vector is propagated (Figure 7-1). Plasmid vectors are detected by the presence of one of a number of available selectable markers, whereas phage vectors are identified by their lysis of the cells they infect, forming plaques. The cloning sites in the early plasmid vectors were positioned within a gene that imparted antibiotic resistance. Insertion of a foreign piece of DNA into the vector could be detected by the inability of cells containing the plasmid with the insertion to grow in the presence of the particular antibiotic. This practice is not commonly used today; instead the sites for inserting DNA are positioned in the gene encoding the enzyme β-galactosidase, which can be screened on plates by a color assay (Figure 7-1).

Basic Strategies for Cloning Involve Three Steps

There are three steps in the process of cloning a gene. The first is to choose the source of DNA for cloning. Sources of DNA for cloning can be the chromosomal DNA itself or a DNA copy of an mRNA, a so-called *complementary DNA,* or *cDNA.* The choice depends on the particular problem to be tackled. If one is interested in the amino acid sequence of a protein, this information can be obtained most readily from the nucleotide sequence of a cloned cDNA. On the other hand, if one is interested in the regions of a gene that regulate its expression or in gene sequences not contained within the mRNA (Chapters 8 and 9), then this information can be obtained only from genes cloned from chromosomal DNA.

The second step is to produce a collection of DNA fragments that can be inserted into appropriate vectors that in turn can be introduced into a population of bacteria. These fragments are prepared from chromosomal DNA by using a restriction endonuclease or from mRNA by using the enzyme *reverse transcriptase* to produce cDNA molecules. Each cDNA molecule is a copy of a single mRNA. The vectors with inserted fragments are introduced into a population of bacteria

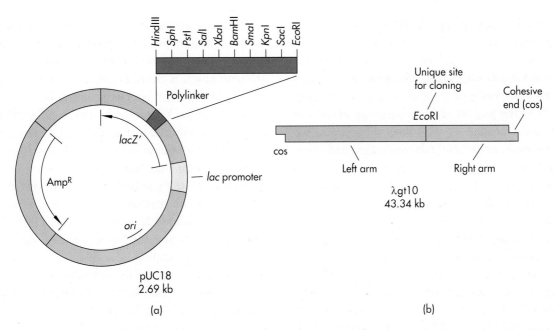

FIGURE **7-1**

Typical cloning vectors. (a) A typical plasmid vector used for propagating cloned DNA fragments. Important features of plasmid vectors include an origin of replication (*ori*) that permits the efficient replication of plasmids to hundreds of copies per cell, a drug-resistance gene (Amp^R for ampicillin resistance) to allow selection of cell clones carrying the plasmid, and restriction enzyme cleavage sites for the insertion of foreign DNA fragments. Modern vectors often carry an array of restriction sites in a *polylinker*. In plasmids of the popular pUC family, the polylinker is situated within a fragment of the *E. coli lacZ* gene (*lacZ'*), which encodes β-galactosidase. The polylinker does not disrupt the *lacZ* reading frame, and in an appropriate *E. coli* host expression of this *lacZ* fragment leads to synthesis of active β-galactosidase enzyme. When these *E. coli* are grown on an agar plate containing a colorless compound called *X-gal*, cleavage of the X-gal by the enzyme produces an insoluble blue product. Thus, these *E. coli* colonies with plasmids that do not have an insert of foreign DNA in their polylinker are blue. However, insertion of a foreign fragment into the polylinker usually shifts or terminates the *lacZ* reading frame in the plasmid, β-galactosidase is not produced, and the resulting plasmid yields colorless colonies which are easily distinguished on the agar plates. (b) A lambda phage vector, λgt10, commonly used for cDNA cloning. Phage structural genes are clustered on the left and right arms of the phage genome. Sequences in the middle are dispensable for purposes of phage replication. In lambda-based cloning vectors, these dispensable sequences are removed and replaced with restriction enzyme sites for cloning of cDNA. In addition to the phage structural genes that permit replication, lambda vectors also retain the single-stranded *cohesive ends* (cos) that are required for phage DNA packaging.

that can be grown on the surface of an agar plate. Because the vectors carry antibiotic-resistance genes, only those bacteria containing plasmids will grow when plated onto agar containing the antibiotic. Each resistant bacterial cell grows to form a colony. Such a collection of cloned DNA fragments propagated in bacteria is called a *library*. Ideally this library should contain representatives of every sequence in chromosomal DNA or of cDNA molecules from every different mRNA.

This library is a library without a catalog to tell us which clone contains a particular sequence, so the third step is to screen all the colonies for the desired sequence. We can screen directly for the sequence by using a nucleic acid probe complementary to that sequence. Part of the sequence may be known already, or it can be deduced from the amino acid sequence of the purified protein. Alternatively, we can screen for the protein produced by a cloned gene by using antibodies to the protein or by using an assay for the function of the protein. In the rest of this chapter we will review some of the methods for cloning and identifying genes.

Choosing the Right Starting Material Is Essential in Cloning

A major difficulty in cloning genes directly from chromosomal DNA is that any one gene represents only a very small fraction of the total DNA in the cell. This problem can be partially circumvented by using mRNA to produce cDNA for cloning, since mRNA represents only the gene sequences *expressed* in a cell. A second consideration is that any given cell type expresses only a subset of its chromosomal genes. Therefore, to clone a cDNA, the library must be prepared from a source that expresses the gene of interest. This may not be easy, since many genes are expressed in only a limited number of cell types or under certain growth conditions or stages in development. However, a variety of methods are available to determine whether a gene is expressed in a given cell. The first eukaryotic gene to be cloned was the rabbit β-globin gene because reticulocytes contain very high levels of globin mRNA; between 50 and 90 percent of their total mRNA is globin mRNA. Nowadays we are able to clone cDNAs from rare mRNAs with concentrations as low as 1 or 2 molecules per cell.

mRNA Is Converted to cDNA by Enzymatic Reactions

The first step in constructing a cDNA library involves isolating total cellular RNA. From this, a fraction that contains mostly mRNA is isolated. As we mentioned in Chapter 4, virtually every eukaryotic mRNA molecule has at its 3′ end a run of adenine

nucleotide residues called a *poly(A) tail*. Whatever its function, the poly(A) tail provides a very convenient way to isolate mRNAs from total cellular RNA, the bulk of which is ribosomal RNA and tRNA (Figure 7-2). Oligonucleotides composed only of deoxythymidine [oligo(dT)] can be linked to cellulose and the oligo(dT)–cellulose packed into small columns. When a preparation of total cellular RNA is passed through such a column, the mRNA molecules bind to the oligo(dT) by their poly(A) tails while the rest of the RNA flows on through the column. The bound mRNAs are eluted from the column. The fraction of poly(A)-containing mRNA is usually about 1 to 2 percent of the total cellular RNA.

The poly(A) tails of the mRNA molecules are used for the next step of cDNA cloning. Short oligonucleotides containing 12 to 20 deoxythymidines are mixed with the purified mRNA and hybridized to the poly(A) tails, where they act as primers for reverse transcriptase (Figure 7-3). This enzyme, which is isolated from certain RNA tumor viruses (Chapter 4), can use RNA as a template to synthesize a DNA strand. (Its name comes from its ability to reverse the normal first step of gene expression.) The product of the reaction is an RNA–DNA hybrid. The use of oligo(dT)-primed cDNA suffers from the disadvantage that because it must begin at the 3′ end the reverse transcriptase may not reach the 5′ end of the mRNA molecule. This is a particular problem for very long mRNAs. To circumvent this difficulty, a second method has been developed termed *randomly primed* cDNA synthesis. Short oligonucleotide fragments 6 to 10 nucleotides long, made up of many possible sequences, are used as primers for the cDNA synthesis. In this way, priming of the mRNA occurs from many positions, not only from the 3′ end as with oligo(dT) primers. Studies have indicated that sequences close to the 5′ end of very long mRNAs are more readily cloned using this method. Again the product of reverse transcription is an RNA–DNA hybrid. From this point, several different procedures are available that convert the RNA–DNA hybrid molecules to double-stranded DNA molecules that can be cloned into appropriate vectors.

One of the first procedures developed to produce cDNA took advantage of a tendency of the newly synthesized DNA strand to form a hairpin loop at its

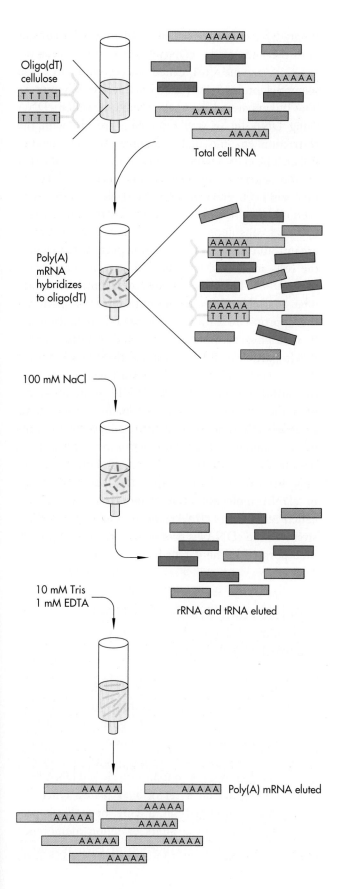

Oligo(dT)
cellulose

Total cell RNA

Poly(A)
mRNA
hybridizes
to oligo(dT)

100 mM NaCl

rRNA and tRNA eluted

10 mM Tris
1 mM EDTA

Poly(A) mRNA eluted

FIGURE **7-2**

Isolation of poly(A) RNA. Most eukaryotic mRNAs carry a poly(A) tail, which can be used to purify the mRNA fraction from the bulk of cellular RNA. Cellular RNA is passed over a column consisting of an inert material, often cellulose or agarose, to which oligonucleotides consisting entirely of deoxythymidine (dT) residues have been attached. The poly(A) tails hybridize to this oligo(dT), causing the mRNA to stick to the column, while the rest of the RNA runs through. After extensive washing of the column to remove the last traces of contaminating material, the column is washed with a buffer of low ionic strength. Under these conditions the poly(A)·oligo(dT) hybrids dissociate, and the purified mRNA washes off the column.

end, apparently as a result of the reverse transcriptase enzyme "turning the corner" and starting to copy the DNA strand. The hairpin loop is probably an in vitro artifact, but it does provide a very convenient primer for the synthesis of the second strand of DNA. The resultant double-stranded cDNA has the hairpin loop intact; this loop can be cleaved by S1 nuclease, a single-strand-specific nuclease. However, the S1 nuclease treatment can trim away much of the desired cDNA sequence as it digests the hairpin loop. This produces a cDNA clone that is missing information from the 5′ end of the mRNA.

A second way to synthesize a double-stranded cDNA molecule from an mRNA–cDNA hybrid makes use of an enzyme from *E. coli*, RNase H, that recognizes RNA–DNA hybrid molecules and digests the RNA strand into many short pieces (see Figure 7-3). These RNA pieces remain hybridized to the first cDNA strand and serve as primers for *E. coli* DNA polymerase I, which uses the original cDNA as a template to synthesize the complementary strand of DNA. Eventually, this process completely replaces the original RNA with DNA, except for a small piece of RNA at the extreme 5′ end. The new DNA strand is not entirely contiguous but rather contains breaks ("nicks"). These breaks are joined (ligated) by the action of DNA ligase, thus forming a double-stranded DNA molecule. The RNase H procedure has an advantage over the S1 nuclease method because it produces longer cDNA molecules that contain all or nearly all the nucleotide sequence from the 5′ end of the mRNA.

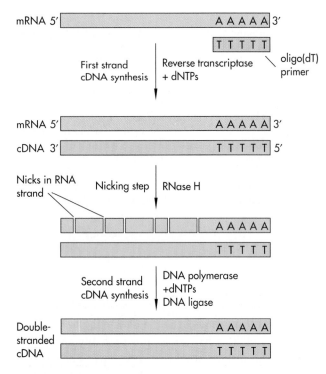

FIGURE **7-3**

One method of converting mRNA to double-stranded cDNA. Poly(A) RNA is incubated with short deoxythymidine-containing oligonucleotides [oligo(dT)], which hybridize to the poly(A) tails, forming primed templates for the enzyme reverse transcriptase. The result of this reaction is a collection of RNA–DNA hybrids. To clone the cDNA molecules, the RNA strands must be destroyed and replaced with DNA. One way this is done is to use an enzyme called *RNase H*, which nicks the RNA–DNA hybrids. These nicks then serve as initiation sites for DNA synthesis by *E. coli* DNA polymerase. Eventually, most of the RNA fragments are replaced by DNA. The double-stranded DNA molecules are finally treated with DNA ligase, which seals up any remaining nicks in the new DNA strand. To prepare cDNA enriched for a specific sequence, cDNA synthesis can be primed with a specific oligonucleotide primer. This method is often used when the goal is to obtain the missing 5' end of a partial cDNA clone.

cDNA Molecules Are Joined to Vector DNA for Propagation in Bacteria

The double-stranded cDNA molecules obtained by these procedures are then inserted into a plasmid or a phage vector, either by tailing with terminal transferase, as described earlier (Figure 5-9), or more com-
monly by attaching artificial restriction enzyme sites onto the ends of the cDNA (Figure 7-4). These restriction enzyme sites, called *linkers,* are 8- to 12-bp oligonucleotides synthesized chemically (Chapter 5). The linkers are added to the double-stranded cDNA using DNA ligase, and then they are cut with the restriction enzyme that recognizes the site on the linker. The cDNA, now carrying sticky ends generated by the restriction enzyme, is inserted into a vector that has been cleaved with the same enzyme. The method by which cDNA vectors are introduced into bacterial cells depends on the vector type. Plasmids are introduced by a transformation procedure in which the cells are heated for a short time to induce uptake of the DNA. Then the cells are spread onto the surface of an agar plate that contains nutrients for growth and the appropriate antibiotic for selection. In contrast, lambda phage vectors are packaged in vitro to form phage particles, which introduce their DNA into an appropriate host by infection of a bacterial lawn on the surface of an agar plate. While plasmid vectors offer the advantage of ease of manipulation of the inserted cDNA, libraries constructed in lambda phage vectors contain a greater number of members and can be screened in much larger numbers. Other methods have been devised to improve the length and yield of cDNA molecules. For example, in the method called *vector-primed cDNA synthesis,* synthesis of the first strand of the cDNAs is primed by a tract of oligo(dT) incorporated in the vector itself.

Nucleic Acid Probes Are Used to Locate Clones Carrying a Desired DNA Sequence

Following packaging into phage particles, the library is plated out on a lawn of bacteria, which is the first step in the screening process (Figure 7-5). The net result of plating out a library is hundreds of thousands to a million phage plaques or (for plasmid vectors) bacterial colonies, each containing a cloned DNA fragment, distributed on a set of agar plates. Once the library has been plated out, a copy, or *replica* is prepared on nitrocellulose filters or nylon membranes. This process transfers a portion of each plaque or

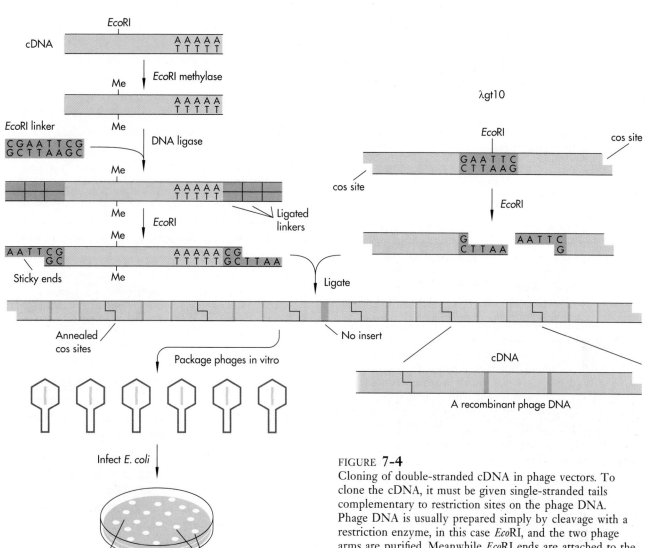

FIGURE **7-4**

Cloning of double-stranded cDNA in phage vectors. To clone the cDNA, it must be given single-stranded tails complementary to restriction sites on the phage DNA. Phage DNA is usually prepared simply by cleavage with a restriction enzyme, in this case EcoRI, and the two phage arms are purified. Meanwhile EcoRI ends are attached to the cDNA. This is done using synthetic oligonucleotide linkers that carry EcoRI sites. First, the cDNA is treated with EcoRI methylase, which methylates any EcoRI sites in the cDNA to protect them from cleavage by EcoRI restriction enzyme. Next, the EcoRI linkers are attached to the ends of the cDNA using DNA ligase; usually, many linkers become ligated to the ends of each cDNA molecule. The linkered cDNA is exhaustively treated with EcoRI to cleave all the linkers, leaving a single EcoRI sticky end at each end of the cDNA molecules. Now the cDNA and phage arms have compatible ends. They are covalently joined with DNA ligase, resulting in a tandem array of recombinant phage molecules flanked by cos sites (see Figure 7-1). This concatenated DNA is the substrate for packaging into infectious phage particles, which is accomplished by adding the DNA to lysates prepared from lambda-infected E. coli. The packaged phages are used to infect fresh E. coli cultures, and the infected cells are spread on agar plates to yield plates carrying thousands of individual phage plaques. Each plaque arises from a single recombinant phage molecule, now propagating as a viable phage.

colony to the nitrocellulose and is done in such a way that the pattern of plaques on the original plates is maintained on the filters. Screening is carried out by incubating these nitrocellulose replicas with a nucleic acid probe or with an antibody.

The most direct method of screening is to use a nucleic acid probe for hybridization, but this does require knowledge of the sequences being sought. In some cases, part of the gene may already have been cloned, and this can be used to search for clones that contain additional sequences flanking the starting clone. In other cases, a closely related gene may have

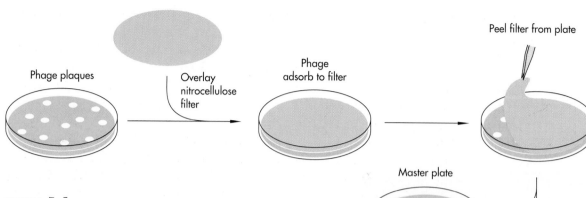

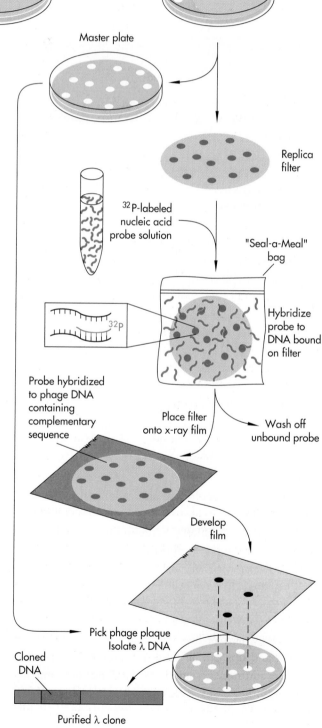

FIGURE **7-5**

Screening a library with a nucleic acid probe to find a clone. Libraries are typically screened by spreading several hundred thousand phages on 10 to 20 large agar plates covered with the host bacterial culture (lawn). After the phage plaques have grown to visible size, nitrocellulose filters are carefully laid onto the surface of the plates. Phage particles from the plate adhere to the filter, creating a precise replica on the filter of the pattern of plaques on the plate. The filters are treated to strip off phage proteins and bind the phage DNA to the filter surface. The filter is incubated in a solution containing a radioactively labeled DNA or RNA probe complementary in sequence to a portion of the gene being sought. This hybridization reaction is often carried out in sealed plastic bags. The filters are carefully washed to remove unbound probe, leaving behind only the probe molecules tightly bound to complementary sequences within phage DNA. The location of the bound probe is determined by exposing the filters to x-ray film (*autoradiography*). The position is represented by a spot of exposure on the film. By orienting the film with the original agar plate, the phage plaque carrying the complementary sequence can be identified and the desired clone can be isolated.

already been cloned and can be used as a probe for the gene of interest using conditions that allow partially matching sequences to hybridize. Hybridization between probe and cloned gene sequences containing mismatched bases will occur if screening is performed at lower temperatures (42°C, for example, instead of 65°C) and lower salt concentrations (low stringency). However, often nothing is known about the nucleotide sequence of the gene. In this situation, nucleic acid probes can be generated from knowledge of the amino acid sequence of a protein. Significant improvements made in peptide sequencing procedures in recent years have made this a feasible and an attractive approach.

Synthetic Oligonucleotide Probes Can Be Designed from Known Protein Sequence

There are several problems associated with using amino acid sequences to derive oligonucleotide probes. First, the oligonucleotide probe has to recognize specifically and uniquely only the cDNA of interest and not unrelated sequences. The minimum-length oligonucleotide that would be unique in all the sequences in a eukaryotic cDNA library is 15 to 16 nucleotides. Generally, oligonucleotides that are 17 to 20 nucleotides long are used. This requires knowledge of a minimum of six contiguous amino acids from the protein sequence. The second difficulty stems from the degeneracy of the genetic code. Some amino acids are coded for by as many as six different codons, and therefore a stretch representing only six amino acids could be encoded by many different DNA sequences. For this reason, oligonucleotide probes are often synthesized as mixtures or pools that contain all possible combinations of codons that translate into the protein sequence (Figure 7-6). Within this pool will be one oligonucleotide that contains the correct sequence of the gene. A disadvantage of this approach is that the correct oligonucleotide is only a small fraction of the total population of molecules in the pool. The remaining oligonucleotides may hybridize to undesired cDNAs, leading to "false positives."

A variation of this method has been developed that uses just a small number of oligonucleotides or even a single long oligonucleotide of 35 to 75 nucleotides (Figure 7-6). The sequence for the oligonucleotide is chosen in the same way as with degenerate probes from the known amino acid sequence. However, a single nucleotide sequence is arrived at by choice of a region of the protein that contains the fewest degenerate codons, by knowledge of codon usage in the particular species, and by educated guessing. Hence, these probes have been called "guessmers" (*mer* from oligo*mer*). The results from a number of cloning experiments indicated that a sequence with only 83 percent of the nucleotides correct successfully hybridized to the desired clone provided that the incorrect nucleotides were clustered and separated by several regions 10 to 12 nucleotides long that perfectly matched the sequence of the cDNA. If the mismatched nucleotides were equally distributed throughout the length of the probe, then the guessmer failed to hybridize to the cDNA of interest. The inherent disadvantage of this method over the short mixed-probe procedure is the requirement for precise knowledge of the amino acid sequence over a much longer portion of the protein of interest.

An interesting method to circumvent the problem of choosing the nucleotide sequence for screening makes use of the polymerase chain reaction (Chapter 6). In one example, a sequence of 32 amino acids from the urate oxidase protein was obtained by conventional peptide sequencing (Figure 7-7). Two regions, one at each end of the sequence, were used to design two pools of degenerate oligonucleotide primers. Single-stranded cDNA was made by reverse transcription of total mRNA, and was amplified by PCR using these primers. The products of the PCR were cloned, and the clones were screened with a unique "guessmer" probe derived from the middle of the known amino acid sequence. A clone was isolated that encoded the entire 32 amino acid peptide. The insert in this clone was subsequently used as a probe to screen a phage cDNA library from which a full-length 2.2-kb cDNA was obtained. Thus, degenerate oligonucleotide primers can be used to generate a probe that contains a unique sequence from the desired gene. The library can then be screened using very stringent conditions to obtain the entire cDNA.

Tissue-Specific cDNAs Are Identified by Differential Hybridization

Thus far we have discussed the isolation of specific cDNAs using information about the particular gene of interest. *Differential hybridization* is a technique used to clone a family of genes that are related by a common mechanism of regulation. The genes themselves can be structurally different, but they are all expressed under the same cellular conditions. Examples include genes expressed in specific tissues, genes expressed at specific stages of the cell cycle, and genes regulated by a growth factor. The basis of this technique relies simply on producing two cell populations, one in which the genes are, and the other in which the genes

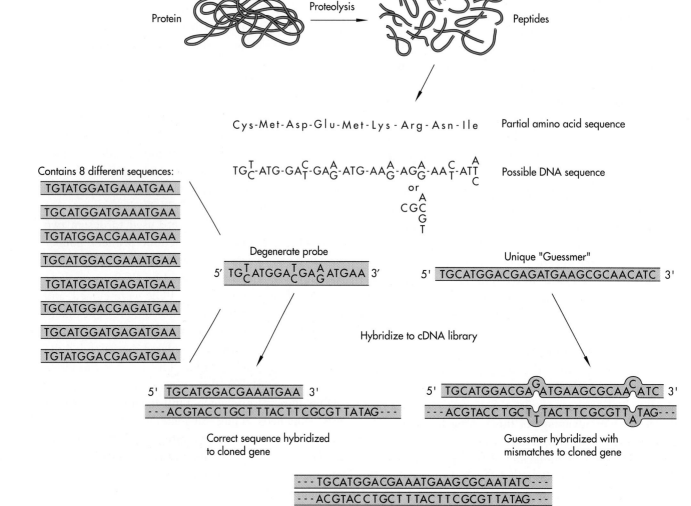

FIGURE **7-6**

Designing oligonucleotide probes based on protein sequence. When no natural nucleic acid probe is available to screen for a clone, one can often be designed and synthesized on the basis of the sequence of the encoded protein. An oligonucleotide is synthesized carrying the codons known to encode the corresponding amino acids. Most amino acids, however, are specified by two or more codons. In this example, Cys, Asp, and Glu are each specified by two codons. Therefore, to ensure that the correct DNA sequence is represented, the oligonucleotide is synthesized using mixtures of nucleotide precursors in the ambiguous positions. Thus, the oligonucleotide synthesized in the figure is actually a pool of eight different oligonucleotides, all the possible sequences that can encode the six amino acids. The complexity of the pool of probes can be reduced by using sequences of 17 rather than 18 nucleotides since the third position of the last codon is omitted because it is degenerate. One of these oligonucleotides must be fully complementary to the gene. An alternative strategy if more than six contiguous amino acids are known is to synthesize a "guessmer," a much longer oligonucleotide of unique sequence. The sequence is chosen on the basis of the frequency of codon usage and other sequence considerations. The guessmer will be largely but incompletely complementary to the target sequence. If patches of complementarity are large, 10 to 12 bp, then hybridization of the guessmer will be sufficiently strong to identify a positive clone.

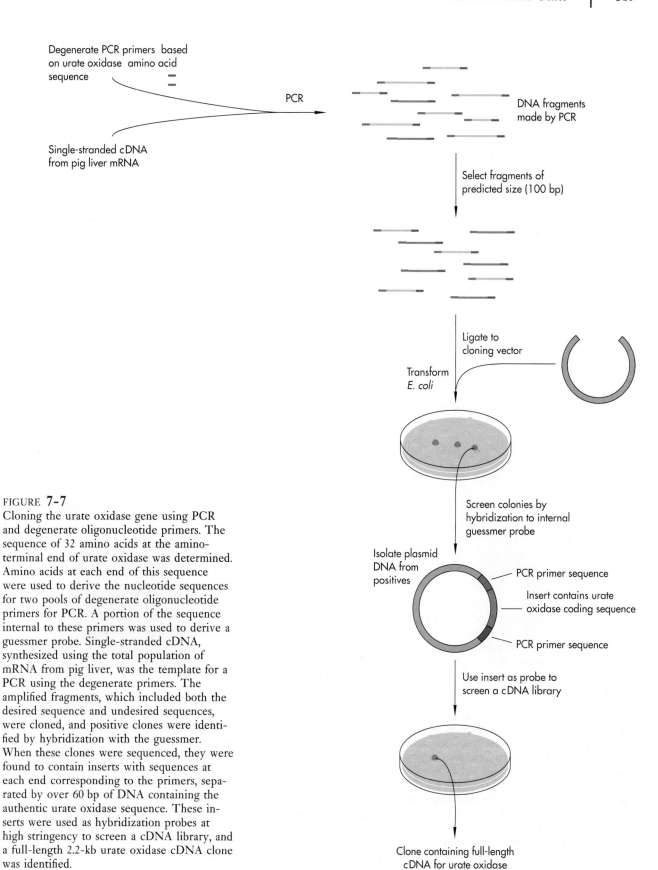

FIGURE **7-7**
Cloning the urate oxidase gene using PCR
and degenerate oligonucleotide primers. The
sequence of 32 amino acids at the amino-
terminal end of urate oxidase was determined.
Amino acids at each end of this sequence
were used to derive the nucleotide sequences
for two pools of degenerate oligonucleotide
primers for PCR. A portion of the sequence
internal to these primers was used to derive a
guessmer probe. Single-stranded cDNA,
synthesized using the total population of
mRNA from pig liver, was the template for a
PCR using the degenerate primers. The
amplified fragments, which included both the
desired sequence and undesired sequences,
were cloned, and positive clones were identified by hybridization with the guessmer.
When these clones were sequenced, they were
found to contain inserts with sequences at
each end corresponding to the primers, separated by over 60 bp of DNA containing the
authentic urate oxidase sequence. These inserts were used as hybridization probes at
high stringency to screen a cDNA library, and
a full-length 2.2-kb urate oxidase cDNA clone
was identified.

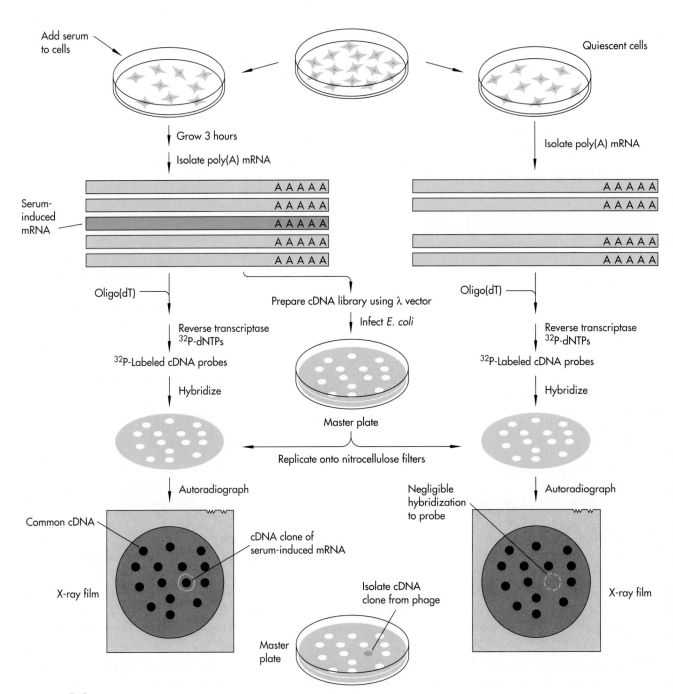

FIGURE **7-8**
Cloning growth factor–regulated genes by differential hybridization. This strategy was used to clone a family of genes that are rapidly induced when quiescent cells are stimulated to grow by the addition of serum (which contains growth factors). Poly(A) RNA was isolated from cultures of quiescent cells and from cells stimulated with serum for 3 hours. RNA from the stimulated cells was used to produce a cDNA library in a lambda phage vector, which was plated out. Phage from the plaques produced were transferred to duplicate filter replicas. Radioactively labeled probes were generated from the mRNA of quiescent cells and stimulated cells by synthesizing the cDNA strands in the presence of ^{32}P-labeled dNTPs. One set of filters was hybridized to each cDNA probe. The resulting autoradiographs were carefully compared to identify the position of phage plaques that hybridized to the probe from stimulated cells but not to the probe from quiescent cells. These phages carried cDNAs from genes turned on in the serum-stimulated cells.

are not, expressed. No specific information about an individual gene is required. Figure 7-8 shows the experimental approach used to isolate a family of genes whose expression is stimulated by addition of serum to quiescent cells. A cDNA library was prepared using poly(A) RNA isolated from cells treated with serum. This phage library was plated out and transferred to duplicate sets of filters. One of the sets of filters was probed with radioactively labeled cDNA prepared by reverse transcription of RNA from cells treated with serum. The other set of filters was hybridized to labeled cDNA prepared from untreated (quiescent) cells. Clones were identified that hybridized more strongly to the first probe. These clones were shown subsequently to encode mRNAs that are serum-inducible.

This method works well in practice for genes that are highly expressed in the stimulated cells. However, mRNAs of low abundance are difficult to isolate by differential screening. The use of absorbed probes and subtracted libraries improves the chances of cloning rare cDNAs. Both of these tricks were used in the very difficult cloning of cDNAs encoding the subunits of the T-cell receptor (Figure 7-9). The strategy assumed that the genes encoding the T-cell receptor were expressed only in T cells and not in B cells. cDNA was prepared from T-cell poly(A) mRNA and hybridized to a large excess of poly(A) mRNA from B cells. T-cell cDNAs from genes expressed in both cell types formed DNA–RNA hybrids with their complementary B-cell mRNA molecules. cDNA molecules expressed only in T cells (about 2 percent of the total mRNA) did not form such a hybrid because a corresponding mRNA was not present in the B cells. The hybridized nucleic acids were passed through a column of hydroxylapatite under conditions in which DNA–RNA hybrids bound to the column and free cDNA passed through. The T-cell-specific cDNA recovered from the column was converted to double-stranded cDNA and cloned into a λ phage vector. This produced a library of 5000 clones highly enriched for T-cell-specific cDNAs. To screen the library, ^{32}P-labeled T-cell cDNA was subtracted with B-cell mRNA in the same fashion and used as a probe. Thirty-five clones reacted with this probe. Among them were clones that encoded the T-cell receptor (see Chapter 16).

Gene Probes from Conserved Segments in Protein Families Identify New Related Genes

Before the availability of recombinant DNA techniques, the primary structure of proteins had to be determined by protein sequencing after classical biochemical purification. With the advent of molecular cloning and our ability to deduce amino acid sequences from gene sequences, the number of new sequences has increased dramatically. This has resulted in a huge database of proteins that can be grouped into families by common structural and functional properties. Careful analysis of the sequences of proteins within families revealed regions of common amino acid sequence. These might be important for catalytic function or might define a particular protein structure. Often a conserved region consists of 6 to 7 contiguous amino acids along the polypeptide chain. This is an appropriate size to construct an oligonucleotide probe to clone cDNAs encoding related proteins. This procedure has been successfully applied to the isolation of cDNAs encoding protein kinases. A cDNA library was probed (by the method shown in Figure 7-6) using a degenerate 17-mer (i.e., an oligo*mer*, 17 nucleotides long) oligonucleotide pool that encoded a sequence found in many protein serine kinases (enzymes that phosphorylate serine residues), Asp-Leu-Lys-Pro-Glu-Asn. In contrast, the sequence of the analogous region in protein tyrosine kinases is different and therefore would not be recognized by the probe. Of 80,000 cDNA clones screened, 89 clones hybridized to the 17-mer probe, and 19 of these also hybridized to another probe from a second region of common sequence. A second probe such as this provides a useful check to eliminate false positives obtained with a first probe. Three of these nineteen clones were characterized by DNA sequencing. All three clones exhibited extensive sequence conservation throughout their coding regions with other members of the protein serine kinase family.

The PCR technique has improved the cloning of gene families by a similar approach. For example, two amino acid sequences common to the members of the protein tyrosine kinase (PTK) family were used to design degenerate primers for a PCR. The primers bracketed a relatively conserved region that was suf-

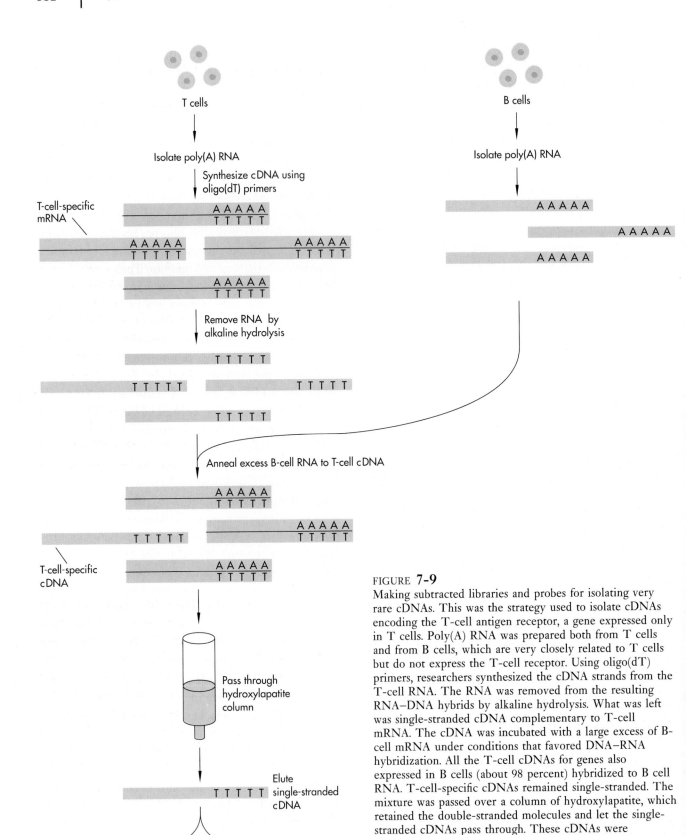

FIGURE **7-9**
Making subtracted libraries and probes for isolating very rare cDNAs. This was the strategy used to isolate cDNAs encoding the T-cell antigen receptor, a gene expressed only in T cells. Poly(A) RNA was prepared both from T cells and from B cells, which are very closely related to T cells but do not express the T-cell receptor. Using oligo(dT) primers, researchers synthesized the cDNA strands from the T-cell RNA. The RNA was removed from the resulting RNA–DNA hybrids by alkaline hydrolysis. What was left was single-stranded cDNA complementary to T-cell mRNA. The cDNA was incubated with a large excess of B-cell mRNA under conditions that favored DNA–RNA hybridization. All the T-cell cDNAs for genes also expressed in B cells (about 98 percent) hybridized to B cell RNA. T-cell-specific cDNAs remained single-stranded. The mixture was passed over a column of hydroxylapatite, which retained the double-stranded molecules and let the single-stranded cDNAs pass through. These cDNAs were recovered, converted to double-stranded cDNAs, and cloned. The library was screened with a subtracted ^{32}P-labeled cDNA probe prepared precisely the same way.

ficiently diverse in the known PTKs that it could be used to identify individual genes. The PCR was performed using pools of 64 and 8 oligonucleotides for the 5′ and 3′ primers, respectively. Amplification produced 210-bp fragments, which were cloned and sequenced. Of 200 clones, 133 contained inserts that were representative of 6 distinct PTK-like sequences. Four of these sequences were identified as PTKs that had already been cloned: the cellular protooncogenes c-*fes*, c-*met* and c-*lyn*, and the insulinlike growth factor receptor. The remaining two sequences were previously unidentified members of the PTK family. It is clear that this simple technique provides a powerful tool to identify new members of a protein family.

Expression Vectors May Be Used to Isolate Specific cDNAs

Often nucleotide sequences may not be available as probes for screening libraries. Instead we can identify specific eukaryotic cDNAs by looking for their gene products in bacteria after cloning the cDNA into appropriately constructed plasmids or phages termed *expression vectors*. cDNAs are inserted into these vectors within regions that promote their expression in *E. coli*. Often regulated bacterial promoters are used. Proteins may be expressed as fusion proteins in which amino acids from a prokaryotic protein are incorporated at one end of the eukaryotic protein. Fusion proteins are often more stable than the corresponding eukaryotic protein in bacteria and are therefore produced at higher levels.

In the preparation of expression libraries, the generation of cDNA from isolated poly(A) mRNA is done in the same way as for conventional nucleic acid screening. However, unlike cloning by nucleic acid hybridization, expression cloning requires that the cDNA fragment be inserted adjacent to the promoter in the expression vector and in the correct orientation and reading frame for the correct protein product to be produced. The simplest procedures ligate an oligonucleotide linker onto each end of the cDNA molecule, which can be inserted into the cloning vector in either of two orientations. Since a cDNA can be inserted in two orientations and there are three reading

frames for each orientation, on average only one out of six clones containing a particular cDNA insert will express the correct gene product. Methods have been developed to reduce this problem by orienting the cDNA with respect to the promoter. In addition, a set of three vectors has been developed, each of which results in the translation of a different reading frame.

As with nucleic acid screening, the vectors for expression cloning can be either plasmid or phage (Figure 7-10). Similarly, the library is plated out and replica nitrocellulose filters are prepared. The filters are first treated to expose the protein in each colony or plaque, then they are mixed with a solution containing an antibody to the desired protein. After an appropriate period, unbound antibody is removed, and a second antibody or staphylococcal protein A is added to locate the position of the first antibody. The second antibody can be radioactively labeled, coupled to a molecule of biotin, or conjugated with an enzyme such as alkaline phosphatase. These agents provide a means to visualize the position of the clone expressing the protein recognized by the specific antibody. The phage DNA (or plasmid) is isolated from the corresponding plaque or colony on the master plate, and the sequence of the cDNA clone is determined.

Cloned Genes Can Be Isolated by Functional Assay in *E. coli*

A second way to locate clones that express a desired protein is to use an assay for the function of the protein. For example, genes encoding proteins that form a tight complex with the calcium-binding protein calmodulin have been isolated using this strategy. Biochemical studies have shown that in the presence of calcium, calmodulin forms a stable complex with a number of enzymes. Radioactively labeled calmodulin was used as a probe to identify clones expressing proteins that are able to associate with the calmodulin in vitro. This method identified a brain-specific cDNA encoding a subunit of a novel Ca^{2+} calmodulin–dependent protein kinase.

A variation of this procedure uses expression cloning to identify cDNAs for proteins that bind to DNA sequences involved in regulating gene expression. A

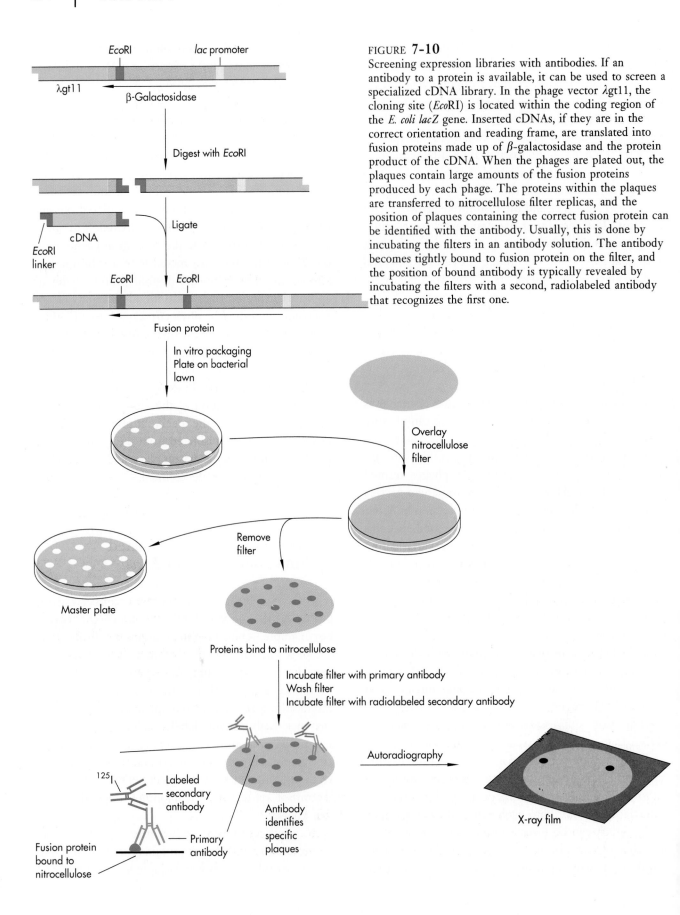

FIGURE **7-10**
Screening expression libraries with antibodies. If an antibody to a protein is available, it can be used to screen a specialized cDNA library. In the phage vector λgt11, the cloning site (*Eco*RI) is located within the coding region of the *E. coli lacZ* gene. Inserted cDNAs, if they are in the correct orientation and reading frame, are translated into fusion proteins made up of β-galactosidase and the protein product of the cDNA. When the phages are plated out, the plaques contain large amounts of the fusion proteins produced by each phage. The proteins within the plaques are transferred to nitrocellulose filter replicas, and the position of plaques containing the correct fusion protein can be identified with the antibody. Usually, this is done by incubating the filters in an antibody solution. The antibody becomes tightly bound to fusion protein on the filter, and the position of bound antibody is typically revealed by incubating the filters with a second, radiolabeled antibody that recognizes the first one.

FIGURE **7-11**

Cloning of the erythropoietin (EPO) receptor by expression screening in mammalian cells. cDNA was prepared from an erythroleukemia cell line that expresses the EPO receptor and inserted in a plasmid vector carrying regulatory sequences that directed expression of the inserted cDNAs in mammalian cells. The library was transformed into *E. coli* and plated out at a density of approximately 1000 colonies per agar dish. Pooled plasmid DNA was prepared from the colonies on each plate, and the pools were separately transfected (Chapter 12) into cultures of mammalian cells that lack EPO receptors. Using a sensitive EPO-binding assay, cultures of cells that were able to bind EPO were identified, suggesting that the plasmid pool transfected into those cells included a plasmid expressing the EPO receptor. That initial pool was then subdivided into several smaller subpools, and each was again tested for the ability to transfer EPO-binding activity to recipient cell cultures. Positive pools were subdivided further and tested and the cycle repeated until an individual plasmid clone was identified that directed high-level expression of the EPO-binding activity. This plasmid clone carried a cDNA for the EPO receptor.

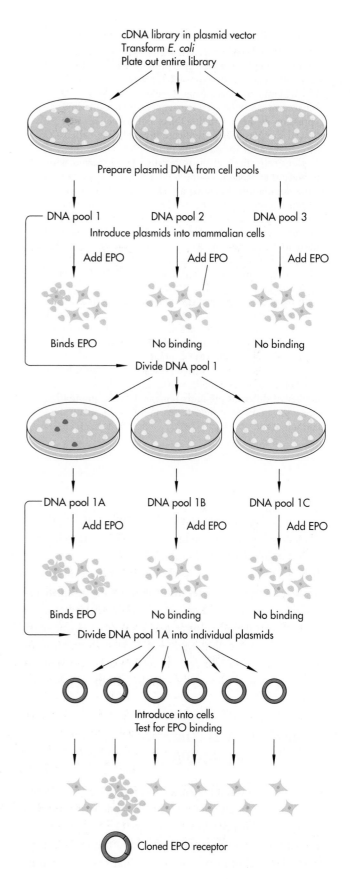

cDNA library is screened with a labeled DNA fragment that contains a protein-binding site. Under appropriate conditions, the oligonucleotide binds only to DNA-binding proteins that specifically recognize it. Using this method, a number of clones have been identified that encode proteins that recognize specific DNA sequences. This method of cloning requires only that the protein domain containing the ligand binding site be correctly expressed and that the association of ligand and protein be stable under the conditions used.

Cloned Genes Can Be Isolated by Functional Assay in Eukaryotic Cells

Although growing eukaryotic cells is more tedious than handling *E. coli*, some proteins are correctly made and functional only in eukaryotic cells. One example is growth factor receptors, which normally span the plasma membrane of eukaryotic cells, displaying a high-affinity ligand-binding site on the cell surface. This property was exploited to clone the receptor for erythropoietin (EPO), a growth and differentiation factor for erythroid cells of the blood (Figure 7-11). A variation on this technique, called *panning*, has been

FIGURE **7-12**

Panning for cDNA clones. This method was used to clone the cDNA for a T-cell surface protein, CD2. Cultures of COS cells that do not express CD2 were transfected with a cDNA library prepared from T cells. The library was prepared in a vector carrying sequences that directed expression of the inserted cDNAs and sequences that allowed the plasmids to replicate in a cell line called *COS*. Three days after introduction of the library into the recipient cells, the cells were harvested and treated with an antibody that recognized the CD2 protein. Only a few cells in the population contained a CD2 cDNA and expressed the protein on their surfaces. These rare cells became coated with antibody. The cultures were transferred to a petri dish coated with a second antibody that recognized and bound the first one. Any cells coated with the CD2 antibody thus adhered to the surface of the dish. The adherent cells were collected, and the replicating plasmid DNA was extracted and reintroduced into *E. coli*. The resulting *E. coli* colonies were pooled and expanded to generate more plasmid DNA. Because each transfected COS cell takes up hundreds of plasmids besides the one carrying the CD2 cDNA, the CD2-encoding plasmids isolated in the first round of panning were contaminated with other plasmids in the library. The CD2-encoding plasmids were enriched by further rounds of panning, until only a small number of different plasmids remained in the population. These plasmids were then tested individually using the panning assay.

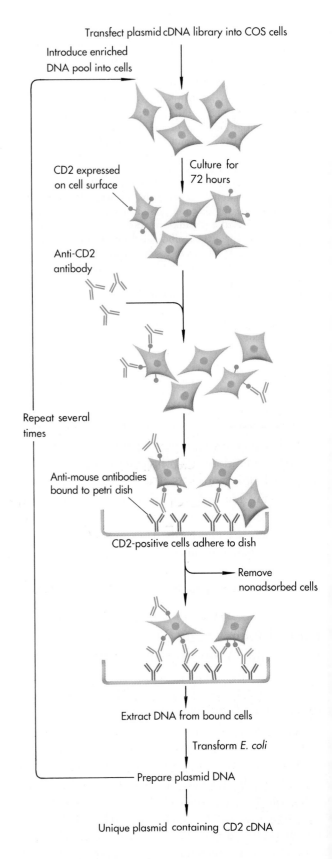

used to clone a cDNA encoding CD2, a protein expressed on the surface of T cells (Figure 7-12). Panning allows direct physical selection of cells expressing the cDNA of interest, and this cDNA can be recovered directly from the cells. Mammalian genes can also be isolated by genetic selection for their function in recipient cells. We will see in Chapter 18 how such a strategy was used to clone an activated oncogene from a human tumor cell line and in Chapter 16 how clever genetic selection allowed cloning of a gene involved in the rearrangement of immunoglobulin genes.

Cloned DNA Is Analyzed by DNA Sequencing

Once an individual plasmid containing a single cDNA fragment has been obtained, how is it determined that it encodes the protein of interest? One method is to determine the nucleotide sequence of the cDNA, now a standard laboratory procedure that can be accom-

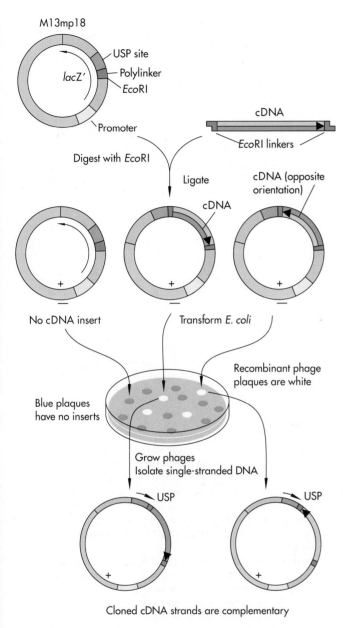

Cloned cDNA strands are complementary

FIGURE **7-13**
Cloning and sequencing in M13 vectors. Double-stranded M13 DNA is isolated from infected cells for use as a cloning vector. The DNA consists of the original phage strand (the + strand) and the strand created during replication (the − strand). Fragments of cDNA are ligated into M13 at a polylinker sequence as with a plasmid vector and transformed into host cells. The cells become chronically infected with M13 and grow more slowly than uninfected cells, so the location of all M13 clones is apparent as "plaques" of poor cell growth. Like the polylinker in the plasmid vectors described in Figure 7-1, the M13 polylinker is situated within a fragment of the *lacZ* gene, providing a screening procedure based on color for clones with inserts. Thus, M13 vectors produce blue "plaques" on an appropriate *E. coli* host. Insertion of a foreign DNA fragment into the polylinker disrupts the *lacZ* sequence, leading to the formation of colorless "plaques," allowing phage clones carrying inserts to be easily identified. "Plaques" are isolated and the infected cells grown up. Infected cells continuously extrude M13 phage particles, and these can be harvested from the growth medium. M13 phages package single-stranded DNA, so phage DNA carries a single strand of the cloned insert. This is the preferred template for DNA sequence analysis and in vitro mutagenesis (see Chapter 11). Insertion of the DNA fragment in either of the two orientations relative to the phage vector yields opposite strands. DNA fragments inserted into M13 vectors can be easily sequenced using a "universal" sequencing primer (USP) that anneals to the M13 sequence at a site just outside the polylinker. The single primer allows sequencing of any insert in the vector.

plished rapidly and unambiguously. In cases in which protein sequence is known, the translation (i.e., decoding) of the nucleotide sequence into the corresponding amino acid sequence will quickly reveal whether the cDNA encodes the desired protein. On the other hand, in cases where no primary amino acid sequence data are known, determination of the nucleotide sequence from a cDNA clone will not unambiguously match it with the protein of interest.

As discussed in Chapter 5, the determination of DNA sequence can be accomplished by either the chemical method developed by Walter Gilbert and Alan Maxam or the enzymatic method of Fred Sanger. A series of extremely useful cloning vectors based on the filamentous phage M13 was developed by Jo Messing. These vectors revolutionized the enzymatic sequencing method (Figure 7-13). M13 is a single-stranded, filamentous DNA bacteriophage. The double-stranded replicative form (RF) can be isolated and used as a cloning vector. DNA fragments are ligated into the vector at unique restriction sites, then the recombinant M13 DNA is transformed into *E. coli* in the same way as with plasmid vectors. Phage particles can be isolated from the culture medium of infected cells. The important feature of filamentous phages is that only one of the DNA strands of the vector is packaged into the phages. M13 cloning vectors were developed to produce single-stranded template DNA, the optimal form of DNA for Sanger dideoxy sequence analyses. By cloning the insert into M13 in each of the two possible orientations, either one strand or the

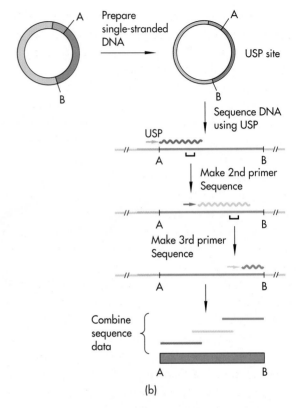

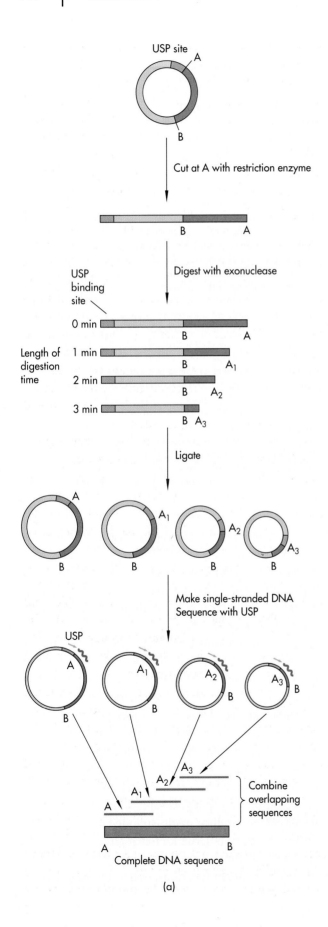

(a)

(b)

FIGURE **7-14**

Long-range sequencing with M13 vectors. Although any insert can be sequenced using the universal sequencing primer, the range of sequencing reactions is short (a few hundred base pairs), whereas the DNA to be sequenced is often much longer. Two general strategies for long-range sequencing are shown. One is to generate a family of M13 clones, each carrying a subfragment of the gene. This can be done by preparing ordered deletions as depicted in (a). Double-stranded DNA carrying the full-length insert is cut open at one end of the insert (labeled A) with a restriction enzyme and insert DNA is digested away with a DNA exonuclease. Several different strategies are employed to limit the extent of exonuclease digestion of the insert. Usually, portions of the exonuclease reaction mixture are removed to stop the reaction at different times and thus create populations of molecules with different extents of digestion. The deleted molecules are recircularized with DNA ligase, bringing the binding site for the universal primer adjacent to the new end created by exonuclease digestion (labeled A_1, A_2, and A_3). The result is a collection of clones in which the universal sequencing primer is poised to sequence different portions of the insert. By sequencing this collection of clones with the single primer, many overlapping sequences are obtained, which can be combined to give the entire contiguous sequence of the insert. A different method for long-range sequencing uses the universal primer only to start the sequence (b). A sequence from near the end of this run is used to make a second primer, which primes sequencing from this position and moves a few hundred nucleotides farther. From the end of this run, a third primer is made, and the cycle is repeated until the entire insert has been covered.

other of the foreign DNA fragment is obtained. DNA is ligated into M13 in a region of the vector termed the *polylinker,* so called because it contains many restriction enzyme recognition sequences that are present only once in the vector. An oligonucleotide primer that anneals adjacent to this polylinker region is used to sequence the inserted DNA fragment. Since the same oligonucleotide can be used to sequence any recombinant M13 clone, this oligomer has been called the *universal sequencing primer (USP).* This can be used to obtain the DNA sequence from one end of the clone to over 400 bases away. In order to sequence clones larger than 400 bases, a collection of M13 clones that contain smaller pieces of the gene is generated (Figure 7-14a). These clones can be created randomly or by various ordered strategies. The entire collection of clones is sequenced using the universal sequencing primer, and the sequence of the entire cloned DNA fragment is assembled from this information. An alternative strategy uses the newly determined sequence for one segment of the cloned DNA fragment to design a new oligonucleotide for use as a primer for the next several hundred base pairs of DNA (Figure 7-14b).

Recently, a new series of vectors, termed *phagemids,* has been developed that consists of hybrids between plasmids and filamentous phage cloning vehicles. These are generally smaller than the M13 vectors and were formed by combining sequences from pBR322 with the M13 origin of replication. These vectors can be propagated normally as double-stranded plasmids within the cell. However, by addition of a "helper phage" to cells that contain the phagemid, single-stranded phagemid DNA is replicated, packaged, then extruded from the cell in a similar fashion to the M13 phage vectors. The advantages of these vectors are their small size (about 3000 bp), allowing the insertion and stable propagation of larger fragments of DNA than M13 can accept, and the ease with which both double-stranded DNA and single-stranded DNA can be prepared.

Computers Have Simplified Translating DNA Sequence into Protein Sequence

The raw data from a DNA sequencing project are the pattern of bands on a piece of x-ray film that result from autoradiography of a sequencing gel (Chapter 5). At first, the data were compiled by reading the position of each band by eye and manually keeping track of the corresponding DNA sequence in a notebook. Computers have now replaced notebooks, and digitizer pads have made it easier to read the sequence data from x-ray films. Each base can be entered manually into the computer through the keyboard or can be entered automatically by touching each band on the x-ray film with a special pen linked to the computer. Recently, *automatic DNA sequencers* have been developed that perform the gel electrophoresis and then automatically determine the sequence using a laser detection system (Chapter 29). Soon, the entire sequencing process, including preparing the DNA and performing the sequencing reactions, will be automated. The ease of conducting the sequencing reactions has been accompanied by improvements in processing and analysis of the vast amounts of sequence data. Computer programs are used to manage each sequencing experiment, merge the data from individual experiments into one contiguous stretch, and analyze the final nucleotide sequence.

In order to identify potential protein coding regions, researchers convert the DNA sequence into protein sequence by translating triplet codons into corresponding amino acids. Without any reference points, the sequence has to be translated in all six of the possible reading frames in the hope that a long open reading frame will be recognized (Figure 7-15). In some cases, the sequence of the cDNA clone can be oriented with respect to the 5′ and 3′ ends of the molecule if the sequence analysis identified which end of the cDNA clone contained the poly(A) tail. The reading frame of a clone obtained through expression screening can be deduced from the known reading frame of the fusion protein. Generally it is not very difficult to find the coding sequence of a cDNA clone using DNA-protein computer translation programs because the coding sequence is contiguous. If the amino acid sequence for the protein of interest is known, then it can be compared with the amino acid sequence predicted from the cDNA. However, if only a little sequence information is known, for example, the portion of a peptide used to make a probe for cloning, then finding this sequence in the cDNA is significant but does not provide conclusive evidence that the cDNA stems from the gene that encodes the desired protein. Sometimes the protein encoded by

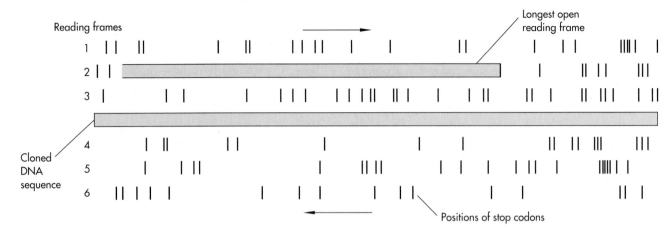

FIGURE **7-15**
Finding an open reading frame. A simple method to spot an open reading frame with the potential to encode a protein is to ask a computer to locate the translation termination codons in all reading frames in both orientations (six frames in all). For example, above and below the bar representing the cDNA sequence of the human c-*FOS* protooncogene (see Chapter 18) are lines that represent the termination codons in each reading frame in the leftward and rightward orientation, respectively. In five of the frames, termination codons are scattered throughout the sequence, so that it is clear these frames cannot be translated into a sizeable protein. In the second frame above the cDNA, however, there is a long gap in which there is no termination codon. It would be expected that the contiguous open reading frame encodes the protein, and in fact it does encode the c-*FOS* protein.

the cloned cDNA is a member of a large protein family (such as the protein kinases) and thus may be similar to but not identical with the protein of interest. In cases in which no amino acid sequence data are available for the protein, the development of databases for gene and protein sequences has provided powerful tools to aid the molecular biologist in discovering the functions of previously unknown proteins.

Searching Sequence Databases to Identify Proteins and Protein Functions

Once the sequencing of proteins and DNA became routine, computerized *sequence databases* were formed to catalog this information so that it could be used by the entire scientific community. Computer programs were developed to explore the sequences in the databases. In addition, these programs can help to identify a newly cloned cDNA sequence. The DNA sequence determined from a cDNA clone can be translated and used to "search" the protein database, in order to determine whether the protein is known or whether the cDNA has been previously cloned. A search may find proteins that are related on the basis

of their primary amino acid sequence (homology). This is important in helping to determine the function of an unknown protein. The information in these databases has expanded our understanding of protein families and has provided clues about how proteins evolve.

Often homology searching can yield surprising results (Figure 7-16). Russell Doolittle performed a computer analysis to compare the amino acid sequence of platelet-derived growth factor (PDGF) determined by peptide sequencing with all the sequences in his protein database. A match was found between PDGF and the amino acid sequence of the protein encoded by an oncogene, v-*sis*, from an RNA tumor virus. This discovery showed that the v-*sis* oncogene was derived from a cellular PDGF gene that had became part of the tumor virus genome (see Chapter 18). For the first time, a physiological function was assigned to an oncogene.

A second example in which computer analysis provided an important hint regarding gene function is the case of the yeast cdc2 gene. As we will learn in Chapter 19, the cdc2 gene is a key controller of the cell cycle in virtually all eukaryotic cells. Cloning of the cdc2

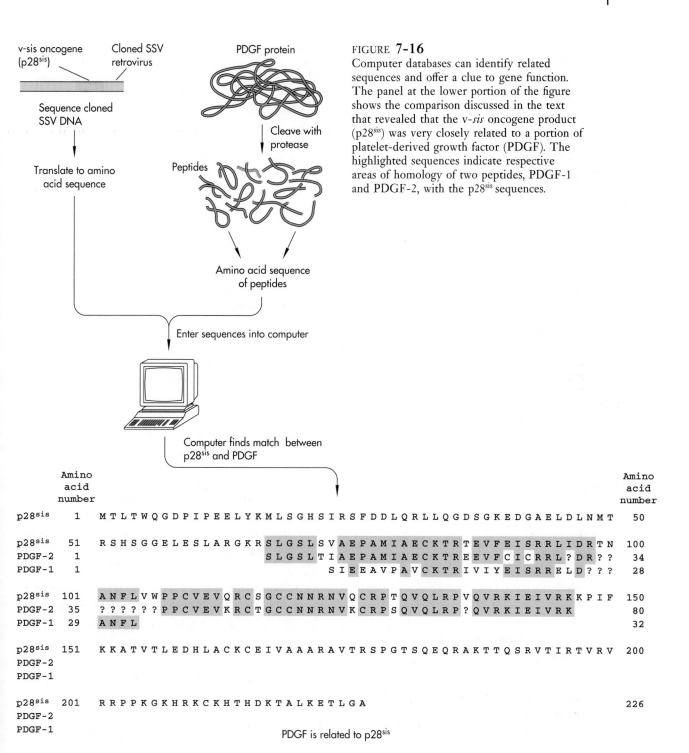

FIGURE **7-16**
Computer databases can identify related sequences and offer a clue to gene function. The panel at the lower portion of the figure shows the comparison discussed in the text that revealed that the v-*sis* oncogene product (p28^{sis}) was very closely related to a portion of platelet-derived growth factor (PDGF). The highlighted sequences indicate respective areas of homology of two peptides, PDGF-1 and PDGF-2, with the p28^{sis} sequences.

gene and comparison of its sequence against a computer database revealed that the protein encoded by the cdc2 gene was closely related to proteins of the protein kinase family. Armed with this information, it was relatively easy to show biochemically that the cdc2 protein is indeed a protein kinase.

Several Procedures Exist to Analyze Proteins Encoded by cDNA Clones

In addition to translating the cDNA sequence by computer, there are several ways to demonstrate experimentally that an isolated clone is structurally or

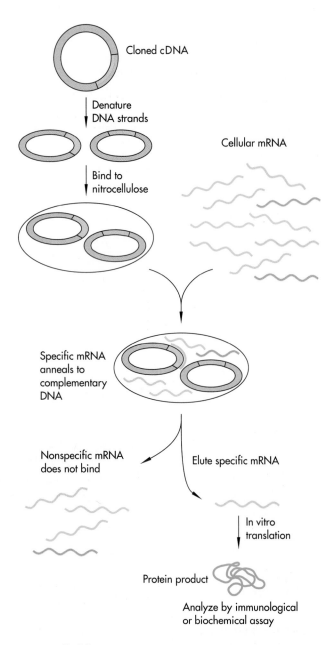

FIGURE **7-17**
Isolating mRNA by hybrid selection. To isolate mRNA for translation in vitro, plasmid DNA carrying the cloned gene is denatured and applied to nitrocellulose filters, which tightly bind single-stranded DNA. The filters are incubated in a solution of cellular RNA. The complementary mRNA hybridizes to gene sequences immobilized on the filter. Noncomplementary RNA is washed off, and the pure mRNA is eluted from the filter. This mRNA can be added to a reticulocyte lysate, which efficiently translates added mRNA into protein, allowing the protein product of the mRNA to be analyzed by a variety of methods.

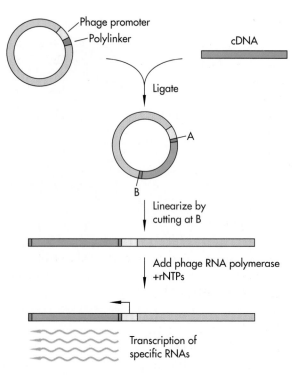

FIGURE **7-18**
In vitro transcription of RNA with bacteriophage RNA polymerase. A widely used procedure for producing biologically active RNA from a DNA sequence is to clone the sequence into a position adjacent to a promoter for an RNA polymerase from bacteriophages such as SP6, T7, and T3, which initiates transcription with a high level of specificity and efficiency. The plasmid DNA is cut with a restriction enzyme at a site ("B") at the end of the inserted gene. The polymerase initiates transcription at its promoter and efficiently transcribes the inserted sequence, producing a large quantity of highly uniform RNA, which is fully active in many types of assays. For example, it can be added to a reticulocyte lysate or microinjected into cells for translation into protein (see, for example, Chapter 20). It can be used as a substrate for studying RNA processing (see Chapter 8). If the RNA is synthesized in the presence of radiolabeled precursors, it can be used as a hybridization probe.

functionally related to the protein of interest. The cDNA can be used to obtain or produce the corresponding mRNA, which can then be translated into protein in vitro using cell lysates prepared from rabbit reticulocytes. There are several ways to prepare mRNA for in vitro translation. In a method called *hybrid selection,* (Figure 7-17) plasmid DNA containing the cDNA insert is denatured and bound to a nitrocellulose filter. The filter is incubated in a solution of cellular RNA under conditions that permit RNA–DNA hybridization. The DNA on the filter captures the complementary mRNA, which can then be simply eluted from the filter. Specific mRNA may then be added to a reticulocyte lysate, and the protein translated from the RNA can be analyzed by gel electrophoresis, using an antibody to the protein, or by biochemical assay for protein function. Alternatively, the mRNA can be prepared synthetically by cloning the cDNA insert into a position adjacent to a promoter for a bacteriophage RNA polymerase (Figure 7-18). The RNA polymerase can be used to transcribe the cDNA into RNA with high efficiency.

The protein encoded by the cDNA can also be characterized by expressing it in *E. coli* or in mammalian cells. A variety of plasmid vectors have been designed to direct high-level expression of foreign proteins in *E. coli*. Generally, these plasmids contain a strong *E. coli* promoter and signals that permit efficient translation by *E. coli* ribosomes. When the plasmid is introduced into *E. coli,* the cell's transcription and translation machinery synthesizes the foreign protein as it would its own. There are many technical variations of this procedure in wide use. Usually the *E. coli* promoter of such vectors is regulated so that it can be kept turned off during growth of the culture, when production of the foreign protein might be toxic to the cells. Expression is induced just before harvesting of the culture (Figure 7-19). The foreign protein can accumulate to a level of 10 percent or more of the total cellular protein. This typically represents an enrichment of 1000-fold or more compared with the natural source of the protein. In some vectors, the

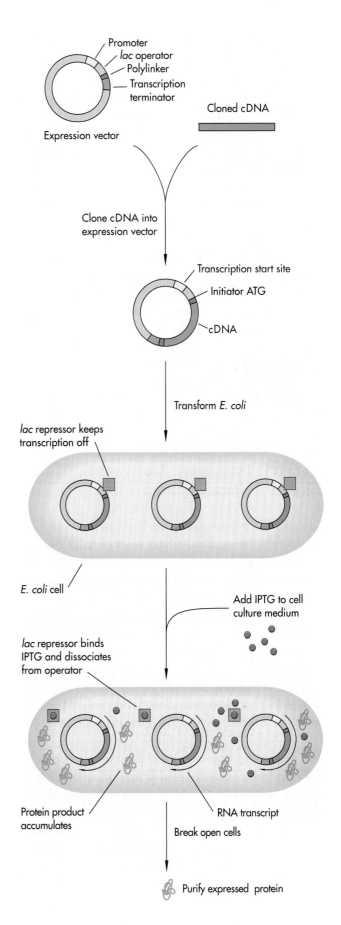

FIGURE **7-19**

The use of a regulated promoter to express a foreign protein in *E. coli.* The expression vector contains a *E. coli* promoter under the control of the *lac* repressor and operator (see Chapter 4). The cDNA is inserted in a restriction site between the promoter and a transcriptional terminator, so that transcripts initiated at the promoter produce mRNA that can be translated into protein. The plasmid is transformed into an *E. coli* strain that overexpresses the *lac* repressor. This keeps transcription of the cDNA tightly repressed in case the foreign protein is toxic to the cells. Once a large enough amount of the *E. coli* culture has been grown, isopropyl-*β*-D-thiogalactopyranoside (IPTG) is added to inactivate the repressor. Now the foreign mRNA is efficiently synthesized and translated into protein.

foreign protein is produced as an in-frame fusion with a bacterial protein. Often fusion proteins are more soluble and therefore easier to work with, and the fused polypeptide sometimes provides a tag or an activity that facilitates purification of the protein from cell extracts. A protein can be targeted for secretion by fusing the coding sequence to a segment of a bacterial protein containing signals for secretion out of the cell. Protein produced in *E. coli* can be used for functional studies if it retains its biochemical activity or as a source of antigen for immunizing animals for the production of specific antisera against the protein. Using similar strategies, cloned cDNAs can be expressed in mammalian cells as well. Mammalian proteins are more likely to be functional when expressed in mammalian cells. For example, expression of a cDNA encoding a protein believed to be a growth factor receptor should result in the expression of binding sites for the growth factor on the surface of the expressing cell, as discussed above for the EPO receptor (Figure 7-12).

Genomic Fragments Are Cloned in Bacteriophage

Cloning of chromosomal genes is relatively easy once the corresponding cDNAs have been isolated and are available as probes. It is essential to analyze genomic DNA if regulatory sequences outside the coding sequences of a gene are to be studied. As we shall see in Chapter 8, analysis of chromosomal genes revealed unforeseen complexities in the organization of eukaryotic genes. Sometimes the abundance of a particular mRNA is so low that a cDNA derived from it is difficult to find in a cDNA library using the available probes. Instead, a genomic clone is first isolated, which serves as a better probe to find the rare cDNA clone in the library. Although genomic clones usually contain noncoding sequences (see Chapter 8), a fragment from the genomic clone containing coding sequences can be isolated and used to screen the library. In addition, the genomic DNA can be introduced into a cell line and expressed using a strong viral promoter. Often under a different promoter, the mRNA is highly expressed, and then its cDNA can be isolated from this cell line.

Early attempts to prepare genomic DNA libraries in plasmids were not successful. While cDNAs are relatively small and of an appropriate size for cloning in plasmids, it soon became clear that plasmids were not suitable for cloning large segments of chromosomal DNA. Because small plasmids replicate more efficiently than large ones, the latter are selected against and portions of the cloned DNA are progressively lost. In contrast, large chromosomal DNA fragments (of about 15,000 bp) are stable when inserted into the DNA of special strains of λ phage. Already at the 1975 Asilomar conference it had been suggested that λ phage could be mutated so that it would be unable to insert its DNA into that of host *E. coli* cells, and thus it would be at least as safe as, if not safer than, disabled plasmid vectors. Such λ vectors exploited the fact that the entire central section of λ phage DNA is not necessary for its replication in *E. coli*, but functions only to ensure the integration of the phage DNA into the host bacterial chromosome during its lysogenic phase. Strains of λ have been created in which recognition sites for a restriction enzyme are located so as to leave intact the left and right end fragments (the *arms*) of the phage DNA that are essential for its replication and packaging (Figure 7-20). After digestion with the enzyme, these end fragments, because of their relatively large sizes, can easily be isolated and can be used to make new λ-like phages containing one left arm, one right arm, and a foreign DNA insert. Maturation of λ phage requires that its DNA chromosome be approximately 45 kb long; thus the only DNAs constructed in vitro that can multiply following such manipulations are recombinant molecules that consist of DNA inserts about 15 kb in length flanked by phage arms.

Genomic libraries were first constructed by cutting genomic DNA to completion with a restriction enzyme such as *Eco*RI and cloning these fragments into a λ vector. But many genomic *Eco*RI fragments are either too large or too small to yield viable phages, and therefore genes carried on such fragments would be missing from a library made this way. Instead, to make a library in which all genomic sequences are represented, genomic DNA is broken in as nearly a random a fashion as possible (Figure 7-20). Random fragments of a size optimal for packaging in λ vectors are selected and cloned. Typically several million independent λ clones carrying genomic fragments are

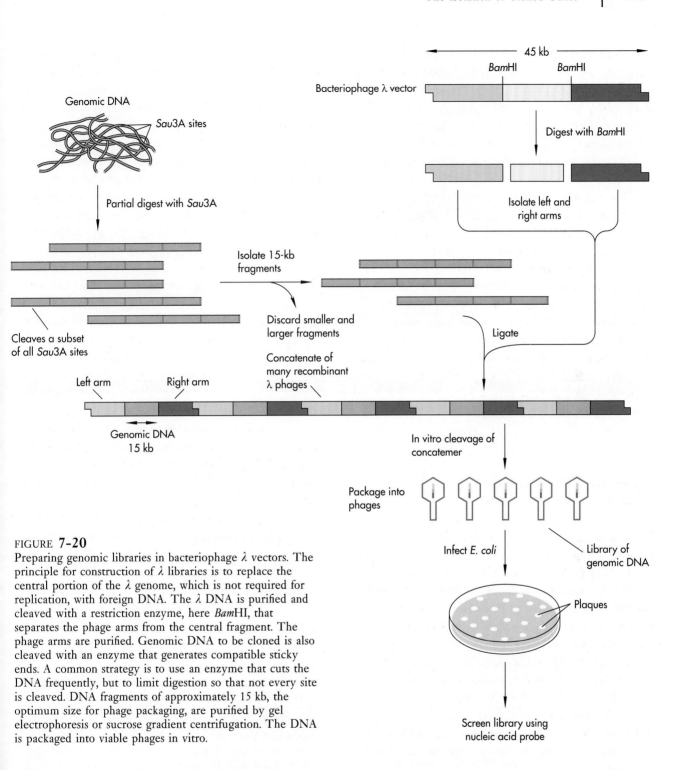

FIGURE **7-20**

Preparing genomic libraries in bacteriophage λ vectors. The principle for construction of λ libraries is to replace the central portion of the λ genome, which is not required for replication, with foreign DNA. The λ DNA is purified and cleaved with a restriction enzyme, here *Bam*HI, that separates the phage arms from the central fragment. The phage arms are purified. Genomic DNA to be cloned is also cleaved with an enzyme that generates compatible sticky ends. A common strategy is to use an enzyme that cuts the DNA frequently, but to limit digestion so that not every site is cleaved. DNA fragments of approximately 15 kb, the optimum size for phage packaging, are purified by gel electrophoresis or sucrose gradient centrifugation. The DNA is packaged into viable phages in vitro.

obtained. Mammalian genomes contain about 3×10^9 bp of DNA. If the average insert size is 15,000 bp, approximately 200,000 phages will carry a genome's worth of DNA. Because, however, the phages carry random DNA fragments, the first 200,000 phages selected will carry some sequences more than once and others not at all. To ensure that all sequences in the

genome are present at least once, simple statistical calculations show that roughly 1 to 2 million phages must be screened (to understand why, simply imagine how many playing cards you would need to draw from a deck to ensure that you have at least one card of each suit; unless you are very lucky, you will need to draw more than four).

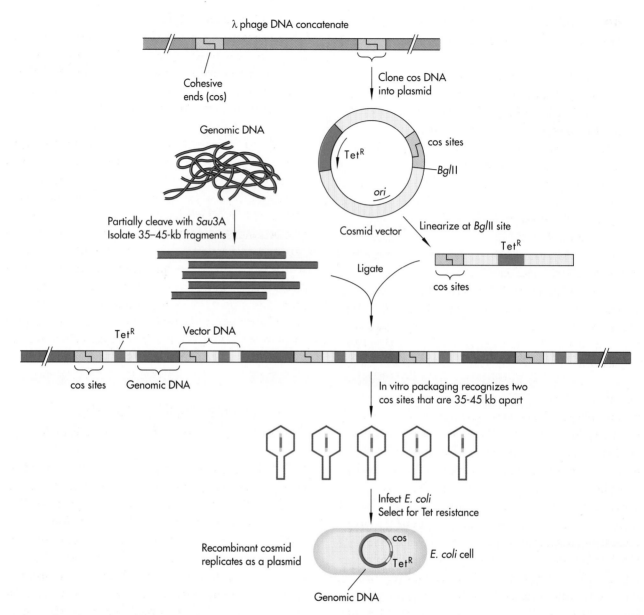

λ phage DNA concatenate

Cohesive ends (cos)

Clone cos DNA into plasmid

Genomic DNA

Tet^R

cos sites

*Bgl*II

ori

Cosmid vector

Partially cleave with *Sau*3A
Isolate 35–45-kb fragments

Linearize at *Bgl*II site

Tet^R

Ligate

cos sites

Tet^R

Vector DNA

cos sites   Genomic DNA

In vitro packaging recognizes two
cos sites that are 35-45 kb apart

Infect *E. coli*
Select for Tet resistance

Recombinant cosmid
replicates as a plasmid

cos

Tet^R

E. coli cell

Genomic DNA

FIGURE **7-21**

Cosmids. Cosmids are plasmid vectors that carry the cos sites from the λ phage as well as a standard plasmid origin of replication and a drug-resistance gene (Tet^R). To clone genomic DNA into a cosmid vector, the vector is linearized with a restriction enzyme, here *Bgl*II, and the genomic DNA is partially digested with *Sau*3A, which leaves *Bgl*II-compatible ends. DNA fragments in the 35 to 45-kb range are isolated and ligated to linearized vector DNA, forming tandem arrays of vector and genomic DNA fragments (not drawn to scale). A λ packaging extract recognizes and packages any ligated DNA that carries two cos sites 35 to 45 kb apart. These segments of DNA are introduced into *E. coli* by infection and replicate as drug-resistance plasmids.

Cosmids Allow the Cloning of Large Segments of Genomic DNA

The size of eukaryotic DNA segments that can be carried in λ phages is limited. It is clear from genomic analysis of a variety of organisms that many genes are larger than this, some as large as 1000 kb. Genes of this size have to be cloned as a set of overlapping genomic fragments. *Cosmid* vectors are hybrids, derived from plasmids and λ phages, that facilitate genomic cloning by being able to carry approximately 45 kb of foreign DNA, three times more than that of phage vectors (Figure 7-21).

The λ phage contains at each end single-stranded complementary stretches of DNA, the so-called *cos* sites. During the normal life cycle of λ, hundreds of copies of newly replicated phage DNA form long chains, or *concatamers*, each λ genome being joined to the next one in the chain through the cos sites. The λ packaging enzymes chop this concatamer into λ-sized pieces by recognizing two cos sites approximately 45 kb apart, cleaving this unit, and packaging it into phages. Thus, the cos sites are all that is necessary for packaging DNA into phage and cosmid vectors. To make a cosmid library, eukaryotic DNA is cleaved with a restriction enzyme under conditions that yield relatively large pieces of DNA. This DNA is then ligated to the cosmid, which has been cleaved with a restriction enzyme that leaves ends complementary to the cleaved genomic DNA. The ligated DNA is packaged in vitro into phages and introduced into *E. coli* by infection. Once inside the *E. coli* cell, the cosmid replicates and can be recovered from the cell as a plasmid. Cosmids represent an important link between λ vectors, which contain up to 15 kb of DNA, and YAC vectors (Chapter 29), which carry over 100 kb of DNA.

Chromosome Walking Is Used to Analyze Long Stretches of Eukaryotic DNA

The isolation of genomic DNA and the analysis of genomic organization must be done in a systematic way. An efficient way of doing this is to use one re-combinant phage or cosmid to isolate another that contains overlapping information from the genome. This technique, known as *chromosome walking*, depends on obtaining a small segment of DNA from one end of the first recombinant and using this piece of DNA to rescreen the library to obtain additional recombinants containing that piece of DNA and the next portion of the genome (Figure 7-22). The second recombinant is used to obtain a third, and so on, to yield a set of overlapping cloned segments. Of course, the small piece of DNA used to rescreen the library must be a single-copy element in the genome; if it is a repeated sequence, many unrelated recombinants will be identified. It may be difficult to clone if it is unstable or toxic in *E. coli*. These situations are molecular roadblocks that may take particular effort to circumvent. Chromosome walking is the only way to search for a gene when its position on a chromosome is only approximately known. We will see in Chapter 26 how chromosome walking is used to clone human disease genes.

Southern and Northern Blotting Procedures Analyze DNA and RNA

Once a cloned cDNA is isolated, it can be used to analyze gene expression and the organization of the genomic DNA. The most widely used method is called *blotting* (Figure 7-23). *Southern blotting*, developed by E. M. Southern, is an extremely powerful tool for analyzing gene structure. To do a Southern blot, genomic DNA is cut with one or several restriction enzymes and the resultant fragments are separated by size on an agarose gel. The gel is then overlaid with a sheet of nitrocellulose filter or nylon membrane, and a flow of buffer is set up through the gel toward the nitrocellulose filter. This causes the DNA fragments to be carried out of the gel onto the filter, where they bind. Thus a "replica" of the DNA in the gel is created on the nitrocellulose. A labeled probe, specific for the gene under study, is then hybridized to the DNA molecules bound on the filter. This probe can be a purified RNA, a cloned cDNA, or a short synthetic oligonucleotide. The labeled probe will hybridize to the specific molecules containing a complementary sequence. Autoradiography of the nitrocellulose filter will result in a pattern of bands indicating the number

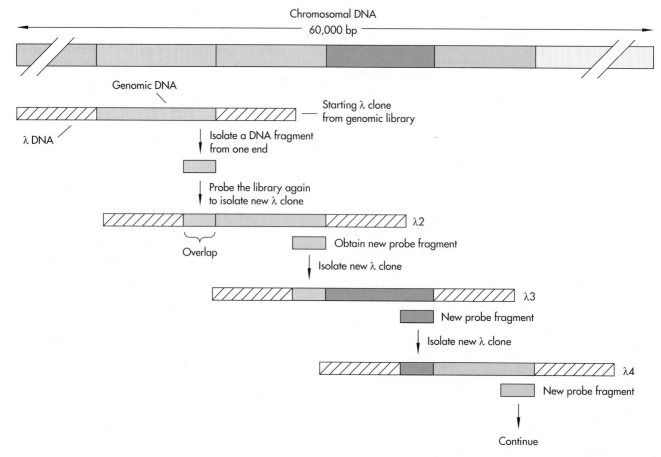

FIGURE **7-22**

Chromosome walking. This method is used to move systematically along a chromosome from a known location. A cloned probe is used to isolate phage clones carrying genomic fragments from that region of the chromosome. A small DNA fragment from the end of the largest phage clone is used to rescreen the same library. Among the clones recovered are new phages (for example, λ2) that carry the probe sequence but whose sequences also extend farther along the chromosome. A new probe is generated from the far end of one of these phages and used to screen the library again to isolate new clones extending still farther. Hundreds of kilobases of contiguous chromosomal DNA can be cloned by repeated cycles of walking. The colored regions in the figure indicate chromosomal DNA, and the hatched regions are λ DNA.

and size of the DNA fragments complementary to the probe. A physical map of the gene, consisting of the positions of landmark restriction sites, can be produced from the sizes of the fragments. Such a *restriction map* can be used to compare the DNA sample with others, allowing, for example, the detection of deletions or other rearrangements involving the gene.

Southern blotting is useful for detecting major gene rearrangements and deletions found in a variety of human diseases (see Chapters 18 and 27). It can also

be used to identify structurally related genes in the same species and homologous genes in other species. Southern blots to a panel of genomic DNAs from a collection of organisms, *zoo blots*, reveal the degree of evolutionary conservation of a gene. For example, Southern blotting identified genes in yeast related to the *RAS* oncogene in human tumors, a remarkable example of evolutionary conservation. In fact, as we will see in Chapter 13, human *RAS* genes are functional in yeast.

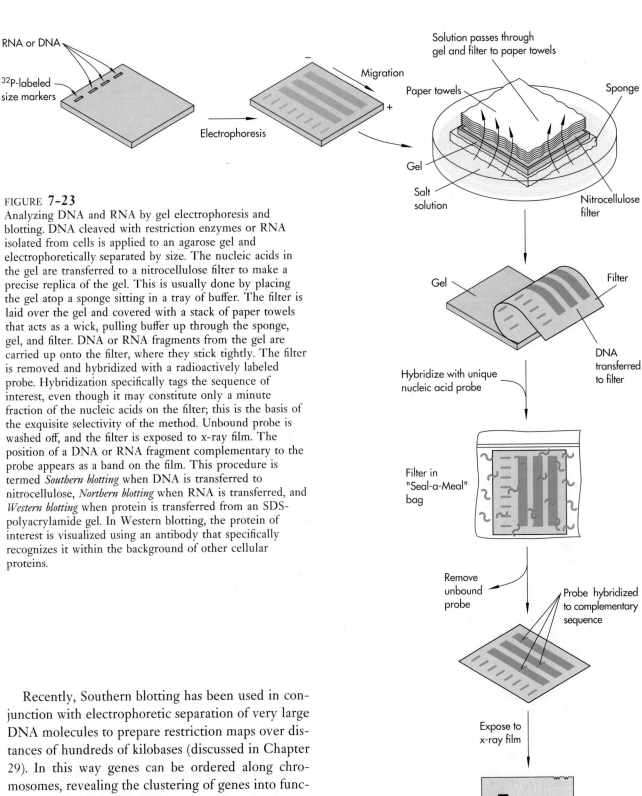

FIGURE **7-23**

Analyzing DNA and RNA by gel electrophoresis and blotting. DNA cleaved with restriction enzymes or RNA isolated from cells is applied to an agarose gel and electrophoretically separated by size. The nucleic acids in the gel are transferred to a nitrocellulose filter to make a precise replica of the gel. This is usually done by placing the gel atop a sponge sitting in a tray of buffer. The filter is laid over the gel and covered with a stack of paper towels that acts as a wick, pulling buffer up through the sponge, gel, and filter. DNA or RNA fragments from the gel are carried up onto the filter, where they stick tightly. The filter is removed and hybridized with a radioactively labeled probe. Hybridization specifically tags the sequence of interest, even though it may constitute only a minute fraction of the nucleic acids on the filter; this is the basis of the exquisite selectivity of the method. Unbound probe is washed off, and the filter is exposed to x-ray film. The position of a DNA or RNA fragment complementary to the probe appears as a band on the film. This procedure is termed *Southern blotting* when DNA is transferred to nitrocellulose, *Northern blotting* when RNA is transferred, and *Western blotting* when protein is transferred from an SDS-polyacrylamide gel. In Western blotting, the protein of interest is visualized using an antibody that specifically recognizes it within the background of other cellular proteins.

Recently, Southern blotting has been used in conjunction with electrophoretic separation of very large DNA molecules to prepare restriction maps over distances of hundreds of kilobases (discussed in Chapter 29). In this way genes can be ordered along chromosomes, revealing the clustering of genes into functionally related groups (Chapter 8). As we shall see, the Southern blotting technique has been important for our understanding of a variety of biological processes such as RNA splicing (Chapter 8) and genomic rearrangements to form antibodies and T cell receptors (Chapter 16) and in the detection of rearranged oncogenes (Chapter 18).

Northern blotting is a technique used to analyze RNA. Total cellular RNA, or poly(A) RNA, is separated by size on an agarose gel. The RNA molecules in the gel are transferred to nitrocellulose or nylon as described above and detected using an appropriate probe. Northern blotting is useful as an adjunct to cDNA cloning because the size of a specific mRNA can be compared with the size of cloned cDNAs, revealing whether the cloned cDNA is full-length. In addition, this simple procedure can indicate which tissues or cell types express a particular gene or the factors that regulate its expression. An example of this is the analysis of regulated genes. The isolation of cDNA clones from serum-regulated genes was previously discussed in the section on differential screening. The first step in characterizing these clones was Northern blotting. Cells were stimulated with serum, then total RNA was isolated at several time points. The RNA obtained at each point was analyzed by Northern blotting using individual cloned cDNAs as a probe (Figure 7-24). The results showed that each mRNA was present in low levels in untreated cells and rapidly accumulated following serum stimulation.

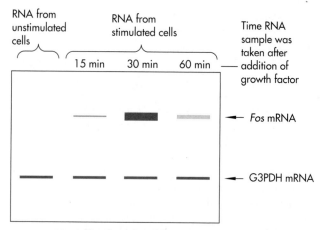

X-ray film of Northern blot

FIGURE **7-24**

Northern blotting. This method is used to examine the size and expression pattern of specific mRNAs. In this experiment, RNA was isolated from cultures of resting fibroblasts stimulated with serum growth factors. Equal amounts of RNA from cells stimulated for the indicated times were applied to an agarose gel and transferred to a nitrocellulose filter by blotting. The filter was hybridized to two radiolabeled probes, one complementary to c-*FOS* (see Chapter 18) known to be regulated by growth factors in serum and one complementary to a nonregulated gene encoding glyceraldehyde 3-phosphate dehydrogenase (G3PDH). The 2.2-kb c-*FOS* mRNA was not found in resting cultures but appeared rapidly and transiently following stimulation. In contrast, the G3PDH mRNA was present at the same level in all samples.

Reading List

General

Gilbert, W., and L. Villa-Komaroff. "Useful proteins from recombinant bacteria." *Sci. Am.,* 242(4): 74–94 (1980). [Review]

Schmidtke, J., and D. N. Cooper. "A comprehensive list of cloned human DNA sequences." *Nuc. Acids Res. Suppl.,* vol. 19: 2111–2126 (1991).

Sambrook, J., E. F. Fritsch, and T. Maniatis. *Molecular Cloning: A Laboratory Manual,* 2nd ed. Cold Spring Harbor Laboratory Press, Cold Spring Harbor, N. Y., 1989.

Glover, D. M., ed. *DNA Cloning: A Practical Approach.* IRL, Oxford, Eng. (A series of volumes.)

Ausubel, F. M., R. Brent, R. E. Kingston, D. D. Moore, J. G. Seidman, J. A. Smith, K. Struhl, eds. *Current Protocols in Molecular Cloning.* Greene Publishing and Wiley Interscience, New York, 1989.

Methods Enzymology, vols. 68, 100, 101, 152, 153, 154, 155. Academic, New York.

Original Research Papers

PLASMID AND PHAGE VECTORS

Blattner, F. R., B. G. Williams, A. E. Blechl, K. Denniston-Thompson, H. E. Faber, L.-A. Furlong, D. J. Grunwald, D. O. Keifer, D. D. Moore, J. W. Schumm, E. L. Sheldon, and O. Smithies. "Charon phages: safer derivatives of bacteriophage lambda for DNA cloning." *Science,* 196: 161–169 (1977).

Bolivar, F., R. L. Rodrigues, P. J. Greene, M. C. Betlach, H. L. Heyneker, H. W. Boyer, J. Crosa, and S. Falkow.

"Construction and characterization of new cloning vehicles, II: a multi-purpose cloning system." *Gene,* 2: 95–113 (1977).

Curtiss, R., III, M. Inoue, D. Pereira, J. C. Hsu, L. Alexander, and L. Rock. "Construction in use of safer bacterial host strains for recombinant DNA research. "In W. A. Scott and R. Werner, eds., *Molecular Cloning of Recombinant DNA: Proceedings of the Miami Winter Symposia,* vol. 13. Academic, New York, 1977, pp. 99–114.

Leder, P., D. Tiemeier, and L. Enquist. "EK2 derivatives of bacteriophage lambda useful in the cloning of DNA from higher organisms: the λgt WES system." *Science,* 196: 175–177 (1977).

Messing, J., and J. Vieira. "The pUC plasmids, an M13mp7-derived system for insertion mutagenesis and sequencing with synthetic universal primers." *Gene,* 19: 259–268 (1982).

Huynh, T. V., R. A. Young, and R. W. Davis. "Constructing and screening cDNA libraries in λgt10 and λgt11." In D. M. Glover, ed., *DNA Cloning: A Practical Approach,* vol. 1. IRL, Oxford, Eng., 1985.

Yanisch-Peron, C., J. Vieira, and J. Messing. "Improved M13 phage cloning vectors and host strains: nucleotide sequences of the M13mp18 and pUC19 vectors." *Gene,* 33: 103–119 (1985).

SYNTHESIS AND CLONING OF cDNA

Rougeon, F., P. Kourilsky, and B. Mach. "Insertion of the rabbit β-globin gene sequence into *E. coli* plasmid." *Nuc. Acids Res.,* 2: 2365–2378 (1975).

Maniatis, T., S. G. Kee, A. Efstratiadis, and F. C. Kafatos. "Amplification and characterization of a β-globin gene synthesized in vitro." *Cell,* 8: 163–182 (1976).

Rabbitts, T. H. "Bacterial cloning of plasmids carrying copies of rabbit globin messenger RNA." *Nature,* 260: 221–225 (1976).

Hohn, B., and K. Murray. "Packaging recombinant DNA molecules into bacteriophage particles in vitro." *Proc. Natl. Acad. Sci. USA,* 74: 3259–3263 (1977).

Scheller, R., R. Dickerson, H. Boyer, A. Riggs, and K. Itakura. "Chemical synthesis of restriction enzyme recognition sites useful for cloning." *Science,* 196: 177–180 (1977).

Land, H., M. Guez, H. Hauser, W. Lindenmaier, and G. Schutz. "5′ terminal sequences of eucaryotic mRNA can be cloned with high efficiency." *Nuc. Acids Res.,* 9: 2251–2266 (1981).

Okayama, H., and P. Berg. "High-efficiency cloning of full length cDNA." *Mol. Cell. Biol.,* 2: 161–170 (1982).

Gubler, U., and B. J. Hoffman. "A simple and very effective method for generating cDNA libraries." *Gene,* 25: 263–269 (1983).

Frohman, M. A., M. K. Dush, and G. R. Martin. "Rapid production of full-length cDNAs from rare transcripts: amplification using a single gene-specific oligonucleotide primer." *Proc. Natl. Acad. Sci. USA,* 85: 8998–9002 (1988).

SCREENING LIBRARIES WITH NUCLEIC ACID PROBES

Grunstein, M., and D. S. Hogness. "Colony hybridization: a method for the isolation of cloned DNAs that contain a specific gene." *Proc. Natl. Acad. Sci. USA,* 72: 3961–3965 (1975).

Benton, W. D., and R. W. Davis. "Screening λgt recombinant clones by hybridization to single plaques in situ." *Science,* 196: 180–182 (1977).

Montgomery, D. L., B. D. Hall, S. Gillam, and M. Smith. "Identification of the yeast cytochrome c gene." *Cell,* 14: 673–680 (1978).

Hanahan, D., and M. Meselson. "Plasmid screening at high colony density." *Gene,* 10: 63–67 (1980).

Suggs, S. V., R. B. Wallace, T. Hirose, E. H. Kawashima, and K. Itakura. "Use of synthetic oligonucleotides as hybridization probes: isolation of cloned cDNA sequences for human b_2-microglobulin." *Proc. Natl. Acad. Sci. USA,* 78: 6613–6617 (1981).

Jaye, M., H. de la Salle, F. Schamber, A. Balland, V. Kohli, A. Findeli, P. Tolstoshev, and J.-P. Lecocq. "Isolation of a human anti-haemophillic factor IX cDNA clone using a unique 52-base synthetic oligonucleotide probe deduced from the amino acid sequence of bovine factor IX." *Nuc. Acids Res.,* 11: 2325–2335 (1983).

Lathe, R. "Synthetic oligonucleotide probes deduced from amino acid sequence data. Theoretical and practical considerations." *J. Mol. Biol.,* 183: 1–12 (1985).

Wood, W. I., J. Gitschier, L. A. Lasky, and R. M. Lawn. "Base composition-independent hybridization in tetramethylammonium chloride: a method for oligonucleotide screening of highly complex gene libraries." *Proc. Natl. Acad. Sci. USA,* 82: 1585–1588 (1985).

Urdea, M. S., B. D. Warner, J. A. Running, M. Stempien, J. Clyne, and T. Horn. "A comparison of non-radioisotopic hybridization assay methods using fluorescent, chemiluminescent and enzyme labeled synthetic oligodeoxyribonucleotide probes." *Nuc. Acids Res.,* 16: 4937–4956 (1988).

CLONING BY DIFFERENTIAL SCREENING AND USING SUBTRACTIVE LIBRARIES

Cochran, B. H., A. C. Zullo, I. M. Verma, and C. D. Stiles. "Expression of the c-fos gene and of a fos-related gene is stimulated by platelet-derived growth factor." *Science,* 226: 1080–1082 (1984).

Hedrick, S. M., D. I. Cohen, E. A. Nielsen, and M. M. Davis. "Isolation of cDNA clones encoding T-cell specific membrane-associated proteins." *Nature,* 308: 149–153 (1984).

Lau, L. F., and D. Nathans. "Identification of a set of genes expressed during the G0/G1 transition of cultured mouse cells." *EMBO J.*, 4: 3145–3151 (1985).

Almedral, J. M., D. Sommer, H. MacDonald-Bravo, J. Burckhardt, J. Perera, and R. Bravo. "Complexity of the early genetic response to growth factors in mouse fibroblasts." *Mol. Cell. Biol.*, 8: 2140–2148 (1988).

Lassar, A. B., J. N. Bushkin, D. Lockshon, R. L. Davis, S. Apone, S. D. Hauschka, and H. Weintraub. "MyoD is a sequence-specific DNA binding protein requiring a region of myc homology to bind to the muscle creatine kinase enhancer." *Cell*, 58: 823–831 (1989).

CLONING cDNAs USING RELATED SEQUENCES

Itoh, H., T. Kozasa, S. Nagata, S. Nakamura, T. Katada, M. Ui, S. Iwai, E. Ohtsuka, H. Kawasaki, K. Suzuki, and Y. Kaziro. "Molecular cloning and sequence determination of cDNAs for a subunit of the guanine nucleotide-binding proteins G_s, G_i, and G_o from rat brain." *Proc. Natl. Acad. Sci. USA*, 83: 3776–3780 (1986).

Hanks, S. "Homology probing: identification of cDNA clones encoding members of the protein-serine kinase family." *Proc. Natl. Acad. Sci. USA*, 84: 388–392 (1987).

CLONING cDNAs BY IMMUNOLOGICAL SCREENING

Broome, S., and W. Gilbert. "Immunological screening method to detect specific translation products." *Proc. Natl. Acad. Sci. USA*, 75: 2746–2749 (1978).

Helfman, D. M., J. R. Fiddes, G. P. Thomas, and S. Hughes. "Identification of clones that encode chicken tropomyosin by direct immunological screening of a cDNA expression library." *Proc. Natl. Acad. Sci. USA*, 80: 31–35 (1983).

Young, R. A., and R. W. Davis. "Efficient isolation of genes by using antibody probes." *Proc. Natl. Acad. Sci. USA*, 80: 1194–1198 (1983).

ISOLATION OF cDNAs BY A FUNCTIONAL SCREEN

Sikela, J. M., and W. Hahn. "Screening an expression library with a ligand probe: isolation and sequence of a cDNA corresponding to a brain calmodulin-binding protein." *Proc. Natl. Acad. Sci. USA*, 84: 3038–3042 (1987).

Singh, S., J. H. LeBowitz, A. S. Baldwin, Jr., and P. A. Sharp. "Molecular cloning of an enhancer binding protein: isolation by screening of an expression library with a recognition site DNA." *Cell*, 52: 415–423 (1988).

Vinson, C. R., K. L. LaMarco, P. F. Johnson, W. H. Landschulz, and S. L. McKnight. "In situ detection of sequence-specific DNA binding activity specified by a recombinant bacteriophage." *Genes and Devel.*, 2: 801–806 (1988).

CLONING cDNAs USING MAMMALIAN EXPRESSION

Seed, B., and A. Aruffo. "Molecular cloning of the CD2 antigen, the T-cell erythrocyte receptor, by a rapid immunoselection procedure." *Proc. Natl. Acad. Sci. USA*, 84: 3365–3369 (1987).

D'Andrea, A. D., H. F. Lodish, and G. G. Wong. "Expression cloning of the murine erythropoietin receptor." *Cell*, 57: 277–285 (1989).

CLONING GENES AND cDNAs BY THE POLYMERASE CHAIN REACTION

Lee, C. C., X. Wu, R. A. Gibbs, R. G. Cook, D. M. Muzny, and C. T. Caskey. "Generation of cDNA probes directed by amino acid sequence: cloning of urate oxidase." *Science*, 239: 1288–1291 (1988).

Wilks, A. F. "Two putative protein-tyrosine kinases identified by application of the polymerase chain reaction." *Proc. Natl. Acad. Sci. USA*, 86: 1603–1607 (1989).

SEQUENCING DNA

Maxam, A. M., and W. Gilbert. "A new method for sequencing DNA." *Proc. Natl. Acad. Sci. USA*, 74: 560–564 (1977).

Messing, J., B. Gronenborn, B. Muller-Hill, and P. H. Hofschneider. "Filamentous *coli* phage M13 as a cloning vehicle: insertion of a *Hin*dII fragment of the lac regulatory region in M13 replicative form in vitro." *Proc. Natl. Acad. Sci. USA*, 74: 3642–3646 (1977).

Sanger, F., A. R. Coulson, B. G. Barrell, A. J. H. Smith, and B. A. Roe. "Cloning in single-stranded bacteriophage as an aid to rapid DNA sequencing." *J. Mol. Biol.*, 143: 161–178 (1980).

Messing, J., Crea, and P. H. Seeburg. "A system for shotgun DNA sequencing." *Nuc. Acids Res.*, 9: 309–321 (1981).

Henikoff, S. "Unidirectional digestion with exonuclease III creates targeted breakpoints for DNA sequencing." *Gene*, 28: 351–359 (1984).

Tabor, S., and C. C. Richardson. "DNA sequence analysis with a modified bacteriophage T7 DNA polymerase." *Proc. Natl. Acad. Sci. USA*, 84: 4767–4771 (1987).

COMPUTERS IDENTIFY HOMOLOGOUS PROTEINS

Doolittle, R. F., M. W. Hunkapillar, L. E. Hood, S. G. Devare, K. C. Robbins, S. A. Aaronson, and H. N. Antoniades. "Simian sarcoma virus *onc*, gene, *v-sis*, is derived from the gene (or genes) encoding a platelet-derived growth factor." *Science*, 221: 275–276 (1983).

Korn, L. J., and C. Queen. "Analysis of biological sequences on small computers." *DNA*, 3: 421–426 (1984).

Lippman, D. J., and W. R. Pearson. "Rapid and sensitive protein similarity searches." *Science*, 227: 1435–1441 (1985).

BACTERIAL EXPRESSION VECTORS

Chang, A. C. Y., J. H. Nunberg, R. J. Kaufman, H. A. Erlich, R. T. Schimke, and S. N. Cohen. "Phenotypic expression in *E. coli* of a DNA sequence coding for mouse dihydrofolate reductase." *Nature*, 275: 617–624 (1978).

Villa-Komaroff, L., A. Efstratiadis, S. Broome, P. Lomedica, R. Tizard, S. P. Nabet, W. L. Chick, and W. Gilbert. "A bacterial clone synthesizing proinsulin." *Proc. Natl. Acad. Sci. USA*, 75: 3727–3731 (1978).

Guarente, L., G. Lauer, T. Roberts, and M. Ptashne. "Improved methods of maximizing expression of a cloned gene: a bacterium that synthesizes rabbit β-globin." *Cell*, 20: 545–553 (1980).

DeBoer, H. A., L. J. Comstock, and M. Vasser. "The tac promoter: a functional hybrid derived from the trp and lac promoters." *Proc. Natl. Acad. Sci. USA*, 80: 21–25 (1983).

Rosenberg, A. H., B. N. Lade, D. S. Chui, J. J. Dunn, and F. W. Studier. "Vectors for selective expression of cloned DNAs by T7 RNA polymerase." *Gene*, 56: 125–135 (1987).

HYBRID SELECTION AND IN VITRO TRANSLATION

Harpold, M. M., P. R. Dobner, R. M. Evans, and F. C. Bancroft. "Construction and identification by positive hybridization translation of a bacterial plasmid encoding a rat hormone structural gene sequence." *Nuc. Acids Res.*, 5: 2039–2053 (1978).

Ricciardi, R. P., J. S. Miller, and B. E. Roberts. "Purification and mapping of specific mRNAs by hybridization selection and all free translation." *Proc. Natl. Acad. Sci. USA*, 76: 4927–4931 (1979).

Melton, D. A., P. A. Kreig, M. R. Rebagliati, T. Maniatis, K. Zinn, and M. R. Green. "Efficient in vitro synthesis of biologically active RNA and RNA hybridization probes from plasmids containing a bacteriophage SP6 promoter." *Nuc. Acids Res.*, 12: 7035–7056 (1984).

SOUTHERN AND NORTHERN BLOTTING

Southern, E. M. "Detection of specific sequences among DNA fragments separated by gel electrophoresis." *J. Mol. Biol.*, 98: 503–517 (1975).

Alwine, J. C., D. J. Kemp, and G. R. Stark. "Method for detection of specific RNAs in agarose gels by transfer to diazobenzyloxymethyl-paper and hybridization with DNA probes." *Proc. Natl. Acad. Sci. USA*, 74: 5350–5354 (1977).

GENOMIC CLONING

Maniatis, T., R. C. Hardison, E. Lacy, J. Lauer, C. O'Connell, D. Quon, G. K. Sim, and A. Efstratiadis. "The isolation of structural genes from libraries of eucaryotic DNA." *Cell*, 15: 687–701 (1978).

COSMIDS

Collins, J., and B. Hohn. "Cosmids: a type of plasmid gene cloning vector that is packageable in vitro in bacteriophage heads." *Proc. Natl. Acad. Sci. USA*, 75: 4242–4246 (1978).

Hohn, B., and J. Collins. "A small cosmid for efficient cloning of large DNA fragments." *Gene*, 11: 291–298 (1980).

CHROMOSOME WALKS

Shimizu, A., N. Takahashi, Y. Yaoita, and T. Honjo. "Organization of the constant region gene family of the mouse immunoglobulin heavy chain." *Cell*, 28: 499–506 (1982).

Steinmetz, M., A. Winoto, K. Minard, and L. Hood. "Clusters of genes encoding mouse transplantation antigens." *Cell*, 28: 489–498 (1982).

8

The Complexity
of the Genome

Examination of the structure of specific genes began by using cloned cDNAs as probes for Southern blotting. Once genomic DNA segments were cloned, the organization of genes could be analyzed in molecular detail using newly developed DNA sequencing methods. This chapter focuses on information about eukaryotic genomes that was obtained by sequencing cloned genes. Initially, attention was focused on the gene sequences corresponding to the region near the 5′ end of the encoded mRNA, since it was thought that these sequences were important for regulation of eukaryotic mRNA synthesis. It was taken for granted that the nucleotide sequence of the cloned genomic DNA ("genes") would be identical with the sequences of the cloned cDNAs. Yet Southern blotting experiments suggested that restriction fragments generated from chromosomal DNA and those from the cloned cDNA frequently differed in size. These observations were finally explained by the surprising finding that genes were "split," and that the intervening sequences (*introns*) were removed from the mRNA. A further demonstration of the complexity of the genome was the finding that some genes produce two or more different proteins by alternative splicing of an mRNA transcript.

As more genes were cloned and analyzed, the basic organization of eukaryotic genes was revealed. For example, eukaryotic genes are far more complex than prokaryotic genes. The genes for histones, ribosomal RNAs, and tRNAs are repeated many times in tandem arrays, and other short repeated sequences are dispersed throughout the genome. The sequences of the α- and β-globin genes suggest that the genes were duplicated from a common precursor and thus provide a clue about the evolution of genes and proteins. Although the first cloned genes were small, now it has been established that some genes span hundreds of kilobases.

Much of the information described in this chapter was obtained in a burst of activity during the late 1970s and early 1980s. Recent work in genomic organization has continued to produce surprises. For example, genes have been identified that are encoded within the introns of other genes in the complementary DNA strand. Other studies showed how a single tropomyosin gene can encode nine distinct tropomyosin polypeptides by alternative mRNA splicing. Although the experiments described in this chapter deal primarily with the cloning and sequencing of genes, this work produced a considerable wealth of knowledge about the complexity of the eukaryotic genome.

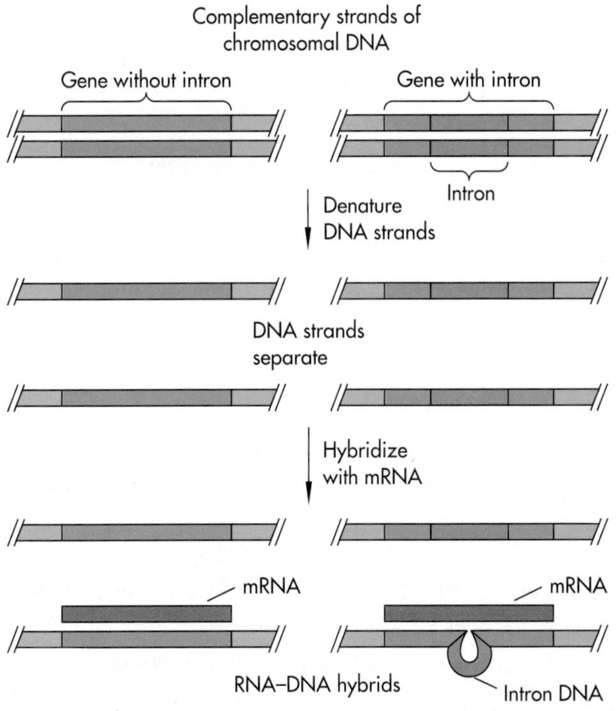

Split Genes Are Discovered

The discovery of mRNA splicing was first announced at the 1977 Cold Spring Harbor Laboratory Symposium on Chromatin. During adenovirus replication, precursors of viral RNA transcripts (pre-mRNAs) within the nucleus of an infected cell were found to be shortened by removal of one or more internal sections to produce smaller mRNA molecules. These moved to the cytoplasm, where they served as templates for viral protein synthesis. The gene segments missing from cytoplasmic mRNAs were identified in electron micrographs in RNA–DNA heteroduplexes between adenovirus DNA and isolated RNA transcripts (Figures 8-1 and 8-2). These segments were present in the long primary transcript and were "spliced out" during formation of the smaller mRNA molecules. Quickly, the generality of splicing was shown to extend to another virus, SV40; and the question immediately arose whether splicing might also be involved in the processing of cellular RNA. For several years it had been known that many eukaryotic mRNAs are first synthesized as large pre-mRNAs that are later processed in the nucleus to much smaller products. But until the announcement of adenovirus splicing, it had always been assumed that this processing necessarily and exclusively involved removal of long sections at the 5' and 3' ends of pre-mRNA.

FIGURE **8-1**
DNA–mRNA hybrids (heteroduplexes) formed by annealing chromosomal DNA with mRNA. On the left side, the segment of a gene that lacks an intron hybridizes entirely with its corresponding mRNA. On the right side, the gene sequences in exons hybridize with the mRNA, whereas the intron DNA does not hybridize but forms a loop (R loop).

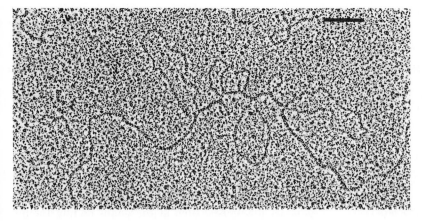

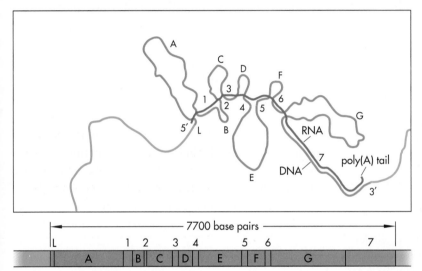

FIGURE **8-2**

Examination of gene structure by electron microscopy of DNA–mRNA hybrids. DNA containing the gene for ovalbumin was hybridized with ovalbumin mRNA. The regions of the gene that hybridize to the mRNA are eight exons (L, 1-7). Genomic DNA that encodes introns does not hybridize to the mRNA but forms seven loops (A-G). The upper portion of the figure shows the actual electron micrograph. Regions where genomic DNA hybridized to mRNA form a thicker line than do the single-stranded genomic DNA loops. The locations and lengths of the introns were estimated simply by plotting the position of each loop along the mRNA molecule. Because the length of the mRNA (in nucleotides) was known, the approximate positions of the introns could be calculated. The middle portion of the figure shows the interpretation of the electron micrograph. The 5′ and 3′ ends of the mRNA are indicated. The bottom portion shows a scheme of the structure of the ovalbumin gene determined subsequently by DNA sequence analysis of the exon-intron boundaries in genomic DNA and comparison with the cDNA sequence. The exons are shown in green and the introns are shown in red (modified from Chambon, 1981).

Introns Are Discovered in Eukaryotic Genes

Excited by these results, people working on the structure of eukaryotic genes searched for splicing of cellular RNAs. Within a very short time after the discovery of adenovirus splicing, the coding sequences of β-globin, ovalbumin, and immunoglobulin genes were also found to be interrupted by noncoding DNA. Proof that chromosomal genes were spliced came initially by electron microscopy. The sizes and locations of these introns were then estimated by a technique called *S1 nuclease mapping* (Figure 8-3). The regions of the chromosomal DNA not present in the mature mRNA were given the name *introns*. The coding sequences were called *exons*, because the processed mRNAs, without the introns, "exit" from the nucleus to the cytoplasm. It should be noted that at the time that the electron microscopy experiments on adenovirus were done, no one had cloned a cellular gene yet. Once the first genes were cloned, introns were identified by comparing the cloned genomic DNA with the corresponding cloned cDNA. For small genes, such as the β-globin gene, the sizes of the introns and the locations of intron-exon boundaries were precisely determined by sequencing cloned genomic DNA and

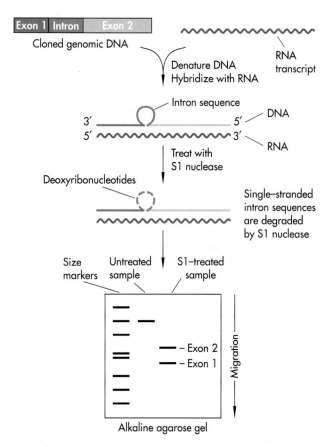

FIGURE **8-3**

S1 nuclease mapping finds introns in a gene. Total cellular RNA is hybridized to a cloned genomic DNA fragment that contains a single intron. An RNA–DNA heteroduplex is formed between the gene DNA and its corresponding mRNA by base-pairing of complementary sequences. Intron sequences in the gene do not hybridize with the mRNA, so these DNA sequences form single-stranded loops. The RNA–DNA hybrid is treated with S1 nuclease, an enzyme that digests single-stranded DNA into mononucleotides. The unpaired intron sequences thus are digested, a process that splits the genomic DNA into two fragments. The sizes of these two DNA fragments are determined by gel electrophoresis. From this information, the length and location of the intron in the gene are deduced.

comparing the sequence with the cDNA sequence and with the protein sequence.

Introns exist in genes from all eukaryotic animals, in plant genes, and, surprisingly, in genes of the *E. coli* phage T4. Often the introns of a gene contain many more nucleotides than do its coding exons, thus ac-

counting for the previously unexplained large sizes of many primary RNA transcripts. The number and size of introns vary widely from one gene to another. Two introns are present in all genes of the *β*-globin family (Figure 8-4). The sizes of introns in the *β*-globin genes from different species differ slightly, but their positions are always the same relative to the coding sequence. The ovalbumin gene is more complicated than the *β*-globin gene and contains seven introns. The length of introns can vary from 31 nucleotides in an SV40 gene to over 210,000 nucleotides in the human dystrophin gene. A few genes, such as the genes coding for the *α* and *β* forms of interferon and most of the genes from the yeast *Saccharomyces cerevisiae* do not contain introns. Intronless mammalian genes can also be generated through recombinant DNA tricks and tested to see how they function in vivo. For certain genes, the complete removal of introns has no consequence. Such genes produce fully active mRNA transcripts. However, for other genes, the removal of their natural introns somehow blocks the exit of functional mRNA products to the cytoplasm. Perhaps in these latter cases the newly made transcripts adopt configurations incompatible with exit from the nucleus.

Specific Base Sequences Are Found at Exon-Intron Boundaries

By the summer of 1978, just a year after the first split genes were discovered, the sequences of many exon-intron boundaries from cellular genes had been determined. It was hoped that such sequence data would be useful in predicting the location of an intron in a gene and would also explain how splicing was accomplished. People expected to find that the sequence at the upstream (5′) and downstream (3′) ends of an intron would be complementary. These sequences would therefore be expected to hybridize and form a stretch of double-stranded RNA, which would be recognized and precisely excised by specific splicing enzymes. However, this idea was soon discounted, since the sequences at the ends of the introns were not complementary. The upstream and downstream splice sites therefore could not be brought together by self-

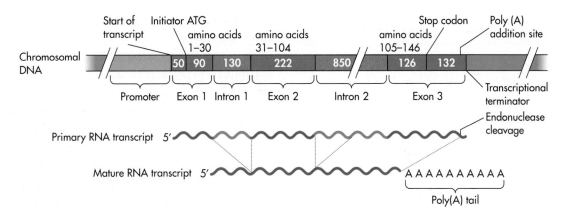

FIGURE **8-4**

Organization of the human β-globin gene, which encodes the adult form of the β chain of hemoglobin. The β-globin gene from all mammals consists of three exons interrupted by two introns. The numbers within the colored segments indicate the number of nucleotides in each exon or intron. The primary transcript (pre-mRNA) contains the exon and intron sequences. The introns are removed by splicing enzymes to form the mature mRNA. The sequence AAUAAA, near the 3′ end of the primary transcript, directs an endonuclease to cleave the RNA 15 to 30 nucleotides farther along the molecule. The end generated by this cleavage is the site of poly(A) addition.

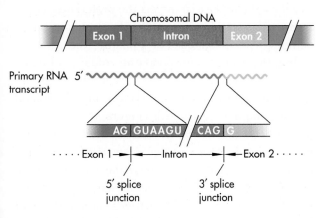

FIGURE **8-5**

Consensus sequences of 5′ and 3′ splice junctions in eukaryotic mRNAs. Almost all introns begin with G-U and end with A-G. From the analysis of many exon-intron boundaries, extended consensus sequences of preferred nucleotides at the 5′ and 3′ ends have been established. In addition to A-G, other nucleotides just upstream of the 3′ splice junction are also important for precise splicing.

complementarity. Yet the base sequences at the boundaries between exons and introns were not random, and after many boundaries had been sequenced, a pattern emerged. For every intron, the sequences at the 5′ end were related to each other. Similarly, the sequences at the 3′ end of every intron were related (Figure 8-5).

These two consensus sequences suggested an alternative mechanism in which removal of an intron was coordinated by interaction with small nuclear RNA molecules (snRNAs), which were present in large numbers in the nucleus of virtually every eukaryotic organism. It was thought that snRNAs might be associated with the splicing enzymes, since it had previously been shown that tRNA processing in *E. coli* was accomplished by a complex of RNA and protein. Inspection of the nucleotide sequence of one such snRNA, U1 RNA, revealed that a sequence at its 5′ end was complementary to the consensus sequence at the 5′ end of splice sites (Figure 8-6). The interaction of this region of U1 snRNA with the 5′ splice site and the involvement of other snRNAs in splicing has now been established experimentally.

As we shall learn in Chapter 26, mutations in the genome are often manifested as genetic diseases. Several examples of β-thalassemia, a disease in humans that affects the β chain of hemoglobin, are caused by mutations in the β-globin gene that disrupt correct splicing of the mRNA transcript. In an unusual finding, it was demonstrated that the *ras* oncogene owes part

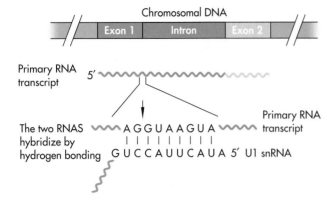

FIGURE **8-6**
Sequences at the 5′ end of U1 snRNA are complementary to the 5′ splice junction consensus sequence. The involvement of snRNAs in splicing was suggested by the observation that U1 could potentially hybridize with the pre-mRNA. Results of experiments support this model.

of its transforming activity to a single nucleotide mutation within an intron. Although this mutation does not affect splicing, it increases the expression of the oncogene 10 times over that of the same gene without this extra mutation. The mechanism by which this mutation exerts its effect is not understood.

Alternative Splicing Pathways Generate Different mRNAs from a Single Gene

Some primary gene transcripts can be spliced in different ways to produce distinct RNA molecules that each encode a different protein. Differential splicing was first demonstrated in mRNAs from adenovirus, SV40, and polyoma virus and subsequently in mRNAs from many cellular genes. Alternative splicing often produces two forms of the same protein that are necessary at different stages of development or in different cell types. For example, immunoglobulins of the IgM class exist either as a membrane-bound protein displayed on the cell surface or as a soluble protein secreted into the blood. The membrane-bound form is expressed first during B-cell development; then as the B cell differentiates into a plasma cell, expression of the membrane-bound form ceases and the secreted form is produced.

As discussed in Chapter 16, an immunoglobulin is a complex of four protein molecules, two "heavy" chains and two "light" chains. By direct analysis of the secreted and membrane-bound antibody proteins, it was demonstrated that the two antibodies contain different heavy chains. Analysis of cloned cDNAs showed that this difference is due to alternative splicing of the heavy-chain gene transcript. The two heavy-chain mRNAs differ only at their 3′ ends (Figure 8-7). The B-cell-specific mRNA contains two exons that encode very hydrophobic amino acids that anchor the protein in the membrane. These exons are missing from the plasma cell IgM heavy-chain mRNA. Instead, this mRNA contains a different exon that is shorter and encodes less hydrophobic amino acids appropriate for secretion of the protein.

The mechanism by which cell-type-specific splicing occurs is not currently understood. A model has been proposed in which the presence (or absence) of specific splicing factors directs the preferential use of a particular set of exons. Alternative splicing can be extremely complicated. For example, the gene encoding the protein α-tropomyosin contains 14 exons. Different combinations of exons are used to form mature tropomyosin mRNAs in skeletal muscle, smooth muscle, and nonmuscle cells (Figure 8-8). This complex process most likely evolved to produce a tropomyosin protein with a particular structure that is necessary for each cell type. Although the overall structure of each tropomyosin protein is similar, the cell-type-specific amino acids may function as binding sites for other proteins.

Introns Sometimes Mark Functional Protein Domains

Through alternative splicing, exons can be combined to form proteins with different functional domains. Introns may also play an important role in the evolution of proteins with new functions. It has been established by x-ray crystallography that proteins are made up of *domains;* each domain is responsible for a particular function of the protein, such as binding of a cofactor or a substrate. When genes for proteins with

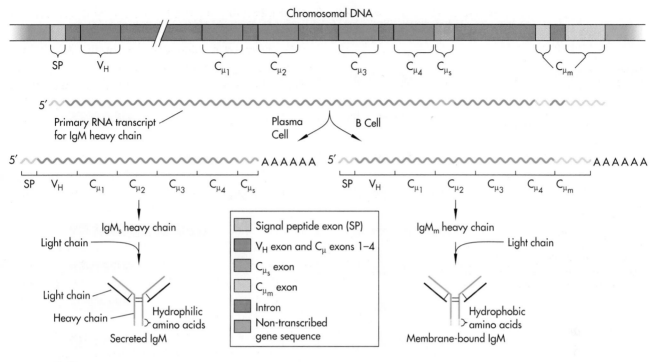

FIGURE **8-7**

Alternative splicing produces secreted and membrane-associated forms of IgM from a single gene. The μ gene encoding the heavy chain for an IgM molecule is shown. As we will discuss in chapter 16, the heavy and light chains of an antibody are composed of a series of structural *domains*. The organization of an immunoglobulin gene parallels this domain structure of the protein. For example, in the heavy chain gene shown here, the coding sequences for the signal peptide (SP)—amino acids at the amino terminus that target the antibody for secretion—are contained within the first exon. Similarly, the sequences encoding the variable (V_H) and constant (C_μ) domains reside within individual exons. The same pre-mRNA is produced in both B cells and plasma cells, but each cell type processes the primary transcript in a different way. In a plasma cell (which secretes immunoglobulin molecules into the blood), the mature mRNA is spliced so that it includes the C_{μ_s} exon, encoding hydrophilic amino acids. In a B cell (which displays immunoglobulin molecules on its surface), the pre-mRNA is spliced so that it includes the two C_{μ_m} exons, encoding hydrophobic amino acids, thus allowing the immunoglobulin to be anchored in the plasma membrane.

known three-dimensional structure were cloned, the positions of the introns were often found to divide the coding sequences in such a way that individual exons encoded well-defined structural domains. For example, the domains in immunoglobulins are each encoded by individual exons. The low-density lipoprotein (LDL) receptor is predicted to have a complex protein structure that appears to have evolved by recruiting functional domains from several other proteins. As we shall see in Chapter 22, the positions of introns and exons that form the LDL receptor gene reflect this domain structure. The mechanism by which these new proteins were formed is difficult to demonstrate experimentally, but they may have been formed by genetic recombination between introns within two different genes. It is conceivable that the long length of many introns helps to ensure that coding sequences are kept intact during genetic crossing over. Imprecise recombination would be tolerated within introns, but would disrupt the coding sequence in exons.

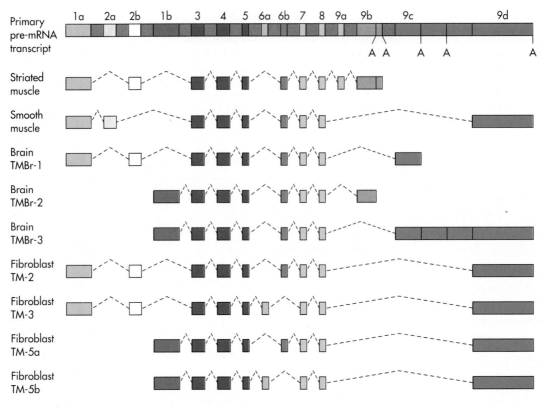

FIGURE **8-8**

Complex patterns of eukaryotic mRNA splicing. The pre-mRNA transcript of the α-tropomyosin gene is alternatively spliced in different cell types. The red boxes represent introns; the other colors represent exons. Polyadenylation signals are indicated by an A. Dashed lines in the mature mRNAs indicate regions that have been removed by splicing. TM, tropomyosin. (After J. P. Lees, et al., 1990.)

Transcriptional Control Regions Occur Throughout Eukaryotic Genes

Eukaryotic genes can be divided into three classes on the basis of the specific RNA polymerase that transcribes them into RNA. RNA polymerase I transcribes only the tandemly repeated ribosomal RNA (rRNA) genes forming the precursor of 18S, 5.8S and 28S rRNAs. RNA polymerase II transcribes all genes encoding proteins as well as most small nuclear RNAs that constitute small nuclear ribonucleoprotein particles (snRNPs). RNA polymerase III transcribes 5S rRNA and all tRNA genes.

It is now clear that any mRNA encoding a protein (and, thus, any cloned cDNA, as well) contains only the information needed for *translation* by ribosomes. In order to identify the specific nucleotides that con-

trolled *gene expression,* regions of the gene flanking the coding region were sequenced. Comparisons of these sequences revealed common patterns near the 5′ and 3′ ends of different genes. These were predicted to be important for proper transcription by RNA polymerase II. The most common motif is the TATA sequence around 30 bp from the transcriptional start site. Other conserved sequences have been found roughly 50 to 100 bp upstream of the transcriptional start site; among these are a GC-rich sequence and the sequence CCAAT. As we shall see in Chapter 9, these sequences are the recognition sites for specific proteins that are transcription factors. Another regulatory element, called an *enhancer,* is often found thousands of base pairs away from the beginning of the mRNA and can exert its effect from the 3′ end of the gene. The nucleotide sequence of an enhancer was first identified in mammalian tumor viruses (Chapter 18), where enhancers are located close to the transcriptional start site. Discovery of the functional importance of these motifs and identification of other regulatory sequences were accomplished experimentally by performing transcription assays and by studying the effects of mutations in promoter sequences (Chapter 9).

At the 3' end of many eukaryotic genes is a sequence transcribed into AAUAAA in the corresponding mRNA. This is believed to act as a signal for the addition of the poly(A) tail to the 3' end of the transcribed mRNA. Although transcription usually continues beyond this six-base sequence (sometimes thousands of bases beyond), the sequence is believed to cause an endonuclease to clip the mRNA strand at a site only 10 to 30 bases away, thus removing most of the additional downstream bases. A second enzyme, poly(A) polymerase, then adds the 100 to 300 bases of the poly(A) tail. There is experimental evidence that the formation of the mature 3' end of eukaryotic mRNAs requires additional proteins and snRNAs. These additional factors will probably be isolated and identified more clearly by fractionating nuclear extracts that are capable of performing the polyadenylation reaction in vitro.

RNA Polymerase III Transcription Is Regulated by Sequences in the Middle of the Gene

RNA polymerase III transcribes 5S RNA, tRNA, and a number of other RNA genes in eukaryotes. The general features of these genes are that they are about 300 bp in length, they encode RNA molecules that perform structural roles (that is, these RNAs do not encode proteins), and they are repeated thousands to millions of times per genome in some organisms. Analysis of the *Xenopus* 5S RNA gene established that the primary region required for transcription is located in the middle of the gene. A series of deletion mutants were constructed in the cloned 5S RNA gene by in vitro mutagenesis (Chapter 11). When these mutant genes were tested in a transcription assay, a 50-bp region in the middle of the gene, termed the *internal control region*, was found to be required for proper expression. By comparison of different polymerase III–dependent genes and by further in vitro mutagenesis studies, the control region has been divided into two

conserved "blocks." Recent experiments have demonstrated that the transcription factor TFIIIA binds to the internal control region of the 5S RNA gene; then other transcription factors, including TFIIIB, TFIIIC, and RNA polymerase III, sequentially bind, forming an active transcription complex. TFIIIA contains a set of distinctive structural domains termed *zinc fingers* that are important for DNA binding (Chapter 9). Transcriptional control regions have been found upstream of other polymerase III–dependent genes. These sequences most likely bind as yet unidentified proteins that are required for proper transcription.

Genes Encoding Abundant Products Are Often Tandemly Repeated

Although most genes are present only once per haploid genome, the genes for histones, rRNAs, and tRNAs are present many times within the genome and are often clustered together. The histone gene family consists of five major genes in most eukaryotes (H1, H2A, H2B, H3, and H4), encoding proteins that maintain and regulate chromosome structure. In lower eukaryotes such as the sea urchin and *Drosophila*, these five genes occur in a cluster of about 5000 to 6000 bp, and each cluster is tandemly repeated between 100 and 1000 times (Figure 8-9). It is thought that these

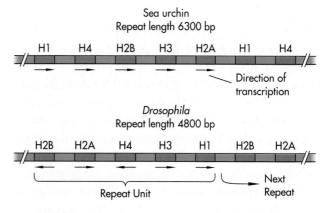

FIGURE **8-9**
The histone gene cluster is tandemly repeated. The genomic organization of the histone gene family is shown for the sea urchin and the fruit fly. The coding sequences are shown in green and the intergenic spacer DNA is shown in blue. The direction of transcription for each gene is designated by an arrow. In sea urchin, the entire 7-kb region is repeated up to 1000 times; in the fruit fly, about 100 times.

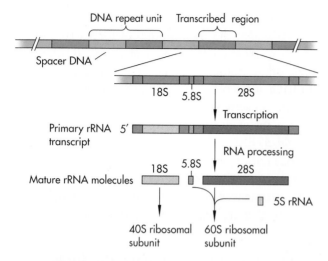

FIGURE **8-10**

Organization and transcription of an rRNA gene repeat. The rRNA gene that encodes the 18S, 5.8S, and 28S ribosomal RNAs is tandemly repeated in all eukaryotes. The basic gene repeat unit consists of a nontranscribed spacer region followed by the transcribed region. In *Xenopus laevis* the rRNA gene is about 13 kb. Initially, an 8.3-kb primary rRNA transcript is made, which is processed to the mature 18S, 5.8S, and 28S rRNA molecules.

organisms require such a large number of genes in order to be able to produce enough histones during periods of rapid DNA replication in embryonic development. A cluster of histone genes exists in higher eukaryotes but at only 10 to 40 copies per genome.

The genes encoding ribosomal RNAs are also clustered and tandemly repeated. The three genes that encode the 18S, 5.8S, and 28S rRNAs are clustered together and are transcribed initially as a large precursor by RNA polymerase I. This precursor is then cleaved into the three individual rRNA molecules (Figure 8-10). Electron microscopy has shown that the rRNA genes are localized in the nucleoli, where rRNAs are processed and the ribosomes are assembled. These electron micrographs show the transcription of rRNA genes from tandemly repeated genes. Between each transcription unit is a region of DNA, known as *spacer DNA,* that is not transcribed. The structure of the rRNA gene repeat is now known in molecular detail following the cloning and sequencing of rRNA genes from many organisms. In the toad *Xenopus laevis* each repeat is approximately 11 to 17 kbp long. The transcribed mRNA is 8.3 kb, and the nontranscribed spacer DNA varies from 2.7 to 9 kbp. In somatic cells,

about 500 copies of the rRNA gene are present in a tandem array. Careful analysis of the nontranscribed sequences has shown that each of these regions is composed of several types of repeated sequences. These repeated sequences can influence the rate of rRNA gene transcription. The rRNA genes in other organisms follow the same organization found in *Xenopus;* however, the length and sequence of the nontranscribed spacer DNA differ in each organism. The remaining rRNA gene encoding the 5S rRNA is also present in tandem repeats but is not associated with the other three rRNA genes.

Clustered Globin Genes Exhibit Coordinated Expression in Development

The α and β subunits of hemoglobin are encoded by two gene families whose members are closely related structurally and whose expression is regulated during development. Different forms of hemoglobin are required in the developing fetus and in the adult. In humans, an embryonic form is produced initially, then a fetal form is expressed during gestation. Shortly before birth, an adult form is produced, and expression of fetal hemoglobin declines until, at 6 months of age,

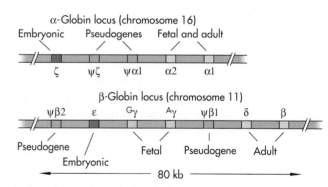

FIGURE **8-11**

The order of human globin genes in the chromosome reflects their expression during development. The human α- and β-globin gene families contain multiple genes that encode the α and β chains of hemoglobin, respectively. The α-globin locus contains the ζ gene, expressed in the embryo, and the $\alpha1$ and $\alpha2$ genes, expressed in both the fetus and the adult. The β-globin locus contains the embryonic ε gene, the fetal $^G\gamma$ and $^A\gamma$ genes, and the δ and β adult genes. The two loci contain pseudogenes ($\psi\beta$, $\psi\zeta$, $\psi\alpha1$) that do not encode functional proteins. The signals that are responsible for the switch in expression during development are not completely understood.

Transcriptionally Inactive Regions of the Genome Are Often Found to Be Methylated

Direct analysis of purified eukaryotic chromosomal DNA found that besides the four bases adenine, guanine, cytosine, and thymine, chromosomal DNA contained a small amount of a modified base, 5-methylcytosine (5-methyl-C). Methylation of cytosine occurs enzymatically after DNA synthesis. The position of 5-methyl-C is not random but, rather, is predominantly found when a C immediately precedes a G, a *CG pair*. About 70 percent of the C's in CG pairs in eukaryotic chromosomal DNA are methylated. (A CG pair is often designated CpG to signify that the two nucleotides are contiguous in DNA and that the C is 5′ to the G.)

For certain genes in higher eukaryotes, methylation affects gene expression. For example, for transcriptionally active genes such as globin, the CG pairs within and near the gene are undermethylated. This is assessed by using restriction endonucleases that contain CG in their recognition sequence. For example, *Msp*I and *Hpa*II both recognize the sequence CCGG. *Msp*I digests either unmethylated (CCGG) or methylated (C^mCGG) DNA, but *Hpa*II cleaves only unmethylated DNA. A Southern blot experiment (Chapter 7) is performed in which chromosomal DNA is cleaved with either *Msp*I or *Hpa*II. The two digests are then electrophoresed on an agarose gel, transferred to nitrocellulose, and the nitrocellulose filter probed using a segment of the cloned gene. If the gene is not methylated, then both enzymes will cleave the DNA at all CCGG sites within the gene. If the gene is methylated, then only *Msp*I will cleave the DNA within the gene. *Hpa*II digestion will yield much larger DNA fragments.

Experiments using cloned genes have strengthened the hypothesized link between methylation and gene expression. Transcription from a methylated globin gene introduced into mammalian cells was lower relative to the expression from the unmethylated gene. The effect of methylation on gene expression was also demonstrated by incorporation into DNA of nucleotide analogs, such as 5-azacytosine, that block methylation and sometimes activate silent genes. The

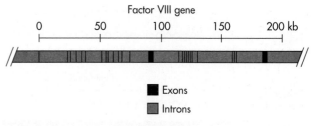

FIGURE **8-13**
Organization of the human Factor VIII gene. The gene spans about 186,000 bp and expresses a 9-kb mRNA transcript that is encoded by 26 exons. The positions of exons are shown by black lines and boxes; introns are in red. The 26 exons range from 69 to 3106 bp, and the introns range from 200 to 32,400 bp.

undermethylation of transcriptionally active genes is thought to function by altering the structure of the DNA so that it is more accessible to activator proteins or, alternatively, so that repressor proteins dissociate from the gene. Although undermethylation generally correlates with gene expression, there are a number of examples in which undermethylation doesn't affect expression at all. In fact, some organisms, such as *Drosophila*, do not contain methylated DNA.

Most Genes Are Much Larger Than Their mRNA

Cloned cDNAs were first used as hybridization probes for Southern blotting to obtain information about the sizes and organization of eukaryotic genes. Although the first cloned genes such as β-globin and ovalbumin contained less than 10 kbp, it is now clear that many genes span lengths often greater than 100 kbp (100,000 base pairs), and several have been reported to be over 1000 kbp. Most of the nucleotides in these large genes are situated in introns. For these genes, only the exon sequences and the portion of the intron sequences at the exon-intron boundaries are determined.

The organization of the human gene encoding the blood clotting protein Factor VIII was recently determined. In humans, this gene resides on chromosome X, spans approximately 186,000 bp, and is divided into 26 exons (Figure 8-13). It has been estimated that this

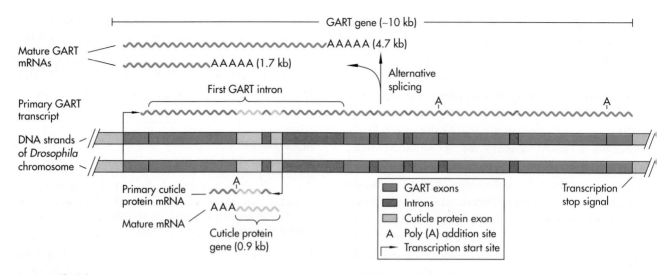

FIGURE **8-14**

A *Drosophila* pupal cuticle gene is nested within an intron of the GART gene. The genomic organization of the GART gene locus is shown. Exons that encode the GART mRNA and exons for the cuticle mRNA are shown. The primary GART mRNA transcript utilizes alternative poly(A) addition signals to yield two mature mRNAs. The cuticle protein mRNA is transcribed from the opposite strand within the first GART intron.

gene makes up about 0.1% of chromosome X. Since the gene is so long, it was isolated from a set of λ and cosmid clones that contained overlapping chromosomal inserts (see Chapter 7). The pre-mRNA transcript produced from this gene is processed to a 9-kb mature mRNA that encodes a 2332 amino acid protein. cDNA fragments were used as probes to map the positions of exons along the gene by Southern blotting of genomic DNA. The cDNA was also used to clone genomic fragments in order to determine the sequences at the intron-exon boundaries. The 26 exons range in size from 69 to 3106 bp, and the introns range from 200 to 32,400 bp. It has been determined from analysis of many cloned genes that most exons contain fewer than 400 bp, though the last exon of a gene is more variable and can contain more than 2000 bp. It is now clear that most of the DNA in the chromosomes of higher eukaryotes is either in introns or in spacer DNA between genes.

Genes Are Sometimes Encoded Within Genes

Since the introns in most eukaryotic genes contain thousands of base pairs, only the DNA sequence of the exons and the sequences near the exon-intron boundaries have been determined. There is a great deal of speculation about the significance of the DNA in introns and the spacer regions between genes. A role for introns in the evolution of new proteins will be discussed in Chapter 22. In some cases, functional genes are embedded within introns. The first example of this was identified in a gene from *Drosophila*. During the analysis of the GART gene, which encodes an enzyme important for the biosynthesis of purines, an unrelated gene for a pupal cuticle protein was found embedded within one of the introns (Figure 8-14). The direction of transcription of the cuticle gene was opposite that of the GART gene transcript. Overlapping coding sequences, or *nested genes,* were previously known to exist in bacteria, bacteriophages, such as φX174, and viruses.

Nested genes encoded on opposite strands have now been characterized in higher eukaryotes. A human cDNA that encodes an 819 amino acid protein was cloned and found to be encoded on the opposite strand of the steroid 21-hydroxylase (P450c21) gene. The largest intron of the human Factor VIII gene encodes a gene that is transcribed in the direction opposite to that of the Factor VIII mRNA. Other examples include Rev-ErbAα, related to the steroid and thyroid hormone receptor gene family and encoded within the thyroid hormone receptor gene; a gene of unknown function transcribed from the strand opposite the one that encodes rat gonadotropin-releasing hormone; and three mRNAs transcribed from an intron of the human NF1 gene, which is associated with neurofibromatosis (NF) disease. For several of these nested genes, it still remains to be shown that they are functionally important.

It should be noted there are very few genes for which the entire DNA sequence including the introns, is known. On the basis of these few experiments, we can predict that as more introns are fully sequenced, it is likely that additional nested genes will be found.

Summary

This chapter reviewed what was learned about the organization of the eukaryotic genome from the cloning and sequencing of genes. In Chapter 7 we learned

that to clone a gene, some information about the encoded protein is required, such as a homologous gene, primary amino acid sequence data, or an antibody that reacts with the protein. In Chapter 26, we will learn about identifying mutant genes that are responsible for genetic diseases. As in bacterial and *Drosophila* genetics, detectable phenotypes in humans caused by a mutant gene can be used to map the chromosomal location of the gene and eventually to clone the gene. The total number of genes in humans has been estimated to be between 100,000 and 300,000. For the vast majority of these, a protein sequence, antibody, or genetic phenotype may not be obtainable. In order to clone every human gene and to understand the complex arrangement and regulation of eukaryotic genes, a project has begun recently to sequence all 3 billion bp that make up the human genome. This endeavor will be discussed further in Chapter 30.

Reading List

General

Cold Spring Harbor Symp. Quant. Biol., vol. 42: *Chromatin.* Cold Spring Harbor Laboratory, Cold Spring Harbor, N.Y., 1978.

Cold Spring Harbor Symp. Quant. Biol., vol. 47: *Structures of DNA.* Cold Spring Harbor Laboratory, Cold Spring Harbor, N.Y., 1983.

Original Research Papers

DISCOVERY OF RNA SPLICING

Witkowski, J. A. "The discovery of split genes: a scientific revolution." *Trends in Biochemical Science*, (1988) [Review]

Berget, S. M., C. Moore, and P. Sharp. "Spliced segments at the 5′ termini of adenovirus-2 late mRNA." *Proc. Natl. Acad. Sci. USA*, 74: 3171–3175 (1977).

Chow, L. T., R. E. Gelinas, T. R. Broker, and R. J. Roberts. "An amazing sequence arrangement at the 5′ ends of adenovirus 2 messenger RNA." *Cell*, 12: 1–8 (1977).

THE ORGANIZATION OF EUKARYOTIC GENES

Chambon, P. "Split genes." *Sci. Am.*, 244(5): 60–71 (1981). [Review]

Breathnach, R., J. L. Mandell, and P Chambon. "Ovalbumin gene is split in chicken DNA." *Nature*, 270: 314–319 (1977).

Jeffreys, A. J., and R. A. Flavell. "The rabbit β-globin gene contains a large insert in the coding sequence." *Cell*, 12: 1097–1108 (1977).

Breathnach, R., C. Benoist, K. O'Hare, F. Gannon, and P. Chambon. "Ovalbumin gene: evidence for a leader sequence in mRNA and DNA sequences at the exon-intron boundaries." *Proc. Natl. Acad. Sci. USA*, 75: 4853–4857 (1978).

Gitschier, J., W. I. Wood, T. M. Goralka, K. L. Wion, E. Y. Chen, D. H. Eaton, G. A. Vehar, D. J. Capon, and R. M. Lawn. "Characterization of the human Factor VIII gene." *Nature*, 312: 326–330 (1985).

Wood, W. I., D. J. Capon, C. C. Simonsen, D. L. Eaton, J. Gitschier, B. Keyt, P. H. Seeburg, D. H. Smith, P. Hollingshead, K. L. Wion, E. Delwart, E. G. D. Tuddenham, G. A. Vehar, and R. M. Lawn. "Expression of active human factor VIII from recombinant DNA clones." *Nature*, 312: 330–337 (1985).

Schmidtke, J., and D. N. Cooper. "A comprehensive list of cloned human DNA sequences." *Nuc. Acids Res.*, 19(Suppl.): 2111–2126 (1991).

SPLICING DEFECTS LINKED TO DISEASE

Treisman, R., N. J. Proudfoot, M. Shander, and T. Maniatis. "A single-base change at a splice site in a β^0-Thalassemic gene causes abnormal RNA splicing." *Cell*, 29: 903–911 (1982).

Treisman, R., S. H. Orkin, and T. Maniatis. "Specific transcription and RNA splicing defects in five cloned β-thalassemia genes." *Nature*, 302: 591–596 (1983).

Cohen, J., and A. D. Levinson. "A point mutation in the last intron responsible for increased expression and transforming activity of the c-Ha-*ras* oncogene." *Nature*, 334: 119–124 (1988).

Gonzalez, F. J., R. C. Skoda, S. Kimura, M. Umeno, U. M. Zanger, D. W. Nebert, H. V. Gelboin, J. P. Hardwick, and U. A. Meyer. "Characterization of the common genetic defect in humans deficient in desbrisoquine metabolism." *Nature*, 331: 442–446 (1988).

ALTERNATIVE SPLICING

Smith, C. W., J. G. Patton, and B. Nadal-Ginard. "Alternative splicing in the control of gene expression." *Ann. Rev. Gen.*, 23: 527–577 (1989). [Review]

Early, P., J. Rogers, M. Davis, K. Calame, M. Bond, R. Wall, and L. Hood. "Two mRNAs can be produced from a single immunoglobulin μ gene by alternative RNA processing pathways." *Cell*, 20: 313–319 (1980).

Amara, S. G., V. Jonas, M. G. Rosenfeld, E. S. Ong, and R. M. Evans. "Alternative RNA processing in calcitonin gene expression." *Nature*, 298: 240–244 (1982).

Crabtree, G. R., and J. A. Kant. "Organization of the rat γ-fibrinogen gene: alternative mRNA splice patterns produce the γA and γB (γ') chains of fibrinogen." *Cell*, 31: 159–166 (1982).

Leff, S. E., R. M. Evans, M. G. Rosenfeld. "Splice commitment dictates neuron-specific alternative RNA processing in calcitonin/CGRP gene expression." *Cell*, 48: 517–524 (1987).

Lees-Miller, J. P., L. O. Goodwin, and D. M. Helfman. "Three novel brain tropomyosin isoforms are expressed from the rat α-tropomyosin gene through the use of alternative promoters and alternative RNA processing." *Mol. Cell. Biol.*, 10: 1729–1742 (1990).

INTRONS MARK STRUCTURAL AND FUNCTIONAL DOMAINS IN PROTEINS

Gilbert, W. "Genes in pieces revisited." *Science*, 228: 823–824 (1985). [Review]

Patthy, L. "Evolution of the proteases of blood coagulation and fibrinolysis by assembly of modules." *Cell*, 41: 657–663 (1985). [Review]

Ny, T., F. Elgh, and B. Lund. "The structure of human tissue-type plasminogen activator gene: correlation of intron and exon structures to functional and structural domains." *Proc. Natl. Acad. Sci. USA*, 81: 5355–5359 (1984).

UPSTREAM (5') AND DOWNSTREAM (3') CONTROL SEQUENCES

Nevins, J. R., and J. E. Darnell. "Steps in the processing of Ad2 mRNA: poly(A)$^+$ nuclear sequences are conserved and poly(A) addition precedes splicing." *Cell*, 15: 1477–1493 (1978).

Benoist, C., K. O'Hare, R. Breathnach, and P. Chambon. "The ovalbumin gene sequence of putative control regions." *Nuc. Acids Res.*, 8: 127–143 (1979).

Fitzgerald, M., and T. Shenk. "The sequence 5'-AAUAAA-3' forms part of the recognition site for polyadenylation of late SV40 mRNAs." *Cell*, 24: 251–260 (1981).

Ghosh, P. K., P. Lebowitz, R. J. Frisque, and Y. Gluzman. "Identification of a promoter component involved in positioning the 5' termini of simian virus 40 early mRNAs." *Proc. Natl. Acad. Sci. USA*, 78: 100–104 (1981).

Hen, R., P. Sassone-Corsi, J. Corden, M. P. Gaub, and P. Chambon. "Sequences upstream from the T-A-T-A box are required in vivo and in vitro for efficient transcription from the adenovirus serotype 2 major late promoter." *Proc. Natl. Acad. Sci. USA*, 79: 7132–7136 (1982).

Kozak, M. "Analysis of 5'-noncoding sequences from 699 vertebrate messenger RNAs." *Nuc. Acids Res.*, 15: 8125–8143 (1987).

RNA POLYMERASE III REGULATORY ELEMENTS

Geiduschek, E. P. "Transcription by RNA polymerase III." *Ann. Rev. Biochem.*, 57: 873–914 (1988). [Review]

Sakonju, S., D. F. Bogenhagen, and D. D. Brown. "A control region in the center of the 5 S gene directs specific initiation of transcription. I: The 5' border of the region." *Cell*, 19: 13–25 (1980).

Galli, G., H. Hofstetter, and M. L. Birnstiel. "Two conserved sequence blocks within eukaryotic tRNA genes are major promoter elements." *Nature*, 294: 626–631 (1981).

Lobo, S. M., and N. Hernandez. "A 7 bp mutation converts a human RNA polymerase II snRNA promoter into an RNA polymerase III promoter." *Cell*, 58: 55–67 (1989).

Kassavetis, G. A., B. R. Braun, L. H. Nguyen, and E. P. Geiduschek. "*S. cerevisiae* TFIIIB is the transcription factor proper of RNA polymerase III, while TFIIIA and TFIIIC are assembly factors." *Cell*, 60: 235–245 (1990).

REPEATED GENES

Petes, T., and G. R. Fink. "Gene conversion between repeated genes." *Nature,* 300: 216–217 (1982). [Review]

Smith, G. P. "Evolution of repeated DNA sequences by unequal crossovers." *Science,* 191: 528–537 (1976).

Wellauer, P. K., and I. B. Dawid. "Isolation and sequence organization of human ribosomal DNA." *J. Mol. Biol.,* 128: 289–303 (1979).

Fritsch, E., R. Lawn, and T. Maniatis. "Molecular cloning and characterization of the human β-like globin gene cluster." *Cell,* 19: 959–972 (1980).

Slauer, J., C. Shen, and T. Maniatis. "The chromosomal arrangement of human α-like globin genes: sequence homology and α-globin gene deletions." *Cell,* 20: 119–130 (1980).

Yen, P. H., and N. Davidson. "The gross anatomy of a tRNA gene cluster at region 42A of the *D. melanogaster* chromosome." *Cell,* 22: 137–148 (1980).

Hentschel, C. C., and M. L. Birnstiel. "The organization and expression of histone gene families." *Cell,* 25: 301–305 (1981).

Worton, R. G., J. Sutherland, J. E. Sylvester, H. F. Willard, S. Bodrug, I. Dube, C. Duff, V. Kean, P. N. Ray, and R. D. Schmickel. "Human ribosomal RNA genes: orientation of the tandem array and conservation of the 5' end." *Science,* 239: 64–68 (1988).

Loomis, W. F., and D. L. Fuller. "A pair of tandemly repeated genes code for gp24, a putative adhesion protein of *Dictyostelium discoideum.*" *Proc. Natl. Acad. Sci. USA,* 87: 886–890 (1990)

PSEUDOGENES

Little, P. F. R. "Globin pseudogenes." *Cell,* 28: 683–684 (1982). [Review]

Proudfoot, N. J., and T. Maniatis. "The structure of a human α-globin pseudogene and its relationship to α-globin gene duplication." *Cell,* 21: 537–544 (1980).

Lemischka, I., and P. A. Sharp. "The sequences of an expressed rat α-tubulin gene and a pseudogene with an inserted repetitive element." *Nature,* 300: 330–335 (1982).

REPETITIVE SEQUENCES

Britten, R. J., and D. E. Kohne. "Repeated sequences in DNA." *Science,* 161: 529–540 (1968).

Jelinek, W. R., T. P. Toomey, L. Leinwand, C. H. Duncan, P. A. Biro, P. V. Choudary, S. M. Weissman, C. M. Rubin, C. M. Houck, P. L. Deininger, and C. W. Schmid. "Ubiquitous interspersed repeated DNA sequences in mammalian genomes." *Proc. Natl. Acad. Sci. USA,* 77: 1398–1402 (1980).

Weiner, A. "An abundant cytoplasmic 7S RNA is partially complementary to the dominant interspersed middle repetitive DNA sequence family in the human genome." *Cell,* 22: 209–218 (1980).

Duncan, C. H., P. Jagadeeswaran, R. R. C. Wang, and S. M. Weissman. "Structural analysis of templates and RNA polymerase III transcripts of Alu family sequences interspersed among the human β-like globin genes." *Gene,* 13: 185–196 (1981).

Schmid, C. W., and W. R. Jelenik. "The Alu family of dispersed repetitive sequences." *Science,* 216: 1065–1070 (1982).

Singer, M. F. "SINEs and LINEs: highly repeated short and long interspersed sequences in mammalian genomes." *Cell,* 28: 433–434 (1982).

Jeffreys, A. J., V. Wilson, and S. C. Thein. "Individual-specific fingerprints of human DNA." *Nature,* 316: 76–78 (1985).

Jurka, J., and T. Smith. "A fundamental division in the *Alu* family of repeated sequences." *Proc. Natl. Acad. Sci. USA,* 85: 4775–4778 (1988).

Korenberg, J. R., and M. C. Rykowski. "Human genome organization: Alu, LINES, and the molecular structure of the metaphase chromosome bands." *Cell,* 53: 391–400 (1988).

METHYLATION AND GENE EXPRESSION

Weissbach, A., C. Ward, and A. Bolden. "Eukaryotic DNA methylation and gene expression." *Curr. Top. Cell. Regul.,* 30: 1–21 (1989). [Review]

McGowan, R., R. Campbell, A. Peterson, C. Sapienza. "Cellular mosaicism in the methylation and expression of hemizygous loci in the mouse." *Gene and Devel.,* 3: 1669–1676 (1989).

Antequera, F., J. Boyes, and A. Bird. "High levels of de novo methylation and altered chromatin structure at CpG islands in cell lines." *Cell,* 62: 503–514 (1990).

GENES WITHIN GENES

Henikoff, S., M. A. Keene, K. Fechtel, and J. W. Fristron. "Gene within a gene: nested *Drosophila* genes encode unrelated proteins on opposite DNA strands." *Cell,* 44: 33–42 (1986).

Lazar, M. A., R. A. Hodin, D. S. Darling, and W. W. Chin. "A novel member of the thyroid/steroid hormone receptor family is encoded by the opposite strand of the rat c-erbAα transcriptional unit." *Mol. Cell. Biol.,* 9: 1128–1136 (1989).

Morel, Y., J. Bristow, S. E. Gitelman, and W. L. Miller. "Transcript encoded on the opposite strand of the human steroid 21-hydroxylase/complement C4 gene locus." *Proc. Natl. Acad. Sci. USA,* 86: 6582–6586 (1989).

Cawthon, R. M., P. O'Connell, A. M. Buchberg, D. Viskochil, R. B. Weiss, M. Culver, J. Stevens, N. A. Jenkins, N. G. Copeland, and R. White. "Identification and characterization of transcripts from the neurofibromatosis 1 region: the sequence and genomic structure of EV12 and mapping of other transcripts." *Genomics,* 7: 555–565 (1990).

Levinson, B., S. Kenwrick, D. Lakich, G. Hammonds, Jr., and J. Gitschier. "A transcribed gene in an intron of the human Factor VIII gene." *Genomics,* 7: 7–11 (1990).

9

Controlling Eukaryotic Gene Expression

As more eukaryotic genes were cloned and sequenced and their structural organization was elucidated, attention turned to their complex patterns of expression. In a multicellular organism, each cell expresses a subset of its genes, and a cell expresses different genes depending on its growth state or environment. Sometimes gene transcripts are spliced differently in different cells. Because these patterns of expression are heritably passed from one organism to its offspring, the patterns, like the genes themselves, must be encoded in DNA. Thus, each gene carries not only structural information but also instructions for its correct expression in space and time. Where are these instructions? And how are they executed? Perhaps the gene sequences offered some clues.

Common Sequence Motifs Are Recognized in 5′-Flanking Regions

In prokaryotes, as we learned in Chapter 4, the sequences that control expression of a gene, the promoter, are reliably found immediately 5′ to the gene. So this is where attention focused first. Indeed, many eukaryotic genes were found to share common sequence motifs in their 5′-flanking regions. One easily spotted

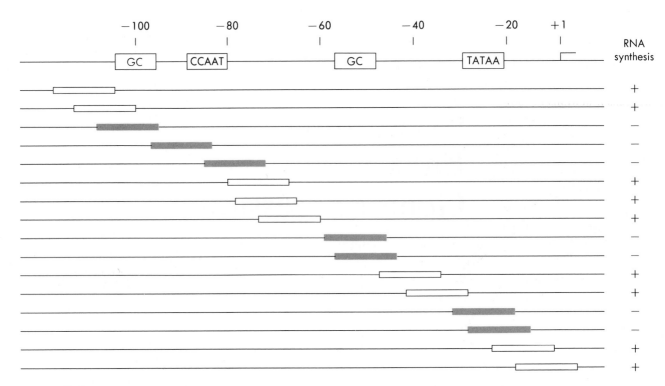

FIGURE **9-1**

Mutational analysis of a eukaryotic promoter demonstrates the functional importance of conserved sequence motifs. The top line shows the organization of the promoter of the thymidine kinase gene of herpes simplex virus, which is transcribed by host enzymes in infected cells. The numbers indicate distance in base pairs from the start of transcription (+1). The boxes indicate the position of the conserved GC, CCAAT and TATAA sequence motifs in the promoter. Below are a series of promoter mutants, created using a technique called *linker-scanning mutagenesis*. Each rectangle represents the position of a clustered region of 6 to 10 nucleotide substitutions. The mutant DNAs were injected into *Xenopus* oocytes, where the gene was transcribed into RNA. The white rectangles indicate promoter mutants that yielded normal amounts of the RNA. The colored rectangles represent mutations that resulted in substantial reduction of promoter activity and consequently reduced levels of this RNA synthesis. All mutations that changed sequences within the conserved motifs reduced promoter activity.

motif was a sequence TATAA, often found about 25 to 30 bp 5′ to the start of transcription. This sequence resembled the −10 element in bacterial promoters (see Chapter 4). Other sequences commonly found upstream of genes included CCAAT and GGGCG (the latter are often called *GC boxes*). Their positions were variable, but they were always farther from the gene than was TATAA. The existence of these sequences more or less where they were expected suggested initially that eukaryotic transcriptional control would be similar to that in bacteria. But, unlike the pattern for bacterial promoters, these sequence motifs were not found in all eukaryotic genes.

Experimental demonstration that sequence elements 5′ to the gene played a role in eukaryotic gene transcription relied on the use of recombinant DNA techniques to perform directed mutagenesis of cloned gene sequences (Chapter 11) and on the ability to return altered genes to the cell (Chapter 12). In one early study of the thymidine kinase gene of herpes virus—a gene organized much like chromosomal genes—a series of clustered point mutations were generated throughout the promoter by a technique called *linker scanning* (see Chapter 11). The mutant genes were microinjected into oocytes of the frog *Xenopus laevis*, where the genes were transcribed by the *Xenopus* transcription machinery. The ability of the injected genes to be transcribed in the oocytes was determined by measuring the level of mRNA transcribed from them (Figure 9-1). These experiments showed that whereas mutations in some areas of the promoter had no effect on transcription, mutations that destroyed the TATA, CCAAT, and GC boxes in the promoter dramatically reduced the level of transcription. These elements therefore constitute important functional units of eukaryotic promoters.

Nowadays, it is a relatively straightforward task to map the sequence elements required for promoter function in eukaryotic genes. Sophisticated and rapid mutagenesis strategies are available for altering promoter sequences and introducing them into living cells (see Chapters 11 and 12). The experimental parameter measured in these types of gene-transfer assays is sometimes the RNA transcribed from the transferred gene, but it is more often the activity of a *reporter gene*. The reporter gene is fused downstream of the promoter of interest, so that transcripts initiating at the promoter proceed through the reporter gene. Reporter genes generally encode some easily measurable protein. One commonly used reporter is the bacterial gene encoding the enzyme chloramphenicol acetyltransferase (CAT), which acetylates the antibiotic chloramphenicol (Figure 9-2). Assays for enzymes such as CAT are simpler than measuring RNA levels and in many cases are more sensitive.

Enhancers Activate Gene Expression over Long Distances

In the beginning, eukaryotic promoters appeared to be like spread-out versions of prokaryotic promoters, but with the same basic principles of operation. However, studies of viruses that infect eukaryotic cells and use host machinery to transcribe their genes produced surprises. The promoter for SV40 virus seemed simple at first glance (Figure 9-3). Just 5′ to the start of transcription lay an AT-rich sequence similar to the TATAA motif. Beyond that, 50 to 100 bp from the start site, was an array of six GC boxes. But initial mutagenesis studies of the regulatory region of the virus suggested that important sequences for transcription of the viral genome were situated farther away in a

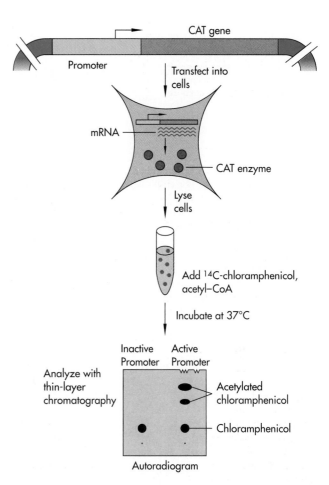

FIGURE **9-2**
Using the CAT gene as a reporter for promoter activity assays. The promoter under study is removed from its gene and cloned adjacent to the coding sequence for the bacterial CAT gene. When this construct is introduced into eukaryotic cells, transcripts initiate at the eukaryotic promoter and proceed through the CAT gene. The resulting transcripts are translated into CAT enzyme by the translation machinery of the host cell. To assay CAT activity in the cells, simple cell lysates are made by cycles of freezing and thawing. The lysates are mixed with isotopically labeled chloramphenicol and acetyl-coenzyme A (acetyl-CoA). The CAT enzyme transfers the acetyl group from acetyl-CoA to the 2- or 3- position of chloramphenicol. The extent of the reaction is monitored by thin-layer chromatography, which separates acetylated chloramphenicol from unreacted material, followed by autoradiography to visualize the reaction products. The 2- and 3-acetylated products (right lane) migrate at slightly different rates, giving two spots of acetylated product. If the promoter were inactive, these two spots would be absent (left lane).

region containing a tandemly duplicated 72-bp sequence. Deletions of this element reduced viral transcription by a factor of 100. Recombinant DNA manipulation of this sequence, termed an *enhancer*,

Late
RNAs
Early
RNAs
Enhancer
GC boxes
TATA
box
Multiple start sites
10 bp

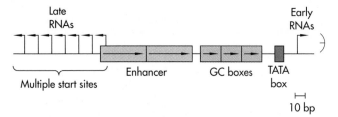

FIGURE **9-3**
Organization of the SV40 promoter region. In SV40, a single regulatory region serves two different sets of genes, encoded on opposite strands and used early and late in infection, respectively. The major functional elements in the promoter are the TATA box, located 25 bp 5′ to the start site for the early viral transcripts; a series of six GC boxes organized into three tandem repeats of 21 to 22 bp; and an enhancer, tandemly repeated 72-bp elements. There is no discernible TATA box for the late viral mRNA promoter. As is often the case for promoters without a TATA box, transcript initiation occurs at many sites for the late viral transcripts.

1
2
3
4

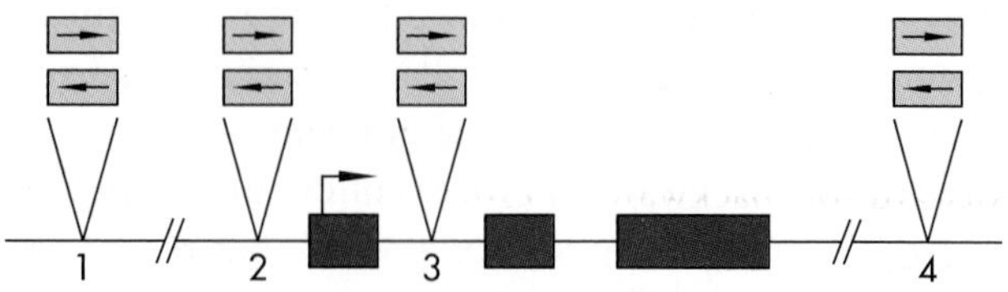

FIGURE **9-4**
The functional properties of an enhancer. The purple boxes indicate the coding exons of an imaginary gene. The light green boxes above the gene indicate positions where enhancers are known to function. In addition to functioning from close to the promoter (position 2, here), as in SV40 (Figure 9-3), enhancers can work from thousands of base pairs upstream (position 1) or downstream (position 4) of a gene. In some genes, the immunoglobulins for example, the enhancer is located within an intron (position 3). In most cases, enhancer sequences function in either orientation relative to the gene.

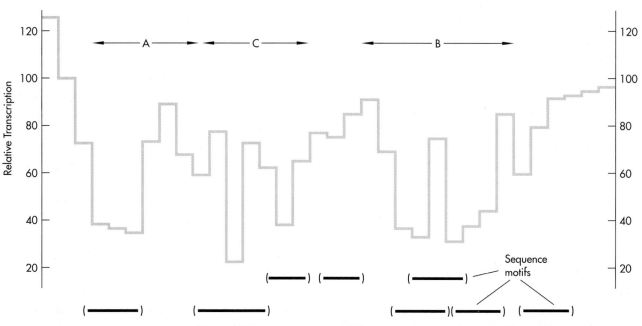

Relative Transcription
120
100
80
60
40
20
A
C
B
Sequence
motifs
5′ AACCAGCTGTGGAATGTGTGTCAGTTAGGGTGTGGAAAGTCCCCAGGCTCCCCAGCAGGCAGAAGTATGCAAAGCATGCATCTCAATTAGTCAGCAACCAG 3′

FIGURE **9-5**
Organization of the SV40 enhancer. The sequence at the bottom is from the enhancer region of SV40. It includes one of the two 72-bp repeats and additional sequences to the late side (see Figure 9-3). Brackets above the sequence indicate sequence motifs recognized in other enhancers or sequences repeated within the SV40 enhancer. As is the case with the conserved motifs found in promoters (Figure 9-1), mutation of these motifs affects enhancer function. This is indicated by the graph of relative transcriptional activity, which shows the effect on enhancer activity of 3-bp mutations at each position shown. No single mutation reduces enhancer activity more than a few fold, because the enhancer comprises many redundant elements. The arrows A, B, and C mark subelements of the enhancer that can by themselves function as enhancers when present in multiple copies.

revealed some startling properties (Figure 9-4). The DNA sequence could be removed from the SV40 genome and placed in a different promoter, and it still worked. Moreover, it worked in either orientation, forward or backward. Perhaps most surprising, it worked from thousands of base pairs away, and it could even work when situated within a gene or beyond the 3′ end of a gene.

Clearly, enhancers, which were soon found in cellular genes as well, were a very different kind of transcriptional element from those previously encountered. Unlike conventional promoter elements, which constituted a single short sequence motif, it became clear that enhancers were more complex. They were larger, 50 to 150 bp; and comparison of different enhancer sequences revealed new sequence motifs common to many of them. The picture that emerged was of a patchwork of distinct and redundant sub-elements. Individual sequence motifs within an enhancer could usually be destroyed without greatly affecting overall enhancer activity. And duplications of one portion of an enhancer could often compensate for mutations in another region. In fact, individual enhancer elements could be synthesized as oligonucleotides of 18 to 20 bp and, when ligated together as a tandem array of several repeats, functioned as fully active enhancers. To date, the best studied enhancer remains the SV40 sequence. A current view of SV40 enhancer structure is shown in Figure 9-5.

Enhancers Can Be Tissue-Specific or Regulated by Signals

One of the outstanding problems in understanding development of complex eukaryotic organisms is explaining how the expression of individual genes can be limited to specific cell types. Part of the answer appears to be that enhancers can function in a cell-type-specific fashion. The immunoglobulin (abbreviated Ig and also called *antibody*) genes are expressed at very high levels in B cells of the immune system and nowhere else (see Chapter 16). These genes carry potent enhancers that work only in B cells. The enhancers for the immunoglobulin genes were localized

FIGURE **9-6**
Organization of the immunoglobulin heavy-chain gene enhancer. The enhancer spans approximately 200 bp within the second intron of the gene. The colored boxes indicate sequence motifs known to be bound by proteins or shared with other enhancers. Mutation of any one of these motifs has little effect on overall enhancer activity, because, like the SV40 enhancer, these elements are redundant. The "E boxes" are bound in vivo by proteins only in B cells, although proteins that bind to these elements are present in all cells. The central portion of the enhancer is active in other cell types. Sequences in the flanking region appear to confer B-cell specificity either by binding transcription factors made only in B cells (for example, at the site marked "OCT") or by inhibiting enhancer function in non-B cells.

to the second intron of the genes. Deletion of the sequence abolished the transcription of the gene in B cells. When the Ig enhancer was removed and placed on an unrelated promoter, it activated that promoter only in B cells and not in other cell types. This property accounts in part for the B-cell-specific expression of Ig genes.

Like the SV40 enhancer, the Ig enhancers are large, approximately 100 to 150 bp. They appear to be made up of many redundant elements, which can be individually eliminated with little consequence (Figure 9-6). They also contain negatively acting elements that suppress transcription in non-B cells. Surprisingly, many of the individual elements within the Ig enhancer are not cell-type-specific at all. It is the particular arrangement of these elements, only a few of which are limited to B cells, that confers B-cell-specific activity to the enhancer.

Whereas the Ig enhancer is active throughout the life of a mature B cell, other enhancers are activated in response to external signals. The enhancer of the metallothionein genes, which encode proteins that detoxify heavy metals, is switched on by the presence of zinc or cadmium. Heat-shock genes are rapidly switched on by a shift to high temperature. Many genes are rapidly activated in response to growth factors and hormones (see Chapters 17 and 18). Each of these genes contains enhancer elements that function only

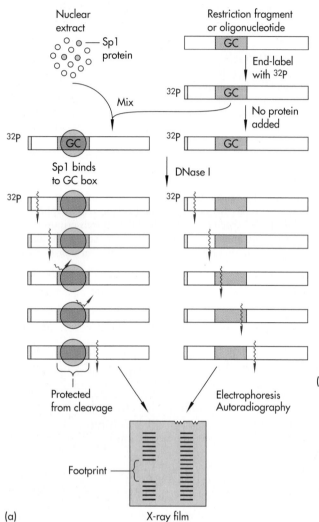

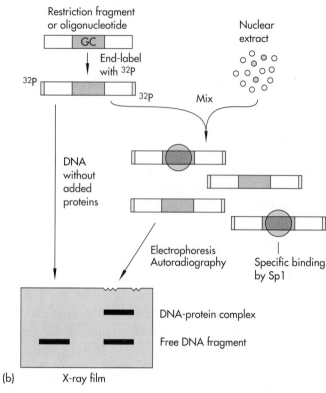

FIGURE 9-7

In vitro DNA-binding assays. (a) DNase I footprinting. A restriction fragment or oligonucleotide carrying a binding site for a protein, here a GC box that binds Sp1, is radioactively labeled at one end. The DNA fragment is incubated with cellular proteins, and any Sp1 protein present binds to the GC box sequence in the fragment. Next the reaction mixture is treated briefly with DNase I, an enzyme that makes single-strand cuts in DNA. The DNase treatment is adjusted so that each DNA strand is on average cut once. In the absence of bound protein, the enzyme cuts throughout the fragment, generating a collection of labeled subfragments of different sizes. When analyzed by gel electrophoresis, this appears as a relatively uniform ladder of bands, each arising from cuts at a single position in the fragment (right lane in the autoradiogram, bottom left). When Sp1 is bound to the GC box, however, the enzyme can't reach the GC box and doesn't cut in that region. When this sample is analyzed on a gel, the ladder is interrupted by a "footprint" representing the sequence protected from DNase cleavage by bound Sp1 (left lane). (b) Mobility-shift assay. In this method, a radioactively labeled DNA fragment is incubated with a protein extract containing the DNA-binding protein of interest. The reaction is analyzed on a polyacrylamide gel run under conditions in which bound proteins remain with the DNA fragment. Free fragment runs to a position characteristic of its size, usually at the bottom of the gel. Fragment with protein bound to it is much larger, migrates more slowly through the gel, and appears as a "shifted" band of DNA higher up in the gel.

in the presence of the particular inducing signal. These elements can often be isolated from their natural genes and placed onto new ones, imparting signal responsiveness to the new genes.

Enhancers Contain Recognition Sites for DNA-Binding Proteins

Although some early models for enhancer function proposed a physical role for enhancers in determining the topology or structure of promoter DNA, it is now clear that the primary function of enhancers is to bind cellular transcription factors. That enhancers are protein-binding sites was revealed by both in vitro and in vivo assays for DNA binding. Two important in vitro assays for DNA binding are illustrated in Figure 9-7. In DNase footprinting, protein bound to DNA is revealed by its ability to protect the DNA sequence to which it is bound from nicking by an endonuclease. In the mobility-shift assay, protein bound to a radiolabeled DNA fragment causes the mobility of the fragment through a polyacrylamide gel to decrease and shifts the DNA band to a higher position in the gel (indicating lower mobility). In both assays, it is simple to demonstrate the sequence specificity of a DNA-protein interaction, by showing that the protein binds to that fragment with a substantially higher affinity (usually 3 to 4 orders of magnitude) than to random DNA. And it is possible to locate the binding site within the fragment at nucleotide resolution.

Related methods have been used to demonstrate protein bound to DNA in intact cells. These methods are much more difficult and considerably less sensitive. But, unlike in vitro binding assays, which provide only circumstantial evidence that a DNA element binds a protein in cells, in vivo footprinting demonstrates this interaction directly. On the other hand, current technology does not allow biochemical identification of the protein that is bound in vivo.

Using a combination of in vitro and in vivo binding assays, we have been able to build maps of protein-binding sites in enhancers (Figure 9-8). These sites generally correspond to functionally defined DNA elements within enhancers. In general, those elements found to function in all cells bind proteins in all cells; those that act in a cell-type-specific fashion bind proteins found only in the active cell type. But the function of an enhancer is not simply the sum of its protein binding sites. For example, all cells contain proteins that bind to the E box sequences in the Ig enhancer (see Figure 9-6), yet it is only in B cells that these sequences are actually bound by these proteins. So other factors must be regulating the accessibility of the enhancer to DNA-binding proteins or the assembly of the protein-DNA complex at the enhancer.

Genes Can Be Transcribed in Cell-Free Extracts

In prokaryotes, simply mixing purified RNA polymerase, a template carrying a promoter, nucleoside triphosphates, and appropriate buffer and salts is sufficient to obtain specific gene transcription in vitro beginning at the correct sites. Purified RNA polymerase from eukaryotes, however, initiates transcription very poorly and essentially at random. This observation suggested that accessory factors required for accurate initiation of transcription do not copurify with RNA polymerase. When crude cellular or nuclear extracts are used in place of purified RNA polymerase, transcription does initiate at the same sites used in intact cells, but initiation is inefficient and elongation of initiated RNA chains is poor. Nevertheless, these extracts provided the starting material for biochemical fractionation schemes that have led to the identification of several important transcription factors. Some of these are general factors required for initiation at all promoters, while others are gene-specific and required only for certain promoters. Among the general

FIGURE **9-8**
Binding sites for proteins in the SV40 enhancer. The colored shapes represent proteins known to bind at these respective positions in the enhancer sequence. This is only a subset of the proteins known to bind the enhancer.

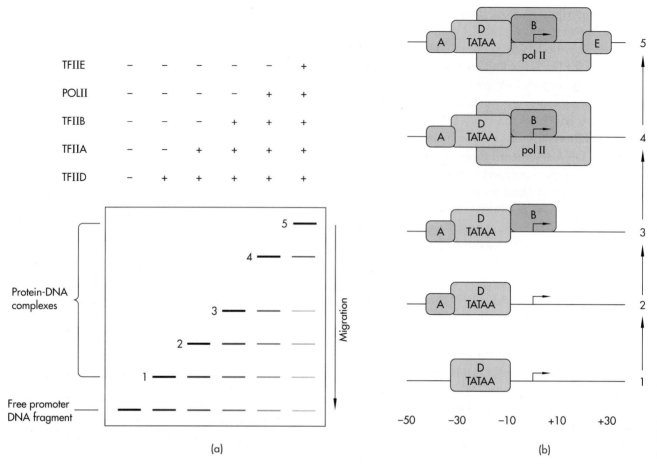

TFIIE	–	–	–	–	–	+
POLII	–	–	–	–	+	+
TFIIB	–	–	–	+	+	+
TFIIA	–	–	+	+	+	+
TFIID	–	+	+	+	+	+

(a)

(b)

FIGURE **9-9**

Assembly of a multiprotein complex at a eukaryotic promoter. Figure 9-9a shows an experiment in which the assembly of a multicomponent transcription complex was monitored using the mobility-shift assay (see Figure 9-7). In the experiment, a radioactively labeled DNA fragment carrying a promoter was incubated with the indicated combination of transcription factor proteins. Binding of the proteins to the DNA fragment was monitored by electrophoresis on a polyacrylamide gel. With addition of each component, a larger complex was formed; experiments showed that the components had to be added in the order shown to assemble an active transcription complex. Note, however, that each new complex is only incompletely formed, so that each reaction mixture contains previously formed complexes. Figure 9-9b depicts the assembly process; the numbers correspond to the complexes observed on the gel.

factors is a protein called *TFIID,* which binds to the TATAA sequence in promoters. Other general factors are involved in the assembly of a multicomponent protein complex at the promoter. Assembly of this complex has been visualized using the mobility-shift assay (Figure 9-9a). Biochemical experiments of this type have led to a general view of the structure of the transcription complex assembled at a eukaryotic promoter (Figure 9-9b).

Gene-Specific Transcription Factors Are DNA-Binding Proteins

Experiments to isolate factors required for the transcription in vitro of the SV40 early promoter identified a protein that activated the transcription of the SV40 promoter but not that of a promoter from a different virus. The action of this protein, called *Sp1,* on the SV40 promoter depends on the presence of the six

GC boxes in the promoter. A promoter from which the GC boxes were deleted did not respond to the factor. When protein fractions containing Sp1 transcription-stimulating activity were assayed by DNase I footprinting, they all contained a protein that bound specifically to the GC boxes (Figure 9-10). Fully purified Sp1 protein retains both DNA-binding and transcriptional stimulatory activities. Thus, the basis for the promoter-specific activity of Sp1 is its sequence-specific DNA-binding activity: Sp1 activates transcription from a promoter by binding to the promoter. Sp1 is the prototype of what is now a large family of sequence-specific DNA-binding transcription factors.

Transcription Factors Are Purified and Cloned

The realization that many transcription factors are sequence-specific DNA-binding proteins has been exploited for both purification of these rare proteins and direct cloning of their encoding genes. Purification from crude nuclear extracts has been facilitated by affinity chromatography on columns containing oligonucleotide binding sites for the proteins (Figure 9-11). To purify Sp1, a synthetic double-stranded oligonucleotide was prepared that carried a GC box sequence from the SV40 promoter. The oligonucleotide was ligated into large tandem arrays using DNA ligase and chemically coupled to an agarose column resin. When a partially purified nuclear fraction enriched for DNA-binding proteins was applied to the column, many of the proteins bound weakly to the column as they would to any random DNA sequence. Sp1, however, because it specifically recognizes this sequence in preference to others, bound much more tightly. By washing the column with buffers of increasing salt concentration, the weakly bound proteins were removed from the column first, whereas Sp1 required much higher salt concentrations to be dissociated from the column. This column effected a significant purification of Sp1 away from other DNA-binding proteins of different specificity. This is now the method of choice for purifying DNA-binding proteins. In many instances, proteins have been purified

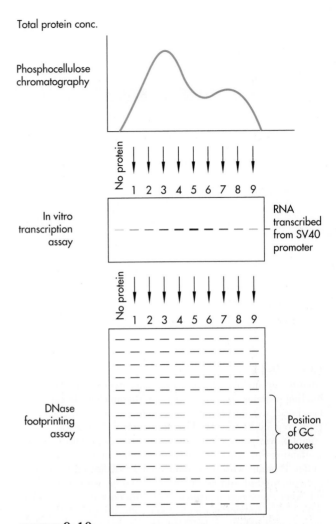

FIGURE **9-10**

Demonstration that transcription factor Sp1 binds to the GC boxes in the SV40 promoter. A protein fraction enriched for Sp1 was fractionated by phosphocellulose chromatography. The top panel shows the amount of protein washed off the column at each concentration of KCl in the chromatography buffer. Each fraction was added to an in vitro transcription reaction containing the SV40 promoter, RNA polymerase, and other accessory factors. Synthesis of RNA from the promoter was monitored by gel electrophoresis, depicted in the middle panel. In the absence of added protein from the phosphocellulose column, only a low level of transcription was obtained. In fractions 4 through 7, increased transcription was observed owing to the activity of Sp1 in those fractions. In the bottom panel, the fractions were assayed for GC box–binding activity using the DNase footprinting assay. This assay showed that the same fractions that contained Sp1 activity in the transcription assay also contained a protein that bound to the GC boxes in the promoter. This experiment showed that Sp1 worked by binding to the GC boxes in the SV40 promoter; this explains why Sp1 has no effect on promoters lacking GC boxes.

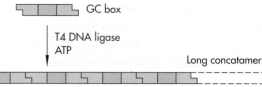

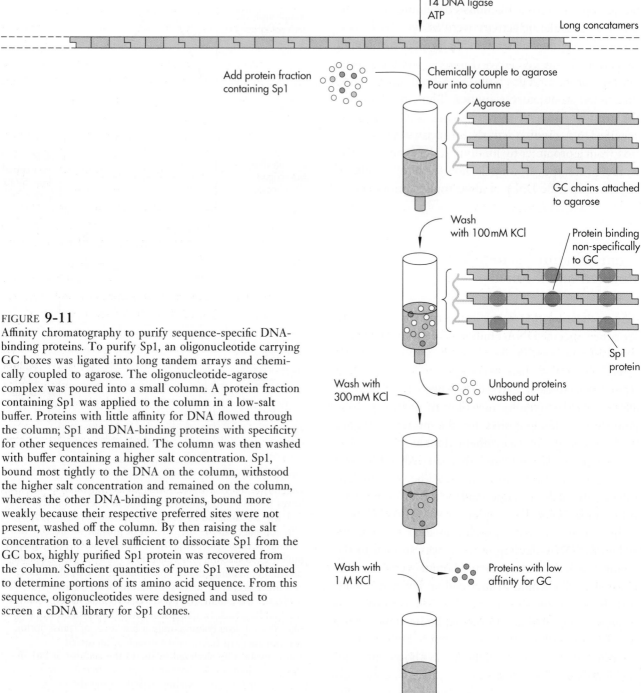

FIGURE **9-11**

Affinity chromatography to purify sequence-specific DNA-binding proteins. To purify Sp1, an oligonucleotide carrying GC boxes was ligated into long tandem arrays and chemically coupled to agarose. The oligonucleotide-agarose complex was poured into a small column. A protein fraction containing Sp1 was applied to the column in a low-salt buffer. Proteins with little affinity for DNA flowed through the column; Sp1 and DNA-binding proteins with specificity for other sequences remained. The column was then washed with buffer containing a higher salt concentration. Sp1, bound most tightly to the DNA on the column, withstood the higher salt concentration and remained on the column, whereas the other DNA-binding proteins, bound more weakly because their respective preferred sites were not present, washed off the column. By then raising the salt concentration to a level sufficient to dissociate Sp1 from the GC box, highly purified Sp1 protein was recovered from the column. Sufficient quantities of pure Sp1 were obtained to determine portions of its amino acid sequence. From this sequence, oligonucleotides were designed and used to screen a cDNA library for Sp1 clones.

using this method in quantities sufficient to obtain a direct amino acid sequence. Synthetic oligonucleotides based on the amino acid sequence were used to isolate cDNA clones encoding the proteins.

Transcription factors have also been cloned directly by expression screening techniques (see Chapter 7). In one method, expression libraries are prepared in the vector λgt11, which expresses inserted cDNAs as β-galactosidase fusion proteins. Nitrocellulose filters carrying replicas of phage plaques are incubated with radiolabeled oligonucleotides or DNA fragments that carry binding sites for specific proteins. The labeled DNA binds to plaques carrying fusion proteins that recognize it with high affinity.

Transcription Factors Fall into Structural Families

The frenzy of factor cloning triggered by these technical developments has led to an explosion in our understanding of the structure and function of these proteins. In general, transcription factors are found to contain two functional domains, one for DNA-binding and one for transcriptional activation. As we will discuss in more detail below, these functions often reside within circumscribed structural domains that retain their function when removed from their natural context. The DNA-binding domains of transcription factors fall into several structural families based on their primary amino acid sequence. Models for these structures are shown in Figure 9-12.

The homeodomain proteins resemble in amino acid sequence several prokaryotic DNA binding proteins whose structures have been extensively studied by x-ray crystallography, and the three-dimensional structures of the first authentic eukaryotic homeodomains have been determined. In these proteins, a protein helix lies across the major groove of the DNA helix and makes contact with the base pairs exposed there. These proteins contain a second helix that lies across the first and contacts other proteins in the transcription apparatus. Homeodomain proteins were first identified as the products of *Drosophila* genes with important roles in embryonic development (see Chapter 20).

Zinc finger proteins are so called because of the presence of repeated motifs of cysteine and histidine that are thought to fold up into a three-dimensional structure coordinated by a zinc ion. This structural motif has been most extensively studied in the RNA polymerase III transcription factor TFIIIA, which is involved in the transcription of various small RNAs, but it is extremely common among transcription factors for mRNA promoters. Notable among the zinc finger proteins are Sp1, the steroid receptor family (Chapter 17), and *Drosophila* embryonic gap gene proteins like Kruppel (Chapter 20). Hundreds of genes encoding zinc finger proteins exist in the mammalian genome. Demonstration that the zinc fingers encode the DNA-binding specificity of these proteins has come from "finger swaps" between different steroid receptors. Recombinant DNA techniques were used to change the sequence of the zinc fingers of one receptor to the sequence of a different one. The resulting hybrid adopted the DNA-binding specificity of the second receptor.

Leucine zipper proteins derive their name from the repeats they share of four or five leucine residues precisely seven amino acids apart. These domains provide hydrophobic faces through which leucine zipper proteins interact to form dimers. These can be dimers of identical proteins (homodimers) or of different ones (heterodimers). Immediately adjacent to the leucine zipper, on the amino-terminal side, is a domain enriched for the positively charged amino acids, arginine and lysine. This region is the DNA-binding domain. Assignment of these functions is based on the results of experiments involving extensive mutagenesis of these proteins. Mutations in the leucine zipper prevent both dimerization and DNA binding, because monomers do not bind DNA. Mutations in the positively charged domain abolish DNA binding but do not prevent dimer formation. The specificity of dimer formation is encoded within the leucine zipper. Swapping leucine zippers between proteins of this class swaps the partner with which the proteins dimerize. The products of the c-*fos* and c-*jun* protooncogenes (Chapter 18) are leucine zipper proteins. The Jun proteins form homodimers and bind specific DNA sequences with modest affinity. Fos proteins are unable to form homodimers and therefore do not bind DNA on their

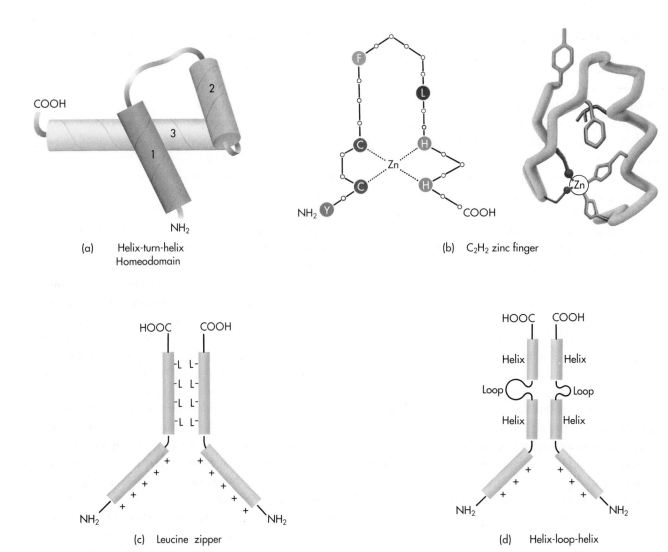

(a) Helix-turn-helix
     Homeodomain

(b) C₂H₂ zinc finger

(c) Leucine zipper

(d) Helix-loop-helix

FIGURE **9-12**
Structure models for DNA-binding domains. (a) The helix-turn-helix motif found in several prokaryotic regulatory proteins and in the homeodomain proteins of eukaryotes. Helix 3 is the "recognition helix" that makes the major contacts with DNA. Helices 1 and 2 lie atop helix 3 and are able to make contact with other proteins. In some homeodomain proteins, there is a fourth helix. (b) The zinc finger. On the left is a diagram of a zinc finger consisting of two cysteine (C) residues and two histidine (H) residues coordinated around a zinc molecule. Other highly conserved amino acids in this class of zinc fingers are also shown (F, phenylalanine; L, leucine; Y, tyrosine). On the right is a structural model of this class of zinc finger; the helix on the right is believed to contact the major groove of the DNA molecule. The zinc fingers of the steroid hormone receptors are of a different class, using four cysteines to coordinate the zinc atom. (c) The leucine zipper. Leucine zipper proteins act as dimers of two subunits. The leucine zipper domain is divided into two regions, both believed to be helical in structure. At the carboxyl terminus is a region containing leucine (L) residues every seventh position, placing each leucine side chain along the same face of the helix. This structure is believed to supply a hydrophobic surface for the interaction of two protein subunits. Amino-terminal to the leucines is a region rich in positively charged amino acids, believed to contact DNA. Results of experiments involving extensive mutagenesis of these proteins support this model. (d) Helix-loop-helix proteins. These proteins contain two helices linked by a loop of unknown structure. Like the leucine zipper, these helices are believed to provide surfaces for protein dimer formation. A third helix rich in positively charged amino acids is responsible for DNA binding.

own. When mixed, however, Jun and Fos proteins form tight heterodimers that bind DNA with high affinity. The ability to mix and match partners gives this family of transcription factors unusual functional flexibility.

A fourth transcription factor family is composed of the helix-loop-helix (HLH) proteins. These proteins are similar to the leucine zipper family in that they bind DNA as dimers, either homodimers or heterodimers, and they have a positively charged domain that recognizes the DNA site. In the helix-loop-helix family are several proteins with important roles in the control of cell growth and differentiation. Among them is a protein called MyoD, which has the remarkable power single-handedly to convert many different cell types into muscle cells. A similar family of HLH proteins appears to control development of the nervous system (see Chapter 21). The protein encoded by the c-*myc* protooncogene (Chapter 18) is also an HLH protein. Although attempts to demonstrate sequence-specific DNA binding by the Myc protein failed for many years, the recognition that it was a member of the HLH protein family sparked a renewed effort to investigate its DNA-binding properties. A clever application of the polymerase chain reaction was used to select a DNA sequence bound by Myc protein from a pool of random oligonucleotides (Figure 9-13). In addition, screening of a λ phage cDNA expression library with bacterially produced Myc protein identified a cDNA encoding a protein that functions as the partner for Myc, generating a heterodimer with strong DNA-binding activity.

Many transcription factors fall into none of these classes, so new structural solutions to the problem of sequence-specific DNA binding remain to be discovered.

The activation domains of transcription factors, the regions required for activating transcription once they are brought to the promoter by the DNA-binding domain, are less well characterized. In some proteins, activation is due to protein domains whose only discernible characteristic is net negative charge (Figure 9-14). Indeed when random protein sequences were fused to a DNA-binding domain, 10 percent of the sequences tested were able to provide an activation function to the protein. Other potent transcription factors, however, lack acidic regions in their activation

domains, so a general view of how these domains function at the biochemical level lags behind our understanding of the DNA-binding domains.

Transcription Factors Are Modular

The observation that DNA binding and transcriptional activation reside within discrete domains in transcription factors has led to the experimental demonstration that transcription factors consist of independently functioning modules. The activation domain of one factor can be joined to the DNA-binding domain of another, and the resulting hybrid protein is fully active in cells. This was first demonstrated in yeast, but it is true of mammalian transcription factors, too. Indeed, this observation solves a thorny experimental problem for investigators studying mammalian transcription factors. Cells often contain many different factors capable of binding to a single DNA sequence. This is the case, for example, for a sequence element responsible for the activation of gene transcription by the second messenger cyclic AMP (Chapter 17). Studying the function of one of these DNA-binding proteins by allowing its cDNA to be expressed in mammalian cells would be difficult because its activity would be masked by the many competing proteins already in the cell. To solve this problem, the DNA-binding specificity of one of these factors, CREB, was reprogrammed by fusing its activation domain to the DNA-binding domain of the yeast GAL4 transcription factor (Figure 9-15). Because there is no GAL4-like protein in mammalian cells, a synthetic promoter carrying only a GAL4-binding site and a TATA box is inactive in normal cells. But in cells expressing the GAL4–CREB fusion protein, the CREB activation domain was brought to the promoter by the GAL4 DNA-binding domain: this promoter then behaved as a regular cAMP-responsive promoter. Thus, the cAMP-regulated transcriptional stimulatory activity of CREB could be expressed through the unrelated DNA-binding domain.

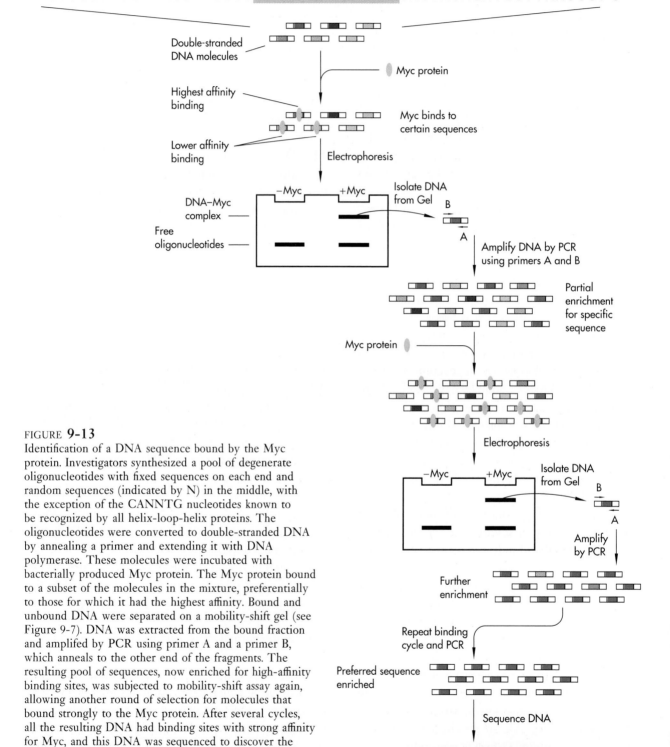

Window of random sequence

5′ AGACGGATCCATTGCATCCCC**NNNNCANNTGNNNN**CTGATCTGTAGGAATTCGGA
  TCTGCCTAGGTAACGTAGGGG**NNNNGTNNACNNNN**GACTAGACATCCTTAAGCCT 5′

Double-stranded DNA molecules

Myc protein

Highest affinity binding

Lower affinity binding

Myc binds to certain sequences

Electrophoresis

−Myc +Myc

DNA−Myc complex

Free oligonucleotides

Isolate DNA from Gel

Amplify DNA by PCR using primers A and B

Partial enrichment for specific sequence

Myc protein

Electrophoresis

−Myc +Myc

Isolate DNA from Gel

Amplify by PCR

Further enrichment

Repeat binding cycle and PCR

Preferred sequence enriched

Sequence DNA

5′.....CCCC**GGGGCACGTGGTNN**CTGAT......3′

Consensus Myc binding sequence

FIGURE **9-13**
Identification of a DNA sequence bound by the Myc protein. Investigators synthesized a pool of degenerate oligonucleotides with fixed sequences on each end and random sequences (indicated by N) in the middle, with the exception of the CANNTG nucleotides known to be recognized by all helix-loop-helix proteins. The oligonucleotides were converted to double-stranded DNA by annealing a primer and extending it with DNA polymerase. These molecules were incubated with bacterially produced Myc protein. The Myc protein bound to a subset of the molecules in the mixture, preferentially to those for which it had the highest affinity. Bound and unbound DNA were separated on a mobility-shift gel (see Figure 9-7). DNA was extracted from the bound fraction and amplifed by PCR using primer A and a primer B, which anneals to the other end of the fragments. The resulting pool of sequences, now enriched for high-affinity binding sites, was subjected to mobility-shift assay again, allowing another round of selection for molecules that bound strongly to the Myc protein. After several cycles, all the resulting DNA had binding sites with strong affinity for Myc, and this DNA was sequenced to discover the preferred site (sequence shown). Surprisingly, Myc exhibited no preference for particular nucleotides at the two positions marked N.

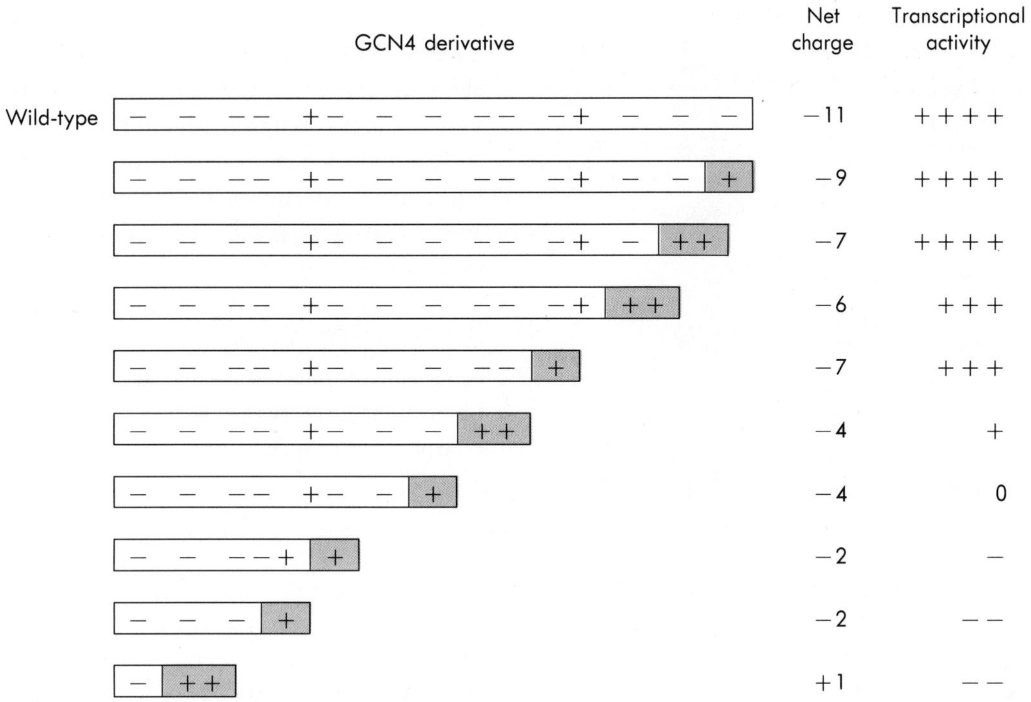

GCN4 derivative	Net charge	Transcriptional activity
Wild-type [− − −− +− − − −− −+ − − −]	−11	+ + + +
[− − −− +− − − −− −+ − − **+**]	−9	+ + + +
[− − −− +− − − −− −+ − **+ +**]	−7	+ + + +
[− − −− +− − − −− −+ **+ +**]	−6	+ + +
[− − −− +− − − −− **+**]	−7	+ + +
[− − −− +− − − **+ +**]	−4	+
[− − −− +− − **+**]	−4	0
[− − −−+ **+**]	−2	−
[− − − **+**]	−2	− −
[− **+ +**]	+1	− −

FIGURE **9-14**

Mapping of GCN4 activation domain to an "acid blob." GCN4 is a yeast transcription factor of the leucine zipper family. The region of the protein responsible for transcriptional activation was localized to a stretch of amino acids rich in negative charges (acidic amino acids) with no other discernible features, hence the term *acid blob*. Fine-resolution mapping was done by making a series of closely spaced, cumulative deletions in the GCN4 gene in the region encoding the acid blob. The deleted GCN4 genes were tested for activity by transformation back into yeast. The top rectangle shows a representation of the acid blob region of GCN4, with the positions of charged amino acids indicated by + and −. Each line below shows a deletion derivative of GCN4. The shaded region represents amino acids encoded by linker DNA placed adjacent to GCN4 sequences in these constructions. Positive signs under "Transcriptional activity" indicate the relative activity of the GCN4 derivative as a transcription factor in yeast. A 0 indicates no increase in activity relative to a yeast strain with no GCN4 protein at all. Negative signs indicate that these deletion derivatives actually inhibited gene expression, presumably because they bound to GCN4 recognition sites and prevented other transcription factors from binding there. Note that there was a steady decrease in activity as the net negative charge was reduced.

How Do Transcription Factors Work?

As transcription factor structure becomes clear and abundant quantities of purified factors become available, thanks to recombinant DNA, a biochemical view of how transcription factors work will eventually emerge. At present, however, most ideas of how these proteins act to stimulate gene transcription are highly speculative. In bacteria, auxiliary transcription factors act at two points in the transcription cycle. They can help RNA polymerase to bind to the promoter, or they can accelerate the rate at which bound RNA polymerase initiates transcription. It is likely that eukaryotic transcription factors are capable of doing the

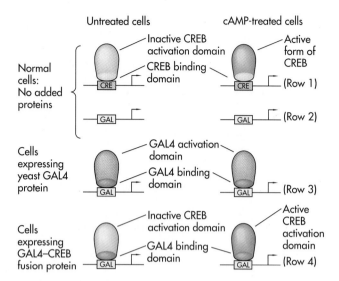

FIGURE **9-15**
The modular nature of transcription factors. This experiment showed that a regulated activation domain can still work when attached to a different DNA-binding domain. When normal mammalian cells are transfected with a promoter containing a cyclic AMP (cAMP)–responsive element (CRE), the promoter becomes activated when the cells are treated with cAMP. This is due to the action of transcription factors [CRE-binding protein (CREB)] in the cell that bind to the CRE and are modified to an active form in the presence of cAMP (row 1). When a promoter carrying a binding site for the yeast transcription factor GAL4 (GAL) is introduced into mammalian cells, it is inactive because mammalian cells have no transcription factors with the DNA-binding specificity of GAL4 (row 2). However, introduction into cells of a plasmid that expresses GAL4 results in activation of the promoter carrying the GAL binding site (row 3); GAL4 activity is not affected by cAMP. In the experiment shown in row 4, the GAL promoter was transfected together with a second plasmid engineered to produce a protein in which the activation domain of CREB was joined to the DNA-binding domain of GAL4. This hybrid protein was able to bind to the GAL4 site in the promoter, bringing along the regulated activation domain of CREB. Under these conditions, the GAL promoter behaved like an authentic CRE-containing promoter and was activated in response to cAMP.

same things. Indeed, evidence for a stabilizing interaction between transcription factors and the TATAA-binding protein TFIID has been obtained.

The organization of the eukaryotic genome adds a few wrinkles, however. Chromosomal DNA is tightly packed into higher-order nucleoprotein structures, the basic unit of which is the nucleosome, a complex of eight histone proteins around which approximately 200 bp of DNA are wound. Transcription factors might act to free DNA from nucleosomes so that the promoter is accessible to the large transcription complex. Recall, too, that in eukaryotes enhancers are able to stimulate transcription from thousands of base pairs away. This fact has led to the proposal that transcription factors bound to DNA at distant sites can physically interact with one another by looping out the intervening DNA (Figure 9-16). In this way, the multiprotein complex assembled at a distant enhancer can be brought directly into contact with the transcription complex at the promoter.

The Cellular Splicing Machinery Is Extraordinarily Selective and Precise

In eukaryotes, as we have learned, merely activating transcription of a gene is not usually sufficient to generate a mature mRNA that can be exported to the cytoplasm for translation. Because most eukaryotic genes are interrupted by introns, the primary transcript must be processed to remove the introns and join the coding sequence into a contiguous mRNA molecule. Splicing of pre-mRNA to mRNA presents the cell with a challenging problem. It must locate every exon within the primary transcript—there may be scores of them interspersed in hundreds of kilobases of DNA (see, for example, Figure 8-13)—and it must join each of them absolutely precisely and in the correct order. If an exon is missed or a splice occurs even one nucleotide away from the correct location, the correct protein will not be translated from the resulting mRNA. How does the splicing machinery locate exons? What guides the precision of a splice? And how does the splicing machinery make decisions when faced with alternative splicing choices? The answers to such questions are beginning to emerge from molecular, genetic, and biochemical characterization of the splicing apparatus.

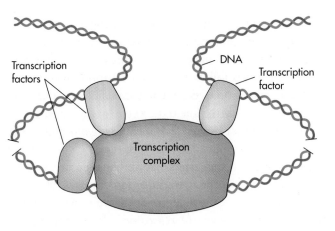

FIGURE **9-16**
How do transcription factors activate transcription when bound to DNA sites far from the promoter? One model is that DNA forms loops, allowing distant factors to contact the transcription complex directly at the promoter.

Synthetic Pre-mRNAs Are Spliced in Oocytes and Cell-Free Extracts

The technical breakthrough that made biochemical analysis of splicing possible was the availability of bacteriophage RNA polymerases and promoters that allowed the enzymatic synthesis of large amounts of any desired RNA molecule (see Chapter 7). In early experiments, pre-mRNA from the β-globin gene was synthesized in vitro in the presence of ^{32}P-labeled nucleoside triphosphates. The resulting labeled RNA was injected into *Xenopus* oocytes. After an incubation period, the labeled RNA was recovered from the oocytes and analyzed by polyacrylamide gel electrophoresis, which revealed that the pre-mRNA had been accurately spliced—the intron removed and the exons joined (Figure 9-17). Pre-mRNA synthesized from a β^0-thalassemia gene with a mutation in the 5'-splice site of the second β-globin intron was not spliced in the oocytes. Although splicing in the oocytes was inefficient, this experiment showed that splicing was not directly coupled to transcription; the splicing machinery could act on exogenously supplied pre-mRNA. This observation set the stage for a biochemical assault on the splicing apparatus.

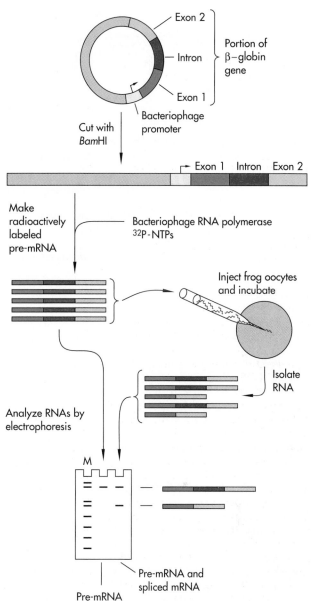

FIGURE **9-17**
Splicing of synthetic pre-mRNAs in *Xenopus* oocytes. A portion of the human β-globin gene was cloned next to a bacteriophage promoter. The resulting plasmid was cleaved with the restriction enzyme *Bam*HI at a position immediately after the globin sequence. Transcription of this linear DNA fragment with bacteriophage RNA polymerase in the presence of ^{32}P-labeled nucleoside triphosphates resulted in the synthesis of labeled globin pre-mRNA containing an intron. The RNA was injected into *Xenopus* oocytes, which spliced it correctly. This was demonstrated by extracting the RNA from the oocytes and analyzing it by gel electrophoresis. In addition to the original pre-mRNA, correctly spliced mature mRNA molecules were observed. M, size markers.

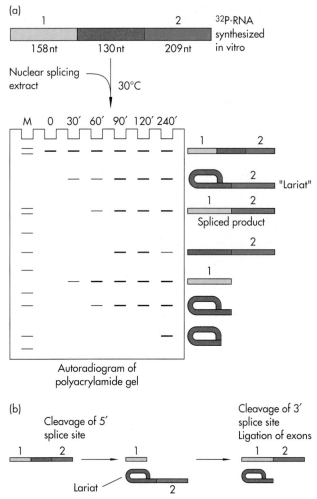

(a)

FIGURE **9-18**
RNA splicing in cell-free extracts. (a) Radioactively labeled globin pre-mRNA was prepared as in Figure 9-17 [the RNA contained two exons and one intron of the sizes indicated in nucleotide (nt) units] and incubated with a nuclear extract of human cells. At various times, samples were removed and analyzed by gel electrophoresis. After a lag of 30 to 45 minutes, in which nothing happened, new RNA species began to appear on the gel, with correctly spliced mRNA first appearing at 60 minutes. The other RNAs on the gel were intermediates and other reaction products as indicated beside the gel. For example, one RNA that appeared late in the reaction (fourth from the top) was a damaged lariat intermediate opened up by enzymes in the extract. These experiments showed that splicing in extracts proceeded in two steps (b). First, the 5' splice site was cleaved, generating free exon 1 and a lariat intermediate containing the intron and exon 2. Next, the 3' splice site was cleaved and the two exons joined, freeing the intron as a lariat. Because this cleavage and the subsequent joining of the two exons occur in concert, free exon 2 molecules never appear on the gels. The tail of the lariat is eventually trimmed off, leading to the bottommost RNA seen on the gel (a). M, size markers.

As with transcription, splicing can take place in cell-free extracts. Synthetic pre-mRNA was added to nuclear extracts of mammalian cells, and processing of the RNA was followed by gel electrophoresis. Under optimal conditions, the intron was efficiently removed from the pre-mRNA, and the two exons were correctly joined together (Figure 9-18a). Splicing required ATP. A 30 to 45 minute lag was observed before splicing began; this lag suggested that perhaps the pre-mRNA had to be incorporated into some larger complex before it could be spliced. Indeed, we now know that splicing takes place in *spliceosomes*, large ribonucleoprotein particles akin to ribosomes.

Intermediates in the splicing reactions were also observed on the gels, but their mobility on the gels was not what was expected for their size. Biochemical analysis of these strange RNAs revealed that they contained an unusual 2'-5' phosphodiester bond not normally found in RNA. This structure arises because in the first step of splicing, the 5' end of the intron loops back on itself, and the guanine nucleotide at the 5' end forms a 2'-5' bond with an adenine nucleotide near the 3' end of the intron. The resulting RNA molecule is a circle with a tail, much like a cowboy's lariat; this unusual structure explains the aberrant mobilities of these molecules on gels. From these biochemical studies, we know that splicing is a two-step process. First, the 5' splice site is cut, freeing the upstream exon and generating a lariat molecule containing the intron still joined to the downstream exon. In the second step, the 3' splice site is cut and the exons are joined, freeing the intron as a lariat (Figure 9-18b).

Cell-Free Extracts Are Fractionated to Identify Splicing Factors

Biochemical fractionation of nuclear extracts coupled with in vitro splicing assays has identified factors required for splicing. One class of splicing factors is the small nuclear ribonucleoproteins (snRNPs), large particles made of snRNAs (Chapter 8) and several protein components. One type of snRNP contains the U1 RNA molecule, which, as we learned in Chapter 8, contains a sequence complementary to the 5' splice site consensus. Considerable experimental evidence points to a role for U1 snRNPs in splicing. Antibodies to proteins in U1 snRNPs block splicing in extracts, presumably by interfering with snRNP function. Enzymatic digestion of U1 RNA in the extracts also blocks splicing. Splicing is blocked at its earliest step, consistent with a role for U1 in recognition of the 5' splice site. Similarly, another snRNP containing the U2 snRNA is involved in recognizing and presumably defining the position near the 3' end of the intron where the lariat forms. The roles of snRNPs in splicing have been further defined by genetic studies in yeast (Chapter 13).

Conventional protein factors have also been identified. One protein, *SF2*, is required for the earliest steps of splicing and appears to play a role in 5' splice site selection. It binds RNA and promotes the annealing of RNA molecules. A second factor, called *U2AF*, aids the U2 snRNP in the recognition of the lariat branch point near the 3' splice site. Many other factors are undoubtedly involved in the recognition of splice sites and the chemistry of RNA cleavage and rejoining. As we will learn in Chapter 22, some RNA molecules have the remarkable ability to splice themselves without need of any protein factors. Conventional splicing may also use this unconventional chemistry.

Trans-Acting Splicing Factors Govern Alternative Splicing

As we learned in Chapter 8, many primary gene transcripts can be spliced in different ways to produce mRNAs encoding different proteins. The different splicing patterns are often carried out in different cell types. How is the choice between alternative splicing patterns made? Current evidence suggests that there

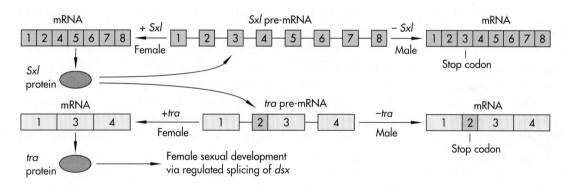

FIGURE **9-19**

Regulated splicing events in the determination of sex in *Drosophila*. Both the *Sxl* and *tra* genes are transcribed into pre-mRNA in males and females, but they are spliced differently. The male, or default, splicing pattern is shown to the right. For both the *Sxl* and *tra* genes in males, an exon (blue) is included in the spliced mRNA that contains a stop codon, terminating translation of the *Sxl* and *tra* proteins before active proteins are made. In females, inclusion of this exon in the spliced end product is blocked by the action of the *Sxl* protein itself, so that female mRNA carries intact open reading frames for active *Sxl* and *tra* proteins. Thus, in females, the *Sxl* protein itself ensures that it will continue to be produced. Its absence in males permits the stop codon to be incorporated into the mRNA, ensuring that the *Sxl* protein will continue *not* to be produced. *Sxl* protein is believed to act by binding to the pre-mRNA in the vicinity of the male exon (the exon containing the stop codon), preventing its inclusion by the splicing machinery. Therefore, the exon is skipped. Active *tra* protein directs female development by regulating the splicing of doublesex (*dsx*) RNA. Spliced one way, in the presence of *tra*, female sex organs develop. Spliced differently in the absence of *tra*, male organs develop. How the first *Sxl* proteins are produced to initiate this cascade is not yet understood, but the mechanism may involve special *Sxl* mRNAs unique to very early embryos.

are specific splicing factors that direct such decisions.

In fruit flies, sexual differentiation is controlled by a cascade of regulated splicing events. Each of the genes that determine whether a fly is male or female is transcribed into RNA in cells of both sexes, but the RNA is spliced differently in males and females. The master gene in this regulatory cascade is Sex-lethal *(Sxl)*. Activation of *Sxl* is triggered by an unknown mechanism that is able to measure the ratio of X chromosomes to autosomes in the developing embryo (in females, which have 2 X chromosomes, this ratio is 1:1, whereas in males it is 1:2). Once activated, the *Sxl* product controls the splicing of its own pre-mRNA and of the next gene in the cascade, transformer *(tra)*. Only the female-specific splices yield functional *Sxl* and *tra* proteins. In males, splicing of *Sxl* and *tra* follows a "default" pattern owing to the action of the cell's regular housekeeping splicing machinery, yielding mRNAs that encode inactive proteins. In females, it is the active *Sxl* protein itself that modifies these splicing choices so that the resulting mRNAs continue to encode active *Sxl* and *tra* proteins (Figure 9-19). By regulating its own splicing, *Sxl* provides each cell with a memory of its sex—once the female-specific splicing pattern is established, active *Sxl* protein continually reinforces the female-specific splice of its own RNA. By controlling the splicing of *tra,* it ensures that only females make active *tra* protein, which in turn directs the development of female sex organs. The *tra* protein itself acts to regulate splicing of another gene, doublesex (see Figure 9-19).

Cloning and sequencing of the *Sxl* and *tra* genes revealed that they encode proteins related to components of snRNPs and other proteins known to bind to RNA. Therefore, *Sxl* and *tra* proteins probably control the splicing choices by binding directly to the pre-mRNA and guiding the splicing machinery one way or the other. Several experimental observations support this model. One class of mutations in the *Sxl* gene results in the expression of the female-specific *Sxl* mRNA in males. All these mutations are near the male-specific *Sxl* exon. These mutations probably interfere with the use of this exon, forcing the cells to make the female-specific splice instead. This is precisely what the *Sxl* protein normally does in females. Therefore, the *Sxl* protein may act by binding to the *Sxl* pre-mRNA to prevent use of the male-specific exon, promoting the female-specific splice instead. Similarly, female-specific splicing of the *tra* RNA can be forced by damaging the splice site normally used in males or by forcing expression of the active female-specific form of *Sxl* mRNA. Again, the implication is that the *Sxl* protein acts at the male splice site, forcing the splicing machinery to use the female site instead. Indeed, there is now direct evidence that *Sxl* protein binds to the *tra* pre-mRNA in the vicinity of the male splice site. It is likely that other alternative splicing choices are regulated by factors that prevent or directly promote one of the two options. The biochemical basis for these mechanisms remains a mystery, but alternative splicing adds yet another level of complexity and flexibility to gene expression in eukaryotic cells.

Reading List

Original Research Papers

EUKARYOTIC PROMOTER STRUCTURE

Maniatis, T., S. Goodbourn, and J. A. Fischer. "Regulation of inducible and tissue-specific gene expression." *Science*, 236: 1237–1245 (1987). [Review]

Benoist, C., and P. Chambon. "In vivo sequence requirements of the SV40 early promoter region." *Nature*, 290: 304–310 (1981).

McKnight, S. L., and R. Kingsbury. "Transcriptional control signals of a eukaryotic protein-coding gene." *Science*, 217: 316–324 (1982).

Myers, R. M., K. Tilly, and T. Maniatis. "Fine structure genetic analysis of a β-globin promoter." *Science*, 232: 613–618 (1986).

ENHANCER STRUCTURE AND FUNCTION

Marriott, S. J., and J. N. Brady. "Enhancer function in viral and cellular gene regulation." *Biochim. Biophys. Acta*, 989: 97–100 (1989). [Review]

Banerji, J., S. Rusconi, and W. Schaffner. "Expression of a β-globin gene is enhanced by remote SV40 DNA sequences." *Cell*, 27: 299–308 (1981).

Herr, W., and J. Clarke. "The SV40 enhancer is composed of multiple functional elements that can compensate for one another." *Cell*, 45: 461–470 (1986).

Zenke, M., T. Grundstrom, H. Matthes, M. Wintzerith, C. Schatz, A. Wildeman, and P. Chambon. "Multiple sequence motifs are involved in SV40 enhancer function." *EMBO J.*, 5: 387–397 (1986).

Ondek, B., L. Gloss, and W. Herr. "The SV40 enhancer contains two distinct levels of organization." *Nature*, 333: 40–45 (1988).

TISSUE-SPECIFIC AND INDUCIBLE ENHANCERS

Chandler, V. L., B. A. Maler, and K. R. Yamamoto. "DNA sequences bound specifically by glucocorticoid receptor in vitro render a heterologous promoter hormone responsive in vivo." *Cell*, 33: 489–499 (1983).

Gillies, S. D., S. L. Morrison, V. T. Oi, and S. Tonegawa. "A tissue-specific transcription enhancer element is located in the major intron of a rearranged immunoglobulin heavy chain gene." *Cell*, 33: 717–728 (1983).

Goodbourn, S., K. Zinn, and T. Maniatis. "Human β-interferon gene expression is regulated by an inducible enhancer element." *Cell*, 41: 509–520 (1985).

Treisman, R. "Transient accumulation of c-*fos* RNA following serum stimulation requires a conserved 5' element and c-*fos* 3' sequences." *Cell*, 42: 889–902 (1985).

ENHANCER-BINDING PROTEINS

Church, G. M., A. Ephrussi, W. Gilbert, and S. Tonegawa. "Cell-type-specific contacts to immunoglobulin enhancers in nuclei." *Nature*, 313: 798–801 (1985).

Davidson, I., C. Fromental, P. Augereau, A. Wildeman, M. Zenke, and P. Chambon. "Cell-type-specific protein binding to the enhancer of SV40 in nuclear extracts." *Nature*, 323: 544–548 (1986).

Sen, R., and D. Baltimore. "Multiple nuclear factors interact with the immunoglobulin enhancer sequences." *Cell*, 47: 705–716 (1986).

GENE TRANSCRIPTION IN CELL-FREE EXTRACTS

Saltzman, A. G., and R. Weinmann. "Promoter specificity and modulation of RNA polymerase II transcription." *FASEB J.*, 3: 1723–1733 (1989). [Review]

Weil, P. A., D. S. Luse, J. Segall, and R. G. Roeder. "Selective and accurate initiation of transcription at the Ad2 major late promoter in a soluble system dependent on purified RNA polymerase II and DNA." *Cell*, 18: 469–484 (1979).

Dynan, W. S., and R. Tjian. "Isolation of transcription factors that discriminate between different promoters recognized by RNA polymerase II." *Cell*, 32: 669–680 (1983).

Buratowski, S., S. Hahn, L. Guarente, and P. A. Sharp. "Five intermediate complexes in transcription initiation by RNA polymerase II." *Cell*, 56: 549–561 (1989).

Lin Y.-S., and M. R. Green. "Mechanism of action of an acidic transcriptional activator in vitro." *Cell*, 64: 971–981 (1991).

PURIFICATION AND CLONING OF TRANSCRIPTION FACTORS

Dynan, W. S., and R. Tjian. "The promoter-specific transcription factor Sp1 binds to upstream sequences in the SV40 early promoter." *Cell*, 35: 79–87 (1983).

Kadonaga, J. T., and R. Tjian. "Affinity purification of sequence-specific DNA binding proteins." *Proc. Natl. Acad. Sci. USA*, 83: 5889–5893 (1986).

Kadonaga, J. T., K. R. Carner, S. R. Masiarz, and R. Tjian. "Isolation of cDNA encoding transcription factor Sp1 and functional analysis of the DNA-binding domain." *Cell*, 51: 1079–1090 (1987).

Singh, H., J. H. LeBowitz, A. S. Baldwin, and P. A. Sharp. "Molecular cloning of an enhancer binding protein: isolation by screening a library with a recognition site DNA." *Cell*, 52: 415–423 (1988).

TRANSCRIPTION FACTOR STRUCTURE

Johnson, P. F., and S. L. McKnight. "Eukaryotic transcriptional regulatory proteins." *Ann. Rev. Biochem.*, 58 797–839 (1989). [Review]

Miller, J., A. D. McLachlan, and A. Klug. "Repetitive zinc-binding domains in the protein transcription factor IIIA from *Xenopus* oocytes." *EMBO J.*, 4: 1609–1614 (1985).

Landschulz, W. M., P. F. Johnson, and S. L. McKnight. "The DNA binding domain of the rat liver nuclear protein C/EBP is bipartite." *Science*, 243: 1681–1688 (1989).

Murre, C., P. S. McCaw, H. Vaessin, M. Caudy, L. Y. Jan, Y. N. Jan, C. V. Cabrera, J. N. Buskin, S. D. Hauschka, A. B. Lassar, H. Weintraub, and D. Baltimore. "Interactions between heterologous helix-loop-helix proteins generate complexes that bind specifically to a common DNA sequence." *Cell*, 58: 537–544 (1989).

Umesono, K., and R. M. Evans. "Determinants of target gene specificity for steroid/thyroid hormone receptors." *Cell*, 57: 1139–1146 (1989).

Blackwell, T. K., L. Kretzner, E. M. Blackwood, R. N. Eisenman, H. Weintraub. "Sequence-specific DNA binding by the c-Myc protein." *Science*, 250: 1149–1151 (1990).

Kissinger, C. R., B. Liu, E. Martin-Blanco, T. B. Kornberg, and C. O. Pabo. "Crystal structure of an engrailed homeodomain-DNA complex at 2.8 Å resolution: a framework for understanding homeodomain-DNA interactions." *Cell,* 63: 579–590 (1990).

TRANSCRIPTION FACTOR FUNCTION

Ptashne, M. "How eukaryotic transcriptional activators work." *Nature,* 335: 683–689 (1988). [Review]

Giniger E., and M. Ptashne. "Transcription in yeast activated by a putative amphipathic α-helix linked to a DNA-binding unit." *Nature,* 330: 670–672 (1987).

Ma, J., and M. Ptashne. "A new class of yeast transcriptional activators." *Cell,* 51: 113–119 (1987).

Hope, I. A., S. Mahadevan, and K. Struhl. "Structural and functional characterization of the short acidic transcriptional activation region of yeast GCN4 protein." *Nature,* 333: 635–640 (1988).

Berkowitz, L. A., and M. Z. Gilman. "Two distinct forms of active transcription factor CREB (cAMP response element binding protein)." *Proc. Natl. Acad. Sci. USA,* 87: 5258–5262 (1990).

MECHANISM OF SPLICING

Maniatis, T., and R. Reed. "The role of small nuclear ribonucleoprotein particles in pre-mRNA splicing." *Nature,* 325: 673–678 (1987). [Review]

Green, M. R., T. Maniatis, and D. A. Melton. "Human β-globin pre-mRNA synthesized in vitro is accurately spliced in *Xenopus* oocyte nuclei." *Cell,* 32: 681–694 (1983).

Krainer, A. R., T. Maniatis, B. Ruskin, and M. R. Green. "Normal and mutant human β-globin pre-mRNAs are faithfully and efficiently spliced in vitro." *Cell,* 36: 993–1005 (1984).

Padgett, R. A., M. M. Konarska, P. J. Grabowski, S. F. Hardy, and P. A. Sharp. "Lariat RNA's as intermediates and products in the splicing of messenger RNA precursors." *Science,* 225: 898–903 (1984).

Frendewey, D., and W. Keller. "Stepwise assembly of a pre-mRNA splicing complex requires U-snRNPs and specific intron sequences." *Cell,* 42: 355–367 (1985).

SPLICING FACTORS (INCLUDING snRNPs AND ALTERNATIVE SPLICING FACTORS)

Padgett, R. A., S. M. Mount, J. A. Steitz, and P. A. Sharp. "Splicing of messenger RNA precursors is inhibited by antisera to small nuclear ribonucleoprotein." *Cell,* 35: 101–107 (1983).

Kramer, A., W. Keller, B. Appel, and R. Luhrmann. "The 5' terminus of the RNA moiety of U1 small nuclear ribonucleoprotein particles is required for the splicing of messenger RNA precursors." *Cell,* 38: 299–307 (1984).

Ruskin, B., P. D. Zamore, and M. R. Green. "A factor, U2AF, is required for U2 snRNP binding and splicing complex assembly." *Cell,* 52: 207–219 (1988).

Krainer, A. R., G. C. Conway, and D. Kozak. "The essential pre-mRNA splicing factor SF2 influences 5' splice site selection by activating proximal site." *Cell,* 62: 35–42 (1990).

ALTERNATIVE SPLICING

Baker, B. S. "Sex in flies: the splice of life." *Nature,* 340: 521–524 (1989). [Review]

Smith, C. W. J., J. G. Patton, and B. Nadal-Ginard. "Alternative splicing in the control of gene expression." *Ann. Rev. Gen.,* 23: 527–577 (1989). [Review]

Bell, L. R., E. M. Maine, P. Schedl, and T. W. Cline. "*Sex-lethal,* a *Drosophila* sex determination switch gene, exhibits sex-specific RNA splicing and sequence similarity to RNA binding proteins." *Cell,* 55: 1037–1046 (1988).

Sosnowski, B. A., J. M. Belote, and M. McKeown. "Sex-specific alternative splicing of RNA from the *transformer* gene results from sequence-dependent splice site blockage." *Cell,* 58: 449–459 (1989).

Inoue, K., K. Hoshijima, H. Sakamoto, and Y. Shimura. "Binding of the *Drosophila Sex-lethal* gene product to the alternative splice site of *transformer* primary transcript." *Nature,* 344: 461–463 (1990).

10

Movable Genes

I n the early 1950s, Barbara McClintock, working at the Cold Spring Harbor Laboratory on the maize plant, began to call attention to the novel behavior of genetic elements she called "controlling elements." These genetic elements, which were first noticed because they inhibited the expression of other maize genes with which they came into close contact, did not have fixed chromosomal locations. Instead, they seemed to move about the maize genome. The controlling elements could be inserted and later excised; after excision, the function of a previously dormant gene often returned. The genes that associated with the controlling elements were rendered unstable and had high mutation rates because of the instability of the controlling element.

For many years, the corn plant provided the only genetic system in which such movable elements were observed. Then there arose some preliminary evidence that several of the highly mutable genetic loci in *Drosophila* might be associated with movable control elements. But most geneticists paid little attention to such loci until the discovery, in the late 1960s, that certain highly pleiotropic mutations (mutations affecting several functions) in *E. coli* resulted from the insertion of large segments of DNA called *insertion sequences* (ISs). By examining heteroduplexes between genes carrying these insertions, it became clear that many independent insertion events involved exactly the same DNA sequences.

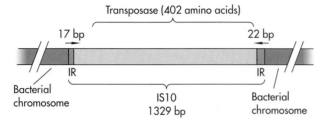

FIGURE **10-1**

Structure of IS10, a simple bacterial insertion sequence. IS10 is a 1329-bp transposable element found in *E. coli*. The element consists of a gene that encodes a 402 amino acid protein, thought to be the transposase enzyme required for IS movement, flanked by short inverted repeats (IR). The IRs are blocks of similar (but not identical) sequence in opposite orientation to one another. The IRs are recognized by the transposase enzyme in the first steps of transposition and therefore define the ends of the sequence to be transposed.

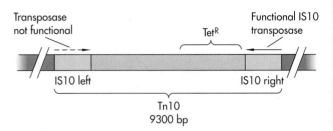

FIGURE **10-2**

Complex transposons are genes flanked by two IS elements. Tn10 is a 9300-bp-long movable element that encodes genes for resistance to the antibiotic tetracycline (Tet^R). Each end contains an IS10 insertion element (oriented in opposite directions). The right IS encodes a functional transposase that is required for movement of the transposon. The left IS10 element has accumulated mutations so that it no longer encodes an active transposase protein. Between the IS elements are genes required for tetracycline resistance. The IS elements can transpose individually or in tandem; in the latter case they carry the intervening DNA with them.

With the arrival of recombinant DNA technology, these insertions could be readily cloned and studied. From these studies flowed a wealth of data on the organization of these elements and how they could "jump" from one place to another, a phenomenon termed *transposition*. Moreover, as larger portions of eukaryotic genomes were explored by cloning and sequencing, it soon became apparent that virtually all genomes were littered with sequences that appeared to be relics of transposition events long past. Cloning also revealed that many organisms used programmed DNA rearrangements as a strategy to solve difficult problems in gene regulation and diversity. But other transposable elements have no clear function. Why are they a part of our DNA? Are they simply molecular parasites, or do they serve a useful function in the evolution of genomes?

Sequencing Reveals the Organization of Bacterial Transposable Elements

As we have mentioned, it was clear from genetic and physical studies of unusual *E. coli* mutations that these events were caused by insertions of new DNA into genes. These bacterial insertion sequences were among the first *transposons*, as they were called, to be sequenced. The organization of a typical insertion sequence is shown in Figure 10-1. ISs range in size from 750 to 2500 bp. They are flanked by inverted repeats—closely related sequences in opposite orientations—from 10 to 40 bp in length. Enclosed by the inverted repeats are one or more open reading frames that encode proteins required for transposition of the element. In the simplest ISs, a single open reading frame encodes a transposase enzyme that recognizes the inverted repeats and catalyzes excision of the element. We will discuss the molecular mechanisms of transposition later in this chapter.

When two IS elements are located close to one another, they can transpose as a unit, mobilizing the host DNA between them. These *complex transposons* (Figure 10-2) were recognized because the mobilized DNA carried genes conferring antibiotic resistance. Complex transposons can easily jump from the bacterial chromosome to phage genomes or conjugative plasmids, and when this happens, the transposons can then spread to other bacteria. Transposons of this sort are the major source of antibiotic resistance in naturally occurring bacteria.

Some Transposable Elements in Eukaryotes Resemble Bacterial Transposons

The "controlling elements" of maize, painstakingly studied by Barbara McClintock, turned out to be remarkably similar to the transposable elements of bacteria (Figure 10-3). Cloning and sequencing of these maize elements also offered a clear explanation for the distinct behavior of the two types of elements identified by McClintock in her genetic experiments, *Ac* and *Ds.* Ac elements caused mutations that reverted at high frequency—Ac mutations in the genes for synthesis of the kernel pigment anthocyanin thus produced mottled kernels with patches of normal color where the mutation had reverted. Ds elements caused stable mutations—the kernels with these mutations stayed colorless unless the plants were crossed to a strain containing Ac. Ac and Ds are derivatives of the same transposable element. Ac is 4500 bp in length and apparently encodes a transposase that catalyzes its movement. Ds elements are simply deleted Ac's that no longer express transposase on their own. Thus, in the absence of Ac, Ds is immobile, and Ds insertion mutations cannot revert. When Ac is present and providing transposase activity, Ds freely hops as a complex transposon, allowing "recovery" (reversion) of the anthocyanin genes.

P elements in *Drosophila* share a similar organization (Figure 10-4). *Drosophila* strains that carry P elements usually carry only a few full-length transposons. Most P sequences are deleted derivatives like Ds, unable to make their own transposase and therefore dependent on full-length elements for movement. P element transposition is limited to the germ line cells of *Drosophila* by an alternative splicing mechanism. Active transposase is encoded by four exons, and the final splice occurs only in germ line cells. In fact, the product synthesized from the first three exons in somatic cells, in addition to being inactive as a transposase, is thought to function as a repressor of transposition. That the inability to make the final splice is what prevents P element transposition in somatic cells was demonstrated elegantly by engineering a P element in which the last intron was precisely deleted, joining

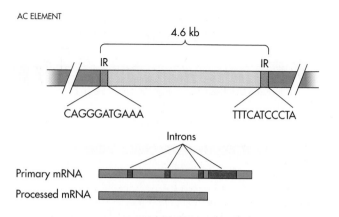

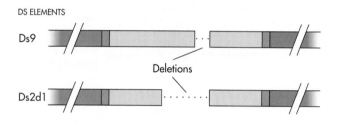

FIGURE **10-3**
The Ac/Ds transposable elements of maize. The Ac element is an intact transposon organized much like bacterial transposons, with short inverted repeats (IR) flanking sequences encoding a transposase enzyme. In Ac the transposon gene carries introns. Because it carries an active transposase gene, Ac moves autonomously. The Ds elements are Ac elements carrying deletions in the transposase gene. They cannot move on their own; thus, Ds insertion mutations are stable in the absence of Ac.

the third and fourth exons as if they were spliced. This engineered P element transposed readily in somatic cells of the fly.

Ty Elements in Yeast Are a Different Class of Transposable Elements

As in maize and *E. coli,* yeast Ty elements were first identified by the genetic properties of the mutations they caused. In addition to extinguishing gene function by insertion into coding or regulatory sequences, Ty

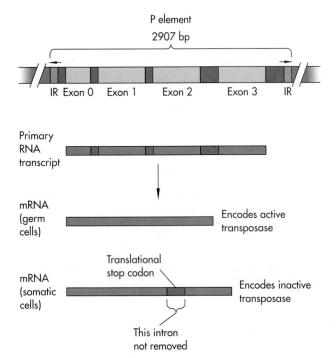

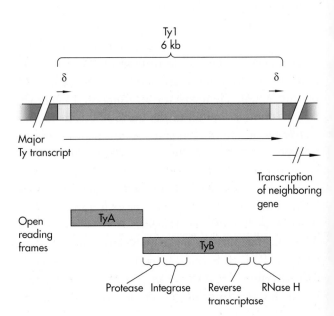

insertions had the unusual property of sometimes activating genes that were previously silent. Once again, cloning and sequencing explained this behavior. Ty elements are organized differently from bacterial transposons (Figure 10-5). Instead of inverted repeats at the ends, the Ty1 element carries 334-bp direct repeats, termed *delta elements*. These repeats, also called *LTRs* (for *long terminal repeats*), carry strong promoters for RNA polymerase (in fact, Ty elements contribute 5 to 10 percent of total yeast mRNA). As is apparent from the structure, each element carries one LTR oriented so that it promotes transcription of chromosomal DNA located downsteam of the insertion. This is how insertion of a Ty1 element can activate silent genes. The organization of Ty elements is shared with the copia element of *Drosophila,* which has many of the same properties, and with the retroviruses of vertebrates. The ability of inserted retroviral genomes to activate the expression of neighboring genes is the basis for their ability to cause tumors (see Chapter 18).

Since the organization of Ty and copia elements was distinct from that of bacterial transposons, it seemed likely that they transpose by a different mechanism. As we will learn shortly, this is indeed the case.

FIGURE **10-4**

The P element of *Drosophila*. Like Ac and the bacterial transposons, the P element consists of a transposase gene flanked by inverted repeats (IR). The transposase is encoded in four exons (exon 0 through exon 3). All four exons are spliced together in cells of the germ line to construct an mRNA that encodes active transposase. In all other tissues of the fly (somatic cells), however, the exon 2–exon 3 splice does not occur, and the protein encoded by the resulting mRNA represses rather than promotes transposition. This form of regulation limits transposition of P elements to germ line cells. Thus, new transposition events occur only from one generation of flies to the next, while each individual fly is stable. If the intron between exons 2 and 3 is removed by genetic engineering, the resulting element transposes frequently in all tissues of the fly.

FIGURE **10-5**

Structure of the Ty1 transposable element of the yeast *Saccharomyces cerevisiae*. The complete element is approximately 6000 bp long and is flanked by direct repeats of 334 bp (the δ elements, or LTRs). The repeats contain a strong promoter for RNA polymerase II, leading to transcription of the complete element into a 5.6-kb mRNA from the left δ and transcription of host DNA from the right δ. The Ty element carries two open reading frames. The TyB gene encodes a long *polyprotein* that is processed to several different enzymes that reverse-transcribe Ty mRNA into DNA and integrate it into the host genome. This gene is related to the POL gene of retroviruses in animals.

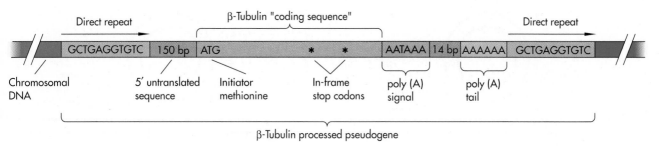

FIGURE **10-6**

A processed human β-tubulin pseudogene. Sequencing of a genomic clone that hybridized to a β-tubulin probe revealed that it carried a processed pseudogene. The sequence lacked all the introns present in the expressed β-tubulin gene. The tubulin sequence in the pseudogene begins in the 5′ untranslated region and ends with a stretch of 17 adenine residues just downstream of the poly(A) addition site. Thus, this sequence appears to be wholly derived from a processed mRNA. None of the untranscribed flanking sequences that occur in the actual β-tubulin gene are present in the corresponding pseudogene. Two nucleotide substitutions in the coding sequence have resulted in in-frame stop codons. The gene is flanked by short direct repeats of the target DNA into which it inserted. Such repeats are characteristic of mobile element insertions (see also Figure 10-7).

Repetitive Elements and Processed Pseudogenes Are Remnants of Transposition Events

Although there is little evidence of active transposable elements in the genomes of animals, cloning and sequencing of animal genes have turned up compelling evidence for transposition events in the genome. As we learned in Chapter 8, animal genomes contain repeated sequence elements dispersed to thousands of locations. These include the LINES and SINES, the best known of which is the *Alu* family. Recall that the *Alu* sequence is closely related to a small RNA molecule called *7SL*. Similar DNA copies of other small RNAs such as tRNAs are found scattered throughout the genome. The most spectacular examples of apparent movement of RNA molecules into genomic DNA are the *processed pseudogenes* (Figure 10-6). These are copies of complete coding genes that precisely lack their introns and the 5′ and 3′ flanking sequences present in the active gene. Many even end with a poly(A) tail. These pseudogenes clearly represent sequences that were processed as RNA molecules before inserting into genomic DNA.

By all appearances, these sequences represent transposition events that involved an RNA intermediate. As we will learn below, transposition via RNA is a common theme in several well-characterized transposons. Because these elements move from RNA to DNA, reversing the normal flow of genetic information, they are sometimes called *retroposons*. The function of retroposons in the genome is far from clear, but many speculate that the presence of mobile genetic elements capable of influencing the pattern of cellular gene expression provides a powerful mechanism for evolution to test novel regulatory strategies for genes (see Chapter 22).

Bacterial Transposons Jump via DNA Intermediates

The bacterial transposon Tn3 does not actually jump at all. Rather it copies itself to a new location, leaving the parental element intact. This was first demonstrated in experiments studying plasmid-to-plasmid

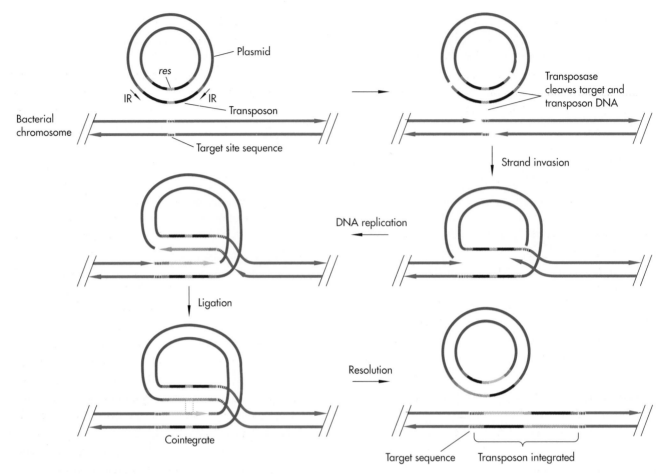

FIGURE **10-7**

Tn3 transposition occurs by a combination of replication and recombination. The diagram shows transposition of Tn3 from a plasmid to the bacterial chromosome. The transposase enzyme initiates transposition by putting single-stranded nicks in the IRs at each end of the transposon. Staggered nicks are also made in the chromosomal DNA, and the free ends of the transposon DNA strands invade the chromosomal molecule. Presumably, this unstable structure is held together by the transposase protein. Note that each strand of the original transposon is now a free single strand. Complementary strands are newly synthesized by DNA polymerase. Ligation of the newly synthesized DNA to the parental molecules yields a cointegrate in which the donor and recipient molecules are covalently joined with a transposon at each junction. Each transposon consists of one strand from the original molecule and one newly synthesized strand. The cointegrate is resolved by recombination to regenerate the donor plasmid molecule and the chromosomal DNA molecule now carrying a transposon insertion. Resolution occurs by alignment of a short DNA sequence *res* in both transposons and recombination catalyzed by the resolvase enzyme encoded by the transposon. Note that transposition also results in a small duplication of the target sequence, into which the transposon is inserted.

transposition: Donor plasmids never lost their transposons. Transposition of Tn3 is a two-step process (Figure 10-7). This was established by isolating Tn3 mutants that got stuck in the intermediate phase. This intermediate structure, termed a *cointegrate,* is basically two full Tn3 elements tandemly repeated. The cointegrate is normally resolved into separate molecules

by recombination between the two elements. Recombination requires the *resolvase* enzyme encoded by the transposon and a target sequence, *res,* in the transposons. Mutants that got stuck in the cointegrate carried mutations in either *res* or the resolvase gene. In the latter case, the mutants could be induced to transpose correctly if resolvase was supplied from another source, whereas mutants lacking *res* never got out of the cointegrate. Together with biochemical studies of the resolvase enzyme, these data suggest that resolvase binds to the two *res* sequences in the adjacent transposons and catalyzes recombination between the elements, generating a new Tn3 insertion in the recipient DNA and leaving the parental element in its original location. Formation of the cointegrate requires activity of the transposase. That replication of the transposon occurs during cointegrate formation was shown by analysis of cointegrates formed in strains carrying two distinguishable forms of Tn3. Cointegrates in these strains always contained two identical elements, indicating that cointegrates are formed by copying of the parental element rather than by joining of two preexisting ones.

In contrast to Tn3, which is replicated to new sites, Tn10 performs a more conventional jump without any replication. This was elegantly demonstrated by constructing two lambda phages carrying Tn10 insertions (Figure 10-8). The Tn10 elements had been engineered to carry a *lacZ* gene encoding β-galactosidase. The only difference between the two phage genomes was three nucleotide substitutions in the *lacZ* gene that inactivated the β-galactosidase enzyme encoded by one Tn10 molecule. The phage genomes were mixed, denatured, and renatured, to form heteroduplex phage molecules consisting of one strand from each parent. The phage molecules carrying heteroduplex transposons were introduced into *E. coli*, and bacterial colonies resistant to tetracycline (also encoded in Tn10) were selected. Because the phages were engineered to neither replicate nor integrate in the bacteria, tetracycline-resistant colonies had to arise by transposition of Tn10 from phage DNA to the bacterial chromosome. If transposition occurred by semiconservative DNA replication as for Tn3, information from only one strand of the heteroduplex transposon would be integrated (the other strand remains at the donor site), and colonies would be uniformly lac⁺ or lac⁻. If, on the other hand, the transposon moved nonreplicatively—cut from the donor site and pasted into the recipient chromosome—then both strands of the heteroduplex would be transferred, and among the resulting colonies would be mixtures of lac⁺ and lac⁻. In this experiment, mixed colonies were recovered, providing strong evidence that transposition of Tn10 is nonreplicative.

Ty Elements Transpose via an RNA Intermediate

We have already learned that Ty elements of yeast are organized differently from bacterial transposons, which transpose as DNA, and that they share a similar organization with integrated retrovirus genomes, which originate as RNA molecules (see Chapter 18). Moreover, sequencing of Ty elements revealed that the open reading frames in Ty elements encode proteins related to retroviral proteins, including the enzyme reverse transcriptase, known to convert retroviral RNA to DNA. But proof that Ty elements transpose through an RNA intermediate came from a series of experiments in which Ty genomes were engineered by recombinant DNA techniques (Figure 10-9). These experiments showed that increasing the rate of transcription of a Ty element increases the rate at which it transposes, and that when an intron is placed in the donor Ty, transposed elements lack the intron. Thus, the newly transposed elements must be DNA copies of spliced RNA molecules transcribed from the donor element. The copia element of *Drosophila* has a very similar structure to the Ty elements and also encodes proteins related to retroviral reverse transcriptase. Like Ty it is transcribed into RNA at a very high level, so it too almost certainly transposes in a similar manner.

As we have mentioned, animal cell genomes also carry sequence elements that appear to have originated from RNA molecules. Unlike Ty and copia, most of these elements do not themselves encode the enzymes able to catalyze their conversion from RNA to DNA and their integration into the genome. Yet such enzymes must have been present in cells in order for these structures to appear. Where the enzymes came from is not really clear, but one possibility is that they were produced by retroviruses.

Mating Type Interconversion in Yeast Occurs by Replicative Transposition

As we will learn in Chapter 13, cells of the yeast *Saccharomyces cerevisiae* can exist either as diploid cells or as haploid cells of two distinct mating types, **a** and α. A single genetic locus, *MAT*, determines whether

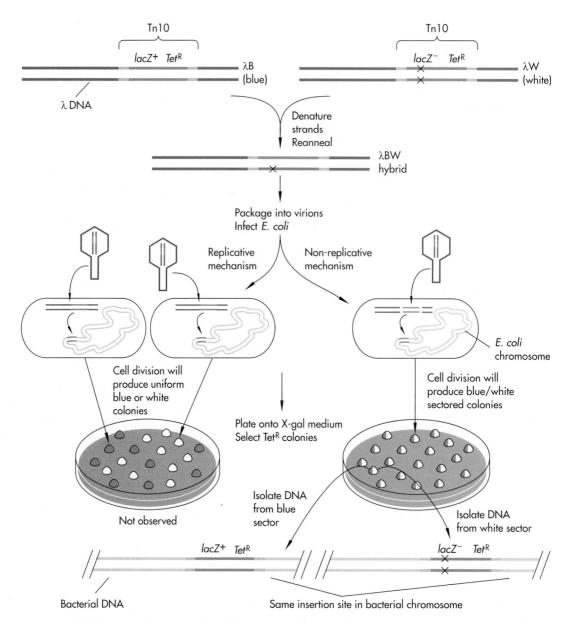

FIGURE **10-8**

Demonstration that movement of Tn10 occurs by a nonreplicative "cut and paste" mechanism. Two lambda phage derivatives were constructed by insertion of a Tn10 molecule carrying a *lacZ* gene (which encodes β-galactosidase). One phage, λBlue, carried a wild-type *lacZ* gene, the other, λWhite, a mutant gene. The two phage DNA molecules were denatured, mixed, and reannealed, yielding heteroduplexes carrying one strand from each parent. The phage DNA was packaged into virion particles and used to infect *E. coli*. The transposon hops to the *E. coli* chromosome, yielding tetracycline-resistant (Tet^R) colonies. The *lac* phenotype of the strains was examined using X-gal, which stains *lac*$^+$ blue. If transposition occurs by a replicative mechanism as seen in Tn3 (see Figure 10-7), only a single strand of the heteroduplex is copied to the chromosome, yielding colonies that are either uniformly blue (if they received the λBlue strand) or white (if they received the λWhite strand). If the transposon is cut from the phage and pasted into the chromosome, both the λBlue and the λWhite strands move to the chromosome. Replication of the chromosome and cell division segregate the two strands, leading to mixed colonies with blue and white cells. This is what happened in the experiment. Comparison of the transposons in blue and white cells from a single colony showed that, as expected, the lac$^+$ and lac$^-$ transposons were integrated at the same location.

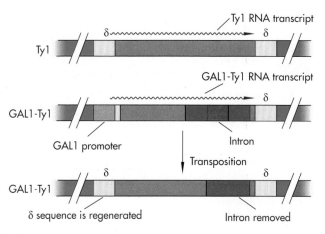

FIGURE **10-9**

Ty1 transposition occurs via an RNA intermediate. An artificial Ty1 derivative was constructed in which the promoter in the left δ was substituted with the galactose-inducible promoter of the yeast *GAL1* gene. Inserted into the body of the element was a fragment of a yeast gene carrying an intron. The engineered Ty was transformed into yeast cells, which were grown on either glucose or galactose. When the cells were grown on galactose, the artificial element was transcribed at a high level. Greatly increased transposition of both this element and endogenous Ty elements in the strain was observed. That endogenous Ty elements also moved shows that the element encodes functions that can act in trans to promote integration of other Ty molecules. Newly transposed copies of the artificial element precisely lacked the intron sequence, showing that a spliced mRNA was the precursor for integration of the new element. Note that the transposed copies also contain an intact left δ sequence. This is because the right δ is duplicated during the transposition process.

a haploid yeast is **a** or α. *MAT* can carry either of two alleles, *MAT***a** or *MAT*α. In homothallic strains of yeast, cells switch their mating type very frequently—cells can switch from **a** to α and back to **a** in consecutive generations. This frequency and reversibility was inconsistent with any known conventional mechanisms of spontaneous mutation and suggested some more unorthodox mechanism for altering the genetic information at the *MAT* locus. A clue came from the discovery of two additional genes flanking the *MAT* locus that carry copies of the *MAT***a** and *MAT*α loci. We now know from cloning and sequencing that these flanking genes, *HML*α and *HMR***a**, serve essentially as storage for the two mating alleles and that information from these loci is transposed to *MAT* when a switch occurs (Figure 10-10). This transposition is similar to

the replicative mechanism of Tn3 in that the sequence at the silent locus is copied into the *MAT* locus concurrently with destruction of the sequence previously at *MAT*. Thus, the *MAT* locus is switched (by transposition) while the donor locus remains unchanged. This mechanism is also termed *gene conversion* because the sequence at the *MAT* locus is converted to that of another gene.

A very similar mechanism is used by a parasite, *Trypanosoma brucei,* in a diabolical strategy to evade destruction by the immune system. The surfaces of these trypanosomes are coated with a highly antigenic glycoprotein (*variant surface glycoprotein,* or *VSG*) that triggers an immune response in the infected host. But, as the name VSG would suggest, the trypanosome is able to switch its glycoprotein to elude destruction by the immune system antibodies, which are blind to each new VSG "disguise." Cloning of a cDNA for the glycoprotein and hybridization of this clone to genomic DNA of the trypanosome showed that, as expected, there was a large family of glycoprotein genes. Yet sequence analysis of these genes (the *basic copies,* or *BCs*) revealed that the sequences at their 3′ ends did not match the sequence found at the 3′ ends of the mRNAs that were actually transcribed from the expressed gene (as found in the cDNA clones). The expressed glycoprotein mRNAs had a common 3′ sequence, which hybridized to a distinct portion of the genome, that could be shown to carry the expressed glycoprotein gene. As with mating type switches in yeast, glycoproteins are switched by replicative transposition of a remote BC sequence to this expression site. Unlike the mating type switch, which can occur at every cell division, glycoprotein switching occurs randomly at a low frequency. Thus, in infected hosts, trypanosome infections occur in waves of destruction and regrowth, with each subsequent wave of trypanosomes carrying a different surface glycoprotein.

Functional Immunoglobulin Genes Are Formed by Ordered Gene Rearrangements

One of the outstanding problems in immunology has been to understand how the immune system is able to generate antibodies to recognize millions of different potential antigens (see Chapter 16). Structural analysis

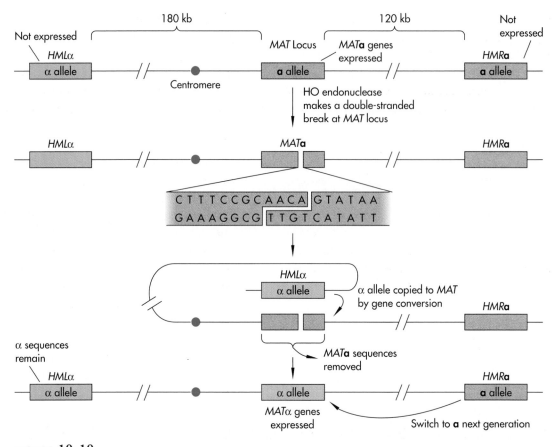

FIGURE **10-10**

Mating type interconversion in yeast. The sequence at the *MAT* locus, either *MAT*a or *MAT*α, determines the haploid mating type of yeast. Unexpressed copies of the *MAT*a and *MAT*α alleles reside at *HMR* and *HML*, respectively, on either side of MAT. Switching is initiated by cleavage of *MAT* DNA by the HO endonuclease. The DNA sequence at *MAT* is degraded and replaced by a copy of one of the unexpressed loci. Transposition occurs by *gene conversion*, a form of replicative transposition (Figure 10-7).

of antibody proteins showed that each B lymphocyte produced a unique antibody molecule. But if each possible antibody molecule were individually encoded in the genome, coding for the immune system alone would account for a large share of the DNA in all of our cells, and limiting expression to one and only one antibody per B cell (*allelic exclusion*) would pose a tricky problem. Antibody genes, because of the interest in this problem and because they are very highly expressed in certain B cells, were among the first genes cloned. And analysis of the structure of antibody genes unveiled a remarkable program of directed gene rearrangements that not only could explain how antibody

diversity was generated, but also showed how B cells practiced allelic exclusion.

Each antibody gene was indeed created from a large family of genes (Figure 10-11). The light-chain locus, for example, contains a single gene encoding the constant region (C) shared by all antibodies and multiple genes encoding different variable regions (V) with distinct sequences. During B-cell development, the genes at the immunoglobulin locus undergo a programmed rearrangement that brings one (and only one) of the V genes to a position immediately upstream of the C gene, generating a full-length coding gene for a functional antibody chain. In one sense, this is akin to mating type interconversion in yeast and glycoprotein switching in trypanosomes—the V genes are stored at a distant location and recruited to the active gene. But as we will learn in more detail in Chapter 16, the molecular mechanism of antibody gene rearrangement is not gene conversion but rather a sloppy cut-and-paste that usually deletes intervening DNA and joins the V and C genes in a somewhat imprecise manner that adds to the diversity of potential antibody sequences. Activated heavy-chain antibody genes also sustain rearrangements during B-cell development that transfer the particular V gene selected to more than one C region. Since it is the C regions that impart differential functions to the various Ig molecules, this *class switching* allows cells to place a single useful recognition specificity on a number of antibody molecules that have different functional properties (see Chapter 16).

Gene Rearrangements Sometimes Go Awry

The purpose of directed gene rearrangements is often to bring silent genetic information to an expressed locus in the genome. Sometimes, however, the wrong gene can be brought in. In many tumors of the immune system, leukemias and lymphomas, this is precisely what happens. As we will learn in Chapter 18, the genome contains a large group of genes, the protooncogenes, that regulate cell proliferation. When these genes are mutated or inappropriately expressed, uncontrolled cell proliferation—cancer—can occur. Chromosomal analysis of B-cell tumors showed that many cells carried chromosomal translocations in which portions of two different chromosomes were fused. Remarkably, these fusions always occurred in

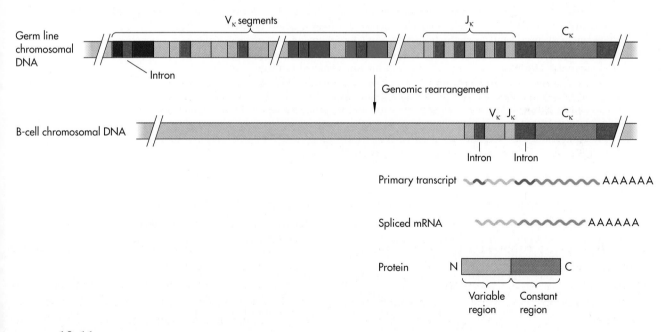

FIGURE **10-11**

Rearrangement of the immunoglobulin (Ig) gene locus to form a functional antibody-coding gene. Antibody molecules consist of a constant region (C) that is shared among all antibodies and a variable region (V) that is unique for each antibody and determines its antigen-recognition specificity. In the κ light-chain gene, about 100 V-region specificities are stored as unexpressed gene segments (V_κ) far upstream of the sequences encoding the constant region (C_κ). A selected V_κ is transposed to a "joining" region (J_κ) immediately upstream of C_κ. This occurs by a cut-and-paste mechanism that joins V_κ to J_κ. There are several J_κ segments in the κ locus, increasing the number of possible junction sequences in the resulting antibody gene. The details of Ig gene rearrangement and expression will be discussed in Chapter 16.

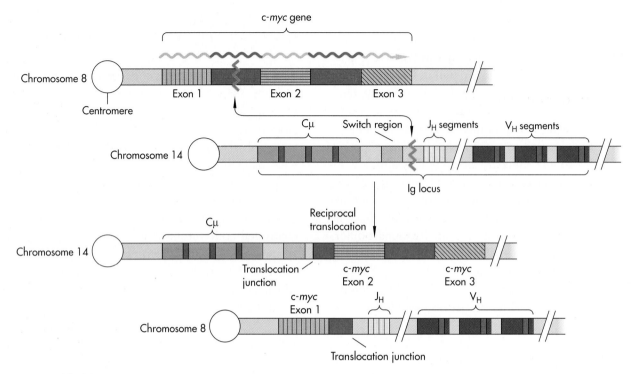

FIGURE **10-12**
Translocation of the c-*myc* oncogene into the immunoglobulin heavy-chain locus. These chromosomal translocations in B-cell tumors are believed to be the result of aberrant Ig gene rearrangement events. In this example, from a Burkitt's lymphoma, a human cancer, the c-*myc* protooncogene on chromosome 8 is translocated to the heavy-chain gene on chromosome 14. In most translocations, c-*myc* coding information, confined to exons 2 and 3, is not affected. Rather, the important consequence is to change the pattern of c-*myc* gene expression, by bringing the coding information into the transcriptionally active Ig locus. In some cases, as shown here, the first non-coding c-*myc* exon and its promoter are also lost. Several variations on Ig–c-*myc* translocations have been observed, involving different Ig genes and different breakpoints in both partners.

the chromosomes that carried the antibody genes. Southern blot hybridization and cloning and sequencing of the antibody genes from these tumors showed that there were indeed physical rearrangements of the antibody genes that joined them to sequences from a different chromosome. And in many tumors, this invading sequence was the same—a protooncogene called c-*myc* (see Chapter 18). The c-*myc* gene is normally turned off in terminally developed B cells. But rearrangement brings the c-*myc* gene under the influence of the powerful transcriptional signals in the immunoglobulin locus, keeping it on and driving proliferation in these cells (Figure 10-12). These tumors and similar ones in T cells, which employ a similar rearrangement strategy for their antigen receptors, appear to be a tax we pay to support the complex machinery that moves DNA for the immune system.

Transposable Elements Are Potent Tools for Identifying and Manipulating Genes

Molecular biologists are quick to exploit advances in biology to further their own experiments. Transposable elements have become powerful tools in the molecular biologist's tool chest. One common use of transposons is in the mapping of cloned genes. The cloned gene is introduced on a plasmid into an *E. coli* strain carrying in its chromosome a transposon with a drug-resistance marker. Occasionally the transposon hops onto the plasmid, and some of these insertions disrupt the cloned gene. Plasmid DNA is recovered and retransformed into a transposon-free strain. Drug-resistant colonies carry plasmids with transposon insertions. These are recovered and tested individually

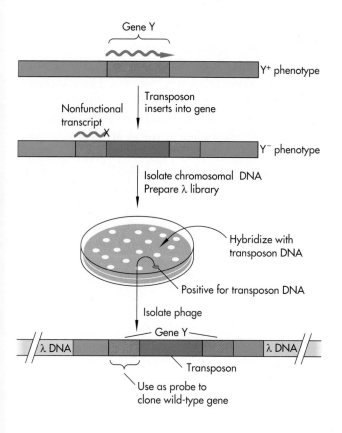

FIGURE **10-13** *(Left)*

Cloning a gene by transposon tagging. Transposons can be used as molecular tags to clone genes for which no hybridization probe exists. In many organisms, mutant genes can be identified that are inactivated by insertion of a natural transposable element. A library is constructed from the genomic DNA of such a mutant organism and screened by hybridization with a transposon DNA probe to identify any phage carrying transposon sequences. Also contained in this clone will be neighboring DNA into which the transposon inserted. Additional genetic manipulations may sometimes be required to separate the tagged gene from other transposons in the genome.

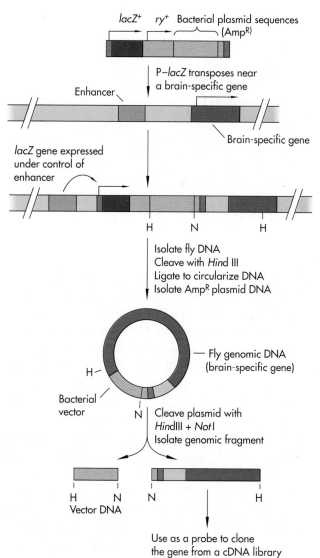

FIGURE **10-14** *(Right)*

Enhancer trapping in *Drosophila* using P element transposition, a variation of transposon tagging. These experiments used a P element derivative carrying an *E. coli lacZ* gene and sequences for propagating the element as a plasmid in *E. coli* (the element also carries a *Drosophila* eye color gene, *ry*, to aid in identification of flies carrying the transposon). To identify *Drosophila* genes expressed in the brain, this element was introduced into flies, where it transposed to many random locations in the fly genome. By appropriate genetic manipulation, it was possible to generate a collection of several hundred fly lines, each with a P-*lacZ* element inserted at a different location. Expression of the *lacZ* gene in these elements can come under the influence of, and adopt the expression pattern of, regulatory elements (i.e., enhancers) in nearby genes. With simple stains for β-galactosidase activity, the flies were examined to find animals that expressed the enzyme only in the brain. To isolate the putative brain-specific gene neighboring the P element, fly DNA was cleaved with restriction enzyme *Hind*III that cuts once in the P element and in the neighboring gene (sites labeled "H"). The resulting fragments were ligated into circles. Among the circles were molecules carrying plasmid sequences from the P element linked to the neighboring fly DNA. These molecules were recovered by transformation into *E. coli*. The fragment of fly DNA in the plasmid was recovered by cleavage with restriction enzymes at sites H and N (*Not*I) and was used as a simple hybridization probe to isolate the brain-specific gene from a library for study.

for gene function. Restriction endonuclease digestion is used to map the position of insertions that inactivate the target gene. Sequencing primers based on the transposon sequence can be used to sequence target DNA adjacent to each insertion.

Once cloned, a natural transposon serves as a mobile sequence tag in the genome of its normal host. And if a transposon insertion occurs within a gene one wishes to clone, then a transposon hybridization probe can be used to isolate the gene from a library (Figure 10-13). This approach could be difficult if there are hundreds of copies of the transposon in the genome, but in many experimental organisms this problem can be solved by repeated crossing of the mutant to a related strain that lacks the transposon. After several crosses, only a few of the transposons remain, one of which is linked to the target gene.

Transposable elements can also be used as vehicles to deliver foreign DNA to organisms. The P element of *Drosophila* has been adapted for this purpose and is now routinely used to put genes into flies (see Chapter 20). A clever adaptation of the P element has been used to place promoterless *E. coli lacZ* genes at many positions in the *Drosophila* genome (Figure 10-14). If these elements land in or near a gene, they can adopt the pattern of expression of the neighboring sequence. Stains for β-galactosidase allow flies with interesting patterns of expression to be readily identified, and the transposon can be used as a tag to clone the neighboring fly gene normally expressed in this pattern.

Reading List

General

Shapiro, J. A., ed. *Mobile Genetic Elements*. Academic, New York, 1983.

Berg, D. E., and M. M. Howe, eds. *Mobile DNA*. American Society for Microbiology, Washington, D. C., 1989.

Original Research Papers

BACTERIAL TRANSPOSONS (Tn3 and Tn10)

Grindley, N. D. F., and R. R. Reed. "Transpositional recombination in prokaryotes." *Ann. Rev. Biochem.*, 54: 863–896 (1985). [Review]

Kleckner, N., R. K. Chan, B.-K. Tye, and D. Botstein. "Mutagenesis by insertion of a drug-resistance element carrying an inverted repetition." *J. Mol. Biol.*, 97: 561–565 (1975).

Gill, R., F. Heffron, G. Dougan, and S. Falkow. "Analysis of sequences transposed by complementation of two classes of transposition-deficient mutants of transposition element Tn3." *J. Bacteriol.*, 136: 742–756 (1978).

Foster, T. J., M. A. Davis, D. E. Roberts, K. Takeshita, and N. Kleckner. "Genetic organization of transposon Tn10." *Cell*, 23: 201–213 (1981).

Reed, R. R., and N. D. F. Grindley. "Transposon-mediated site-specific recombination in vitro: DNA cleavage and protein-DNA linkage at the recombination site." *Cell*, 25: 721–728 (1981).

Bender, J., and N. Kleckner. "Genetic evidence that Tn10 transposes by a nonreplicative mechanism." *Cell*, 45: 801–815 (1986).

DROSOPHILA (COPIA AND P ELEMENTS)

Rio, D. C. "Molecular mechanisms regulating *Drosophila* P element transposition." *Ann. Rev. Gen.,* 24: 543–578 (1990). [Review]

Rubin, G. M., M. G. Kidwell, and P. M. Bingham. "The molecular basis of P-M hybrid dysgenesis: the nature of induced mutations." *Cell,* 29: 987–994 (1982).

Rubin, G. M., and A. C. Spradling. "Genetic transformation of *Drosophila* with transposable element vectors." *Science,* 218: 348–353 (1982).

Mount, S. M., and G. M. Rubin. "Complete nucleotide sequence of the *Drosophila* transposable element copia: homology between copia and retroviral proteins." *Mol. Cell Biol.,* 5: 1630–1638 (1985).

Laski, F. A., D. C. Rio, and G. M. Rubin. "Tissue specificity of *Drosophila* P element transposition is regulated at the level of mRNA splicing." *Cell,* 44: 7–19 (1986).

Wilson, C., R. Kurth-Pearson, H. Bellen, C. J. O'Kane, U. Grossniklaus, and W. J. Gehring. "P-element mediated enhancer detection: an efficient method for isolating and characterizing developmentally regulated genes in *Drosophila.*" *Genes and Devel.,* 3: 1301–1313 (1989).

MAIZE (Ac AND Ds)

Federoff, N. V. "About maize transposable elements and development." *Cell,* 56: 181–191 (1989). [Review]

McClintock, B. "Controlling elements and the gene." *Cold Spring Harbor Symp. Quant. Biol.,* 21: 197–216 (1957).

Fedoroff, N., S. Wessler, and M. Shure. "Isolation of the transposable maize controlling elements Ac and Ds." *Cell,* 35: 243–251 (1983).

Dooner, H., J. English, E. Ralston, and E. Weck. "A single genetic unit specifies two transposition functions in the maize element Activator." *Science,* 234: 210–211 (1986).

YEAST Ty ELEMENTS

Kingsman, A. J., and S. M. Kingsman. "Ty: a retroelement moving forward." *Cell,* 53: 333–335 (1988). [Review]

Cameron, J. R., E. Y. Loh, and R. W. Davis. "Evidence for transposition of dispersed repetitive DNA families in yeast." *Cell,* 16: 739–751 (1979).

Boeke, J., D. J. Garfinkel, C. A. Styles, and G. R. Fink. "Ty elements transpose through an RNA intermediate." *Cell,* 40: 491–500 (1985).

Garfinkel, D., J. Boeke, and G. R. Fink. "Ty element transposition: reverse transcriptase and virus-like particles." *Cell,* 42: 507–517 (1985).

MAMMALIAN RETROPOSONS

Weiner, A. M., P. L. Deininger, and A. Efstradiatis. "Non-viral retroposons: genes, pseudogenes, and transposable elements generated by the reverse flow of genetic information." *Ann. Rev. Biochem.,* 55: 631–661 (1986). [Review]

Haynes, S. R., and W. R. Jelinek. "Low molecular weight RNAs transcribed in vitro by RNA polymerase III from Alu-type dispersed repeats in Chinese hamster DNA are also found in vivo." *Proc. Natl. Acad. Sci. USA,* 78: 6130–6134 (1981).

Lee, M. G.-S., S. A. Lewis, C. D. Wilde, and N. J. Cowan. "Evolutionary history of a multigene family: an expressed human beta-tubulin gene and three processed pseudogenes." *Cell,* 33: 477–487 (1983).

Ullu, E., and C. Tschudi. "Alu sequences are processed 7SL RNA pseudogenes." *Nature,* 312: 171–172 (1984).

YEAST MATING TYPE INTERCONVERSION

Herskowitz, I. "A regulatory hierarchy for cell specialization in yeast." *Nature,* 342: 749–757 (1989). [Review]

Hicks, J. B., J. N. Strathern, and I. Herskowitz. "The cassette model of mating type interconversion." In A. I. Bukhari, J. A. Shapiro, and S. L. Adhya, eds., *DNA Insertion Elements, Plasmids, and Episomes.* Cold Spring Harbor Laboratory, Cold Spring Harbor, N. Y., 1977, pp. 457–462.

Hicks, J. B., J. N. Strathern, and A. J. S. Klar. "Transposable mating-type genes in *Saccharomyces cerevisiae.*" *Nature,* 282: 478–483 (1979).

Strathern, J. N., A. J. S. Klar, J. B. Hicks, J. A. Abraham, J. M. Ivy, K. A. Nasmyth, and C. McGill. "Homothallic switching of yeast mating type cassettes is initiated by a double-stranded cut in the MAT locus." *Cell,* 31: 183–192 (1982).

TRYPANOSOME SURFACE GLYCOPROTEIN SWITCHING

Pays, E., and M. Steinert. "Control of antigen gene expression in African trypanosomes." *Ann. Rev. Gen.,* 22: 107–126 (1988). [Review]

Hoeijmakers, J., A. Frasch, A. Bernards, P. Borst, and G. Cross. "Novel expression-linked copies of the genes for variant surface antigens in trypanosomes." *Nature,* 284: 78–80 (1980).

Bernards, A., L. H. T. V. d. Ploeg, A. C. C. Frasch, P. Borst, J. C. Boothroyd, S. Coleman, and G. A. M. Cross. "Activation of trypanosome surface glycoprotein genes involves a duplication-transposition leading to an altered 3' end." *Cell,* 27: 497–505 (1981).

Thon, G., T. Baltz, C. Giroud, and H. Eisen. "Trypanosome variable surface glycoproteins: composite genes and order of expression." *Genes and Devel.,* 9: 1374–1383 (1990).

IMMUNOGLOBULIN GENE REARRANGEMENTS

Yancopoulos, G., and F. Alt. "The regulation of variable region gene assembly." *Ann. Rev. Immunol.,* 4: 339–368 (1986). [Review]

Hozumi, N., and S. Tonegawa. "Evidence for somatic rearrangement of immunoglobulin genes coding for variable and constant regions." *Proc. Natl. Acad. Sci. USA,* 73: 3628–3632 (1976).

Brack, C., M. Hirama, R. Lenhard-Schuller, and S. Tonegawa. "A complete immunoglobulin gene is created by somatic recombination." *Cell,* 15: 1–14 (1978).

Early, P., H. Huang, M. Davis, K. Calame, and L. Hood. "An immunoglobulin heavy-chain variable region gene is generated from three segments of DNA: V_H, D and J_H." *Cell,* 19: 981–992 (1980).

Kataoka, T., T. Kawakami, N. Takahashi, and T. Honjo. "Rearrangement of immunoglobulin γ1-chain gene and mechanisms for class switching." *Proc. Natl. Acad. Sci. USA,* 77: 919–923 (1980).

Sheng-Ong, G. L. C., E. J. Keath, S. P. Piccoli, and M. D. Cole. "Novel *myc* oncogene RNA from abortive immunoglobulin gene recombination in mouse plasmacytomas." *Cell,* 31: 443–452 (1982).

Taub, R., I. Kirsch, C. Morton, G. Lenoir, D. Swan, S. Tronick, S. Aaronson, and P. Leder. "Translocation of the *c-myc* gene into the immunoglobulin heavy chain locus in human Burkitt lymphoma and murine plasmacytoma cells." *Proc. Natl. Acad. Sci. USA,* 79: 7838–7841 (1982).

11

In Vitro Mutagenesis

Recombinant DNA technology and DNA sequencing provided the tools to clone and characterize genes. As we learned in Chapter 8, simple inspection of gene sequences told us much about genomic organization. Functional sequences, such as transcriptional control elements, could often be identified by comparing sequences of a number of genes. However, to delve deeply into the structure and function of genes required the ability to change the DNA sequence and examine the effect of the change on gene function. For decades before the advent of recombinant DNA, this was done by classical genetics, the identification of mutant organisms with new properties. From the genetic properties of mutants, information about the structure and function of the underlying genes could often be inferred. This approach, however, was limited to organisms in which simple genetic analysis was possible—bacteria, yeast, fruit flies. Genetic analysis of more complex, longer-lived organisms like mice and men was slow and difficult.

Recombinant DNA changed all that. The ability to isolate genes as molecular clones, the development of tools to modify gene sequences in the test tube, and the power to return altered genes to the organism to test their function have revolutionized the way genetics is done in higher organisms. Because we now often work "backwards" from gene sequence to gene function, in contrast to

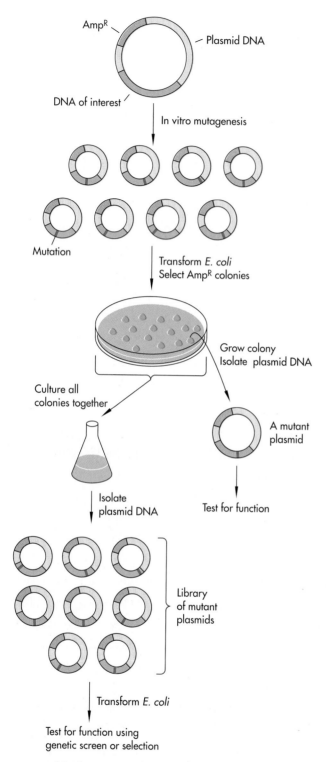

FIGURE **11-1**

General strategy for an in vitro mutagenesis experiment. Most procedures for in vitro mutagenesis follow the same basic scheme: Plasmid DNA is "mutagenized" in vitro, then introduced into *E. coli* by transformation. Depending on the method, mutant clones can be isolated and tested individually, or a library of mutant plasmids can be obtained, which are tested using a genetic screen.

classical genetics, this new approach spawned by recombinant DNA is called *reverse genetics.* In this chapter we will learn ways to alter the sequence of a cloned gene at will and how these methods are used to understand the structure and function of genes and gene products.

In Vitro Mutagenesis Is Used to Study Gene Function

In vitro mutagenesis of cloned genes has become a standard tool in the functional analysis of nucleic acids and proteins. Most procedures follow the same basic scheme (Figure 11-1). Plasmid DNA containing the gene of interest is treated in vitro by some mutagenesis procedure that alters the DNA either chemically or enzymatically. The mutagenized plasmid DNA is introduced into *E. coli* by transformation, and colonies containing plasmid molecules are selected by antibiotic resistance. Mutants can be made one at a time, or hundreds of different mutants can be created in a single mutagenesis experiment. Mutant plasmids can be isolated from single colonies and tested individually. Alternatively, plasmid DNA can be prepared from pooled colonies and the resulting library tested en masse to identify mutant plasmids.

The various approaches to mutagenesis can be grouped broadly into random and site-directed methods. Random methods put mutations anywhere in a plasmid. They are best used to identify the location and boundaries of a particular function within a cloned DNA fragment and are most readily used for this purpose when a simple genetic screen (or selection) is available. A genetic screen or selection consists of a system to test the function of the DNA of interest in cells without having to isolate each plasmid individually. Random mutagenesis is often used as a first step, when little is known about the function encoded by particular DNA fragment. Analysis of random mutants generally provides only a simple identification of the functional region but does not explain how things work on a molecular level. The value of such a strategy is that it quickly helps to narrow down the focus of attention from a large DNA fragment to a smaller region that can be studied subsequently in greater detail. As we will learn, random mutagenesis can be accomplished by several different methods, such as altering the sequences within restriction endonuclease sites, inserting an oligonucleotide linker randomly into a plasmid, damaging plasmid DNA in vitro with chemicals, or incorporating incorrect nucleotides during in vitro DNA synthesis.

Once an important functional domain in a gene has been identified by random mutagenesis, site-directed methods—putting mutations precisely where they are needed—are used to define the role of specific sequences. In addition, directed mutagenesis provides a powerful tool for the analysis of protein function, by allowing researchers to make specific and subtle changes in the structure of the protein. A number of strategies have been developed to construct site-directed mutants in vitro, but this type of mutagenesis is best accomplished using synthetic oligonucleotides. With an oligonucleotide the desired sequence is simply built into the wild-type framework. Nowadays, oligonucleotide-directed mutagenesis reactions are relatively straightforward, and oligonucleotides are cheap and easy to obtain. The limitation of site-directed mutagenesis is that you must already have enough information to know what you wish to change. There are two standard ways of using oligonucleotides to construct site-directed mutants: mutagenesis by gene synthesis and mutagenesis by enzymatic extension of a mutagenic oligonucleotide. By using degenerate oligonucleotides (see Chapter 7) a set of "random" mutations at a specific site can also be made.

Restriction Endonuclease Sites Provide the Simplest Access for Mutagenesis

One of the first experiments done with a cloned DNA fragment is to map the positions of restriction endonuclease cleavage sites in the DNA by using a battery of different enzymes. Although this information could be precisely obtained from the DNA sequence, mapping restriction sites can be accomplished rapidly and is often done in conjunction with sequencing. Restriction endonuclease recognition sites provide the simplest way to modify a DNA clone in vitro (Figure 11-2). Cleaving plasmid DNA with a restriction enzyme that recognizes only one site produces a linear molecule. This serves as an entry point for modifying the DNA sequence in the vicinity of the restriction site. For example, the enzyme EcoRI recognizes the sequence GAATTC and produces ends with 5′ overhangs. The ends can be made even (blunt) by treating

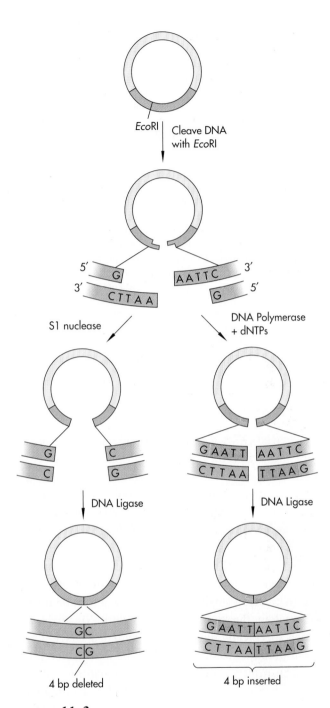

FIGURE **11-2**

Creating a mutation by manipulation of a restriction site. Plasmid DNA is cleaved with EcoRI restriction endonuclease, which generates a linear fragment with 5′ ends that have four unpaired nucleotides (so-called sticky ends). Treatment with S1 nuclease (left) removes these nucleotides, and the linear fragment is then treated with DNA ligase. The resulting circular molecule contains a deletion of 4 bp. Alternatively, addition of DNA polymerase and deoxyribonucleotide triphosphates (dNTPs) to the plasmid cleaved by EcoRI extends the 3′ ends by DNA synthesis (right). After ligation, the resulting molecule contains an insertion of 4 bp. In both cases, the EcoRI site has been destroyed.

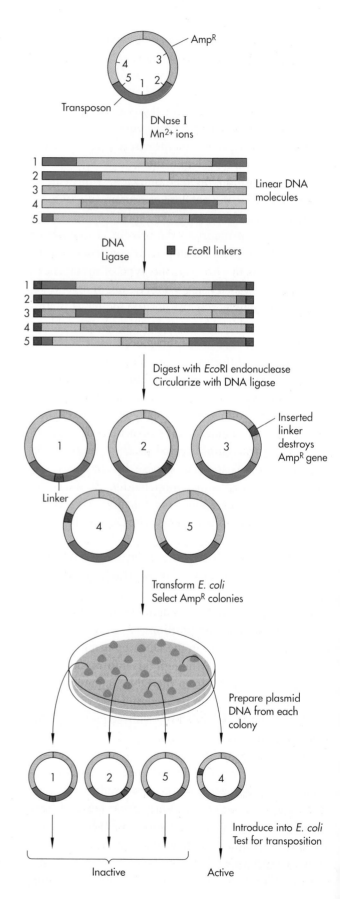

FIGURE **11-3**
Linker insertion mutagenesis to map functional domains of a bacterial transposable element. The starting plasmid contains an intact transposon, an ampicillin-resistance gene for selection in *E. coli*, and sequences for plasmid replication. The DNA is treated with a low concentration of deoxyribonuclease I in the presence of Mn^{2+}. Under these conditions, the enzyme makes double-stranded cuts at random positions in the plasmid, generating a collection of linear DNA molecules broken at different positions. Oligonucleotide linkers encoding an *Eco*RI restriction site are added to the ends with DNA ligase, the linear molecules are treated with *Eco*RI endonuclease to create sticky ends on the linkers, and the molecules are recircularized. The circular molecules are transformed into *E. coli*, and ampicillin-resistant colonies are selected. Plasmid DNA is isolated from individual colonies, introduced into another strain of *E. coli*, and tested for activity of the transposon. The positions of the inserted linkers are mapped by restriction digestion. Linkers inserted in one region (blue) of the plasmid inactivated the transposon. No linker insertions in the ampicillin-resistance gene were recovered, because these plasmids would fail to yield a drug-resistant colony in the original selection of transformed *E. coli*.

the cleaved DNA with DNA polymerase in the presence deoxyribonucleotide triphosphates. The two blunt ends can then be linked together again (ligated) by incubating the linear plasmid molecule with DNA ligase. A few nanograms of DNA from the in vitro ligation reaction is used to transform *E. coli*, and the new modified plasmid is isolated from one of the resulting colonies. The net result of these manipulations is to insert 4 bp into the plasmid at the *Eco*RI site. Alternatively, a small deletion mutation can be made by treating the linearized DNA with S1 nuclease, which specifically digests single-stranded DNA. This creates blunt ends by removal of the four nucleotides that constitute the 5′ overhang generated by *Eco*RI at each end. Subsequent ligation of the DNA into a covalently closed circular molecule thus results in the deletion of 4 bp from the DNA. In each example, the new sequence no longer encodes the *Eco*RI recognition site. These types of manipulations, if done to a protein-coding sequence, would change the translational reading frame, resulting in production of a grossly altered protein. The major limitation of using restriction sites to make mutations is that there simply may not be sites in regions of the gene the experimenter wishes to alter.

Linker Insertion Is Used to Map a Bacterial Transposon

We have learned that it is a simple matter to cleave a plasmid with a restriction enzyme, blunt the ends by treatment with a DNA polymerase, and rejoin them by ligation. A variation on this technique is to rejoin the ends in the presence of a synthetic oligonucleotide "linker," often one that encodes a restriction site. Insertion of the linker disrupts the gene sequence; the position of the inserted linker can be easily mapped by cleavage of the plasmid with the restriction enzyme that cuts the linker.

A similar method was used to define the functional regions of a bacterial transposable element (a "jumping gene," see Chapter 10), by inserting linkers at many alternative positions throughout the element. To place linker insertions in the transposon, a plasmid carrying a clone of the transposon was treated with a nuclease that cleaved the plasmids at random positions (Figure 11-3). Cleavage conditions were adjusted so that each plasmid was cut just once on average. The linearized molecules were isolated and ligated into circles again in the presence of an 8-bp linker oligonucleotide containing an EcoRI restriction site, resulting in insertion of the linkers into random sites, one in each plasmid. The resulting plasmids were transformed into E. coli and, using a genetic screen, examined to see if the transposon could jump. Insertion of a linker into a region of the transposon critical for its function inactivates it, presumably by putting a protein-coding sequence out of frame. By mapping the positions of the inserted linkers by restriction analysis, the locations of functional regions of the transposon were deduced.

Construction of Nested Deletions Maps the Boundaries of a Transcriptional Control Region

Transcription of the gene encoding the 5S ribosomal RNA molecule is carried out by RNA polymerase III (pol III, see Chapter 8). To identify the sequences within the 5S gene required for transcription by pol III, a series of deletion mutations was made and tested for their ability to support accurate transcription. Two sets of deletions were made. One was made by cutting a plasmid carrying a cloned 5S gene at a restriction site on the 5' side of the gene. The linearized plasmid was treated with a combination of nucleases that digested away DNA from the ends of the molecule (Figure 11-4). The amount of DNA removed was controlled by varying the time, temperature, or enzyme concentration in the reaction. A second set of deletions was generated from plasmid DNA cleaved at a site on the 3' side of the gene. The result was two sets of plasmids with progressively larger deletions toward the gene from both directions. Testing these genes revealed that only deletions entering a 35-bp region within the transcribed region of the 5S gene abolished transcription by pol III. Therefore, this deletion analysis mapped the transcriptional regulatory element to this 35-bp stretch, which has subsequently been analyzed in much greater detail by site-directed mutagenesis.

Several different types of enzymes can be used to produce deletions. Generally, these enzymes delete DNA from both ends of a linearized plasmid molecule. Often, however, one end of the molecule contains sequences that need to be retained in the plasmid because, for example, they are required for plasmid replication. In the 5S gene deletion experiment, this limitation was accommodated by isolating the deleted gene fragments and recloning them into a new vector. Alternatively, a strategy can be used that limits deletion to one end of a linearized plasmid molecule (Figure 11-5). This method is widely used to generate nested deletions for DNA sequencing (see Chapter 7).

Linker-Scanning Mutagenesis Permits Systematic Analysis of Promoters

Deletion mutagenesis of the 5S gene mapped the boundaries of the transcriptional control region in the gene. But not all the nucleotides within the boundaries of that 35-bp region are necessarily critical for function. Therefore, methods were needed to change individual nucleotides in a target without generating gross deletions or other rearrangements. This was accomplished for a viral promoter using an elegant ad-

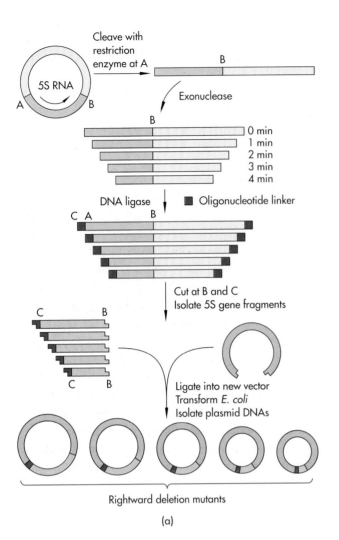

FIGURE **11-4**

Construction of a nested set of deletion mutants to map the transcription control region of a 5S ribosomal RNA gene. (a) A plasmid clone was linearized with a restriction enzyme at a position (A) on the 5′ side of the gene. The linear fragments were treated with an exonuclease, which digests DNA from both ends of the molecule. Portions of the reaction were removed at different times to recover populations of molecules with progressively larger deletions. Linkers were added to the ends, and the molecules were cleaved with restriction enzymes specific for sites B and C to separate the 5S gene fragments from the remnants of the vector. The fragments were recloned into a new vector, generating the set of rightward deletion mutants. To create the leftward deletion mutants, this process was repeated after cleaving the plasmid at restriction site B. (b) Individual plasmids were isolated after transformation, their deletion endpoints determined by DNA sequencing, and their ability to support transcription by RNA polymerase III tested with an in vitro assay. As can be seen by comparing transcription activity with the extent of deletion, transcription is inhibited when the rightward (5′) deletions enter the +40 region and when the leftward (3′) deletions pass the +80 point. This suggests that the transcription control region lies between +40 and +80.

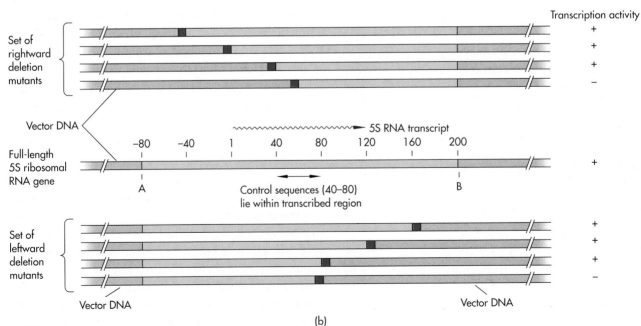

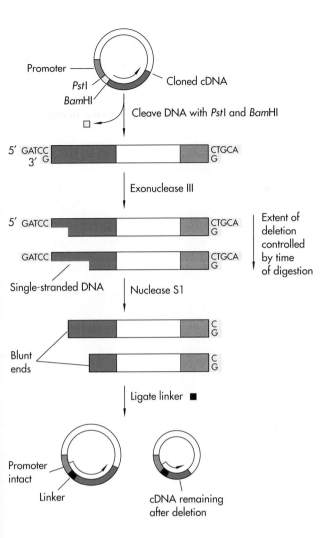

FIGURE **11-5**
Construction of unidirectional deletions using exonuclease III. Exonuclease III attacks preferentially the 3' end of a linear DNA molecule with 5' protruding nucleotides. Therefore, by cleaving a plasmid molecule at adjacent sites with *Bam*HI, which leaves a 5' overhang, and *Pst*I, which leaves a 3' overhang, only the end generated by *Bam*HI is attacked by exonuclease III. After exonuclease III treatment, the remaining single-stranded tail (along with the overhang at the other end) is removed with S1 nuclease, which digests only single-stranded DNA. An oligonucleotide linker is attached, and the fragments are ligated to form closed circular molecules. In the experiment shown here, deletions are being used to map the functional domains of a cloned gene inserted in an expression vector. This strategy allows deletions to be made only in the cloned gene, without damaging the promoter sequence.

aptation of deletion mutagenesis called *linker scanning.* Using the methods outlined in Figure 11-4, two sets of plasmids were constructed that contained deletions within the promoter. One set of deletions started from a site beyond the 5' end and proceeded toward the gene, leaving the 3' end intact; the other set started at a point within the gene and proceeded in the opposite direction, leaving the 5' end intact. Each deletion terminated with a 10-bp *Bam*HI linker. The extent of the deletion in the DNA was determined for each plasmid by DNA sequencing. Pairs of plasmids from the two deletion sets with endpoints 10 bp apart were recombined at their *Bam*HI sites (Figure 11-6). The effect was to preserve the length and organization of the promoter—thought to be important for promoter function—but to replace various 10-bp segments of wild-type promoter sequence with the sequence in the linker. Thus, this experiment created a library of promoter mutants of similar structure but with nucleotide substitutions clustered within 10-bp windows located at various sites in the promoter. This collection of mutants spanned the length of the promoter. The results of this analysis were discussed in Chapter 9. At the time, this experiment represented the most thorough analysis of a promoter in a mammalian gene.

Random Nucleotide Substitutions Are Obtained by Chemical Modification of DNA or by Enzymatic Misincorporation

While linker scanning allows the creation of nucleotide substitutions, each mutant generally contains several substitutions, and the positions of the mutations depend on the availability of appropriately placed deletions. Therefore, several strategies have been developed for placing *single* nucleotide substitutions at random positions in a DNA molecule. The simplest methods employ chemicals that modify or damage DNA. Generally, plasmid DNA or DNA fragments are treated with chemicals, transformed into *E. coli*, and propagated as a library of mutant plasmids. Chemicals most commonly used for in vitro mutagenesis include sodium bisulfite, which deaminates cytosine residues to uracil, and reagents that damage or remove

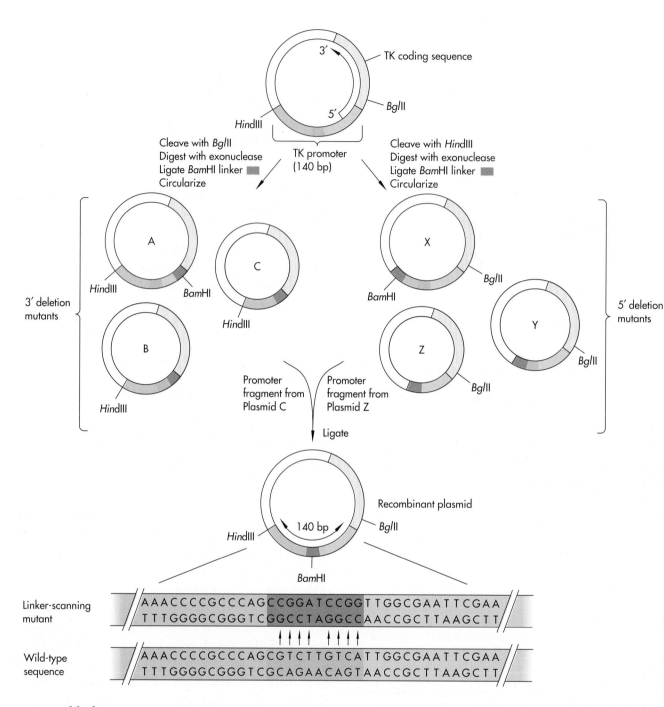

FIGURE **11-6**

Linker-scanning mutagenesis of the viral promoter for the thymidine kinase (TK) gene. Two sets of deletion mutants were made beginning from restriction sites on the 5′ and 3′ sides of the promoter by the method described in Figure 11-4. (The promoter is divided into three colors to make the extent of deletion more obvious.) Approximately one hundred plasmids were sequenced to determine their deletion endpoints. Pairs of deletion fragments, where the 5′ deletion of one fragment ended precisely 10 bp downstream from the endpoint of the 3′ deletion of the other fragment, were identified. The *Hin*dIII–*Bam*HI fragment of the 3′ deletion mutant and the *Bam*HI–*Bgl*II fragment of the 5′ deletion mutant were joined via their *Bam*HI sticky ends and cloned into a new plasmid. This strategy yields molecules like the one shown at the bottom: they are wild-type in sequence except for the substitution of the 10-bp *Bam*HI linker in place of the sequence between the two deletion endpoints. In the example shown, this results in a cluster of eight nucleotide substitutions (arrows).

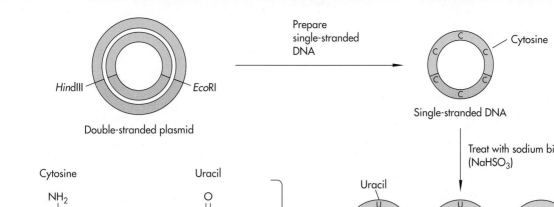

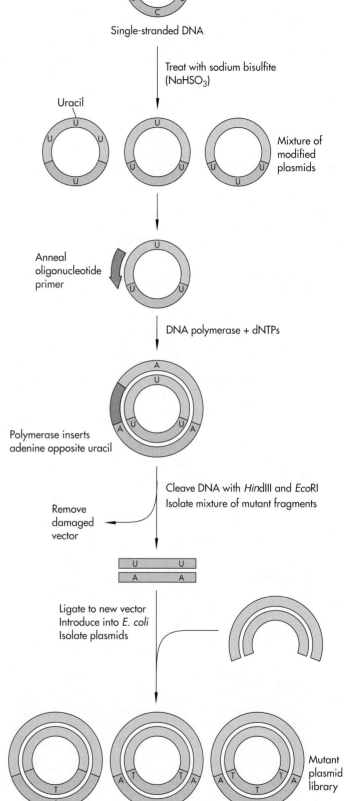

FIGURE **11-7**

Chemical mutagenesis using sodium bisulfite. Sodium bisulfite reacts with cytosine bases of single-stranded DNA to convert them to uracil, a thymine analog that base-pairs with adenine. Single-stranded DNA is treated with sodium bisulfite to modify a small number of cytosine residues in each molecule. An oligonucleotide primer is annealed to the DNA and serves as a primer for synthesis by DNA polymerase. When the polymerase encounters a uracil in the template strand, it incorporates an adenine into the newly synthesized DNA. Since the vector sequences are also damaged by bisulfite treatment, it is necessary to excise the double-stranded DNA fragment by restriction endonuclease cleavage and reclone it into an undamaged vector. Following transformation into *E. coli*, a library of mutant plasmids can be isolated or individual plasmids can be purified and tested. The average number of substitutions in the DNA fragment can be controlled by altering the conditions of bisulfite treatment.

bases, thereby preventing normal Watson-Crick base-pairing (these include hydrazine and formic acid, which are used in Gilbert-Maxam DNA sequencing, Chapter 5). Most often, chemical mutagenesis is performed on single-stranded DNA and followed by in vitro synthesis of the complementary strand using a DNA polymerase (Figure 11-7). This synthesis incorporates the mutation into the new strand. In DNA treated with bisulfite, an adenine nucleotide is incorporated opposite the uracil; after transformation into

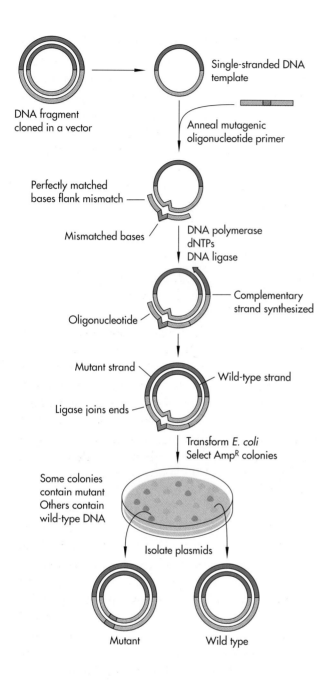

DNA fragment
cloned in a vector

Single-stranded DNA
template

Anneal mutagenic
oligonucleotide primer

Perfectly matched
bases flank mismatch

Mismatched bases

DNA polymerase
dNTPs
DNA ligase

Oligonucleotide

Complementary
strand synthesized

Mutant strand

Wild-type strand

Ligase joins ends

Transform *E. coli*
Select Amp^R colonies

Some colonies
contain mutant
Others contain
wild-type DNA

Isolate plasmids

Mutant

Wild type

FIGURE **11-8**
Oligonucleotide-directed mutagenesis by enzymatic primer extension. A "mutagenic" oligonucleotide encoding the desired mutation embedded in wild-type flanking sequence is annealed to a single-stranded DNA template. The sequence of the oligonucleotide is complementary to the template except for the nucleotides that define the mutation. Generally, the mutagenic oligomer is designed so that the mismatched nucleotides are positioned in the middle and there are at least 8 to 12 nucleotides on either side that base-pair with the template DNA. The mutagenic oligonucleotide serves as a primer for DNA synthesis by DNA polymerase. Once the entire template has been copied, the ends of the newly synthesized strand are covalently linked by DNA ligase. The heteroduplex DNA is transformed into *E. coli*. Theoretically, both strands can replicate, segregating into separate mutant and wild-type plasmids. In practice, however, most colonies contain only one or the other, because enzymes in the cell recognize and repair mismatched nucleotides in the heteroduplex before replication. Plasmid DNA is isolated from the resulting colonies and is screened to identify mutants.

E. coli, the wild-type C·G base pair becomes a T·A pair. In DNA treated with reagents that eliminate bases, any nucleotide can be incorporated opposite the "abasic" site, which still retains its deoxyribose backbone although it has lost its base. The major limitation of chemical mutagenesis is the specificity of the individual reagents: bisulfite mutagenesis, for example, changes only cytosines.

All possible nucleotide substitutions can be generated using enzymatic misincorporation. Here the

strategy is to perform in vitro DNA synthesis under nonideal conditions—suboptimal ionic conditions, unbalanced concentrations of nucleotide precursors—that encourage DNA polymerase occasionally to incorporate the wrong nucleotide during synthesis. For example, synthesis is carried out in the presence of high concentrations of three of the precursors and a very low concentration of the fourth. At positions that normally call for the fourth (scarce) nucleotide, one of the others is sometimes incorporated instead. These methods also exploit DNA polymerases that lack a proofreading activity—a 3′ to 5′ exonuclease mechanism that checks each base pair after incorporation and removes nucleotides that are mismatched. *Thermus aquaticus (Taq)* DNA polymerase, used in the polymerase chain reaction (Chapter 6), lacks such an activity. Though this is a problem when accuracy of synthesis is required, the PCR is a very simple and efficient way to introduce random nucleotide substitutions into a DNA fragment.

A general problem with random mutagenesis approaches is that they often produce mutants with more than one substitution. Multiple substitutions in a single mutant complicate the interpretation of an experiment, because it isn't clear which substitution (or which combination of substitutions) is responsible for observed changes in the properties of the mutant. Extraordinary methods have been used to circumvent this problem—essentially, significantly reducing the extent of mutagenesis and using enrichment protocols to find rare mutants—but almost all these procedures have been supplanted by new methods that use synthetic oligonucleotides.

Synthetic Oligonucleotides Facilitate Mutagenesis

Most of the methods for mutagenesis we have discussed so far have some significant shortcoming—they rely on fortuitous access to a sequence via a restriction site, forced entry through deletion strategies, or tedious screens to find randomly generated mutations in the region of interest. To be most powerful, mutagenesis must allow the experimenter to place *any* modification at *any* position desired in cloned DNA. This has become not only possible, but simple and cheap, with the advent of synthetic DNA oligonucleotides. Oligonucleotides provide the means to design a particular mutation and then to place it precisely where you want it.

The simplest method for doing oligonucleotide-directed mutagenesis is by enzymatic primer extension (Figure 11-8). In this method, an oligonucleotide is designed that carries the mutation flanked by 10 to 15 nucleotides of wild-type sequence. This "mutagenic" oligonucleotide is hybridized to its complementary sequence in single-stranded wild-type DNA prepared from a phage or phagemid clone, forming a heteroduplex with mismatched nucleotides at the site of the mutation. Although the oligonucleotide is not perfectly complementary, it will anneal if the hybridization conditions are not very stringent. The oligonucleotide serves as a primer for in vitro enzymatic DNA synthesis by a DNA polymerase that converts the single-stranded DNA into double-stranded form, using the wild-type strand as template. In this way, all regions of the plasmid except the region containing the mutagenic oligonucleotide will be wild-type in sequence. Once the primer has been extended completely around the template, the ends of the newly synthesized strand are ligated, forming a double-

stranded circular DNA molecule. This heteroduplex DNA—one strand has the wild-type sequence and the other strand has the mutant sequence—is transformed into *E. coli*, where either strand can be replicated. By the time a colony grows up, however, it usually contains only one type of plasmid, wild-type or mutant. The types of mutations that can be made by this approach range from single nucleotide substitutions to deletions or insertions, limited only by the size of the oligonucleotide needed.

Mutant Clones Can Be Identified by Hybridization and DNA Sequencing

Theoretically, half the daughter molecules of a mutagenesis reaction will be wild-type and half mutant. In practice, however, the percentage of mutant plasmids is often much lower. This is due to a variety of technical factors, but the consequence is that methods for identifying or enriching mutant clones are vital. Mutant molecules can be distinguished from wild-type if there is gain or loss of a restriction site. Alternatively, the oligonucleotide that was originally used to make the mutation can be used as a hybridization probe to distinguish mutant from wild-type molecules (Figure 11-9). The mutagenic oligonucleotide is radioactively labeled with ^{32}P-ATP and hybridized to DNA from bacterial colonies on nitrocellulose filters, as described in Chapter 7. If the temperature of the hybridization is raised in 5 or $10°C$ increments, a point can usually be reached at which the labeled oligonucleotide will hybridize only to the mutant molecules (to which it is perfectly complementary) and not to the wild-type molecules, because the hybrid is destabilized by the mismatched nucleotides. Plasmid DNA is isolated from an *E. coli* colony that strongly hybridizes to the probe. Verification that the desired mutation was made is accomplished by sequencing the DNA of this putative mutant clone. This technique can identify one mutant clone among several hundred wild-type clones.

Several clever methods enrich for mutant clones so that the tedious task of screening by hybridization is not necessary. In one of these techniques, the template DNA is biologically marked so that it is destroyed after transformation into *E. coli* and the mutant strand

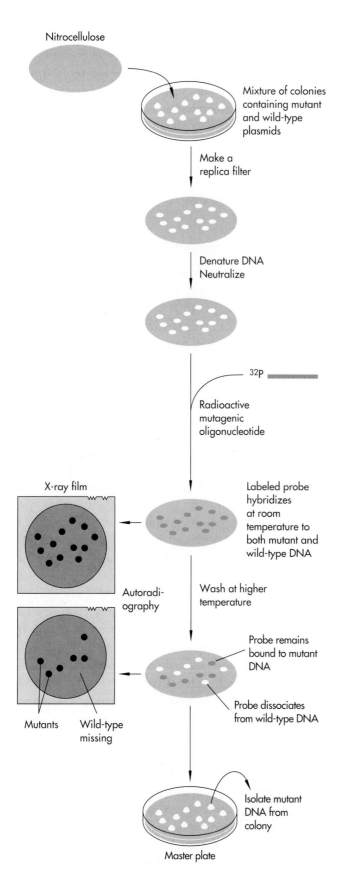

FIGURE **11-9**
Searching for mutant plasmids using the mutagenic oligonucleotide as a probe. Colonies (or plaques) resulting from
transformation by mutagenized plasmids (see Figure 11-8)
are prepared for colony hybridization on nitrocellulose filters using methods described in Chapter 7. The mutagenic
oligonucleotide is radioactively labeled by phosphorylating
its 5′ end using ^{32}P-ATP and polynucleotide kinase. The labeled oligonucleotide is hybridized to the plasmid DNA on
the nitrocellulose filters. At low temperature, the oligonucleotide will hybridize to both mutant and wild-type DNAs.
As the temperature is increased, the mismatched oligonucleotide hybridized to the wild-type plasmid DNA begins to
dissociate from the wild-type clones. Eventually a temperature is reached at which the mismatched oligomers completely dissociate from the wild-type clones but remain
hybridized to the mutants. Since the oligonucleotide is radioactively labeled, the nitrocellulose filter is exposed to
x-ray film and mutant clones are identified by the presence
of a strong signal on the autoradiograph. Mutant plasmid
DNA is then isolated from the corresponding colony on the
master plate, using the replica filter as a guide.

is preferentially replicated (Figure 11-10). In a second
method, the template strand is enzymatically destroyed before transformation. Both methods can yield
mutants at a frequency of greater than 50 percent, so
that plasmid DNA is simply isolated from three or
four randomly picked colonies and analyzed by DNA
sequencing with the expectation that a mutant will be
found among the DNA selected.

Oligonucleotide Cassettes Provide a Simple Method for Introducing Directed Mutations

We learned earlier that restriction enzyme sites provide access to a cloned DNA for mutagenesis. If two
restriction sites are close together, the intervening
fragment can be removed and replaced with a synthetic
double-stranded fragment (a *cassette*) made from two
complementary single-stranded oligonucleotides carrying any desired sequence. Often, however, convenient restriction sites are not available; fortunately, it
is a simple matter to create them using the oligonucleotide-directed mutagenesis procedures described in
the previous sections. Once the sites are in place, any

FIGURE **11-10**

Enrichment for oligonucleotide-directed mutants by using a uracil-containing template. Single-stranded template DNA is prepared in a strain of *E. coli* that lacks the enzyme uracil deglycosidase (ung⁻), so that it contains several uracil residues in place of thymines. (Although uracil is not usually incorporated into DNA, it is not actually mutagenic and it does form a base-pair with adenine.) The mutagenic oligonucleotide is annealed and primes the synthesis of a strand that extends around the template in a reaction using the four standard dNTPs (as in Figure 11-8). Following ligation, the heteroduplex DNA molecules are introduced into an ung⁺ strain of *E. coli*. Once in the cell, the wild-type (template) strand is attacked by uracil deglycosidase, which causes breaks in the DNA strand, and the DNA strand is degraded before it can be replicated. Since the strand containing the mutagenic oligonucleotide does not contain uracil, it is not attacked and is replicated normally. When this procedure is used, 50 percent or more of colonies contain mutant plasmids.

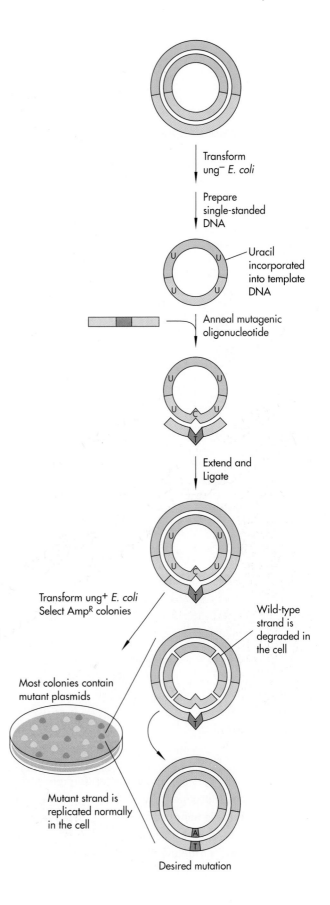

Transform ung⁻ *E. coli*

Prepare single-standed DNA

Uracil incorporated into template DNA

Anneal mutagenic oligonucleotide

Extend and Ligate

Transform ung⁺ *E. coli* Select Ampᴿ colonies

Wild-type strand is degraded in the cell

Most colonies contain mutant plasmids

Mutant strand is replicated normally in the cell

Desired mutation

number of new mutants can be made by inserting synthetic fragments into the plasmid (Figure 11-11), just as different cassettes can be inserted into a tape player.

This method of cassette mutagenesis was the basis for an elegant experiment that verified a structural model for DNA recognition by phage repressors. The repressors of the λ-like phages 434 and P22 contain a helix-turn-helix structure (see Chapter 9) that recognizes the operator DNA in the phage genome. It was hypothesized that amino acid side chains on one face of an α helix in the repressor protein make sequence-specific contacts with operator DNA. To test this hypothesis, a *helix swap* was performed (Figure 11-12). Oligonucleotides were synthesized that encoded the amino acids of the helix in the 434 repressor, with the five positions thought to contact DNA changed to those found in the P22 repressor. This synthetic fragment was swapped for the natural fragment in the 434 gene. The resulting hybrid protein gained the recognition specificity of the P22 repressor, demonstrating that this helix indeed contacts the DNA.

Cassette mutagenesis with degenerate oligonucleotides can be used to create a large collection of random mutations in a single experiment. This method was

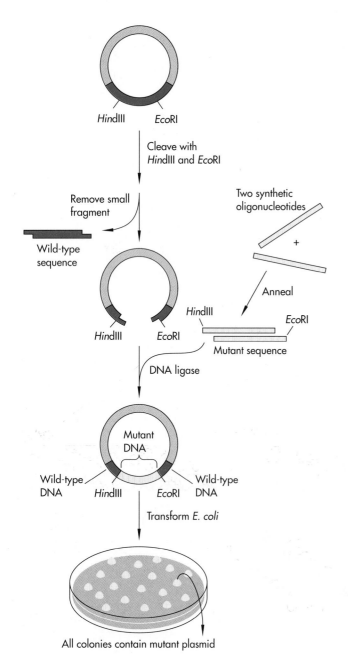

FIGURE 11-11
Mutagenesis by cassette replacement. Plasmid DNA is cleaved with restriction enzymes *Eco*RI and *Hin*dIII, which cut at sites that flank the sequence to be mutated. The small cleaved DNA fragment containing a portion of the wild-type sequence is removed, and a DNA fragment (cassette) containing the desired mutation is ligated into the plasmid. This mutant DNA fragment is composed of two complementary synthetic oligonucleotides that have *Eco*RI and *Hin*dIII sticky ends when annealed. Because there is no heteroduplex intermediate—the mutant cassette is simply swapped for the wild-type fragment—the recombinant plasmids are all mutants. A mutant cassette can be composed of degenerate oligonucleotides (see Chapter 7), resulting in a library of mutant plasmids containing different sequences.

GRE function precisely, single point mutations throughout the 30-bp region were generated and tested in cells for inducibility by glucocorticoid hormone. Two complementary oligonucleotides were synthesized that carried the 30-bp GRE, but synthesis was performed under conditions in which incorrect nucleotides were incorporated at a low frequency (Figure 11-13). These "doped" oligonucleotides (that is, oligonucleotides produced by doping; see Figure 11-13) were annealed and inserted as a cassette into a promoter that lacked a GRE. Using this method, most single-nucleotide substitutions at the 30 positions were obtained. Such a collection of mutants would have been unthinkable before oligonucleotides revolutionized in vitro mutagenesis.

Gene Synthesis Facilitates Production of Normal and Mutant Proteins

The oligonucleotide-directed mutagenesis methods we have described use a single oligonucleotide or a pair of complementary oligonucleotides to insert mutant sequences into an otherwise natural DNA fragment. With the increasing availability of longer oligonucleotides, it is now feasible to assemble an entire gene from synthetic units. This is done by synthesizing a set of oligonucleotides, typically 40 to 80 nucleotides in length, that can be annealed and ligated in vitro to assemble an entire double-stranded DNA

used to study the structure of the glucocorticoid response element (GRE), an enhancer sequence that activates a family of genes in response to certain steroid hormones. The element had been mapped by deletion mutagenesis to a 30-bp region in a glucocorticoid-regulated gene. To define the sequence required for

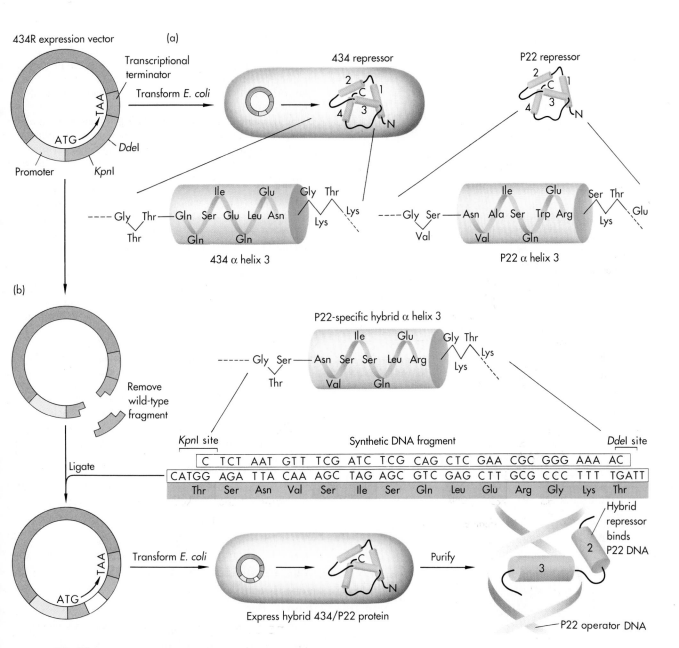

FIGURE **11-12**
The helix swap experiment. Amino acids in the phage 434 repressor protein believed responsible for recognition of the 434 operator were changed by cassette mutagenesis (Figure 11-11) of 434 DNA to the amino acids believed to perform the same function in an analogous region of phage P22 repressor protein. (a) Expression in *E. coli* of the 434 repressor protein (left), with an enlargement of the site believed to bind the 434 operator; (right) the corresponding section of the P22 repressor protein. (b) A cassette was synthesized resembling the 434 domain, but with P22-type substitutions at positions thought to be essential for recognizing P22 operator DNA. This was ligated into the digested 434 plasmid, and the recombinant vector was introduced into *E. coli* to produce the hybrid protein, which then recognized P22 operator DNA but not the 434 operator.

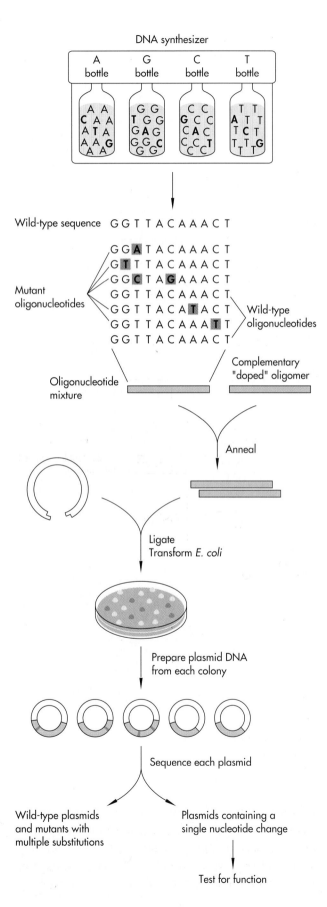

DNA synthesizer

Wild-type sequence GGTTACAAACT

Mutant oligonucleotides

Wild-type oligonucleotides

Oligonucleotide mixture

Complementary "doped" oligomer

Anneal

Ligate Transform *E. coli*

Prepare plasmid DNA from each colony

Sequence each plasmid

Wild-type plasmids and mutants with multiple substitutions

Plasmids containing a single nucleotide change

Test for function

FIGURE 11-13

Cassette mutagenesis using doped oligonucleotides to generate numerous mutants in a single experiment. An oligonucleotide cassette encoding the glucocorticoid response element (GRE) was synthesized by a DNA synthesis machine. Synthesis was done under conditions in which each bottle containing a particular nucleotide precursor was "contaminated" (doped) with small amounts of the other three precursors. In the example above, the DNA synthesizer was instructed to make an oligonucleotide with the sequence GGTTACAAACT. Thus, when a nucleotide precursor is called for—a C for example—the machine adds an aliquot of the solution from the C bottle, and a C base is coupled to the end of most of the oligonucleotide chains. However, because the C bottle contains a small amount of A, G and T, an incorrect base is sometimes added instead. Since the concentration of C is roughly 30 times that of A, G and T, an incorrect base will be added to about 1 out of 30 molecules. This results in a doped collection of oligonucleotides, which actually consists of many different sequences, some wild-type and some with substitutions. The level of contamination was adjusted to favor synthesis of oligonucleotides with only one substitution, but because substitutions occur randomly, some molecules in the collection had none and others had two or more. Cassettes were formed by annealing complementary doped oligonucleotides and ligated into a vector. Plasmid DNA was isolated from 546 individual *E. coli* transformants and analyzed by sequencing. Of these, 224 were wild-type, 218 contained one substitution (for the 30 bases, of interest, 74 of the 90 possible single substitutions were recovered), and the rest contained two or more.

molecule (Figure 11-14). In gene synthesis, the experimenter has total control over the sequence of the gene. It can be wild-type or mutant in any way required. Because most amino acids are encoded by multiple triplet codons, genes encoding wild-type proteins can be constructed using different codons. Codons can be chosen to place unique restriction sites throughout the sequence so that mutant cassettes can be easily swapped in. This was done with the bacterial rhodopsin gene. Replacing a fragment of the synthetic gene with a new synthetic fragment identified the amino acid that is linked to the photon-absorbing chromophore that initiates photosynthesis. Other fragments can be exchanged as cassettes to study other important structural features of the protein.

Codons can also be changed by gene synthesis to allow production of proteins at high levels in other organisms. Studies of the biochemistry of the Fos protein, encoded by a cellular protooncogene in animal

FIGURE **11-14**

Gene synthesis by ligation of complementary oligonucleotides. To synthesize a gene that encodes a protein of interest, a set of overlapping complementary oligonucleotides are
designed that can be combined to form a double-stranded
DNA molecule that encodes the entire protein. The oligonucleotides are mixed together, heated at 90°C for a few
minutes to denature the strands, and then cooled slowly to
room temperature. During this period the oligonucleotides
anneal through complementary base pairs. The oligonucleotides are designed so that each one anneals to two adjacent
oligonucleotides from the opposite strand, bridging them.
Generally, oligonucleotides ranging in length from 40 to 80
nucleotides are used in gene synthesis. The annealed oligonucleotides are covalently linked by DNA ligase, producing two contiguous DNA strands. This synthetic gene is
usually purified from a gel before ligation into a vector. The
resultant recombinant plasmid is obtained following transformation into *E. coli* and is sequenced to check that the
correct sequence was synthesized. The sequence of the synthetic gene can be designed to place restriction sites at convenient locations for cassette mutagenesis.

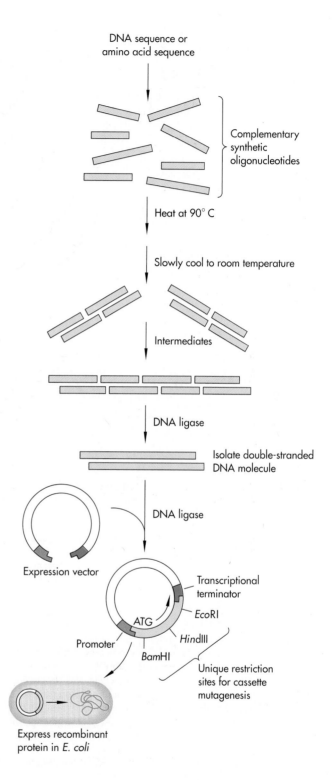

cells (Chapter 18), have been severely hampered by
the inability to produce the protein in *E. coli*. This
problem was finally solved by synthesizing a portion
of the *fos* gene entirely from oligonucleotides, changing natural *fos* codons to the codons used most efficiently in *E. coli*. Insertion of this synthetic gene into
an *E. coli* expression vector allowed for the first time
the production of large quantities of active Fos protein.
The gene was also designed with several unique restriction sites so that efficient cassette mutagenesis can
now be coupled to the biochemical assays for Fos
function.

The PCR Can Be Used to Construct Genes Encoding Chimeric Proteins

The ease with which mutations can be made in a
protein coding sequence has revolutionized the study
of protein function. A functional domain can be identified by making a series of mutant proteins, then
testing which substitutions cause a change in function.
However, it is not often easy to decide where to make
a mutation. In the example of the helix swap experiment (Figure 11-12), the domain that bound DNA
had been previously identified. And the design of the
experiment was guided by having a model for the
three-dimensional structure of the repressor protein.

However, for most proteins, little structural information is available. Identifying a functional domain—
for example, a region of the protein that may interact
with another protein—is difficult to do by inspecting
the primary amino acid sequence. A simple strategy

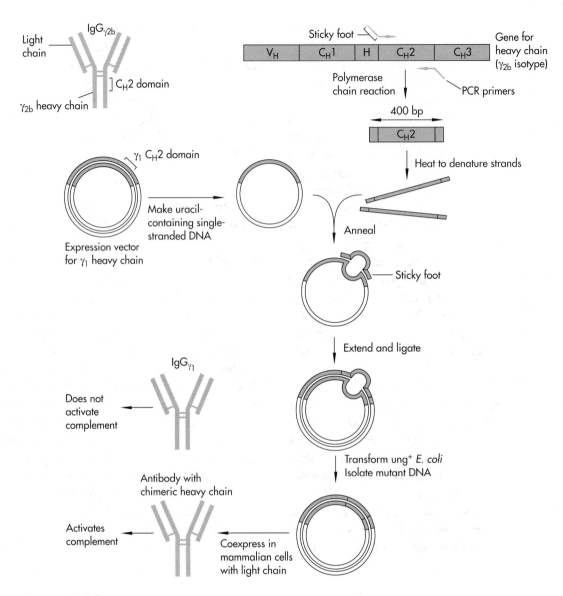

FIGURE **11-15**

Construction of a chimeric antibody heavy-chain-encoding gene by "sticky feet–directed" mutagenesis. Antibodies containing a γ2b heavy chain are known to participate in complement-dependent cell lysis, whereas antibodies containing γ1 heavy chains do not. In order to identify which domain of the γ2b heavy chain is responsible for this property, an antibody containing a chimeric heavy chain was produced. To construct a gene encoding the chimeric heavy chain, a 400-bp fragment encoding the C_H2 domain from a γ1 heavy chain was replaced with the homologous segment from a γ2b gene. Since there were no convenient restriction sites at the ends of the C_H2 segments, the 400-nucleotide-long γ2b DNA was prepared by PCR. The PCR primers were complementary to the ends of the γ2b DNA but contained additional nucleotides (the sticky feet) that were complementary to γ1 DNA at the boundaries of the γ1 C_H2 domain. The strands of the PCR-generated fragment were separated by heating, then one strand was used as the primer in a mutagenesis experiment using a uracil-containing single-stranded γ1 DNA template by the method shown in Figure 11-10. The resulting chimeric heavy-chain gene was coexpressed with a light chain gene in mammalian cells to form an antibody that now activated complement. Since only the C_H2 domain came from the γ2b heavy-chain, this result demonstrated that the γ2b C_H2 domain contains the information necessary to activate complement-dependent cell lysis. Sticky feet–directed mutagenesis provided a simple means for constructing this complicated gene.

that helps to narrow down important amino acids in a protein is the analysis of chimeras between related proteins. We have previously discussed the use of computer programs to identify related proteins by comparison of their amino acid sequences (Chapter 8). Chimeric proteins are constructed by replacing a segment of one protein with the *homologous* segment from another protein. Although the two proteins have functional differences, their sequence similarity often indicates that they share a common overall structure. A striking example of this was in the analysis of human growth hormone (hGH). A series of chimeric proteins were made in which most of the amino acids were derived from hGH but which contained segments from related hormones, such as human prolactin. Using this strategy, regions of hGH that interact with the hGH receptor were identified. In Chapter 17, we will see how functional regions of a receptor which spans the membrane seven times were identified by the study of chimeras.

The 434/P22 repressor (Figure 11-12) and hGH chimeras were constructed by ligation of short oligonucleotide cassettes into the coding sequence. A different strategy (Figure 11-15) was used to prepare a chimeric antibody in which a 400-bp segment from a γ1 heavy-chain gene was replaced by the homologous segment from a γ2b gene. A 400-bp DNA fragment was generated by PCR that encoded the new sequence to be inserted and two 30-base "sticky feet" on each end. The double-stranded PCR fragment was heated to denature the two strands, and then one of the single-stranded molecules was utilized in a primer-extension experiment (as in Figure 11-8). Had the gene synthesis method been employed, construction of the chimeric gene would have required twenty 40-nucleotide-long oligomers. Instead, the sticky feet method used only two oligonucleotide primers for PCR.

Mutagenesis Is the Gateway to Gene Function and Protein Engineering

It would be difficult to overestimate the importance of in vitro mutagenesis techniques to biology and biotechnology. The harnessing of enzymes that operate on DNA and the refinement of oligonucleotide synthesis have made changing gene sequences an almost trivial task. And the ability to operate on DNA lets us also change the structure of the products of genes—RNA and, most importantly, proteins. Thus, the impact of this technology is twofold. It has revolutionized how research is done in molecular biology by creating the entirely new concept of "reverse genetics"—changing gene sequence first, then examining gene function. And it opens the door to sophisticated protein engineering (see Chapter 23), the ability to make changes in natural gene products that make them do their jobs better. The impact of protein engineering on medicine and industry will be substantial.

Reading List

General

Wu, R., L. Grossman, and K. Moldave, eds. *Methods in Enzymology*, vol. 100: *Recombinant DNA, Part B*. Academic, New York, 1983.
Gait, M. *Oligonucleotide Synthesis: A Practical Approach*. IRL, Oxford, Eng., 1984.
Wu, R., and L. Grossman, eds. *Methods in Enzymology*, vol 154: *Recombinant DNA, Part E*. Academic, New York, 1987.

Original Research Papers

STRATEGIES FOR IN VITRO MUTAGENESIS

Botstein, D., and D. Shortle. "Strategies and applications of in vitro mutagenesis." *Science*, 229: 1193–1201 (1985). [Review]
Smith, M. "In vitro mutagenesis." *Ann. Rev. Gen.*, 19: 423–462 (1985). [Review]

Zoller, M. J. "New molecular biology methods for protein engineering." *Curr. Opin. Struct. Biol.,* 1: 605–610 (1991). [Review]

INSERTION MUTAGENESIS

Heffron, F., M. So, and B. J. McCarthy. "In vitro mutagenesis of a circular DNA molecule using synthetic restriction sites." *Proc. Natl. Acad. Sci. USA,* 75: 6012–6016 (1978).

Abraham, J., J. Feldman, K. A. Nasmyth, J. N. Strathern, A. J. S. Klar, J. R. Broach, and J. B. Hicks. "Sites required for position-effect regulation of mating type information in yeast." *Cold Spring Harbor Symp. Quant. Biol.,* 47: 989–998 (1983).

Rees-Jones, R. W., and S. P. Goff. "Insertional mutagenesis of the Abelson murine leukemia virus genome: identification of mutants with altered kinase activity and defective transformation ability." *J. Virol.,* 62: 978–986 (1988).

DELETION MUTAGENESIS

Lai, C. J., and D. Nathans. "Deletion mutants of SV40 generated by enzymatic excision of DNA segments from the viral genome." *J. Mol. Biol.,* 89: 179–193 (1974).

Mertz, J. E., J. Carbon, M. Herzberg, R. W. Davis, and P. Berg. "Isolation and characterization of individual clones of simian virus 40 mutants containing deletions, duplications and insertions in their DNA." *Cold Spring Harbor Symp. Quant. Biol.,* 39: 69–84 (1974).

Sakonju, S., D. Bogenhagen, and D. D. Brown. "A control region in the center of the 5S gene directs specific initiation of transcription, I: the 5' border of the region." *Cell,* 19: 13–26 (1980).

McKnight, S. L., and R. Kingsbury. "Transcriptional control signals of a eukaryotic protein-coding gene." *Science,* 217: 316–324 (1982).

Struhl, K. "The yeast his3 promoter contains at least two distinct elements." *Proc. Natl. Acad. Sci. USA,* 79: 7385–7389 (1982).

CHEMICAL MUTAGENESIS

Chu, C. T., D. S. Parris, R. A. F. Dixon, F. E. Farber, and P. A. Schaffer. "Hydroxylamine mutagenesis of HSV DNA and DNA fragments: introduction of mutations into selected regions of the viral genome." *Virol.,* 98: 168–181 (1979).

Shortle, D., and D. Nathans. "Regulatory mutants of simian virus 40: constructed mutants with base substitutions at the origin of viral DNA replication." *J. Mol. Biol.,* 131: 801–817 (1979).

Weiher, H., and H. Schaller. "Segment-specific mutagenesis: extensive mutagenesis of a lac promoter/operator element." *Proc. Natl. Acad. Sci. USA,* 79: 1408–1412 (1982).

Myers, R. M., L. S. Lerman, and T. Maniatis. "A general method for saturation mutagenesis of cloned DNA fragments." *Science,* 229: 242–247 (1985).

ENZYMATIC MISINCORPORATION MUTAGENESIS

Shortle, D. P. Grisafi, S. J. Benkovic, and D. Botstein. "Gap misrepair mutagenesis: efficient site-directed introduction of transition, transversion, and frameshift mutations in vitro." *Proc. Natl. Acad. Sci. USA,* 79: 1588–1592 (1982).

Zakour, R. A., and L. A. Loeb. "Site-specific mutagenesis by error-directed DNA synthesis." *Nature,* 295: 708–710 (1982).

Abarzua, P., and K. J. Marians. "Enzymatic techniques for the isolation of random single-base substitutions in vitro at high frequency." *Proc. Natl. Acad. Sci. USA,* 81: 2030–2034 (1984).

Leung, D. W., E. Chen, and D. V. Goeddel. "A method for random mutagenesis of a defined DNA segment using a modified polymerase chain reaction." *Technique,* 1: 11–15 (1989)

OLIGONUCLEOTIDE-DIRECTED MUTAGENESIS

Hutchinson, C. A., Phillips, M. H. Edgell, S. Gillam, P. Jahnke, and M. Smith. "Mutagenesis at a specific position in a DNA sequence." *J. Biol. Chem.,* 253: 6551–6560 (1978).

Wallace, R. B., M. Schold, M. J. Johnson, P. Dembek, and K. Itakura. "Oligonucleotide-directed mutagenesis of the human β-globin gene: a general method for producing specific point mutations in cloned DNA." *Nuc. Acids Res.,* 9: 3642–3656 (1981).

Zoller, M. J., and M. Smith. "Oligonucleotide-directed mutagenesis using M13-derived vectors: an efficient and general procedure for production of point mutations in any fragment of DNA." *Nuc. Acids Res.,* 10: 6487–6500 (1982).

Kramer, W., V. Drutsa, H. W. Jansen, M. Pflugfelder, and H.-J. Fritz. "The gapped duplex DNA approach to oligonucleotide-directed mutation construction." *Nuc. Acids Res.,* 12: 9441–9456 (1984).

Kunkel, T. A. "Rapid and efficient site-specific mutagenesis without phenotypic selection." *Proc. Natl. Acad. Sci. USA,* 82: 477–492 (1985).

Taylor, J. W., J. Ott, and F. Eckstein. "The rapid generation of oligonucleotide-directed mutations at high frequency using phosphorothioate-modified DNA." *Nuc. Acids Res.,* 13: 8765–8785 (1985).

CASSETTE MUTAGENESIS

Lo, K.-M., S. S. Jones, N. R. Hackett, and H. G. Khorana. "Specific amino acid substitutions in bacterioopsin: replacement of a restriction fragment in the structural gene by synthetic DNA fragments containing altered codons." *Proc. Natl. Acad. Sci. USA,* 81: 2285–2289 (1984).

Wells, J. A., M. Vasser, and D. B. Powers. "Cassette mutagenesis: an efficient method for generation of multiple mutations at defined sites." *Gene,* 34: 315–323 (1985).

Wharton, R., and M. Ptashne. "Changing the binding specificity of a repressor by redesigning an alpha-helix." *Nature,* 316: 601–605 (1985).

Reidhaar-Olson, J. F., and R. T. Sauer. "Combinatorial cassette mutagenesis as a probe of the informational content of protein sequences." *Science,* 241: 53–57 (1988).

DOPED OLIGONUCLEOTIDE MUTAGENESIS

McNeil, J. B., and M. Smith. "*Saccharomyces cerevisiae* CYC1 mRNA 5'-end positioning: analysis by in vitro mutagenesis, using synthetic duplexes with random mismatch base pair." *Mol. Cell. Biol.,* 5: 3545–3551 (1985).

Hutchison, C. A., S. K. Nordeen, K. Vogt, and M. H. Edgell. "A complete library of point substitution mutations in the glucocorticoid response element of mouse mammary tumor virus." *Proc. Natl. Acad. Sci. USA,* 83: 710–714 (1986).

GENE SYNTHESIS

Ferreti, L., S. S. Karnık, H. G. Khorana, N. Nassal, and D. D. Oprian. "Total synthesis of a gene for bovine rhodopsin." *Proc. Natl. Acad. Sci. USA,* 83: 599–603 (1986).

Caruthers, M. H., A. D. Barone, S. L. Beaucage, D. R. Dodds, E. F. Fisher, L. J. McBride, M. Matteucci, Z. Stabinsky, and J.-Y. Tang. "Chemical synthesis of deoxyoligonucleotides by the phosphoramidite method." *Meth. Enzymol.,* 154: 287–313 (1987).

Abate, C., D. Luk, R. Getz, F. J. Rauscher, III, and T. Curran. "Expression and purification of the leucine zipper and DNA-binding domains of Fos and Jun: both Fos and Jun contact DNA directly." *Proc. Natl. Acad. Sci. USA,* 87: 1032–1036 (1990).

CONSTRUCTING CHIMERIC GENES BY PCR

Higuchi, R., B. Krummel, and R. K. Saiki. "A general method of in vitro preparation and specific mutagenesis of DNA fragments: study of protein and DNA interactions." *Nuc. Acids Res.,* 16: 7351–7367 (1988).

Clackson, T., and G. Winter. "'Sticky feet'-directed mutagenesis and its application to swapping antibody domains." *Nuc. Acids Res.,* 17: 10163–10170 (1989).

PROTEIN STRUCTURE AND FUNCTION

Winter, G., A. R. Fersht, A. J. Wilkinson, M. Zoller, and M. Smith. "Redesigning enzyme structure by site-directed mutagenesis." *Nature,* 299: 756–758 (1982).

Cunningham, B. C., P. Jhurani, P. Ng, and J. A. Wells. "Receptor and antibody epitopes in human growth hormone identified by homolog-scanning mutagenesis." *Science,* 243: 1330–1336 (1989).

Gibbs, C. S., and M. J. Zoller. "Rational scanning mutagenesis of a protein kinase identifies functional regions involved in catalytic and substrate interactions." *J. Biol. Chem.,* 266: 8923–8931 (1991).

CHAPTER

12

Transferring Genes into Mammalian Cells

Three essential tools form the basis for studying the function of mammalian genes. First, we must be able to isolate a gene by cloning. Second, we must be able to manipulate the sequence of a gene in the test tube. Third, we must be able to return an altered gene to cells to determine how it functions. In many ways, this last step has presented the greatest obstacles to progress. Methods needed to be developed to persuade cells to take up foreign DNA, despite millions of years of evolutionary barriers to DNA uptake. Although this problem was solved relatively quickly, investigators soon discovered that once they got DNA into cells, they lost all control over what happened to it. Thus, much work has gone into developing methods for retaining the correct functioning of transferred genes. Methods are also being designed to inactivate genes already present in cells, so that their functions in an intact cell or animal can be evaluated.

With many of these methods now in place, gene transfer has become a routine tool for studying gene structure and function. It is used to identify the regulatory sequences that control gene expression. Transfer of new or altered genes into new cellular environments provides a means to determine gene function. Gene transfer provides the basis for high-level protein expression, a capability used by researchers to understand the proteins they study and by the biotechnology industry to produce new protein drugs. And sophisticated methods allow investigators to introduce new genes or alter existing ones in intact animals, making possible the promise of useful animal models of difficult human diseases.

Establishment of Immortal Cell Lines Makes Gene Transfer Practical

Gene transfer into mammalian cells is an inefficient process, so an abundant source of starting cells is necessary to end up with a workable number of *transfected* cells (that is, cells containing the transferred gene). Thus, the availability of mammalian cell lines that grow indefinitely in culture has made gene transfer experiments practical. As we will learn in Chapter 18, most tissues explanted directly from animals will not grow indefinitely in culture. However, many permanently growing cell lines originate from tumors, cells which have escaped normal limits to growth. Not only can tumor cell lines grow indefinitely in chemically defined media, they grow quickly, usually doubling in number in less than a day. But, because they are tumor cells, their properties are not always the same as those of the normal body cells from which they arise. Thus, efforts continue to develop cell lines that behave like their normal counterparts.

Nevertheless, much of the basic business of the cell is transacted properly in tumor cells and other immortal cell lines. Most of what we know about the function and regulation of mammalian genes comes from transferring genes into cell lines rather than primary tissue samples. And the availability of batteries of cell lines representing different tissues of the body has given us experimental access to many of the specialized processes of the body. For example, we can easily study cells of the blood, skin, muscle, liver, and, to a limited extent, brain.

Gene Transfer Was First Used in the Study of Tumor Viruses

Much of the landmark work in the study of mammalian genes came from working with tumor viruses. Tumor viruses infect mammalian cells, insert their own genetic material, and hijack the cellular biosynthetic machinery to manufacture more viruses. Often viral multiplication kills the cell. Sometimes the presence of tumor virus genes permanently alters the growth properties of an infected cell, transforming it into a tumor cell. Tumor viruses attracted the attention of researchers for several compelling reasons. Their genomes, which program both viral life cycle in cells and the structural components of the virus itself, are often astonishingly small—the smallest DNA tumor viruses comprise a handful of genes packed into only 5000 bp. Within this small stretch of DNA lies the information for switching viral genes on and off in an ordered fashion, for replicating the viral genome, for assembling new virus particles, and, most tantalizing, for converting a normal cell into a cancer cell. Before recombinant DNA, these viruses, which could be purified in large quantities, provided the only source of purified genetic material—DNA for some viruses, RNA for others—that functioned in mammalian cells. Thus, it was the study of tumor viruses and their genomes that provided experimenters with the impetus to develop methods to transfer purified genetic material into mammalian cells.

The earliest gene transfer experiments were done with DNA tumor viruses, viruses whose genes are encoded in DNA just as cellular genes are. DNA was isolated from purified viruses and introduced into cultures of uninfected cells. These cells eventually produced fully infectious viruses. This DNA-mediated transfer of infectious virus was dubbed *transfection,* to distinguish it from *infection,* the natural route of entry for viruses. These early experiments were critical ones in the development of our understanding of the biology of mammalian cells, for they established clearly that the entire program of viral reproduction (and by inference reproduction of animals, too) is encoded in pure DNA. The rest of the virus functions just to get the DNA into cells.

How were these experiments done? Cells do not naturally take up DNA. Indeed, throughout evolution cells have gone to great lengths to protect themselves from invading genetic material. Thus, a variety of technical tricks were used to bypass natural barriers to gene transfer. Perhaps the simplest to understand, though the toughest to do, is microinjection, in which DNA is injected directly into the nucleus of cells through painstakingly constructed fine glass needles. This is an efficient process on a per cell basis—that is, a large fraction of the injected cells actually get the DNA—but only a few hundred cells can be injected in a single experiment. Nevertheless, some of the early tumor virus experiments were done by microinjection, and it remains the method of choice for getting DNA, RNA, and proteins into cells in certain kinds of experiments.

FIGURE **12-1**

Transfection of DNA by calcium phosphate coprecipitation. The figure depicts an updated version of the original Graham and van der Eb experiment, which demonstrated that transfer of naked adenovirus DNA yields infectious virus. Many variations of this basic protocol are now in use. Adenovirus particles are purified, and their DNA is extracted. The DNA is mixed with a carefully buffered solution containing phosphate. Addition of calcium chloride forms a fine precipitate of calcium phosphate and DNA (which contains phosphate groups in its backbone). The precipitate is pipetted onto a monolayer of cells growing in a petri dish and left on the cells for several hours, during which time many of the cells take up the precipitated DNA. The precipitate is removed from the cells, which are then incubated in fresh growth medium for several days. Infectious virus particles soon emerge from the cells.

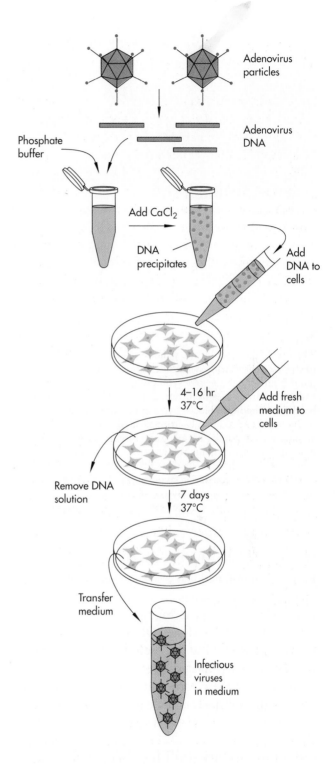

The earliest method for introducing DNA into cells en masse was to incubate the DNA with an inert carbohydrate polymer (dextran) to which a positively charged chemical group (DEAE, for diethylamino-ethyl) has been coupled. The DNA sticks to the DEAE–dextran via its negatively charged phosphate groups. These large DNA-containing particles stick in turn to the surfaces of cells, which are thought to take them in by a process known as *endocytosis,* a normal consequence of membrane turnover. Some of the DNA evades destruction in the cytoplasm of the cell and escapes to the nucleus, where it can be transcribed into RNA like any other gene in the cell.

The DEAE–dextran method, while relatively simple, was very inefficient for many types of cells, so it was not a reliable method for the routine assay of the biological activity of a purified DNA preparation. The breakthrough that eventually made gene transfer a routine tool for workers studying mammalian cells was the discovery that cells efficiently took in DNA in the form of a precipitate with calcium phosphate (Figure 12-1). This discovery arose from prior work showing that divalent cations such as calcium and magnesium promoted the uptake of DNA into bacteria. With this new method, the yield of virus from cells transfected with viral DNA was a hundred times greater than with the DEAE–dextran method. It was by using the calcium phosphate coprecipitation method that researchers found that purified tumor virus DNA could transform normal cells into cancer cells, a compelling demonstration that cancerous growth can be encoded in DNA. This was a watershed

observation that formed the basis for our advanced understanding of the genes underlying cancer in our own cells (see Chapter 18).

Once the tools of recombinant DNA became available, increasingly sophisticated gene transfer experiments were possible. Restriction enzymes were used to cut tumor virus DNA into fragments, which could

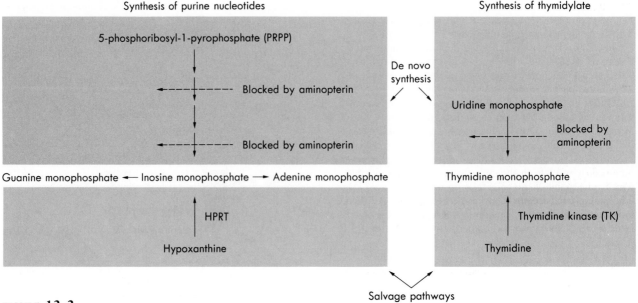

FIGURE **12-2**
The metabolic basis for the genetic selection for *tk*. Mammalian cells possess two distinct routes for synthesizing deoxynucleoside triphosphates for DNA synthesis. They can make them from scratch (*de novo* synthesis, shown in blue boxes), or they can salvage free purine and pyrimidine bases from intracellular or extracellular sources (pathways shown in pink boxes). Aminopterin blocks two steps in the biosynthesis of purines and one in the biosynthesis of thymidine. If cells are provided with hypoxanthine and thymidine, they can survive aminopterin treatment by using the salvage pathways. The key salvage enzyme for thymidine nucleotides is thymidine kinase (TK), which phosphorylates free thymidine to thymidine monophosphate. Cells that carry mutations in the gene encoding this enzyme, *tk*, cannot grow in medium containing hypoxanthine, aminopterin, and thymidine (*HAT medium*). Cells that have gained a *tk* gene by transfection can grow in HAT. This selection can also be used for *hprt*, the gene that encodes the key salvage enzyme (hypoxanthine phosphoribosyltransferase—HPRT) for purines.

be tested individually by transfection, allowing the localization of functional genes to specific DNA fragments. Ultimately, when it became possible to clone tumor virus DNA, gene transfer experiments of this type established that cloned DNA, propagated exclusively in bacteria, had precisely the same biological activity as DNA isolated directly from viruses. The significance of these early tumor virus gene transfer experiments for modern biology was profound.

Selectable Markers That Work in Mammalian Cells Allow Gene Transfer by Cotransformation

One of the basic principles of recombinant DNA technology is the use of biological markers to identify cells carrying recombinant DNA molecules. In bacteria, these are commonly drug-resistance genes. We use

drug resistance to select bacteria that have taken up cloned DNA from the much larger population of bacteria that have not. In the early mammalian gene transfer experiments involving viral genes, the transfer of exogenous DNA into cells was detected because the DNA had a biological activity—it led to production of infectious virus or produced stable changes in the growth properties of the transfected cells. But what if we want to study genes whose function we cannot detect simply by looking at the transfected cells? Finding the transfected cells is not a trivial problem, because even the best transfection methods result in stable transfer to only one cell in a thousand.

The solution came from studies of a more complex DNA tumor virus, herpes simplex virus (HSV). HSV was found to contain a gene (*tk*) encoding an enzyme, thymidine kinase, which catalyzes a step in the synthesis of thymidine triphosphate (TTP), one of the four precursor nucleotides for DNA synthesis (Figure

FIGURE 12-3

Transfer of a nonselectable gene by cotransfection with a marker gene. A herpes simplex virus DNA fragment carrying the *tk* gene was mixed with a fragment carrying the rabbit β-globin gene. The two fragments were applied to dishes of mouse *tk⁻* cells as a calcium phosphate precipitate. The precipitate was removed, and the cells were incubated in HAT medium, which supports only the growth of *tk⁺* cells (see Figure 12-2). After 2 to 3 weeks, colonies of *tk⁺* cells grew up on the dishes. The colonies were picked and grown up. Chromosomal DNA was isolated, digested with the restriction enzyme *Kpn*I, (which cleaves at the sites indicated by "K" in the β-globin gene), and analyzed by Southern blotting, using a radioactive probe that detected the rabbit β-globin sequence in the fragment generated by cleavage with *Kpn*I. Most of the *tk⁺* colonies had also acquired the rabbit-β-globin gene. Detailed analysis of the structure of the transfected DNA in these clones showed that it integrated as large arrays carrying multiple copies of both transfected fragments, usually at one or a few chromosomal sites. M, molecular size markers.

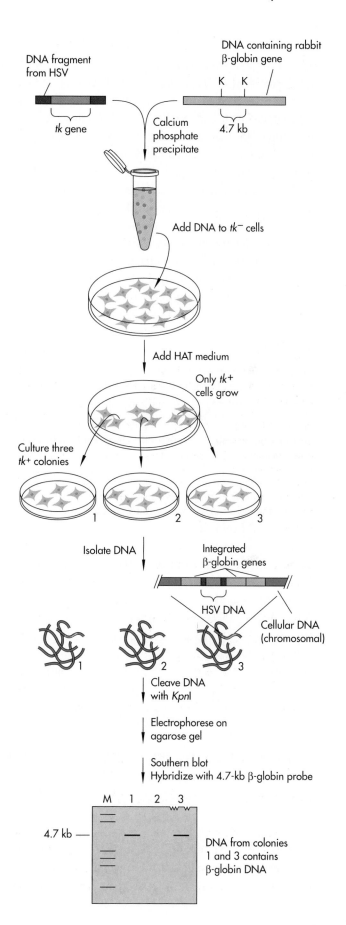

12-2). Years earlier, cell lines had been isolated that were deficient in this activity, presumably owing to inactivation of the cellular genes encoding this enzyme. Infection of these *tk⁻* cells with HSV and, in later experiments, transfection with naked HSV DNA were able to correct the defect in thymidine kinase activity in these cells, allowing them to grow in a specially concocted medium in which thymidine kinase is absolutely required for the cells to make DNA (and thus, to multiply). It was quickly realized that the HSV *tk* gene was functioning as a selectable genetic marker in much the same way that drug-resistance genes worked in bacteria, to allow rare transfected cells to grow up out of a much larger population that did not take up any DNA.

Could such a marker be used to identify cells that have taken up a gene for which there is no simple selection? The answer came from experiments like the one depicted in Figure 12-3. Researchers isolated a fragment of HSV DNA carrying the *tk* gene, simply mixed the *tk* DNA with a another DNA fragment carrying a rabbit globin gene, and applied the mixture of DNA fragments to cultures of *tk⁻* cells as a calcium phosphate precipitate. The cells were transferred to *selective* growth medium, which permits growth only of cells that took up a functional *tk* gene, and after a few weeks, the resulting colonies of growing cells were examined by Southern blotting to determine whether the cells had taken up the rabbit globin DNA as well. The vast majority had done so.

These experiments led to the view that while few cells in the culture took up DNA, those that did took up a great deal of it (perhaps 1000 kb or more). And physically unlinked fragments became covalently joined once inside cells, presumably through the action of cellular DNA ligases. The complexes somehow integrated into chromosomal DNA at only one or a few locations. These large complexes of transfected DNA usually contained many copies of each fragment. Thus, cells selected for the presence of one fragment (for example, the *tk* gene) were almost always found also to contain the fragment for which no selection was applied. Moreover, when cells from a single transfected colony were examined, they could be shown to be clonal in the same way a transformed bacterial colony is—all cells in the colony appeared to contain the same arrangement of transfected DNA integrated at the same site in the host cell genome, meaning that they were all descendants of a single transfected cell. Thus, at cell division, the transfected DNA must be faithfully replicated along with the rest of the DNA in the cell. The upshot of all this was that any gene could now be stably transferred into mammalian cells simply by mixing it with a *tk* gene. The cells that took up DNA could be easily selected via the *tk* marker, and surviving colonies could be examined for the presence of the other gene with the expectation that, with high probability, it would be there.

But *tk* is a marker of limited utility, because it can be used only in cells carrying mutations in their own *tk* genes. The ideal marker would permit a simple dominant genetic selection that could operate in any cell the experimentor wished to use. Many such markers have now been developed (Table 12-1). The most commonly used marker in mammalian cells is, ironically, a bacterial drug-resistance gene that confers resistance to a neomycin-related drug, G418, which kills mammalian cells by blocking protein synthesis. The marker gene encodes an enzyme that destroys the drug.

Exogenous DNA Is Transiently Expressed in Many Cells Immediately Following Transfection

Thus far, we have discussed the use of gene transfer to incorporate new DNA *stably* into the genome of the transfected cell. This is a rare event, occurring in one in a thousand to one in a million cells. Many more cells, up to half, actually take up DNA in a transfection but fail to integrate it. In these cells, the DNA persists in the nucleus for several days before disappearing (Figure 12-4). During this period, however, the transfected DNA is subject to many of the regulatory activities that control the expression of endogenous genes in the chromosomes.

Researchers have exploited this window of *transient expression* in several ways. Transient expression is used

TABLE **12-1**

Dominant Selectable Markers Used in Transfection Experiments

ENZYME (abbreviation)	DRUG FOR SELECTION	SELECTION MECHANISM
Aminoglycoside phosphotransferase (APH)	G418 (inhibits protein synthesis)	APH inactivates G418
Dihydrofolate reductase (DHFR): Mtx-resistant variant	Methotrexate (Mtx; inhibits DHFR)	Variant DHFR resistant to Mtx
Hygromycin-B-phosphotransferase (HPH)	Hygromycin-B (inhibits protein synthesis)	HPH inactivates hygromycin-B
Thymidine kinase (TK)	Aminopterin (inhibits de novo purine and thymidylate synthesis)	TK synthesizes thymidylate
Xanthine-guanine phosphoribosyltransferase (XGPRT)	Mycophenolic acid (inhibits de novo GMP synthesis)	XGPRT synthesizes GMP from xanthine
Adenosine deaminase (ADA)	9-β-D-xylofuranosyl adenine (Xyl-A; damages DNA)	ADA inactivates Xyl-A

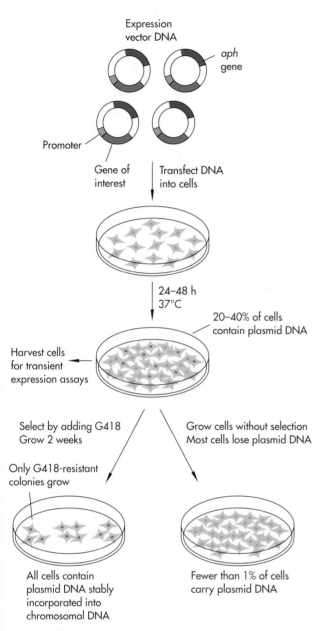

Expression
vector DNA

aph
gene

Promoter

Gene of
interest

Transfect DNA
into cells

24–48 h
37°C

20–40% of cells
contain plasmid DNA

Harvest cells
for transient
expression assays

Select by adding G418
Grow 2 weeks

Grow cells without selection
Most cells lose plasmid DNA

Only G418-resistant
colonies grow

All cells contain
plasmid DNA stably
incorporated into
chromosomal DNA

Fewer than 1% of cells
carry plasmid DNA

FIGURE **12-4**

Transfected DNA is maintained transiently in many cells, but stably integrates in only a few. During the first 48 hours following transfection, as many as 50 percent of the cells in the culture carry the transfected DNA. This is termed the *transient phase*, and for many kinds of experiments, the cultures are used at this point. As the cells continue to incubate, the transfected DNA is progressively lost from most of them (owing to a combination of degradation and dilution, because unintegrated DNA is not usually replicated along with the host cell chromosomes). In only a few cells on the dish does the DNA become stably integrated into a chromosome. Isolating those cells and their descendants requires the application of a *selection* for a transfected marker gene (left). The selection kills off all the cells that failed to integrate transfected DNA, allowing the rare integrants to grow up into large isolated colonies (clones), which can be recovered for further analysis.

as a rapid assay to map the regulatory elements in a gene that control transcription and RNA processing. For example, in the most commonly used assay for studying the activity of mammalian promoters and enhancers, plasmids carrying cloned promoters (often with different sequences deleted or mutated) are transfected into cells, and the cells are harvested 48 to 72 hours later, while the cells are in the transient expression phase. During this period, active promoters are recognized and used by the transcription machinery of the cells, resulting in transcription of downstream sequences. Often these downstream sequences carry a *reporter gene*, such as the bacterial *CAT* gene, that encodes an enzyme that is simple to assay (see Figure 9-2). Or the amount of RNA transcribed from the transfected genes can be measured directly by hybridization to radiolabeled gene probes. Another application of transient expression is the production of large amounts of protein or RNA from a cloned gene. The amount of protein and RNA produced from a transiently expressed transfected gene can be high.

From our studies of viral and cellular genes, we have learned many of the strategies used by these genes to achieve high-level expression. Modern expression vectors are designed on the basis of what we have learned and incorporate strong viral or cellular promoters and efficient translational initiation signals. In addition, most vectors incorporate an intron, since splicing increases the efficiency of export of mRNA from the nucleus for translation. Vectors can be engineered with protein coding sequences that direct the expressed protein to particular locations in the cell, such as secretory vesicles, or that tag the protein with some additional peptide sequence that can be used to purify it. An example of such a vector is shown in Figure 12-5.

Gene Amplification Is Used to Achieve High-Level Protein Expression

Gene amplification is a special application of stable transfection, used to obtain very high levels of expression of a transfected gene. When cell cultures are treated with methotrexate (Mtx), an inhibitor of a critical metabolic enzyme, dihydrofolate reductase (DHFR), most cells die, but eventually some Mtx-resistant cells grow up. Upon examination, the resistant cells are found to have *amplified* their *dhfr* genes; that is, they

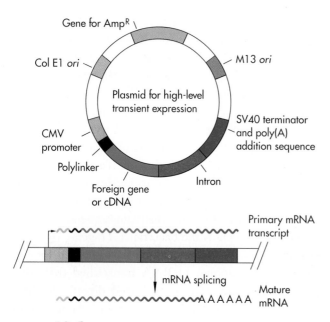

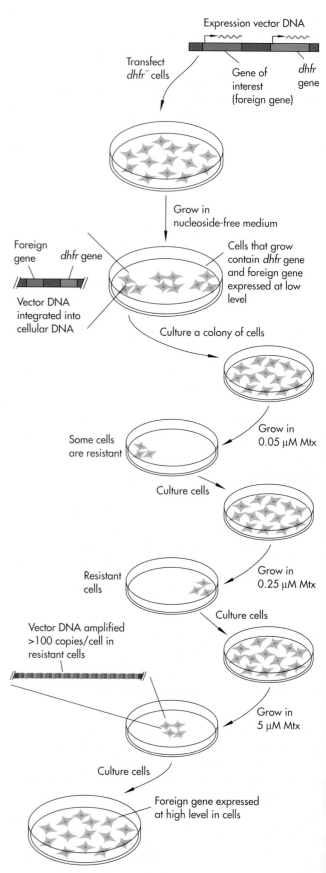

FIGURE **12-5** *(Above)*

A typical mammalian expression vector for high-level protein expression. Modern expression vectors are stitched together from several distinct units that provide the different components of a functional gene. The vector shown here carries a potent promoter from cytomegalovirus (CMV), an intron from a globin gene, and the poly(A) addition signal from SV40 virus. Such plasmids usually contain several restriction sites (collectively called a polylinker) for inserting the foreign gene to be expressed. The plasmid backbone carries the usual sequences for replication (Col E1 *ori*) and drug-resistance (Amp^R) in *E. coli,* including in this case an M13 origin of replication (*ori*), allowing this vector to be recovered from bacteria as single-stranded phages for easy mutagenesis.

FIGURE **12-6** *(Right)*

Obtaining high-level expression by coamplification with *dhfr.* The gene to be expressed is transfected together with a *dhfr* gene into a cell line that lacks endogenous DHFR enzyme activity. The cells are grown in a special nucleoside-free medium in which only cells that have acquired a *dhfr* gene can grow. Clonal colonies are picked, grown up, and shifted to medium containing 0.05 μM methotrexate (Mtx), an inhibitor of the DHFR enzyme. Most cells fail to grow at this concentration of Mtx. But a few will have amplified their *dhfr* genes, allowing them to produce sufficient amounts of the enzyme to escape inhibition by Mtx. After a few weeks of growth, cells with amplified *dhfr* genes will have multiplied enough to predominate in the culture. At this point, the concentration of MTX is raised, and the cycle is repeated. The Mtx concentration is steadily raised in small increments until it reaches about 50 μM. Cells growing at this concentration have amplified their *dhfr* genes along with the linked foreign gene several hundredfold.

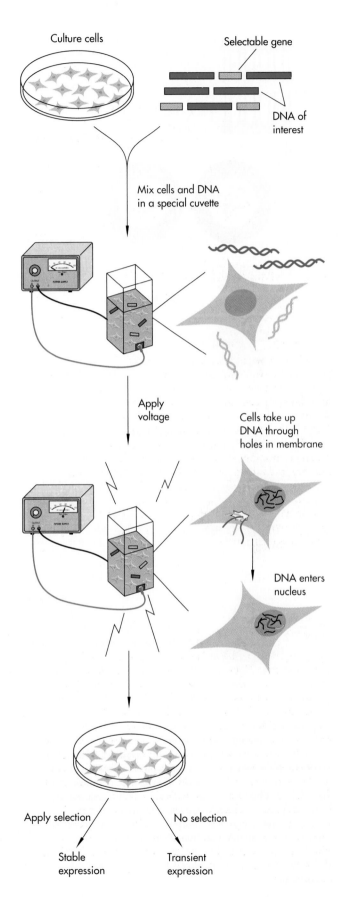

FIGURE **12-7**

Gene transfer by electroporation. Cells are concentrated, mixed with the DNA to be transfected, and placed in a small chamber with electrodes connected to a specialized power supply. A brief electric pulse is discharged across the electrodes, transiently opening holes in cell membranes. DNA enters the cells, which are removed and plated in fresh medium. The cultures can be harvested during the transient expression phase, or selection can be applied to isolate stably transfected clones.

have managed somehow to significantly increase the number of *dhfr* genes, leading to expression of higher enzyme levels, allowing the cells in turn to escape inhibition by the drug. Study of the amplification process revealed that the unit of DNA that is amplified is much larger than the *dhfr* gene itself, so that neighboring DNA becomes amplified as well.

This observation has been exploited as shown in Figure 12-6. A gene to be expressed in cells is cotransfected with a cloned *dhfr* gene, and the transfected cells are subjected to selection with a low concentration of Mtx. Resistant cells that have taken up the *dhfr* gene (and, in most cases, the cotransfected gene) multiply. Increasing the concentration of Mtx in the growth medium in small steps generates populations of cells that have progressively amplified the *dhfr* gene, together with linked DNA. Although this process takes several months, the resulting cell cultures capable of growing in the highest Mtx concentrations will have stably amplified the DNA encompassing the *dhfr* gene a hundredfold or more, leading to significant elevation of the expression of the cotransfected gene.

Specialized Methods Are Developed to Transfect Difficult Cell Types

Although calcium phosphate coprecipitation is the most widely used method for introducing DNA into mammalian cells, in some cells it doesn't work. Cells such as lymphocytes, which grow in suspension, are especially resistant to transfection by calcium phosphate precipitates. In another method, *electroporation* (Figure 12-7), cells are placed in a solution containing DNA and subjected to a brief electrical pulse that causes holes to open transiently in their membranes. DNA enters through the holes directly into the cy-

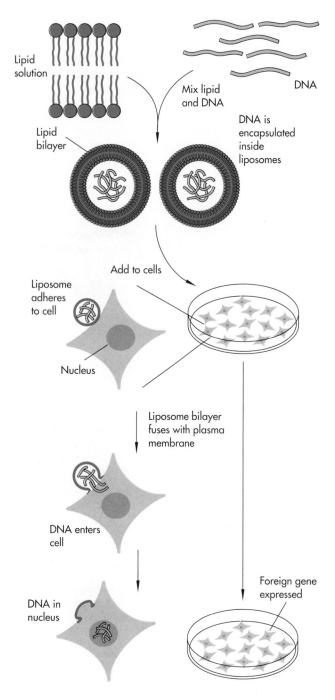

FIGURE **12-8**
Liposome-mediated gene transfer (*lipofection*). A number of
methods have been developed to encapsulate DNA in
synthetic lipid bilayers resembling cell membranes. These
liposomes are essentially spheres of synthetic membrane filled
with DNA. They fuse spontaneously to cell membranes,
disgorging their contents into the cytoplasm. Making
liposomes from scratch is complicated, but commercial
reagents are available that considerably simplify the
procedure.

toplasm, bypassing the endocytotic vesicles through
which they pass in the DEAE–dextran and calcium
phosphate procedures (passage through these vesicles
may sometimes destroy or damage DNA). DNA can
also be incorporated into artificial lipid vesicles—
liposomes—which fuse with the cell membrane,
delivering their contents directly into the cytoplasm
(Figure 12-8). Microinjection, the surest way to get
DNA into cells, can now be performed with a
computer-assisted apparatus that increases by 10-fold
or more the number of cells that can be injected in
one experiment. And in a stunningly direct approach,
used primarily with plant cells and tissues, DNA is
absorbed to the surface of tungsten microprojectiles
and fired into cells with a device resembling a shotgun
(see Figure 15-9).

Several of these methods, microinjection, electroporation, and liposome fusion, have been adapted to
introduce proteins into cells. This is a powerful approach for studying protein function in living cells.
Pure proteins can be directly introduced into cells to
evaluate their function in their natural environment,
or antibodies to a protein can be introduced to prevent
the endogenous protein from functioning.

Viral Vectors Introduce Foreign DNA into Cells with High Efficiency

Although naked DNA introduced by transfection can
be transiently expressed in up to half the cells in a
culture, more frequently the fraction of transiently
transfected cells in much lower. In fact, some cells are
almost completely refractory to transfection by the
artificial methods we have discussed in this chapter.
Many applications of recombinant DNA technology
require introducing foreign genes into recalcitrant cell
types. Potential gene therapy strategies, for example,
require efficient means for transferring genes into
normal human cells (see Chapter 28). For some applications in research and biotechnology, such as high-
level production of protein from a cloned gene, it is
paramount to get the cloned DNA into as many cells
as possible.

To solve these problems, researchers have turned
to viruses. Viral growth depends on the ability to get

FIGURE **12-9**

SV40 viral vectors. The gene to be expressed (foreign DNA) is cloned into a site after the viral late promoter, in place of the SV40 late genes. The recombinant viral genome is excised from the plasmid (using *Bam*HI restriction enzyme which cuts at "B") and circularized with DNA ligase. The circularized molecules are cotransfected into cells with DNA from a defective SV40 helper virus that carries the functional late genes that are missing from the recombinant viral genome (this helper virus, however, is defective because it lacks the early genes). Cells that take up both plasmids begin to produce viral proteins that replicate both of the DNAs introduced and package them into infectious viral particles. The cells soon lyse, releasing virus. The resulting virus stock is harvested and can be used to infect a new culture of monkey cells. At the height of this second infection, the infected cells produce high levels of protein from the cloned gene.

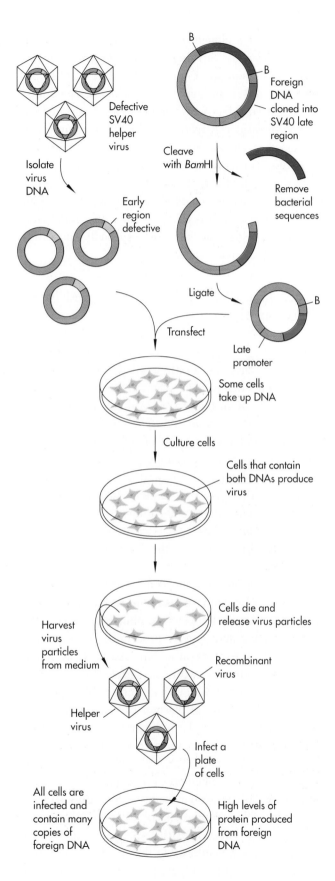

the viral genome into cells, and viruses have devised devilishly clever and efficient methods for doing it. Once they get in, viruses seize the cell's biosynthetic machinery to produce viral RNA, DNA, and protein. Viruses are sensationally souped-up gene transfer vectors. Ironically, it was the development of DNA-mediated gene transfer that allowed researchers to work out the organization of viral genomes so that they could be harnessed for more efficient gene transfer strategies. The earliest viral vectors, based on the monkey tumor virus SV40, simply substituted some of the viral genes with the foreign gene (Figure 12-9). These recombinant molecules, prepared as bacterial plasmids, were transfected into monkey cells together with a second plasmid that supplied the missing viral genes. Once inside the cells, viral gene products produced from the two plasmids cooperate to replicate both plasmids and package each into virus particles. The virus *stock* that emerges from the cell is a mixture of two viruses, each of which is by itself defective (that is, it cannot replicate on its own because it is missing necessary viral genes). Nevertheless, this virus stock can now be used to infect new cells, efficiently introducing and expressing the foreign gene in the recipient cells.

A hybrid method that uses transfection to get DNA into cells and a viral protein to replicate it once inside is now commonly used for high-level production of protein from a cloned gene. This procedure uses a cell line, *COS cells,* carrying a stably integrated portion of the SV40 genome (Figure 12-10). These cells produce the viral T antigen protein, which triggers replication of viral DNA by binding to a DNA sequence termed the *origin of replication.* The foreign gene to be expressed is cloned into a plasmid that carries the SV40 origin of replication. After transfection into COS cells, the plasmid is replicated to a very high number of copies, increasing the expression level of the foreign gene. This system was the basis for many expression cloning strategies and has been especially successful for cloning cytokine receptors.

Vaccinia and Baculovirus Are Used for High-Level Protein Production

Use of SV40-based viral vectors is limited for a number of reasons—they infect only monkey cells, the size of foreign gene that can be inserted is small, and the genomes are often rearranged or deleted. Other viral vectors are more commonly used now, either because they can infect a wider range of cells or because they accept a wider range of foreign genes. Vaccinia virus is a large DNA-containing virus that replicates entirely in the cytoplasm. Early vaccinia vectors incorporated the foreign gene directly into a nonessential region of the viral genome. Recombinant viruses are viable and upon infection transcribe the foreign gene from a nearby viral promoter. Because the viral genome is large (185,000 bp), foreign genes cannot be inserted into vaccinia by standard recombinant DNA methods; instead, it must be done by recombination inside cells, a cumbersome and lengthy procedure. A more versatile vaccinia expression system uses a ready-made recombinant virus that expresses a bacteriophage RNA polymerase (Figure 12-11). The gene to be expressed is simply cloned into a plasmid carrying a bacteriophage promoter. The plasmid is transfected into cells that have been previously infected with the vaccinia virus that expresses the RNA polymerase. The gene on the plasmid is efficiently transcribed by the bacteriophage polymerase, accounting for up to 30 percent of the RNA in the cell. An additional feature of vaccinia virus infection is that the virus shuts down host cell protein synthesis so that viral mRNA (and mRNA from the plasmid) are preferentially translated into protein.

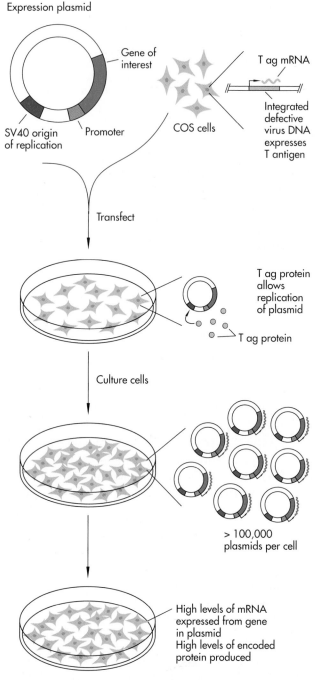

FIGURE **12-10**
Use of COS cells for high-level protein expression. The gene to be expressed is cloned into a plasmid that carries a small piece of SV40 DNA that includes the viral origin of replication. The plasmid is transfected into a special cell line, COS, that contains an integrated defective SV40 genome that expresses the viral T antigen (T ag), required for replication of viral DNA. The plasmid, because it contains the viral origin of replication, is recognized by T ag, causing the plasmid to be massively replicated. Before the transfected cells finally die a few days later, protein expressed from the gene on the plasmid accumulates to high levels.

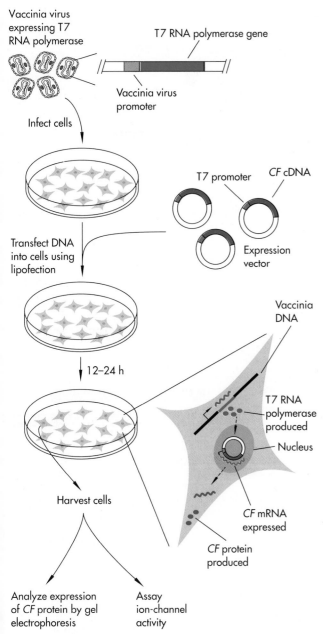

Vaccinia virus expressing T7 RNA polymerase

T7 RNA polymerase gene

Vaccinia virus promoter

Infect cells

Transfect DNA into cells using lipofection

T7 promoter

CF cDNA

Expression vector

12–24 h

Vaccinia DNA

T7 RNA polymerase produced

Nucleus

CF mRNA expressed

CF protein produced

Harvest cells

Analyze expression of *CF* protein by gel electrophoresis

Assay ion-channel activity

FIGURE **12-11**

Expression of the cystic fibrosis gene (*CF*) in a vaccinia virus vector. To prove that a wild-type *CF* gene could correct the ion-transport defect in cells from a cystic fibrosis patient, the wild-type gene was expressed in these cells using a two-component vaccinia virus expression system. First, the cells were infected with a specially engineered vaccinia virus that carries the gene encoding bacteriophage T7 RNA polymerase. The infected cells began to produce the RNA polymerase. Next, the cells were transfected with a plasmid carrying the wild-type *CF* gene downstream of a bacteriophage T7 promoter. The T7 RNA polymerase efficiently transcribed the *CF* gene into mRNA, and the mRNA was translated into protein. The resulting cells were analyzed for their ion-transport activity and found to have normal activity, proving that the cloned *CF* gene could correct the defect in the patient's cells.

Another virus widely used for protein production is an insect virus, baculovirus. Baculovirus attracted the attention of researchers because during infection, it produces one of its structural proteins (the coat protein) to spectacular levels. If a foreign gene were to be substituted for this viral gene, it too ought to be produced at high level. Baculovirus, like vaccinia, is very large, and therefore foreign genes must be placed in the viral genome by recombination (Figure 12-12).

Retroviruses Provide High-Efficiency Vectors for Stable Gene Transfer

All of the viruses we have discussed so far are *lytic* viruses—they enter cells, take over, replicate massively, and get out, killing the cell in the process. So these vectors cannot be used to introduce a gene into cells in a stable fashion. This task is most ably performed by *retroviruses*. Retroviruses are RNA viruses with a life cycle quite different from that of the lytic viruses. When they infect cells, their RNA genomes are converted to a DNA form (by the viral enzyme *reverse transcriptase*, now a key reagent in cloning). The viral DNA is efficiently integrated into the host genome, where it permanently resides, replicating along with host DNA at each cell division. This integrated *provirus* steadily produces viral RNA from a strong promoter located at the end of the genome (in a sequence called the *long terminal repeat* or *LTR*). This viral RNA serves both as mRNA for the production of viral proteins and as genomic RNA for new viruses. Viruses are assembled in the cytoplasm and bud from the cell membrane, usually with little effect on the cell's health. Thus, the retrovirus genome becomes a permanent part of the host cell genome, and any foreign gene placed in a retrovirus ought to be expressed in the cells indefinitely.

So, retroviruses make attractive vectors because they can permanently express a foreign gene in cells. Moreover, they can infect virtually every type of mammalian cell, making them exceptionally versatile. Retroviruses are widely used to study the effects of a foreign gene on cells. Oncogenes, genes that cause cancer, and their counterparts, the antioncogenes, which act to retard cell growth, are commonly studied by placing them in retrovirus vectors. Retroviruses carrying oncogenes are used to infect unusual cells taken directly from animals to establish permanent cell lines

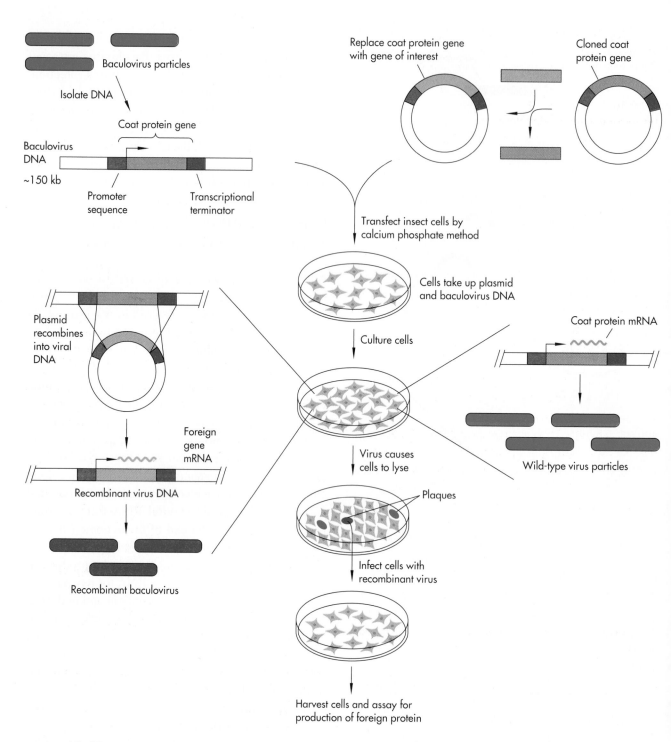

FIGURE **12-12**
Baculovirus is a very large DNA virus (genome of about 150 kb) that infects insect cells. To express a foreign gene in baculovirus, the gene of interest is cloned in place of the viral coat protein gene in a plasmid carrying a small portion of the viral genome. The recombinant plasmid is cotransfected into insect cells with wild-type baculovirus DNA. At a low frequency, the plasmid and viral DNAs recombine through homologous sequences, resulting in the insertion of the foreign gene into the viral genome. Virus plaques develop, and the plaques containing recombinant virus look different because they lack the coat protein. The plaques with recombinant virus are picked and expanded. This virus stock is then used to infect a fresh culture of insect cells, resulting in high expression of the foreign protein.

for detailed study (this is a promising approach for studying brain function, as we will see in Chapter 21). Retroviruses carrying the *E. coli lacZ* gene, encoding β-galactosidase, can be used to mark individual cells in living animals, so that they and their progeny can be located later in development. Descendants of the infected cells can be easily identified in tissue samples by treating them with X-gal, a molecule turned blue by β-galactosidase (this is the same reaction used in the blue-white screens for cloning that we described in Chapter 7). Because of their versatility, retroviruses are also the vector of choice for gene therapy (see Chapter 28).

The design and use of retroviral vectors are summarized in Figure 12-13. The vectors usually contain a selectable marker as well as the foreign gene to be expressed. Most of the viral structural genes are gone, so these vectors cannot replicate as viruses on their own. To prepare virus stocks, cloned proviral DNA is transfected into a *packaging cell*. These cells usually contain an integrated provirus with all its genes intact, but lacking the sequence recognized by the packaging apparatus. Thus, the packaging provirus produces all the proteins required for packaging of viral RNA into infectious virus particles but it cannot package its own RNA. Instead, RNA transcribed from the transfected vector is packaged into infectious virus particles and released from the cell. The resulting virus stock is termed *helper-free* because it lacks wild-type replication-competent virus. This virus stock can be used to infect a target cell culture. The recombinant genome is efficiently introduced, reverse-transcribed into DNA (by reverse transcriptase deposited in the virus by the packaging cells), and integrated into the genome. Thus, the cells now express the new virally introduced gene, but they never produce any virus, because the recombinant virus genome lacks the necessary viral genes.

The Experimenter Has Little Control over the Fate of Transferred DNA

The most significant limitation of current gene transfer methodologies is that once foreign DNA is taken up by cells, it is frequently lost; and in the rare instances that it integrates stably into the host cell genome, the integration occurs at random. Random integration

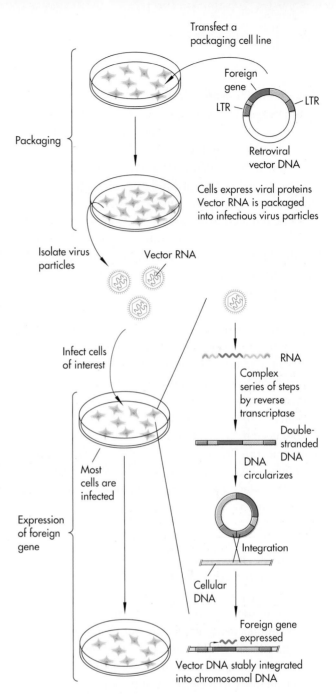

FIGURE **12-13**
Use of a retrovirus vector for stable, long-term expression of a foreign gene. The gene is cloned into a retrovirus vector that lacks most viral genes. The gene is usually expressed under the control of the strong viral promoter in the LTR. The recombinant plasmid is transfected into a special packaging cell line that harbors an integrated provirus. The provirus has been crippled so that, although it produces all the proteins required to assemble infectious viruses, its own RNA cannot be packaged into virus. Instead, RNA produced from the recombinant virus is packaged. The virus stock released from the packaging cells thus contains only recombinant virus. The virus can be used to infect virtually any other cell type, resulting in the integration of the viral genome and the stable production of the foreign gene product.

means wide fluctuations in the expression levels of the transferred gene and in the pattern of its regulation. In some cases, the foreign gene integrates in the vicinity of a highly expressed cellular gene and comes under the influence of that gene's regulatory apparatus, leading to high levels of foreign gene expression. In other cases, foreign genes insert into quiet areas of the genome, and their expression is likely to be suppressed. For the experimenter, this often means tedious and lengthy sifting of dozens of clones of transfected cells to find the cell clone with the desired level of foreign gene expression.

A second limitation of gene transfer in mammalian cells is that it can be used only to bring new genetic information into cells. For example, one can study the effects of a adding a gene to cells, but not the effects of removing a gene. In addition, studying the function of introduced genes is sometimes complicated by the presence of the normal counterpart of the gene in the recipient cells. Several strategies have been used to overcome this limitation—for example, by hunting for mutant cell lines lacking the endogenous activity— but recent developments promise more general and powerful approaches for dealing with these problems.

Antisense RNA and DNA Can Extinguish Gene Function

A versatile new approach for shutting off an endogenous cellular gene targets the gene's mRNA rather than the gene itself. The idea is to introduce into cells an RNA or single-stranded DNA molecule that is complementary to the mRNA of the target gene. (Figure 12-14). This *antisense* molecule can base-pair with the mRNA, preventing translation of the mRNA into protein. Several strategies have been used to introduce antisense nucleic acids into cells. In the earliest antisense experiments, antisense RNA was synthesized in vitro using bacteriophage RNA polymerase and then microinjected into cells. Injection of antisense RNA for the gene encoding actin, a major constituent of the cytoskeleton, which maintains cell shape, caused the cytoskeleton to disintegrate and the cells to change their shape. Injection of fruit fly embryos with antisense RNAs corresponding to important regulatory genes caused the embryos to develop as if they carried mutations in those genes.

Expression vectors can be constructed to produce high levels of antisense RNA in transfected cells (Figure 12-14). In several cases, this strategy has led to reduced expression of the corresponding cellular gene. Particularly dramatic results have been achieved with antisense constructs for oncogenes, the genes responsible for the aberrant growth properties of tumor cells. In several instances, antisense oncogene constructs have reverted the growth properties of tumor cells to near normal or slowed their growth. An antisense construct was used to make a transgenic mouse with a neurological disease. This construct produced antisense RNA for the gene encoding a major component of the myelin sheath that surrounds nerve fibers. The mice were therefore defective in the production of myelin and had neurological defects. This approach offers particular promise for generating animal models to study important human diseases.

A third strategy for introducing antisense nucleic acids into cells is to use synthetic single-stranded DNA oligonucleotides. When short oligonucleotides complementary to the sequence around the translational initiation site (the AUG codon) of an mRNA enter cells, they hybridize to the mRNA and prevent initiation of translation. Simply adding high concentrations of such oligonucleotides to cell cultures is sometimes sufficient to prompt cells to take them in. Chemically modifying the oligonucleotides can greatly increase the efficiency with which they enter cells and their stability once inside. This method has also been used to extinguish oncogene function, and antisense oligonucleotides directed against viral RNA can protect cells from viral infection. Although significant technical hurdles remain to be cleared, antisense technology offers hope for new therapies for cancer and for viral diseases, such as AIDS.

Homologous Recombination Is Used to Inactivate Cellular Genes

We have learned that once a gene is introduced into mammalian cells, investigators have little control over where the gene *recombines* into the host genome. In the vast majority of cases, this occurs by *heterologous recombination*, when the gene recombines into an unrelated sequence. But in some instances, a gene can recombine precisely into the identical sequence in the

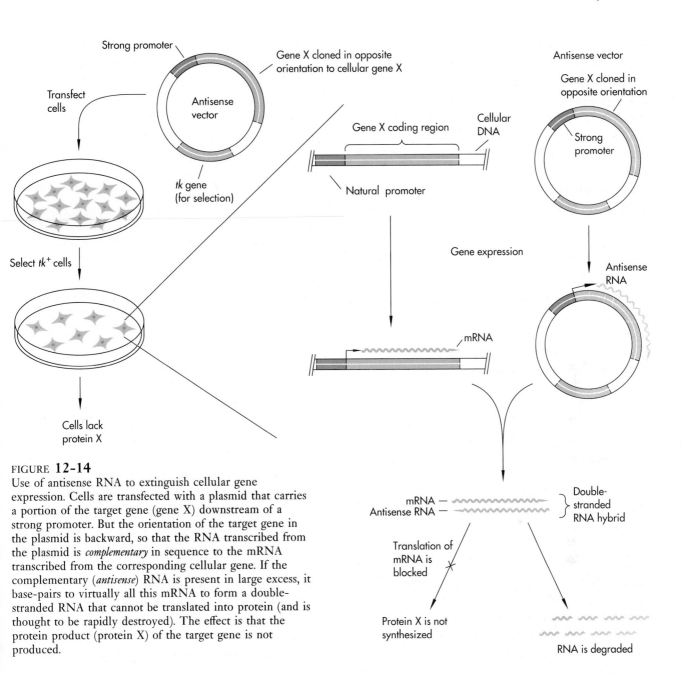

FIGURE **12-14**

Use of antisense RNA to extinguish cellular gene expression. Cells are transfected with a plasmid that carries a portion of the target gene (gene X) downstream of a strong promoter. But the orientation of the target gene in the plasmid is backward, so that the RNA transcribed from the plasmid is *complementary* in sequence to the mRNA transcribed from the corresponding cellular gene. If the complementary (*antisense*) RNA is present in large excess, it base-pairs to virtually all this mRNA to form a double-stranded RNA that cannot be translated into protein (and is thought to be rapidly destroyed). The effect is that the protein product (protein X) of the target gene is not produced.

genome by *homologous recombination*. Homologous recombination is frequently observed in bacteria, yeast, and certain viruses, but it is exceedingly rare in mammalian cells. Homologous recombination is potentially a powerful tool for studying gene function because it can be used for *gene targeting*, placing foreign DNA at a precise locus in the genome. Gene targeting is now

used to inactivate mammalian genes by homologous recombination of a disrupted gene construct into the gene of interest. In the future, it may become possible to perform gene *transplacement*, by which an endogenous gene is precisely replaced with an engineered derivative, a procedure commonly done in yeast (Chapter 13).

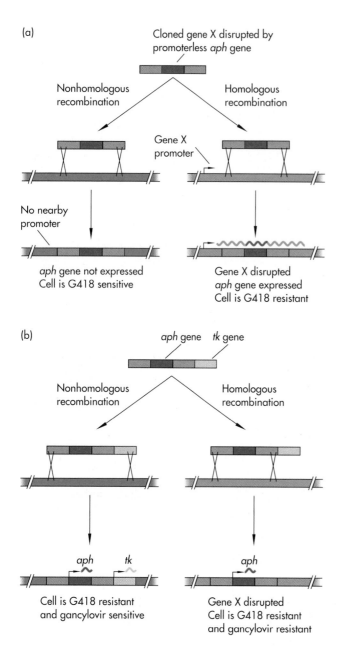

(a)

Cloned gene X disrupted by promoterless *aph* gene

Nonhomologous recombination · Homologous recombination

Gene X promoter

No nearby promoter

aph gene not expressed
Cell is G418 sensitive

Gene X disrupted
aph gene expressed
Cell is G418 resistant

(b)

aph gene · *tk* gene

Nonhomologous recombination · Homologous recombination

aph · *tk*

aph

Cell is G418 resistant
and gancyclovir sensitive

Gene X disrupted
Cell is G418 resistant
and gancyclovir resistant

FIGURE **12-15**

Two strategies for enriching for gene targeting or homologous recombination events. (a) The cloned gene is disrupted by insertion of an *aph* gene (encoding resistance to G418), rendering it nonfunctional. This *aph* gene, however, lacks a promoter of its own, so most random integration events do not lead to transcription of the gene, and these cells are usually G418-sensitive. In rare targeted integrations, the *aph* gene becomes part of the targeted gene's mRNA, leading to a G418-resistant cell and disruption of the targeted gene. (b) In this scheme, termed *positive-negative selection*, the gene is disrupted with an *aph* gene that *does* have its own promoter. But downstream of the gene is cloned a *tk* marker. In most random integrations, which occur near the ends of the transfected fragments, both markers become incorporated into the genome. A homologous recombination, however, requires recombination within the homologous cellular DNA fragments. These homologous recombination events retain the *aph* gene but not the *tk* marker. Thus, after transfection, a double selection is applied with G418 and gancyclovir. G418 ensures that only cells with an integrated *aph* gene grow up, and gancyclovir kills the cells that also express *tk*. Thus, only *aph⁺tk⁻* cells survive this selection, and these colonies frequently carry the desired targeted insertion.

Because homologous recombination is such a rare event in mammalian cells, occurring roughly once per thousand heterologous insertions, the feasibility of gene targeting relies on the ability to enrich a population of transfected cells for those with homologous recombination, or on sensitive screening methods to find the rare cells carrying the desired insertion. One way to enrich for homologous recombination events is depicted in Figure 12-15(a). The gene to be knocked out is interrupted with a selectable marker gene that

lacks its own promoter. After transfection, selection for the marker is applied. Few heterologous insertions result in the incorporation of the marker gene into a genomic sequence encoding an mRNA, so the marker is rarely expressed. Homologous recombination results in the incorporation of the marker into the transcription unit of the target gene, allowing marker expression and the survival of the cell during the selection. A second enrichment strategy, shown in Figure 12-15(b), employs a double selection—for expression of the marker that disrupts the gene and against expression of a *tk* gene placed immediately downstream. Heterologous recombination usually results in the insertion of both marker genes, so the cells do not survive the double selection. In a homologous recombination, only the first marker is retained, and the cells survive double selection. Using methods of this type, the enrichment for homologous recombinants in the small number of colonies recovered can be a hundredfold or more.

Gene targeting can also be performed without the use of selection, if there is a sensitive method for detecting homologous recombinants. A mouse devel-

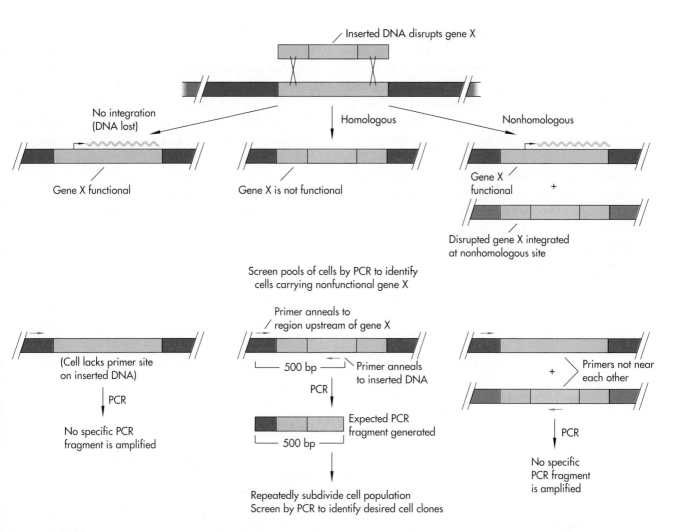

FIGURE **12-16**
Using PCR to detect gene targeting events without the use of a selectable marker. With this method, genes can be disrupted with an insertion of as little as a few base pairs. The disrupted gene fragment is microinjected directly into cells, and the injected cells are collected and grown up. To determine whether a homologous recombination event has occurred in this population, DNA is extracted from the cells and a polymerase chain reaction (PCR) is performed. One of the primers is designed to anneal and prime only at the site of the small insertion (pink), whereas the other anneals and primes in an endogenous sequence beyond the boundary of the injected fragment. In cells with no insertions or with nonhomologous recombinations, no fragments are amplified by PCR, because the primers are nowhere near one another (in fact, they likely anneal to different chromosomes). In a targeted insertion, however, the primers anneal closely to one another and in the proper configuration to yield an amplified fragment of a predicted size. If this fragment is detected in the PCR, then there must be homologously recombined genes within the cell population. The cells are split into smaller subpopulations and retested by PCR, this cycle is repeated until individual cell clones carrying the targeted gene are isolated.

opmental gene was successfully knocked out by microinjecting a few hundred cells with a DNA fragment containing a copy of the gene with a 20-bp insertion disrupting the reading frame. The cells were grown up, and DNA was extracted and analyzed by the polymerase chain reaction (Figure 12-16). The PCR used two primers, one that annealed to the inserted sequence and one that annealed to natural mouse DNA beyond the end of the injected fragment. Only in the event of a homologous recombination did the PCR yield a fragment of the expected size. Because of the sensitivity of the PCR, the presence of a homologous insertion in approximately 1 cell in 150 was detectable.

By repeated rounds of subdivision and PCR testing of the original culture, pure cell clones in which the targeted gene was knocked out were recovered.

As we will learn in Chapter 14, the development of gene targeting, combined with our increased knowledge of mouse embryology, has made possible the prospect of introducing any desired gene mutation into a living mouse, where it can be passed on to offspring like any other genetic locus. These methods will contribute substantially to our understanding of cell growth and development and will have the practical benefit of generating animal models for human genetic diseases.

Reading list

General

Gluzman, Y. *Eukaryotic Viral Vectors.* Cold Spring Harbor Laboratory, Cold Spring Harbor, N. Y., 1982.

Original Research Papers

DEVELOPMENT OF GENE TRANSFER TECHNOLOGY

Pellicer, A., D. Robins, B. Wold, R. Sweet, J. Jackson, I. Lowy, J. M. Roberts, G. K. Sim, S. Silverstein, and R. Axel. "Altering genotype and phenotype by DNA-mediated gene transfer." *Science,* 209: 1414–1422 (1980). [Review]

Graham, F. L., and A. J. van der Eb. "A new technique for the assay of infectivity of human Adenovirus 5 DNA." *Virol.,* 52: 456–467 (1973).

Goff, S. P., and P. Berg. "Construction of hybrid viruses containing SV40 and λ phage DNA segments and their propagation in cultured monkey cells." *Cell,* 9: 695–705 (1976).

Graessmann, M., and A. Graessmann. "Early simian-virus-40-specific RNA contains information for tumor antigen formation and chromatin replication." *Proc. Natl. Acad. Sci. USA,* 73: 366–370 (1976).

Wigler, M., S. Silverstein, L.-S. Lee, A. Pellicer, Y.-C. Cheng, and R. Axel. "Transfer of purified Herpes virus thymidine kinase gene to cultured mouse cells." *Cell,* 11: 223–232 (1977).

Wigler, M., A. Pellicer, S. Silverstein, and R. Axel. "Biochemical transfer of single-copy eucaryotic genes using total cellular DNA as donor." *Cell,* 13: 725–731 (1978).

Mulligan, R. C., B. H. Howard, and P. Berg. "Synthesis of rabbit β-globin in cultured monkey kidney cells following infection with an SV40 β-globin recombinant genome." *Nature,* 277: 108–114 (1979).

Wigler, M., R. Sweet, G. K. Sim, B. Wold, A. Pellicer, E. Lacy, T. Maniatis, S. Silverstein, and R. Axel. "Transformation of mammalian cells with genes from procaryotes and eucaryotes." *Cell,* 16: 777–785 (1979).

Lowy, I., A. Pellicer, J. F. Jackson, G. K. Sim, S. Silverstein, and R. Axel. "Isolation of transforming DNA: cloning of the hampster *aprt* gene." *Cell,* 22: 817–823 (1980).

Perucho, M., D. Hanahan, and M. Wigler. "Genetic and physical linkage of exogenous sequences in transformed cells." *Cell,* 22: 309–317 (1980).

Gluzman, Y. "SV40-transformed simian cells support the replication of early SV40 mutants." *Cell,* 23: 175–182 (1981).

Mulligan, R. C., and P. Berg. "Selection for animal cells that express the *Escherichia coli* gene coding for xanthine-guanine phosphoribosyltransferase." *Proc. Natl. Acad. Sci. USA,* 78: 2072–2076 (1981).

Robins, D. M., S. Ripley, A. S. Henderson, and R. Axel. "Transforming DNA integrates into the host chromosome." *Cell,* 23: 29–39 (1981).

Southern, P. J., and P. Berg. "Transformation of mammalian cells to antibiotic resistance with a bacterial gene under control of the SV40 early region promoter." *J. Mol. Appl. Gen.* 1: 327–342 (1982).

GENE AMPLIFICATION

Schimke, R. T. "Gene amplification in cultured animal cells." *Cell,* 37: 705–713 (1984).

SPECIALIZED METHODS OF TRANSFECTION

Mannino, R. J., and S. Gould-Fogerite. "Liposome mediated gene transfer." *BioTechniques,* 6: 682–690 (1988). [Review]

Shigekawa, K., and W. J. Dower. "Electroporation of eukaryotes and prokaryotes: a general approach to the introduction of macromolecules into cells." *BioTechniques,* 6: 742–751 (1988). [Review]

Capecchi, M. "High efficiency transformation by direct microinjection into cultured mammalian cells." *Cell,* 22: 479–488 (1980).

Felgner, P. L., T. R. Gadek, M. Holm, R. Roman, H. W. Chan, M. Wenz, J. P. Northrop, G. M. Ringold, and M. Danielson. "Lipofection: a highly efficient, lipid-mediated DNA-transfection procedure." *Proc. Natl. Acad. Sci. USA,* 84: 7413–7417 (1987).

Klein, T. M., E. D. Wolf, R. Wu, and J. C. Sanford. "High-velocity microprojectiles for delivering nucleic acids into living cells." *Nature,* 327: 70–73 (1987).

DNA VIRAL VECTORS

Berkner, K. L. "Development of adenovirus vectors for the expression of heterologous genes." *BioTechniques,* 6: 616–629 (1988). [Review]

Miller, L. K. "Insect baculoviruses: powerful gene expression vectors." *Bioessays,* 11: 91–95 (1989). [Review]

Moss, B., O. Elroy-Stein, T. Mizukami, W. A. Alexander, and T. R. Fuerst. "New mammalian expression vectors." *Nature,* 348: 91–92 (1990). [Review]

DiMaio, D., R. Treisman, and T. Maniatis. "Bovine papilloma virus vector that propagates as a plasmid in both mouse and bacteria cells." *Proc. Natl Acad Sci USA,* 79: 4030–4034 (1982).

Rich, D. P., M. P. Anderson, R. J. Gregory, S. H. Cheng, S. Paul, D. M. Jefferson, J. D. McCann, K. W. Klinger, A. E. Smith, and M. J. Welsh. "Expression of cystic fibrosis transmembrane conductance regulator corrects defective chloride channel regulation in cystic fibrosis airway epithelial cells." *Nature,* 347: 358–363 (1990).

RNA VIRAL VECTORS

Cepko, C. "Retrovirus vectors and their application in neurobiology." *Neuron,* 1: 345–353 (1988). [Review]

Eglitis, M. A., and W. F. Anderson. "Retroviral vectors for introduction of genes into mammalian cells." *BioTechniques,* 6: 608–614 (1988). [Review]

Shimotohno, K., and H. M. Temin. "Formation of infectious progeny virus after insertion of Herpes Simplex thymidine kinase gene into DNA of an avian retrovirus." *Cell,* 26: 67–77 (1981).

Mann, R., R. C. Mulligan, and D. Baltimore. "Construction of a retrovirus packaging mutant and its use to produce helper-free defective retrovirus." *Cell,* 33: 153–159 (1983).

Cepko, C. L., B. E. Roberts, and R. C. Mulligan. "Construction and applications of a highly transmissible murine retrovirus shuttle vector." *Cell,* 37: 1053–1062 (1984).

Lemischka, I. R., D. H. Raulet, and R. C. Mulligan. "Developmental potential and dynamic behavior of hematopoietic stem cells." *Cell,* 45: 917–927 (1986).

Sanes, J. R., J. L. R. Rubenstein, and J.-F. Nicolas. "Use of a recombinant retrovirus to study post-implantation cell lineage in mouse embryos." *EMBO J.,* 5: 3133–3142 (1986).

Price, J., D. Turner, and C. Cepko. " Lineage analysis in the vertebrate nervous system by retrovirus-mediated gene transfer." *Proc. Natl. Acad. Sci. USA,* 84: 156–160 (1987).

ANTISENSE RNA AND DNA

Helene, C., and J.-J. Toulme. "Specific regulation of gene expression by antisense, sense, and antigene nucleic acids." *Biochim. Biophys. Acta,* 1049: 99–125 (1990). [Review]

Weintraub, H. M. "Antisense RNA and DNA." *Sci. Am.,* 40–46 (1990). [Review]

Izant, J. G., and H. Weintraub. "Inhibition of thymidine kinase gene expression by antisense RNA: a molecular approach to genetic analysis." *Cell,* 36: 1007–1015 (1984).

Rozenberg, U. B., A. Preiss, E. Seifert, H. Jackle, and D. C. Knipple. "Production of phenocopies by *Kruppel* antisense RNA injection into *Drosophila* embryos." *Nature,* 313: 703–706 (1985).

Heikkila, R., G. Schwab, E. Wickstrom, S. L. Loke, D. H. Pluznick, R. Watt, and L. M. Neckers. "A c-*myc* antisense oligodeoxynucleotide inhibits entry into S phase but does not progress form G0 to G1." *Nature,* 328: 445–449 (1987).

Agarwal, S., J. Goodchild, M. P. Civeira, A. H. Thornton, P. S. Sarin, and P. C. Zamecnik. "Oligodeoxynucleoside phosphoramidates and phosphorothioates as inhibitors of human immunodeficiency virus." *Proc. Natl. Acad. Sci. USA,* 85: 7079–7083 (1988).

HOMOLOGOUS RECOMBINATION

Bollag, R. J., A. S. Waldman, and R. M. Liskay. "Homologous recombination in mammalian cells." *Ann. Rev. Gen.* 23: 199–225 (1989). [Review]

Capecchi, M. R. "Altering the genome by homologous recombination." *Science,* 244: 1288–1292 (1989). [Review]

Smithies, O., R. G. Gregg, S. S. Boggs, M. A. Koralewski, and R. S. Kucherlapati. "Insertion of DNA sequences into the human chromosomal β-globin locus by homologous recombination." *Nature,* 317: 230–234 (1985).

Thomas, K. R., and M. R. Capecchi. "Site-directed mutagenesis by gene targeting in mouse-embryo-derived stem cells." *Cell,* 51: 503–512 (1987).

Doetschman, T., N. Maeda, and O. Smithies. "Targeted mutation of the *hprt* gene in mouse embryonic stem cells." *Proc. Natl. Acad. Sci. USA.,* 85: 8583–8587 (1988).

Mansour, S. L., K. R. Thomas, and M. R. Capecchi. "Disruption of the proto-oncogene *int*-2 in mouse embryo-derived stem cells: a general strategy for targeting mutations to non-selectable genes." *Nature,* 336: 348–352 (1988).

Zimmer, A., and P. Gruss. "Production of chimaeric mice containing embryonic stem (ES) cells carrying a homeobox Hox 1.1 allele mutated by homologous recombination." *Nature,* 338: 150–153 (1989).

13

Using Yeast to Study Eukaryotic Gene Function

Although we now have the ability to study gene function in mammalian cells thanks to the gene transfer methods discussed in Chapter 12, our power to manipulate the genome of mammalian cells remains severely limited. For example, although we can introduce modified genes into mammalian cells with ease, we cannot necessarily ensure that their regulation will be correct. And the endogenous wild-type gene remains present, confounding study of the modified gene. With current technology, eliminating a wild-type gene from the genome of a mammalian cell is exceedingly difficult, preventing us from asking the simplest question about a gene: Is it required for cell growth (or any other cell function we can examine)? These limitations, in part due to lack of technology and in part due simply to the biology of mammalian cells, are significant and have slowed progress in the study of many of the complex systems in these cells.

Fortunately, this gap in our experimental arsenal has been filled by studies of an unlikely experimental organism, the baker's yeast *Saccharomyces cerevisiae.* The study of yeast such as *S. cerevisiae* and the distantly related species *Schizosaccharomyces pombe* has been favored by microbiologists because of several attractive aspects of their biology that make them almost as easy to work with as bacteria. First, they grow quickly, doubling in about 2 hours. This means that thousands of clonal yeast colonies can be cultured on petri dishes in 2 days. Colonies of mammalian cells, which double in 16 to 24 hours at best, take 2 to 3 weeks of careful tending to culture enough cells for experiments. Second, the yeast genome is very small, only a few times larger than that of *E. coli* and 200-fold smaller than that of mammalian cells (Figure 13-1), which greatly simplifies both genetic and molecular analysis. For example, a yeast genomic library can be carried on only a few thousand plasmids or phages, whereas a complete mammalian library requires about a million. Third, yeast is especially suited for genetic analysis because cells can be maintained either as haploids (one genome complement) or diploids (two genome complements). Thus, genetically recessive mutations can be easily obtained by working with haploid cells, and genetic complementation, the geneticist's fundamental tool, can be done simply by mating two different haploid mutants to form a diploid. Mammalian cells are diploid (or sometimes even more complex), making recessive mutations virtually impossible to detect.

The interest in studying yeast has been greatly spurred by the realization that, as eukaryotes, yeast cells are organized much like cells in more complex organisms and that many yeast proteins are closely related in structure and function to their mammalian counterparts. Thus, by studying yeast, which is so much easier, we also learn about what our own cells do. In this chapter we will learn about how the yeast genome is experimentally manipulated, and we will examine some cases in which the study of yeast has contributed to our understanding of complex cellular processes.

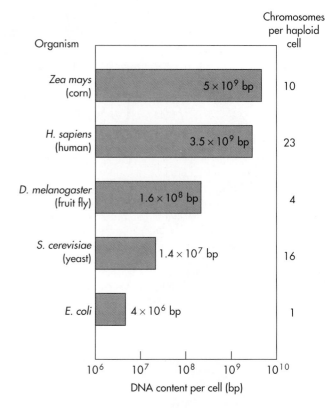

FIGURE **13-1**
The size of the genome in selected organisms. Each bar represents the total size in base pairs of the genome of the indicated organism (note that it is a logarithmic scale). The number of chromosomes in a haploid genome is shown. The compact size of the yeast genome relative to the human genome is due to both lower genetic complexity (yeast live a much simpler lifestyle) and a higher density of genetic information (few yeast genes contain introns and the genes are tightly packed). Note also that despite the small amount of DNA that makes up the yeast genome, it is nevertheless distributed to 16 chromosomes.

Yeast Biosynthetic Genes Are Cloned by Complementation of *E. coli* Mutations

We have already learned that genetic markers (drug resistance, for example) are used to identify the rare cells in a population into which new DNA has been introduced. Obtaining genetic markers that work in yeast was one of the breakthroughs that permitted the use of recombinant DNA to manipulate the yeast genome. The first yeast genetic markers were genes encoding enzymes involved in amino acid and nucleotide biosynthesis. These could be obtained because they

functionally complemented well-characterized *E. coli* mutations in these biosynthetic pathways. For example, the yeast *LEU2* gene, which encodes the enzyme β-isopropylmalate dehydrogenase in the leucine biosynthesis pathway, can complement *E. coli leuB* mutants, which are deficient in the same enzyme. Several yeast genes of this type were cloned by transforming *E. coli* mutants with libraries of plasmids carrying yeast DNA and simply selecting for plasmids that restored the missing function. In the case of *LEU2*, the yeast gene permitted the bacteria to grow on plates lacking leucine (Figure 13-2).

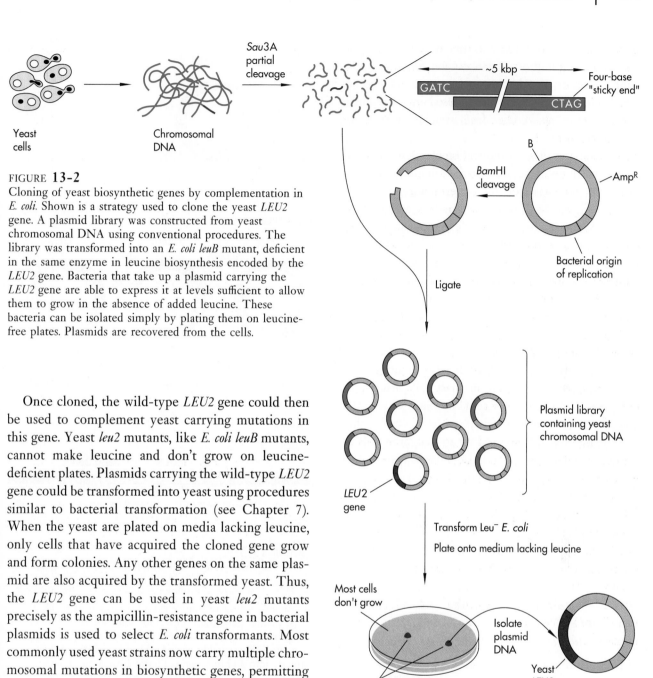

FIGURE **13-2**
Cloning of yeast biosynthetic genes by complementation in *E. coli.* Shown is a strategy used to clone the yeast *LEU2* gene. A plasmid library was constructed from yeast chromosomal DNA using conventional procedures. The library was transformed into an *E. coli leuB* mutant, deficient in the same enzyme in leucine biosynthesis encoded by the *LEU2* gene. Bacteria that take up a plasmid carrying the *LEU2* gene are able to express it at levels sufficient to allow them to grow in the absence of added leucine. These bacteria can be isolated simply by plating them on leucine-free plates. Plasmids are recovered from the cells.

Once cloned, the wild-type *LEU2* gene could then be used to complement yeast carrying mutations in this gene. Yeast *leu2* mutants, like *E. coli leuB* mutants, cannot make leucine and don't grow on leucine-deficient plates. Plasmids carrying the wild-type *LEU2* gene could be transformed into yeast using procedures similar to bacterial transformation (see Chapter 7). When the yeast are plated on media lacking leucine, only cells that have acquired the cloned gene grow and form colonies. Any other genes on the same plasmid are also acquired by the transformed yeast. Thus, the *LEU2* gene can be used in yeast *leu2* mutants precisely as the ampicillin-resistance gene in bacterial plasmids is used to select *E. coli* transformants. Most commonly used yeast strains now carry multiple chromosomal mutations in biosynthetic genes, permitting the use of several different genetic markers (Table 13-1).

Shuttle Vectors Replicate in Both *E. coli* and Yeast

The second breakthrough that advanced our ability to apply recombinant DNA technology to yeast was the construction of plasmid vectors that could replicate in both *E. coli* and yeast. Such vectors allowed plasmids to be manipulated by conventional recombinant DNA methods, propagated in bacteria like any other plasmid, and then returned to yeast for study. Shuttle vectors must contain selectable markers and origins of DNA replication that work in each organism. Although, as we discussed, many yeast markers will work in *E. coli*, modern shuttle vectors usually carry the workhorse ampicillin-resistance gene found in most cloning vectors, in addition to a yeast marker gene. The bacterial replication origin used in all modern shuttle vectors also comes from standard laboratory cloning vectors.

Several types of yeast replication elements are used in shuttle vectors, depending on the number of plasmid molecules per yeast cell the experimenter wishes to achieve (Figure 13-3). Vectors that lack a yeast replication origin, and thus are unable to replicate in yeast, are termed *integrating vectors,* because the only way that cells can stably express the marker and form a colony is by integrating the plasmid directly into a yeast chromosome. Yeast replication origins have been isolated from a naturally occurring yeast plasmid, the *2μ circle,* and from chromosomal DNA, where they are termed *autonomously replicating sequence,* or *ARS,* elements. The presence of the *2μ* or ARS sequence in a vector allows it to replicate in yeast cells, usually to a dozen or more copies per cell. Plasmids carrying the *2μ* sequence are usually fairly stable in yeast—that is, they are efficiently passed on to daughter cells at mitosis. ARS-containing plasmids, however, are not so

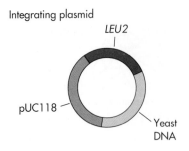

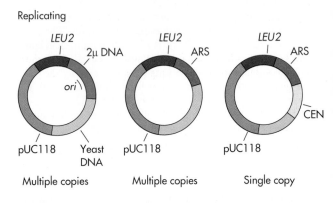

FIGURE **13-3**

Classes of yeast vectors. The simplest yeast vector is an *integrating plasmid*. It carries bacterial plasmid sequences (here from the plasmid pUC118) that provide a replication origin and a drug-resistance gene for growth in *E. coli,* a yeast marker gene (here *LEU2*) for selection of transformants in yeast, and restriction sites for inserting new sequences. Yeast stably acquire this plasmid by integrating it directly into a chromosome. Figure 13-6 shows how linearizing the plasmid before transformation directs integration to a particular location. *Replicating vectors* exist as free plasmid circles in yeast. One of these uses a replication origin (*ori*) from a natural yeast plasmid, the *2μ* circle. Others use cellular replication origins called *autonomously replicating sequence (ARS)* elements. ARS-containing vectors are stabilized by the addition of a centromere (CEN) sequence, which confers stable partitioning of the plasmid at cell division. A CEN sequence also holds the plasmid copy number to 1 to 2 per cell. *Yeast artificial chromosomes* (YACs) contain both a centromere and two telomeres which function as stable chromosome ends, allowing YACs to replicate as small linear chromosomes. YACs can carry several hundred thousand base pairs of DNA, making them appropriate for specialized genome mapping procedures.

TABLE **13-1**

Yeast Selectable Markers

GENE	ENZYME	SELECTION
HIS3	Imidazole glycerolphosphate dehydratase	histidine
LEU2	β-Isopropylmalate dehydrogenase	leucine
LYS2	α-Aminoadipate reductase	lysine
TRP1	*N*-(5′-phosphoribosyl)-anthranilate isomerase	tryptophan
URA3	Orotidine-5′-phosphate decarboxylase	uracil

Yeast labs usually have a series of mixes, termed *drop-out media,* that contain all pertinent nutrients but one. Thus, *HIS3* transformants are selected on media lacking histidine. These are all positive selections; that is, they select for yeast that have acquired a new gene. The *URA3* gene provides additional flexibility because a selection can also be applied for loss of the gene (the selection for loss of *URA3* is resistance to 5-fluoroorotic acid).

efficiently partitioned at mitosis and are often lost if genetic selection for the markers on the plasmid is not continued.

ARS-containing plasmids can be greatly stabilized by the addition of a *centromere*, or *CEN*, sequence. Centromeres are sequences found on each chromosome that mediate the stable replication and precise partitioning of chromosomes at mitosis and meiosis. Not only are plasmids containing a CEN element extraordinarily stable, but they are tightly maintained in yeast at a copy number equivalent to that of a chromosome, usually one or two. Thus, by choosing either a 2μ or a CEN plasmid, an experimenter can achieve a gene copy number equivalent to a single-copy chromosomal gene or a highly elevated copy number, usually leading to higher expression of the gene product. The ability to overexpress yeast genes simply by placing them on a high-copy plasmid is the basis for many of the genetic experiments we will study shortly.

A variation of the CEN vector is the *yeast artificial chromosome (YAC)*. YACs contain additional sequences called *telomeres*, which are found at the ends of chromosomes. The presence of telomeres on the ends of YACs permits YACs to replicate as linear chromosomelike molecules. Although YACs are not used for routine cloning experiments, their unprecedented capacity for foreign DNA (hundreds of kilobase pairs) makes them an important tool in the physical characterization of large genomes (see Chapter 29).

Yeast Genes Can Be Cloned by Simple Complementation Strategies

Although yeast genes are sometimes cloned by conventional hybridization procedures when probes are available, often a gene can be directly cloned based on its ability to correct a defect in a characterized mutant strain. This procedure exploits genetic complementation in precisely the same way as use of a *LEU2* marker. The availability of flexible cloning vectors that replicate in yeast, coupled with the organism's small genome, makes it relatively simple to clone yeast genes by complementation (Figure 13-4). Once a yeast mutant with an interesting defect is isolated, the mutant can be transformed with a plasmid library carrying yeast genomic DNA fragments. Since the yeast gen-

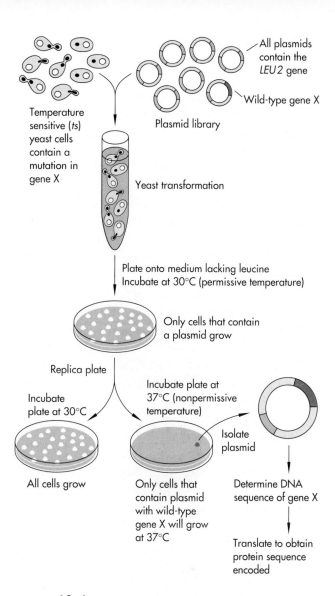

FIGURE **13-4**

Cloning a yeast gene by genetic complementation. Because inactivating mutations in essential genes are lethal, such genes are often identified by *conditional-lethal mutations*. One common type is a temperature-sensitive (*ts*) mutant, which grows at a low (*permissive*) temperature but not at a high (*nonpermissive*) temperature. Temperature-sensitive phenotypes usually arise when a gene sustains a missense mutation encoding a protein that is unstable or otherwise inactivated at the higher temperature. The wild-type gene damaged in a *ts* mutant can be easily cloned by genetic complementation as shown here for a *ts* mutation in gene X. A plasmid library is transformed into the *ts* mutant strain, which also has a mutation in the *LEU2* gene. Transformed cells are plated at the permissive temperature (30°C) on plates lacking leucine to select colonies that have acquired plasmids (all of the plasmids carry the *LEU2* gene). The colonies are transferred by replica plating (Figure 13-5) to a second plate, which is incubated at 37°C. Only colonies with plasmids carrying the wild-type gene X can survive when the mutant gene product is inactivated. Plasmids are isolated from the surviving colonies, yielding the cloned wild-type gene. Occasionally, this strategy will lead to the isolation of different genes, termed *suppressors*, which suppress the mutant phenotype because they have a related function.

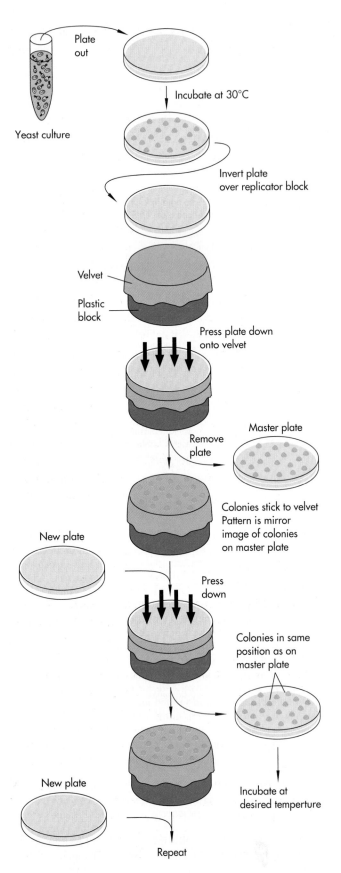

Yeast culture

Plate out

Incubate at 30°C

Invert plate over replicator block

Velvet

Plastic block

Press plate down onto velvet

Master plate

Remove plate

Colonies stick to velvet Pattern is mirror image of colonies on master plate

New plate

Press down

Colonies in same position as on master plate

New plate

Incubate at desired temperture

Repeat

FIGURE **13-5**

Replica plating, one of the basic tools of the yeast geneticist. Replica plating allows the experimenter to transfer yeast colonies quickly from one plate to another for genetic analysis. Yeast colonies are grown up on a plate and pressed onto a sterile piece of velvet. Yeast adhere to the velvet nap, yielding a print of all the colonies. The original plate is removed and can be reincubated to grow the colonies up again (usually this serves as the master plate on which the colonies are stored). A fresh plate is placed on the velvet and the transfer is reversed generating a *replica* of the colonies from the master plate onto the new plate. Several new plates can by imprinted from a single velvet. Replica plating can be used to screen colonies, selected for one marker, for presence of another. Or it can be used to generate sets of plates for incubation at different temperatures, as in Figure 13-4.

ome can be carried on only a few thousand plasmids, it is usually fairly simple to devise a genetic selection or screen, using replica plating (Figure 13-5), to identify yeast transformants in which the defect has been corrected by the presence of the wild-type gene (or sometimes a related one) on a plasmid. Once such a transformant has been found, the plasmid is recovered directly from those cells and transformed into *E. coli* for amplification.

To locate the gene within the yeast genomic sequence carried on the plasmid, individual restriction enzyme fragments (or sometimes randomly generated deletions or insertions) are cloned into new plasmids and tested individually for their ability to correct the genetic defect in the mutant strain. Once the gene's location is pinpointed, its DNA sequence is determined. This feature—the ability to move swiftly from a genetic phenotype to the sequence of the responsible gene—is one of yeast's greatest strengths as an experimental organism. In no other eukaryote is it so simple to clone genes.

Homologous Recombination Is a Relatively Frequent Event in Yeast

One of the most attractive features of working with yeast is the ease with which mutations can be targeted to specific sites in the genome. Gene targeting is extremely difficult in mammalian cells, because homologous recombination between a transferred gene and its chromosomal counterpart is rare. In yeast, however, because of the smaller genome size and differences in the recombination machinery, homologous recombination is a much more frequent event. So gene targeting is a simple procedure for yeast geneticists. Gene targeting is a powerful tool for genetic analysis because a gene modified by mutagenesis in the test tube can be placed back into the yeast genome with precision. Once incorporated into the genome, the modified sequence is stably passed to daughter cells with the fidelity of any natural gene. But most important, gene targeting allows the most stringent test of function for a modified gene, because the modified gene can be directly substituted for the natural gene.

Gene targeting is facilitated in yeast by transforming them with linear DNA molecules rather than circles, because DNA ends are favored substrates for the recombination enzymes. Thus, linearizing a plasmid within a particular yeast sequence targets the integration to that gene in the intact genome (Figure 13-6). These simple integrations result in duplication of the target gene and insertion of plasmid sequences. The plasmid sequence, which may include a marker gene, can be used to follow the integration site in genetic crosses. This strategy is sometimes used in genetic mapping of a cloned gene.

To *replace* a natural gene precisely with a modified gene involves a slightly more complex process, termed *transplacement*, which requires two closely linked homologous recombination events (Figure 13-7). Nevertheless, transplacement is experimentally quite simple, and the efficiency of the reaction makes its use routine. In the most extreme form of transplacement, the natural gene can be replaced with a completely inactive, or *null*, derivative. By performing this type of transplacement, often called a *gene knockout*, an experimenter can sometimes determine the function of the gene simply by asking if there is anything the knocked out yeast (that is, the yeast containing the knocked out gene) cannot do normally. In the case of a gene essential for growth, a knockout is lethal in a haploid cell, which normally carries only a single copy of the gene. Consequently, gene knockouts are often done in diploids—if one copy of the gene is destroyed, the yeast still survive on the remaining wild-type copy. The resulting diploid is induced to undergo meiosis or *sporulation*, leading to the generation of four haploid spores. If the gene knockout is lethal, half the resulting spores will be dead (Figure 13-8).

Cloning Genes Required for Mating Reveals a Signaling Pathway Similar to That Seen in Higher Organisms

We have mentioned that genetic analysis of yeast mutants is facilitated by the yeast sexual cycle—that two haploid cells mate to form a diploid. But the mating process itself has proved an especially fertile area for study because it involves regulatory processes common to all higher eukaryotes. Yeast cells can exist in either of two haploid mating types, called **a** and α. One cell of each type mates to form a diploid. Mating is coordinated by the release of pheromones, small peptide hormones secreted by each partner. Reception of a pheromone signal triggers a series of events that prepare the cell for mating. These include changes in cell shape, activation of genes required for mating, and arrest of cell growth so that mating can proceed. How these events are triggered by the pheromone signal is of great interest to biologists because similar signaling events control most aspects of the development and function of multicellular organisms more complex than yeast.

The components of this signaling pathway have been identified using a simple genetic selection. If **a** cells are incubated in the presence of α factor, the pheromone produced by α cells, they stop growing, because they are expecting to mate. Thus, **a** cells won't form colonies on plates containing α factor. To identify mutations in the signaling pathway activated by α factor, mutant **a** cells were selected that failed to arrest their growth and, therefore, formed colonies on α factor plates. These mutations identified a number of genes, termed *STE* genes, because mutants are sterile (unable to mate). The wild-type *STE* genes were then cloned by complementation of the mating defect (this is accomplished using a simple replica plating assay). Sequence analysis of the cloned *STE* genes suggested in many cases what their functions might be. The *STE2* gene, for example, encoded a protein similar to hormone receptors in mammalian cells, which suggested that the protein might be the α factor receptor.

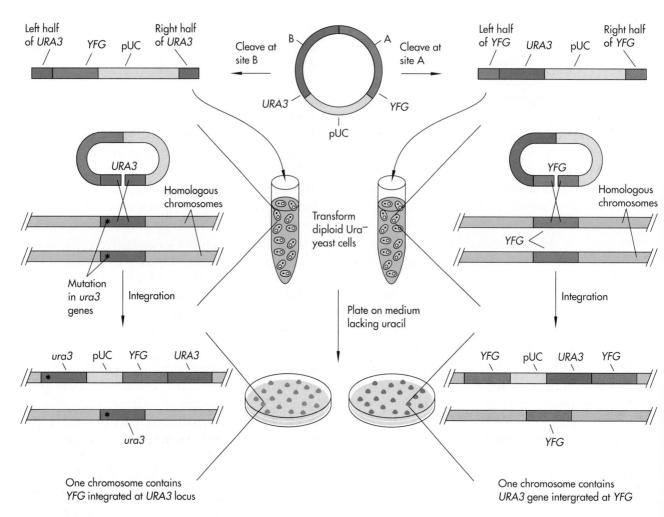

FIGURE **13-6**

Gene targeting by homologous recombination. Integration is targeted by linearizing the plasmid with a restriction enzyme. Because free ends are preferred substrates for recombination, the plasmid integrates preferentially into the chromosomal sequence homologous to the plasmid ends. Thus, a plasmid carrying a marker gene (*URA3*) and your favorite gene (*YFG*) can be directed to either gene by cleaving the plasmid at sites B or A, respectively. In either case, integration is done in a yeast strain carrying a chromosomal *ura3* mutation (marked with an *), and transformants are selected on plates lacking uracil. Note that the target gene is duplicated and the entire plasmid becomes integrated. Usually, no genetic information is lost at the target locus.

Additional biochemical and genetic experiments confirmed this assignment. Other *STE* gene products were readily identified as enzymes involved in the production of the **a** cell pheromone; protein kinases, key regulatory enzymes known to mediate signaling processes in other organisms and subunits of G proteins, also known to work as regulatory switches in other systems (see Chapters 17, 18, and 19).

Based on the sequence information provided by cloning of the *STE* genes, a preliminary view of the pheromone signaling pathway has emerged (Figure 13-9). Genetic and biochemical experiments have provided further information about the function of these genes and the order in which they work along the pathway. For example, biochemical studies of the *STE12* protein suggest that it is a DNA-binding protein

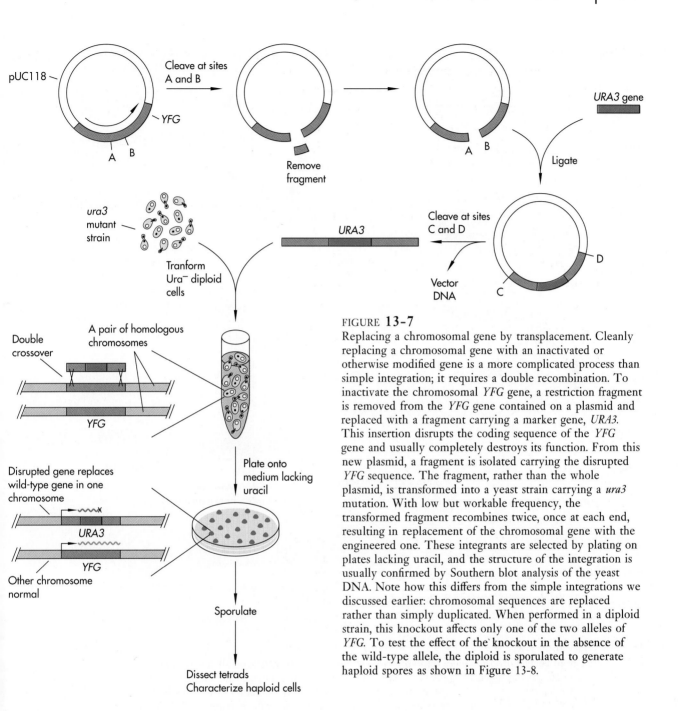

FIGURE **13-7**
Replacing a chromosomal gene by transplacement. Cleanly replacing a chromosomal gene with an inactivated or otherwise modified gene is a more complicated process than simple integration; it requires a double recombination. To inactivate the chromosomal *YFG* gene, a restriction fragment is removed from the *YFG* gene contained on a plasmid and replaced with a fragment carrying a marker gene, *URA3*. This insertion disrupts the coding sequence of the *YFG* gene and usually completely destroys its function. From this new plasmid, a fragment is isolated carrying the disrupted *YFG* sequence. The fragment, rather than the whole plasmid, is transformed into a yeast strain carrying a *ura3* mutation. With low but workable frequency, the transformed fragment recombines twice, once at each end, resulting in replacement of the chromosomal gene with the engineered one. These integrants are selected by plating on plates lacking uracil, and the structure of the integration is usually confirmed by Southern blot analysis of the yeast DNA. Note how this differs from the simple integrations we discussed earlier: chromosomal sequences are replaced rather than simply duplicated. When performed in a diploid strain, this knockout affects only one of the two alleles of *YFG*. To test the effect of the knockout in the absence of the wild-type allele, the diploid is sporulated to generate haploid spores as shown in Figure 13-8.

and likely functions as a transcription factor that turns on genes in response to the pheromone signal. Consistent with this hypothesis, simply overexpressing the *STE12* gene on a high-copy 2μ plasmid is sufficient to partially correct the mating defect in mutants for genes that function earlier in the pathway (including receptor mutations).

Genetic analysis has also clarified our view of how the G protein controls the signaling pathway. G proteins usually contain three subunits, α, β, and γ. The *STE4* and *STE18* genes, which encode the β and γ subunits, respectively, were identified because mutations in these genes make cells α factor–resistant and sterile. Thus, these gene products must be positive activators of the signaling pathways. In contrast, the gene encoding the α subunit was not identified in the

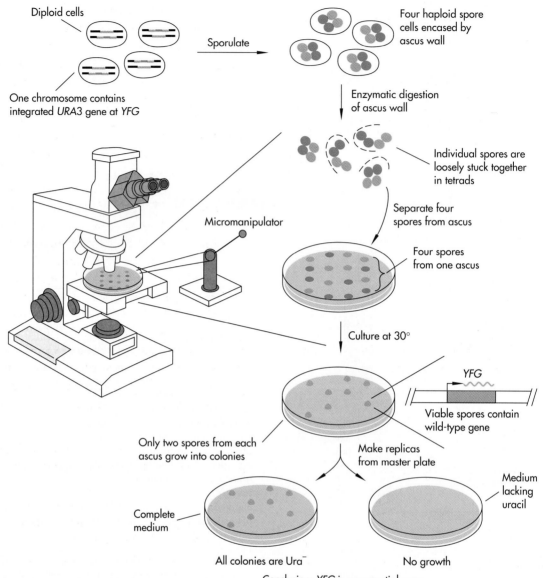

Diploid cells

One chromosome contains
integrated *URA3* gene at *YFG*

Sporulate

Four haploid spore
cells encased by
ascus wall

Enzymatic digestion
of ascus wall

Individual spores are
loosely stuck together
in tetrads

Separate four
spores from ascus

Micromanipulator

Four spores
from one ascus

Culture at 30°

YFG

Viable spores contain
wild-type gene

Only two spores from each
ascus grow into colonies

Make replicas
from master plate

Medium
lacking
uracil

Complete
medium

All colonies are Ura⁻

No growth

Conclusion: *YFG* is an essential gene

FIGURE **13-8**

Tetrad analysis: sporulating a diploid to separate mutant and wild-type alleles. Diploids
generated as described in Figure 13-7 are induced to undergo meiosis and sporulation (by
depriving them of a nitrogen source). Sporulation results in the creation of four haploid
spores tightly encased in a structure called an *ascus*. The ascus wall is stripped off with an
enzyme, and the spore clusters (*tetrads*) are placed on a plate. Through the use of a joy
stick–operated device called a *micromanipulator*, each tetrad is dissected into four individual
spores. The spores are each allowed to germinate and grow into a colony. Of the four
spores, two will carry the wild-type allele (*YFG*), and two the *YFG* allele disrupted by
integration of the *URA3* gene (*yfg::URA3*). If the knocked out gene is essential for growth,
only two of the four spores from each tetrad will grow into colonies, and both will be unable
to grow on plates lacking uracil (since the *URA3* genes are associated only with the knocked
out allele).

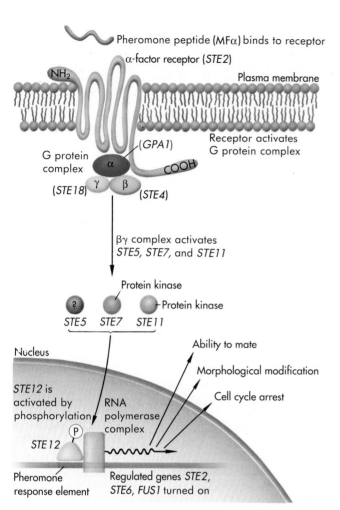

FIGURE **13-9**
The pheromone signaling pathway, as determined by molecular and classical genetics. The pheromones act through highly specific cell surface receptor proteins, encoded by the *STE2* and *STE3* genes, found on **a** and α cells, respectively. The receptors act through a three-subunit G protein (see Chapter 17) encoded by the *GPA1, STE4,* and *STE18* genes. Genetic analysis suggests that the *STE4* and *STE18* proteins activate the pathway, whereas the *GPA1* protein inhibits the pathway. Three genes—*STE5, STE7, STE11*—encode proteins that function downstream of the G protein. Unknown is whether the activation of these proteins is due to direct interaction between one of these proteins and the G protein or is due to a small molecule (*second messenger*) produced in response to G protein activation. *STE7* and *STE11* encode protein kinases; the product of the *STE5* gene is not known. The proteins encoded by *STE5, STE7,* and *STE11* act on the protein encoded by *STE12,* a transcription factor that exhibits increased activity in response to pheromone signaling. The result is activation of genes required for mating. This pathway shares many similarities with *signal transduction* pathways in higher organisms, which we discuss in detail in Chapter 17.

original α-factor resistance screen. This was because the α subunit inhibits the pathway, so mutants fire the pathway constantly and arrest their growth even in the absence of pheromone (and therefore don't grow). The lethality of these mutations is reversed, as would be expected, by breaking the pathway downstream (by mutations in *STE4, 5, 7, 11, 12,* or *18*). The gene encoding the α subunit was cloned by low-stringency hybridization with a mammalian G protein probe. Another laboratory used a genetic approach, selecting for plasmids carrying yeast genes that reversed a mutation that made cells hypersensitive to pheromone, so that even low concentrations of pheromone caused them to arrest (Figure 13-10). Overexpression of the α subunit (encoded by the gene called *GPA1*) on a high-copy 2μ plasmid reverses this hypersensitivity simply because it acts to inhibit the signaling pathway. Interestingly, expression of the same gene on a low-copy CEN plasmid has no effect. The ease of cloning yeast genes by complementation and the ability to modulate the expression of genes both up and down in yeast cells is a powerful combination for dissecting regulatory pathways of this type.

Genetic Experiments in Yeast Can Answer Precise Biochemical Questions

The ability to move genes in and out of yeast easily makes possible the design of genetic experiments that illuminate the biochemical events that occur in the cell. In the case of the pheromone signaling pathway, sequence analysis of the *STE2* and *STE3* genes suggested that they encoded the α factor and **a** factor receptors, respectively. Biochemical studies were consistent with this model as well. But proof came from genetic experiments with mutants that carried *ste2* or *ste3* mutations. These mutants do not respond to pheromone. Providing them with a wild-type *STE2* gene allowed them to respond to α factor, whereas giving them wild-type *STE3* gene allowed them to respond to **a** factor (Figure 13-11). This receptor swap experiment clearly established two facts—first, that these genes indeed encoded the receptor proteins and, second, that both receptors activated precisely the same signaling pathway in the cells.

Another example of genetic confirmation came from studies of RNA splicing in yeast. Biochemical

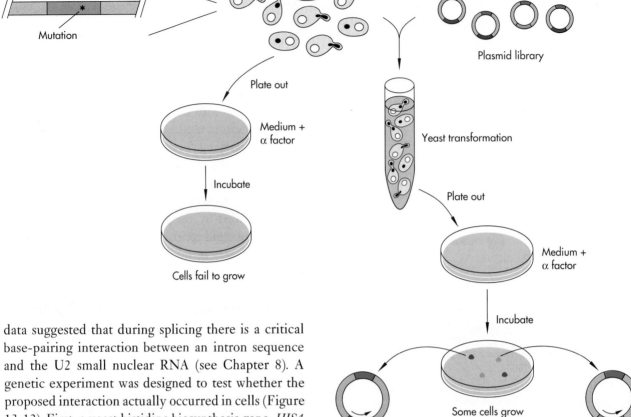

data suggested that during splicing there is a critical base-pairing interaction between an intron sequence and the U2 small nuclear RNA (see Chapter 8). A genetic experiment was designed to test whether the proposed interaction actually occurred in cells (Figure 13-12). First, a yeast histidine biosynthesis gene, *HIS4*, was engineered so that its expression required splicing of an intron. This modified gene was integrated into the genome of a yeast strain carrying a *his4* mutation. These cells grew on plates lacking histidine so long as they could make the engineered splice. A derivative was made carrying a single nucleotide substitution in the intron sequence thought to be recognized by U2 RNA. Cells carrying this gene didn't grow on plates lacking histidine, because the mutant RNA was not spliced. To test whether the U2 RNA recognized the intron sequence, a mutant U2 gene was constructed that carried a single substitution predicted to restore the base-pairing interaction with the mutant intron sequence containing a single substitution. Adding the mutant U2 gene to the strain carrying the mutant intron restored the ability to grow on plates lacking histidine by restoring splicing of the mutant intron. Examining several different combinations of intron and U2 mutations showed that only combinations that allowed correct base-pairing allowed expression of *HIS4*. Thus, this recognition must be direct.

FIGURE **13-10**

Cloning of the *GPA1* gene as a high-copy suppressor of pheromone sensitivity. Yeast mutants have been identified that are supersensitive (*sst*) to pheromone, probably because of a defect in adaptation and recovery in the presence of pheromone. These cells cannot grow in even low concentrations of pheromone that are inconsequential to normal cells. To clone genes that could overcome this defect, a yeast library in a high-copy 2μ plasmid was transformed into *sst2* cells, and cells were selected for their ability to grow in the presence of a low concentration of the α factor pheromone. Two plasmids were isolated. One carried the wild-type *SST2* gene. This fragment also corrected the defect when placed on a low-copy CEN plasmid, because it was acting simply by complementing the *sst2* mutation in the strain. The other gene worked only when it was present on the high-copy plasmid, suggesting that it was acting to *suppress* the defect in the mutant by acting elsewhere in the pathway. This gene, which was called *SCG1*, is identical to *GPA1*, which encodes the α subunit of the G protein in the pheromone signaling pathway. We now know that this protein inhibits the pathway, and this is why it reverses supersensitivity to pheromone.

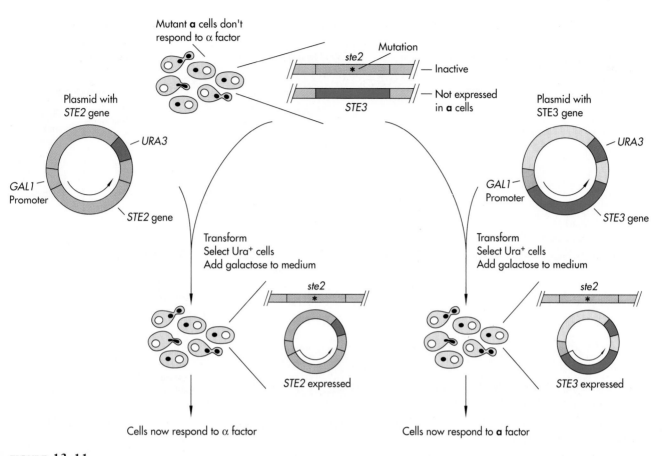

FIGURE **13-11**

The receptor swap experiment. Haploid **a** cells normally express only the *STE2* gene, which encodes the α factor receptor, but not the *STE3* gene, which encodes the **a** factor receptor. Therefore, **a** cells respond only to α factor. If the *STE2* gene carries a mutation (*ste2*), **a** cells respond to neither factor. To determine whether the putative receptor genes actually specify the response to pheromone, plasmids were constructed that carried either gene expressed under the control of the *GAL1* promoter, which can be turned on or off depending on the energy source provided to the cells. Not surprisingly, transforming the *ste2* mutants with a wild-type *STE2* gene restored response to α factor. But transforming them with a wild-type *STE3* gene switched their response, now to **a** factor. This experiment established that these genes indeed encode the pheromone receptors and that the receptors alone determine the specificity of the signaling process.

These two examples illustrate how genetic experiments in yeast have allowed the resolution of specific questions about the biochemical interactions of macromolecules inside cells. Indeed, a general strategy has been devised to use yeast genetics for direct study of protein-protein interactions, previously the exclusive domain of protein biochemists. The idea, which is illustrated in Figure 13-13, is based on the observation that the yeast GAL4 protein, like many other transcription factors, comprises two separable functional domains, one for binding DNA and one for activation of the transcription complex. Thus, it is possible to test whether any two proteins of interest interact in cells by fusing one to the GAL4 DNA-binding domain and one to the activation domain. If the proteins associate inside the yeast cell, an active GAL4-like transcription factor is assembled, and the *GAL1* promoter (which is activated by GAL4 protein) is turned on. By placing a *lacZ* reporter gene under the control of the *GAL1* promoter, association of the proteins can be

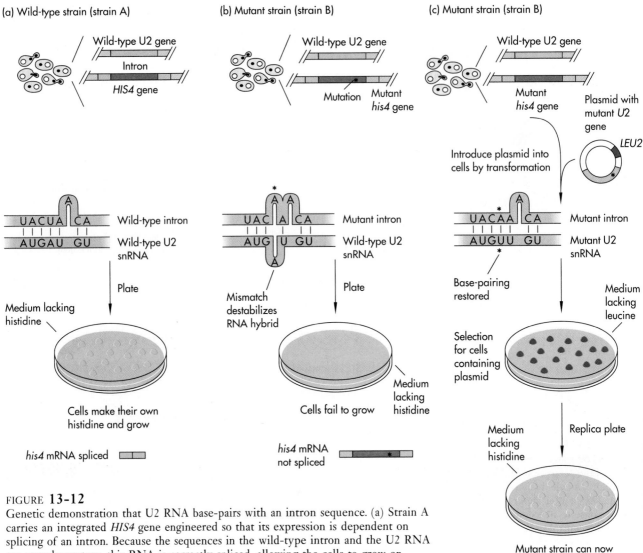

(a) Wild-type strain (strain A) **(b) Mutant strain (strain B)** **(c) Mutant strain (strain B)**

FIGURE **13-12**
Genetic demonstration that U2 RNA base-pairs with an intron sequence. (a) Strain A
carries an integrated *HIS4* gene engineered so that its expression is dependent on
splicing of an intron. Because the sequences in the wild-type intron and the U2 RNA
are complementary, this RNA is correctly spliced, allowing the cells to grow on
histidine-free plates. (b) If a single substitution is introduced into the intron sequence
(strain B), it is not stably bound by the U2 RNA, interfering with splicing, and the
cells can't grow on these plates. (c) Strain B is transformed with a plasmid carrying a
U2 gene with a nucleotide substitution that restores base-pairing with the mutant
intron. This new match restores splicing of the *HIS4* RNA, and cells can grow once
again on histidine-free medium. This experiment established that U2 RNA base-pairs
directly with the intron during splicing.

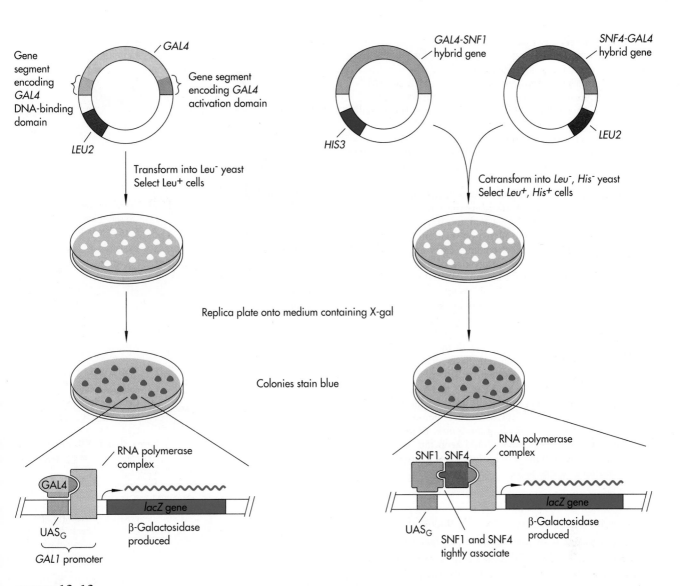

FIGURE **13-13**
A genetic assay for protein-protein interactions. The GAL4 transcription factor is composed of two functional domains, an amino-terminal domain that binds to the *GAL1* promoter upstream-activating sequence (UAS_G) and a carboxy-terminal domain that activates transcription (Chapter 9). To demonstrate that this system can be used to identify an interaction between two foreign proteins, the gene segment encoding the DNA-binding domain of GAL4 was fused to the *SNF1* gene, and the gene segment encoding the activation domain of GAL4 was fused to the *SNF4* gene. The SNF1 and SNF4 proteins are known to associate, forming a complex that has protein kinase activity. The two fusion genes, on individual plasmids, were transformed into a strain of yeast carrying the *E. coli lacZ* gene, encoding β-galactosidase, under control of the *GAL1* promoter. The cells that received both fusion genes stained blue when cultured in the presence of X-gal, a colorless substrate that is converted enzymatically to a blue dye by β-galactosidase. Control cells that received a plasmid that constitutively produces the wild-type GAL4 protein also stained blue. In contrast, cells that received only one of the fusion genes remained white in the presence of X-gal (not shown). The results show that a functional transcription factor was formed by association of the SNF1 and SFN4 proteins. This simple plate screening assay can thus be used to clone genes for proteins that interact with each other.

FIGURE **13-14**

Plasmid shuffle. This is a general technique for studying the function of mutant or foreign genes. The experiment shown here was used to demonstrate that the mammalian cAMP-dependent protein kinase could substitute for its yeast equivalent, encoded by *TPK1*. Yeast have three functionally equivalent *TPK* genes, *TPK1*, *TPK2*, and *TPK3*. A strain was constructed in which the three chromosomal *TPK* genes were knocked out by integration with *URA3*, *TRP1*, and *HIS3*. Yeast require at least one of the *TPK* genes in order to grow, so this strain was kept alive by a *TPK1* gene carried on an *ADE8* plasmid (the ADE8 gene allows growth in medium lacking adenine). A cDNA encoding the mouse protein kinase was placed on a *LEU2* plasmid and transformed into the same strain. Cells were plated on medium lacking both leucine and adenine to select colonies carrying both plasmids. The next step was to culture one of these colonies nonselectively in a complete medium (YPD) that contains leucine and adenine. Under these growth conditions, the cells no longer need to preserve the markers on the plasmids and therefore are free to lose the plasmids. Since at least one protein kinase gene is required for the cells to grow, they must keep one of the two plasmids. To determine which plasmid the cells have retained, they were plated first onto a nonselective YPD plate and then replicaplated to a plate lacking leucine and a plate lacking adenine. The result that some cells grew on the plate lacking leucine but not on the plate lacking adenine indicated that the mouse cDNA encoding the mouse protein kinase (carried on the *LEU2* plasmid) could keep the cells alive. That these cells contained only the mouse protein kinase was confirmed biochemically.

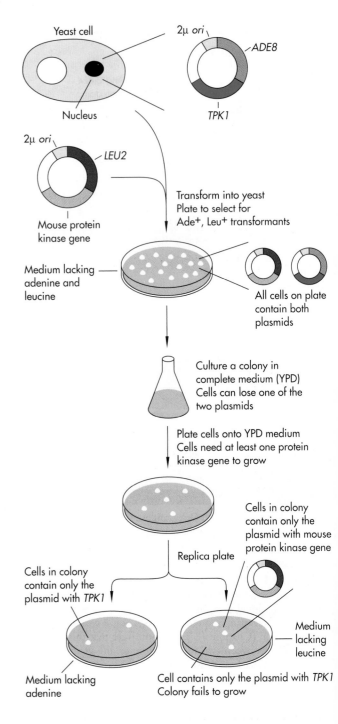

detected simply by measuring β-galactosidase enzyme activity directly or with a dye (X-gal) incorporated into the yeast plates. This type of assay can be used to hunt for the partner of a known protein by fusing the protein-coding sequence of the known protein to the sequence encoding the GAL4 DNA-binding domain and inserting a library of genes or cDNAs in frame with the sequence encoding the GAL4 activation domain. Plasmids encoding protein domains that associate with the known protein will activate the *lacZ* gene.

Genetic Analysis in Yeast Can Be Exploited to Identify and Study Genes from Higher Organisms

We have emphasized in this chapter that the study of complex processes in yeast has consistently provided insights into the function of similar processes in cells more difficult to work with. Several examples, including transcription, splicing, and cell signaling, have already been mentioned, and others will come up later in this book. But more direct approaches to studying mammalian genes in yeast have emerged. Some genes are so similar in mammals and yeast that they function normally, or nearly so, in the other organism. If this is the case, then the power of yeast genetics can be harnessed to study the mammalian gene directly. One simple tool for testing mammalian gene function in yeast is the plasmid shuffle procedure, outlined in Figure 13-14.

The study of mammalian genes in yeast has been particularly useful for the human *RAS* genes, oncogenes involved in the development of tumors (see Chapter 18). Yeast also contain *RAS* genes, very closely related in structure and function to the human genes. Indeed, a suitably engineered human *RAS* gene will substitute for the yeast genes. In mammalian cells, Ras proteins interact with a protein called GAP (for *GTPase activating protein*). Yeast also contain GAP-related genes, *IRA1* and *IRA2*. Both human GAP and another GAP-like protein involved in an inherited human disease, neurofibromatosis (see Chapter 26), will substitute for the proteins produced by the yeast *IRA* genes. These observations have greatly accelerated our understanding of these crucial human proteins.

The most dramatic use of yeast to study human genes is for cloning new human genes by direct genetic complementation of yeast mutations. A major development in our understanding of how cells regulate growth was the isolation of a human gene that complemented a mutation in a cell cycle control gene in the yeast *Schizosaccharomyces pombe*. How this experiment was done and its impact on our understanding of cell cycle control are described in Chapter 19. It is only one of the many spectacular examples of how working with a deceptively simple fungus has taught us fundamental lessons about how our own cells work.

Reading List

General

Strathern, J. N., E. W. Jones, and J. R. Broach, eds. *The Molecular Biology of the Yeast* Saccharomyces: *Life Cycle and Inheritance.* Cold Spring Harbor Laboratory, Cold Spring Harbor, N. Y., 1981.

Strathern, J. N., E. W. Jones, and J. R. Broach, eds. *The Molecular Biology of the Yeast* Saccharomyces: *Metabolism and Gene Expression.* Cold Spring Harbor Laboratory, Cold Spring Harbor, N. Y., 1982.

Sherman, F., G. R. Fink, and J. B. Hicks, eds. *Methods in Yeast Genetics.* Cold Spring Harbor Laboratory, Cold Spring Harbor, N. Y., 1986.

Guthrie, C., and G. R. Fink, eds. *Guide to Yeast Genetics and Molecular Biology.* Academic, New York, 1991.

Original Research Papers

DEVELOPMENT OF YEAST CLONING METHODS

Beggs, J. D. "Transformation of yeast by a replicating hybrid plasmid." *Nature,* 275: 104–109 (1978).

Hinnen, A., J. B. Hicks, and G. R. Fink. "Transformation of yeast." *Proc. Natl. Acad. Sci. USA,* 75: 1929–1933 (1978).

Struhl, K., D. T. Stinchcomb, S. Scherer, and R. W. Davis. "High-frequency transformation of yeast: autonomous replication of hybrid DNA molecules." *Proc. Natl. Acad. Sci. USA,* 76: 1035–1039 (1979).

Nasmyth, K. A., and S. I. Reed. "Isolation of genes by complementation in yeast: molecular cloning of a cell cycle gene." *Proc. Natl. Acad. Sci. USA,* 77: 2119–2123 (1980).

Stinchcomb, D. T., M. Thomas, J. Kelly, E. Selker, and R. W. Davis. "Eukaryotic DNA segments capable of autonomous replication in yeast." *Proc. Natl. Acad. Sci. USA,* 77: 4559–4563 (1980).

Orr-Weaver, T. L., J. W. Szostak, and R. J. Rothstein. "Yeast transformation: a model system for the study of recombination." *Proc. Natl. Acad. Sci. USA,* 78: 6354–6358 (1981).

Bloom, K. S., and J. Carbon. "Yeast centromere DNA is in a unique and highly ordered structure in chromosomes and small circular minichromosomes." *Cell,* 29: 305–317 (1982).

Shortle, D., J. E. Haber, and D. Botstein. "Lethal disruption of the yeast actin gene by integrative DNA transformation." *Science,* 217: 371–373 (1982).

Szostak, J. W., and E. H. Blackburn. "Cloning yeast telomeres on linear plasmid vectors." *Cell,* 29: 245–255 (1982).

Guarente, L. "Yeast promoters and lacZ fusions designed to study expression of cloned genes in yeast." *Meth. Enzymol.,* 101: 181–191 (1983).

Ito, H., Y. Fukuda, K. Murata, and A. Kimura. "Transformation of intact yeast cells treated with alkali cations." *J. Bacteriol.,* 153: 163–168 (1983).

Rothstein, R. J. "One-step gene disruption in yeast." *Meth. Enzymol.,* 101: 202–210 (1983).

Boeke, J., F. LaCroute, and G. R. Fink. "A positive selection for mutants lacking orotidine-5'-phosphate decarboxylase activity in yeast: 5-fluoroorotic acid resistance." *Mol. Gen. Genet,* 197: 345–346 (1984).

Burke, D. T., G. F. Carle, M. V. Olsen. "Cloning of large segments of exogenous DNA into yeast by means of artificial chromosome vectors." *Science,* 236: 806–812 (1987).

Sikorski, R. S., and P. Hieter. "A system of shuttle vectors and yeast host strains designed for efficient manipulation of DNA in *Saccharomyces cerevisiae.*" *Genetics,* 122: 19–27 (1989).

PHEROMONE RESPONSE IN SACCHAROMYCES CEREVISIAE

Herskowitz, I. "A regulatory hierarchy for cell specialization in yeast." *Nature,* 342: 749–757 (1989). [Review]

Fields, S. "Pheromone response in yeast." *Trends Biochem.,* 15: 270–273 (1990). [Review]

Hartwell, L. H. "Mutants of *Saccharomyces cerevisiae* unresponsive to cell division control by polypeptide mating hormone." *J. Cell. Biol.,* 85: 811–822 (1980).

Kurjan, J., and I. Herskowitz. "Structure of yeast pheromone gene (MFα): a putative α-factor precursor contains four tandem copies of mature α-factor." *Cell,* 30: 933–943 (1982).

Jenness, D. D., A. C. Burkholder, and L. H. Hartwell. "Binding of α-factor pheromone to yeast **a** cells: chemical and genetic evidence for an α-factor receptor." *Cell,* 35: 521–529 (1983).

Burkholder, A. C., and L. H. Hartwell. "The yeast alpha-factor receptor: structural properties deduced from the sequence of the STE2 gene." *Nuc. Acids Res.,* 13: 8463–8475 (1985).

Nakayama, N., A. Miyajima, and K. Arai. "Nucleotide sequences of STE2 and STE3: cell type-specific sterile genes from *Saccharomyces cerevisiae.*" *EMBO J.,* 4: 2643–2648 (1985).

Bender, A., and G. F. Sprague, Jr. "Yeast peptide pheromones, **a**-factor and α-factor, activate a common response mechanism in their target cells." *Cell,* 47: 929–937 (1986).

Hagen, D. C., G. McCaffrey, and G. F. Sprague, Jr. "Evidence the yeast STE3 gene encodes a receptor for the peptide pheromone **a** factor: gene sequence and implications for the structure of the presumed receptor." *Proc. Natl. Acad. Sci. USA,* 83: 1418–1422 (1986).

Dietzel, C., and J. Kurjan. "The yeast SCG1 gene: a G_α-like protein implicated in the **a**- and α-factor response pathway." *Cell,* 50: 1001–1010 (1987).

Miyajima, I., M. Nakafuku, N. Nakayama, C. Brenner, A. Miyajima, K. Kaibuchi, K. Arai, Y. Kaziro, and K. Matsumoto. "GPA1, a haploid-specific essential gene, encodes a yeast homolog of mammalian G protein which may be involved in mating factor signal transduction." *Cell,* 50: 1011–1019 (1987).

Nakayama, N., A. Miyajima, and K. Arai. "Common signal transduction system shared by STE2 and STE3 in haploid cells of *Saccharomyces cerevisiae:* autocrine cell-cycle arrest results from forced expression of STE2." *EMBO J.,* 6: 249–254 (1987).

Jahng, K.-Y., J. Ferguson, and S. I. Reed. "Mutations in a gene encoding the α subunit of a *Saccharomyces cerevisiae* G protein indicate a role in mating pheromone signaling." *Mol. Cell. Biol.,* 8: 2484–2493 (1988).

Nakayama, N., Y. Kaziro, K. Arai, and K. Matsumoto. "Role of STE genes in the mating factor signaling pathway mediated by GPA1 in *Saccharomyces cerevisiae.*" *Mol. Cell. Biol.,* 8: 3777–3783 (1988).

Blinder, D., S. Bouvier, and D. D. Jenness. "Constitutive mutants in the yeast pheromone response: ordered function of the gene products." *Cell,* 56: 479–486 (1989).

Dolan, J. W., and S. Fields. "Overproduction of the yeast STE12 protein leads to constitutive transcriptional induction." *Genes and Devel.,* 4: 492–502 (1990).

Tanaka, K., M. Nakafuku, T. Satoh, M. S. Marshall, J. B. Gibbs, K. Matsumoto, Y. Kaziro, and A. Toh-e. "*S. cerevisiae* genes IRA1 and IRA2 encode proteins that may be functionally equivalent to mammalian ras GTPase activating protein." *Cell,* 60: 803–807 (1990).

SIGNAL TRANSDUCTION BY THE RAS/cAMP PATHWAY

Cross, F., L. H. Hartwell, C. Jackson, and J. B. Konopka. "Conjugation in *Saccharomyces cerevisiae.*" *Ann. Rev. Cell Biol.,* 4: 429–457 (1988). [Review]

Broach, J. R. "RAS genes in *Saccharomyces cerevisiae:* signal transduction in search of a pathway." *Trends Gen.,* 7: 28–33 (1991). [Review]

DeFeo-Jones, D., E. M. Scolnick, R. Koller, and R. Dhar. "ras-related gene sequences identified and isolated from *Saccharomyces cerevisiae.*" *Nature,* 306: 707–709 (1983).

Powers, S., T. Kataoka, O. Fasano, M. Goldfarb, J. Strathern, J. Broach, and M. Wigler. "Genes in *S. cerevisiae* encoding proteins with domains homologous to the mammalian ras proteins." *Cell,* 36: 607–612 (1984).

DeFeo-Jones, D., K. Tatchell, L. C. Robinson, I. S. Sigal, W. C. Vass, D. R. Lowy, and E. M. Scolnick. "Mammalian and yeast ras gene products: biological function in their heterologous systems." *Science,* 228: 179–184 (1985).

Kataoka, T., S. Powers, S. Cameron, O. Fasano, M. Goldfarb, J. Broach, and M. Wigler. "Functional homology of mammalian and yeast RAS genes." *Cell*, 40: 19–26 (1985).

Toda, T., S. Cameron, P. Sass, M. Zoller, and M. Wigler. "Three different genes in *S. cerevisiae* encode the catalytic subunits of the cAMP-dependent protein kinase." *Cell*, 50: 277–287 (1987).

Ballester, R., T. Michaeli, K. Ferguson, H. P. Xu, F. McCormick, and M. Wigler. "Genetic analysis of mammalian GAP expressed in yeast." *Cell*, 59: 681–686 (1989).

Colicelli, J., C. Nicolette, C. Birchmeier, L. Rodgers, M. Riggs, and M. Wigler. "Expression of three mammalian cDNAs which interfere with RAS function in *S. cerevisiae*." *Proc. Natl. Acad. Sci. USA*, 88: 2913–2917 (1991).

Zoller, M. J., W. Yonemoto, S. S. Taylor, and K. E. Johnson. "Mammalian cAMP-dependent protein kinase functionally replaces its homolog in yeast." *Gene*, 99: 171–179 (1991).

EXAMPLES OF YEAST AS AN EXPERIMENTAL SYSTEM TO STUDY OTHER CELLULAR PROCESSES

Guarente, L. "UASs and enhancers: common mechanism of transcriptional activation in yeast and mammals." *Cell*, 52: 303–305 (1988). [Review]

Silver, P. "How proteins enter the nucleus." *Cell*, 64: 489–497 (1991) [Review]

Novick, P., C. Field, and R. Schekman. "Identification of 23 complementation groups required for post-translational events in the yeast secretory pathway." *Cell*, 21: 205–215 (1980).

Guarante, L., and M. Ptashne. "Fusion of *Escherichia coli lacZ* to the cytochrome c gene of *Saccharomyces cerevisiae*." *Proc. Natl. Acad. Sci. USA*, 78: 2199–2203 (1981).

Brent, R., and M. Ptashne. "A eukaryotic transcriptional activator bearing the DNA specificity of a prokaryotic repressor." *Cell*, 43: 729–736 (1985).

Lee, M. G., and P. Nurse. "Complementation used to clone a human homologue of the fission yeast cell cycle control gene cdc2." *Nature*, 327: 31–35 (1987).

Parker, R., P. G. Siliciano, and C. Guthrie. "Recognition of the TACTAAC box during mRNA splicing in yeast involves base pairing to the U2-like snRNA." *Cell*, 49: 229–239 (1987).

Fields, S., and O.-K. Song. "A novel genetic system to detect protein-protein interactions." *Nature*, 340: 245–246 (1989).

Rothblatt, J. A., R. J. Deshaies, S. L. Sanders, G. Daum, R. Schekman. "Multiple genes are required for proper insertion of secretory proteins into the endoplasmic reticulum." *J. Cell. Biol.*, 109: 2641–2652 (1989).

King, K., H. G. Dohlman, J. Thorner, M. G. Caron, and R. J. Lefkowitz. "Control of yeast mating signal transduction by a mammalian β_2-adrenergic receptor and G_s α subunit." *Science*, 250: 121–123 (1990).

Kranz, J. E., and C. Holm. "Cloning by function: an alternative approach for identifying homologs of genes from other organisms." *Proc. Natl. Acad. Sci. USA*, 87: 6629–6633 (1990).

14

The Introduction of Foreign Genes into Mice

We have seen in previous chapters how eukaryotic gene expression can be studied by introducing genes into yeast cells and mammalian cells in tissue culture. But what if we want to study the genetic control of development of higher organisms, where cell interactions play a crucial role? Fortunately technical developments in recent years have provided mouse geneticists with a set of tools that they can use to perform in mice experiments previously restricted to lower organisms. They can now manipulate the genes of developing embryos and introduce foreign genes into embryos. Once these genes have integrated into the genome of the recipient embryo, the embryos or the adult animals that develop from them can be analyzed to determine the pattern of expression and the phenotypic effect of the introduced gene. This provides an opportunity not only to identify genetic elements that determine tissue-specific gene expression, but also to examine the phenotypic effects of the targeted expression of one gene or several genes acting in concert. Experiments can be designed to study embryogenesis by cell ablation, nuclear transfer, or tagging of genomic sequences important for development. Transgenic mice can also provide model systems that can be used to broaden our understanding of human disease conditions, including developmental defects and cancer.

Foreign Genes Become Integrated in the Chromosomes of Recipient Animals

Before the advent of recombinant DNA techniques, the only sources of large amounts of pure genes were viruses. In the early studies designed to introduce DNA into mouse embryos, purified SV40 DNA was microinjected into the blastocoels of mouse embryos at the blastocyst stage. When the injected blastocysts were reimplanted into the uteri of foster mothers and allowed to develop, about 40 percent of the progeny had SV40 DNA in some of their cells. The mice were mosaic: in each individual tissue, some cells had SV40 DNA in their chromosomes and other cells did not. This proved that foreign DNA injected into very early embryos can be incorporated into the chromosomes of some embryonic cells and maintained in them as they both proliferate and differentiate into adult tissues. Similar experiments using the Moloney murine leukemia virus (MoMLV) showed that the provirus could become integrated into the germ line.

Once cloned genes became available, they could be microinjected into early mouse embryos. The best technique has proved to be microinjection of the cloned genes into fertilized eggs, which contain two pronuclei, one from the sperm (male), and one from the egg (female), that ultimately form the nucleus of the one-celled embryo. A few hundred copies of the foreign DNA in about 2 picoliters of solution are microinjected directly into one of the two pronuclei; the injected embryos are then transferred to the oviduct of a foster mother, and upon subsequent implantation in the uterus, many develop to term (Figure 14-1). The percentage of eggs that survive the manipulation and develop to term varies, but it is usually between 10 and 30 percent. Of the survivors, the number that have the foreign DNA integrated into their chromosomes is between a few percent and 40 percent. The introduced DNA appears to integrate randomly without preference for a particular chromosomal location, usually in a tandem array of many copies at a single locus. Mice that carry the foreign gene are referred to as *transgenic*, and the foreign DNA is termed a *transgene*.

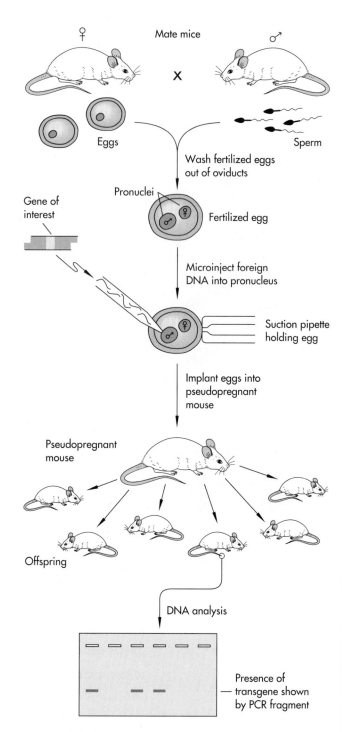

FIGURE **14-1**

Producing transgenic mice by microinjection. Fertilized eggs are collected by washing out the oviducts of mated females, and the gene of interest is injected into one of the two pronuclei. The injected eggs are transferred to foster mothers, female mice made pseudopregnant by mating with vasectomized males. Three weeks after the birth, the offspring are checked for the presence of the transgene by Southern blotting of DNA extracted from a small piece of the tail. Screening can be performed rapidly using the polymerase chain reaction if suitable primers are available. In the example shown, three of the offspring carry the transgene.

Foreign DNA Can Become Stably Integrated into Germ Line Cells

It was quickly shown that DNA microinjected into one-celled embryos could be stably integrated in both somatic cells and germ line cells. Mice derived from embryos injected with cloned human interferon DNA or with rabbit β-globin DNA transmitted these transgenes to their offspring as a Mendelian trait just as with their own genes (Figure 14-2). All the mice derived from such a single founder mouse form a *line* of mice; every member of a line of mice has the same transgene at the identical position in its genome. The transmission pattern implies that the integration event occurs very early in development, before the first cell division of the zygote, and certainly before the germ cell population that gives rise to eggs or sperm is segregated from the primordial somatic cells. Once it was shown that foreign genes introduced into embryos are present in a stably integrated form in the somatic and germ cells of adult mice, it was necessary to determine whether these genes are expressed and, if so, whether this expression is appropriately regulated.

Embryonic Stem Cells Can Carry Foreign Genes

Direct injection of DNA into the pronuclei of fertilized mouse eggs is an efficient way of producing transgenic mice, but there is no opportunity to manipulate or otherwise control DNA integration. However, this can be done by introducing the DNA into special cells called *embryonic stem cells (ES cells)* and then injecting the transfected cells into embryos, where they become incorporated into the developing embryo (Figure 14-3). The ES cells are obtained by culturing the inner cell mass of mouse blastocysts. They are grown in tissue culture just like other cells except that the ES cells must be prevented from differentiating by growing them on a feeder layer of fibroblasts, or by adding leukemia inhibitory factor (LIF) to the culture medium. Under these conditions, ES cells can be grown for many weeks but still retain a remarkable capacity for differentiation—myocardium, blood vessels, myoblasts, cartilage, and nerve cells can be obtained. These extraordinary ES cells can be regarded as the equivalent of unicellular mice, and when they are injected into mouse blastocysts, they are able to participate in the formation of all tissues.

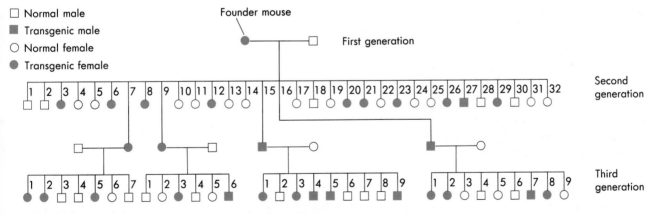

FIGURE **14-2**

Microinjected foreign DNA can be transmitted through the germ line as with a normal Mendelian gene. It is important to determine whether the transgene has been integrated into the germ line as well as the somatic cells of transgenic mice because mice with the transgene integrated into germ line cells can be used to establish lines of transgenic mice. To do this, transgenic mice are mated with normal, nontransgenic mice and the inheritance of the transgene determined. In this case, a female mouse transgenic for the herpes simplex virus thymidine kinase gene linked to the promoter for the metallothionein gene was mated to a normal male. Approximately half the offspring in each generation carried the transgene, as would be expected for a transgenic mouse with a germ line integration.

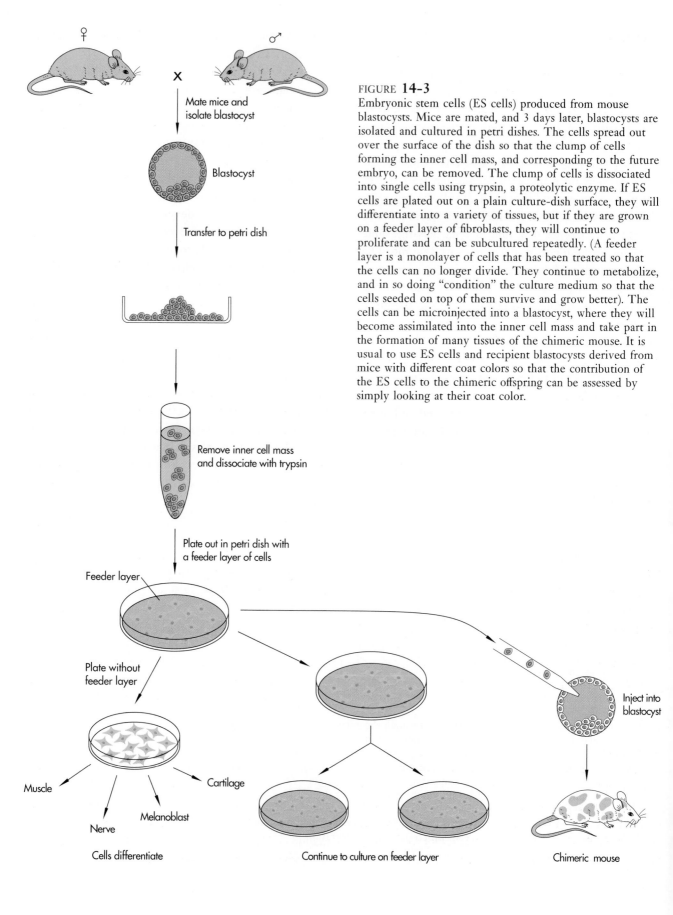

FIGURE **14-3**
Embryonic stem cells (ES cells) produced from mouse blastocysts. Mice are mated, and 3 days later, blastocysts are isolated and cultured in petri dishes. The cells spread out over the surface of the dish so that the clump of cells forming the inner cell mass, and corresponding to the future embryo, can be removed. The clump of cells is dissociated into single cells using trypsin, a proteolytic enzyme. If ES cells are plated out on a plain culture-dish surface, they will differentiate into a variety of tissues, but if they are grown on a feeder layer of fibroblasts, they will continue to proliferate and can be subcultured repeatedly. (A feeder layer is a monolayer of cells that has been treated so that the cells can no longer divide. They continue to metabolize, and in so doing "condition" the culture medium so that the cells seeded on top of them survive and grow better). The cells can be microinjected into a blastocyst, where they will become assimilated into the inner cell mass and take part in the formation of many tissues of the chimeric mouse. It is usual to use ES cells and recipient blastocysts derived from mice with different coat colors so that the contribution of the ES cells to the chimeric offspring can be assessed by simply looking at their coat color.

DNA can be introduced into ES cells by transfection, retroviral infection, or electroporation (Chapter 10). Their most important advantage for gene transfer into mice is that cells carrying the transgene can be selected for before being injected into a blastocyst. In early experiments, ES cells were infected with retroviral vectors, or transfected with plasmids, carrying the *neo* gene. This gene confers resistance to the antibiotic G418. Only ES cells that have taken up the *neo* gene grow in medium containing G418, and these G418-resistant cells were introduced into mouse blastocysts. Not only did the resulting mice have *neo* integrated into their genomes, as shown by Southern blotting, but also the gene was transmitted to the offspring of the mice, and cell lines from the F2 generation were G418-resistant. Because ES cells can be manipulated in vitro before injection into the embryo, mouse geneticists can use homologous recombination to produce transgenic mice with mutations in specific genes, or to replace a mutant gene with the normal equivalent (Chapter 28).

Transgenes Can Be Regulated in a Tissue-Specific Pattern

Although a transgene integrates in a chromosomal location different from that of its endogenous counterpart, it is often expressed in a manner that mimics the expression of the endogenous gene. To determine the pattern of expression, various tissues are analyzed for the presence of RNA or protein products encoded by the transgene. Species differences may be capitalized on to distinguish the transgene product from its endogenous counterpart. For example, the RNA encoded by the human insulin gene can easily be differentiated from the RNA transcribed from the mouse insulin gene.

Insulin is a polypeptide hormone involved in the regulation of glycogen metabolism. This protein is normally produced in the *β cells,* which are found in discrete clusters of endocrine cells, called the *islets of Langerhans,* in the pancreas. When transgenic mice harboring the human insulin gene were analyzed, human insulin RNA was found in the pancreas but not in other tissues. Transcription of the human insulin transgene was induced by the same signals that induced the endogenous mouse insulin genes. Thus, not only can a foreign transgene be expressed in the correct tissue, but it may be subject to the same regulatory signals as the endogenous genes. Similar results have been obtained using many genes introduced into transgenic mice by ES cell transfer.

Transgene Expression Can Be Targeted to Specific Tissues

If the sequences responsible for tissue-specific regulation of a gene are known, they can be used to target expression of a gene product to a tissue in which it is not normally expressed. For example, the islets of Langerhans in the pancreas are composed of four cell types—α, β, δ, and PP—characterized by the hormone that they produce—glucagon, insulin, somatostatin, and pancreatic polypeptide, respectively. The promoter-enhancer control region of the insulin gene was used to target expression of SV40 large T antigen, a viral oncogene, to the β cells of the islets. The recombinant DNA molecule consisted of 660 bp from the 5′ region of the rat insulin gene linked to the region of the SV40 genome encoding T antigen (Figure 14-4). Mice carrying this transgene died at 9 to 12 weeks of age, and pathological analysis showed hyperplasia (abnormal proliferation of cells) and tumors of the islets of Langerhans. All other tissues of the transgenic mice were normal despite containing the transgene, showing that tissue-specific expression was occurring in the islets. All the mice carrying this transgene developed pancreatic tumors, although only a minority of islets in each mouse was affected. Immunohistochemical analysis of tissue sections from the transgenic mice showed that the tumor cells expressed T antigen and that they were exclusively β cells. That is, the insulin gene control region had directed expression of the T antigen to precisely the appropriate cell type. These β-cell tumors are similar to a human tumor called an *insulinoma* except that the naturally occurring tumors also involve the δ cells of the islets.

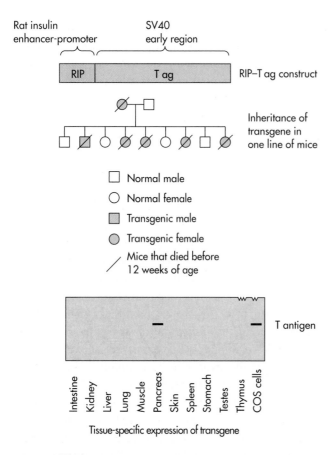

Rat insulin enhancer-promoter

SV40 early region

RIP — T ag — RIP-T ag construct

Inheritance of transgene in one line of mice

☐ Normal male

○ Normal female

▨ Transgenic male

◉ Transgenic female

╱ Mice that died before 12 weeks of age

T antigen

Intestine Kidney Liver Lung Muscle Pancreas Skin Spleen Stomach Testes Thymus COS cells

Tissue-specific expression of transgene

FIGURE **14-4**

Gene targeting to specific tissues. The insulin gene is expressed exclusively in the β cells of the pancreas. The enhancer-promoter region of the insulin gene was linked to the gene for the large T antigen (T ag) of the DNA tumor virus SV40. It was expected that the resulting transgene would be expressed only in the pancreas of transgenic animals. The transgene integrated into the germ line, as shown by the inheritance pattern. The most striking phenotypic expression of the transgene was death of all transgenic animals 9 to 12 weeks after birth with tumors of the islets of Langerhans. The tissue specificity of large T antigen expression was determined by using an antibody to large T antigen to immunoprecipitate tissue homogenates, carrying out electrophoresis, and detecting large T antigen with a second, radio-labeled antibody. Large T antigen was found only in the pancreas and in extracts of COS cells that constitutively express large T antigen.

Transgenes Can Be Used to Kill Specific Cell Types

Targeting of genes to specific cell types is also a powerful tool for studying mouse development. In another type of experiment, the control region for a gene expressed in a specific differentiating cell type is used to restrict expression of a toxin gene to those cells

alone. Only the specific cell type is killed, and the effects of the absence of that cell type on subsequent mouse development can be followed. For example, the elastase 1 gene is expressed only in the exocrine pancreas, and its regulatory sequences were used to direct expression of a potent toxin, diphtheria toxin A chain, to these cells. Twenty-four transgenic mice were obtained, and seven of these lacked a normal pancreas. Microscopic examination showed that some of the mice had a rudimentary pancreas in which the exocrine pancreas was highly abnormal. Similar *cell ablation* experiments have been carried out using other cell-specific enhancer-promoter sequences. Those from the crystallin genes (crystallins are lens proteins) have been used to direct expression of diptheria toxin A chain and the lectin ricin to the lens cells of the developing eye.

One difficulty with this approach is that the toxin gene is turned on as soon as the cell switches on the endogenous gene. If this happens early in development, the ablation of some cell lineages might be lethal for subsequent development. Instead an inducible system for selective killing of cells has been devised by linking the herpes virus 1 thymidine kinase gene (*tk*) to cell-specific enhancer-promoter sequences. Cells containing such a transgene are not killed until synthetic nucleosides are injected into the mice. The nucleosides are not toxic, but the herpes virus thymidine kinase metabolizes the nucleosides into products that kill dividing cells.

This technique has been used to determine cell lineage relationships in the anterior pituitary. Cells synthesizing prolactin are thought to be derived from a precursor cell that synthesizes growth hormone (GH). The GH- and prolactin-synthesizing cells of the anterior pituitary were made the target cells in transgenic mice by using *tk* constructs with either the rat growth hormone or the prolactin promoter, respectively. Following injection of the synthetic nucleosides, animals with GH promoter–*tk* transgene grew up to be dwarfs and were devoid of both GH- and prolactin-synthesizing cells, but animals with the prolactin–TK transgene were normal. The first conclusion of this study is that cells synthesizing prolactin do not divide; otherwise they would have been killed by the nucleoside metabolites. The second conclusion is that a GH-synthesizing cell must be the precursor of both cell types, since the mice in which only the GH-synthesizing cells had been killed experimentally had also lost their prolactin-synthesizing cells.

Retroviruses Can Be Used to Trace Cell Lineages

Cell lineage studies have also been performed by introducing marker genes into newborn mice. Retroviruses are efficient vectors for delivering genes to cells (Chapter 10), and recombinant Moloney murine leukemia viruses have been made in which the *gag*, *env*, and *pol* genes of the virus are replaced by the *E. coli lacZ* gene. This codes for the β-galactosidase enzyme, which can convert the chemical X-gal to a blue dye. Cells containing this recombinant virus integrated in their DNA can be identified in microscopic sections of tissues by this blue staining. The aim of these experiments is to deliver the marker virus to a particular site in an embryo or young animal so that a limited number of cells is infected. Each infected cell gives rise to a clone of cells containing the integrated marker virus. The distributions and types of stained cells in a clone provide information about the cell lineages that gave rise to them. This has been particularly effective in studies of the developing retina. Marker virus was injected between the retina and the pigment epithelium on the day of birth or on days 2, 4, and 7 after birth. Tissue sections of the retina were prepared four to six weeks later and stained with X-gal. A small amount of virus was injected such that each cluster of blue-stained cells would be derived from only a single infected cell and was therefore clonal. Remarkably, many clones contained more than one cell type, showing that a single precursor cell could give rise to rods, bipolar cells, and Müller glial cells. There are some problems with this technique. For example, some cells carrying the marker virus may not express β-galactosidase, and the virus cannot be directed to a particular cell. Alternative methods, including direct injection of markers like fluorescent dextrans into cells, have been used with *Xenopus*, but this is a difficult technique to apply to mammalian embryos.

Transgenes Can Disrupt the Functioning of Endogenous Genes

On rare occasions, transgene insertion itself disrupts the functioning of some endogenous gene whose product is required for normal embryogenesis. The resulting abnormalities can give clues to developmental pathways involving the disrupted gene. One line of transgenic mice (S) was created using a recombinant DNA molecule comprising the mouse mammary tumor virus (MMTV) LTR and the mouse c-*myc* gene (Figure 14-5). When heterozygous mice were bred together, some of the offspring had forelimb and hind limb abnormalities, including single bones in place of the radius and ulna and the tibia and fibula, and absence of digits. Breeding experiments showed cosegregation of the MMTV-*myc* integration site with the mutant phenotype, confirming that integration of the transgene had damaged an endogenous cellular gene.

The phenotype of the homozygous mice resembled that of a known mutation at a locus called *ld* for "limb deformity." Experiments were carried out to confirm that the new transgene mutation, called *ld*[Hd] (for "limb deformity[Harvard]") was an allele of *ld*. First, the chromosomal location of the *ld*[Hd] insertion was determined and compared with the location of the *ld* locus. The integrated MMTV-*myc* DNA was used as a tag to clone mouse DNA from the region of the *ld*[Hd] insertion. This DNA was used as a probe and hybridized against DNA from a panel of mouse-hamster cell hybrids containing different mouse chromosomes. The *ld*[Hd] DNA hybridized against DNA from the hybrid with mouse chromosome 2, the same chromosome that carries the *ld* locus. Second, breeding experiments mating *ld* and *ld*[Hd] mice showed that *ld*/*ld*[Hd] offspring had limb deformities identical with those in *ld*[Hd]/*ld*[Hd] mice; that is, the two loci are allelic. Similar results have been obtained with an insertional mutation that is allelic to the known mouse mutation, Purkinje cell degeneration (*pcd*).

Thus, transgene insertion can be used not only to identify, but also to clone, developmentally important genes.

Knocking Out Genes by Homologous Recombination Can Elucidate Complex Gene Systems

Insertional mutagenesis creates *loss-of-function mutations*, in which expression of a gene is turned off. However, there is no control over which gene is affected, because the plasmid or retrovirus does not have a preferred site for integration. Homologous recombi-

(a)

MMTV LTR   *myc*

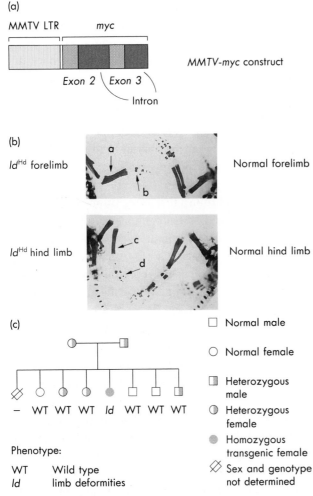

MMTV-*myc* construct

Exon 2   Exon 3

Intron

(b)

Id^Hd forelimb   Normal forelimb

Id^Hd hind limb   Normal hind limb

(c)

□ Normal male

○ Normal female

◧ Heterozygous male

◐ Heterozygous female

● Homozygous transgenic female

◇ Sex and genotype not determined

—   WT   WT   WT   *Id*   WT   WT   WT

Phenotype:

WT   Wild type
Id   limb deformities

FIGURE **14-5**

Insertional mutations occur when an integrated transgene disrupts the function of a cellular gene. (a) Experiments were performed to examine the effects of c-*myc* oncogene expression in transgenic mice by using part of the c-*myc* oncogene linked to the control region [long terminal repeat (LTR)] of the mouse mammary tumor virus (MMTV). (b) When mice heterozygous for the transgene were mated to produce homozygotes, one line of mice was found to carry a mutation (*Id*^Hd) that caused severe limb abnormalities. These included fusion of the long bones and digits of both forelimbs and hind limbs. Arrows *a* and *c* show radius-ulna and tibia-fibula fusions, respectively. Arrows *b* and *d* show fusion of the digits. The transgene in this mouse line has integrated within and disrupted a host gene involved in limb development. (c) Insertional mutations show characteristic recessive inheritance because integration of the transgene disrupts only one copy of a gene which is compensated by a normal copy of the gene on the sister chromosome. (Woychick et al., 1985)

nation (described in Chapter 10) has provided the mouse geneticist with a tool that can be used to manipulate a specified gene in the mouse genome.

The class I molecules of the major histocompatability complex (MHC) are found on the surfaces of cells and are the most important of the molecules that are involved in immunological cell recognition. MHC class I molecules also have important nonimmunological functions. One way to try to dissect the immunological and other functions of the MHC class I molecules is to produce mice lacking these surface molecules. This was done by disrupting the gene for β_2-microglobulin, one of the two components that make up the MHC class I molecule. The targeting was done in ES cells by electroporation to introduce a vector that contained 10 kb of the mouse β_2-microglobulin gene, including the first three exons, with the *neo* gene inserted into the second exon. The 10-kb segment of the β_2-microglobulin gene provides the sequence for homologous recombination of the vector with the endogenous gene. Cells containing the vector are selected using G418, and PCR is used to determine in which cells the vector has undergone homologous recombination with the endogenous $\beta2$-microglobulin gene. After injection of these cells into blastocysts, chimeric mice were produced that had a disrupted β_2-microglobulin gene and transmitted the mutation to their offspring. Mice homozygous for the mutation were produced by mating, and they were indistinguishable from wild-type mice despite a total absence of β_2-microglobulin and cell surface MHC class I molecules! This result clearly rules out a crucial role for MHC class I molecules in development. Further studies of these animals will examine the immunological roles of the MHC class I molecules. The mice lack the CD4$^-$8$^+$ cytotoxic lymphocytes that are involved in killing infected cells, and while the mice should be able to resist viral infection, they should be very poor at recovering from an infection.

Targeting of the kind just described for the β_2-microglobulin locus results in *null mutations,* in which no protein at all is produced. However, null mutations of so-called housekeeping genes, which are responsible for general functions, are likely to be lethal. Now small mutations that produce an altered protein can be inserted into genes using the so-called *hit-and-run* procedure. Homologous recombination inserts the complete vector, carrying a mutation in the target gene and the *neo* and *tk* genes, into the endogenous gene, and cells are selected with G418. Because integration produces a duplication of the target sequence, intrachromosomal recombination occurs in which the duplicated sequence is lost. If it is the endogenous sequence that is lost, the net effect is to replace it with the vector sequence containing the engineered mutation. In addition, the plasmid and selectable marker sequences are lost, so these cells can be selected by using gancyclovir. The advantage of the technique is that the mutation is the only change made to the cell DNA. The ability to make precise mutations will make gene targeting an even more valuable tool for dissecting the genetics of complex processes.

Studying Genetic Control of Mouse Development

As we will learn in Chapter 20, the fruit fly *Drosophila* is the organism of choice for studying developmental genetics. However, the techniques described in this chapter are being used to explore the parallels between genetic control of development in the fly and the mouse. For example, there is a class of genes called the *homeobox* genes that plays a crucial role in *Drosophila* morphogenesis. Mutations in the Antennapedia and bithorax clusters of genes lead to dramatic changes because these genes determine how the fly's body segments will develop. These genes, called *homeotic* genes, are characterized by the presence of a short sequence called the *homeobox*, and vertebrate genes containing the same sequence have been cloned. The homologous vertebrate genes are called *Hox* and exist in four clusters, *Hox-1* through *Hox-4*. The expression of these genes has been studied in mice by in situ hybridization, and it is clear from the spatial distribution and timing of their expression that they play an important part in mouse development. The fascinating question is whether the *Hox* genes play the same role in vertebrate development as the genes of the Antennapedia and bithorax complexes do in *Drosophila* development. This has been explored by creating transgenic mice that overexpress the *Hox-1.1* gene, and mice in which the *Hox-1.5* gene has been knocked out.

Hox-1.1 expression has a clearly defined anterior boundary and is restricted during development to cells of the neural tube, spinal ganglia, and somites. Transgenic mice expressing *Hox-1.1* in all cells were produced by injecting a construct containing the *Hox-1.1* gene under control of the chicken β-actin promoter. One mouse carried the transgene in the germ line, and all her transgenic offspring died within a few days after birth. The only significant abnormalities were in the head, including cleft palate and incomplete eyelid fusion at birth, changes that may reflect effects on neural crest cells. More detailed analysis of the skeleton showed that the transgenic mice seem to have retained the proatlas vertebra that disappears in normal development and to have developed an extra intervertebral disk. In one mouse an extra rib had developed at the level of the seventh cervical vertebra. These findings suggest that *Hox-1.1* does act as a homeotic gene—changing the level of anterior expression of the gene makes an anterior structure resemble a more posterior structure.

The *Hox-1.5* gene is expressed early during mouse development, and destroying both copies of the *Hox-1.5* gene by homologous recombination produces severe developmental abnormalities. Mice homozygous for the mutation ($Hox\text{-}1.5^-/Hox\text{-}1.5^-$) die at birth with a variety of defects. Among the most interesting are an absence of the thymus, a rearrangement of the vertebrae of the neck, alterations of the heart and its associated blood vessels, and abnormalities of the throat. There is a striking resemblance of the phenotype of these $Hox\text{-}1.5^-/Hox\text{-}1.5^-$ mice to a human congenital disorder, the DiGeorge syndrome, suggesting that further analysis of the functions of the *Hox* genes may lead to an understanding of the mechanisms of such disorders.

Transgenes Can Be Used to Detect Host Genes

The classical approach to identifying genes regulating developmental processes is to screen for animals showing phenotypic abnormalities presumed to result from mutations in such genes. This has been a powerful approach, as demonstrated by the many lines of mutant mice that have been developed. A limitation of this

Gene trap construct

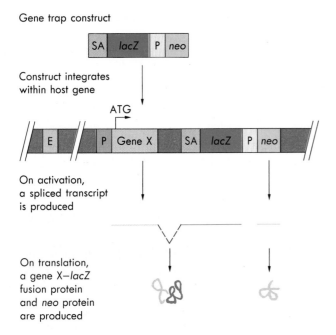

FIGURE **14-6**

Transgenes as reporters for host gene expression. A so-called gene trap construct consists of three parts. First, there is a splice acceptor site from the mouse *En-2* homeobox gene (SA), positioned so that the reporter gene mRNA can be correctly spliced to the host gene mRNA. Second, there is the reporter gene, in this case the *lacZ* gene whose protein product, β-galactosidase, can be detected in cells. Third, there is the neomycin gene (*neo*) under the control of its own promoter (P; yellow). ES cells containing the gene trap can be selected for by their survival in G418, prior to their injection into blastocysts. When this reporter gene integrates within the introns (red) of an active gene (gene X), and provided it does so in the correct orientation and in frame with the transcription of the host gene, a fusion protein of the product of the upstream exons of gene X and β-galactosidase is produced. The expression of this fusion protein is under the control of the enhancer (E) and promoter (P; orange) of gene X. Cells that express the gene X–β-galactosidase fusion protein can be stained blue with X-gal.

approach is that it is restricted to genes that produce recognizable phenotypes. Mutations in genes that kill the embryo or produce only subtle changes in the adult animal will not be detected. An alternative approach made possible by recombinant DNA technology uses *reporter transgenes* that can be detected only when they are expressed under the control of active endogenous genes. Developmentally important endogenous genes may be revealed by looking for transgenes expressed only in specific tissues or at particular times during development.

One kind of reporter transgene, the *gene trap,* uses the *E. coli lacZ* gene as a marker (Figure 14-6). The gene trap contains a *lacZ* gene with a splice acceptor site at its 5' end. If it integrates within the intron of a gene, and in the correct orientation and reading frame, the upstream exons of the host gene are spliced to the *lacZ* gene. The end result is a fusion protein that includes part of the host gene product fused to β-galactosidase. Expression of the transgene, which now depends on expression of the host gene containing it, is detectable by X-gal staining. The advantages of this method are that the gene trap acts as an insertion mutation so that animals may reveal interesting phenotypes, and cloning of the interrupted gene is easy because the construct integrates within the gene.

The actual process for introducing these transgenes and analyzing their effects is shown in Figure 14-7. The transgenes, which contain the *neo* gene, are electroporated into ES cells, and G418-resistant cell lines are selected. ES cell lines expressing β-galactosidase are introduced into mouse blastocysts, and tissues in the resulting chimeric embryos are analyzed for β-galactosidase expression. Embryos have been found in which expression was restricted to particular tissues, for example to the hindbrain and the spinal cord. One chimeric mouse transmitted the transgene in the germ line, and homozygous mice survived to birth. The interrupted genes in these animals will be cloned and sequenced to try to determine the functions of the gene products. Similar results have been obtained in transgenic animals produced by injecting reporter constructs directly into oocytes. These experiments demonstrate the general utility of using this approach for identifying and cloning developmentally important genes in mice.

A Single Gene Turns Female Mice into Males

Transgenic mice have also been used to analyze one of the most fascinating events in development, *sex determination.* In the absence of a Y chromosome, the cells of the genital ridge always give rise to ovaries. If, however, the embryo has a Y chromosome, testes develop. The gene responsible for testes development has been defined genetically; it is called *Tdy* in mice

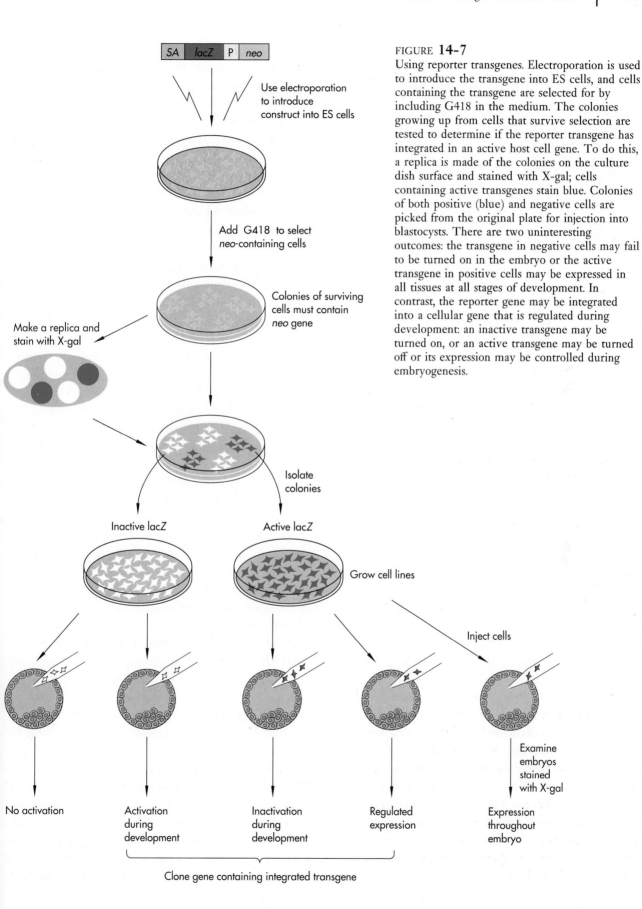

FIGURE **14-7**
Using reporter transgenes. Electroporation is used to introduce the transgene into ES cells, and cells containing the transgene are selected for by including G418 in the medium. The colonies growing up from cells that survive selection are tested to determine if the reporter transgene has integrated in an active host cell gene. To do this, a replica is made of the colonies on the culture dish surface and stained with X-gal; cells containing active transgenes stain blue. Colonies of both positive (blue) and negative cells are picked from the original plate for injection into blastocysts. There are two uninteresting outcomes: the transgene in negative cells may fail to be turned on in the embryo or the active transgene in positive cells may be expressed in all tissues at all stages of development. In contrast, the reporter gene may be integrated into a cellular gene that is regulated during development: an inactive transgene may be turned on, or an active transgene may be turned off or its expression may be controlled during embryogenesis.

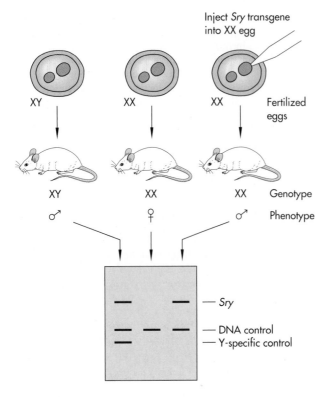

FIGURE **14-8**
Under normal circumstances, genetically male embryos (XY) develop into males (♂), and genetically female embryos (XX) develop into females (♀). However, an XX embryo injected with the *Sry* transgene develops a male phenotype. Analysis of the DNA from these mice shows that both males contain the *Sry* gene, but only the XY mouse is positive for another Y chromosome probe. The female XX mouse has neither *Sry* nor the Y-specific sequence, but the presence of the DNA control band serves as a positive control to verify that the DNA from the female XX embryo was not lost during the analysis.

and *TDF* in humans. A gene, *Sry*, has been cloned from the same region of the chromosome as *Tdy*. Genetic studies and analyses of *Sry* expression have shown that *Sry* is involved in testes determination, but the direct way to demonstrate that *Sry* is *Tdy* is to show that *Sry* alone can turn an XX mouse into a male. A 14-kb fragment containing the *Sry* gene was used to produce transgenic mice (Figure 14-8). Although the *Sry* transgene did not invariably produce testes in transgenic XX females, two embryos were found that were genetically female and had testes. Interestingly, one of the adult sex-reversed mice studied had normal copulatory behavior but was sterile. Histologically the testes of this mouse were normal except that there

was no spermatogenesis. The *Sry* transgene was the only Y-chromosome DNA in the sex-reversed mice. These results show that *Sry* is the testes-determining gene on the Y chromosome, and that genes on other chromosomes must be responsible for turning on *Sry*, and for implementation of the developmental program initiated by *Sry*.

Transgenic Mice Provide Models of Human Diseases

Transgenic mice can be used to model human diseases. Many viruses have a restricted host range; that is, they infect cells of only one species. This is a serious problem because the pathogenesis of a human viral disease is very difficult to study if the virus does not infect animals. However, transgenic techniques provide a way of sidestepping this problem; the gene sequences of human viruses carried in a transgene can be microinjected directly into oocytes. For example, a human papovavirus called *JC* may cause multifocal leukoencephalopathy, a disease characterized by destruction of the myelin-producing cells of the brain, in patients who already have deficiencies in their immune system. When transgenes consisting of the regulatory and coding sequences of the JC virus were used to produce transgenic mice, a specific pathology developed in the animals. The offspring of two of the founder mice developed tremors that became progressively worse as the mice aged. Analysis of the nervous system of these mice revealed that the affected animals produced the JC virus proteins and had decreased myelination in the central nervous system. This study suggested that demyelination can result from the expression of JC virus in infected brain tissue, and that these mice may serve as a model for studying demyelinating disease.

Knowledge of the molecular defect responsible for an inherited disease can be used to make a more accurate animal model. Osteogenesis imperfecta type II affects the extracellular matrix and causes prenatal or neonatal death. The disease is caused by abnormal type I collagen. Normal type I collagen is a triple helical molecule made up of two pro-α1(I) chains, and one pro-α2(I) chain. The osteogenesis imperfecta mutation is in the pro-α1(I) chain, and a similar mutation was made in the mouse gene by site-directed mutagenesis. When the mutant gene was injected into fertilized eggs, embryos carrying the transgene died around the time of birth. The mutant form of pro-$\alpha 1(I)$ collagen mRNA accounted for between 4 and 30 percent of the total pro-$\alpha 1(I)$ collagen mRNA in the transgenic animals, and as little as 10 percent of the mutant pro-$\alpha 1(I)$ collagen mRNA was sufficient to reduce the amount of normal collagen type I to 46 percent of control values. Perhaps abnormal collagen chains have more than one effect, not only leading to reduced stability and increased degradation of collagen molecules containing mutant chains, but also affecting interactions of collagen with other extracellular matrix materials.

Another impressive transgenic mouse model for a human genetic disease is that for the childhood cancer of the eye, retinoblastoma. The transgene used in these experiments included the promoter for the β subunit of the luteinizing hormone, linked to the coding sequence for SV40 large T antigen. Expression of the transgene was limited to the anterior pituitary, and 14 of the transgenic lines did not develop tumors for many months. However, one line of mice derived from a single founder male showed an unusual pattern in which eye tumors developed at about 5 months of age. Transgene mRNA was expressed at high levels in the eyes of these mice, and the tumors bore a striking resemblance to those found in retinoblastoma. Interestingly, the molecular mechanism of tumor formation in these mice is *not* the same as in human retinoblastoma. Studies of human retinoblastoma tumors have shown that they arise when both copies of the retinoblastoma susceptibility gene are mutated. It might be expected that in these mice, the transgene had integrated into one copy of the gene and that a second mutation had occurred that inactivated the copy of the gene on the other chromosome. However, in situ hybridization experiments showed that the transgene was integrated at a single site on chromosome 4, whereas the mouse homologue of the human retinoblastoma susceptibility gene is on chromosome 14. T antigen has been shown to bind to the protein product of the retinoblastoma susceptibility gene, p105^{Rb}, presumably inhibiting its activity. When proteins from the mouse tumors were precipitated with an antiserum to the p105^{Rb} protein and analyzed by Western blotting using an anti−T antigen antibody, it was clear that the p105^{Rb} and T antigen were as-

sociated, just as the model predicted. It appears then, that the transgene in these mice is producing so much T antigen that this is able to "mop up" the normal levels of p105^{Rb} in the cells. Even though the mutation is not the same as the one that produces the disease in children, this transgenic retinoblastoma model is likely to prove very useful in studying the behavior of these tumors and their spread to other tissues.

However, mutations in mice that are *identical* with those in human beings do not necessarily have the same phenotype. For example, the Lesch-Nyhan syndrome is an X-linked disorder characterized by mental retardation and bizarre behavior of affected children, including self-mutilation. The syndrome is caused by mutations in the gene for hypoxanthine guanine phosphoribosyltransferase (HPRT), an enzyme involved in the pathway for the salvage of hypoxanthine and guanine. HPRT$^-$ ES cells were used to make germ line chimeras, and males totally deficient in the HPRT enzyme were bred using chimeric mice that transmitted the HPRT$^-$ mutation in their germ line. These mice are genetically equivalent to patients with the Lesch-Nyhan syndrome, but remarkably the mice exhibit no symptoms whatsoever. It seems that mice can use urate oxidase as well as HPRT for salvaging these nucleotides, and both urate oxidase and HPRT may have to be knocked out to create a mouse equivalent of the Lesch-Nyhan syndrome. It is probable that other biochemical pathways differ between human beings and the mouse, and that mice will be found to respond differently than human beings for other mutations.

The converse of these experiments, that is, correcting a mutation in a mouse by introducing the wild-type gene, has been achieved for the *shiverer* mutation and for one of the HPRT$^-$ lines produced by targeted gene knockout. These experiments will be described in Chapter 28 on gene therapy.

Imprinting—Males and Females Make Unequal Genetic Contributions to Their Offspring

Each parent contributes one haploid set of chromosomes to their offspring, and so it seems logical to conclude that they make an equal genetic contribution. However, experiments transferring nuclei between fertilized eggs show otherwise. Microsurgery can be

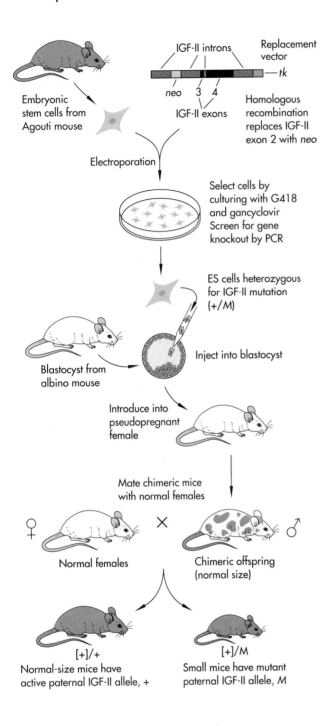

Embryonic stem cells from Agouti mouse

Replacement vector

IGF-II introns

neo 3 4

IGF-II exons

Homologous recombination replaces IGF-II exon 2 with *neo*

Electroporation

Select cells by culturing with G418 and gancyclovir Screen for gene knockout by PCR

ES cells heterozygous for IGF-II mutation (+/M)

Blastocyst from albino mouse

Inject into blastocyst

Introduce into pseudopregnant female

Mate chimeric mice with normal females

Normal females

Chimeric offspring (normal size)

[+]/+
Normal-size mice have active paternal IGF-II allele, +

[+]/M
Small mice have mutant paternal IGF-II allele, M

FIGURE **14-9**

Imprinting in transgenic mice. The insulinlike growth factor (IGF-II) gene in embryonic stem cells derived from a black, agouti strain of mice was knocked out by homologous recombination. The replacement vector contained IGF-II exons 3 and 4, the *neo* gene (conferring resistance to G418), and the *tk* gene (conferring susceptibility to gancyclovir). When the vector undergoes homologous recombination (see Figure 12-15) with the endogenous gene, the mice acquire a *neo* gene, which replaces exon II of the mouse gene. At the same time, the *tk* gene is eliminated from the transgene. A double selection eliminates cells that have simply taken up the vector but have not undergone homologous recombination—they contain an active *tk* gene and are killed by the gancyclovir—and cells that do not contain the vector at all—they lack a *neo* gene and are killed by the G418. The resulting cells (+/M) were injected into blastocysts from albino mice, which were introduced into pseudopregnant mice. Chimeric mice were recognized by their agouti hair. Males whose germ line was derived entirely from the embryonic stem cells were mated with females. The litters were composed equally of small and normal-size mice. The IGF-II genes from the females are inactive because of imprinting ([+]); the normal-size offspring have a normal (+) and the small offspring a mutant (M) gene from their fathers.

used to remove the male or female pronucleus from a fertilized egg, and to replace it with the male or female pronucleus from another fertilized egg. In this way, fertilized eggs can be produced readily that contain only maternal (gynogenetic) or paternal (androgenetic) genomes. Of over 200 control embryos in which pronucleus transfer resulted in embryos containing the genomes of both parents, 7 embryos developed normally. In contrast, none of similar numbers of gynogenetic and androgenetic embryos survived. Thus, in order to complete embryogenesis, a fertilized egg must contain one female and one male pronucleus.

As both a maternal and a paternal pronucleus are required to complete development, they must make functionally distinct contributions to embryogenesis. This phenomenon of differential genetic contributions of parents to their offspring is called *imprinting*. Its most obvious consequence is that the activity of a gene in an embryo depends on the sex of the parent from which a particular allele was inherited. One example of imprinting was discovered during research on the function of the insulinlike growth factor II (IGF-II). Homologous recombination was used to knock out one allele of IGF-II in embryonic stem cells derived from an agouti strain of mice (Figure 14-9). The stem cells were injected into blastocysts from albino mice, and mice with chimeric coat colors were obtained. These chimeric mice looked normal, presumably because the normal cells from the host blastocyst compensated for the reduction of IGF-II in the cells with the knocked out gene. When the chimeric males were

MATERNAL INHERITANCE OF TRANSGENE

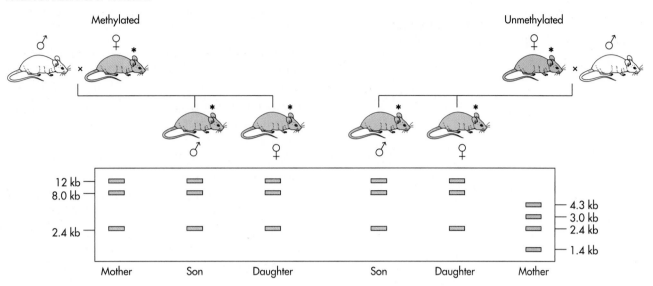

PATERNAL INHERITANCE OF TRANSGENE

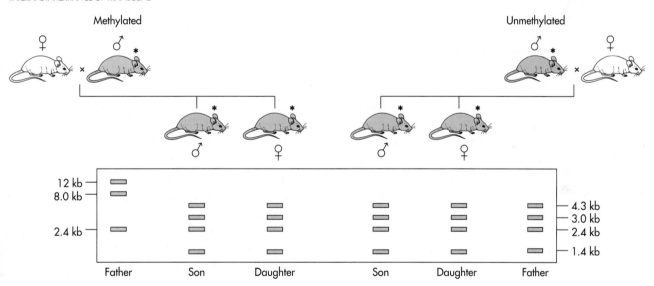

FIGURE **14-10**

Gene imprinting depends on methylation of the imprinted gene. A line of mice contains a transgene made up of a fragment of the c-*myc* gene joined to the immunoglobulin locus. The transgene is subject to imprinting; it is expressed in the offspring of transgenic males, and not in the offspring of transgenic females. This pattern of expression parallels the methylation state of the transgene. The latter can be determined by digesting DNA from the mice with the enzyme *Hpa*II and probing Southern blots with a probe from the immunoglobulin locus. The restriction site for *Hpa*II is CCGG, and the enzyme fails to cut when either of the cytosines is methylated. Restriction fragments of 12 kb, 8 kb, and 2.4 kb are produced from the methylated transgene (pink), and fragments of 4.3 kb, 3 kb, 2.4 kb, and 1.4 kb are produced from the unmethylated transgene (aqua). (Strictly speaking, the transgene is *under*methylated, not totally unmethylated in these animals). Matings are set up between transgenic (*) and normal mice. The methylation state of the transgene in the offspring depends on the sex of the transgenic parent. A transgenic female *always* passes on a methylated transgene to her offspring, even if her transgene is unmethylated. A transgenic male *always* passes on an unmethylated transgene to his offspring, even if his transgene is methylated.

TABLE 14-1

Expression of the c-*myc* Transgene in Mice Depends on the Parental Origin of the Transgene

PARENT		ACTIVITY OF TRANSGENE IN TRANSGENIC OFFSPRING (NUMBER OF MICE)	
FEMALE	MALE	NOT	
		EXPRESSED	EXPRESSED
Transgenic	Not transgenic	0	42
Not transgenic	Transgenic	20	0
Transgenic	Transgenic	6	9

mated with normal females, all the offspring of some males had agouti hair, showing that their germ lines were derived entirely from the injected ES cells. Surprisingly, half these offspring were small. When the agouti mice were analyzed, the normal-sized progeny were homozygous for IGF-II, while the small mice were heterozygous, containing one wild-type and one disrupted allele of IGF-II. These effects might have been due to gene dosage, the small mice simply having one-half the levels of IGF-II of the normal-sized mice. However, analysis of the small mice showed that the amount of IGF-II mRNA produced from their one intact allele was about one-tenth, rather than one-half,

that found in wild-type embryos. What happened to the maternally derived allele in these small heterozygous mice? The answer appears to be imprinting. The maternally derived IGF-II gene is subject to imprinting, so the only IGF-II alleles active in the offspring are paternally derived, producing normal-sized mice (in those inheriting the wild-type paternal allele), or small mice (if they inherited the mutant paternal IGF-II allele).

DNA methylation has been suggested to be one mechanism by which imprinting is regulated. This can be assessed by using restriction enzymes like *Hpa*II that cannot cut DNA sequences once they have been methylated. Systematic breeding experiments showed that a c-*myc* transgene in mice was expressed only in the offspring of matings between male carriers of the transgene and noncarrier females, showing that the transgene was active only when inherited from a male (Table 14-1). Cutting DNA from these transgenic mice with *Hpa*II, and hybridizing with a c-*myc* probe, produced three patterns (Figure 14-10). Analysis showed that females transmitted a methylated transgene, and males transmitted a nonmethylated transgene always, no matter the state of methylation in the parents. The results show clearly that gametogenesis in males and females differs in the way certain genes are turned off or on. This may be important in some human diseases, for example, spinocerebellar ataxia and Wilm's tumor, where there is evidence that inheritance of these disorders depends on which parent contributes the mutant gene.

Reading List

General

Gordon, J. W. "Transgenic animals." *Intl. Rev. Cytol.,* 115: 171–229 (1989).

Jaenisch, R. "Transgenic animals." *Science,* 240: 1468–1474 (1989).

Rossant, J. "Manipulating the mouse genome: implications for neurobiology." *Neuron,* 2: 323–334 (1990).

Original Research Papers

INTEGRATION AND GERM LINE TRANSMISSION OF TRANSFERRED GENES

Jaenisch, R., and B. Mintz. "Simian virus 40 DNA sequences in DNA of healthy adult mice derived from preimplantation blastocysts injected with viral DNA." *Proc. Natl. Acad. Sci. USA,* 71: 1250–1254 (1974).

Gordon, J. W., G. A. Scangos, D. J. Plotkin, J. A. Barbosa, and F. H. Ruddle. "Genetic transformation of mouse embryos by microinjection of purified DNA." *Proc. Natl. Acad. Sci. USA*, 77: 7380–7384 (1980).

Costantini, F., and E. Lacy. "Introduction of a rabbit β-globin gene into the mouse germ line." *Nature*, 294: 92–94 (1981).

Gordon, J. W., and F. H. Ruddle. "Integration and stable germ line transmission of genes injected into mouse pronuclei." *Science*, 214: 1244–1246 (1981).

EMBRYONIC STEM CELLS

Evans, M. J., and M. H. Kaufman. "Establishment in culture of pluripotential cells from mouse embryos." *Nature*, 292: 154–156 (1981).

Martin, G. R. "Isolation of a pluripotent cell line from early mouse embryos cultured in medium conditioned by teratocarcinoma stem cells." *Proc. Natl. Acad. Sci. USA*, 78: 7634–7638 (1981).

Gossler, A., T. Doetschman, R. Korn, E. Serfling, and R. Kemler. "Transgenesis by means of blastocyst-derived embryonic stem cell lines." *Proc. Natl. Acad. Sci. USA*, 83: 9065–9069 (1986).

Robertson, E., A. Bradley, M. Kuehn, and M. Evans. "Germ-line transmission of genes introduced into cultured pluripotential cells by retroviral vector." *Nature*, 323: 445–448 (1986).

Beddington, R. S. P., and E. J. Robertson. "An assessment of the developmental potential of embryonic stem cells in the midgestation mouse embryo." *Development*, 105: 733–737 (1989).

TISSUE-SPECIFIC EXPRESSION AND TARGETING OF TRANSGENES

Swift, G. H., R. E. Hammer, R. J. MacDonald, and R. L. Brinster. "Tissue-specific expression of the rat pancreatic elastase I gene in transgenic mice." *Cell*, 38: 639–646 (1984).

Hanahan, D. "Heritable formation of pancreatic β-cell tumours in transgenic mice expressing recombinant insulin/simian virus 40 oncogenes." *Nature*, 315: 115–122 (1985).

Edwards, R. H., W. J. Rutter, and D. Hanahan. "Directed expression of NGF to pancreatic β cells in transgenic mice leads to selective hyperinnervation of the islets." *Cell*, 58: 161–170 (1989).

KILLING SPECIFIC CELL TYPES

Bernstein, A., and M. Breitman. "Genetic ablation in transgenic mice." *Mol. Biol. Med.*, 6: 523–530 (1989). [Review]

Palmiter, R. D., R. R. Behringer, C. J. Quaife, F. Maxwell, I. H. Maxwell, and R. L. Brinster. "Cell lineage ablation in transgenic mice by cell-specific expression of a toxin gene." *Cell*, 50: 435–443 (1987).

Borrelli, E., R. Heyman, M. Hsi, and R. M. Evans. "Targeting of an inducible toxic phenotype in animal cells." *Proc. Natl. Acad. Sci. USA*, 85: 7572–7576 (1988).

Borrelli, E., R., A. Heyman, C. Arias, P. E. Sawchenko, and R. M. Evans. "Transgenic mice with inducible dwarfism." *Nature*, 339: 538–541 (1989).

ANALYZING CELL LINEAGES

Sanes, J. R. "Analysing cell lineage with a recombinant retrovirus." *Trends Neurosci.*, 12: 21–28 (1989). [Review]

Turner, D. L., and C. L. Cepko. "A common progenitor for neurons and glia persists in rat retina late in development." *Nature*, 328: 131–136 (1987).

Walsh, C., C. L. Cepko. "Clonally related cortical cells show several migration patterns." *Science*, 241: 1342–1345 (1988).

TRANSGENES AS INSERTIONAL MUTAGENS

Woychik, R. P., T. A. Stewart, L. G. Davis, P. D'Eustachio, and P. Leder. "An inherited limb deformity created by insertional mutagenesis in a transgenic mouse." *Nature*, 318: 36–40 (1985).

Krulewski, T. F., P. E. Neumann, and J. W. Gordon. "Insertional mutation in a transgenic mouse allelic with Purkinje cell degeneration." *Proc. Natl. Acad. Sci. USA*, 86: 3709–3712 (1989).

GENE KNOCKOUT BY HOMOLOGOUS RECOMBINATION

Mansour, S. L., K. R. Thomas, and M. R. Capecchi. "Disruption of the proto-oncogene *int-2* in mouse embryo-derived stem cells: a general strategy for targeting mutations to nonselectable genes." *Nature*, 336: 348–352 (1988).

Zijlstra, M., E. Li, F. Sajjadi, S. Subramani, and R. Jaenisch. "Germ-line transmission of a disrupted β₂-microglobulin gene produced by homologous recombination in embryonic stem cells." *Nature*, 342: 435–438 (1989).

Zijlstra, M., M. Bix, N. E. Simister, J. M. Loring, D. H. Raulet, and R. Jaenisch. "β₂-Microglobulin deficient mice lack CD4⁻8⁺ cytolytic T cells." *Nature*, 344: 742–746 (1990).

Hasty, P., R. Ramires-Solis, R. Krumlauf, and A. Bradley. "Introduction of a subtle mutation into the *Hox-2.6* locus in embryonic stem cells." *Nature*, 350: 243–246 (1991).

STUDYING GENETIC CONTROL OF MOUSE DEVELOPMENT

Holland, P. W. H., and B. L. M. Hogan. "Expression of homeo box genes during mouse development: a review." *Genes and Devel.,* 2: 773–782 (1988). [Review]

Hunt, P., and R. Krumlauf. "Deciphering the Hox code: clues to patterning branchial regions of the head." *Cell,* 66: 1075–1078 (1991). [Review]

Balling, R., G. Mutter, P. Gruss, and M. Kessel. "Craniofacial abnormalities induced by ectopic expression of the homeobox gene *Hox-1.1* in transgenic mice." *Cell,* 58: 337–347 (1989).

Kessel, M., R. Balling, and P. Gruss. "Variations of cervical vertebrae after expression of a *Hox-1.1* transgene in mice." *Cell,* 61: 301–308 (1990).

Chisaka, O., and M. R. Capecchi. "Regionally restricted developmental defects resulting from targeted disruption of the mouse homeobox gene *hox-1.5.*" *Nature,* 350: 473–479 (1991).

Koopman, P., J. Gubbay, N. Vivian, P. Goodfellow, and R. Lovell-Badge. "Male development of chromosomally female mice transgenic for *Sry.*" *Nature,* 351: 117–121 (1991).

REPORTER TRANSGENES

Allen, N. D., D. G. Cran, S. C. Barton, S. Hettle, W. Reik, and M. A. Surani. "Transgenes as probes for active chromosomal domains in mouse development." *Nature,* 333: 852–855 (1988).

Kothary, R., S. Clapoff, A. Brown, R. Campbell, A. Peterson, and J. Rossant. "A transgene containing *lacZ* inserted into the *dystonia* locus is expressed in neural tube." *Nature,* 335: 435–437 (1988).

Gossler, A., A. L. Joyner, J. Rossant, and W. C. Skarnes. "Mouse embryonic stem cells and reporter constructs to detect developmentally regulated genes." *Science,* 244: 463–465 (1989).

ANIMAL MODELS OF HUMAN DISEASES

Small, J. A., G. A. Scangos, L. Cork, G. Jay, and G. Khoury.

"The early region of human papovavirus JC induces dysmyelination in transgenic mice." *Cell,* 46: 13–18 (1986).

Hooper, M., K. Hardy, A. Handyside, S. Hunter, and M. Monk. "HPRT-deficient (Lesch-Nyhan) mouse embryos derived from germline colonization by cultured cells." *Nature,* 326: 292–295 (1987).

Kuehn, M. R., A. Bradley, E. J. Robertson, and M. J. Evans. "A potential animal model for Lesch-Nyhan syndrome through introduction of HPRT mutations into mice." *Nature,* 326: 295–298 (1987).

DeCaprio, J. A., J. W. Ludlow, J. Figge, J.-Y. Shew, C.-M. Huang, W.-H Lee, E. Marsilio, E. Paucha, and D. M. Livingston. "SV40 large tumor antigen forms a specific complex with the product of the retinoblastoma susceptibility locus." *Cell,* 54: 275–283 (1988).

Stacey, A., J. Bateman, T. Choi, T. Mascara, W. Cole, and R. Jaenisch. "Perinatal lethal osteogenesis imperfecta in transgenic mice bearing an engineered mutant pro-α1(I) collagen gene." *Nature,* 332: 131–136 (1988).

Erickson, R. P. "Why isn't a mouse more like a man?" *Trends Gen.,* 5: 1–3 (1989).

Windle, J. J., D. M. Albert, J. M. O'Brien, D. M. Marcus, C. M. Disteche, R. Bernards, and P. L. Mellon. "Retinoblastoma in transgenic mice." *Nature,* 343: 665–669 (1990).

IMPRINTING

McGrath, J., and D. Solter. "Completion of mouse embryogenesis requires both the maternal and paternal genomes." *Cell,* 37: 179–183 (1984).

Swain, J. L., T. A. Stewart, and P. Leder. "Parental legacy determines methylation and expression of an autosomal transgene: a molecular mechanism for parental imprinting." *Cell,* 50: 719–727 (1987).

Reik, W. "Genomic imprinting and genetic disorders in man." *Trends Gen.,* 5: 331–336 (1989).

DeChiara, T. M., A. Efstratiadis, and E. J. Robertson. "A growth-deficiency phenotype in heterozygous mice carrying an insulin-like growth factor II gene disrupted by targetting." *Nature,* 345: 78–80 (1990).

15

Genetic Engineering of Plants

Genetic manipulation of plants has been practiced for many hundreds of years with great success by plant breeders, and plant breeding has become a very sophisticated branch of applied genetics. Breeders have developed elegant schemes for crossing plants to introduce and maintain desirable traits in inbred lines, and the yields of crops like maize and wheat have steadily increased over the past 60 years. However, the methods of classical plant breeding are slow and uncertain. To introduce a desired gene or set of genes by conventional methods requires a sexual cross between two lines, and then repeated back-crossing between the hybrid offspring and one of the parents until a plant with the desired characteristics is obtained. This process, however, is restricted to plants that can sexually hybridize, and genes in addition to the desired gene will be transferred.

Recombinant DNA techniques promise to circumvent these limitations by enabling plant geneticists to identify and clone specific genes for desirable traits, such as resistance to an insect pest, and to introduce these genes into already useful varieties of plants. Sexual compatibility becomes irrelevant, and the process becomes faster because transgenic plants expressing the gene can be selected directly. Plants have a number of unique biological features that can be explored with recombinant DNA techniques. These features include their pattern of growth, the means plants have devised to cope with the challenges of a changing environment from which they cannot escape, and, of course, photosynthesis. In this chapter we will discuss the basic techniques being used to genetically manipulate plants, especially plants that are important in agriculture. The progress in developing agriculturally important plants by using recombinant DNA methods will be reviewed in Chapter 24.

Plants Have Advantages and Disadvantages for Genetic Engineering

Plants present advantages and disadvantages for the genetic engineer. The long history of plant breeding means that plant geneticists have a wealth of strains carrying genetically characterized mutations that can be exploited at the molecular level. Plants are particularly amenable to genetic manipulation because many can be self-fertilized or *selfed*. When a plant heterozygous for a mutation is selfed, the progeny include wild-type plants, plants homozygous for the mutation, and also heterozygotes, in which the mutation is maintained. In addition, because plants produce very large numbers of progeny, rare mutations and recombinations can be found. Genetic manipulation of some plants is particularly refined because of the many years scientists have spent analyzing plant transposable elements, which can be exploited as vectors and as insertion mutagens. Plant geneticists are also helped by the regenerative capabilities of plants. As we will describe in a moment, genetic manipulations can be performed on plant cells in culture, and then, with various degrees of difficulty, whole plants can be regenerated from single cells. The relative immobility of plant cells is important for experimental genetics; it is easy to observe clones of cells, because the descendants of a cell remain associated.

However, although plants are attractive subjects for genetic research, they do have some disadvantages. For the molecular geneticist, one disadvantage is that many plants have very large genomes, often because of *polyploidy*, the presence of many genomes in the cell. Many groups of plants have polyploid species; for

example, about two-thirds of the grasses are polyploid, and species in the group that includes the potato have chromosome numbers ranging from 24 to 144. Polyploidy may contribute to the phenomenon of *somaclonal variation* exhibited by plants cells in tissue culture. In other words, plants regenerated from single cells are not genetically homogeneous, for it appears that plant cells growing in tissue culture are genetically unstable. This is a potentially serious problem in gene transfer experiments. A final difficulty arises because of our preoccupation with plants like maize, rice, and wheat, which have great agricultural importance. These are *monocotyledonous* plants ("monocots"), whose seeds have a single cotyledon (meaning "seed leaf"). These monocots are proving to be very difficult to transform with the DNA vector systems that are very efficient with *dicotyledonous* plants (those with two cotyledons). As we will describe later in this chapter, some novel methods are being devised to overcome this limitation.

Whole Plants Can Be Grown from Single Cells

An extraordinary phenomenon, and one that is very useful to the geneticist, is that whole plants can be regenerated from single cells. (Recall, however, that the usefulness of this technique may be limited by somaclonal variation.) When a plant is wounded mechanically, a patch of soft cells called a *callus* grows over the wound. If a piece of young callus is removed and placed in a culture medium containing the appropriate nutrients and plant growth hormones, the cells will continue to grow and divide as a suspension culture (a culture of individual cells or small clumps of cells suspended in the liquid medium). These cells can be plated out and they will grow to form new calli. It is sometimes necessary to use other cells as a *nurse culture*, equivalent to the feeder layers of cells sometimes used in mammalian cell culture. The callus will then redifferentiate into shoots and roots, and ultimately a whole flowering plant will be produced. Studies by Skoog and Miller showed that the differentiation of the cells in a callus depends on the relative concentrations of the plant hormones (phytohormones), *auxins* and *cytokinins*. If the ratio of auxins to

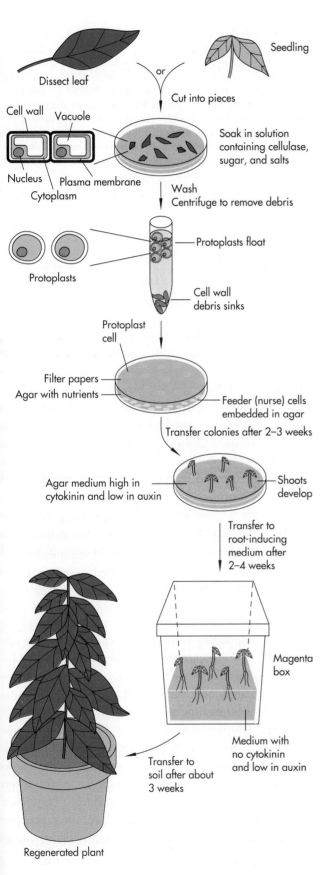

Regenerated plant

FIGURE **15-1**
Regeneration of plants from protoplasts. Leaf cells are characterized by a cytoplasmic compartment containing numerous chloroplasts, a large vacuole, and a nucleus. The plasma membrane is surrounded by a tough cellulose cell wall that can be removed by incubating pieces of plant tissue in a solution containing cellulase. Sugars and salts are added to the solution to maintain osmotic balance and prevent the protoplasts from lysing. Once the cell debris is removed, the protoplasts are placed on filter paper covering a layer of nurse cells. The filter paper is impervious to the cells, but growth factors and other molecules produced by the nurse cells can diffuse into the protoplasts, which divide and grow to form microcolonies. For most plant cells nurse cell feeder layers are not needed. The microcolonies are carefully transferred to a medium high in cytokinin and low in auxin. Shoots appear in about two to four weeks. Then the cultured cells are transferred to a container called a Magenta box, which contains root-inducing medium lacking cytokinin and low in auxin. Once the roots appear, the plantlets can be placed in soil, where they develop into regenerated plants.

cytokinins is high, then roots develop; shoots develop when the ratio is low.

These cells are not very useful for uptake of DNA because like all plant cells, they are surrounded by a cellulose wall. However, this cellulose wall can be removed by treating the cells with fungal cellulase enzymes (Figure 15-1). The resulting *protoplast* is enclosed only by a plasma membrane and is much more amenable to experimental manipulation. Protoplasts will take up macromolecules like DNA, and they are capable of regenerating whole plants via the formation of calli.

Even though protoplasts have been used successfully for many species, the most important group of agricultural plants, cereals, are very difficult to regenerate from protoplasts. Recently, there have been some successes in growing both maize and rice from genetically engineered protoplasts derived from embryonic cell cultures, that is, cultures containing cells from plant embryos. Plant embryos have been grown in vitro for over 100 years and are a very important source of cells for culture. Fertile rice and maize plants have now been produced from genetically manipulated cells from embryogenic suspension cultures.

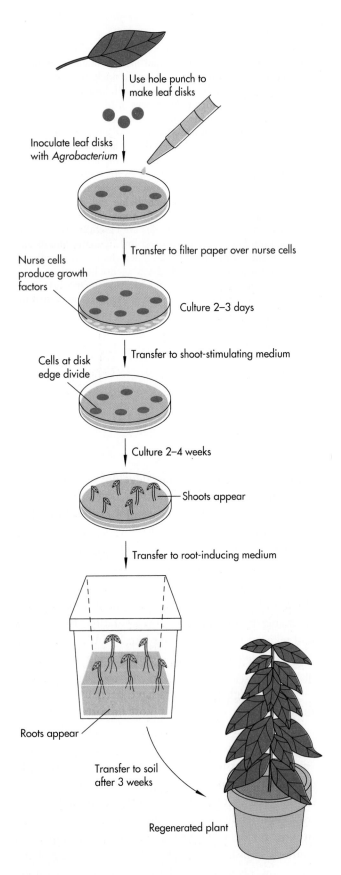

Use hole punch to
make leaf disks

Inoculate leaf disks
with *Agrobacterium*

Transfer to filter paper over nurse cells

Nurse cells
produce growth
factors

Culture 2–3 days

Transfer to shoot-stimulating medium

Cells at disk
edge divide

Culture 2–4 weeks

Shoots appear

Transfer to root-inducing medium

Roots appear

Transfer to soil
after 3 weeks

Regenerated plant

FIGURE **15-2**

Regeneration of leaf disks infected by *Agrobacterium*. Leaf disks are cut out and placed in a shallow dish. A solution of agrobacteria is added, and after a few minutes, the leaf disks are transferred onto nurse cell medium. Wounded cells at the edge of the disk release factors that induce the agrobacteria to infect the cells. The plant disks are cultured in a fashion similar to that described for protoplasts in a medium containing an antibiotic such as cefotaxime that kills *Agrobacterium* but does not harm plant cells (Figure 15-1), to yield a regenerated plant.

Leaf Disks Are an Important Target for Gene Transfer

Growing whole plants from protoplasts is not easy, even for the most amenable species of plants. A simple but very significant improvement came with the development of the *leaf disk technique* (Figure 15-2). The technique is so important because it can be used with the most effective system for transferring genes into plants, a system using the Ti plasmid carried by the bacterium *Agrobacterium tumefaciens*. Plant cells must be wounded to be targets for Ti gene transfer, and pieces of roots and stems have been used as targets. Leaves are a good source of regenerating cells, the cells coming from small disks cut from a leaf. The cells at the edge of the disk begin to regenerate, and when these disks are cultured briefly in a medium containing agrobacteria, these cells are efficiently exposed to the transfecting agent (Figure 15-2). The disks are then transferred for several days to nurse cultures containing medium that stimulates shoot development. Cells carrying the plasmid are selected by culturing in shoot-stimulating medium with an appropriate antibiotic, such as kanamycin, and an antibiotic like cefotaxime to kill the *Agrobacterium*. Shoots develop within a few weeks, and these shoots are transferred to medium that induces root formation. The whole process, from cutting out the leaf disk to having rooted plants, takes between four and seven weeks. This process is extraordinarily fast compared with protoplast cultures. Furthermore, the technique is applicable to a wide variety of dicotyledons and is now used routinely.

FIGURE 15-3
Agrobacteria cause crown gall tumors in plants. When a wounded plant is infected by *Agrobacterium,* the agrobacteria cells do not enter the plant cell but transfer a DNA segment called the T-DNA from the circular extrachromosomal *tumor-inducing* (Ti) plasmid. The T-DNA becomes stably incorporated into the plant cell chromosomal DNA. Genes within T-DNA from natural Ti plasmids are expressed and their products stimulate the cells to divide uncontrollably. The structure formed by the rapidly dividing cells is called a *crown gall tumor.*

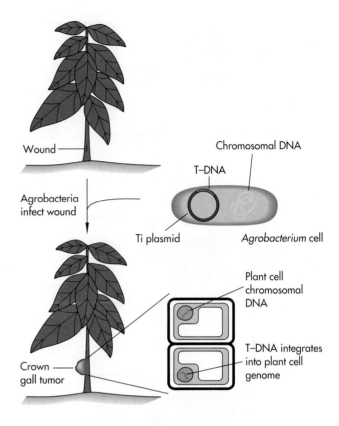

Ti Plasmid of *Agrobacterium* Causes Crown Gall Tumors

Crown galls are tumors of plants that arise at the site of infection by some species of the bacteria *Agrobacterium* (Figure 15-3). The cells of crown galls have acquired the properties of independent, unregulated growth (that is, they are transformed). In culture, these cells grow in the absence of the plant hormones that are necessary for the culture of normal plant cells, and the cells retain this phenotype even in the absence of the bacterium. The tumor-inducing agent in *Agrobacterium* is a plasmid that integrates some of its DNA into the chromosome of its host plant cells. Ti plasmids are large, circular double-stranded DNA molecules of about 200 kb, and like other bacterial plasmids, they exist in *Agrobacterium* cells as independently replicating genetic units.

Ti plasmids are maintained in *Agrobacterium* because a part of the plasmid DNA, called *T-DNA,* carries the genes coding for the synthesis of unusual amino acids called *opines.* The infected plant cell is induced to synthesize these amino acids, but the plant cannot utilize them. Instead, the Ti plasmid is believed to carry genes coding for enzymes that can degrade opines, so the opines may act as a nutrient for the *Agrobacterium.* This subversion of the plant's metabolism could provide a selective advantage for *Agrobacterium.* A second set of genes in T-DNA causes the unregulated growth of the plant cell. Two of these genes, *iaaM* and *iaaH,* code for enzymes that lead to the production of an auxin. The third gene, *iptZ,* codes for an enzyme that causes production of a second

phytohormone. These two hormones cause the infected plant cell to divide; they also affect the neighboring cells.

T-DNA, Part of the Ti Plasmid, Is Transferred to Plant Cells

There are three components involved in Ti plasmid tumor induction (Figure 15-4). One is T-DNA, which is transferred to the host cell and is a form of mobile element. In addition, genes called *vir* (for virulence), present elsewhere on the Ti plasmid, are needed for the production of trans-acting proteins that are essential for, or at least enhance, plant cell transformation. A third set of genes is indirectly involved in transformation. These genes are carried on the *Agrobacterium* chromosome and are responsible for binding the bacterial cells to the plant.

The virulence genes in *Agrobacterium* are switched on by chemicals produced by wounded plant cells (Figure 15-4). Following activation of the *vir* genes, the T-DNA element is excised from the plasmid DNA.

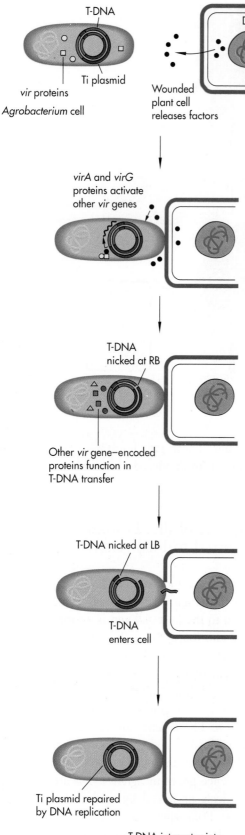

vir proteins

Agrobacterium cell

Wounded plant cell releases factors

Plant cell

virA and **virG** proteins activate other **vir** genes

T-DNA nicked at RB

Other **vir** gene–encoded proteins function in T-DNA transfer

T-DNA nicked at LB

T-DNA enters cell

Ti plasmid repaired by DNA replication

T-DNA integrates into plant chromosomal DNA

FIGURE **15-4**

Transfer of T-DNA from *Agrobacterium* into a plant cell. When a plant cell is wounded, it releases factors that stimulate transcription of the *vir* genes on the Ti plasmid that function in the transfer of the T-DNA into the plant cell. Only the T-DNA region of the Ti plasmid is transferred to the plant cell. T-DNA is bounded by 25-bp imperfect repeats termed the *left border* (LB) and the *right border* (RB). Transfer begins with a nick in the DNA strand in the RB, then a nick occurs at the LB producing a single-stranded T-DNA molecule. By a mechanism that is still not completely worked out, the T-DNA molecule enters the plant cell, where it integrates randomly into the chromosomal DNA. The single-stranded T-DNA region of the Ti plasmid is repaired by DNA replication, so the *Agrobacterium* has not lost any information by transfering DNA to the plant cell.

The T-DNA is flanked by Ti plasmid sequences, each 25 bp long. These flanking sequences are called *borders,* and they are involved in excision of the T-DNA sequence. Excision is a two-stage process in which the right-hand border is nicked between the third and fourth bases of the 25-bp repeat. A second nick in the left-hand border releases the T-DNA as a single strand. The process of transfer from the bacterial cell to the plant cell is analogous to the process of bacterial conjugation; it is as though the *Agrobacterium* is mating with a plant cell! The functions of the *vir* proteins in the transfer process are still being explored. Incorporating extra copies of one of the *vir* genes into *Agrobacterium* leads to increased production of T-DNA and enhanced transformation. Other *vir* genes are associated with the single-stranded T-DNA itself and may be involved in the transfer process. However, this is not the whole story, because once inside the plant cell, the T-DNA has to enter the nucleus and integrate into the plant cell DNA. Usually, multiple copies of T-DNA integrate at a single random site in the plant chromosome, but little is known of the mechanism.

T-DNA Has Been Modified to Act as a Gene Vector

A method called *cointegration* was first used for gene transfer with the T-DNA, Ti plasmid, and *Agrobacterium* system (Figure 15-5). This method was

FIGURE **15-5**

Transferring genes into plant cells by cointegration using T-DNA, Ti plasmids, and *Agrobacterium*. A cloned gene can be introduced into plant cells by first inserting it into the cloning site of a plasmid that can replicate in *E. coli* and contains a segment of T-DNA. The resulting intermediate shuttle vector is introduced into *E. coli* cells, and transformants are selected by resistance to ampicillin, encoded within the pBR322 sequences. Next, the plasmid is transferred from the *E. coli* cell to an *Agrobacterium* cell by mating. Once inside the *Agrobacterium*, the plasmid *integrates* into the Ti plasmid by means of homologous recombination of the T-DNA sequences on the two plasmids. This process places the entire integrative plasmid (the plasmid integrated into the Ti plasmid) between the left and right boundaries of the T-DNA. Plasmids that fail to integrate do not accumulate because they lack an origin of replication for *Agrobacterium*. Agrobacteria containing the recombinant Ti plasmid are selected and used to infect plant cells. Plant cells that have taken up the T-DNA are identified by the plant selectable marker NPTII, which confers resistance to kanamycin. These cells also contain the cloned gene of interest.

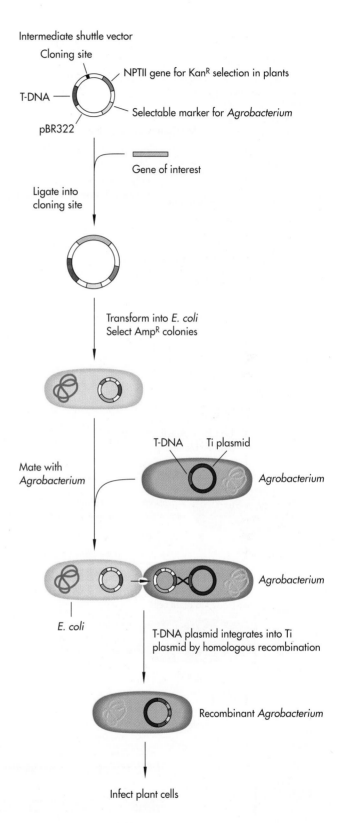

developed to avoid the problems associated with manipulating large pieces of DNA the size of the Ti plasmid. T-DNA was first cloned into a standard *E. coli* cloning vector, and the plant gene subsequently cloned into a second cloning site carried in the vector. This intermediate vector was introduced into *Agrobacterium* containing intact Ti plasmids. Recombination occurs between the homologous regions of the intermediate vector and the wild-type Ti plasmid, and on infection of a plant with the *Agrobacterium*, the recombinant plasmid is transferred to the plant cells. The *E. coli* plasmid used in this process is called an *integrative plasmid* because it becomes part of the Ti plasmid.

The standard method for T-DNA transfer is now the *binary system*. This method was devised when investigators realized that the essential functions for transfer are supplied separately by the T-DNA itself and by the Ti plasmid, and that the components can be carried on separate vectors. The *binary vector* contains the 25-bp borders of the T-DNA that are needed for excision and integration. The phytohormone genes of the T-DNA can be removed to create room for the insertion of foreign DNA, which will be transferred to the plant cell. At the same time deleting the phytohormone genes prevents the uncontrolled growth of the recipient cells. The other essential genes are the

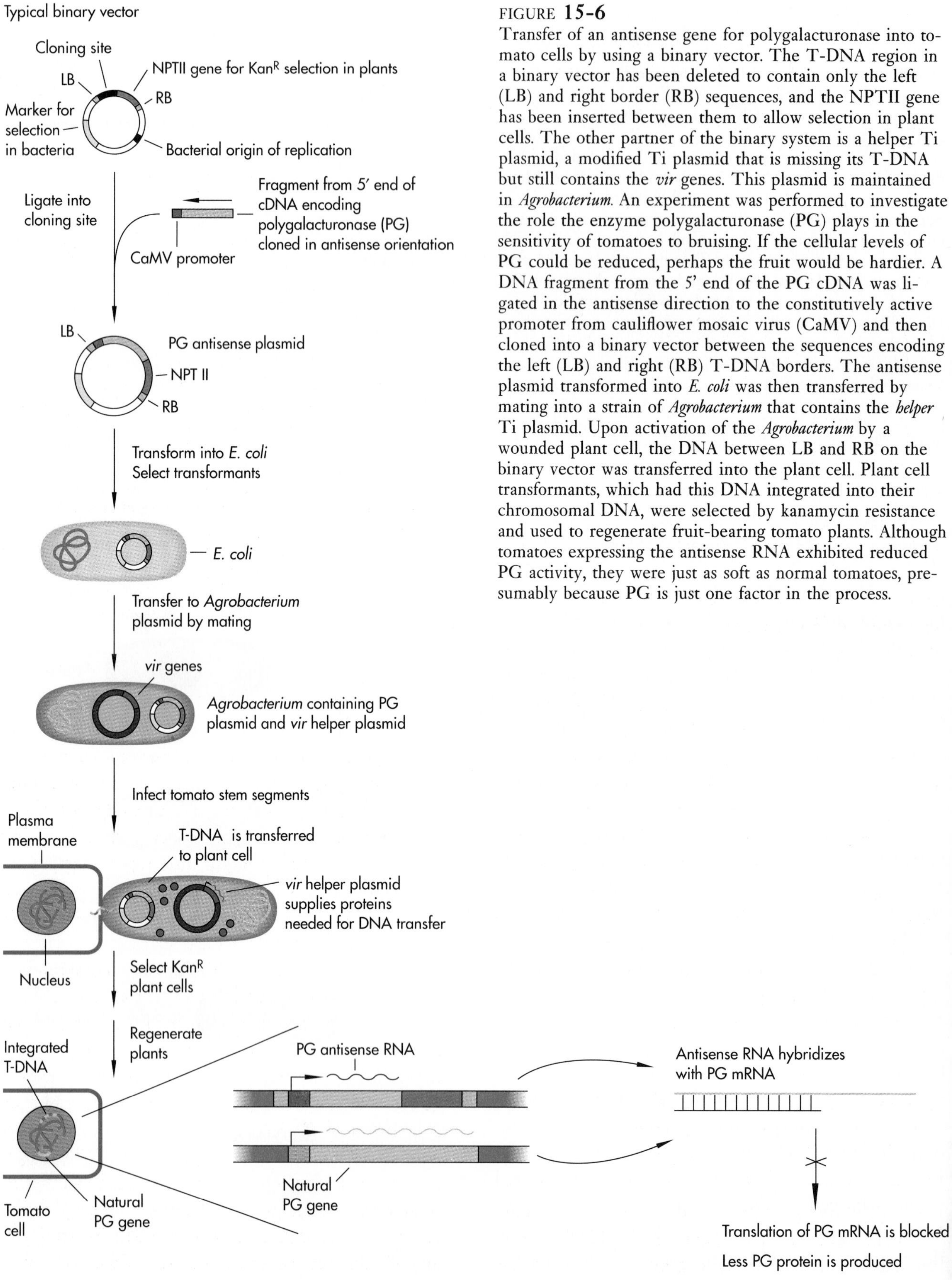

FIGURE **15-6**
Transfer of an antisense gene for polygalacturonase into tomato cells by using a binary vector. The T-DNA region in a binary vector has been deleted to contain only the left (LB) and right border (RB) sequences, and the NPTII gene has been inserted between them to allow selection in plant cells. The other partner of the binary system is a helper Ti plasmid, a modified Ti plasmid that is missing its T-DNA but still contains the *vir* genes. This plasmid is maintained in *Agrobacterium*. An experiment was performed to investigate the role the enzyme polygalacturonase (PG) plays in the sensitivity of tomatoes to bruising. If the cellular levels of PG could be reduced, perhaps the fruit would be hardier. A DNA fragment from the 5' end of the PG cDNA was ligated in the antisense direction to the constitutively active promoter from cauliflower mosaic virus (CaMV) and then cloned into a binary vector between the sequences encoding the left (LB) and right (RB) T-DNA borders. The antisense plasmid transformed into *E. coli* was then transferred by mating into a strain of *Agrobacterium* that contains the *helper* Ti plasmid. Upon activation of the *Agrobacterium* by a wounded plant cell, the DNA between LB and RB on the binary vector was transferred into the plant cell. Plant cell transformants, which had this DNA integrated into their chromosomal DNA, were selected by kanamycin resistance and used to regenerate fruit-bearing tomato plants. Although tomatoes expressing the antisense RNA exhibited reduced PG activity, they were just as soft as normal tomatoes, presumably because PG is just one factor in the process.

vir genes of the Ti plasmid, and these can act in trans if they are supplied on a separate plasmid, called the *helper* plasmid. The binary system is illustrated in Figure 15-6. A very important factor in the development of T-DNA–based vectors is the availability of selectable markers such as neomycin phosphotransferase II (NPTII), and dihydrofolate reductase. These markers are included within the 25-bp repeats of the binary vector, so they too are transferred into the plant cell. The vectors carry a second selectable marker so that they can be manipulated easily in *E. coli*. Binary vectors (Figure 15-6) differ from integrative vectors (Figure 15-5) in that the binary plasmid containing the DNA to be transferred to the plant cell is maintained as a separate replicating vector in *Agrobacterium*.

An example of the use of the binary system to introduce functional genes in plants comes from experiments using antisense RNA to control plant gene expression. Polygalacturonase (PG) is an enzyme that solubilizes the walls of plant cells by digesting pectin. Reducing expression of polygalacturonase could lead to fruit that bruise less easily. A segment from the 5' end of the PG gene was inserted in the reverse orientation, together with the promoter from the cauliflower mosaic virus (CaMV), into a binary vector (Figure 15-6). This vector was transferred to *Agrobacterium* and subsequently to tomato plants. Ripe fruit from transformed plants expressing the antisense construct had significantly reduced levels of PG enzyme activity. Disappointingly, these fruits were as soft as wild-type tomatoes, presumably because PG activity is just one of several factors that contribute to fruit softening. Another factor is ethylene, and successful inhibition of tomato ripening has now been achieved by expressing an antisense RNA for an enzyme in the metabolic pathway of ethylene. The grower can ship such tomatoes without bruising them and then use ethylene to ripen them.

Reporter Genes Demonstrate Transgene Expression in Plant Tissues

A very interesting and important aspect of plant biology is the interaction of the plant with its environment. Plants are particularly sensitive to environmental factors such as light, gravity, and so on, and

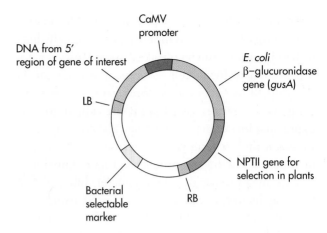

FIGURE **15-7**
A GUS vector. This plasmid is a binary vector used to identify DNA sequences that encode regulatory functions, such as a tissue-specific enhancer. Genomic DNA is cloned adjacent to the *gusA* gene, which encodes *E. coli* β-glucuronidase. The DNA between the T-DNA LB and RB is transferred to plant cells, as described in Figure 15-6. Transformed plant cells are selected by kanamycin resistance encoded by the NPTII gene. The promoter activity of the cloned fragment can be tested by assaying β-glucuronidase activity either histochemically in cells or quantitatively in a fluorometric assay. GUS vectors are also used in a screen for plant cells that have taken up plasmid DNA.

these factors induce changes in gene expression. Rapid, quantitative assays are used to measure gene expression, while morphological analyses using histochemical methods provide data on tissue specificity. We saw in the previous chapter how reporter genes are being used in studies of mouse development, but such genes, inserted into T-DNA vectors, have been used much more extensively with plants.

The *E. coli* gene for the enzyme *β-glucuronidase* (GUS) is especially useful as a reporter gene (Figure 15-7). This enzyme is similar to the *E. coli* β-galactosidase used with animal cells but instead uses glucuronides as substrates. Its advantage as a reporter gene for plants is that plants have virtually undetectable levels of GUS, the enzyme being restricted to vertebrates and their microorganisms. When plant cells expressing β-glucuronidase are incubated with X-glucuronide (X-gluc, analogous to X-gal), a blue color is produced that can be detected histochemically; and, if a different substrate is used, GUS can be measured quantitatively with a fluorometer. The marker β-glucuronidase has one disadvantage: the cells have to be killed for histochemical analysis. However, the distribution of living cells expressing a gene can be determined by using the *luciferase* gene as a reporter gene. When luciferin and ATP are added to cells expressing luciferase, the enzyme catalyzes a reaction in which the luciferin is oxidized, and AMP, carbon dioxide, and light are produced. Light emission is usually measured with a luminometer, but cells expressing luciferase can also be detected using photographic film. The surface of a leaf, for example, is abraded with carbide paper and then soaked in buffer containing luciferin and ATP. When this leaf is overlaid with photographic film, luciferase-expressing cells expose the film, analogous to the way radioactive labels expose x-ray film.

Viruses Can Be Used as Vectors for Whole Plants

Viruses are attractive for use as vectors for introducing genes into plants. Their principal advantage is that they are evolutionarily adapted to do just that, to distribute their own genome throughout an infected plant. If the viral genome includes a foreign gene, then that too should be spread systemically throughout the plant. This method has the potential for greatly simplifying delivery systems. Another advantage is that viral vectors may circumvent the problems of delivering genes to monocotyledonous plants like maize. For example, geminiviruses have a wide host range and experiments have been performed to assess their potential as vectors. These are DNA viruses with genomes made up of two single-stranded DNA molecules that each go through a double-stranded replicative form. The A molecule alone can replicate in plant cells, but the B molecule is required for infectivity. Both A and B genes must be present in a cell for productive viral infection. Because the double-stranded replicative DNA is infectious in the absence of the protein coat, much of the coat protein–coding region can be deleted from the A component to make room for a transgene. A recombinant version of

DNA A of tomato golden mosaic virus (TGMV) was made with an NPTII gene in place of the viral coat protein gene (DNA A–neo). This fragment was inserted between the borders of T-DNA cloned in a plasmid and introduced into a binary *Agrobacterium* system (Figure 15-8). When these bacteria were injected directly into the stems of transgenic tobacco plants that already contained integrated DNA B, replicating DNA A containing and expressing the NPTII gene was found. Similar results have been obtained by using GUS as a reporter gene. Promising though these results are, there appear to be limitations on the size of the DNA that can be inserted into DNA A. In experiments with another strain of tobacco plants, the only DNA A molecules found systemically after infection had lost the GUS insert. There appears to be a strong selective pressure for variants that resemble the wild-type DNA A in size. Clearly we need to know more about the biology of such viruses in vivo so that they can be used more effectively as vectors.

Guns and Electric Shocks Transfer DNA into Plant Cells

The *Agrobacterium*-Ti plasmid vector system is very effective for introducing DNA into dicotyledonous plants, but what can be done about the monocots? These, remember, are of tremendous agricultural importance. One way to sidestep the species restrictions imposed by *Agrobacterium's* host range is to introduce DNA directly into cells by using physical rather than biological means. We learned in Chapter 12 that mammalian cells can be made to take up "naked" DNA presented as a calcium phosphate precipitate, or by electroporation. The calcium phosphate method works poorly for plant transformation, but electroporation is being used successfully with plant cells. Typically, a high concentration of plasmid DNA containing a cloned gene is added to a suspension of protoplasts and the mixture shocked with an electrical field of 200 to 600 V/cm. Following electroporation, the protoplasts are grown in tissue culture for one or two weeks before beginning selection for cells that have taken up the DNA. Both maize and rice protoplasts have been

FIGURE **15-8**

Systemic infection of a plant with recombinant geminivirus vectors. Tomato golden mosaic virus (TGMV) is composed of two 2.5-kb single-stranded DNA molecules packaged together in a protein coat. DNA A encodes the coat protein and replication functions. DNA B is required for cell-to-cell infection. The naked DNAs are infections in plants. These DNA molecules have been engineered into plant expression vectors that transfer cloned DNA of interest into all cells of a plant by infection. This technique eliminates the need to regenerate plants from a transformed cell. To test this vector system, the NPTII gene for resistance to kanamycin was ligated into a cloned fragment of the DNA A molecule, replacing the coat protein gene. This DNA fragment, positioned between T-DNA border sequences (LB and RB) in a binary vector, was then transferred to *Agrobacterium*, as described in Figure 15-6. The DNA was transferred to tobacco plants by injecting the *Agrobacterium* directly into the stems. These plants had previously been transformed with T-DNA containing two copies of the geminivirus DNA B. Within three weeks, the infection had spread through the plant. Single-stranded DNA B and DNA A–NPTII recombinant molecules were detected in leaves from infected plants by Southern blot DNA hybridization. Extracts prepared from infected leaves contained high levels of NPTII activity. An alternative binary vector that has been developed contains both DNA A and DNA B sequences. With this vector, plants do not have to have be transformed first with DNA B, and any plant capable of participating in *Agrobacterium*-mediated transfer can be infected.

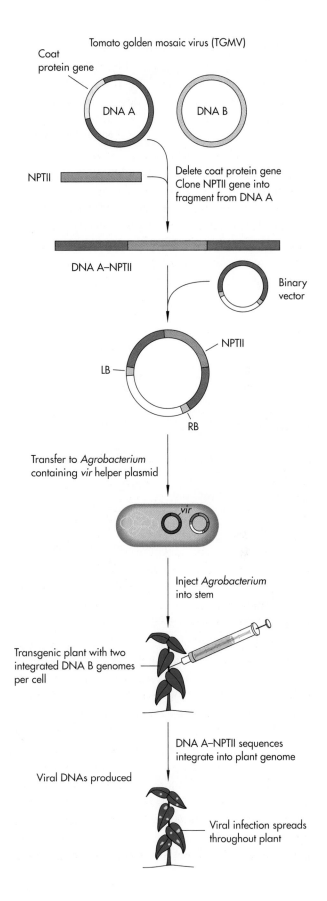

successfully transformed, with efficiencies of between 0.1 and 1 percent.

Electroporation still requires the use of protoplasts, with the concomitant difficulties of regenerating whole plants from protoplasts and the extra complication of somaclonal variation resulting from prolonged periods of culture. It would be much more satisfactory to inject DNA directly into whole cells with intact cell walls; and this has been done, although with limited success. Instead, plant geneticists have developed what must be the most extraordinary delivery system so far devised: minute metal beads coated with DNA are shot directly into cells (Figure 15-9). The tungsten beads are one micrometer in diameter. The DNA is simply precipitated onto the surfaces of the beads, which are fired from the "gun" with velocities of about 430 meters per second. The targets have included suspension cultures of embryonic cells plated on filters and intact

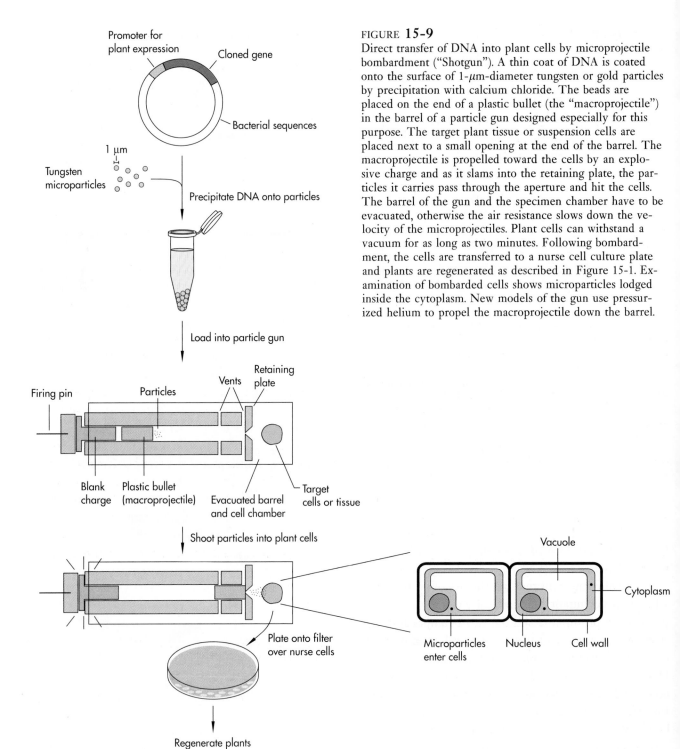

FIGURE **15-9**
Direct transfer of DNA into plant cells by microprojectile bombardment ("Shotgun"). A thin coat of DNA is coated onto the surface of 1-μm-diameter tungsten or gold particles by precipitation with calcium chloride. The beads are placed on the end of a plastic bullet (the "macroprojectile") in the barrel of a particle gun designed especially for this purpose. The target plant tissue or suspension cells are placed next to a small opening at the end of the barrel. The macroprojectile is propelled toward the cells by an explosive charge and as it slams into the retaining plate, the particles it carries pass through the aperture and hit the cells. The barrel of the gun and the specimen chamber have to be evacuated, otherwise the air resistance slows down the velocity of the microprojectiles. Plant cells can withstand a vacuum for as long as two minutes. Following bombardment, the cells are transferred to a nurse cell culture plate and plants are regenerated as described in Figure 15-1. Examination of bombarded cells shows microparticles lodged inside the cytoplasm. New models of the gun use pressurized helium to propel the macroprojectile down the barrel.

leaves and maize kernels. Cells in the direct line of fire are killed, but there is a concentric zone of cells where the projectiles penetrate the cells without killing them. Morphological analysis of leaves bombarded with a GUS vector showed that the tungsten particles can penetrate at least one layer of tissue, the leaf epidermis, to reach the mesophyll.

Visually convincing evidence that the transferred genes are expressed comes from studies of anthocyanin development in maize kernels. The *Lc* gene is a member of the *R* gene family that encodes proteins that determine the expression of anthocyanin pigmentation (the red color) in maize. Sequence analysis showed that the protein encoded by the *Lc* gene had characteristics of DNA-binding proteins and transcriptional activators. That it worked in this way was demonstrated by making a construct in which the *Lc* gene was under the control of a constitutive promoter from cauliflower mosaic virus. This promoter ensures that the *Lc* protein is produced at high levels. Microprojectiles were coated with the plasmid and blasted into kernels. There is an *R* gene expressed in the aleurone (outer layer of cells) of the maize kernel, but the kernels used in this experiment were mutant for this gene and were nonpigmented. However, within 36 hours, individual cells within the aleurones of bombarded kernels were red, because anthocyanin synthesis had been turned on by the *Lc* gene.

Bombardment has been used successfully to transform embryogenic maize cells with reporter genes and, what is more interesting, with the *bar* gene for the enzyme phosphinothricin acetyltransferase (PAT). This enzyme inactivates phosphinothricin (PPT), a component of herbicides, and so protects the plant. The gene-coated beads were shot into embryogenic cells, and transformed cells were selected by culturing them in medium containing PPT. Whole plants regenerated from these cells were resistant to a commercial herbicide applied directly to the leaves. So bombardment has tremendous potential for performing recombinant DNA experiments on maize.

Bombardment with DNA-Coated Beads Can Produce Transgenic Organelles

Nuclear DNA is not the only DNA targeted by plant geneticists, for chloroplasts have their own DNA genome. Not surprisingly, many of these genes are concerned with photosynthesis. If recombinant DNA techniques are to be used for genetic analysis of photosynthesis, then ways have to be devised for introducing DNA into chloroplasts. That this can be done with microprojectiles was first shown by experiments performed with *Chlamydomonas*, a unicellular alga with a single, large chloroplast that occupies a large part of the cell. Transformation of plant chloroplasts is a different matter, because plant cells have numerous chloroplasts and these are very small. Nevertheless, transformation of tobacco plant chloroplasts has been achieved, using a gene for 16S rRNA with mutations conferring resistance to spectinomycin and streptomycin (Figure 15-10). On the appropriate selection medium, plants carrying chloroplasts with spectinomycin resistance are green, whereas non resistant plants are white. Leaves were bombarded, cut into smaller pieces, and cultured. Green, spectinomycin-resistant calli on white leaves were selected, but only three transgenic clones were found. The investigators estimated that transformation of chloroplasts was about 100-fold less efficient than transformation of nuclear genomes. Nevertheless, breeding experiments showed that seeds from selfed plants were all spectinomycin-resistant and as expected, resistance is maternally inherited through the ovum. It is not at all clear why this method works. Chloroplasts are small disk-shaped organelles, and the tungsten beads are probably too large to lodge in a chloroplast. Provided the efficiency of transformation can be improved, this may be a very important approach for genetic manipulation of chloroplasts.

Plant Genes Can Be Cloned by Using Transposable Elements

As we learned in Chapter 10, transposable elements have been used by molecular biologists as tools to detect and clone interesting genes. It is not surprising, given that the first detailed descriptions of transposable elements came from Barbara McClintock's work on maize, that plant geneticists have made extensive use of transposons for cloning. The first work was done in maize itself, because maize transposons were the first to be cloned and to be available for experimental manipulation. The most popular of these transposons is *Activator* (*Ac*). First cloned because of a mutation it

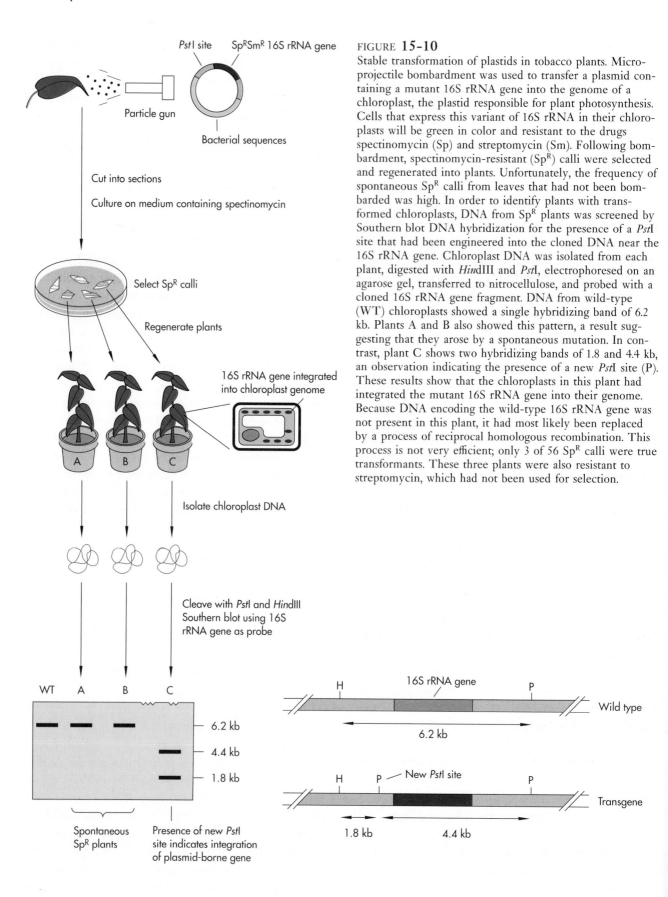

FIGURE **15-10**

Stable transformation of plastids in tobacco plants. Micro-projectile bombardment was used to transfer a plasmid containing a mutant 16S rRNA gene into the genome of a chloroplast, the plastid responsible for plant photosynthesis. Cells that express this variant of 16S rRNA in their chloroplasts will be green in color and resistant to the drugs spectinomycin (Sp) and streptomycin (Sm). Following bombardment, spectinomycin-resistant (Sp^R) calli were selected and regenerated into plants. Unfortunately, the frequency of spontaneous Sp^R calli from leaves that had not been bombarded was high. In order to identify plants with transformed chloroplasts, DNA from Sp^R plants was screened by Southern blot DNA hybridization for the presence of a *Pst*I site that had been engineered into the cloned DNA near the 16S rRNA gene. Chloroplast DNA was isolated from each plant, digested with *Hin*dIII and *Pst*I, electrophoresed on an agarose gel, transferred to nitrocellulose, and probed with a cloned 16S rRNA gene fragment. DNA from wild-type (WT) chloroplasts showed a single hybridizing band of 6.2 kb. Plants A and B also showed this pattern, a result suggesting that they arose by a spontaneous mutation. In contrast, plant C shows two hybridizing bands of 1.8 and 4.4 kb, an observation indicating the presence of a new *Pst*I site (P). These results show that the chloroplasts in this plant had integrated the mutant 16S rRNA gene into their genome. Because DNA encoding the wild-type 16S rRNA gene was not present in this plant, it had most likely been replaced by a process of reciprocal homologous recombination. This process is not very efficient; only 3 of 56 Sp^R calli were true transformants. These three plants were also resistant to streptomycin, which had not been used for selection.

caused in the *Waxy* gene, *Ac* sequences were quickly used as probes to identify *Ac*-containing clones in libraries made from plants with *Ac*-induced mutations. Plants are particularly well suited for transposon mutagenesis experiments because mutations can be maintained in heterozygous plants and plants homozygous for a mutation can be recovered in the F2 generation simply by selfing the F1 progeny. These experiments can also be performed on a large scale. For example, a transposon mutation in the *floricaula* gene of the snapdragon (*Antirrhinum*) was found in experiments with 13,000 plants in the F1 generation, and 40,000 plants in the F2 generation!

Among the most interesting of the mutations revealed in transposon mutagenesis experiments are those that affect plant morphogenesis. Flower development is a process unique to plants and involves complex temporal and spatial changes in differentiation. The *infloresecence* of the growing plant is the structure that gives rise to flowers. In many plant species, the *inflorescence meristem* produces small leaves called *bracts*. A *floral meristem* differentiates at the point each bract joins the stem. It is the floral meristem that gives rise to the flower. Some of the transposon-induced mutations in plants affect flowering in a way that suggests that the genes involved are *homeotic* genes. Homeotic genes control morphogenesis, and mutations in homeotic genes transform one structure into another. They have been intensively studied in *Drosophila* (Chapter 20).

The *floricaula* (*flo*) mutation in *Antirrhinum* plants transforms the floral meristem into another inflorescence meristem, and this in turn gives rise to yet another inflorescence meristem, and so on. The *Flo* gene was cloned using a *Flo* allele, *flo*-613, discovered in a mutagenesis experiment using the Tam3 transposon (Figure 15-11). The involvement of a transposon was suggested by the finding that wild-type flowers were produced occasionally by *flo*-613 plants, suggesting that a transposon had excised itself. If the *flo* locus does contain a Tam3 transposon, then an abnormal fragment should be found on a Southern blot of DNA from a *flo*-613 plant hybridized with a probe for Tam3. Such fragment was found and cloned. It contained part of the Tam3 transposon together with some of the plant DNA flanking the insertion. This plant DNA

was used to clone the wild-type gene. A key observation linking this gene with the mutation was made on a revertant flower that, by self-fertilization, produced offspring with either wild-type or mutant flowers. All the former plants contained a normal-size fragment, a result showing that loss of the Tam3 insert leads to reversion to wild type. In situ hybridization revealed, as might be expected from the phenotype of the *flo* mutation, that *flo* is expressed in the early floral meristems, an observation suggesting that it regulates the transition from inflorescence meristem to floral meristem. It is expressed in sepal, carpel, and petal but not stamen primordia; this observation indicates that *flo* also plays an important role in tissue-specific gene expression.

T-DNA Is Used as an Insertion Mutagen

Although gene cloning in maize and *Antirrhinum* illustrates the power of transposon tagging for finding genes, this approach is limited to plants with active transposons. Unfortunately, active transposable elements are not found in plants such as *Arabidopsis*, which is being studied intensively as a model plant genetic system, and tomato, which is an important cash crop. However, the maize transposons *Ac* and *Spm* (also called *Enhancer*) and the *Antirrhinum* transposon Tam3 are active when introduced into other plants, including potato, tomato, tobacco, and *Arabidopsis*.

With *Arabidopsis*, another strategy is available to geneticists; T-DNA from the Ti plasmid is proving to be a very efficient insertion mutagen. You will recall that after infection by *Agrobacterium*, T-DNA transfers from the bacterial cell to the plant cell. In so doing, the T-DNA carries into the plant cell genes inserted between the 25-bp borders that flank the T-DNA in the Ti plasmid. The NPTII gene, conferring resistance to kanamycin, was cloned into T-DNA, and germinating *Arabidopsis* seeds were incubated with *Agrobacterium*. T-DNA transforms seeds with low efficiency, but selection is efficient; and this procedure has the advantage of avoiding the complications arising from somaclonal variation in cultured cells. In the first study,

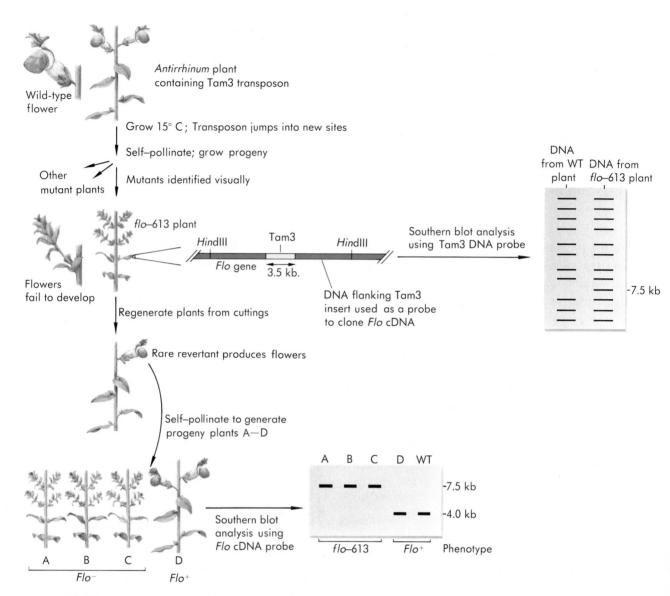

FIGURE **15-11**
Cloning a plant gene by transposon tagging. *Antirrhinum* plants containing a 3.5-kb transposon, Tam3, were grown under conditions that induced the transposon to move to new sites. Forty thousand F2 plants were visually inspected for phenotypes that affected flower development and morphology. One of these, *flo*-613, failed to develop flowers. DNAs from wild-type parental (*Flo*⁺) and *flo*-613 (*Flo*⁻) plants were digested with *Hind*III and probed with cloned Tam3 DNA by Southern blotting. The *flo*-613 mutant contained a new band about 7.5 kb in size. (Because these plants contain many copies of Tam3, there were many hybridizing bands.) This 7.5-kb *Hind*III fragment was cloned by screening a genomic library from *flo*-613 with Tam3. The genomic DNA flanking the inserted Tam3 was used as a probe to clone a cDNA from a wild-type plant cDNA library. The predicted protein sequence from the wild-type cDNA was not related to any known protein. Occasionally, a revertant flower would arise on a mutant plant. Progeny from these flowers usually showed only the Flo⁻ phenotype, but one family produced both Flo⁻ and Flo⁺ plants. (This reversion event must have happened in a germ cell.) Southern blot analysis of DNA from this family progeny showed that the Flo⁻ plants (A, B, and C) still had a 7.5-kb *Hind*III band, but the Flo⁺ revertant (plant D) now had a normal 4.0-kb band. Examination of DNA from the Flo⁺ revertant showed that the Tam3 has been excised in a way that maintained the Flo reading frame.

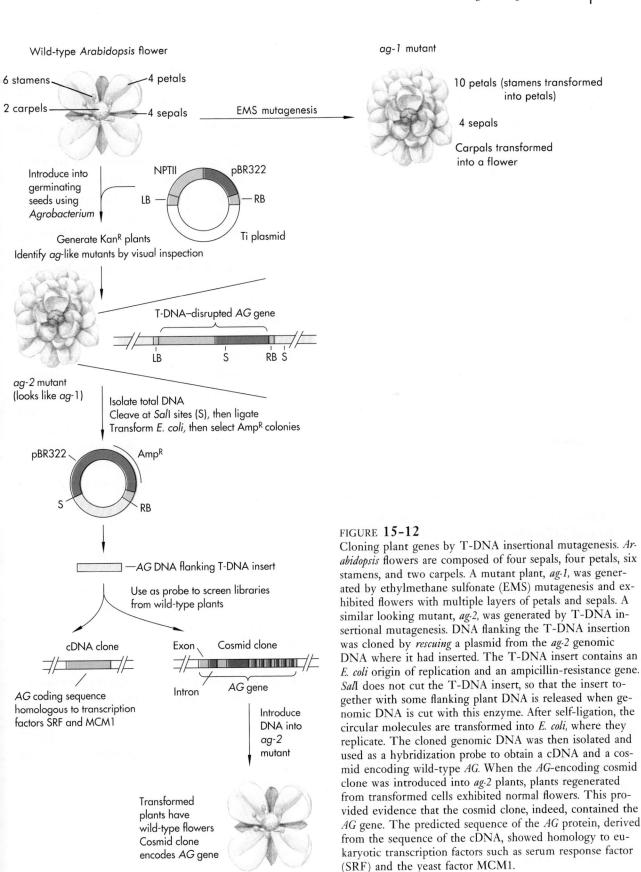

FIGURE 15-12

Cloning plant genes by T-DNA insertional mutagenesis. *Arabidopsis* flowers are composed of four sepals, four petals, six stamens, and two carpels. A mutant plant, *ag-1*, was generated by ethylmethane sulfonate (EMS) mutagenesis and exhibited flowers with multiple layers of petals and sepals. A similar looking mutant, *ag-2*, was generated by T-DNA insertional mutagenesis. DNA flanking the T-DNA insertion was cloned by *rescuing* a plasmid from the *ag-2* genomic DNA where it had inserted. The T-DNA insert contains an *E. coli* origin of replication and an ampicillin-resistance gene. *Sal*I does not cut the T-DNA insert, so that the insert together with some flanking plant DNA is released when genomic DNA is cut with this enzyme. After self-ligation, the circular molecules are transformed into *E. coli*, where they replicate. The cloned genomic DNA was then isolated and used as a hybridization probe to obtain a cDNA and a cosmid encoding wild-type *AG*. When the *AG*-encoding cosmid clone was introduced into *ag-2* plants, plants regenerated from transformed cells exhibited normal flowers. This provided evidence that the cosmid clone, indeed, contained the *AG* gene. The predicted sequence of the *AG* protein, derived from the sequence of the cDNA, showed homology to eukaryotic transcription factors such as serum response factor (SRF) and the yeast factor MCM1.

as many as 36 of 136 kanamycin-resistant plant lines exhibited a mutant phenotype. One of these, for example, was a dwarf plant, one-fifth the size of a wild-type *Arabidopsis* plant.

Another of these insertion mutations resembled *agamous* (*ag*), a mutation previously studied after it was induced by the mutagen ethylmethane sulphonate (Figure 15-12). The *ag-1* allele is a homeotic mutation of the flower, in which the six stamens of the normal flower are replaced with six petals. The carpels are replaced by another flower, to create a flower within a flower. The mutation (*ag-2*) caused by the insertion of T-DNA was shown to be an allele of the *AG* locus by breeding experiments—crossing the *ag-2* plants with *ag-1* plants. The cellular DNA flanking the integrated T-DNA was cloned by plasmid rescue. The most convincing evidence that this cloned DNA was from the *AG* locus came from complementation studies. The cloned DNA was used to isolate the gene from a wild-type *Arabidopsis* cosmid library. When this uninterrupted gene was introduced into *ag-2* plants, a normal phenotype was produced. And in another tribute to the power of comparative genome analysis, the sequence of *AG* was found to be similar to those of a human transcription factor, the serum response factor (SRF), the yeast factor MCM1, and a homeotic gene from *Antirrhinum*! It has now been shown that the expression of *AG* is controlled by another *Arabidopsis* homeotic gene, *ap2,* and it is probable that the genetic control of flower morphogenesis will prove to be as complex as the development of segmentation in the *Drosophila* embryo (Chapter 20).

Arabidopsis Is Being Used as a Model Organism for Molecular Genetic Analysis of Plants

The history of genetics shows that progress in understanding the functions and functioning of genes has depended on framing the right questions and using the right organism to answer them. Geneticists moved from *Drosophila,* through *Neurospora,* to bacteria and bacteriophage in search of organisms in which genetic analysis could be carried out faster and with finer resolution. Similarly, geneticists have exploited plants with special genetic features, like the transposable elements of maize, and are now beginning to use a flowering plant called *Arabidopsis thaliana* as a model to explore the molecular biology and genetics of plants.

Arabidopsis has all the advantages for genetic analysis that other plants have—it can be selfed and very large numbers of progeny can be obtained (up to 10,000 seeds from a single plant). *Arabidopsis* also has a short generation time of five to six weeks, so crosses can be set up and the progeny analyzed within reasonable periods of time. *Arabidopsis* seeds can be mutagenized easily, and, because the plant is physically small, many individuals can be handled in each experiment (a very important attribute for large-scale mutagenizing projects). Mutation screens have identified thousands of mutations affecting many aspects of basic plant biology, including morphogenesis, photosynthesis, fertility, starch and lipid metabolism, mineral nutrition, and so on. Cloning and sequencing all these genes is also proceeding rapidly because of a cooperative effort going on in laboratories throughout the world. Indeed, *Arabidopsis* has special advantages for the molecular geneticist. Its haploid genome is only 10^8 base pairs. This is about one-half that of *Drosophila,* and similar to that of the nematode worm *Caenorhabditis elegans,* which is also the subject of intense genetic analysis. In 1946, *Arabidopsis* was called the "botanical *Drosophila*"; the "botanical worm" is probably more appropriate now!

The application of the techniques of recombinant DNA to plant biology is producing a revolution every bit as profound as that in other organisms. Indeed, the revolution will likely be of global significance as the new biological knowledge is put to use in the genetic engineering of agriculturally important plants (Chapter 24).

Reading List

General

Gasser, C. S., and R. T. Fraley. "Genetically engineering plants of crop improvement." *Science,* 244: 1293–1299 (1989). [Review]

Steeves, T. A., and I. M. Sussex. *Patterns in Plant Development.* Cambridge University Press, Cambridge, 1989.

Walden, R., and J. Schell. "Techniques in plant molecular biology." *Eur. J. Biochem.,* 192: 563–576 (1990). [Review]

Fraley, R., and J. Schell, eds. "Plant Biotechnology." *Curr. Opinion Biotechnol.,* 2: 145–210 (1991). [Review]

Original Reseach Papers

PLANT CELL CULTURE

Horsch, R. B., J. E. Fry, N. L. Hoffman, D. Eichholts, S. G. Rogers, and R. T. Fraley. "A simple and general method for transferring genes into plants." *Science,* 227: 1229–1231 (1985).

Datta, S. W., A. Peterhans, K. Datta, and I. Potrykus. "Genetically engineered fertile indica-rice recovered from protoplasts." *Bio/Technology,* 8: 736–740 (1990).

AGROBACTERIUM, Ti PLASMIDS, AND T-DNA

Potrykus, I. "Gene transfer to cereals: an assessmet." *Trends Biotechnol.,* 7: 269–273 (1989). [Review]

Zambryski, P., J. Tempe, and J. Schell. "Transfer and function of T-DNA genes from *Agrobacterium* Ti and Ri plasmids in plants." *Cell,* 56: 193–201 (1989). [Review]

Jefferson, R. A., T. A. Kavanaugh, and M. W. Bevan. "GUS fusions: glucuronidase as a sensitive and versatile gene fusion marker in higher plants." *EMBO J.,* 6: 3901–3907 (1987).

Smith, C. J. S., C. F. Watson, J. Ray, C. R. Bird, P. C. Morris, W. Schuch, and D. Grierson. "Antisense RNA inhibition of polygalacturonase gene expression in transgenic tomatoes." *Nature,* 334: 724–726 (1988).

Jefferson, R. A. "The GUS reporter gene system." *Nature,* 342: 837–838 (1989).

Fromm, M. E., F. Morrish, C. Armstrong, R. Williams, J. Thomas, and T. M. Klein. "Inheritance and expression of chimeric genes in the progeny of transgenic maize plants." *Bio/Technology,* 8: 833–839 (1990).

Oeller, P. W., L. Min-Wong, L. P. Taylor, D. A. Pike, and A. Theologis. "Reversible inhibition of tomato fruit senescence by antisense RNA." *Science,* 254: 437–439 (1991).

VIRAL VECTORS FOR WHOLE PLANTS

Hayes, R. J., I. T. D. Petty, R. H. A. Coutts, and K. W. Buck. "Gene amplification and expression in plants by a replicating geminivirus vector." *Nature,* 334: 179–182 (1988).

Hayes, R. J., R. H. A. Coutts, and K. W. Buck. "Stability and expression of bacterial genes in replicating geminivirus vectors in plants." *Nucl. Acids Res.,* 17: 2391–2403 (1989).

Elmer, S., and S. G. Rogers. "Selection for wild type size derivatives of tomato golden mosaic virus during systemic infection." *Nucl. Acids Res.,* 18: 2001–2006 (1990).

GUNS AND ELECTRIC SHOCKS

Klein, T. M., E. D. Wolff, R. Wu, and J. C. Sanford. "High-velocity microprojectiles for delivering nucleic acids into living cells." *Nature,* 327: 70–73 (1987).

McCabe, D. E., W. F. Swain, B. J. Martinell, and P. Cristou. "Stable transformation of soybean (*Glycine max*) by particle acceleration." *Bio/Technology,* 6:923–926 (1988).

Rhodes, C. A., D. A. Pierce, I. J. Mettler, D. Mascarenhas, and J. J. Detmer. "Genetically transformed maize plants from protoplasts." *Science,* 240: 204–207 (1988).

Klein, T. M., B. A. Roth, and M. E. Fromm. "Regulation of anthocyanin genes introduced into intact maize tissues by microprojectiles." *Proc. Natl. Acad. Sci. USA,* 86: 6681–6685 (1989).

Shimamoto, K., R. Terada, T. Izawa, and H. Fujimoto. "Fertile transgenic rice plants regenerated from transformed protoplasts." *Nature,* 338: 274–276 (1989).

Gordon-Kamm, W. J., T. M. Spencer, M. L. Mangano, T. R. Adams, R. J. Daines, W. G. Start, J. V. O'Brien, S. A. Chambers, W. R. Adams, Jr., N. G. Willetts, T. B. Rice, C. J. Mackey, R. W. Krueger, A. P. Kausch, and P. G. Lemaux. "Transformation of maize cells and regeneration of fertile transgenic plants." *Plant cell,* 2: 603–618 (1990).

Ludwig, S. R., B. Bowen, L. Beach, and S. R. Wessler. "A regulatory gene as a novel visible market for maize transformation." *Science,* 247: 449–450 (1990).

PRODUCING TRANSGENIC CELL ORGANELLES

Boynton, J. E., N. W. Gillham, E. H. Harris, J. P. Hosler, A. M. Johnson, A. R. Jones, B. L. Randolph-Anderson, D. Robertson, T. M. Klein, K. B. Shark, and J. C. Sanford. "Chloroplast transformation in *Chlamydomonas* with high velocity microprojectiles." *Science,* 240: 1534–1538 (1988).

Johnston, S. A., P. Q. Anziano, K. Shark, J. C. Sanford, and R. A. Butow. "Mitochondrial transformation in yeast by bombardment with microprojectiles." *Science,* 240: 1538–1541 (1988).

Svab, Z., P. Hajdukiewicz, and P. Maliga. "Stable transformation of plastids in higher plants." *Proc. Natl. Acad. Sci. USA,* 87: 8526–8530 (1990).

CLONING GENES BY TRANSPOSON TAGGING

McClintock, B. *The Discovery and Characterization of Transposable Elements: The Collected Papers of Barbara McClintock.* Garland Publishing, New York, 1987.

Balcells, L., J. Swinburne, and G. Coupland. "Transposons as tools for the isolation of plant genes." *Trends Biotechnol.,* 9: 31–36 (1991). [Review]

Hake, S., E. Vollbrecht, and M. Freeling. "Cloning *Knotted,* the dominant morphological mutant in maize using *Ds2* as a transposon tag." *EMBO J.,* 8: 15–22 (1989).

Carpenter, R., and E. S. Coen. "Floral homeotic mutations produced by transposon-mutagenesis in *Antirrhinum majus.*" *Genes Dev.,* 4: 1483–1493 (1990).

Coen, E. S., J. M. Romero, S. Doyle, R. Elliott, G. Murphy, and R. Carpenter. "*floricaula*: A homeotic gene required for flower development in Antirrhinum majus." *Cell,* 63: 1311–1322 (1990).

Vollbrecht, E., B. Veit, N. Sinha, and S. Hake. "The developmental gene *Knotted-1* is a member of a maize homeobox gene family." *Nature,* 350: 241–243 (1991).

CLONING GENES IN *ARABIDOPSIS* BY T-DNA INSERTIONAL MUTAGENESIS

Feldmann, K. A., and M. D. Marks. "*Agrobacterium*-mediated transformation of germinating seeds of *Arabidopsis thaliana*: a non-tissue culture approach." *Mol. Gen. Genet.,* 208: 1–9 (1987).

Feldmann, K. A., M. D. Marks, M. L. Christianson, and R. S. Quatrano. "A dwarf mutant of *Arabidopsis* generated by T-DNA insertion mutagenesis." *Science,* 243: 1351–1354 (1989).

Koncz, C., N. Martini, R. Mayerhofer, Z. Koncz-Kalman, H. Korber, G. P. Redei, and J. Schell. "High-frequency T-DNA-mediated gene tagging in plants." *Proc. Natl. Acad. Sci. USA,* 86: 8467–8471 (1989).

Yanofsky, M. F., H. Ma, J. L. Bowman, G. N. Drews, K. A. Feldmann, and E. M. Meyerowitz. "The protein encoded by the *Arabidopsis* homeotic gene *agamous* resembles transcription factors." *Nature,* 346: 35–39 (1990).

Drews, G. N., J. L. Bowman, and E. M. Meyerowitz. "Negative regulation of the Arabidopsis homeotic gene *AGAMOUS* by the *APETALA2* product." *Cell,* 65: 991–1002 (1991).

ARABIDOPSIS AS A MODEL ORGANISM FOR GENETIC ANALYSIS OF PLANTS

Myerowitz, E. M. "Arabidopsis, a useful weed." *Cell,* 56: 263–269 (1989). [Review]

CHAPTER

16
Molecules of Immune Recognition

For a single-cell organism, distinguishing itself from its environment is not difficult—everything inside its membrane is self, everything outside is non-self. For multicellular organisms, distinguishing self from non-self is much more challenging and crucial to survival. Our bodies consist of hundreds of distinct cell types that differ in appearance and act in unique ways. And we provide a particularly hospitable environment for a menagerie of inventive and aggressive invading organisms. Therefore, we need a mechanism to identify each cell in our body as ours, we need a way to recognize the complete range of invaders as foreign, and we need mechanisms to dispatch invaders and repair the damage they have caused.

This task falls to the immune system. The immune system consists of a number of highly specialized cells that are in close communication with one another and possess a variety of highly sensitive molecules on their surfaces dedicated to the recognition of other cells. Among the most important products of the immune system are the *antibody* molecules (also called *immunoglobulins* and abbreviated Ig), proteins secreted into the bloodstream that seek out and mark foreign agents for destruction. Before recombinant DNA, we knew a great deal about the structure of antibodies but had little idea of how the body is able to produce antibodies that recognize millions of possible *antigens*, the molecular configurations the immune system recognizes as foreign. With the advent of cloning, not only did the basis of antibody diversity become apparent, but the way quickly opened to identification of dozens of additional genes that provide the basis for self versus non-self recognition, cellular immunity, and cell-cell communication.

The Basic Structure of Antibody Molecules Is Established

For decades the fundamental mystery of the immune system was this: How was the body able to produce antibodies that recognize millions of different foreign antigens? The answer began to emerge from the biochemical analysis of antibody structure. These studies were facilitated by the fact that antibodies constitute a significant fraction of the protein in the blood. From these studies, it was known that antibody molecules consisted of four individual protein chains: two light (L) chains of molecular weight 17,000 and two heavy (H) chains of molecular weight 35,000 held together by disulfide bonds (Figure 16-1). There are five classes of H chains, μ, δ, γ, ε, and α, and two classes of L chains, κ and λ. It is the class of H chain that characterizes an individual Ig as an IgM, IgD, IgG, IgE, or IgA, respectively. How antibodies achieved their unique specificities became clear from the study of myelomas, tumors of the immune system derived from normal antibody-producing *B cells*. These tumor cells secrete enormous quantities of a single antibody molecule, with each different tumor producing a distinct antibody. Comparing the amino acid sequences of antibodies secreted by different myelomas revealed that each antibody had a unique sequence. Antibody chains were especially variable in their amino terminal domains but nearly identical in their carboxy terminal domains. Thus each antibody chain could be divided into a *variable* (V) and *constant* (C) region.

But if each of the millions of antibodies has a unique amino acid sequence, each must be encoded by a unique gene, and antibody genes would have to account for a majority of the DNA in the mammalian genome. In 1965, Dreyer and Bennett proposed a dra-

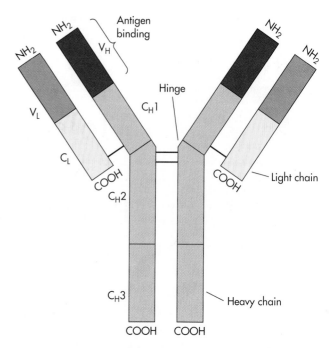

FIGURE **16-1**

The structure of an antibody protein. Two light chains (yellow and orange) and two heavy chains (purple and light blue) are held together by disulfide bonds (black lines). Each chain consists of a variable region (V_L [orange] and V_H [purple]) at its amino-terminal end that is different for each antibody molecule and constant regions (C_L [yellow] and C_H [light aqua] that differ little. The light chain contains a single constant region domain, while the heavy chain contains three related constant region domains and a so-called *hinge* domain that provides conformational flexibility to the molecule. The specificity of antigen binding is determined by the amino acid sequence of the V regions of the L and H chains. Thus, each antibody molecule binds two molecules of antigen.

matic hypothesis to explain how antibody diversity was encoded in the genome. They suggested that the constant regions for each type of antibody chain were encoded by single genes but that the short variable regions were each encoded by thousands of unique genes. Genetic recombination, they proposed, brought V and C genes together to form a functional antibody gene. Few accepted this model, because it was believed that the arrangement of the genome never changed. Ten years later, however, some of the first recombinant DNA experiments proved Dreyer and Bennett's hypothesis correct.

Fledgling Recombinant DNA Technology Verifies the Dreyer and Bennett Hypothesis

The radical idea that functional antibody genes were formed by DNA rearrangements could not be put to the test until the development of tools to manipulate DNA and RNA in the test tube. In 1976, Susumu Tonegawa was able to purify the mRNA encoding an antibody light chain from a myeloma cell line. This RNA preparation could be radioactively labeled and hybridized to DNA to determine the structure and abundance of the gene from which the mRNA arose. The then recent discovery of restriction enzymes provided Tonegawa with the perfect tool to investigate whether the light-chain gene was arranged differently in the myeloma cell genome than in other cells of the mouse that do not express the antibody genes. He simply digested myeloma and embryo cell DNA into fragments with a restriction enzyme, sorted the fragments out by size by agarose gel electrophoresis, and determined which size fragments in each DNA sample hybridized to the mRNA probe. He found that in embryo DNA there were two different fragments that hybridized to the probe, but in the myeloma cell DNA there was only one and its size differed from both embryo fragments (Figure 16-2). Thus, the DNA encoding the antibody gene had been rearranged in the myeloma cell.

The existence and nature of these rearrangements were dramatically confirmed when the first antibody genes were cloned. Antibody genes were among the first mammalian genes cloned, thanks to the availability of myeloma mRNA probes. Indeed, the labs studying antibody genes were at the forefront of the developing technology of gene cloning. Physical characterization and, eventually, direct DNA sequence analysis of these early clones showed that a functional antibody gene in a B cell consisted of a segment encoding the V region close to a separate segment encoding the C region. In the DNA of other cells, however, the V gene segment was not to be found anywhere in the vicinity of the C gene segment. Indeed, we now know that hundreds of different V segments lie clustered in a region of the same chromosome up to a million base pairs away from the C segment.

Somehow, during development of a B cell, a single V gene segment is selected and brought by gene rearrangement to a position just upstream of the C gene segment, and a functional antibody gene is born.

Rearrangement of Antibody Genes Generates Additional Diversity at the V-C Junction

Detailed comparisons of cloned antibody genes from B cells (rearranged) and non-B cells showed that complete L and H chain genes are actually assembled from three and four discrete gene segments, respectively, and that variable joining of the segments creates further diversity in the encoded amino acid sequence (Figures 16-3 and 16-4). In the L-chain genes, the component that provides additional diversity is called the J (for "joining") region. In the κ light-chain gene of the mouse, for example, there are five J sequences located 2.4 kb upstream of the C gene. Rearrangement can join a V gene to any one of the J regions, resulting in a fivefold increase in the number of potential L-chain sequences (Figure 16-3). In the H-chain locus of the mouse, there are 4 J segments and in addition about 20 D (for "diversity") gene segments. Assembly of a complete H-chain gene therefore requires two discrete rearrangements. First, a D segment is joined to a downstream J, and then a V gene is joined to the rearranged D-J unit (Figure 16-4). As with L-chain genes, generating an H-chain gene from these discrete units dramatically increases the number of possible antibody sequences arising from recombination.

Further diversity in antibody sequence arises because these joining events are imprecise. Careful comparison of nucleotide sequences at the joints with the sequences in the donor segments revealed that joining sometimes resulted in the loss of a few base pairs. As can be seen in Figure 16-5, changes in the nucleotide sequence at the joint result in changes in the amino acids encoded at these positions in antibody molecules. Indeed, it was already known that these positions in purified antibodies were "hypervariable." Of course, the enzymes in the cell that catalyze these rearrangements cannot know the eventual translational reading

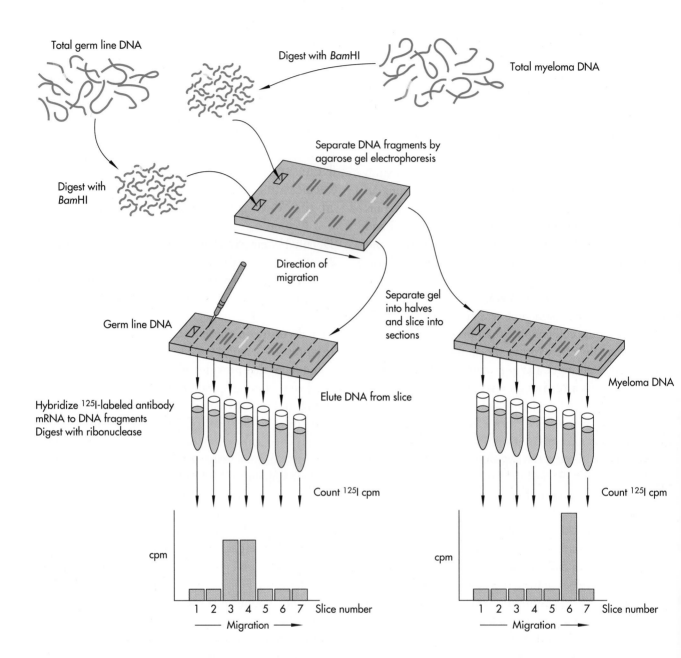

FIGURE **16-2**

Experimental demonstration that antibody genes are rearranged in myeloma cells. DNA was isolated from a myeloma cell line and from mouse embryos (germ line DNA), which contain almost no B cells. The DNA was digested with the restriction enzyme *Bam*HI, and each sample was fractionated by agarose gel electrophoresis, which separated the many fragments according to size. To determine which fragments carried the antibody genes, the agarose gel was cut into small slices. Each slice contained DNA fragments of a different size. The DNA was extracted from each slice and hybridized to radioactively labeled antibody mRNA, painstakingly extracted from the myeloma cells. DNA fractions that contained antibody gene sequences formed DNA–RNA hybrids with the mRNA probe, resulting in the protection of the radiolabeled RNA from digestion with ribonuclease. After treatment with ribonuclease, the remaining radioactivity in the hybrids was measured. Two slices from the embryo DNA hybridized to the probe, showing that the antibody gene sequences were contained on two different fragments of DNA in these cells. In contrast, only a single fragment of myeloma DNA hybridized, showing that in these cells, all antibody gene sequences lay within a single *Bam*HI restriction fragment.

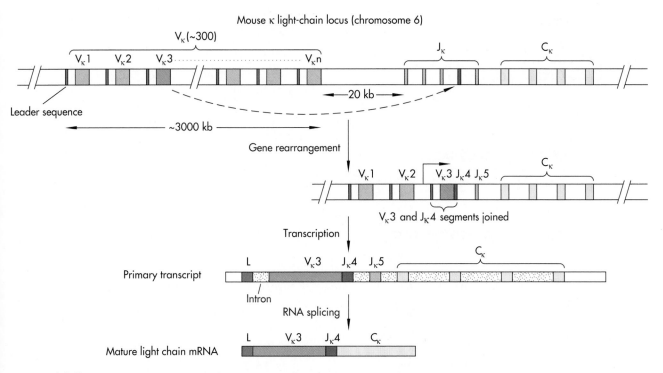

FIGURE **16-3**
Rearrangement of antibody light chain (κ) gene. The mouse κ locus consists of 300 V segments, 4 J segments (an additional J segment is not functional), and a C segment. Rearrangement joins one of the V gene segments to one of the J gene segments. This rearrangement of the genome constructs a transcription unit that, when spliced as shown, yields an mRNA encoding a complete κ light-chain protein. The segments marked L encode a leader peptide found at the amino terminus of each chain that specifies that the protein be secreted. The leader peptide is cleaved off during secretion.

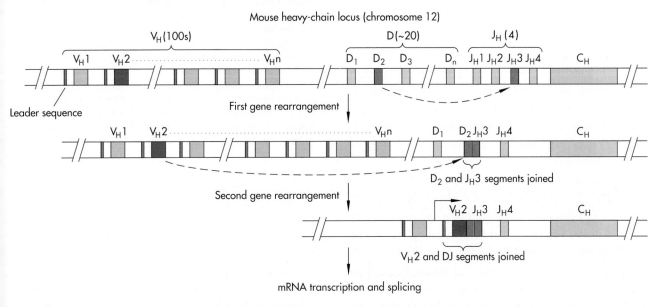

FIGURE **16-4**
Rearrangement of the heavy-chain gene. In addition to V, J, and C segments, the heavy-chain locus also contains 20 different D segments, which provide additional sequence diversity. Assembly of a heavy-chain gene proceeds in two steps: first, one of the D segments is fused to one of the J segments, and, second, a V gene is joined to the fused D-J segment. The rearranged gene is then transcribed, and the primary transcript is spliced.

Productive rearrangements

Non-productive rearrangements

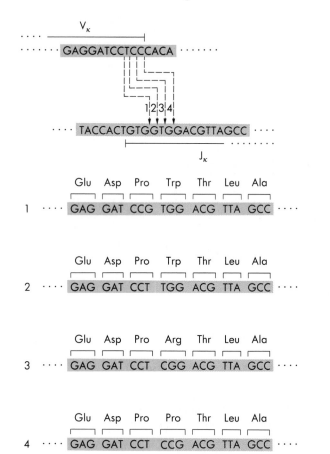

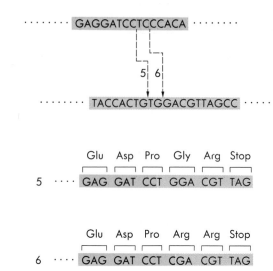

FIGURE **16-5**

Further antibody diversity is generated by imprecise V-J joining. The machinery that catalyzes V-J joining can fuse the two segments at any of several nearby nucleotide positions. Shown are the sequences at the ends of a V and J segment before rearrangement. The last complete codon in the V segment, CCT, encodes a proline at position 95 in the L chain. The first complete codon of the J segment encodes a tryptophan at position 96. However, the V and J sequences can be fused in any of several ways shown. On the left are four "productive" rearrangements that preserve the reading frame but generate codons encoding three different amino acids at position 96. Because there are three possible reading frames, only one of three V-J joinings manages to preserve the reading frame across the junction. Two-thirds of the rearrangements cause the J segment to be translated out of frame; these are termed *nonproductive rearrangements*, because these genes do not encode full-length antibody chains. Two examples are shown on the right.

frame of the gene, so two-thirds of these "sloppy" joints result in a nonfunctional gene in which the reading frame is lost (Figure 16-5). Thus, two-thirds of all rearrangements are nonproductive; this can be viewed as a "tax" the immune system pays for this further diversity in antibody specificity.

Gene rearrangement presents a tricky problem for controlling the transcription of antibody genes. As we learned in Chapter 9, the promoter for a eukaryotic gene is located on the 5' side of the gene. Because the 5' end of the functional antibody gene comes from the V segment, each of the several hundred V segments must carry its own promoter. How can the cell avoid tying up precious transcription factors at these unnecessary promoters and at the same time ensure that only the promoter associated with the selected V gene functions? The cell's solution to this problem is to place the control of antibody gene transcription under the influence of a potent complex enhancer located in the intron between the J and C gene segments (Figure 16-6). The antibody gene promoters are simple, binding a single B-cell-specific transcription factor. Only the promoter brought in with the selected V gene becomes activated by the enhancer. More distant V gene promoters are not affected.

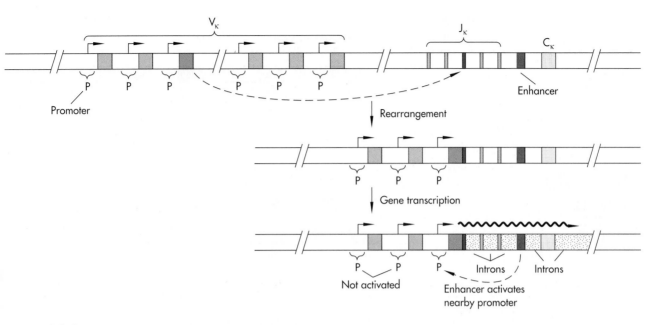

FIGURE **16-6**
Transcriptional regulation of the antibody genes. Each of the 300 V genes in the κ locus has its own promoter (P). These promoters are very simple, however, and promote little if any transcription on their own. When a V segment becomes fused to a J segment, its promoter comes under the influence of a potent transcriptional enhancer (E) located in the intron between the J and C segments.

Class Switching Places Useful Recognition Specificities on Antibody Molecules with Different Functional Properties

The immune system produces five different classes of antibodies, each class defined by a different heavy chain. Each class of antibodies has a distinct set of *effector functions* that it activates in the immune system. IgM and IgG, which carry μ and γ heavy chains, respectively, are secreted into the bloodstream, where they recognize circulating antigens. Complexes of antigen with IgM or IgG activate the complement system, a set of proteins that kills cells to which the antibody is bound. They also activate cells called *macrophages* that actively engulf and destroy bacteria and other antigens. IgA antibodies (α H chains) are found in various secreted fluids of the body, such as saliva, tears, and intestinal secretions, where they provide the first line of defense against invasion. IgE antibodies (ε H chains), when complexed with antigens, prompt *mast cells* to pump out histamine, which triggers allergic

reactions; its helpful function may be to protect against parasitic infections. IgD antibodies (δ H chains) are expressed solely on the surface of B cells. Their function is unknown.

The gene segments encoding these different heavy chains are clustered on a single chromosome (Figure 16-7). The C$_\mu$ gene is located at the 5' end of the cluster, and μ is the first heavy chain made after V-D-J rearrangement. The C$_\delta$ gene is located immediately downstream of C$_\mu$ and is expressed in the same cells by differential processing of a long RNA molecule that includes both genes, resulting in the splicing of the V-D-J sequence to either the C$_\mu$ or the C$_\delta$ sequence. Formation of the other antibody classes requires a further genetic rearrangement, termed a *class switch*. In class switching, the entire V-D-J segment including promoter and enhancer is transferred to a downstream C gene, with deletion of the intervening DNA. The effect is to allow antibody cells to use the same antigen recognition element, encoded by the V-D-J segment, on antibodies with different functions in the body.

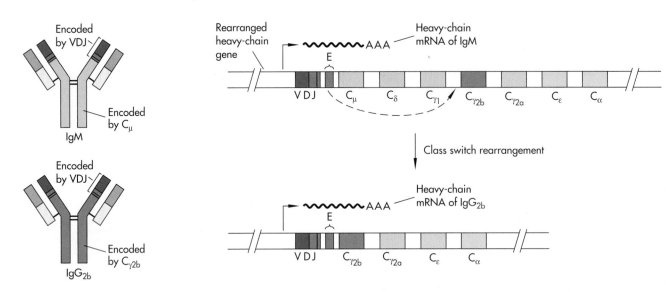

FIGURE **16-7**

Heavy-chain class switching. The top line shows a rearranged heavy-chain gene in which V-D-J joining has occurred. It is transcribed into an mRNA encoding an IgM heavy chain. After class switching, the V-D-J unit becomes joined to a new heavy-chain gene, here $C_{\gamma 2b}$, encoding the heavy chain of an IgG$_{2b}$ antibody. All the sequences between the V-D-J and $C_{\gamma 2b}$ genes in the top line are lost from the genome. Thus, the same V-D-J sequence can be part of different types of antibodies.

The Mechanisms of Antibody Gene Rearrangement Can Be Studied by Introducing Artificial Genes into Cells

The first clue to the mechanism of antibody gene rearrangement came from inspection of the DNA sequences flanking V, D, and J genes. The rearrangement of the sequences flanking the V and J segments of the light-chain gene is shown in Figure 16-8. On the immediate 3' boundary of each V segment is a 7-bp sequence, identical in all V genes, followed by 12-bp of random sequence, and then a 9-bp sequence, also identical in all V genes. These same *heptamer* and *nonamer* sequences are found immediately 5' to each J gene, but in the opposite orientation and separated by a 23-bp spacer of random DNA. Recombination occurs only between an element with the 12-bp spacer and one with the 23-bp spacer.

The role of these elements in recombination and the molecular mechanisms of the reaction have been studied by introducing artificial recombination substrates into B-cell lines that are actively rearranging their antibody genes. These unusual cell lines became available from studies with a mouse tumor virus that caused B-cell cancers consisting of cells blocked early in the B-cell lineage, at a stage when they are still undergoing antibody gene rearrangement. Thus, these cells contained the enzymes that catalyzed the recombination reaction. The design of such an artificial recombination experiment is shown in Figure 16-9. Basically, a DNA molecule was created in which a selectable marker such as a drug-resistance gene was separated from a promoter by a stretch of spacer DNA flanked by the heptamer-nonamer signals. Appropriate rearrangement of the DNA resulted in the expression of the drug-resistance gene from the promoter and the consequent formation of a drug-resistant colony of cells. The rearranged DNA could be recovered from these cells and the precise nature of the joint determined by restriction mapping and DNA sequence analysis. From these studies, it was learned that rearrangement of antibody genes could proceed by deletion of the sequence between two elements in the same orientation (as seen in Figure 16-8), or by inversion of an entire DNA segment to join two elements initially sited in opposite orientations (Figure 16-9).

(a) κ light-chain DNA
(germ line)

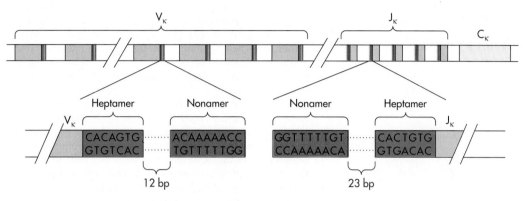

(b) Heavy-chain DNA
(germ line)

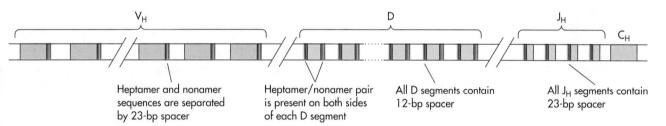

FIGURE **16-8**

The heptamer-nonamer signals for antibody gene rearrangement. (a) The top line shows a nonrearranged, or germline, κ light-chain gene. Immediately 3′ to each V gene is the sequence shown below it, consisting of conserved heptamer and nonamer sequences separated by a 12-bp spacer. Immediately 5′ to each J gene is the same heptamer and nonamer (in reversed orientation) with a 23-bp spacer. Recombination occurs exclusively between an element with a 12-bp spacer and an element with a 23-bp spacer. Thus, a V is always joined to a J but never to another V. (b) This 12/23 rule is especially important in the heavy-chain gene, where the additional D segments complicate recombination considerably. Here the V and J segments are both flanked by 23-bp elements, and the D segments, which recombine at both ends, are flanked on both sides by 12-bp elements. This rearrangement ensures that V segments always recombine to D segments and not directly to J segments and that V's, D's, and J's cannot recombine with other V's, D's, and J's, respectively.

Perhaps the most spectacular use of artificial recombination substrates was in a hunt for the genes that actually control the recombination reaction. When a molecule of this type was introduced into fibroblast cells, it never rearranged and no drug-resistant cells arose, because fibroblasts lack the necessary enzymes. As an assay for a gene that activates recombination of antibody genes, DNA from an antibody-expressing B-cell line was transferred into a population of fibroblasts carrying a previously introduced recombination substrate. The cultures were treated with a drug that killed off all of the fibroblasts, except for any rare cells that could activate a drug-resistance gene by rearranging the recombination substrate. Analysis of the rearranged molecules showed that the rearrangement was precisely as seen in B cells. Indeed, it was soon established that these rare fibroblasts contained a novel activity that permitted them to recombine antibody gene substrates. This activity could be transferred to new fibroblast cultures by sub-

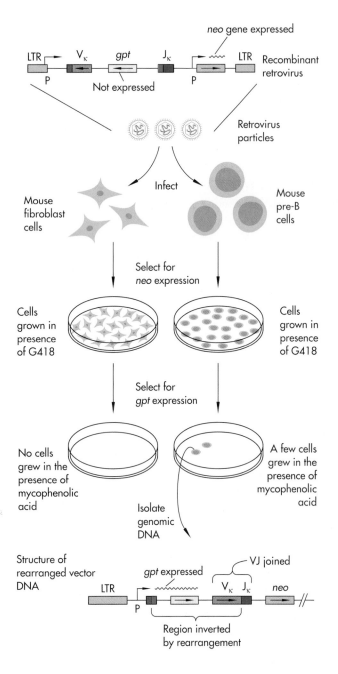

FIGURE **16-9**

Artificial recombination substrates are used to study antibody gene rearrangement. A retrovirus vector was designed as a recombination substrate with the structure shown at the top of the figure. It contained V and J gene segments together with their recombination signals (red and blue boxes) in opposite orientations to one another. Between the segments was placed a drug-resistance gene (*gpt*) in the reverse orientation relative to the strong viral promoter located in the LTR. Because the V and J are oriented in opposite direction, V-J joining requires inversion of the intervening DNA segment. This inversion also flips the *gpt* gene, placing it in the correct orientation for transcription from the LTR promoter. Expression of the *gpt* gene allows cells to grow in the presence of mycophenolic acid. The experiment was performed by packaging the substrate molecule into retrovirus particles, which were used to infect cultures of recipient cells. Infected cells were selected for resistance to the drug G418, a property encoded by a second gene in the vector (*neo*) that is expressed even in the absence of recombination. Next, the cells were treated with mycophenolic acid. The only cells that grew were those able to flip the *gpt* gene by performing V-J recombination. Resistant cells were obtained only in cell lines of the B lineage known to be undergoing antibody gene rearrangements. Introduction of the construct into fibroblast cells, for example, never yielded mycophenolic acid–resistant cells. Fibroblast cells bearing a recombination substrate construct like this served as the recipient in an experiment to identify the genes responsible for recombination (see text).

The Study of Cellular Immunity Was Greatly Advanced by Gene Cloning

Thus far, we have discussed the mechanisms of *humoral immunity*, which is mediated by B cells secreting antibody molecules. The targets of humoral immunity are generally free foreign antigens, like bacteria and viruses. But a second immune system, termed *cellular immunity*, exists to recognize and destroy cells of the body displaying foreign antigens on their surface (owing, for example, to viral infection). An additional function of this system is to regulate and coordinate cell killing and antibody production. These actions are carried out by *T cells*. T cells are closely related to B cells, arising from a common precursor cell. Like B cells, T cells have the capacity to recognize specific antigens, but unlike B cells, T cells do not recognize free antigens. Instead, they recognize only antigens displayed on the surface of other cells of the body. For many years, the nature of the T-cell surface molecule that recognized and bound antigen was only

sequent rounds of DNA transfer (a similar method had been used to isolate a cancer-inducing gene from a tumor cell line, see Chapter 18). Eventually, the segment of human B-cell DNA responsible for this recombination activity was cloned and found to contain two closely linked genes, RAG-1 and RAG-2, which cooperate to activate recombination with high efficiency.

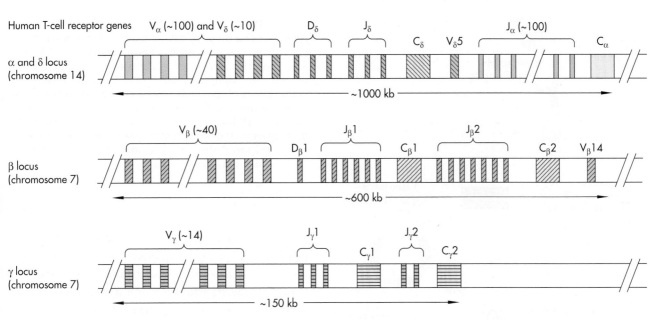

FIGURE **16-10**
The organization of the human T-cell receptor genes. The T-cell α and δ genes are located
in a single complex locus that spans over a million base pairs on chromosome 14. Interspersed V_α and V_δ genes are followed by D, J, and C segments for the δ gene. Downstream
of the δ gene are the J and C segments for the α gene. Note that assembly of a complete α
gene eliminates the δ gene. This is consistent with the observation that T cells express the
γδ receptor only early in development. Mature T cells usually express αβ. The β and γ loci
are both located on chromosome 7, but at opposite ends. They are somewhat smaller than
the αδ locus, but are organized in generally the same fashion. Like the antibody light-chain
genes, the α and γ genes do not use D segments. Although T-cell receptors are constructed
from fewer component V gene segments than are antibody genes, the total diversity of possible T-cell receptors is thought to exceed that of antibodies because of substantially greater
heterogeneity in V-J and V-D-J joining.

sketchily known. But with the cloning of the genes encoding the T-cell antigen receptor, our knowledge of their structure and function grew enormously.

Cloning of the T-cell receptor genes was considerably more difficult than cloning antibody genes and came 6 years later. Unlike antibodies, which are secreted in huge quantities by plasma cells, the T-cell receptor is confined to the surface of T cells and is present in minute quantities. So there was no ready source of either T-cell receptor protein or mRNA to derive probes for cloning. Instead, the first T-cell receptor gene was cloned by an arduous technique known as *subtraction,* described in detail in Chapter 7. Briefly, a technique was developed to isolate cDNA clones of mRNAs found in T cells but not in B cells. Assuming that, like antibodies, the T-cell receptor would be formed by gene rearrangements, the handful of clones isolated were tested individually in Southern

blotting experiments to determine whether any of them hybridized to DNA fragments that underwent rearrangement in T cells. This strategy resulted first in the isolation of the gene encoding the β subunit of the heterodimeric receptor molecule, followed shortly by the α subunit. Along the way, two additional receptor genes were isolated, encoding subunits (γ and δ) not previously identified by biochemical and immunological methods. An explosion of cloning and sequencing ensued, and soon the structural organization of the T-cell receptor genes was apparent (Figure 16-10).

The T-cell receptor genes proved remarkably similar to the antibody genes. The sequences of the proteins themselves were clearly related, helping to define the so-called *immunoglobulin superfamily,* which we discuss later in this chapter. Assembly of a complete T-cell receptor subunit gene entails the recombination

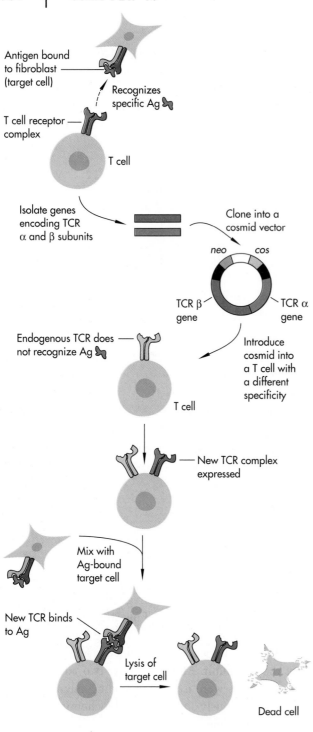

Antigen bound
to fibroblast
(target cell)

Recognizes
specific Ag

T cell receptor
complex

T cell

Isolate genes
encoding TCR
α and β subunits

Clone into a
cosmid vector

neo   *cos*

TCR β
gene

TCR α
gene

Introduce
cosmid into
a T cell with
a different
specificity

Endogenous TCR does
not recognize Ag

T cell

New TCR complex
expressed

Mix with
Ag-bound
target cell

New TCR binds
to Ag

Lysis of
target cell

Dead cell

FIGURE **16-11**

Demonstration by gene transfer that the T-cell receptor (TCR) genes encode antigen-recognition specificity. From a T-cell line that recognized a known specific antigen (red), the genes encoding the α and β subunits of the T-cell receptor were isolated. Both genes were inserted into a single cosmid vector, and the 50-kb cosmid was transfected into a second T-cell line that recognized a different antigen. Cells that expressed the newly introduced α and β genes now recognized and destroyed cells bearing the antigen recognized by the donor T-cell line.

recognition specificity were transfected into other T-cell lines with either a different specificity or with no receptors at all. Analysis of recipient cell lines that expressed the transfected α and β genes showed that these cell lines now recognized the same antigen as the cell line from which the α and β genes were cloned (Figure 16-11).

Not only did the transfected cells recognize the antigen, but they also responded biologically by killing cells bearing this antigen. Thus, the proteins encoded by the transferred receptor genes fulfilled a second receptor function, which is to communicate to the interior of the cell that antigen has been bound. We will discuss in detail in Chapter 17 the mechanisms by which such signals are generated (a process termed *signal transduction*). In T cells, the receptor is found in a large complex with several other proteins that carry out these signaling functions.

T Cells Recognize Antigens Only on Cells from the Same Individual

We have learned that T-cell receptors recognize antigens presented on the surface of cells. Immunological experiments with mice established the surprising fact that T cells can recognize antigen only on cells from the same individual. The same antigen on a cell from a different mouse is ignored. What are the molecules that specify the identity of an individual, and how do T cells combine antigen recognition with self recognition? The answers were to be found in a large complex of genes known as the *major histocompatibility complex (MHC)*. Within the MHC are genes encoding glycoproteins found on the surface of almost every cell of the body. These genes were initially defined in tissue transplantation experiments, in which skin

of V, D, J, and C segments. Recombination even uses the same nonamer-heptamer signals identified in the antibody genes.

Proof that the T-cell receptor α and β genes actually encode the proteins responsible for antigen-recognition specificity came from gene transfer experiments. The α and β genes cloned from a T cell with one

grafts from one individual to another were recognized as foreign and rejected by the host immune system. These experiments showed that MHC molecules on the surface of cells function as signposts to establish the identity of a given individual.

Cloning of the MHC Genes Reveals That Self Identity is Determined by a Few Polymorphic Genes

To establish the unique identities of individuals, many different MHC molecules must be made in a given population. How is this diversity achieved? Surprisingly, the answer turned out to be quite different for the MHC proteins than for antibodies and T-cell receptors. We have learned that most of the diversity in antibody and T-cell receptor structure is encoded directly in the genome and further amplified by physical shuffling of these gene segments. Cloning of the MHC genes revealed that, in contrast, self identity is hard-wired, directly encoded in only a handful of genes that differ in their primary DNA sequence from one individual to another. In other words, whereas a newly born individual inherits the potential for the complete diversity of antibody and T-cell receptor specificities, he inherits only his unique identity at the MHC locus.

The organization of the MHC was determined by gene cloning. The first foothold in the locus was achieved by cloning cDNAs for MHC molecules based on known protein sequence (see Chapter 7). The cDNAs were used to clone genomic DNA from the MHC locus, and most of the 2000 kb in the locus have now been cloned by chromosome walks performed in several labs. The genes of the MHC are grouped into three classes. Class I genes encode the classic transplantation antigens that reside on all cells and are the self molecules recognized in conjunction with antigen by the *cytotoxic T cells* (the type that kill infected self cells). Class II genes encode surface proteins found mainly on B cells and macrophages, cells that bind free antigens. Macrophages process the antigens into peptides which are then presented on class II molecules to *helper T cells,* which in turn regulate the antibody response. The remaining genes in the locus encode a variety of blood proteins (proteins of the complement cascade, for example) and other cell surface proteins whose functions are poorly understood.

After the genes of the MHC were cloned, it became possible to compare directly the sequences of the class I genes in different individuals. Indeed, these genes differ in sequence quite dramatically between individuals to a much greater extent than do any other genes in the body. And the resulting differences (or *polymorphisms*) in amino acid sequence in the MHC proteins form the basis for self recognition.

T-Cell Receptors Recognize Antigens Only in Association with an MHC Molecule

The MHC proteins play a special role in antigen recognition by T cells. We now know that antigen proteins are broken down by intracellular proteases into peptides which are then transported to the cell surface bound tightly in a cleft on the MHC molecule. T cells recognize these antigens only when they are bound within this cleft. How does a T cell recognize the self MHC molecule? Is it recognized by the T-cell receptor or by some other recognition molecule yet to be found in the T cells? The receptor-gene transfer experiment (Figure 16-11) provided the answer: the newly introduced receptor genes specified both antigen and self recognition. Thus the T-cell receptor recognizes the MHC molecule *and* the antigenic peptide as a single complex (Figures 16-12 and 16-13).

The structural basis for such specificity was first suggested by the polymorphisms found in MHC genes. The functional consequences of these differences in amino acid sequence were clarified by the determination of the three-dimensional structure of a class I MHC molecule. Most of the sequence differences encoded amino acids in the cleft predicted to bind the peptide antigen. Thus, the characteristics of the particular amino acids in the peptide-binding pocket determine which peptides will associate with a specific MHC molecule. Polymorphisms within the peptide-binding cleft probably account for variation among individuals in the ability of the MHC proteins to bind different antigens and present them to the immune system for action. Other polymorphic amino acids are not part of the peptide-binding pocket and are thought to make specific contacts with the T-cell receptor. This accounts for the self-recognition displayed by the T-cell receptor.

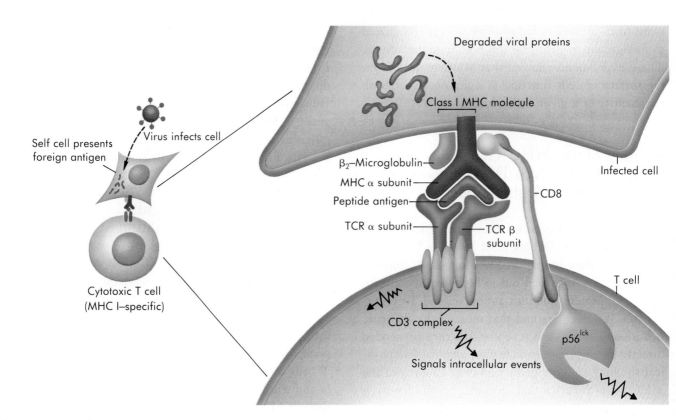

FIGURE **16-12**
Recognition of a virally infected cell by an MHC class I–specific cytotoxic T cell. Antigenic peptides derived from viral proteins appear on the surface of the infected cell in a complex with a class I MHC molecule. The class I molecule is made up of two polypeptides, the polymorphic α chain and β_2-microglobulin, an invariant protein. The class I MHC–peptide complex is specifically recognized by a T-cell receptor (TCR), which is itself composed of two protein chains, α and β. Another T cell–specific surface glycoprotein, CD8, binds to nonpolymorphic regions on the class I MHC molecule. The carboxy-terminal portion of the TCR is associated with a group of transmembrane proteins called the *CD3 complex*. The carboxy-terminal portion of CD8 is associated with an intracellular protein tyrosine kinase, p56^{lck}. Formation of the TCR-antigen-MHC-CD8 complex relays signals to the interior of the cell and activates p56^{lck} and proteins associated with CD3. In the case of cytotoxic T cells, the cells respond to these signals by synthesizing and secreting enzymes that bore holes into the infected cell.

Two other T-cell surface glycoproteins, CD4 and CD8, are important in the formation of the T-cell receptor–MHC complex. CD8 is a 34,000-dalton protein found on class I MHC–specific T cells and CD4 is a 55,000-dalton protein associated with class II MHC–specific T cells. These proteins are thought to strengthen the affinity of the T-cell receptor–MHC complex by contacting nonpolymorphic regions of the MHC (Figures 16-12 and 16-13). CD4 and CD8 also have a role in transmitting a signal, via a protein kinase associated with their carboxy-termini, to intracellular components involved in T-cell activation. And, CD4 is the receptor through which HIV, the virus that causes AIDS, enters T cells (Chapter 25).

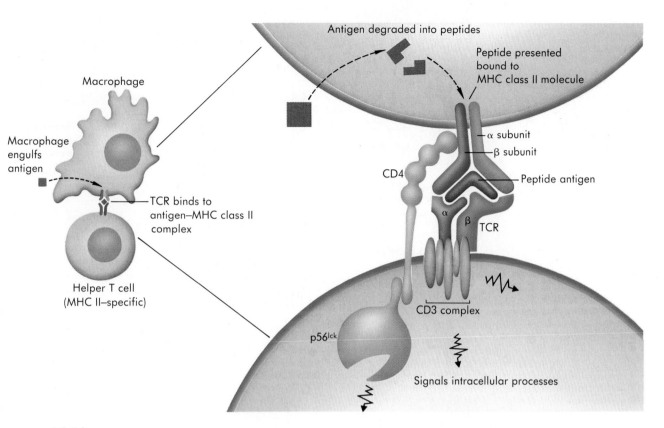

FIGURE **16-13**

Recognition of processed antigen presented on a macrophage by an MHC class II–specific helper T cell. Macrophages engulf an antigen and degrade it to small peptides. The peptides reappear on the cell surface in complex with class II MHC proteins, which are composed of two polymorphic chains termed α and β. Class II–specific T cells express CD4 instead of the CD8 protein present on class I–specific T cells. The class II MHC–peptide complex is bound by the TCR and the CD4 protein present on helper T cells. Upon formation of the complex, the helper T cell synthesizes and secretes lymphokines that cause antibody-producing B cells to divide and differentiate into plasma cells, which secrete antibodies. Helper cell activation occurs through the CD3 complex and by turning on the p56^{lck} protein tyrosine kinase.

Intercellular Communication Regulates Immune System Function

The immune response is like a military operation. Outlying scouts spot the first intruders and inform their commanding officers. The commanding officers activate the troops to attack and destroy the intruders. As in a military operation, communication is the key to a successful outcome. B cells and macrophages function as the scouts, hunting for foreign antigens. When a macrophage encounters an antigen, it presents the antigen via its MHC proteins to a helper T cell that recognizes the complex of MHC and antigen peptide. The helper T cell is the commanding officer, issuing orders in the form of secreted molecules called *lymphokines.* The lymphokines recruit and activate the cells

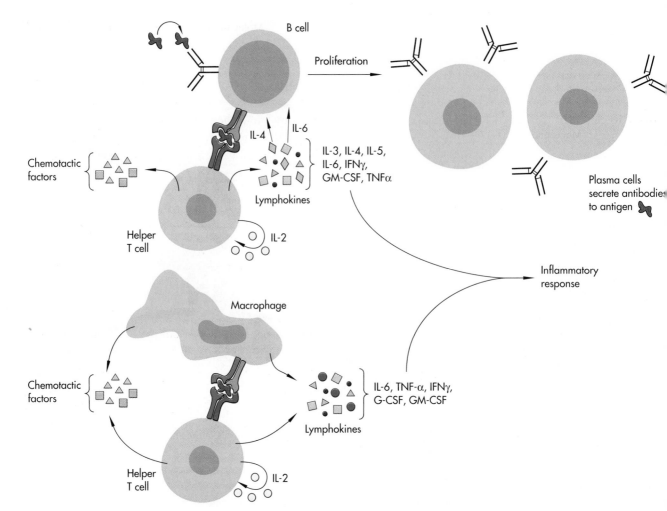

FIGURE **16-14**
Lymphokines exchanged among cells of the immune system coordinate the immune response. When a helper T cell is presented with an antigen by a B cell or macrophage, it responds by synthesizing and secreting a potpourri of small peptide signaling molecules termed *lymphokines*. One lymphokine, IL-2, feeds back to the T cell, causing it to grow and multiply. Other lymphokines act on the antigen-presenting B cell, ordering it to proliferate and develop into plasma cells, which secrete massive quantities of antibody. Other factors summon help in the form of macrophages, platelets, and other blood cells. Still others trigger the proliferation of dormant stem cells, the progenitors of all the cells in the blood, to replenish the ranks. The total coordinated effort of all these cell types results in the inflammatory response, which attacks and destroys invading organisms and repairs damage caused by them. IL, interleukin; TNF, tumor necrosis factor, IFN, interferon; G, granulocyte; M, macrophage; CSF, colony-stimulating factor.

that attack and destroy the invader. For example, some lymphokines signal B cells to differentiate into plasma cells, the factory cells that synthesize and secrete antibodies into the bloodstream (Figure 16-14). Others act as *chemotactic factors* that attract macrophages, platelets, and other cells to the scene of the battle to participate in cleanup and repair. Still others act on the helper T cell itself and cause it to multiply, further amplifying the response. Additional signals activate

precursor cells to replenish the immune cell population.

All the activities of the various components of the immune system are coordinated by lymphokines. Their power to modulate immune function has attracted the attention of the biotechnology industry, which has made the cloning of lymphokines and their receptors a major priority (see Chapter 23). One of the first lymphokines produced by the biotechnology industry was IL-2, which is the autostimulatory factor produced by helper T cells after stimulation by antigen. Recombinant IL-2 is being used to supercharge T cells in cancer patients, to help the immune system attack and destroy tumor cells.

The Immunoglobulin Superfamily Encodes Proteins That Participate in Cell-Cell Communication

Inspection of the sequences of cloned genes for antibodies, T cell receptors, and MHC proteins revealed that these proteins are actually quite closely related to each other. This similarity was first noted when the first antibody genes available were sequenced. A conserved sequence of about 100 amino acids was discerned in the constant regions of both the heavy and the light chains. This sequence was marked by two cysteine residues, about 60 amino acids apart, that form an intrachain disulfide bond. A similar sequence could

also be found in the variable regions of antibody molecules. Studies of the three-dimensional structure of antibody molecules confirmed that these regions of the protein constitute a structural domain folded very similarly in all molecules. Subsequent cloning and sequencing has predicted a similar domain in scores of other proteins, most with a role in cell-cell communication. Members of the so-called *immunoglobulin superfamily* (Figure 16-15) include the T-cell receptor subunits, both the polymorphic and invariant chains of the MHC proteins, various proteins involved in cell-cell adhesion (proteins like the T-cell CD4 protein, which helps tie helper T cells to antigen-presenting cells, and N-CAM [neural cell adhesion molecule], which helps neurons in the brain contact each other), and receptors for lymphokines and growth factors. In the genes encoding members of the immunoglobulin superfamily, this domain is almost always encoded within a single exon, suggesting a role for exon shuffling and duplication in the creation of these genes (see Chapter 22).

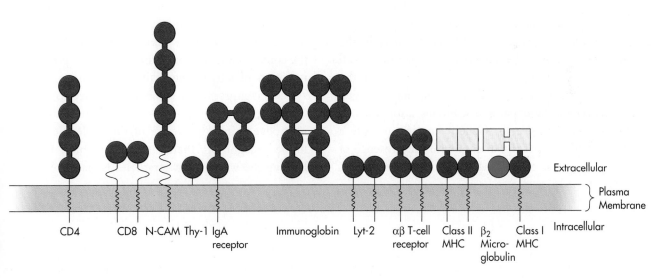

FIGURE **16-15**

The immunoglobulin (Ig) superfamily. Many cellular genes that encode proteins involved in intercellular communication contain a common structural domain, the Ig fold (designated by circles), first identified in antibodies. This figure illustrates the structural organization of several of the proteins that share this domain. It is thought that all these genes have arisen from a single ancestral gene in the distant evolutionary past. For a protein to be a member of the Ig superfamily, its predicted amino acid sequence from cloned cDNA data must exhibit similarities with an immunoglobulin V or C domain sequence. It is expected that the Ig domain in such proteins will be structurally similar to an Ig domain of an antibody. This prediction was confirmed with the solution of the three-dimensional structure of CD4.

Malfunctions of the Immune System Underlie Many Diseases

The immune system is charged with a nearly impossible task. It has to recognize millions of molecules as foreign and destroy them. But millions more must be ignored because they are normal constituents of the body. When the immune system malfunctions, the consequences can be serious. The most common form of immune malfunction is the allergic reaction, triggered by the overproduction of the IgE class of antibodies in response to common antigens like dust and pollen. The IgE-antigen complexes activate mast cells to release histamine, a signaling molecule that triggers events leading to the symptoms we experience as allergies. Allergic reactions are normally treated with antihistamines or with epinephrine, which blocks the release of histamines from mast cells.

A much more severe breakdown of the immune system is autoimmune disease, when the immune system mistakes a self molecule for foreign. Autoimmune diseases range from the rare systemic lupus erythematosus, in which the body can make antibodies to such common self molecules as DNA and RNA, to more common diseases such as arthritis and diabetes. Inspection of the pancreas in patients with type 1 (insulin-dependent) diabetes shows that the β cells, the insulin producers of the pancreas, are destroyed, and the organ is infiltrated by the shock troops of the immune system—B cells, T cells, and macrophages. The blood of these patients contains antibodies to proteins made in β cells. Genetic studies indicate a strong correlation of the disease with the presence of certain alleles of class II MHC genes. Indeed, pre-disposition to the disease correlates with certain amino acids at a single position in the peptide-binding cleft of the β chain of a class II molecule. This suggests that the development of diabetes might require the presentation of β-cell antigens to helper T cells; certain class II proteins might do this better than others.

Different diseases arise when the immune system loses its ability to recognize and attack foreign antigens. Some of these *immunodeficiencies* are inborn. Each year, for example, a few children are born lacking a functional gene encoding the metabolic enzyme adenosine deaminase (ADA). For mysterious reasons, these children have no working immune system whatsoever and must live in a totally sterile environment with no physical contact with the outside world (the "boy in the bubble" disease). These children were the subjects of the first sanctioned gene therapy experiment, in which precursor immune cells were removed, provided with a cloned functional ADA gene, and returned to the body (see Chapter 28). The most widespread and feared immunodeficiency is AIDS (*acquired immunodeficiency syndrome*). AIDS is characterized by increased susceptibility to viral and microbial infections. As we shall learn in Chapter 25, AIDS is itself caused by a virus, *HIV*, with a particularly insidious target of infection. It infects and destroys the helper T cells of the immune system. Because these cells are the commanding officers of the immune response, HIV infection effectively neutralizes the entire immune system, compromising the body's ability to fight off invaders, including the AIDS virus itself. But our rapidly developing knowledge of the molecules of immune recognition, thanks to recombinant DNA, is leading the way to new strategies to battle AIDS.

Reading List

General

Williams, A. F., and A. N. Barclay. "The immunoglobulin superfamily—domains for cell surface recognition." *Ann. Rev. Immunol.,* 6:381–406 (1988).
Hood, L. E., I. L. Weissman, and W. B. Wood. *Immunology.*
Benjamin-Cummings, Menlo Park, Calif., 1984.
Immunological Recognition. Cold Spring Harbor Symp. Quant. Biol., vol. 54, Cold Spring Harbor Laboratory, Cold Spring Harbor, N. Y., 1989.

Original Research Papers

MATURE IMMUNOGLOBULIN GENES ARE FORMED BY RECOMBINATION

Tonegawa, S. "Somatic generation of immune diversity." (Nobel lecture in physiology or medicine, 1987.) *In Vitro Cell. Dev. Biol.,* 24: 253–265 (1988). [Review]

Dreyer, W. J., and J. D. Bennett. "The molecular basis of antibody formation: a paradox." *Proc. Natl. Acad. Sci. USA,* 54: 864–869 (1965).

Hozumi, N., and S. Tonegawa. "Evidence for somatic rearrangement of immunoglobulin genes coding for variable and constant regions." *Proc. Natl. Acad. Sci. USA,* 73: 3628–3632 (1976).

Brack, C., M. Hirama, R. Lenard-Schuller, and S. Tonegawa. "A complete immunoglobulin gene is created by somatic recombination." *Cell,* 15: 1–14 (1978).

Seidman, J. G., E. E. Max, and P. Leder. "A κ-immunoglobulin gene is formed by site-specific recombination without further somatic mutation." *Nature,* 280: 370–375 (1979).

Davis, M. M., K. Calame, P. W. Early, D. L. Livant, R. Joho, I. L. Weissman, and L. E. Hood. "An immunoglobulin heavy-chain gene is formed by at least two recombinational events." *Nature,* 283: 733–739 (1980).

Early, P., H. Huang, M. Davis, K. Calame, and L. Hood. "An immunoglobulin heavy chain variable region gene is generated from three segments of DNA: V_H, D and J_H." *Cell,* 19: 981–992 (1980).

ALTERNATIVE RNA SPLICING IN THE FORMATION OF ANTIBODIES

Early, P., J. Rogers, M. Davis, K. Calame, M. Bond, R. Wall, and L. E. Hood. "Two mRNAs can be produced from a single immunoglobulin μ gene by alternative RNA processing pathways." *Cell,* 20: 313–319 (1980).

Moore, K. W., J. Rogers, T. Hunkapiller, P. Early, C. Nottenburg, I. Weissman, H. Bazin, R. Wall, and L. E. Hood. "Expression of IgD may use both DNA rearrangement and RNA splicing mechanism." *Proc. Natl. Acad. Sci. USA,* 78: 1800–1804 (1981).

HEAVY-CHAIN CLASS SWITCHES

Esser, C., and A. Radbruch. "Immunoglobulin class switching: molecular and cellular analysis." *Ann. Rev. Immunol.,* 8: 717–735 (1990). [Review]

Sledge, C., D. S. Fair, B. Black, R. G. Krueger, and L. E. Hood. "Antibody differentiation: apparent sequence identity between variable regions shared by IgA and IgG immunoglobulins." *Proc. Natl. Acad. Sci. USA,* 73: 923–927 (1976).

Kataoka, T., T. Kawakami, N. Takahashi, and T. Honjo. "Rearrangement of immunoglobulin $\gamma 1$-chain gene and mechanisms for heavy-chain class switch." *Proc. Natl. Acad. Sci. USA,* 77: 919–923 (1980).

UNDERSTANDING THE MECHANISM OF IMMUNOGLOBULIN GENE REARRANGEMENT

Yancopoulos, G. D., T. K. Blackwell, H. Suh, L. Hood, and F. W. Alt. "Introduced T cell receptor variable region gene segments recombine in pre-B cells: evidence that B and T cells use a common recombinase." *Cell,* 44: 251–259 (1986).

Hesse, J. E., M. R. Lieber, K. Mizuuchi, and M. Gellert. "V(D)J recombination: a functional definition of the joining signals." *Genes and Devel.,* 3: 1053–1061 (1989).

Matsunami, N., T. Hamaguchi, Y. Yamamoto, K. Kuze, K. Kanawa, H. Matsuo, M. Kawaichi, and T. Honjo. "A protein binding to the J_κ recombination sequence of immunoglobulin genes contains a sequence related to the integrase motif." *Nature,* 342: 934–937 (1989).

Schatz, D. G., M. A. Oettinger, and D. Baltimore. "The V(D)J recombination activating gene, RAG-1." *Cell,* 59: 1035–1048 (1989).

Oettinger, M. A., D. G. Schatz, C. Gorka, and D. Baltimore. "RAG-1 and RAG-2, adjacent genes that synergistically activate V(D)J recombination." *Science,* 248: 1517–1523 (1990).

CLONING THE GENES FOR T-CELL RECEPTORS

Marrack, P., and J. Kappler. "The T cell receptor." *Science,* 238: 1073–1079 (1987). [Review]

Davis, M. M., and P. J. Bjorkman. "T cell antigen receptor genes and T cell recognition." *Nature,* 334: 395–402 (1988). [Review]

Hedrick, S. M., D. I. Cohen, E. A. Nielsen, and M. M. Davis. "Isolation of cDNA clones encoding T cell-specific membrane associated proteins." *Nature,* 308: 149–153 (1984).

Saito, H., D. M. Kranz, Y. Takagaki, A. C. Hayday, H. N. Eisen, and S. Tonegawa. "Complete primary structure of a heterodimeric T-cell receptor deduced from cDNA sequences." *Nature,* 309: 757–762 (1984).

Yanagi, Y., Y. Yoshikai, K. Leggett, S. P. Clark, I. Aleksander, and T. W. Mak. "A human T cell-specific cDNA clone encodes a protein having extensive homology to immunoglobulin chains." *Nature,* 308: 145–149 (1984).

Brenner, M., J. McLean, D. Dialynas, J. Strominger, J. Smith, F. Owen, J. Seidman, S. Ip, F. Rosen, and M. Krangel. "Identification of a putative second T-cell receptor." *Nature,* 322: 145–149 (1986).

Loh, E. Y., L. L. Lanier, C. W. Turck, D. R. Littman, M. M. Davis, Y. H. Chien, and A. Weiss. "Identification and sequence of a fourth human T cell antigen receptor chain." *Nature*, 330: 569–572 (1987).

STUDYING THE FUNCTION OF T-CELL RECEPTORS

Dembic, C., W. Haas, S. Weiss, J. McCubrey, H. Kiefer, and M. Steinmetz. "Transfer of specificity by murine α and β T cell receptor genes." *Nature*, 320: 232–238 (1986).

Saito, T., A. Weiss, J. Miller, M. A. Norcross, and R. N. Germain. "Specific antigen-Ia activation of transfected human T cells expressing Ti αβ-human T3 receptor complexes." *Nature*, 325: 125–130 (1987).

Berg, L. J., B. Fazekas de St. Groth, F. Ivars, C. C. Goodnow, S. Gilfillan, H. -J. Garchon, and M. M. Davis. "Expression of T-cell receptor alpha-chain genes in transgenic mice." *Mol. Cell. Biol.*, 8: 5459–5469 (1988).

STRUCTURE AND FUNCTION OF MAJOR HISTOCOMPATIBILITY COMPLEX PROTEINS

Flavell, R. A., H. Allen, L. C. Burkly, D. H. Sherman, G. L. Waneck, and G. Widera. "Molecular biology of the H-2 histocompatibility complex." *Science*, 233: 437–443 (1986). [Review]

Bjorkman, P. J., and P. Parham. "Structure, function and diversity of class I major histocompatibility complex molecules." *Ann. Rev. Biochem.*, 59: 253–288 (1990). [Review]

Ploegh, H. L., H. T. Orr, and J. L. Strominger. "Molecular cloning of a human histocompatibility antigen cDNA fragment." *Proc. Natl. Acad. Sci. USA*, 77: 6081–6085 (1980).

Sood, A. K., D. Pereira, and S. M. Weissman. "Isolation and partial nucleotide sequence of a cDNA clone for human histocompatibility antigen HLA-B by use of an oligodeoxynucleotide primer." *Proc. Natl. Acad. Sci. USA*, 78: 616–620 (1981).

Steinmetz, M., A. Winoto, K. Minard, and L. E. Hood. "Clusters of genes encoding mouse transplantation antigens." *Cell*, 28: 489–498 (1982).

Bjorkman, P. J., M. A. Saper, B. Samraoui, J. L. Strominger, and D. C. Wiley. "Structure of the human histocompatibility antigen, HLA-A2." *Nature*, 329: 506–512 (1987).

CD4 AND CD8

Littman, D. R. "The structure of CD4 and CD8 genes." *Ann. Rev. Immunol.*, 5: 561–584 (1987). [Review]

Salter, R. D., R. J. Benjamin, P. K. Wesley, S. E. Buxton, T. P. J. Garrett, C. Clayberger, A. M. Krensky, A. M. Norment, D. R. Littman, and P. Parham. "A binding site for the T-cell co-receptor CD8 on the α_3 domain of HLA-A2." *Nature*, 345: 41–46 (1990).

Wang, J., Y. Youwei, T. P. J. Garrett, J. Liu, D. W. Rodgers, R. L. Garlick, G. E. Tarr, Y. Husain, E. L. Reinherz, and S. C. Harrison. "Atomic structure of a fragment of human CD4 containing two immunoglobulin-like domains." *Nature*, 348: 411–418 (1990).

LYMPHOKINES

Paul, W. E. "Pleiotropy and redundancy: T cell-derived lymphokines in the immune response." *Cell*, 57: 521–524 (1989). [Review]

Arai, K. -i., F. Lee, A. Miyajima, S. Miyatake, N. Arai, and T. Yokota. "Cytokines: coordinators of immune and inflammatory responses." *Ann. Rev. Biochem.*, 59: 783–836 (1990). [Review]

DISEASES OF THE IMMUNE SYSTEM

Schuler, W., I. J. Weller, A. Schuler, R. A. Phillips, N. Rosenberg, T. W. Mak, J. F. Kearney, R. P. Perry, and M. J. Bosma. "Rearrangement of antigen receptor genes is defective in mice with severe combined immune deficiency." *Cell*, 46: 963–972 (1986).

Baekkeskov, S., H. -J. Aanstoot, S. Christgau, A. Reetz, M. Solimena, M. Cascalho, F. Folli, H. Richter-Olesen, and P. -De Camilli. "Identification of the 64K autoantigen in insulin-dependent diabetes as the GABA-synthesizing enzyme glutamic acid decarboxylase." *Nature*, 347: 151–156 (1990).

Romain, L. M., L. F. Simons, R. E. Hammer, J. F. Sambrook, and M. -J. Gething. "The expression of influenza virus hemagglutinin in the pancreatic β-cells of transgenic mice results in autoimmune diabetes." *Cell*, 61: 383–396 (1990).

17

Moving Signals
Across Membranes

Cells of all organisms have evolved mechanisms to sense environmental stimuli and to respond to them appropriately. Single-cell organisms must seek out nutrients and avoid noxious substances to survive. Higher organisms must respond to an even greater diversity of signals, mounting coordinated responses to nutrients, growth factors, hormones, neurotransmitters, and sensory input. These signals fulfill many functions. They control the growth state of cells, mobilizing resting cell populations into active growth and telling proliferating cells to cease growing. And they modulate the functions of nongrowing differentiated cells, controlling, for example, glycogen breakdown in muscle and cell-cell communication in the brain. In each case, however, the receiving cell faces the same problem: the primary signal must be received, transmitted across the cell (plasma) membrane, and in some cases, transmitted across the nuclear membrane as well.

Nearly a century ago, scientists postulated that the detection of extracellular signals would involve *receptors*—structures that bind chemical stimuli specifically and somehow *transduce* the message into the intracellular environment. Although receptors had been characterized physiologically using bioassay methods, their direct detection and measurement became possible in the 1970s with the development of highly specific and sensitive binding assays that exploited

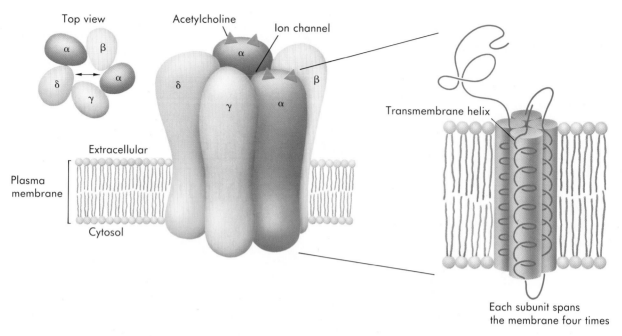

Top view

Acetylcholine

Ion channel

α β

δ α

γ

δ α β

γ α

Transmembrane helix

Extracellular

Plasma membrane

Cytosol

Each subunit spans
the membrane four times

FIGURE **17-1**
Ligand-gated ion channels. The acetylcholine receptor is a ligand-gated ion channel composed of multiple protein subunits, designated α, β, γ, and δ (center and left). Each of these subunits contains four transmembrane segments (right). Amino acid residues from one or more transmembrane segments line the channel and determine its pore size and ion selectivity. The nicotinic acetylcholine receptor is composed of five of these subunits, whereas the receptors for GABA (gamma-aminobutyric acid) and glycine, two other neurotransmitters, each consist of four subunits.

radioactively labeled *ligands* (the molecules that selectively bind to receptors). With the introduction of powerful affinity-chromatography methods, small quantities of highly purified receptor preparations could be characterized biochemically for the first time, proving the receptors to be proteins, and revealing their structural diversity.

By the 1960s, we had discovered how some receptors transduced their signals into the cell. Investigators found that the receptor for epinephrine (also known as adrenaline), when bound by epinephrine, activates an intracellular enzyme, *adenylate cyclase,* which catalyzes the synthesis of cyclic AMP (cAMP) from ATP. cAMP is a *second messenger,* which carries the information encoded by extracellular signals to targets within the cell that produce the biological response. Indeed, simply adding cAMP to epinephrine-responsive cells

mimics the effect of authentic ligand. Many ligand receptors are now known to modulate production of cAMP, whereas others are coupled to the production of other second messengers, including cyclic GMP, calcium ions, and phospholipids. Not all receptors, however, act via second messengers. The nicotinic type of acetylcholine receptor is a membrane ion channel, which triggers changes in the electrical properties of the cell, whereas the receptors for steroid hormones are intracellular transcription factors that interact directly with the genome to induce or repress the expression of certain genes.

In this chapter we will see how the application of recombinant DNA technology has revealed the structures of receptors, clarified their evolutionary relationships, and provided the sophisticated tools required to dissect their signaling mechanisms.

The Acetylcholine Receptor Is a Ligand-Gated Ion Channel

The acetylcholine receptor located at the neuromuscular junction (where nerve impulses stimulate muscle contraction) historically received close attention because of its pharmacological importance, ease of manipulation, and anatomical sequestration from other types of receptors. Many of the basic principles of receptor physiology have arisen from studies of the acetylcholine receptor. Binding of acetylcholine, the ligand, to the receptor results in a rapid influx of sodium ions across the plasma membrane into the cell. The influx of sodium ions results in *depolarization* of the membrane (decrease in the voltage difference between the outside and inside of the cell) within microseconds of ligand binding, permitting rapid muscle contraction in response to the nerve impulse. The receptor protein itself forms the ion channel. In the absence of acetylcholine, the channel is closed and ions cannot pass through. Upon binding acetycholine, it undergoes a conformational change that opens the ion pore to allow passage of sodium ions (but not others). Because this type of acetycholine receptor responds to nicotine, it is referred to as a *nicotinic cholinergic receptor.*

The unusually large amounts of nicotinic receptor present in the electric organ of the eel facilitated its purification and allowed investigators to derive portions of the amino acid sequence of the protein. From this sequence, oligonucleotide probes were synthesized and used to screen cDNA libraries (Chapter 7), and the nicotinic receptor became the first receptor to be cloned and sequenced. In vertebrate muscle the receptor is encoded by four different genes (α, β, γ, and δ) of similar sequence, each of which encodes a unique receptor subunit of approximately 500 amino acids. Each subunit contains four protein helices that each span the membrane; these are connected by three loops, two inside the cell and one outside (Figure 17-1). Five subunits assemble noncovalently in the plasma membrane to produce a single receptor molecule with the structure $\alpha_2\beta\gamma\delta$ in muscle. The receptor complex forms a central ion channel that is controlled by acetylcholine binding to each of the α subunits. From gene cloning studies, we now know that there are several different genes encoding each of the receptor subunits and that further diversity is produced by alternative splicing of their RNA transcripts. Thus, the receptor can exist in many forms, differing in their kinetic properties and sensitivity to drugs and toxins. In Chapter 21, we discuss further the diversity and organization of these receptor channels and their roles in the brain.

Some Receptors Are Coupled to Second Messenger Systems via GTP-Binding Proteins

Whereas the acetylcholine receptor is directly responsible for passing a signal across the membrane, in the form of sodium ions, many hormone receptors seem to work differently. These receptors do not physically pass molecules across the plasma membrane. Instead the effect of hormone binding is activation of an enzyme, adenylate cyclase, inside the cell. Adenylate cyclase catalyzes the synthesis of cAMP, and cAMP becomes the second messenger that ferries the hormone's message to intracellular recipients. How do the receptors communicate with adenylate cyclase, to which they are not physically linked?

An early clue was that adenylate cyclase activation by hormones is dependent on the presence of the nucleotide GTP, and that GTP also influences hormone binding. These observations led to the *G protein hypothesis*, that a GTP-binding protein, a *G protein*, connects the hormone receptors to adenylate cyclase. By 1980, such a protein was identified and isolated. The protein consisted of three subunits, with the largest, or α, subunit being involved in GTP binding. In a series of classical biochemical experiments, this G protein was shown to exchange GTP for bound GDP when activated by hormone receptors. The GTP-carrying form then bound to and activated adenylate cyclase. Hydrolysis of GTP to GDP, catalyzed by the G protein itself, returned the G protein to its basal, inactive form (Figure 17-2). Thus, the G protein serves a dual role, as an intermediate that relays the signal from receptor to effector, and as a clock that controls the duration of the signal.

The G protein was purified and a partial amino acid sequence was obtained, allowing cDNA clones

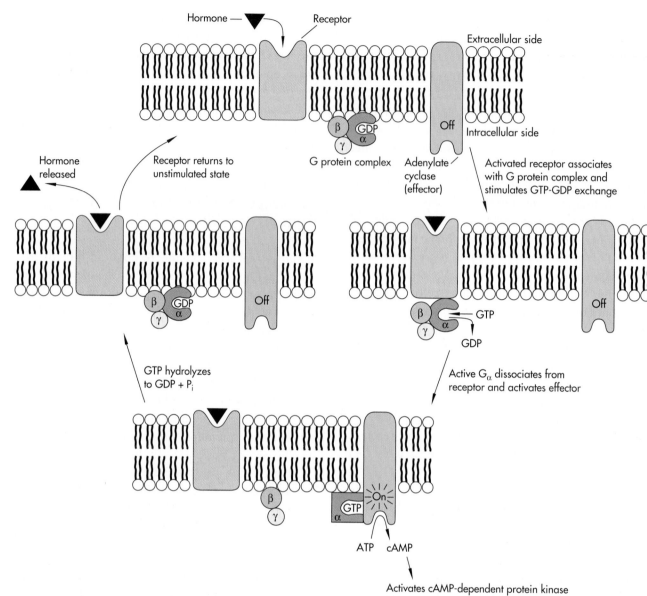

FIGURE **17-2**

The G protein cycle, coupling hormone receptors to adenylate cyclase. The receptor, the inactive G protein complex—consisting of α, β, and γ subunits—and inactive adenylate cyclase are all associated with the plasma membrane. Binding of hormone to receptor triggers the association of the G protein complex with the cytoplasmic face of the hormone receptor. The α subunit of the G protein expels its bound GDP molecule, which is quickly replaced by GTP (which is more abundant in the cell). Binding of GTP in turn triggers a conformational change in the α subunit, which unlinks from β and γ and associates with a neighboring adenylate cyclase molecule. The cyclase switches on, continuing to synthesize cAMP until the G_α subunit spontaneously hydrolyzes its GTP to GDP, returning itself to the inactive state. G_α dissociates from adenylate cyclase, rejoins β and γ, and the inactive G protein trimer reassembles. Hormone dissociates from the receptor, restoring the unstimulated configuration. Thus, G proteins act as molecular switches, toggling between the inactive GDP-bound and the active GTP-bound state. Its built-in GTPase activity ensures that the G protein remains active for only a short time. Mutations in the α subunit that slow or block its GTPase prolong the active state, sometimes indefinitely. In cells in which cAMP is a signal for growth, such G protein mutations contribute to the deregulated growth of tumors.

to be isolated and the entire G protein sequence to be determined. These studies revealed that the receptor-coupled G proteins were similar in structure to *transducin*, a GTP-binding protein that fulfills a similar stimulus-response coupling function in the visual system (see below). We now know that a whole family of G proteins couple many different receptors to their effector proteins. In fact, G proteins serve a general role as switchable molecular couplers (that is, they can be switched on and off) throughout the plant and animal kingdoms, and even in unicellular organisms such as yeast. They are involved in many functions unrelated to signal transduction, such as protein synthesis, intracellular transport, and exocytosis. Members of the G protein family play a critical role in the development of cancer (see Chapter 18). Dozens of new members of this family have now been identified by homology screening and genetic methods. In many cases, however, the partners coupled by the newly isolated G proteins are not yet known.

G Protein–Linked Receptors Span the Membrane Seven Times

Pharmacologists knew for a long time that the diverse effects of epinephrine on target tissues were mediated by two different types of *adrenergic* receptors, termed α and β. Several subtypes of α and β receptors are now known; each receptor consists of a single polypeptide chain and is coupled to a second messenger system through a G protein. Some of the receptors stimulate adenylate cyclase, but others inhibit the cyclase enzyme or activate a different effector enzyme, phospholipase.

The development of subtype-specific affinity chromatography allowed purification of each receptor in amounts sufficient for partial protein sequencing, and the receptor genes were then cloned and sequenced. Surprisingly, the deduced amino acid sequence of the first adrenergic receptor turned out to resemble that of the visual pigment rhodopsin, sequenced a few years earlier by protein chemists. These *integral membrane proteins*, that is, proteins that are inserted in the membrane, consist of a single polypeptide chain of approximately 450 amino acids that is predicted to fold into a structure containing seven membrane-spanning helices (Figure 17-3). The loops that connect the helices protrude into the cytoplasm (loops C1, C2, and C3) or into the extracellular space (loops E1, E2, and E3), and the amino and carboxyl ends of the proteins reside outside and inside the cell, respectively.

Cloning of the adrenergic receptors revealed that all shared this seven-helix structure. Moreover, the cloned receptors for many neurotransmitters were also found to be members of the rhodopsin-adrenergic receptor family. We now know that the seven-helix motif is a common feature of G protein–coupled receptors, regardless of the nature of their ligand. Like the G proteins, this family of receptors is large and evolutionarily ancient; they are used, for example, by yeast to detect mating pheromones and they also form the basis for the senses of sight and smell in higher organisms.

Domain Swapping Reveals the Structural Basis of Receptor-Effector Coupling

Our understanding of the molecular basis of receptor-effector coupling is a triumph of classical biochemistry. Proof of the G protein hypothesis came from painstaking reconstitution experiments in which purified proteins were inserted into lipid vesicles that mimic the plasma membrane. But this approach is prohibitive for detailed structure-function studies, and the tools of recombinant DNA offer a potent alternative to biochemical reconstitution.

The power of this technology to characterize the organization of complex proteins was illustrated in studies of the adrenergic receptors. Each of these receptors must distinguish among different ligands and G protein partners. What structural features impart this specificity? This question has been addressed by constructing chimeric receptor "genes" from cDNA clones of different adrenergic receptor subtypes (Figure 17-4), transcribing them into mRNA with bacteriophage RNA polymerases and injecting the RNA into *Xenopus* oocytes. Because the oocyte supplies the

(a)

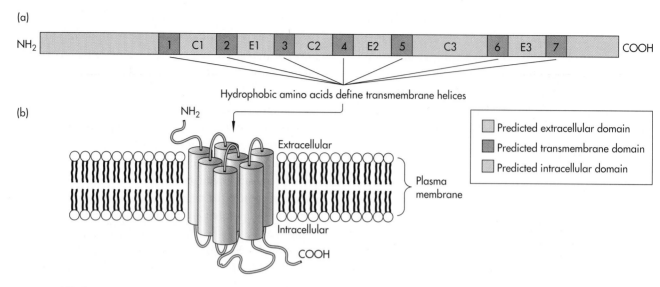

FIGURE **17-3**

Structural organization of seven-helix receptors. (a) A linear representation of the receptor structure. Seven blocks of hydrophobic amino acids (1–7) form helices that are predicted to span the membrane. The helices are linked by loops that are alternatively outside (E) or inside (C) the cell. Similarly, the amino-terminal tail is outside, whereas the carboxy-terminal tail is inside. (b) A view of the proposed folded structure of the receptors.

adenylate cyclase and G proteins for functional assays, but lacks any adrenergic receptor of its own, each chimera could be expressed and its protein product assayed for its ability to bind ligands and to modulate adenylate cyclase. With this simple assay, investigators could reconstitute receptor-effector coupling with receptor proteins engineered to their specifications.

The initial chimeras were built by splicing together restriction fragments encoding different structural domains of the α_2 and β_2 subtypes of adrenergic receptors. These receptors were chosen because they differ in their sensitivities to *agonists* (compounds that activate receptors) and *antagonists* (compounds that block receptor activation). Moreover, the β_2 receptor is coupled to a G protein that stimulates adenylate cyclase, whereas the α_2 receptor is coupled to a G protein that inhibits the cyclase. When oocytes producing the chimeric receptors were tested for ligand binding and changes in cyclase activity, they were indeed found to function as adrenergic receptors and displayed characteristics that combined features of the two parental receptors. For example, replacement of the carboxy-terminal 100 residues of the β_2 receptor (encoding loop E3, the seventh membrane-spanning helix, and the carboxyl terminus) with α_2 receptor sequences

produced a hybrid receptor that activated cyclase like the β_2 receptor, yet bound antagonists like an α_2 receptor. These studies revealed that the specificity for G protein interaction is localized between the fifth and sixth membrane-spanning helices, including loop C3, whereas the seventh membrane-spanning helix and carboxy-terminal tail determine the specificity of ligand binding.

Visual Pigments Are Signaling Receptors

Rhodopsin, the first seven-helix-motif receptor sequenced, is still the most completely characterized member of the G protein–linked receptor superfamily. Localized in the light-sensitive rod cells of the retina, rhodopsin converts light energy into a hyperpolarization of the rod cell, which transmits this signal to the brain via the optic nerve. The mechanism through which rhodopsin elicits rod cell hyperpolarization is a paradigm for understanding how all receptors of this class work (Figure 17-5). Rhodopsin activation is coupled to hyperpolarization by a G protein, transducin. Active (GTP-bound) transducin activates an enzyme called *phosphodiesterase*, which destroys cyclic GMP (a

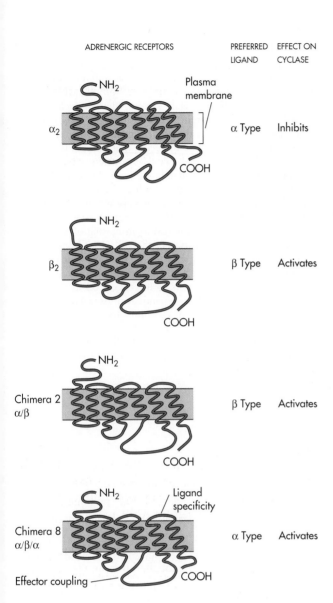

ADRENERGIC RECEPTORS	PREFERRED LIGAND	EFFECT ON CYCLASE
α_2	α Type	Inhibits
β_2	β Type	Activates
Chimera 2 α/β	β Type	Activates
Chimera 8 $\alpha/\beta/\alpha$	α Type	Activates

FIGURE **17-4**

Construction of chimeric receptors (domain swapping) is used to localize functional domains in receptors. The α_2 (red) and β_2 (blue) adrenergic receptors share a similar overall organization, yet they prefer different ligands and exert opposite effects on adenylate cyclase (because they interact with different G proteins). To map the regions of the receptors that specify these functions, receptor cDNAs were recombined so that they encoded receptor chimeras composed of pieces derived from each parent. The recombinant cDNAs were transcribed into RNA in vitro, and the RNA was injected into *Xenopus* oocytes, which produced the chimeric receptor proteins. Chimera 2, in which the amino-terminal half is α_2 and the carboxy-terminal half is β_2, behaves indistinguishably from an authentic β_2 receptor, showing that both ligand-binding and G protein interaction take place in the carboxy-terminal half of the protein. The more complex chimera 8, which is all α_2 except for the fifth and six transmembrane helices, the extracellular loop preceding them, and the cytoplasmic loop linking them, exhibits recombinant properties: it binds ligand like an α_2 receptor but activates the G protein normally used by β_2 (resulting in cyclase activation). Thus, the cytoplasmic loop between helices 5 and 6 must specify G protein binding (and therefore the type of effector action triggered by the receptor), while ligand binding is specified by the carboxy-terminal extracellular loop and the seventh helix.

close relative of cAMP). A drop in the cGMP concentration in the rod cell leads to closure of cGMP-dependent ion channels and subsequent hyperpolarization. A key feature of the visual system is the enormous amplification achieved: photolyzed rhodopsin can activate several hundred transducin molecules, and each activated phosphodiesterase enzyme destroys many cGMP molecules. This amplification cascade endows each rod cell with the sensitivity to respond to a single photon of light.

If this system is used to sense light intensity in rod cells, what mechanism confers the quality of color sensitivity to retinal cone cells? Cloning and sequencing of the *opsin* genes responsible for color sensitivity

revealed protein structures related closely to rhodopsin. Although opsins employ the same chromophore as rhodopsin, 11-*cis*-retinal, amino acid sequence differences in the pocket of the protein that holds the chromophore result in a unique absorption maximum for each opsin. These maxima correspond to the primary colors we see as blue, green, and red. Individual cone cells express only a single species of opsin, which functions as a receptor for a single primary color. Mutations affecting these opsin genes cause color blindness.

Growth Factor Receptors Have Intrinsic Enzymatic Activity

We have already encountered two different strategies for moving signals across membranes—use of receptors that act as gates, admitting ions to the cell interior, and use of receptors that activate intracellular enzymes that produce second messenger molecules. A third strategy is widely employed by receptors for polypeptide growth factors. These receptors themselves carry an intrinsic enzymatic activity that is triggered

(a)

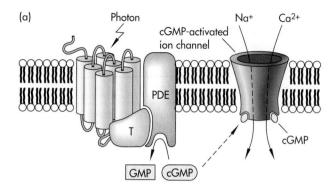

(b)

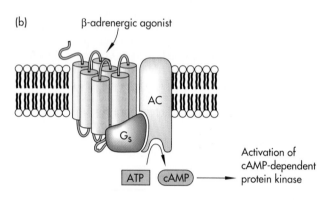

(c)

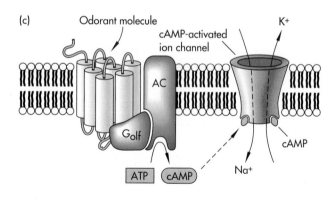

FIGURE **17-5**

Signal transduction by G protein-coupled receptors. Seven-helix receptors can be coupled to several different second messenger systems, depending on the G protein with which they associate. (a) In the visual system, the first such system studied, seven-helix proteins (rhodopsin in the rods, and three different opsins for the color-sensitive cones) serve as the receptors for light. The molecule that absorbs the photon is actually 11-*cis*-retinal, which is covalently bound to a lysine in the seventh transmembrane helix deep inside the protein. The receptor associates with a G protein termed transducin (T), which couples the receptor to phosphodiesterase (PDE), an enzyme that hydrolyzes cyclic GMP. A drop in cGMP concentration results in closing of an ion channel and hyperpolarization of the cell. (b) The β-adrenergic receptor associates with the G_s protein, which couples it to adenylate cyclase (AC). Adenylate cyclase produces cAMP, which exerts its biological effects in this system by activation of cAMP-dependent protein kinase. (c) In the olfactory system, odorant receptors associate with a unique G protein termed G_{olf} (see Chapter 21). This G protein couples the receptor to a form of adenylate cyclase unique to olfactory neurons. Generation of cAMP opens a Na^+-K^+ ion channel.

by ligand binding (Figure 17-6). One of the surprising discoveries of recent years arose from the cloning of the receptor for atrial natiuretic peptide (ANP), a hormone that controls blood volume. Sequencing of the cDNA showed that the receptor is a large protein with a domain situated outside the cell to bind its ligand, a single short helix that spans the membrane, and a large domain that resides in the cytoplasm. The cytoplasmic domain contains a region closely related in sequence to guanylate cyclase, the enzyme that synthesizes cGMP. This unexpected finding suggested that ANP, like other hormones, acts through a cyclic

nucleotide second messenger. But unlike hormones that activate a distinct adenylate cyclase enzyme through a G protein–coupled receptor, the guanylate cyclase activity triggered by ANP is resident in the receptor protein itself.

This paradigm of receptors with intrinsic enzymatic activity is most firmly established for the receptors for polypeptide growth factors that control cell proliferation and differentiation. These receptors were isolated and characterized in a variety of ways. Some were snared as the cellular counterparts of viral *oncogenes* (see Chapter 18), showing that, when altered, these receptors have the power to derail normal cell growth control. Some were purified by conventional biochemistry. And others were isolated by homology cloning using probes from other receptor genes. These receptors share a similar overall structure with the ANP receptor and with some of the molecules of the immune system we discussed in Chapter 16. They possess a large extracellular ligand-binding domain, a single membrane-spanning helix, and a sizeable cytoplasmic domain. The cytoplasmic domain is an enzyme; in the case of the growth factor receptors, it is a ligand-activated *protein kinase* that transfers phosphate to tyrosine residues of other proteins. As we will

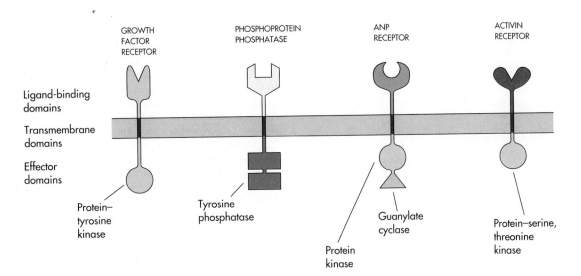

FIGURE **17-6**
The organization of receptors with intrinsic enzymatic activity. Many polypeptide growth factors, which control the growth of cells, act through cell-surface receptors that possess an extracellular ligand-binding domain linked by a single transmembrane helix to a large cytoplasmic domain endowed with tyrosine-specific protein kinase activity. Binding of ligand turns on the receptor kinase activity, which adds phosphate groups to tyrosine residues on itself and other proteins. Some receptors carry cytoplasmic domains that act in opposition to protein-tyrosine kinases, encoding tyrosine-specific protein phosphatases, which remove phosphates from tyrosine residues (most of these receptors actually carry two such domains). Still others, notably the receptor for atrial natiuretic peptide (ANP), a hormone that controls blood volume, have guanylate cyclase activity. These receptors also possess a domain that is similar in sequence to protein kinases, but whose precise function is not yet known. The receptor for activin, a polypeptide growth factor important in embryonic development, has a serine/threonine-specific protein kinase–like domain.

learn repeatedly in the next few chapters, protein kinases are the chief regulators of cellular activity, and tyrosine-specific kinases are of particular importance in the control of cell proliferation.

Recombinant DNA technology has been especially helpful in the identification of receptors, the localization of their functional domains, and the testing of hypotheses about how they work. For example, a candidate cDNA believed to encode the receptor for platelet-derived growth factor (PDGF) was evaluated by placing it in an expression vector and transfecting the resulting plasmid into a cell line that lacked PDGF receptors of its own (Figure 17-7). The transfected cells acquired measurable binding sites for PDGF on their cell surfaces, showing that the cDNA indeed encodes a PDGF-binding protein. But, in addition, the cells gained the ability to grow in response to added

PDGF, proving that the cDNA encodes the receptor's signaling function as well and that the machinery to respond to the receptor must exist in the recipient cells. To prove that the protein kinase activity contained in the receptor is required for its signaling activity, in vitro mutagenesis was used to alter a lysine residue in this domain that was known to be part of the active site in other protein kinases. Mutation of this lysine to alanine abolished the kinase activity of the receptor and also extinguished its signaling function (although, as we learn below, it is also possible to extinguish signaling with other kinds of mutations). Other types of experiments, including receptor domain swaps similar to those done with the adrenergic receptors, have established the location of specific functional domains in receptors, such as the ligand-binding domain.

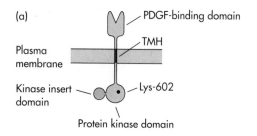

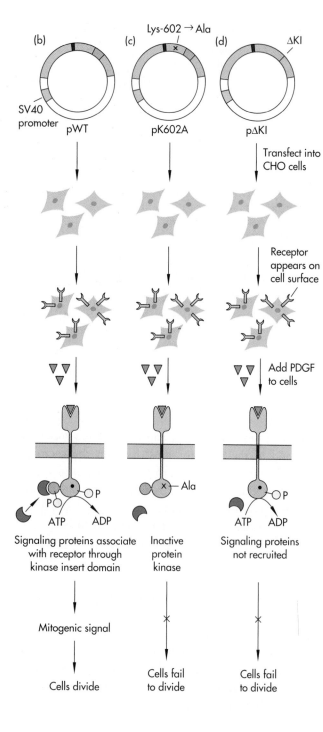

(a)

PDGF-binding domain

TMH

Plasma membrane

Kinase insert domain — Lys-602

Protein kinase domain

(b) (c) Lys-602 → Ala (d) ΔKI

SV40 promoter

pWT   pK602A   pΔKI

Transfect into CHO cells

Receptor appears on cell surface

Add PDGF to cells

Signaling proteins associate with receptor through kinase insert domain

Inactive protein kinase — Ala

Signaling proteins not recruited

Mitogenic signal

Cells divide

Cells fail to divide

Cells fail to divide

FIGURE **17-7**

Expression of normal and mutant receptors to study receptor structure and function. (a) The receptor for platelet-derived growth factor (PDGF) contains an extracellular PDGF-binding domain, a single transmembrane helix (TMH), and a tyrosine-specific protein kinase domain interrupted by a stretch of polypeptide termed the *kinase insert domain*. To prove that a receptor cDNA clone encoded a functional protein, researchers transfected an expression vector containing the cDNA into CHO cells, which do not contain PDGF receptors of their own. (b) When cDNA for a normal (wild-type, pWT) receptor was used, the transfected cells produced on their surfaces a protein that bound PDGF; when these cells were treated with PDGF, tyrosine residues on this receptor became phosphorylated and the cells were prompted to divide. These observations established that the cDNA encoded a functional receptor with both PDGF-binding and signaling properties. (c) To prove that the protein kinase activity was required for signaling, researchers altered the receptor cDNA so that a lysine residue predicted to reside in the kinase active site (Lys-602) was changed to an alanine (pK602A). The receptor produced from this plasmid still bound PDGF, but it failed to become phosphorylated. Moreover, the cells did not divide when treated with PDGF. Thus, kinase activity is required for signaling. (d) Researchers also deleted the kinase insert domain (pΔKI). Now the receptor produced bound PDGF and became phosphorylated normally, but still the cells did not divide. This experiment established that the kinase insert domain, which we now know is the site of recruitment of signaling proteins, is also required for signaling.

Receptors Can Be Associated with Intracellular Protein Kinases

A number of cell-surface proteins that function as receptors share a similar structural organization with the protein kinase class of growth factor receptors, except that their cytoplasmic domains are exceptionally short, too short by far to harbor any enzymatic activity. How do these proteins communicate with the cell interior? A clue came from biochemical studies of one of these cell-surface proteins, the CD4 protein, a receptor that helps T cells bind to antigen-presenting cells (Chapter 16) and serves as the receptor and port-of-entry for the AIDS virus (Chapter 25). When CD4 protein was recovered from T cells with specific anti-CD4 antibodies, a tyrosine-specific protein kinase was recovered with it. This kinase was recognized as the product of a gene called *lck*, a member of a large family of genes that encode intracellular protein kinases and are known to function as oncogenes. By placing the CD4 and *lck* cDNAs in expression vectors and thus producing these two proteins in cells that don't normally express either one, the basis for this interaction was determined (Figure 17-8). The carboxy-terminal cytoplasmic tail of the CD4 protein is bound tightly to the amino-terminal portion of the *lck* protein. Deleting either of these domains abolished the ability of these proteins to form a complex. Transferring these two domains to completely unrelated proteins allowed the modified proteins to form a tight complex. Presumably this non-covalent CD4-*lck* kinase connection allows the *lck* protein kinase activity to be regulated by CD4-mediated cell-cell interactions.

Receptors Activate Common Second Messenger Systems

Once a ligand binds to a receptor and the signal that binding has occurred is received inside the cell, how is that signal distributed to the molecules in the cell that need to respond to it? As we have learned, this task is usually performed by second messengers, small molecules produced inside the cell in response to receptor activation. Many receptors, activate or inhibit adenylate cyclase, the enzyme that synthesizes cAMP, a second messenger. cAMP exerts its effects on cells through the action of a cAMP-dependent protein kinase, which we will discuss in more detail shortly.

But receptors can activate other second messenger systems. Some receptors activate an enzyme called a *phospholipase*, which hydrolyzes *inositol phospholipids*, minor lipid constituents of the plasma membrane. The products of the phospholipase reaction are two molecules with second messenger properties. One, *inositol trisphosphate*, binds to intracellular vesicles that store calcium ions, causing them to dump calcium into the cytoplasm. Calcium ions bind to a small protein called *calmodulin*, which acts as a regulatory subunit for other enzymes in the cell. One of the principal enzymes activated by the calcium-calmodulin complex is a protein kinase. The other product of the phospholipase reaction is *diacylglycerol*, a lipid molecule that remains in the membrane, where it activates yet another protein kinase called *protein kinase C* (cloning of protein kinase C cDNAs has shown that protein kinase C

actually consists of a large family of closely related enzymes). This phospholipid second messenger system is a crucial one in many cells. The combination of artificially raising the calcium concentration with molecules called *ionophores* (which transport ions across the cell membrane) and activating protein kinase C with drugs can often fool cells into thinking they have received a natural stimulus. For example, in response to these drugs, T cells will initiate their response to antigen even though they have not bound antigen, and egg cells will begin their program of cell division without ever encountering sperm. Because of the powerful responses elicited in cells by second messengers, the biochemistry of these signaling systems is under active study, and new candidate second messengers are continually being uncovered.

How the seven-helix receptors are coupled to second messenger–producing enzymes is relatively well understood (they use G proteins). But for the growth factor receptors—those that act through an intrinsic or associated protein-tyrosine kinase activity—how the ligand-binding signal is transmitted to targets within the cell is less clear. We know that the kinase activity is required for receptor signaling, leading many to believe that the receptor transmits its signal by phosphorylating proteins on tyrosine. But which proteins? One answer is the receptor itself. Binding of ligand to receptor leads to rapid phosphorylation of the receptor on one or more tyrosine residues in its cytoplasmic domain, a phenomenon termed *autophosphorylation*. We now believe that this is not a truly autocatalytic reaction, but that instead one receptor molecule is phosphorylated by another neighboring receptor molecule. This was most convincingly demonstrated by expressing in the same cells two distinguishable forms of the same growth factor receptor, one with an active kinase and one without. Treatment of these cells with the growth factor produced tyrosine phosphorylation of the receptor with the inactive kinase, ruling out true autophosphorylation (Figure 17-9). So, one result of growth factor binding is to induce receptor molecules to cluster and phosphorylate one another. This in turn appears to result in the recruitment of several cellular enzymes to the cytoplasmic domain of the receptor. These recruits include enzymes involved in the production of phospholipid second messengers, and, perhaps, other pro-

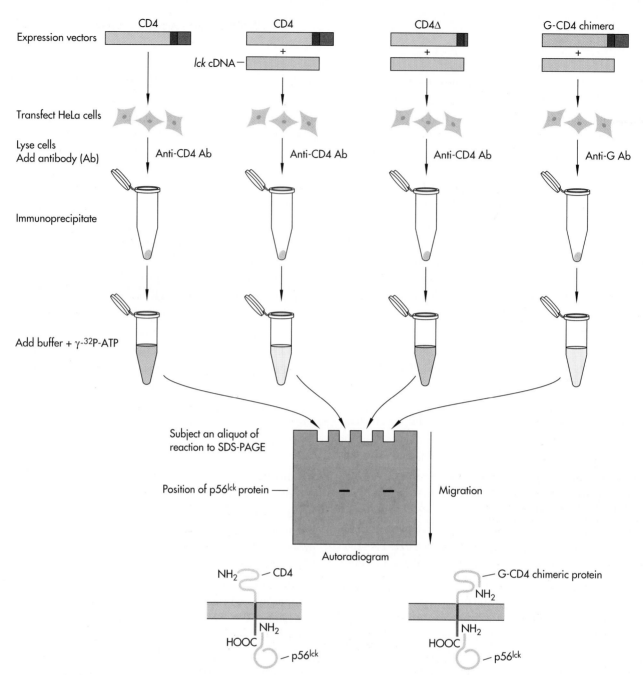

FIGURE **17-8**

Demonstration of association between the cytoplasmic domain of the CD4 protein and the amino-terminal domain of the p56[lck] protein kinase, product of the *lck* gene. (The p56[lck] designates a *lck*-encoded polypeptide of MW 56,000.) CD4 and *lck* cDNAs were placed into expression vectors, and the plasmids were transfected into HeLa cells, human tumor cells that do not normally express either protein. Anti-CD4 antibodies (Ab) were used to capture CD4 and any proteins associated with it from a cell lysate. When radiolabeled ATP was added to the captured proteins, any p56[lck] protein present phosphorylated itself, transferring the labeled phosphate from ATP to a tyrosine residue in its polypeptide chain. The labeled p56[lck] was detected by separating the proteins on a gel and exposing the gel to x-ray film. When the CD4 protein is expressed alone in HeLa cells, no labeled p56[lck] is detected on the film (first lane). When both expression vectors are present, however, antibodies to CD4 recover bound p56[lck] (detectable in second lane of gel). Using this assay, investigators could produce mutant CD4 and p56[lck] proteins and ask whether they associate. They found that deleting a portion of the cytoplasmic tail of CD4 (CD4Δ expression vector) eliminated its ability to bind p56[lck] (third lane), and that appending this tail to an unrelated protein (G-CD4 chimera) allowed p56[lck] to bind to this new partner (fourth lane). Similar modifications to p56[lck] (not shown) localized its CD4-binding domain to the amino-terminal tail, leading to the model shown at the bottom.

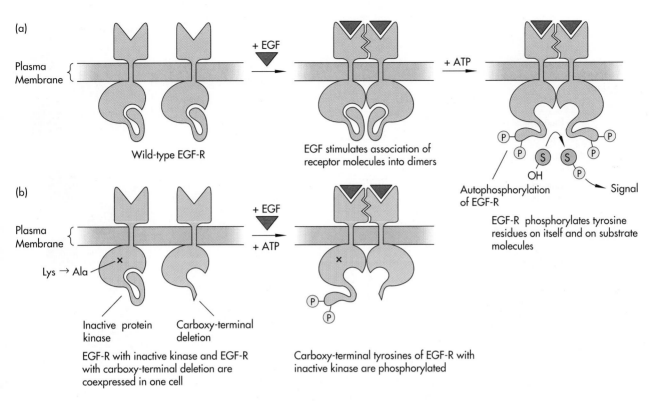

(a)

Plasma Membrane

Wild-type EGF-R

+ EGF

EGF stimulates association of receptor molecules into dimers

+ ATP

Autophosphorylation of EGF-R

OH

Signal

EGF-R phosphorylates tyrosine residues on itself and on substrate molecules

(b)

Plasma Membrane

Lys → Ala

Inactive protein kinase

Carboxy-terminal deletion

EGF-R with inactive kinase and EGF-R with carboxy-terminal deletion are coexpressed in one cell

+ EGF

+ ATP

Carboxy-terminal tyrosines of EGF-R with inactive kinase are phosphorylated

FIGURE **17-9**

Demonstration that growth factor receptors pair and phosphorylate one another. (a) Model for activation of epidermal growth factor receptor (EGF-R) by EGF. Researchers noted early that binding of growth factors to their receptors caused the receptors to become phosphorylated on tyrosine, a phenomenon termed *autophosphorylation.* Subsequent work suggested that receptor signaling required the association of two receptor proteins. Left unresolved was the mechanism of autophosphorylation: Does each receptor phosphorylate itself (as well as other substrates, S), or does one receptor phosphorylate its neighbor? (b) To test for intermolecular phosphorylation, researchers expressed two different EGF receptor mutants in cells. One was full length but carried a single amino acid substitution that inactivated its kinase (a lysine to alanine substitution akin to the one shown in Figure 17-17). The other had an active kinase but lacked the carboxy-terminal amino acids normally autophosphorylated. If autophosphorylation were truly autocatalytic, neither mutant protein should become phosphorylated, because neither can phosphorylate itself: one receptor lacks the kinase activity and the other lacks the sites of phosphorylation. If, on the other hand, a receptor phosphorylates its neighbor, the active tailless receptor should be able to phosphorylate the kinase-inactive protein. The result: the inactive receptor was phosphorylated. This key experiment showed that receptor phosphorylation is not autocatalytic and suggested that formation of receptor dimers in the presence of ligand is an important step in receptor activation.

tein kinases. Recruitment of these enzymes appears to be a key step in receptor signaling, because a mutant receptor engineered to retain its intrinsic kinase activity but lose its ability to recruit the cellular enzymes is inactive for signaling. The ability to produce wild-type and mutant receptor proteins at will, thanks to cloning methodologies, is greatly accelerating receptor research.

Protein Phosphorylation Is a Principal Mechanism for Signaling

A recurring theme in studies of signal transduction is the involvement of protein kinases. Each of the second messengers we have discussed, cAMP, calcium, diacylglycerol, exerts all or part of its effects on cells through the action of a protein kinase. And many

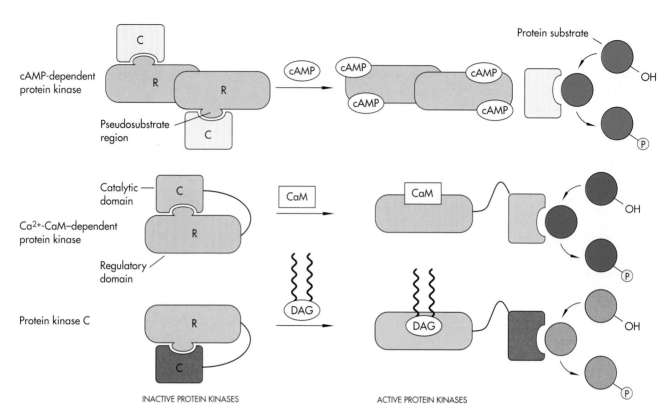

FIGURE **17-10**
The pseudosubstrate model for protein kinase regulation. Each of these protein kinases possesses a regulatory domain (R) carrying a pseudosubstrate, which mimics an actual substrate and occupies the active catalytic site (C) of the kinase. In each case, the ligands that activate these kinases cause an unfolding of the pseudosubstrate domain from the active site. In cAMP-dependent protein kinase, the pseudosubstrate domain resides in a separate regulatory protein subunit, which upon binding cAMP dissociates from the catalytic subunit. In calcium-calmodulin (Ca²⁺-CaM)–dependent protein kinase and protein kinase C, the pseudosubstrate domain is an integral part of the kinase polypeptide. Binding of ligand triggers a refolding of the enzyme that pulls the pseudosubstrate domain out of the active site. DAG, diacylglycerol (ligand for protein kinase C).

receptors have intrinsic or tightly associated protein kinase activity. Protein kinases represent critical nodes for the amplification and distribution of signals, since a single protein kinase can phosphorylate many different target proteins, altering their activities in significant ways. Some protein kinases, however, have very limited substrate specificities, phosphorylating only one or a few proteins. But among these targets can be other protein kinases, leading to further amplification and distribution of the signal.

How are protein kinases turned on by second messengers? Answers to this question have come from biochemical studies coupled with directed mutagenesis and expression of mutant proteins. Although at

first glance the strategies appear different for each kinase, the underlying mechanisms are remarkably similar (Figure 17-10). For the calcium-calmodulin–dependent protein kinase, the enzyme is inactive unless bound by the calcium-calmodulin complex, which as a consequence of conformational changes, causes the enzyme to unfold an inhibitory domain (called a *pseudosubstrate*) from its active site. Protein kinase C is activated when diacylglycerol binds to a regulatory region of the protein, triggering a similar conformational change that exposes the active site. For cAMP-dependent protein kinase, the inhibitory domain resides in a separate regulatory subunit polypeptide. This polypeptide forms a tight complex with the catalytic subunit, which carries the active site. Binding of cAMP to the regulatory subunit disassembles the complex and frees the catalytic subunit to phosphorylate substrate proteins.

If protein phosphorylation is important in signaling, cells must also possess enzymes that remove phosphate groups from proteins. The identification and characterization of such enzymes, *protein phosphatases,* has been greatly spurred by cloning studies. When the first tyrosine-specific protein phosphatase was purified and its protein sequence determined, a screening of computer sequence databases revealed that it was closely related to the cytoplasmic domain of a receptor-like protein, termed *CD45,* found on the surface of lymphocytes. Thus, like protein kinase activity, phosphatase activity may also be directly regulated by extracellular ligands.

The explosion of sequence information on protein kinases and phosphatases has in turn set off echoing explosions of homology cloning (isolation of new genes using probes from existing ones) that have unearthed dozens of new kinase- and phosphatase-encoding genes. This is one of the most striking examples of how recombinant DNA methods have reversed the direction of biological experimentation. Now instead of searching for the genes that encode identified proteins, more and more investigators are getting the genes first and then hunting for their encoded proteins and the roles they perform.

Steroid Receptors Are Transcription Factors

In the past several sections, we have traced hormone and growth factor signals across the plasma membrane to cytoplasmic second messengers. But to effect long-term changes in cellular behavior (for example, from resting to growing), the signals must ultimately reach the nucleus and affect the pattern of cellular gene expression. There are several routes for getting extracellular signals to the nucleus. The most direct route is employed by small lipid-soluble hormones such as the steroids. These molecules, because they can freely pass through the cell membrane, bypass cell-surface receptors entirely and bind directly to soluble receptor proteins inside the cell. While biochemists began to purify these receptors proteins using ligand binding

as an assay, other investigators turned their attention to the genes that are turned on by these hormones.

Hormone-activated gene expression is a rapid and selective phenomenon. Only a few genes are turned on in response to hormone treatment, and they switch on within minutes. Using a combination of in vitro mutagenesis and gene transfer assays (as described in Chapter 9), investigators mapped the ability to respond to specific hormones to short stretches of DNA (approximately 15 bp) flanking the regulated genes. Indeed, transfer of these DNA sequences, termed *hormone-response elements* (HREs), to unregulated genes converted them to hormone-inducible genes. Remarkably, when the hormone receptor proteins were purified, they were found to bind to HREs. But DNA binding by these receptor proteins was only seen in the presence of hormone. Thus, the hormone receptors do double duty, serving both as the cytoplasmic receptor for the signal and as the nuclear effector, activating (or, in some cases, repressing) the transcription of target genes.

How do these receptors work? Answers to this question flowed in rapidly with the isolation of cDNAs encoding the receptors. As we have learned many times before, the ability to mutate cDNAs at will and study their function after returning them to cells has been one of the most powerful gifts of the recombinant DNA revolution. This has been an especially effective approach for studying the hormone receptors. The general strategy employs a *cotransfection* assay, in which a plasmid that expresses a receptor protein is transfected into cells together with a plasmid carrying a reporter gene whose transcription is dependent on the HRE of the corresponding receptor. Using this assay, investigators quickly mapped the receptor domains responsible for hormone binding, DNA-binding, and transcriptional activation. As with the adrenergic receptors we discussed earlier in this chapter, these domains could be swapped among hormone receptors, allowing the construction of chimeric receptors that bound one hormone but regulated genes normally controlled by another.

A revealing experiment was the grafting of the cDNA for the hormone-binding domain of one receptor to the cDNA for a completely unrelated DNA-binding domain (from a bacterial repressor protein; Figure 17-11). Surprisingly, the resulting protein with this new DNA-binding domain also fell under hormone control. This observation rather resoundingly

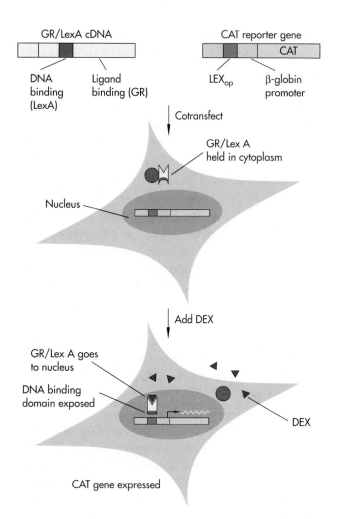

FIGURE 17-11

The ligand-binding domain in steroid hormone receptors is an intraprotein switch. Binding of ligand to a steroid hormone receptor unmasks the DNA-binding activity of the receptor, allowing it to bind to regulatory sequences in target genes and activate their transcription. How does the ligand-binding domain work? In this experiment, recombinant DNA methods were used to replace the DNA-binding domain of the glucocorticoid receptor (GR) with a completely unrelated DNA-binding domain from a bacterial repressor, LexA, to give a cDNA encoding a GR/LexA chimeric protein. To determine whether this foreign DNA-binding domain could be regulated by hormone, investigators used a cotransfection assay, in which an expression vector that produces the chimeric receptor protein was transfected into cells together with a reporter gene that produces a measurable enzymatic activity (CAT, chloramphenicol acetyltransferase; see Chapter 9) in response to receptor activation. The reporter gene contains a neutral promoter derived from the β-globin gene linked to a sequence normally bound by the LexA repressor (the *lex*A operator, Lex$_{op}$). In the experiment, the GR/LexA protein activated the reporter, but only in the presence of the receptor ligand, dexamethasone (DEX), a synthetic steroid. This experiment established that the hormone-binding domain exerts a general control over the activity of the DNA-binding domain, rather than acting as a specific lock-and-key inhibitor. It does this, we now know, by anchoring the receptor to a cytoplasmic protein complex, sequestering it from its DNA targets. Binding of ligand to the receptor uncouples it from its cytoplasmic anchor, allowing it to migrate to the nucleus to bind to specific sites in target genes.

ruled out one hypothesis for hormone regulation of DNA binding, namely that the hormone-binding domain fit snugly, lock-and-key style, into the DNA-binding domain of the receptor until released by hormone. Instead, the effect of the hormone-binding domain on receptor structure evidently must be a more general one. We now know that in the absence of hormone, the receptor is part of a much larger protein complex confined to the cytoplasm of the cell. Binding of hormone to the receptor causes it to unfold and escape from this complex, allowing it to move to the nucleus.

Molecular biologists, eager to exploit this remarkably tight biological switch, have now grafted the cDNA for hormone-binding domains onto the cDNAs of a variety of unrelated proteins, generating novel hormone-regulated proteins. For example, fusion of the sequence for a hormone-binding domain to an oncogene generated a gene that could convert normal cells to cancerous ones, merely by the addition of the appropriate hormone to the cells' growth medium. From studying what occurs in these cells as they transform into cancer cells over the course of a few hours, investigators hope to be able to examine the cellular events that trigger cancer.

cAMP Signals Reach the Nucleus via Transcription Factor Phosphorylation

Steroid hormones get their signal to the nucleus in a very direct fashion—the receptor protein for the hormone is the sole intermediate between the hormone and the target genes. But what of signaling molecules that cannot freely enter the cell and must act instead through cell-surface receptors? Here, the best-understood example is the control of gene expression by cAMP. We have already learned that treating cells with certain hormones leads to activation of adenylate cyclase, which produces cAMP, and that cAMP acts on cells by its ability to activate a cAMP-dependent protein kinase (termed protein kinase A, PKA). Among the many effects of cAMP is the rapid and selective activation of gene transcription. This, too, is an effect of PKA.

As with genes responsive to steroid hormones, investigators were able to map the regulatory elements in cAMP-inducible genes that were responsible for induction. And again, these elements proved to be short DNA sequences that could transfer the cAMP response to other genes. Not surprisingly, these CREs (cAMP-response elements) were binding sites for transcription factors, and several of these transcription factor proteins have now been purified and cloned. One of these proteins, called CREB (CRE-binding protein), appeared to be phosphorylated by PKA. Evidence that phosphorylation of CREB by PKA is one mechanism for transmitting signals to genes came from cotransfection experiments slightly more complicated than those done with the steroid receptors. Here, researchers performed a series of triple transfections involving introduction of a reporter gene under the control of a CRE, an expression plasmid that produced CREB, and one that produced the catalytic subunit of PKA (Figure 17-12). In the presence of both the CREB and PKA plasmids, expression of the reporter gene was greatly elevated, compared with transfections in which either CREB or PKA was eliminated. This observation established that CREB and PKA worked together to regulate gene expression. Most important, when the single serine residue in CREB known to be phosphorylated by PKA was mutated to an alanine (which cannot be phosphorylated), expression of the reporter gene was lost. Thus, this nuclear signaling pathway is a two-component system: cAMP activates a protein kinase (PKA), which in turn phosphorylates a transcription factor (CREB), converting it from an inactive to an active form. Unlike the situation with steroid hormone receptors, however, activation of CREB does not affect its DNA-binding activity; instead the transcriptional activation function of CREB is turned on. This was established by fusing the transcriptional activation domain of CREB to an unrelated DNA-binding domain (see Chapter 9).

Studying Target Genes Reveals the Organization of Signal Transduction Pathways

Other signaling pathways are more difficult to trace from the cell surface to the nucleus. The tyrosine kinase growth factor receptors, for example, generate several potential second messengers, each of which may eventually activate a protein kinase. But, unlike for the cAMP pathway, in this case we don't know whether any of these protein kinases phosphorylate transcription factors or, for that matter, what the transcription factors are. For these pathways, it has proved fruitful to track them in reverse, starting from a gene that is known to be a target for the signals. First, investigators locate the DNA elements required to turn the gene on (or off) when cells are exposed to the growth factor. Then, using methods detailed in Chapter 9, they identify transcription factors that bind to those elements. Finally, how these transcription factors are modified in response to signals can be studied.

The c-*fos* gene, a cellular proto-oncogene, is rapidly activated within minutes (and perhaps seconds) of exposure of cells to signals of many different types. Indeed, c-*fos* transcription is induced by activation of virtually every known signal transduction pathway, including those involving cAMP, calcium, diacylglycerol (which activates protein kinase C), and growth factor–triggered signals not yet biochemically defined. To identify the regulatory elements in the c-*fos* gene that respond to each of these signals, an extensive collection of mutations in the c-*fos* promoter was prepared, and the mutant promoters were transfected into cells and tested for inducibility by different signals (Figure 17-13). Mutations in an element located close to the start of the gene prevented induction by cAMP, but the gene still responded to growth factors such as platelet-derived growth factor. Mutations in a different part of the sequence abolished response to growth factors but spared the cAMP response. Finer mutagenesis of this growth factor–responsive element generated mutant genes that responded to some growth factor signals but not others. These studies showed that each signaling pathway communicates with a different nuclear target and suggested that the pattern of regulatory DNA sequences in a gene specifies the spectrum of signals to which the gene can respond.

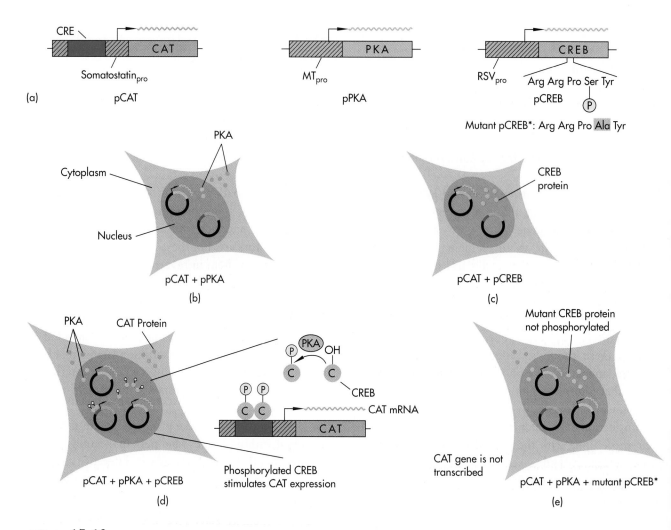

FIGURE **17-12**
Demonstration that cAMP regulation of gene transcription involves the phosphorylation of a transcription factor (CREB) by cAMP-dependent protein kinase (PKA). (a) Using a strategy similar to that in Figure 17-11, investigators transfected a reporter gene (pCAT) carrying a cAMP-responsive element (CRE) into cells together with expression vectors that produced PKA (pPKA) and CREB (pCREB). (b, c) Transfection of pCAT with either pPKA or pCREB alone resulted in no activation of the reporter gene. (d) Transfection of all three plasmids together led to strong activation, detected by production of the CAT enzyme. The CREB protein (C), phosphorylated by PKA, stimulated CAT expression by acting on the CRE. (e) But when the single serine residue in CREB that is phosphorylated by PKA was changed to an alanine (mutant pCREB*), transfection of all three plasmids did not result in gene activation. Thus, activation of target genes by cAMP requires the phosphorylation of CREB by PKA. Pro, promoter; MT, metallothionein; RSV, Rous sarcoma virus.

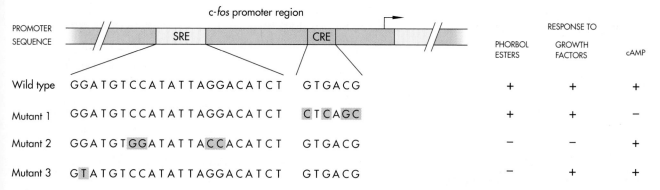

FIGURE **17-13**
Different signal transduction pathways communicate with distinct nuclear targets to regulate gene expression. The c-*fos* gene is rapidly transcribed after activation of many signal transduction pathways. Extensive mutagenesis of the c-*fos* promoter (mutated bases are shown in colored boxes) revealed that each pathway communicates with a distinct target sequence in the promoter. For example, Mutant 1, which carries mutations in the CRE, lost response to cAMP but preserved response to growth factors and phorbol esters, which activate protein kinase C. Mutant 2, which carries substitutions that inactivate the SRE (serum-response element, a regulatory element that responds to serum growth factors) had the reciprocal effect. Detailed mutagenesis of the SRE yielded mutant promoters, such as Mutant 3, that lost response to phorbol esters, but still responded to other growth factor signals. Thus, each of these signals must communicate with a distinct transcription factor.

Thus, we can begin to understand how it is that individual growth factors can have unique effects on cells, because the collection of genes they activate can be different.

Immediate Early Genes Are Third Messengers

What are the genes that are turned on by signaling molecules such as growth factors? We learned in Chapter 7 how growth factor-induced genes have been cloned by differential hybridization. All told, approximately 100 *immediate early genes* are activated within the first hour after treatment of cells with growth factors. A large fraction of these genes encode transcription factors, whose job is to activate or repress the next set of genes in the biological program initiated by the extracellular signal. Thus, some of these immediate early gene products may be thought of as *third messengers,* because they propagate signals from second messengers. Third messenger status has been most clearly bestowed on the c-*fos* gene, which encodes a transcription factor. Experimental strategies that block

the production of the Fos protein (expression of antisense RNA or microinjection of anti-Fos antibodies, for example) prevent cells from responding fully to signals.

The immediate early response is a stereotypic response of cells to signals. Many genes, like c-*fos,* are activated in most cells by most signals. Other genes are induced only in particular cell types or by specific signals. In the rich complexity of this response resides much of the information processing by which cells decide what signal they have received and how they will respond. How these gene products control subsequent cellular behavior is still not understood. But their potential power to control cell proliferation is evident: among them are several proto-oncogenes, genes with the potential to cause the uncontrolled cell growth we see as cancer. The immediate early genes may also prove to be the key links between the growth factor signals that initiate cell growth and the clocklike machinery that times and operates the cell cycle in proliferating cells (Chapter 19). We will also encounter the immediate early genes in Chapter 21, because some investigators believe that these genes provide the basis for learning and memory in the brain.

Reading List

General

Signal Transduction. Cold Spring Harbor Symp. Quant. Biol., vol. 53. Cold Spring Harbor Laboratory, Cold Spring Harbor, N.Y. 1988.

Original Research Papers

ACETYLCHOLINE RECEPTOR STRUCTURE

Stroud, R. M., and J. Finer-Moore. "Acetylcholine receptor structure, function, and evolution." *Ann. Rev. Cell Biol.,* 1: 317–351 (1985). [Review]

Noda, M., H. Takahashi, T. Tanabe, M. Toyosato, M. Furutani, T. Hirose, M. Asai, S. Inayama, T. Miyata, and S. Numa. "Primary structure of α-subunit precursor of *Torpedo californica* acetylcholine receptor deduced from cDNA sequence." *Nature,* 299: 793–797 (1982).

Noda, M., Y. Furutani, H. Takahashi, M. Toyosato, T. Tanabe, S. Shimizu, S. Kiyotani, T. Kayano, T. Hirose, S. Inayama, and S. Numa. "Cloning and sequence analysis of calf cDNA and human genomic DNA encoding the α-subunit precursor of muscle acetylcholine receptor." *Nature,* 305: 818–823 (1983).

G PROTEINS

Gilman, A. G. "G proteins." *Ann. Rev. Biochem.,* 56: 615–649 (1987). [Review]

Bourne, H. R., D. A. Sanders, and F. McCormick. "The GTPase superfamily: conserved structure and molecular mechanism." *Nature,* 349: 117–127 (1991). [Review]

Harris, B. A., J. D. Robishaw, S. M. Mumby, and A. G. Gilman. "Molecular cloning of complementary DNA for the alpha subunit of the G protein that stimulates adenylate cyclase." *Science,* 229: 1274–1277 (1985).

Medynski, D. C., K. Sullivan, D. Smith, C. Van Dop, F.-H. Chang, B. K-K Fung, P. H. Seeburg, and H. R. Bourne. "Amino acid sequence of the subunit of transducin deduced from the cDNA sequence." *Proc. Natl. Acad. Sci. USA,* 82: 4311–4315 (1985).

Strathmann, M., and M. I. Simon. "G protein diversity: a distinct class of alpha subunits is present in vertebrates and invertebrates." *Proc. Natl. Acad. Sci. USA,* 87: 9113–9117 (1990).

G-PROTEIN-COUPLED RECEPTORS

Lefkowitz, R. J. "Variations on a theme." *Nature,* 351: 353–354 (1991). [Review]

Kobilka, B. K., T. S. Kobilka, D. Daniel, J. W. Regan, M. G. Caron, and R. J. Lefkowitz. "β2-adrenergic receptors:

delineation of domains involved in effector coupling and ligand binding specificity." *Science,* 240: 1310–1316 (1988).

Ashkenazi, A., E. G. Peralta, J. W. Winslow, J. Ramachandran, and D. J. Capon. "Functionally distinct G proteins selectively couple different receptors to PI hydrolysis in the same cell." *Cell,* 56: 487–493 (1989).

Buck, L., and R. Axel. "A novel multigene family may encode odorant receptors: a molecular basis for odor recognition." *Cell,* 65: 175–187 (1991).

RECEPTORS THAT ENCODE ENZYMES

Williams, L. T. "Signal transduction by the platelet-derived growth factor receptor." *Science,* 243: 1564–1570 (1989). [Review]

Ullrich, A., and J. Schlessinger. "Signal transduction by receptors with tyrosine kinase activity." *Cell,* 61: 203–212 (1990). [Review]

Fischer E. H., H. Charbonneau, and N. K. Tonks. "Protein tyrosine phosphatases: a diverse family of intracellular and transmembrane enzymes." *Science,* 253: 401–406 (1991). [Review]

Tonks, N. K., H. Charbonneau, C. D. Diltz, E. H. Fischer, and K. A. Walsh. "Demonstration that the leukocyte antigen CD45 is a protein tyrosine phosphatase." *Biochemistry,* 27: 8695–8701 (1988).

Chinkers, M., D. L. Garbers, M. S. Chang, D. G. Lowe, H. M. Chin, D. V. Goeddel, and S. Schulz. "A membrane form of guanylate cyclase is an atrial natriuretic peptide receptor." *Nature,* 388: 78–83 (1989).

Honegger, A. M., A. Schmidt, A. Ullrich, and J. Schlessinger. "Evidence for epidermal growth factor (EGF)-induced intermolecular autophosphorylation of the EGF receptors in living cells." *Mol. Cell. Biol.,* 10: 4035–4044 (1990).

Mathews, L. S., and W. W. Vale. "Expression cloning of an activin receptor, a predicted transmembrane serine kinase." *Cell,* 65: 973–982 (1991).

RECEPTOR-ASSOCIATED PROTEINS

Kazlauskas, A., and J. A. Cooper. "Autophosphorylation of the PDGF receptor in the kinase insert region regulates interactions with cell proteins." *Cell,* 58: 1121–1133 (1989).

Meisenhelder, J., P. G. Suh, S. G. Rhee, and T. Hunter. "Phospholipase C-gamma is a substrate for the PDGF and EGF receptor protein-tyrosine kinases in vivo and in vitro." *Cell,* 57: 1109–1122 (1989).

Varticovski, L., B. Druker, D. Morrison, L. Cantley, and T. Roberts. "The colony stimulating factor-1 receptor associates with and activates phosphatidylinositol-3 kinase." *Nature,* 342: 699–702 (1989).

Anderson, D., C. A. Koch, L. Grey, C. Ellis, M. F. Moran, and T. Pawson. "Binding of SH2 domains of phospholipase C gamma 1, GAP, and Src to activated growth factor receptors." *Science,* 250: 979–82 (1990).

Morrison, D. K., D. R. Kaplan, S. G. Rhee, and L. T. Williams. "Platelet-derived growth factor (PDGF)-dependent association of phospholipase C-gamma with the PDGF receptor signaling complex." *Mol. Cell. Biol.,* 10: 2359–2366 (1990).

SECOND MESSENGERS

Levitzki, A. "From epinephrine to cyclic AMP." *Science,* 241: 800–806 (1988). [Review]

Rasmussen, H. "The cycling of calcium as an intracellular messenger." *Sci. Am.,* 261: 66–73 (1989). [Review]

Berridge, M. J., and R. F. Irvine. "Inositol phosphates and cell signalling." *Nature,* 341: 197–205 (1990). [Review]

PHOSPHORYLATION AND SIGNAL TRANSDUCTION

Hunter, T. "A thousand and one protein kinases." *Cell,* 50: 823–829 (1987). [Review]

Hardie, G. "Pseudosubstrates turn off protein kinases." *Nature,* 335: 592–593 (1988). [Review]

Schulman, H., and L. L. Lou. "Multifunctional Ca^{+2}/calmodulin-dependent protein kinase: domain structure and regulation." *Trends Biochem.,* 14: 62–66 (1989). [Review]

Taylor, S. S., Buechler, JA., and W. Yonemoto. "cAMP-dependent protein kinase: framework for a diverse family of regulatory enzymes." *Ann. Rev. Biochem.,* 59: 971–1005 (1990) [Review]

House, C., and B. E. Kemp. "Protein kinase C contains a pseudosubstrate prototope in its regulatory domain." *Science,* 238: 1726–1728 (1987).

Shaw, A. S., K. E. Amrein, C. Hammond, D. F. Stern, B. M. Sefton, and J. K. Rose. "The *lck* tyrosine protein kinase interacts with the cytoplasmic tail of the CD4 glycoprotein through its unique amino-terminal domain." *Cell,* 59: 627–636 (1989).

Anderson, N. G., J. L. Maller, N. K. Tonks, and T. W. Sturgill. "Requirement for integration of signals from two distinct phosphorylation pathways for activation of MAP kinase." *Nature,* 334: 651–653 (1990).

SIGNAL TRANSDUCTION TO THE NUCLEUS

Beato, M. "Gene regulation by steroid hormones." *Cell,* 56: 335–344 (1989). [Review]

Almendral, J. M., D. Sommer, H. MacDonald-Bravo, J. Burckhardt, J. Perera, and R. Bravo. "Complexity of the early genetic response to growth factors in mouse fibroblasts." *Mol. Cell. Biol.* 8: 2140–2148 (1988).

Godowski, P. J., D. Picard, and K. R. Yamamoto. "Signal transduction and transcriptional regulation by glucocorticoid receptor-lexA fusion proteins." *Science,* 241: 812–816 (1988).

Picard, D., S. J. Salser, and K. R. Yamamoto. "A movable and regulable inactivation function within the steroid binding domain of the glucocorticoid receptor." *Cell,* 54: 1073–1080 (1988).

Gonzalez, G. A., and M. R. Montminy. "Cyclic AMP stimulates somatostatin gene transcription by phosphorylation of CREB at serine 133." *Cell,* 59: 675–680 (1989).

Graham, R., and M. Gilman. "Distinct protein targets for signals acting at the c-*fos* serum response element." *Science,* 251: 189–192 (1991).

18

Oncogenes and Anti-oncogenes

I n no area of medicine has recombinant DNA had a greater impact than in our understanding of cancer. Although this disease has been inscrutable for decades, the seed for the recent explosive growth in our knowledge of cancer was planted in 1911. In that year, Peyton Rous discovered that filtered cell-free extracts from chicken tumors could cause new tumors when inoculated into healthy chickens. It was eventually recognized that what Rous had discovered was a tumor virus. The realization that a tumor virus with an exceedingly small genome could trigger the entire panoply of traits that characterize a cancer cell was an intellectual watershed: the task of understanding how a cell's entire growth program is rewritten in cancer now seemed a feasible one. What was not anticipated was that tumor viruses would provide a window on our own genomes, a window opened wide by recombinant DNA technology.

What we have learned since the tools of recombinant DNA were unleashed on the cancer problem is that our own genome is littered with genes having the potential to cause cancer and other genes having the power to block it. From the study of these genes, we have derived a greater understanding of the molecular events that occur in tumor cells. And as a bonus, we have discovered many of the players that orchestrate normal cell growth and differentiation. The implications for cancer as a disease are significant. Already recombinant DNA is a valuable tool for diagnosing cancer. Soon it may offer us new ways to treat the disease.

Cancer Is a Genetic Disease

A fundamental feature of cancer cells is that when they divide, both daughters are also cancer cells. Cancer is stably inherited during cell division. Indeed, many tumors can be shown to be clonal, derived from a single common progenitor cancer cell that divides incessantly to generate a tumor of identical sibling cells. That cells within a tumor are genetically related suggests that the disease phenotype is genetically determined, that is, encoded within tumor cell DNA. Considerable circumstantial evidence for a genetic basis for cancer has built up over decades of study of chemical carcinogenesis. These studies showed that virtually all chemicals that cause cancer in experimental animals are *mutagens,* chemicals that damage DNA. This common property of carcinogens suggested that their ability to inflict genetic damage is the basis for their carcinogenic properties. The role for genetic alterations in cancer was confirmed with the discovery that normal cells can be converted to cancer cells by the direct introduction of exogenous genetic information. As we will see, this exogenous genetic information can be introduced experimentally by DNA transfection or by tumor viruses.

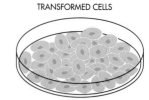

NORMAL CELLS	TRANSFORMED CELLS
Flat and organized	Piled up and disorganized
Contact inhibited	Not contact inhibited
Serum dependent	Growth factor independent
Anchorage dependent	Anchorage independent
Nontumorigenic	Form tumors in nude mice

FIGURE **18-1**
Normal and transformed cells. Normal, nontumorigenic cells grow in a flat and organized pattern until they cover the surface of the culture dish. Then, when each cell is touching its neighbors, they stop, a phenomenon termed *contact inhibition.* In addition, cell growth requires a cocktail of growth factors, usually supplied in the form of *serum* (a cell-free preparation from animal blood), and the cells grow only when attached to a solid surface (*anchorage dependence*). In contrast, cancerous cells usually have a different shape. They disregard their neighbors and grow to high densities in disorganized piles. Their growth often does not require serum growth factors, and they can grow simply suspended in growth medium, without being attached to a surface. When transformed cells are injected into *nude* mice (mutant animals with a compromised immune system), they rapidly form tumors.

Tumor Cells Have Aberrant Growth Properties in Cell Culture

When normal tissues are explanted from adult animals and placed into culture in the laboratory, they seldom grow. Tissues from fetal or newborn animals will often grow steadily for a few weeks and then cease. Rarely, cell lines arise from these cultures that are immortal. They grow indefinitely in culture but otherwise retain the growth properties of their normal counterparts. Cultures like this can be treated with carcinogenic agents such as chemicals, radiation, and tumor viruses, and from these treated cultures, *transformed* cells with novel growth properties arise. These cells often have a distinctive shape, they are less dependent on specific factors in their growth medium, they continue to grow long after their neighbors have stopped—piling atop one another—and when injected into animals they form tumors (Figure 18-1). These traits are stably passed from mother cell to daughter. Thus, these transformed cells behave very much like real tumor cells. The ability to recapitulate the transformation process in cell culture has been the cornerstone for building the molecular view of cancer.

Tumor Viruses Opened the Study of Cancer to Molecular Methods

The discovery that certain viruses, when inoculated into animals, elicited tumors considerably simplified our ideas about cancer, for somewhere within the tiny genomes of these viruses—nearly a millionth the size of the genome of an animal cell—lurked the genetic instructions to derail normal cell growth. So, these viruses came under intense scrutiny. And while they did not give up their secrets easily, study of the tumor viruses uncovered many of the basic principles of carcinogenesis as well as those of eukaryotic gene organization and regulation.

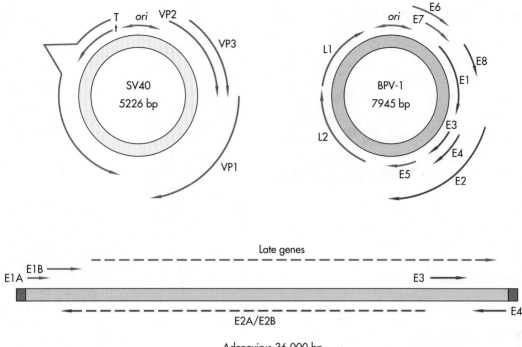

FIGURE **18-2**

Organization of three DNA tumor viruses. SV40, a monkey virus, and BPV-1, a bovine pap-
illomavirus, have small circular double-stranded DNA genomes. Adenovirus, which causes
colds and conjunctivitis in humans, has a longer linear double-stranded genome. Each virus
contains genes responsible for their transforming activity (red). These genes presumably are
important regulators of the viral life cycle. Other regulatory genes with no direct transforming properties are indicated in purple. These genes are generally expressed early in the viral
infection. The late genes (green) encode structural proteins involved in the assembly of new
virus particles. The sequences in blue represent the regions required for initiation of viral
DNA replication. The linear genome of adenovirus is replicated from the ends. The adenovirus gene map has been considerably simplified for clarity.

Tumor viruses come in two types, depending on
whether their genome is encoded in DNA or RNA.
The genomes of DNA tumor viruses vary in size from
5 to 200 kbp (Figure 18-2). The smallest viruses encode only a few genes and depend heavily on host
cell enzymes for transcription and replication of their
genomes. In some cells, the viruses grow lytically: they
enter, multiply rapidly, and kill the host cell by rup-
turing its cell membrane. Certain cells, however, resist
lytic infection, and, with a low frequency, become
transformed into cancer cells. Invariably, transformation is associated with integration of the viral genome (or portions thereof) into host cell DNA and its
stable passage to daughter cells. Thus, tumor viruses
induce genetic alterations in transformed cells.

Tumor viruses carry discrete genetic elements,
called *oncogenes,* that are responsible for their ability
to transform cells. These genes include the T antigen
genes of SV40 and polyoma virus, *E1A* and *E1B* of
adenovirus, and *E6* and *E7* of papillomaviruses. Oncogenes encode proteins, often termed *oncoproteins,* that
play a number of important roles in the viral life cycle,
such as initiation of DNA replication and transcriptional control of viral genes, and the ability of oncogenes to transform cells appears to be a consequence
of these activities. Viruses normally infect nongrowing
cells, which represent the vast majority of cells in an
animal, and because they require host enzymes to
replicate their DNA, one of the critical things a virus
has to do after infecting a cell is to prod the cell into

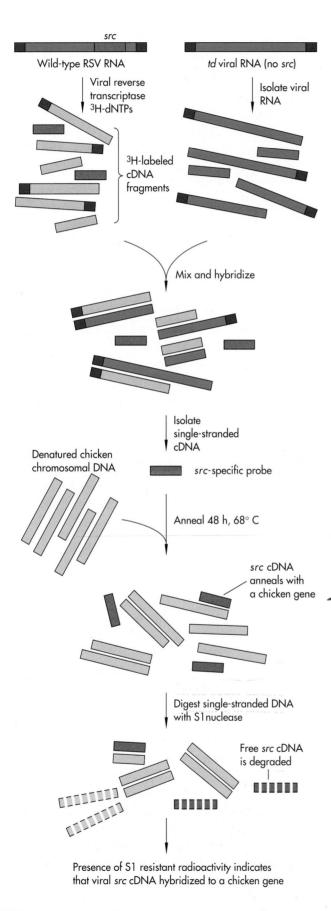

src

Wild-type RSV RNA

Viral reverse transcriptase
³H-dNTPs

³H-labeled cDNA fragments

td viral RNA (no *src*)

Isolate viral RNA

Mix and hybridize

Isolate single-stranded cDNA

Denatured chicken chromosomal DNA

src-specific probe

Anneal 48 h, 68° C

src cDNA anneals with a chicken gene

Digest single-stranded DNA with S1 nuclease

Free *src* cDNA is degraded

Presence of S1 resistant radioactivity indicates that viral *src* cDNA hybridized to a chicken gene

FIGURE **18-3**

Retroviral oncogenes have cellular counterparts. To determine the origin of the *src* transforming gene of Rous sarcoma virus (RSV), investigators needed to prepare a radiolabeled probe specific for *src* sequences. Cloning of tumor virus genomes was not possible at the time, as a result both of technical limitations and of a voluntary moratorium on such experiments. Therefore, the investigators isolated RSV particles and, using the endogenous viral reverse transcriptase, transcribed the viral RNA genome into fragments of radiolabeled cDNA. They hybridized this cDNA to excess amounts of RNA from transformation-defective (*td*) RSV derivatives that lacked the *src* oncogene and isolated the labeled cDNA that failed to hybridize, thereby generating a probe highly enriched for *src* sequences. This *src* probe was in turn annealed to denatured chicken DNA (chicken is the host species for RSV). After annealing, the mixture was treated with S1 nuclease, an enzyme that degrades only single-stranded DNA. Some of the *src* DNA was resistant to S1 degradation, a finding indicating that it had formed double-stranded molecules with chicken DNA fragments. Thus, chicken DNA must possess a gene very similar to *src*.

synthesizing the cellular DNA replication machinery. Activating dormant cells so viral DNA can be replicated is the chief role of the oncoproteins of the DNA tumor viruses. Thus, when the viral genome integrates and becomes a permanent part of the cell genome, the viral oncogene keeps the cell growing indefinitely—and a tumor forms.

Although the oncogenes of the DNA tumor viruses have been studied in intimate detail for nearly two decades, a thorough understanding of how they work has been elusive. They encode complex, multifunctional proteins that have been streamlined by centuries of evolution for one specific purpose, efficient replication of viral DNA. Unlike the oncogenes of the RNA tumor viruses, which we will discuss shortly, the DNA tumor virus oncogenes have no obvious cellular counterparts, and scientists began to wonder whether these oncogenes were leading them on a wild-goose chase that would tell them little about nonviral cancer. But recent discoveries have dramatically justified the study of these genes. We will learn later in this chapter that they act by specifically targeting a newly discovered class of cellular genes whose role is to retard cell growth.

Retroviral Oncogenes Are Captured from Cellular DNA

While researchers struggled with the refractory oncogenes of the DNA tumor viruses, those working with RNA tumor viruses, or *retroviruses,* made the astonishing discovery that their potent oncogenes originated from cellular DNA. During their evolution, these viruses captured cellular genes that gave them their dramatic growth-transforming properties. The demonstration that retroviral oncogenes are cellular in origin arose from studies of the virus originally discovered by Peyton Rous, now called Rous sarcoma virus (RSV). The isolation of mutant strains of RSV that were deficient in transformation provided the first evidence that the virus carried a specific genetic locus, termed *src,* that encoded its ability to transform cells. One class of these transformation-defective (*td*) viruses carried extensive deletions in the viral genome. These deletions marked the site of the *src* oncogene. By laboriously isolating the RNA molecules from large

quantities of normal and transformation-defective viruses and converting the former to isotopically labeled cDNA, it was possible to remove from the normal viral cDNA the genetic information also present in the transformation-defective viruses (Figure 18-3). Left behind was isotopically labeled cDNA containing only sequences specific to the *src* oncogene. When this *src*-specific DNA preparation was annealed to DNA from uninfected cells, the cDNA annealed stably, suggesting that cellular DNA contained sequences very closely related to *src.* These *src*-related sequences were found not only in chicken, the natural host for RSV, but in all vertebrates, including humans.

Thus, the *src* oncogene carried by RSV is actually a derivative of a normal cellular gene, captured and stably incorporated into the viral genome by a poorly understood process called *transduction.* With the ability to generate molecular clones of retroviral genomes, it soon became apparent that each of the acutely transforming retroviruses, those that cause rapidly growing tumors within a few weeks of inoculation, carried on-

---

TABLE **18-1**

A Selection of Well-Characterized Oncogenes and the Proteins They Encode.

ONCOGENE		PROTEIN	
CLASSIFICATION*	HOW ISOLATED[†]	LOCATION	FUNCTION
CLASS 1: GROWTH FACTORS			
sis	Retrovirus	Secreted	Growth factor (PDGF)
CLASS 2: GROWTH FACTOR RECEPTORS			
*erb*B	Retrovirus	Membrane	Receptor for epidermal growth factor (EGF)
fms	Retrovirus	Membrane	Receptor for colony-stimulating factor–1 (CSF-1)
trk	Tumor	Membrane	Receptor for nerve growth factor (NGF)
CLASS 3: INTRACELLULAR TRANSDUCERS			
src	Retrovirus	Cytoplasm	Protein-tyrosine kinase
abl	Retrovirus	Cytoplasm	Protein-tyrosine kinase
raf	Retrovirus	Cytoplasm	Protein-serine kinase
gsp	Tumor	Cytoplasm	G protein α subunit
ras	Retrovirus, tumor	Cytoplasm	GTP/GDP-binding protein
CLASS 4: NUCLEAR TRANSCRIPTION FACTORS			
jun	Retrovirus	Nucleus	Transcription factor (AP-1)
fos	Retrovirus	Nucleus	Transcription factor (AP-1)
myc	Retrovirus, tumor	Nucleus	DNA-binding protein
*erb*A	Retrovirus	Nucleus	Member of steroid receptor family

* Oncogenes can be divided into four classes on the basis of the function of the proteins they encode. Virtually all known oncogenes encode constituents of growth factor signal transduction pathways.
[†] Most oncogenes were found first as retroviral oncogenes, but some were first isolated or extensively characterized as oncogenes in human tumors.

cogenes derived from cellular DNA, and these cellular genes were quickly isolated. In all, more than 25 distinct cellular genes, now referred to as *proto-oncogenes,* have been transduced and employed as oncogenes by retroviruses that infect birds, rodents, cats, and monkeys (see Table 18-1). By convention, the viral form of such oncogenes is labeled with the prefix "v" (e.g., v-*src*), whereas the cellular gene is labeled with "c" (e.g., c-*src*). Although retroviruses of this class do not appear to be involved in human cancer, study of these viruses has revealed a family of genes in the human genome that probably play a very direct role in the disease.

Viruses Can Activate Cellular Proto-Oncogenes by Insertional Mutagenesis

Acutely transforming retroviruses are quite rare. More common are retroviruses causing tumors that take a long time to develop. These viruses lack oncogenes of their own. Nevertheless, the tumors they generate are stable and clonal, an observation suggesting that the viruses trigger genetic events in the tumor cells. With the availability of cloned viral DNA probes, the modus operandi of these viruses became apparent. In the case of lymphomas in chickens infected with avian leukosis virus (ALV), hybridization of viral probes to DNA isolated from tumor cells showed that, as expected, these tumors harbor integrated viral genomes. Closer examination revealed that the integrated *proviruses* in many different tumors are located within the same gene. Indeed, some of these tumors contain mRNA molecules in which viral sequences are linked to sequences from the adjoining cellular gene. Thus, these viruses are causing cells to express abnormally large amounts or novel versions of a normal cellular message (Figure 18-4). The identity of the cellular gene disturbed by the integrated ALV soon became known: it is the cellular progenitor (proto-oncogene) of a previously identified retroviral oncogene, *myc.* An important model emerged to explain carcinogenesis by these oncogene-less retroviruses. As the viruses spread through the animal, they infect cells and integrate into cellular DNA essentially at random. In a rare cell, the virus drops into the c-*myc* gene and

perturbs the expression of this gene in a way that confers a growth advantage to the infected cell, which eventually multiplies to form a tumor. The clear implication of this important model is that other, nonviral, agents could in principle elicit the same changes in cellular proto-oncogenes. Thus, we can begin to understand how genetic damage in the cell can lead to cancer.

Analysis of Tumor Chromosomes Reveals Rearrangements of Proto-Oncogenes

Tumors have long been recognized to contain chromosomal abnormalities—gain and loss of chromosomes, translocations, amplifications. Is this chromosomal instability of tumors a contribution to or a consequence of transformation? That retroviruses could contribute to tumor development by physically disrupting cellular genes strongly suggested that physical rearrangement of chromosomes could do the same thing. B cell tumors in rodents and humans often contain chromosomal translocations in which the highly expressed immunoglobulin genes are joined to a new chromosome. As with the ALV-induced lymphomas, it was soon recognized that these translocations planted the immunoglobulin genes into a common genetic locus. Cloning of these translocations by using immunoglobulin gene probes and analysis of the new neighboring sequence revealed that the target was again the c-*myc* proto-oncogene. Thus, this rearrangement of cellular DNA accomplished the same end as ALV integration, perturbing the expression of the c-*myc* gene by placing it under the influence of the powerful transcriptional signals in the immunoglobulin locus. The result of the disruption is similar in both cases: the expression but not the coding potential of the c-*myc* gene is affected (Figure 18-5a).

A similar chromosomal translocation with quite different consequences occurs in human chronic myelogenous leukemia. These tumors very often carry a hallmark chromosome rearrangement called the *Philadelphia chromosome,* in which the end of chromosome 9 is attached to chromosome 22. Located on the piece of chromosome 9 that moves is the cellular homologue

of another viral oncogene, *abl*. Cloning of the c-*abl* genes from tumor DNAs showed that chromosome 9 always breaks within the first intron of the large c-*abl* gene. And the position on chromosome 22 to which c-*abl* is joined is also always the same, within a previously unknown gene, *bcr* (Figure 18-5b). The fused genes are expressed as a hybrid mRNA that produces a novel protein jointly encoded by both genes. The expression of this *bcr-abl* fusion protein is clearly important in the development of leukemia, because introduction of exogenously supplied *bcr-abl* fusions causes a similar disease in animals.

Other chromosomal translocations are common in tumors, but the genes affected by the rearrangements are not yet known. With modern cloning technologies, these chromosomal breakpoints are rapidly being cloned and the genes adjacent to the breaks are being studied. These studies will be fertile ground for the identification of new cellular proto-oncogenes.

A different type of chromosomal alteration often seen in tumors is *DNA amplification,* in which portions of a chromosome are overreplicated and maintained in many copies. Amplified DNA is often visible microscopically as HSRs (*homogeneously staining regions*) or DMs (*double minute chromosomes*). In some neuroblastomas, these cytogenetic abnormalities contain many copies of a cellular gene closely related to c-*myc*, called N-*myc*, which was first isolated from these tumors by hybridization to a c-*myc* probe. Amplification raises the level of gene expression and also increases the target size for mutations with the potential to influence proto-oncogene activity. Surveys of proto-oncogene structure in tumors now routinely reveal amplification of proto-oncogene sequences. Such surveys are proving useful as diagnostic tools and in designing treatments for cancer patients.

The Transformed Phenotype Can Be Passed via DNA-Mediated Transfection

While retrovirologists were rounding up suspects in the hunt for cellular oncogenes, others were hunting directly for the culprits by using the newly developed methods for gene transfer into mammalian cells described in Chapter 12. Earlier, it had been possible to

pass the transformed phenotype from virally transformed cells to normal recipients. Passage of the phenotype was accompanied by transfer of the integrated viral genome. It was reasoned that if cells transformed by chemical carcinogens carry cellular oncogenes activated by carcinogen-induced DNA damage, it should be possible to pass the transformed phenotype from these cells to normal cells via genomic DNA from the transformed cells. These experiments were difficult, requiring careful isolation of donor DNA to ensure that genes occupying large fragments of DNA remained intact. The DNA was added as a calcium phosphate precipitate to monolayers of immortal mouse fibroblasts with relatively normal properties (Figure 18-6). The cultures were nurtured for several weeks, and on some of the dishes foci of cells developed that overgrew their neighbors. It was difficult to distinguish foci that arose as a result of gene transfer from those that occasionally arose spontaneously in these cultures. But clearly some foci of transformed cells were being generated as a result of gene transfer. And when these cells were grown and their DNA extracted and transferred to new cultures of normal fibroblasts, new foci appeared. This cycle could be repeated indefinitely. Moreover, when the DNA was cleaved with restriction enzymes prior to transfection, some enzymes destroyed the transforming activity and others spared it. The investigators concluded that there was a single defined segment of DNA, probably a single gene, that was responsible for passing the transformed phenotype from one culture to the next.

Very quickly these experiments were repeated with donor DNA from tumors of various tissues and species. About a third of these tumor DNAs were capable of transforming recipient fibroblast cultures. It was evident that this gene transfer assay was detecting powerful cellular oncogenes present in a wide range of tumors.

An Activated Human Oncogene Is Cloned

And then the race was on to clone the oncogenes from human tumors. With no knowledge of the structure or sequence of the oncogenes, more arduous methods

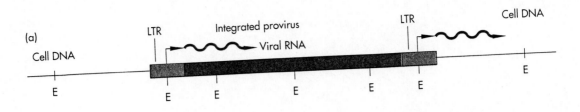

(a)

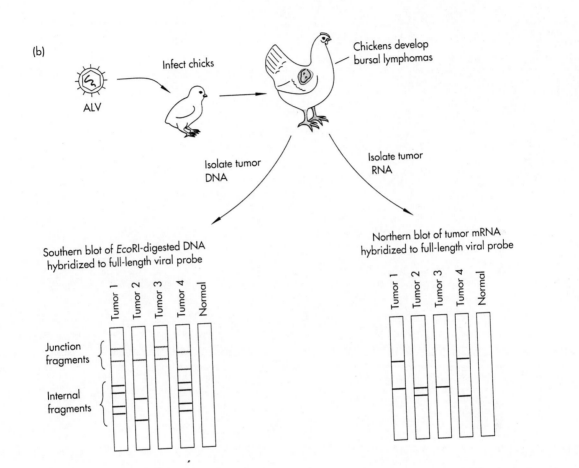

(b)

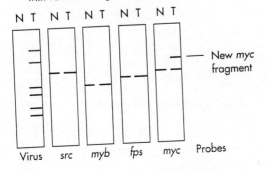

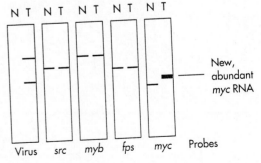

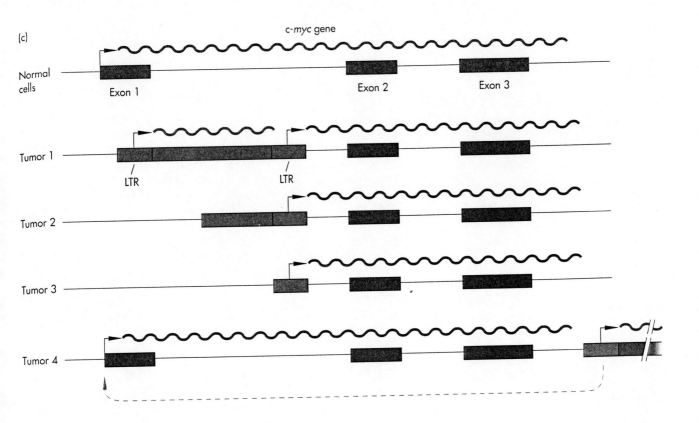

FIGURE **18-4** (*Above and facing page*)

Retroviruses can cause tumors by activating cellular proto-oncogenes. (a) When integrated into chromosomal DNA, a provirus contains repeated sequences at its ends, termed *long terminal repeats* or *LTRs* (orange), which carry the strong viral enhancer and promoter. The left LTR promotes the transcription of the viral genome (blue), whereas the right LTR promotes transcription of flanking cellular DNA. "E" indicates *Eco*RI restriction sites. (b) When chicks are infected with avian leukosis virus (ALV) they develop lymphoid tumors localized to an avian organ called the bursa. Analysis of DNA from the tumors by using Southern blotting and a viral probe indicated that all tumors had acquired viral DNA fragments and by virtue of a unique pattern of bands for each, that each tumor arose from a single infected cell. *Eco*RI digestion generated six fragments that hybridized to the viral probe, four carrying internal viral fragments and two carrying *junction fragments* containing both viral DNA and flanking cellular sequences. Some of the tumors (2 and 3, for example) lacked some or all of the internal fragments, which means that expression of viral genes is not required for tumor formation. Moreover, many of the tumors (1, 2, and 3 here) had a junction fragment in common, an observation suggesting that each of the viruses had integrated at the same location in the cell genome. Analysis by Northern blotting showed the presence in the tumors of RNAs carrying entirely viral sequences (blue) and other RNAs carrying flanking cellular sequences (green). Screening of blots of DNA and RNA from these tumors with a panel of oncogene probes revealed that tumors acquired a new fragment hybridizing to the c-*myc* gene and now expressed a new c-*myc* mRNA at high levels. Thus (c), the viruses had integrated into the c-*myc* gene (exons 1–3), either allowing direct transcription of the c-*myc* gene from a viral LTR, as in tumors 1–3, or enhancing from a distance the use of the natural c-*myc* promoter, as in tumor 4. Since exon 1 of c-*myc* does not encode any of the protein, all tumors produce a normal protein.

(a)

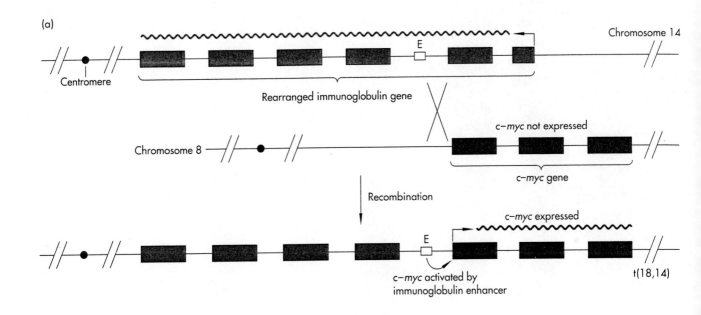

(b)

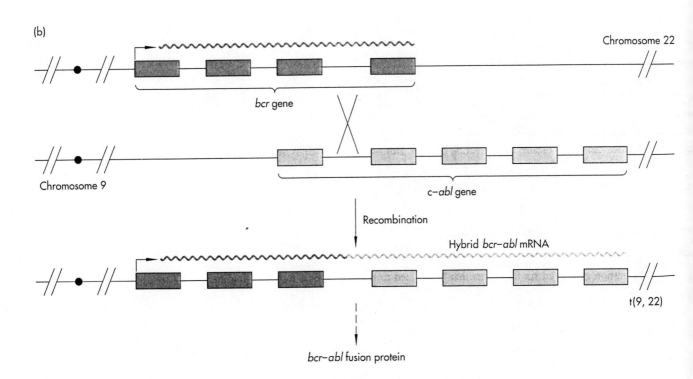

FIGURE **18-5**

(a) Many B cell tumors carry chromosomal translocations that join the c-*myc* proto-oncogene to one of the immunoglobulin genes. The c-*myc* gene, normally off in terminally differentiated B cells, is kept on by the immunoglobulin gene enhancer (E). Like the integrated viral RNA shown in Figure 18-4, these translocated genes usually produce a normal c-*myc* protein. (b) Chronic myelogenous leukemia cells carry a hallmark chromosome translocation, which joins the c-*abl* proto-oncogene on chromosome 9 to chromosome 22 in the middle of a gene termed *bcr*. The translocation produces a novel *bcr-abl* protein kinase no longer subject to its normal control. These translocations are *reciprocal*, meaning that a second aberrant chromosome carrying the other pieces of each translocated chromosome is also formed. Boxes represent gene exons.

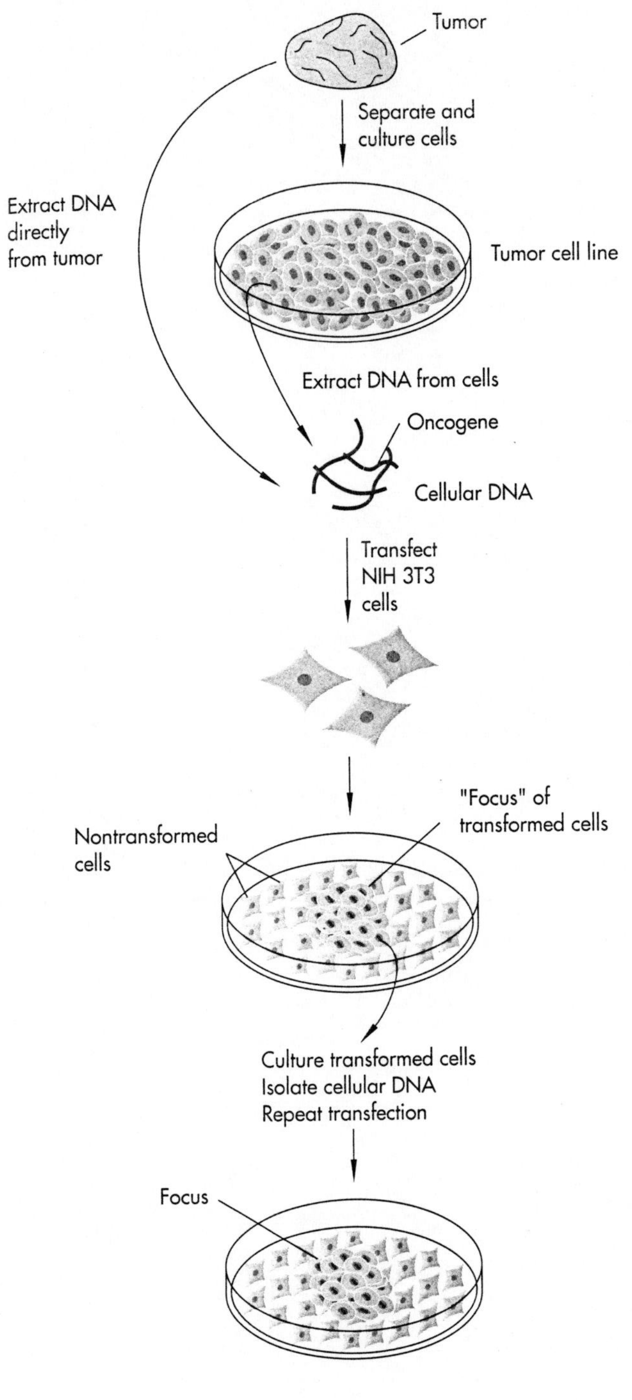
Tumor
Separate and culture cells
Extract DNA directly from tumor
Tumor cell line
Extract DNA from cells
Oncogene
Cellular DNA
Transfect NIH 3T3 cells
"Focus" of transformed cells
Nontransformed cells
Culture transformed cells
Isolate cellular DNA
Repeat transfection
Focus

FIGURE **18-6**
Passage of the transformed phenotype via genomic DNA. DNA isolated either directly from tumors or from tumor cells in culture is transfected into mouse fibroblasts, NIH 3T3 cells (see Chapter 12 for an explanation of transfection procedures). The cells take up tumor DNA, and a few acquire an activated oncogene from the tumor. The descendants of these cells grow as a *focus* of transformed cells with altered growth properties (see Figure 18-1). These transformed cells can be cultured and their DNA used in a second round of transfection.

were required to clone them. The method that eventually worked was the tagging of the oncogenes with markers that could be used to retrieve oncogene clones from a library.

In one case, a natural tag was exploited. As we learned in Chapter 8, human DNA carries a widely dispersed family of repetitive DNA sequences termed *Alu repeats.* These repeats occur about every few thousand base pairs, conveniently placing a repeat within or near every human gene. Sequences closely related to the *Alu* repeats are not found in the mouse genome, so the presence of human DNA in transfected mouse cells can be monitored by hybridizing a Southern blot of restriction enzyme-digested transfected cell DNA with an *Alu* probe (Figure 18-7). Investigators attempting to isolate the oncogene from a human bladder tumor cell line (EJ) passed the oncogene through two rounds of transfection into mouse fibroblasts, following the human DNA with an *Alu* probe. After the second round a single common *Alu*-reactive fragment remained in the transformed cells. Because the same fragment was found in several independently derived transformants, it must be closely linked to, or perhaps even carry, the active oncogene. It was then a relatively simple matter to make a library from transfected fibroblasts and screen the library with an *Alu* probe to find the clones carrying human DNA. In the case of the oncogene from the EJ bladder carcinoma cell line, DNA from the phage clone isolated in this manner yielded thousands of transformed foci when transfected into mouse fibroblasts—a dramatic demonstration that a powerfully active human oncogene had been cloned.

The Human Bladder Carcinoma Oncogene Is an Activated *ras* Gene

But what was this EJ oncogene? The frenzy of tumor virus cloning in the late 1970s and early 1980s enabled leading oncogene labs to develop libraries of DNA probes that reacted with the two dozen or so known retroviral oncogenes. Using this library of oncogene probes, researchers examined the DNA of recipient cultures transformed by human tumor DNA for the presence of new DNA fragments related to the oncogenes. When genomic DNA from normal mouse fibroblasts is hybridized to a given oncogene probe in

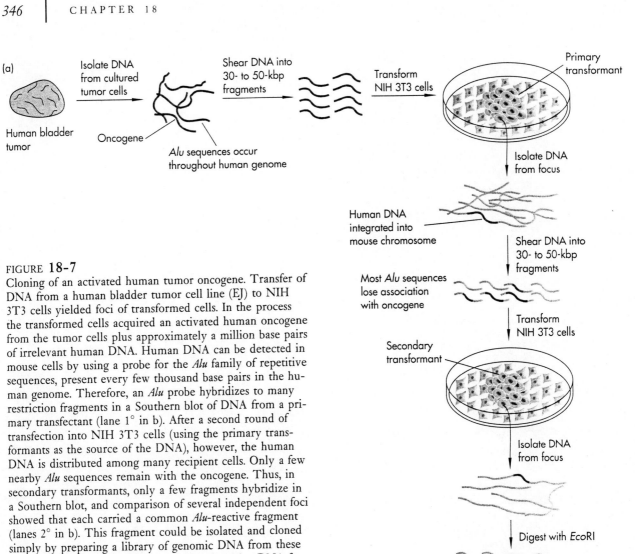

FIGURE **18-7**
Cloning of an activated human tumor oncogene. Transfer of DNA from a human bladder tumor cell line (EJ) to NIH 3T3 cells yielded foci of transformed cells. In the process the transformed cells acquired an activated human oncogene from the tumor cells plus approximately a million base pairs of irrelevant human DNA. Human DNA can be detected in mouse cells by using a probe for the *Alu* family of repetitive sequences, present every few thousand base pairs in the human genome. Therefore, an *Alu* probe hybridizes to many restriction fragments in a Southern blot of DNA from a primary transfectant (lane 1° in b). After a second round of transfection into NIH 3T3 cells (using the primary transformants as the source of the DNA), however, the human DNA is distributed among many recipient cells. Only a few nearby *Alu* sequences remain with the oncogene. Thus, in secondary transformants, only a few fragments hybridize in a Southern blot, and comparison of several independent foci showed that each carried a common *Alu*-reactive fragment (lanes 2° in b). This fragment could be isolated and cloned simply by preparing a library of genomic DNA from these cells and screening the library with an *Alu* probe. DNA from the phage isolated from this library carried the intact oncogene and very efficiently transformed NIH 3T3 cells.

a Southern blot, usually only a small number of DNA fragments of characteristic size hybridize. These fragments represent the endogenous proto-oncogene in mouse DNA. When DNA from normal cells was compared to DNA from fibroblasts transformed with DNA from the EJ cell line, the patterns of hybridization were the same for most oncogene probes. When examined with a probe derived from the *ras* oncogene of Harvey sarcoma virus (a mouse retrovirus), however, new fragments were detected in the transformed fibroblasts (Figure 18-8). Several fibroblast clones independently transformed by the human tumor DNA each carried new *ras*-related DNA fragments. Indeed, the transformed phenotype correlated perfectly with

FIGURE **18-8**
The human bladder carcinoma oncogene is a derivative of
the c-*ras* proto-oncogene. DNA was prepared from normal
NIH 3T3 cells and from NIH 3T3 cells transformed with
the human bladder carcinoma oncogene. The DNA was
cleaved with a restriction enzyme, applied to alternate lanes
of an agarose gel, and then transferred to nitrocellulose
(Southern blot). Strips of nitrocellulose carrying one lane
each of normal and transformed cell DNA were hybridized
to different oncogene probes. Most probes showed that normal and transformed cells carried precisely the same fragments, which arise from the resident proto-oncogene in the
NIH 3T3 cells. A *ras* probe, however, showed that the
transformant carried a new hybridizing fragment, indicating
that it had acquired a new *ras* gene. This result showed that
a *ras* gene derivative was responsible for the transforming
activity in the bladder carcinoma DNA. This conclusion
was confirmed by analysis of the cloned gene.

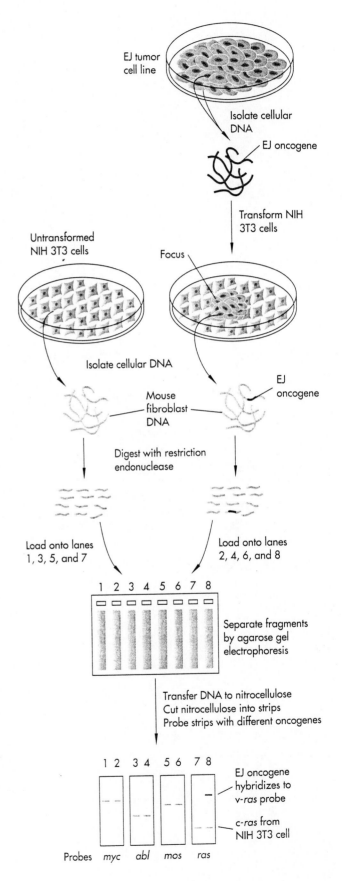

the presence of new *ras* DNA in the cells. And the
cloned bladder oncogene also hybridized to the *ras*
probe. The conclusion: the oncogene in the human
tumor passed from cell to cell by transfection is the
cellular counterpart of the retroviral *ras* gene. At last
there was a smoking gun. A cellular proto-oncogene,
found in every cell, was responsible for the transformed phenotype being passed from the tumor to the
fibroblasts.

The Bladder Carcinoma *ras* Oncogene Is Activated by Point Mutation

The identification of this active human oncogene as
a form of the c-*ras* gene greatly simplified the task of
learning how the *ras* proto-oncogene became a vir-
ulent oncogene. By this time, the normal c-*ras* gene
had been cloned and studied in some detail. However,
preliminary comparison of the activated *ras* gene from
the tumor with its normal counterpart showed no evidence of the gross gene rearrangements many expected
might be involved in oncogene activation. Researchers
realized that the structural difference between the active and inactive *ras* genes must be very subtle, even
though the difference in their biological activities is
dramatic.

Finding the difference was another demonstration
of the power of recombinant DNA technology. Re-

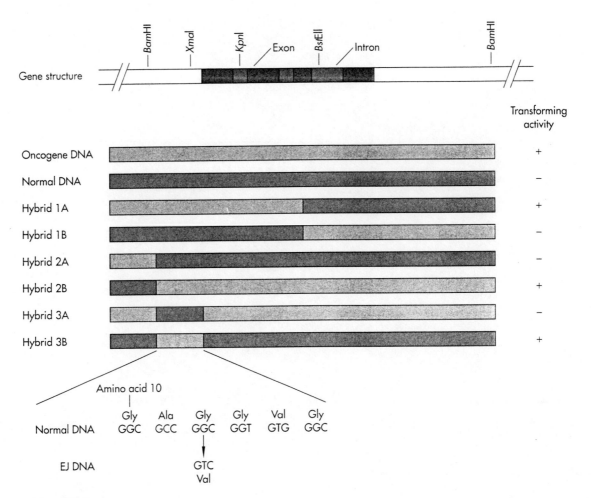

FIGURE **18-9**

Locating the mutation that activated the *ras* gene—the mix-and-match experiment. Whereas a normal *ras* gene (orange) does not transform NIH 3T3 cells, the bladder carcinoma *ras* oncogene (green) does. To locate the sequence difference that accounted for the dramatic difference in the biological activity of these genes, investigators stitched together chimeric genes from the two parent genes, using restriction sites common to both genes. Some chimeras transformed NIH 3T3 cells (indicated by +), whereas others did not (indicated by −). By making progressively finer chimeras, researchers quickly homed in on a short restriction fragment (*Xma*I to *Kpn*I), which, upon sequencing, yielded a single nucleotide difference, shown at the bottom, for the DNA strand corresponding to the sequence in the mRNA.

searchers performed a simple mix-and-match experiment similar to the chimera constructions used to study the β-adrenergic receptors (Chapter 17). Investigators recombined pieces of the normal and transforming *ras* genes, assaying the chimeras for their ability to transform cells (Figure 18-9). This procedure allowed them to rapidly map the transforming mutation to a 350-bp restriction fragment. And within this sequence, they found a *single* nucleotide substitution. A guanine nucleotide within the normal gene

had been converted to a thymine in the oncogene. This mutation changed the twelfth codon of the *ras* gene from a codon encoding glycine to one encoding valine. The startling climax of this remarkable drama of exploration was that a single base-pair change within the thousands of base pairs in this gene among thousands of genes in the recipient cells was the force that drove these cells to form tumors. We will learn shortly about the biochemical significance of the mutations that activate *ras* genes.

Proto-Oncogenes Encode Components of Signal Transduction Pathways

The identification of cellular proto-oncogenes told us that our genomes carried genes with the potential to kill us. What are these genes? What are their normal functions? How do viruses and direct DNA damage unleash their oncogenic potential? The fact that these genes have the potential to dramatically perturb cell growth and the observation that cellular proto-oncogenes have been extraordinarily conserved in evolution suggested that in their normal context, these genes must have important functions in growth control. Indeed, most proto-oncogenes encode proteins that participate in signal transduction pathways through which signals to grow (or not to grow) are relayed from outside the cell to the regulatory machinery within (Table 18-1; see also Chapter 17). Thus, the proto-oncogene proteins are links in a molecular bucket brigade that pass growth signals into the cell. As we will see, the oncogenic versions of these proteins are somehow able to pass buckets down the line even if they are not receiving them from their neighbors.

Proto-Oncogenes Can Encode Growth Factors or Their Receptors

Some proto-oncogenes encode growth factors, the molecules that are themselves the signals to grow. One example is the *sis* oncogene, first identified in a monkey retrovirus. The *sis* oncogene encodes a form of platelet-derived growth factor (PDGF), a potent mitogen for mesenchymal cells such as fibroblasts. Cells infected with the *sis*-carrying virus become transformed via *autocrine stimulation*. They secrete a growth factor to which they can also respond. Thus the cells bathe themselves constantly in a factor that makes them grow. Although the form of PDGF encoded by the v-*sis* oncogene is a slightly altered form of the natural factor, cells artificially engineered to overexpress a natural growth factor behave exactly the same way *sis*-transformed cells do (Figure 18-10).

Other proto-oncogenes encode growth factor receptors. As we learned in Chapter 17, many growth factors act on cells through specific high-affinity cell surface receptor proteins endowed with protein-

tyrosine kinase activity. Binding of growth factors to their receptors triggers a series of growth-promoting signals inside the cell. Some oncogenes encode altered receptors that trigger these signals even in the absence of ligand (Figure 18-11). The viral *erb*B gene, for example, encodes a form of the epidermal growth factor (EGF) receptor that is shortened at both ends. In particular, its entire extracellular (EGF-binding) domain has been lopped off. This decapitated receptor acts as if it is constantly bound to ligand and therefore sends growth-promoting signals constitutively.

Many Proto-Oncogenes Act as Intracellular Signal Transducers

The largest class of oncogenes is in many ways the most mysterious. These encode proteins that occupy the shadowy area just inside the plasma membrane, where, it is believed, the proteins associate with growth factor receptors and pass their signals to downstream targets. The single largest family of oncogenes are those that encode protein-tyrosine kinases found on the inner surface of the membrane. This family includes *src* (the first identified oncogene), *abl* (the protein affected on the Philadelphia chromosome), and *lck* (the protein kinase associated with the CD4 protein; Chapter 17). The very large number of oncogenic protein-tyrosine kinases argues very persuasively that tyrosine phosphorylation is a critical event in growth control. Indeed, phosphotyrosine is vanishingly rare in normal cells. It is found only when cells are stimulated with growth factors or in cells transformed with this class of oncogenes. Protein-tyrosine kinase oncogenes are frequently found in avian retroviruses. The activation of these oncogenes is manifested by a combination of mechanisms, including overexpression of the oncogene product, mutations that increase its enzymatic activity, and changes in its location within the cell.

A second class of oncogenes with a similar function is the *ras* family. There are three very closely related *ras* genes in the mammalian genome. You will recall that an activated *ras* gene was the oncogene in a human bladder carcinoma. Indeed, roughly 20 percent of human tumors possess activated *ras* genes detectable by gene transfer. The *ras* genes are exceptionally conserved in evolution; there are functionally homologous

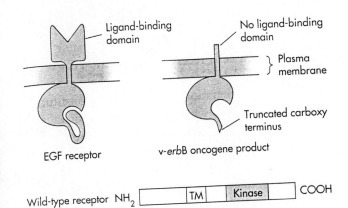

FIGURE 18-11

Alteration of a growth factor receptor can transform cells. The viral *erb*B oncogene is a derivative of the gene encoding the epidermal growth factor (EGF) receptor. The viral gene is truncated at both ends, however, an alteration leading to the production of a receptor protein that signals even in the absence of EGF. Some animal tumors arise from viral integrations into the c-*erb*B gene that lead to the production of a similarly truncated receptor protein. TM, portion of the receptor protein lying within the membrane (transmembrane segment).

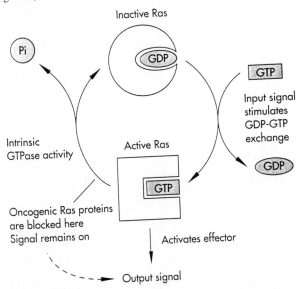

FIGURE 18-12

The *ras* cycle. Ras proteins are related to the G proteins we encountered in Chapter 17. When bound to a molecule of GDP (top), Ras proteins are inactive for signaling. In response to an unknown input stimulus, Ras evicts GDP in favor of GTP. This exchange puts Ras in its active state, in which it can drive cell growth. Normally, Ras proteins remain in this state only briefly, hydrolyzing the GTP molecule to GDP and returning to the inactive state. The mutations that convert *ras* to an oncogene eliminate the ability of its product, Ras protein, to hydrolyze GTP. Thus, the protein is locked in its active conformation.

hydrolysis activity, locking the protein in its active conformation (Figure 18-12). Similar mutations have now been found in the conventional G proteins.

In the case of Ras, we have an unprecedented opportunity to see the precise details of oncogene activation. The three-dimensional structure of the Ras protein has been determined by x-ray crystallography, and therefore we can see the precise location of the amino acids mutated in the products of active *ras* oncogenes. Each of these amino acids lies within the pocket that binds the guanine nucleotide. It is easy to appreciate how substitutions at these positions might interfere with GTP hydrolysis. Moreover, the structures of active (GTP-bound) and inactive (GDP-bound) Ras have now been compared. These studies showed that an entire face of the protein is rearranged when the protein is activated by GTP. It is ironic that despite having such a detailed atomic view of how the Ras protein works, we still have few clues to how active Ras protein drives cell growth. From what does it receive its signal? To what does it pass the signal?

Some Proto-Oncogenes Are Transcription Factors

To change the growth state of cells, stable changes in the pattern of cellular gene expression must occur. Therefore, gene transcription must be considered as an ultimate target for oncogene action. Indeed, several oncogene proteins act directly in the nucleus, where it is assumed they control the expression of cellular genes required for proliferation. These proteins bind DNA and very likely function as transcription factors (Chapter 9). A striking feature of many nuclear proto-oncogenes (that is, proto-oncogenes whose products are found in the nucleus) is that their expression is highly regulated. Genes like c-*fos*, c-*myc*, and c-*rel* are expressed at very low levels in quiescent cells but are quickly turned on in response to mitogenic signals. The current view is that these proteins are nuclear signal transducers, or *third messengers*, that act to convert short-term signals occurring within a few minutes of the primary stimulus to the long-term cellular responses that occur over hours and days. Because the regulation of expression of nuclear proto-oncogenes is intrinsic to their function, activation of nuclear oncogenes often occurs by deregulated expression. We

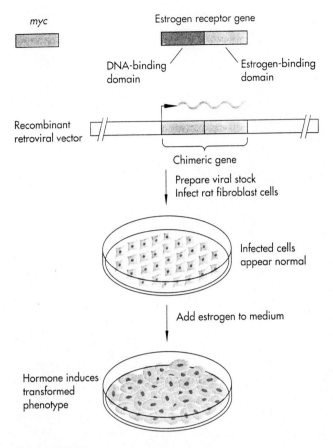

FIGURE **18-13**

Deregulated production of a normal Myc protein is sufficient to transform cells. To study transformation by the Myc protein, investigators constructed a hybrid gene in which *myc* coding sequences were fused to the sequence encoding the hormone-binding domain of the estrogen receptor. As we learned in Chapter 17, the hormone-binding domain acts as an intracellular switch, keeping linked protein domains inactive in the absence of ligand, active in its presence. The chimeric gene was placed in a retroviral vector, and the resulting virus stock was used to infect a culture of normal cells. In the absence of estrogen, the Myc protein was inactive, and the cells continued to grow normally. Estrogen treatment allowed active Myc protein to accumulate in the cell nuclei, and the cells became fully transformed.

have already seen, for example, that the c-*myc* gene is a frequent target for integration of proviruses and for chromosomal rearrangements, both events that promote increased expression of c-*myc* or expression at an inappropriate time. Indeed, placing a normal *myc* gene in an expression vector that cannot be shut off by the cell can be sufficient to transform cells (see, for example, Figure 18-13).

In virtually every case where their functions are known, oncogene proteins lie along the signaling pathways by which cells receive and execute growth instructions. The mutations that activate these genes are either structural mutations that lead to the constitutive activity of a protein without an incoming signal (for example, the protein kinases and Ras) or regulatory mutations that lead to the production of the protein at the wrong place or time (for example, the nuclear oncogene products and the growth factors themselves). Damage to oncogenes gives a cell a persistent internal growth signal in the absence of any external stimuli.

Experimentally Modeling the Complexities of Cancer: Multiple Hits

It is axiomatic for those who study the disease that cancer is a multiple-hit phenomenon, that a tumor results from several independent events occurring sequentially within a single cell. Yet according to the findings of molecular biologists, adding a single active oncogene to apparently normal cells is sufficient to drive the cells to full-blown tumorigenicity. The explanation offered for this paradox was that the cells used in the oncogene transfection assays were not really normal, that as an immortal cell line they were already teetering on the brink of transformation and that addition of an activated *ras* oncogene was just enough to push them over. So attention turned to developing transformation assays with cell cultures directly explanted from animals. Indeed, in these cells transfection of an activated *ras* gene was not sufficient to transform the cells to a fully tumorigenic state. Instead *two* oncogenes needed to be introduced simultaneously. And, interestingly, only certain combinations worked—*ras* and *myc*, for example, or *ras* and the *E1A* oncogene of adenovirus (Figure 18-14). In general, transformation of these primary cells required both a nuclear oncogene and a cytoplasmically acting oncogene. The system was not perfect; many oncogenes could not be made to fit into these two categories. Nevertheless, these experiments established some important points. Single oncogenes could not alone transform normal cells; multiple events were in fact required. And, because only certain combinations worked, these oncogenes clearly had comple-

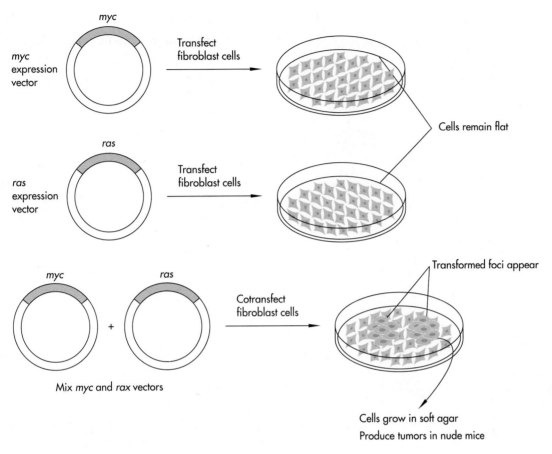

FIGURE **18-14**

The two-hit hypothesis. When investigators transfected *myc* or *ras* oncogenes into primary cultures of rat embryo fibroblasts, neither gene was sufficient to produce transformed foci. When both were transfected together, however, foci arose. The investigators concluded that at least two different genetic events are required to transform these cells.

mentary functions. Two oncogenes that do the same thing would not transform the cells. At the very least, the system had to be perturbed at two distinct points.

Oncogenes Cause Cancer in Transgenic Mice

Development of the ability to generate transgenic mice (see Chapter 14) provided an opportunity to critically test the oncogene hypothesis in intact animals. And it quickly became apparent that mice born with an activated oncogene in their genome reproducibly developed cancer at a young age. This was a dramatic demonstration of the power of dominantly acting oncogenes in a real animal, not just in cells on a cul-

ture dish. In general, experiments with transgenic mice carrying active oncogenes strongly support the multiple-hit model for cancer. Although the oncogene is present in all cells, all cells expressing the oncogene were not transformed. Instead tumors arose sporadically in the affected tissue. And the tumors were clonal. That the tumors were sporadic and clonal means that additional, relatively rare events had to occur to allow oncogene-containing cells to grow out into tumors.

With two lines of transgenic mice bearing different oncogenes, it was possible to breed animals with two hits. Mice expressing activated *ras* in their mammary tissues were bred to mice expressing deregulated *myc* in their mammary tissues. Mice in these individual lines developed clonal mammary tumors at roughly

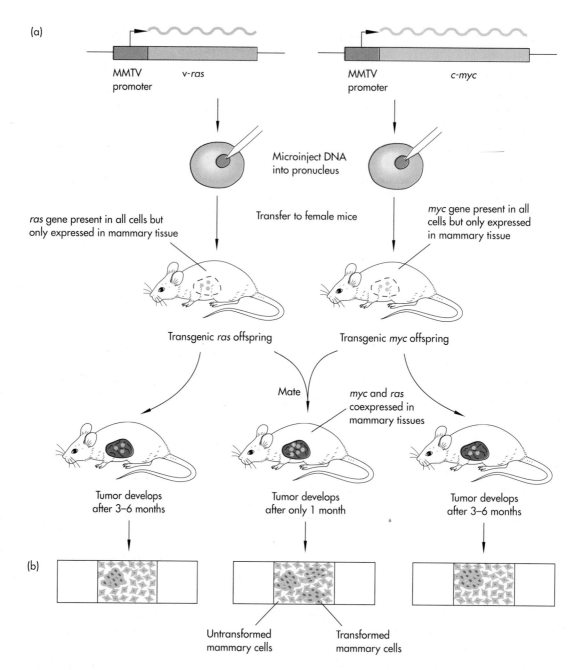

FIGURE **18-15**
Demonstration that more than two hits are required for tumor formation in animals. (a) Researchers developed two lines of transgenic mice that express the *ras* and *myc* oncogenes, respectively, under the control of the LTR (promoter) of mouse mammary tumor virus (MMTV). This LTR targets expression of the oncogenes to mammary tissue. Mice bearing either of these oncogenes predictably develop mammary tumors at about three months of age. These tumors are focal; that is, as shown on the microscope slides (b), they grow out of a background of apparently normal mammary epithelium, an observation indicating that they arise from single progenitor cells that have undergone one or more additional events. When *ras* mice were bred to *myc* mice, one quarter of their offspring carried both oncogenes (a). These mice developed mammary tumors more rapidly than their parents, after only one month. But microscopic examination of tissue sections showed that the tumors were still focal; surrounding mammary tissue remained relatively normal (b). Thus, the presence of both activated *ras* and *myc* oncogenes is not sufficient for tumor formation; other things must happen.

three months of age. What happens to mice expressing both *ras* and *myc* in their mammary cells? If two hits are sufficient for transformation, the mammary tissue of these animals should be uniformly transformed. But this was not observed. Instead these mice still developed clonal mammary tumors, but much more quickly (Figure 18-15). These observations argued that each oncogene contributes significantly to tumor formation, but that still additional events are required. We will see how multiple events can be documented during the development of a human cancer later in this chapter.

The Fusion Paradox: Normal Growth Is Dominant to Transformation

In retrospect, working with oncogenes was easy because of the powerful influence they exert on cell growth. Simply adding an oncogene to cells accelerated their growth rate and allowed them to grow under conditions prohibitive for normal cells. And, importantly, the effect of oncogenes on cells is genetically *dominant*—they exert their biological effect even in the presence of their normal counterpart. Yet, researchers using somatic cell genetics to study cancer reached the opposite conclusion, that tumorigenicity is *recessive* to normal growth.

The type of experiments these geneticists performed is shown in Figure 18-16a. They took cultures of two cell lines, one a tumor cell line and one normal, and fused them to form cell *hybrids* that retained the genomes of both parents. If the tumorigenic phenotype is dominant, as oncogene researchers believed, the hybrids should all be tumorigenic because they carry the oncogenes from the tumorigenic parent. But in fact the hybrids were all normal. Therefore, normal cells must possess genes that can overpower the oncogenes and keep them in check.

Evidence that there were specific *tumor suppressor genes*, or *anti-oncogenes*, in normal cells came from examining revertants of these hybrid cells that regained the tumorigenic phenotype. Careful analysis of the chromosomes carried by these cells revealed that the revertants always lost the same chromosomes from the set acquired from the normal parent (Figure 18-16a). In human cells, this was commonly chromosome 11

or 13. Demonstration that chromosome 11 indeed carried an anti-oncogene came from sophisticated chromosome transfer experiments in which isolated chromosomes from normal human cells were inserted into tumor cells (Figure 18-16b). Insertion of chromosome 11 was sufficient to reprogram the tumor cell into normal growth.

Other evidence for the existence of genes that block tumor growth is the frequent observation of chromosome deletions in tumors, detectable microscopically or by hybridization to molecular probes. These observations support the idea that cells have growth-suppressing genes that must be inactivated before tumors can develop. Identifying, mapping, and cloning the genes deleted from these chromosomes has been a very difficult task, requiring sophisticated and tedious molecular mapping methods, which we describe in detail in Chapters 26 and 27. But, as we discuss shortly, several such genes have now been cloned. And they provide a critical piece of the cancer puzzle: some of the steps in the development of a tumor entail the loss of these growth-suppressing genes.

Susceptibility to Cancer Can Be Inherited

Further evidence for the existence of tumor suppressor genes came from studies of rare cancers that run in families. Members of affected families appear to inherit *susceptibility* to cancer and develop certain kinds of tumors at rates much higher than the normal population. A dramatic example of an inherited tumor susceptibility is *retinoblastoma*, a tumor of the eye that strikes young children. Retinoblastoma can occur sporadically, an isolated event in a family with no history of the disease. This form of the disease is the most common. But about one-third of retinoblastoma patients develop multiple tumors, usually in both eyes. And the children and siblings of such patients often develop the same disease, an observation suggesting that the susceptibility to retinoblastoma is inborn. The data are consistent with a mechanism involving a single responsible gene.

In 1971, Alfred Knudson suggested a prescient model to account for the sporadic and inherited forms of retinoblastoma. He supposed that the development

(a)

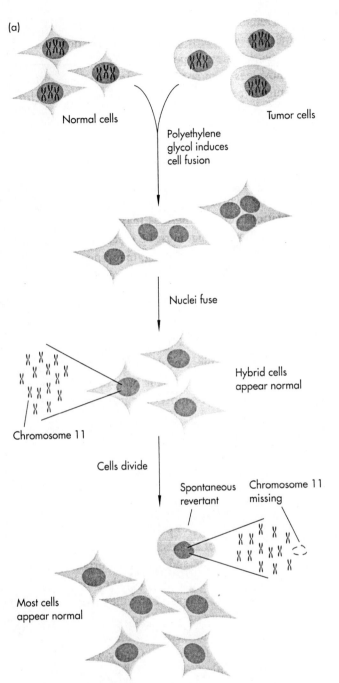

Normal cells

Tumor cells

Polyethylene glycol induces cell fusion

Nuclei fuse

Hybrid cells appear normal

Chromosome 11

Cells divide

Spontaneous revertant

Chromosome 11 missing

Most cells appear normal

(b)

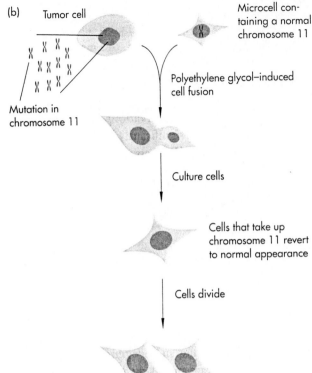

Tumor cell

Microcell containing a normal chromosome 11

Mutation in chromosome 11

Polyethylene glycol–induced cell fusion

Culture cells

Cells that take up chromosome 11 revert to normal appearance

Cells divide

FIGURE **18-16**

Normal growth phenotype is dominant to the transformed phenotype. (a) Investigators mixed normal cells and tumor cells and induced them to fuse, by using the chemical polyethylene glycol (PEG). The resulting hybrids carry chromosomes from both parents, and they were uniformly normal in phenotype. Thus, genes from the normal partner are able to suppress the growth defect encoded in the genes of the transformed cell. But these hybrids are unstable, and they randomly shed chromosomes. By examining hybrids that spontaneously regained the transformed phenotype, investigators could identify the normal chromosome that was always lost from such cells. In certain crosses, it was chromosome 11. Indeed, transfer of chromosome 11 alone to tumor cells containing a mutated chromosome 11 was sufficient to revert their growth to normal (b). This chromosomal transfer was done by *microcell transfer*, which involves treating cells with drugs that disrupt the cytoskeleton and generate small cell fragments (microcells) containing single (or a few) chromosomes in their nuclei.

of retinoblastoma required two rare mutations. In the sporadic form of the disease, both mutations would have to occur within a single retinoblast cell, an exceedingly infrequent circumstance that explains why these children never develop more than one tumor. In the inherited form of the disease, however, he suggested that one of the mutations is already present (inherited) in all retinal cells. Therefore, only a single additional mutations is required for a full-blown tu-

mor, and tumors are much more frequent in these individuals.

A decade later, analysis of the chromosomes in tumor cells and normal tissues of retinoblastoma patients resoundingly confirmed Knudson's hypothesis (Figure 18-17). Many retinoblastoma patients carried deletions in chromosome 13. And the gene inherited by afflicted children, termed *RB*, was mapped to this same chromosome. Most important, while unaffected

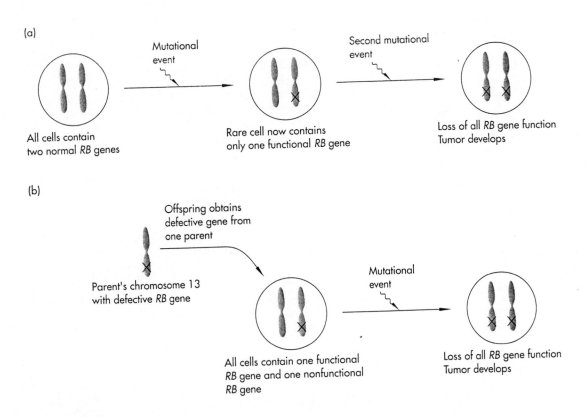

(a)

Mutational event

All cells contain two normal *RB* genes

Rare cell now contains only one functional *RB* gene

Second mutational event

Loss of all *RB* gene function Tumor develops

(b)

Offspring obtains defective gene from one parent

Parent's chromosome 13 with defective *RB* gene

All cells contain one functional *RB* gene and one nonfunctional *RB* gene

Mutational event

Loss of all *RB* gene function Tumor develops

FIGURE **18-17**
The development of sporadic and inherited retinoblastoma. (a) In the sporadic form of the disease, all cells in the body carry two functioning *RB* genes, one on each copy of chromosome 13. In a rare cell, a mutational event destroys one of the active genes. If one of the descendants of this cell suffers a second blow to its remaining *RB* gene, it develops into a tumor. The likelihood that two rare events will strike in the same cell is very low, hence sporadic retinoblastoma is a very uncommon tumor. (b) Some children, however, are born with one *RB* gene already damaged. This mutation can be inherited from a parent or can occur during development of the sperm or egg cell that gives rise to the child. Now, only a single event—damage to the remaining *RB* gene—is all that is required for a tumor to develop. These children often develop multiple tumors in both eyes.

tissues in these children could be shown to carry one mutant *RB* allele and one normal one, tumor DNA carried only mutant *RB* alleles. Thus, development of retinoblastoma appeared to require that both copies of the *RB* gene be mutated; these are Knudson's two hits. Because in many cases the mutant *RB* alleles had large deletions, it was clear that the mutations eliminated the function of this gene. The *RB* gene must therefore be a tumor suppressor gene or anti-oncogene that normally functions to arrest the growth of retinal cells. Even one copy is sufficient to keep growth in check. But loss of both copies of *RB* eliminates the block, and a tumor develops.

Genetically, mutations in tumor suppressor genes behave differently from oncogene mutations. Whereas activating oncogene mutations are dominant to wild

type—they emit their proliferative signals regardless of the presence of wild-type gene product—tumor suppressor mutations are recessive. Mutation in one gene copy usually has no effect, as long as a reasonable amount of wild-type protein remains. Thus, tumor suppressor genes are sometimes called *recessive oncogenes*.

The Cloning of the Retinoblastoma Susceptibility Gene

With the recognition that the *RB* gene was a member of the elusive family of tumor suppressor genes, another cloning race was on. In contrast to the human bladder oncogene, whose presence was easily detected by the tumorigenic properties it conferred on cells,

the *RB* gene expressed its effect only by its absence. So its presence could not be followed experimentally, nor could its activity be easily measured. Thus, the *RB* gene had to be hunted for by exploring, blindly at first, the region of chromosome 13 to which it had been mapped (see Chapter 26). Investigators isolated DNA probes from chromosome 13 and tested them for features that they might expect for the *RB* gene— expression as a messenger RNA, presence of closely related sequences in other animal genomes, and, most important, the ability to detect deletions or other rearrangements in the DNA of retinoblastoma cells.

A DNA probe that met these criteria was found and used to isolate a cDNA clone and then genomic DNA clones for the candidate gene. That this gene was indeed *RB* came from analysis of the sequence in a large panel of retinoblastoma tumors. In many of these tumors, the *RB* gene was completely or partially deleted. In others, the gene was intact, but molecular analysis revealed simple point mutations, often in splice junctions. These splice junction mutations caused exons to be skipped in the assembly of the *RB* mRNA, thus leading to the production of an aberrant RB protein. Finally, it was shown that introducing the *RB* gene within an expression vector into tumor cells could revert their growth properties to normal—compelling evidence that the cloned gene had the biological activity expected of a tumor suppressor gene.

The RB Protein Is the Target of DNA Tumor Virus Oncogenes

What does the RB protein do? How does it put the brakes on cell growth? The availability of an *RB* clone made a number of new experiments possible. For example, the protein (or fragments of it) could be produced and used to make antibodies. The antibodies showed that the RB protein was located in the nucleus, where presumably it influences cellular gene expression. And the protein was phosphorylated cyclically as cells grew: in newly divided cells the protein was only lightly phosphorylated, but as cells began to duplicate their DNA, the protein became very heavily phosphorylated. This observation suggested some connection to the clock mechanism by which cells regulate their growth cycles (see Chapter 19).

But answers to the mystery of *RB* function began to flow in earnest from an unexpected direction, the classic DNA tumor viruses. Over the years, investigators had accumulated much information about the oncogenes of these viruses. They had generated many mutations in the oncogenes. They knew what parts of the genes were required for their oncogenic activities. And they knew, by sight if not by name, some of the cellular proteins with which the oncogene proteins consorted. In one of the key discoveries in cancer biology, investigators realized that one of the proteins that associated with the DNA tumor virus oncoproteins was the RB protein. Now investigators could go back to the dozens of mutant viral oncogenes they had generated over the years. There was an excellent correlation with the ability of these genes to transform cells and their ability to associate with RB (Figure 18-18). Thus, the oncoproteins of the DNA tumor viruses, whose functions had proved so difficult to comprehend compared to their cousins in the RNA tumor viruses, were in fact working in a completely different fashion. They drove cell growth, not by pressing on the accelerator like conventional oncogenes, but by eliminating the brakes. Moreover, the RB protein is leading oncogene researchers beyond the primary signal transduction pathways they had been studying to the internal clock that regulates cell growth after the initial growth signal is received.

The Strange Case of p53: An Oncogene Crosses to the Other Side

The search for cellular proteins that interact with tumor virus oncoproteins and thereby led to the RB connection really began with another protein, p53. Named for its size, p53 (53,000 daltons) was first discovered as a cellular nuclear protein associated with and stabilized by the T antigen protein of SV40. There seemed to be a strict correlation between the abundance of p53 and the oncogenic activity of the tumor virus. Uninfected cells and cells infected with nontumorigenic virus mutants did not accumulate much p53. Strikingly, in cells transformed with a mutant SV40 virus carrying a temperature-sensitive T antigen, shifting the cells to the nonpermissive temperature resulted in simultaneous reversion of the cells to normal and loss of p53 protein. Driven by the emerging paradigm of dominantly acting oncogenes, researchers fingered p53 as a cellular oncogene used by the tumor viruses to transform cells. This view was

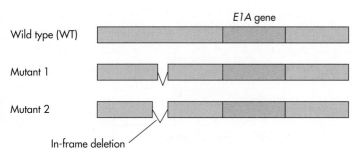

	E1A gene	Deleted amino acids	Transforming activity
Wild type (WT)		None	+
Mutant 1		77–85	+
Mutant 2		30–85	−

In-frame deletion

FIGURE **18-18**

The ability to bind to the RB protein correlates with the ability of the *E1A* gene to transform cells. The adenovirus *E1A* gene is a potent oncogene in certain cells. Extensive mutagenesis and mapping of the gene identified the domains that encode its transforming potential. Two mutants, one transforming (mutant 1) and one nontransforming (mutant 2), are shown. To determine whether the ability of the E1A proteins to associate with the RB protein is required for transformation, investigators infected cells with recombinant adenoviruses that carried the *E1A* mutations. The infected cells were lysed, and antibodies against the E1A proteins were used to recover E1A along with any cellular proteins with which it associated. Analysis of the E1A-associated proteins by gel electrophoresis revealed that all transforming versions of E1A (WT and mutant 1) associated with the RB protein, $p105^{RB}$. Nontransforming versions (mutant 2) did not. In reciprocal experiments, investigators mapped the regions of the RB protein required to associate with E1A. They found that these were precisely the regions deleted in many retinoblastoma tumors. Both E1A and the RB protein associate with several cellular proteins that investigators believe are critical regulators of cell growth.

reinforced when several groups showed that expression vectors carrying p53 cDNAs could transform primary cells in cooperation with *ras* genes, the litmus test for a nuclear oncogene.

But with time the p53 story unraveled. Some p53 cDNAs did not transform cells. Moreover, investigators were beginning to find tumors that lacked p53 entirely, because their p53 genes were deleted or otherwise inactivated. Disappearance in tumors is not acceptable behavior for an oncogene. It took 10 years for investigators to realize that the original p53 cDNAs, which had been isolated from tumor cell lines, were mutants, carrying point mutations in their coding sequence. Wild-type p53 genes, finally isolated from normal cells, had the opposite effect of the original clones: they suppressed cell growth.

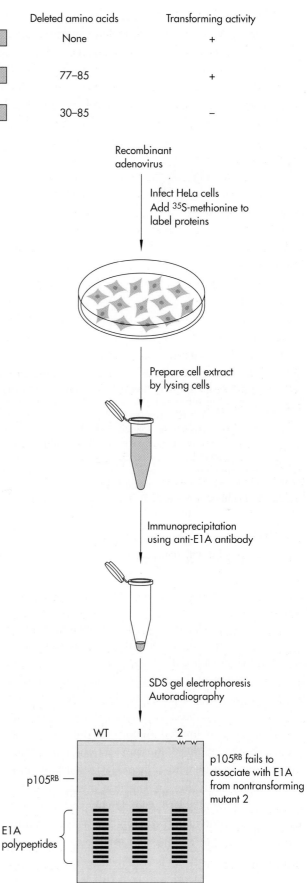

Recombinant adenovirus

Infect HeLa cells
Add ^{35}S-methionine to label proteins

Prepare cell extract by lysing cells

Immunoprecipitation using anti-E1A antibody

SDS gel electrophoresis
Autoradiography

WT 1 2

$p105^{RB}$ —

E1A polypeptides

$p105^{RB}$ fails to associate with E1A from nontransforming mutant 2

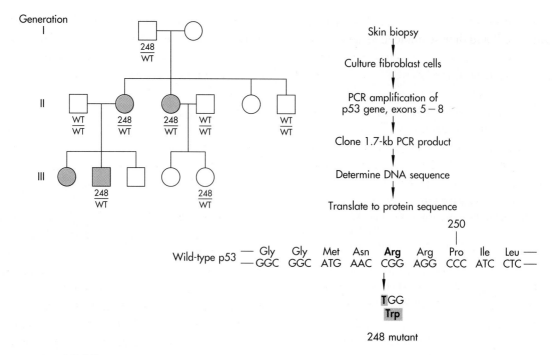

FIGURE **18-19**

A mutant p53 gene is responsible for inherited cancers in the Li-Fraumeni syndrome. A pedigree for a family of patients afflicted with the syndrome is shown. Those marked in blue had tumors. To determine whether mutations in the p53 gene were responsible for the greatly elevated frequency of cancer in these patients, skin samples were taken and grown in culture. DNA was isolated from these samples and subjected to PCR with primers specific for a portion of the p53 gene. The amplified p53 gene fragments were cloned and sequenced. All affected patients carried both the wild-type gene (WT) and a p53 gene with a single nucleotide substitution in the codon for amino acid 248 (marked "248" in the pedigree). This mutation could be traced back through the generations (note, however, that not all family members were tested). The grandfather (generation I) and granddaughter (generation III) carry the mutant p53 gene but have not yet developed tumors. Because 90 percent of persons with this gene develop tumors by age 70, doctors will closely monitor these individuals for early signs of tumor development. In other Li-Fraumeni families, mutations were detected in different p53 codons.

So p53 changed—in the minds of cancer researchers—from a full-fledged cellular oncogene to a member of the rival camp, the tumor suppressor genes. In fact, inactivation of the p53 gene is now known to be the most common genetic defect detectable in tumors. The gene is inactivated by deletion or by point mutations in certain limited regions of the coding sequence. Some of these mutants act in a dominant fashion to inactivate wild-type p53, which is how some of the initial confusion arose: these mutant proteins are stable, accumulate to high levels, and sequester the wild-type protein in the cytoplasm where it cannot perform its growth-suppression function.

Inheriting a damaged p53 gene leads to dramatically elevated frequencies of cancer, a situation resembling the inheritance of retinoblastoma. The Li-Fraumeni syndrome is an inherited susceptibility to a variety of cancers, in which 50 percent of affected individuals develop cancer by age 30 and 90 percent by age 70. Conventional genetic mapping of the locus responsible for inheritance of the syndrome was not successful. But on the basis of the retinoblastoma precedent, in which patients are born with one inactivated allele of *RB*, investigators checked the DNA of Li-Fraumeni patients for changes in p53, the only other well-characterized tumor suppressor gene at that time.

They amplified portions of the p53 gene from patient DNA by using PCR and then sequenced these samples. Indeed, in all five families tested, inheritance of the syndrome correlated with the presence of point mutations in the p53 gene (Figure 18-19). The rapid linking of this rare and mysterious syndrome to single nucleotide changes in the genomes of affected individuals is yet another striking illustration of the power of modern technology to find the precise molecular defect underlying a complex disease. And, for p53, its contribution to the development of cancer and its role as a bona fide tumor suppressor were dramatically confirmed.

The Several Steps to Colorectal Cancer: A Real-Life Tale of Oncogenes and Anti-oncogenes

The epidemiology of cancer has long suggested that cancer is a multistep disease. Statistical calculations based on the increased frequency of cancer with age estimate the number of steps as four to six. Now that we suspect what some of these steps might be (for example, activation of oncogenes and loss of tumor suppressor genes), can we detect these events in naturally occurring cancers? Thanks to the power of today's recombinant DNA technology—the increasing availability of molecular probes for candidate genes and the exquisite sensitivity of detection methods such as PCR—investigators have begun a molecular dissection of one of the most common human cancers, colorectal cancer.

Colorectal cancer occurs both sporadically and in an inherited form. The inherited form, a disease called familial adenomatous polyposis (FAP), is manifested as hundreds of precancerous polyps forming on the colon, some of which progress to fully malignant carcinomas. Like retinoblastoma and Li-Fraumeni syndrome, FAP is inherited as an autosomal dominant lesion—a single copy of the FAP gene from either parent is sufficient for polyp formation and the susceptibility to colon carcinoma—suggesting that the FAP locus encodes yet another tumor suppressor gene. Inheritance of this disease has been mapped to a region of chromosome 5, and a candidate gene from this region was recently cloned. The chromosome 5 gene is important in the development of sporadic colorectal tumors as well: almost half of such tumors have detectable allelic losses at this locus. Thus loss of this putative tumor suppressor gene is one of the steps toward colorectal cancer.

Conventional oncogenes are involved as well. Fully half of colon carcinomas and polyps larger than 1 cm carry activating mutations in one of the *ras* genes. The presence of these mutations in precancerous polyps suggest that activation of a *ras* gene is an early step in tumor development.

A variety of chromosomal losses, signaling the loss of a tumor suppressor gene, are found in colorectal tumors. Most common are losses in chromosomes 17 and 18. The chromosome 17 mutations delete the p53 gene. Deletion of p53 is a relatively late step in tumor development, found in 70 percent of carcinomas but rarely in polyps. And frequently, the remaining p53 allele carries a point mutation. One hypothesis is that mutation of one p53 allele, into a form that interferes with the normal protein, is actually a relatively early step in tumor development. Cells carrying this mutation develop a growth advantage over their neighbors, which is further amplified when the remaining wild-type allele is lost.

The chromosome 18 region consistently lost in colorectal tumors (detectable in over 70 percent of cases) did not correspond to any known gene, and a heroic PCR-based strategy had to be employed to locate and clone the candidate gene in this region (Figure 18-20). This gene, termed *DCC*, encodes a protein related to the cell adhesion molecules that mediate contact between cells and with the extracellular material within which cells grow. Perhaps loss of this gene allows tumor cells to escape their normal confines and invade surrounding tissue, a cardinal property of tumors.

The proposed order of events in the development of colorectal cancer is shown in Figure 18-21. These are probably not the sole events involved. Colorectal tumors possess an average of one or two additional chromosomal losses, which may correspond to unidentified tumor suppressor genes. And although there is a preferred order for these events—*ras* mutations occur early, for example, and p53 deletions late—it is the accumulation of these events rather than their order that matters in tumor development. What is

(a)

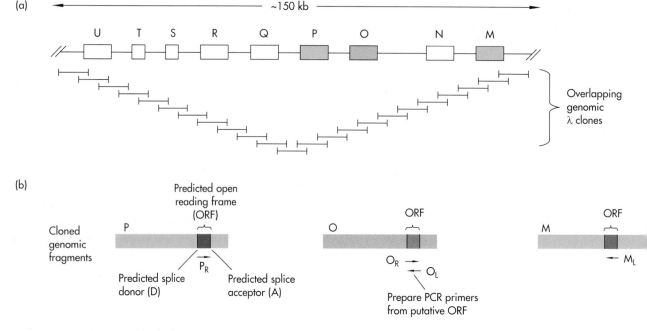

(b)

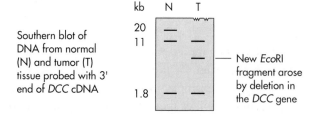

(c) Subfragments from fragments O and P were used as probes to isolate a 1.65-kb *DCC* cDNA

Southern blot of DNA from normal (N) and tumor (T) tissue probed with 3' end of *DCC* cDNA

New *Eco*RI fragment arose by deletion in the *DCC* gene

FIGURE **18-20**

Cloning of the *DCC* gene by exon-connection. (a) Having realized that a gene often deleted in colorectal tumors was located on chromosome 18, investigators used a probe to perform a chromosome walk (Chapter 7) in both directions along chromosome 18, eventually obtaining a set of 140 overlapping λ phage clones encompassing nearly 400 kb of DNA. Restriction fragments from the clones were systematically screened for their ability to cross-hybridize to DNA from other species, a hallmark for coding sequences in genes. Several such fragments, represented by boxes M through U, were found. But none of these fragments hybridized to mRNAs on Northern blots or to cDNAs in several different libraries. (b) Upon sequencing the cross-hybridizing regions and inspecting the sequences for features—such as splice acceptors ("A"), donors ("D"), and open reading frames ("ORF")—that signify gene exons, researchers found sequences that looked convincingly like exons. As a more sensitive probe for expression of these sequences in an mRNA, the investigators devised a PCR-based assay they termed *exon-connection.* They prepared RNA from several different cell lines, converted it to cDNA, and performed PCR with primer pairs from adjacent suspected exons. If these sequences were indeed found in an mRNA, PCR should amplify a fragment of a predictable size. One pair of primers, from regions P and O, yielded the expected 233-bp fragment from several of the RNA samples. The sequence of this fragment showed that it contained the expected splice product of the exons in P and O. (c) The fragment was used to screen a cDNA library from one of the PCR positive cell lines to isolate cDNA clones of this gene. When this *DCC* cDNA was used to probe genomic DNA from colorectal tumors, deletions in this sequence were frequently detected, a result indicating that a tumor suppressor gene had indeed been cloned.

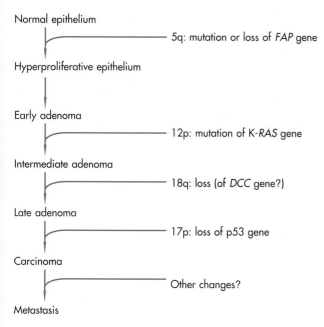

Normal epithelium

— 5q: mutation or loss of *FAP* gene

Hyperproliferative epithelium

Early adenoma

— 12p: mutation of K-*RAS* gene

Intermediate adenoma

— 18q: loss (of *DCC* gene?)

Late adenoma

— 17p: loss of p53 gene

Carcinoma

— Other changes?

Metastasis

FIGURE **18-21**
The road to colorectal cancer. The arrows indicate the stages in the development of a colorectal tumor, from normal colon epithelium through benign adenoma to malignant and metastatic carcinoma. Most steps are associated with a defined genetic change, indicated alongside each arrow. This order of genetic events does not appear to be obligatory. Rather, it is the cumulative effect of all of them that leads to development of colorectal cancer.

especially striking about this remarkable exploration of the molecular steps to colorectal cancer is that the number of events unearthed so far correlates perfectly with the prediction of epidemiologists based purely on the age-dependent frequency of the disease.

Cancer Results from Accumulation of Dominant and Recessive Mutations

With remarkable speed and steadily increasing clarity, the molecular defects in cancer have come into view. And the molecular picture provides a solid foundation for understanding decades of cancer research. Cancer

is a genetic disease, and now we know many of the genes involved and, in some cases, what they do. Cancer often takes decades to develop, requiring multiple independent events to systematically dismantle the complex and redundant regulatory circuits that maintain normal growth in a cell. Now we know what many of these events are.

Cancer results when a single cell sustains a mutation, to an oncogene or a tumor suppressor gene, that gives it a growth advantage over its neighbors. As the number of descendants of the original mutant increases, so does the likelihood that one of them will sustain a second mutation that in turn allows its immediate descendants to grow even faster. And the terrible cycle continues as cells accumulate additional mutations that accelerate their growth further and allow them to invade surrounding tissue. Further events that abet tumor growth may also arise from mutations. Some tumors secrete abundant quantities of *angiogenesis factors,* peptide hormones that direct the growth of blood vessels into the tumor (an adequate blood supply is critical for tumor growth). This change may be the result of a regulatory mutation. Additional mutations may account for *metastasis,* the deadly property of certain tumors that enables them to escape to seed new tumors elsewhere in the body. The individual events on the road to cancer are rare, but we live for a long time, and chance dictates that occasionally this lethal combination of events will unfold.

The increase in our understanding of cancer has been breathtaking, and finally this knowledge is beginning to reach the cancer clinic. Molecular analysis of the oncogenes and anti-oncogenes affected in tumors can predict the course of the disease and suggest the appropriate treatment. The benefits of simply detecting cancer at an earlier stage should not be underestimated. In colorectal cancer, mortality can be prevented if the disease is caught any time prior to the last step; we now have four or five different molecular probes to do that. And many of the same genes will undoubtedly be involved in other cancers. In coming years, cancer diagnosis will be revolutionized by recombinant DNA.

But what about cancer treatment? Cancer is a complex disease, and our recent progress tells us that it is perhaps even more complex than we had appreciated.

But recombinant DNA technology is pushing toward new cancer treatments in several significant ways. First, it is uncovering new targets for cancer therapy, the genes and proteins that drive tumor growth or restore growth to normal. Second, molecular research into cancer has unerringly brought to our immediate attention the systems that regulate cell growth, and it is an axiom of medicine that the deeper our understanding of the underlying biology of disease, the more effective our therapies. And, third, our ability to produce critical cellular proteins at will and manipulate them to change their properties promises new therapeutic drugs and strategies unlike any available at the present time.

Reading List

General

Weiss, R., N. Teich, H. Varmus, and J. Coffin, eds. *RNA Tumor Viruses,* 2nd ed. Cold Spring Harbor Laboratory, Cold Spring Harbor, N.Y., 1985.

Botchan, M., T. Grodzicker, and P. A. Sharp, eds. *"Cancer Cells: DNA Tumor Viruses: Control of Gene Expression and Replication."* Cold Spring Harbor Laboratory, Cold Spring Harbor, N.Y., 1986.

Weinberg, R. A., ed. *Oncogenes and the Molecular Biology of Cancer.* Cold Spring Harbor Laboratory Press, Cold Spring Harbor, N.Y., 1989.

Cooper, G. *Oncogenes.* Jones & Bartlett, Boston, 1990.

Original Research Papers

EARLY CHARACTERIZATION OF DNA TUMOR VIRUSES

Smith, A. E., R. Smith, and E. Paucha. "Characterization of different tumor antigens present in cells transformed by simian virus 40." *Cell,* 18: 335–346 (1979).

Hassell, J. A., W. C. Topp, D. B. Rifkin, and P. Moreau. "Transformation of rat embryo fibroblasts by cloned polyoma virus DNA fragments containing only part of the early region." *Proc. Natl. Acad. Sci. USA,* 77: 3978–3982 (1980).

Houweling, A., P. J. van den Elsen, and A. J. van der Eb. "Partial transformation of primary rat cells by the left most 4.5% fragment of adenovirus 5 DNA." *Virology,* 105: 537–550 (1980).

RETROVIRAL ONCOGENES AND THEIR CELLULAR ORIGINS

Bishop, J. M. "Viral oncogenes." *Cell,* 42: 23–38 (1985). [Review]

Duesberg, P. H., and P. K. Vogt. "Differences between the ribonucleic acids of transforming and non-transforming avian tumor viruses." *Proc. Natl. Acad. Sci. USA,* 67: 1673–1680 (1970).

Martin, G. S. "Rous sarcoma virus: a function required for the maintenance of the transformed state." *Nature,* 227: 1021–1023 (1970).

Stehelin, D., H. E. Varmus, J. M. Bishop, and P. K. Vogt. "DNA related to the transforming gene(s) of avian sarcoma viruses is present in normal avian DNA." *Nature,* 260: 170–173 (1976).

Spector, D. H., H. E. Varmus, and J. M. Bishop. "Nucleotide sequences related to the transforming gene of avian sarcoma virus are present in the DNA of uninfected vertebrates." *Proc. Natl. Acad. Sci. USA,* 75: 4102–4106 (1978).

Shilo, B.-Z., and R. A. Weinberg, "DNA sequences homologous to vertebrate oncogenes are conserved in *Drosophila melanogaster."* *Proc. Natl. Acad. Sci. USA,* 78: 6789–6792 (1981).

INSERTIONAL ACTIVATION OF CELLULAR PROTO-ONCOGENES

Hayward, W. S., B. G. Neel, and S. M. Astrin. "Activation of a cellular *onc* gene by promoter insertion in ALV-induced lymphoid leukosis." *Nature,* 290: 475–480 (1981).

Nusse, R., and H. E. Varmus. "Many tumors induced by the mouse mammary tumor virus contain a provirus integrated in the same region of the host genome." *Cell,* 31: 99–109 (1982).

Payne, G. S., J. M. Bishop, and H. E. Varmus. "Multiple arrangements of viral DNA and an activated host oncogene in bursal lymphomas." *Nature,* 295–214 (1982).

ONCOGENE ACTIVATION BY CHROMOSOME TRANSLOCATION

Dalla-Favera, R., M. Bregni, J. Erickson, D. Patterson, R. C. Gallo, and C. M. Croce. "Human c-myc onc gene is located on the region of chromosome 8 that is translocated in Burkitt lymphoma cells." *Proc. Natl. Acad. Sci. USA,* 79: 7824–7827 (1982).

Sheng-Ong, G. L. C., E. J. Keath, S. P. Piccoli, and M. D. Cole. "Novel myc oncogene RNA from abortive immunoglobulin gene recombination in mouse plasmacytomas." *Cell,* 31: 443–452 (1982).

Taub, R., I. Kirsch, C. Morton, G. Lenoir, D. Swan, S. Tronick, S. Aaronson, and P. Leder. "Translocation of the c-myc gene into the immunoglobulin heavy chain locus in human Burkitt lymphoma and murine plasmacytoma cells." *Proc. Natl. Acad. Sci. USA,* 79: 7838–7841 (1982).

Collins, S. J., I. Kubonishi, I. Miyoshi, and M. T. Groudine. "Altered transcription of the c-*abl* oncogene in K-562 and other chronic myelogenous leukemia cells." *Science,* 225: 72–74 (1984).

Groffen, J., J. R. Stephenson, N. Heisterkamp, A. de Klein, C. R. Bartram, and G. Grosveld. "Philadelphia chromosome breakpoints are clustered within a limited region, bcr, on chromosome 22." *Cell,* 36: 93–99 (1984).

PASSAGE OF THE TRANSFORMED PHENOTYPE BY GENOMIC TRANSFECTION

Hill, M., and J. Hillova. "Recovery of the temperature-sensitive mutant of Rous sarcoma virus from chicken cells exposed to DNA extracted from hamster cells transformed by the mutant." *Virology,* 49: 309–313 (1972).

Shih, C., B.-Z. Shilo, M. P. Goldfarb, A. Dannenberg, and R. A. Weinberg. "Passage of phenotypes of chemically transformed cells via tansfection of DNA and chromatin." *Proc. Natl. Acad. Sci. USA,* 76: 5714–5718 (1979).

Murray, M. J., B.-Z. Shilo, C. Shih, D. Cowing, H. W. Hsu, and R. A. Weinberg. "Three different human tumor cell lines contain different oncogenes." *Cell,* 25: 355–361 (1981).

Perucho, M., M. Goldfarb, K. Shimizu, C. Lama, J. Fogh, and M. Wigler. "Human-tumor-derived cell lines contain common and different transforming genes." *Cell,* 27: 467–476 (1981).

CLONING OF AN ACTIVATED HUMAN ONCOGENE

Der, C. J., T. G. Krontiris, and G. M. Cooper. "Transforming genes of human bladder and lung carcinoma cell lines are homologous to the ras genes of Harvey and Kirsten sarcoma viruses." *Proc. Natl. Acad. Sci. USA,* 79: 3637–3640 (1982).

Goldfarb, M., K. Shimizu, M. Perucho, and M. Wigler. "Isolation and preliminary characterization of a human transforming gene from T24 bladder carcinoma cell." *Nature,* 296: 404–409 (1982).

Parada, L. F., C. J. Tabin, C. Shih, and R. A. Weinberg. "Human EJ bladder carcinoma oncogene is homologue of Harvey sarcoma virus ras gene." *Nature,* 297: 474–478 (1982).

Shih, C., and R. A. Weinberg. "Isolation of a transforming sequence from a human bladder carcinoma cell line." *Cell,* 29: 161–169 (1982).

Tabin, C. J., S. M. Bradley, C. I. Bargmann, R. A. Weinberg, A. G. Papageorge, E. M. Scolnick, R. Dhar, D. R. Lowy, and E. H. Chang. "Mechanism of activation of a human oncogene." *Nature,* 300: 143–149 (1982).

Taparowsky, I., Y. Suard, O. Fasano, K. Shimizu, M. Goldfarb, and M. Wigler. "Activation of the T24 bladder carcinoma transforming gene is linked to a single amino acid change." *Nature,* 300: 762–765 (1982).

PROTO-ONCOGENE FUNCTION

Cantley, L. C., K. R. Auger, C. Carpenter, B. Duckworth, A. Graziani, R. Kapeller, and S. Soltoff. "Oncogenes and signal transduction." *Cell,* 64: 281–302 (1991). [Review]

Cross, M., and T. M. Dexter. "Growth factors in development, transformation, and tumorigenesis." *Cell,* 64: 271–280 (1991). [Review]

Lewin, B. "Oncogenic conversion by regulatory changes in transcription factors." *Cell,* 64: 303–312 (1991). [Review]

Collet, M. S., and R. L. Erikson. "Protein kinase activity associated with the avian sarcoma virus src gene product." *Proc. Natl. Acad. Sci. USA,* 75: 2021–2024 (1978).

Hunter, T., and B. M. Sefton. "Transforming gene product of Rous sarcoma virus phosphorylates tyrosine." *Proc. Natl. Acad. Sci. USA,* 77: 1311–1315 (1980).

Shih, T. Y., A. G. Papageorge, P. E. Stokes, M. O. Weeks, and E. M. Scolnick. "Guanine nucleotide-binding and autophosphorylating activities associated with the p21 protein of Harvey murine sarcoma virus." *Nature,* 287: 686–691 (1980).

Doolittle, R. F., M. W. Hunkapiller, L. E. Hood, S. G. Deuare, K. G. Robbins, S. A. Aaronson, and H. N. Antoniades. "Simian sarcoma virus *onc* gene, v-*sis,* is derived from the gene (or genes) encoding a platelet-derived growth factor." *Science,* 221: 275–276 (1983).

Downward, J., Y. Yarden, E. Mayes, G. Scrace, N. Totty, P. Stockwell, A. Ullrich, J. Schlessinger, and M. D. Waterfield. "Close similarity of epidermal growth factor receptor and v-*erb*B oncogene protein sequences." *Nature,* 307: 521–527 (1984).

Leal, F., L. T. Williams K. C. Robbins, and S. A. Aaronson. "Evidence that the v-*sis* gene product transforms by interaction with the receptor for platelet-derived growth factor." *Science,* 230: 327–330 (1985).

Bohmann, D., T. J. Bos, A. Admon, T. Nishimura, P.K. Vogt, and R. Tjian. "Human proto-oncogene c-*jun* encodes a DNA-binding protein with structural and functional properties of transcription factor AP-1." *Science,* 238: 1386–1392 (1987).

Stern, D. F., D. L. Hare, M. A. Cecchini, and R. A. Weinberg. "Construction of a novel oncogene based on synthetic sequences encoding epidermal growth factor." *Science,* 235: 321–324 (1987).

de Vos, A. M., L. Tong, M. V. Milburn, P. M. Matias, J. Jancarik, S. Noguchi, S. Nishimura, K. Miura, E. Ohtsuka, and S.-H. Kim. "Three dimensional structure of an oncogene protein: catalytic domain of human c-H-ras p21." *Science,* 239: 888–893 (1988).

Franza, B. R., Jr., F. J. Rauscher III, S. F. Josephs, and T. Curran. "The Fos complex and Fos-related antigens recognize sequence elements that contain AP-1 binding sites." *Science,* 239: 1150–1153 (1988).

Ellers, M., D. Picard, K. R. Yamamoto, and J. M. Bishop. "Chimeras of Myc oncoprotein and steroid receptors cause hormone-dependent transformation of cells." *Nature,* 340: 66–68 (1989).

Hempstead, B. L., D. Martin-Zanco, D. R. Kaplan, L. F. Parada, and M. V. Chao. "High-affinity NGF binding requires coexpression of the trk proto-oncogene and the low-affinity NGF receptor." *Nature,* 350: 678–683 (1991).

ONCOGENE COOPERATION

Land, H., L. F. Parada and R. A. Weinberg. "Cellular oncogenes and multistep carcinogenesis." *Science,* 222: 771–778 (1983). [Review]

Hunter, T. "Cooperation among oncogenes." *Cell,* 64: 249–270 (1991). [Review]

Rassoulzadegan, M., A. Cowie, A. Carr, N. Glaichenhaus, R. Kamen, and F. Cuzin. "The roles of individual polyomavirus early proteins in oncogenic transformation." *Nature,* 300: 713–718 (1982).

Land, H., L. F. Parada, and R. A. Weinberg. "Tumorigenic conversion of primary embryo fibroblasts requires at least two cooperating oncogenes." *Nature,* 304: 596–602 (1983).

Ruley, H. E. "Adenovirus early region 1A enables viral and cellular transforming genes to transform primary cells in culture." *Nature,* 304: 602–606 (1983).

ONCOGENES IN TRANSGENIC MICE

Cory, S. J., and J. M. Adams. "Transgenic mice and oncogenesis." *Annu. Rev. Immunol.,* 6: 25–48 (1988). [Review]

Hanahan, D. "Dissecting multistep tumorigenesis in transgenic mice." *Annu. Rev. Genet.,* 22: 479–521 (1988). [Review]

Brinster, R. L., H. Y. Chen, A. Messing, T. van Dyke, A. J. Levine, and R. D. Palmiter. "Transgenic mice harboring SV40 T-antigen genes develop characteristic brain tumors." *Cell,* 37: 367–379 (1984).

Stewart, T. A., P. K. Pattengale and P. Leder. "Spontaneous mammary adenocarcinomas in transgenic mice that carry and express MTV/myc fusion genes." *Cell,* 38: 627–637 (1984).

Hanahan, D. "Heritable formation of pancreatic beta cell tumors in transgenic mice expressing recombinant insulin/SV 40 oncogenes." *Nature,* 315: 115–122 (1985).

Sinn, E., W. Muller, P. Pattengate, I. Tepler, R. Wallace, and P. Leder. "Coexpression of MMTV/v-Ha-ras and MMTV/c-myc genes in transgenic mice: synergistic action of oncogenes in vivo." *Cell,* 49: 465–475 (1987).

TUMOR SUPPRESSOR GENES

Stanbridge, E. J. "Human tumor suppressor genes." *Annu. Rev. Genet.,* 24: 615–657 (1990). [Review]

Marshall, C. J "Tumor suppressor genes." *Cell,* 64: 313–326 (1991). [Review]

Stanbridge, E. J., C. J. Der, C. J. Doersen, R. Y. Nishimi, D. M. Peehl, B. E. Weissman, and J. Wilkinson. "Human cell hybrids: analysis of transformation and tumorigenicity." *Science,* 215: 252–259 (1982).

Pereira-Smith, O. M., and J. R. Smith. "Evidence for the recessive nature of cellular immortality." *Science,* 221: 964–966 (1983).

Weissman, B. E., P. J. Saxon, S. R. Pasquale, G. R. Jones, A. G. Geiser, and E. J. Stanbridge. "Introduction of a normal human chromosome 11 into a Wilms' tumor cell line controls its tumorigenic expression." *Science,* 236: 175–180 (1987).

RETINOBLASTOMA

Haber, D. A., and D. E. Housman. "Rate-limiting steps: the genetics of pediatric cancers." *Cell,* 64: 5–8 (1991) [Review]

Knudson, A. G. "Mutation and cancer: statistical study of retinoblastoma." *Proc. Natl. Acad. Sci. USA,* 68: 820–823 (1971).

Cavenee, W. K., M. F. Hansen, M. Nördenskjold, E. Kock, I. Maumenee, J. A. Squire, R. A. Phillips, and B. L. Gallie. "Genetic origin of mutations predisposing to retinoblastoma." *Science,* 228: 501–503 (1985).

Friend, S. H., R. Bernards, S. Rogelj, R. A. Weinberg, J. M. Rapaport, D. M. Albert, and T. P. Dryja. "A human DNA segment with properties of the gene that predisposes to retinoblastoma and osteosarcoma." *Nature,* 323: 643–646 (1986).

Huang, H.-J. S., J.-K. Lee, J.-Y. Shew, P.-L. Chen, R. Bookstein, T. Friedmann, E. Y.-H. P. Lee, and W.-H. Lee. "Suppression of the neoplastic phenotype by replacement of the RB gene in human cancer cells." *Science,* 242: 1563–1566 (1988).

Whyte, P., K. Buchkovich, J. M. Horowitz, S. H. Friend, M. Raybuck, R. A. Weinberg, and E. Harlow. "Association between an oncogene and an anti-oncogene: the adenovirus E1A proteins bind to the retinoblastoma gene product." *Nature,* 334: 124–129 (1988).

Whyte, P., N. M. Williamson, and E. Harlow. "Cellular targets for transformation by the adenovirus E1A proteins." *Cell,* 56: 67–75 (1989).

P53 GENE

Levine, A. J., J. Momand, and C. A. Finlay. "The p53 tumour suppressor gene." *Nature,* 351: 453–456 (1991). [Review]

Lane, D. P., and L. V. Crawford. "T-antigen is bound to host protein in SV40-transformed cells." *Nature,* 278: 261–263 (1979).

Parada, L. F., H. Land, R. A. Weinberg, D. Wolf, and V. Rotter. "Cooperation between gene encoding p53 tumour antigen and ras in cellular transformation." *Nature,* 312: 649–651 (1984).

Mowat, M. A., A. Cheng, N. Kimura, A. Bernstein, and S. Benchimol. "Rearrangements of the cellular p53 gene in erythroleukemic cells transformed by Friend virus." *Nature,* 314: 633–636 (1985).

Finlay, C. A., P. W. Hinds, and A. J. Levine. "The p53 proto-oncogene can act as a suppressor of transformation." *Cell,* 57: 1083–1093 (1989).

Nigro, J. M., S. J. Baker, A. C. Presinger, J. M. Jessup, R. Hostetter, K. Clearly, S. H. Bigner, N. Davidson, S. Baylin, P. Devilee, T. Glover, F. S. Collins, A. Weston, R. Modali, C. C. Harris, and B. Vogelstein. "Mutations in the p53 gene occur in diverse human tumour types." *Nature,* 342: 705–708 (1989).

Malkin, D., F. P. Li, L. C. Strong, J. F. Fraumeni, Jr., C.E. Nelson, D. H. Kim, J. Kassel, M. A. Gryka, F. Z. Bischoff, M. A. Tainsky, and S. H. Friend. "Germ line p53 mutations in a familial syndrome of breast cancer, sarcomas, and other neoplasms." *Science,* 250: 1233–1238 (1990).

MULTISTEP TUMORIGENESIS

Weinberg, R. A. "Oncogenes, anti-oncogenes, and the molecular bases of multistep carcinogenesis." *Cancer Res.,* 49: 3713–3721 (1989). [Review]

Fearon, E. R., and B. Vogelstein. "A genetic model for colorectal tumorigenesis." *Cell,* 61: 759–767 (1990). [Review]

Bishop, J. M. "Molecular themes in oncogenesis." *Cell,* 64: 235–248 (1991). [Review]

Baker, S. J., E. R. Fearon, J. M. Nigro, S. R. Hamilton, A. C. Preisinger, J. M. Jessup, P. van Tuinen, D. H. Ledbetter, D. F. Barker, Y. Nakamura, R. White, and B. Vogelstein. "Chromosome 17 deletion and p53 gene mutation in colorectal carcinomas." *Science,* 244: 217–221 (1989).

Fearon, E. R., K. R. Cho, J. M. Nigro, S. E. Kern, J. W. Simons, J. M. Ruppert, S. R. Hamilton, A. C. Preisinger, G. Thomas, K. W. Kinzler and B. Vogelstein. "Identification of a chromosome 18q gene that is altered in colorectal cancers." *Science,* 247: 49–56 (1990).

CHAPTER

19

Molecular Analysis of the Cell Cycle

The ability of one cell to become two is a fundamental property of living organisms, and so the mechanism by which cells divide has always fascinated biologists. Beyond its intrinsic interest as a basic feature of living things, cell division is at the root of many problems of both academic and practical importance. For instance, cancer is a problem of badly controlled cell division—cells divide persistently when they shouldn't divide at all. Development of an adult organism from a zygote requires a particular pattern of cell division, with some cell types and lineages dividing much more often than others, and with all divisions occurring at the right times.

The key to our understanding of the cell cycle was the cloning of cell cycle control genes. With our first look at these genes, it was apparent that cell cycle control was practiced virtually identically by all eukaryotic organisms, from humans to frogs to flies to clams to yeast. With this understanding, decades of data from these diverse organisms almost instantly coalesced to provide our first molecular view of the machinery that controls the cell cycle clock. This chapter describes how the cloning of cell cycle control genes catalyzed this dramatic consolidation of knowledge.

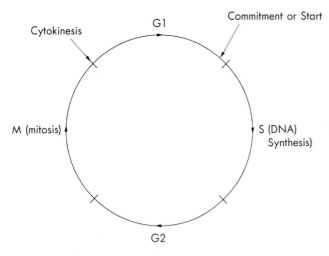

FIGURE **19-1**
The cell cycle. The cycle is traditionally divided into four phases, G1 (gap 1), S (DNA synthesis), G2 (gap 2), and M (mitosis). These phases were defined by light microscopy and DNA synthesis measurements. Other events are cytokinesis, the final separation of the two daughter cells, and commitment, or Start, an event that commits the cell to a full round of division.

There Are Four Stages in the Cell Cycle

Eukaryotes from yeast to humans have similar cell division cycles. Early studies using light microscopy defined four cell cycle stages: G1, S, G2, and M (Figure 19-1). *G1* stands for "gap 1," so called because it is a gap in the action—no special cell cycle events were seen. Most adult, differentiated, nondividing cells are in G1.

S stands for "synthesis" (of DNA). The cell's chromosomal DNA replicates during S phase. The replication is extremely accurate, in that the sequence of the new strands is exactly complementary to that of the template, and also in that every portion of the chromosomal DNA is copied exactly once, no more and no less. Some genes whose products are needed for DNA replication are highly expressed during S phase. For instance, synthesis of histone proteins is limited to S.

G2 stands for "gap 2," the second break in the visible action. At the boundary of S and G2, the replication machinery somehow signals that DNA replication is completed. The cell then enters G2 and prepares for mitosis. The nature of the signal is not known, but its existence has been inferred from the behavior of mu-

tant cells that enter mitosis before DNA synthesis has finished. This premature mitosis kills the cells.

M is "mitosis," the most visually spectacular part of the cell cycle (Figure 19-2). Chromosomes condense and become visible as discrete bodies. The two microtubule organizers, or spindle pole bodies, move apart to opposite sides of the nucleus. Arrays of microtubules grow from the two spindle pole bodies to form the mitotic spindle. Some of these microtubules become attached to the kinetochores of the chromosomes. The attached chromosomes become aligned on the "metaphase plate," the plane halfway between the spindle pole bodies. When all the kinetochores are attached to microtubules and are aligned on the metaphase plate, a signal is sent, and anaphase begins. The chromosomes start moving toward the poles, and the poles move apart from each other. Finally, the cell pinches into two (cytokinesis), producing two G1 phase daughter cells.

There Are Two Types of Control of Cell Division

Some of the processes mentioned above—DNA replication, formation of the mitotic spindle, movement of chromosomes—have been studied in detail, and there is a large body of knowledge about the mechanisms of cell division. However, this chapter is concerned with the *control* of cell division.

Controls on cell division are of two general kinds. First, there are controls that regulate cell metabolism and enlargement. An adult neuron, for instance, does not grow in mass or size and does not divide. In fact, most adult, differentiated cells in a human neither grow nor divide, making only enough protein to replace what is lost. The lack of net growth in size, mass, or protein content, despite an abundance of available nutrients, shows that cells have precise mechanisms for regulating their metabolism to prevent inappropriate enlargement.

Second, for growing cells, there are controls that coordinate enlargement and division. If cells doubled in mass and size slightly faster than they divided, they would become larger and larger with each cycle. If they doubled in mass and size more slowly than they divided, they would shrink. Since most cell types maintain the same average size over many generations, there must be some mechanism that ties division to

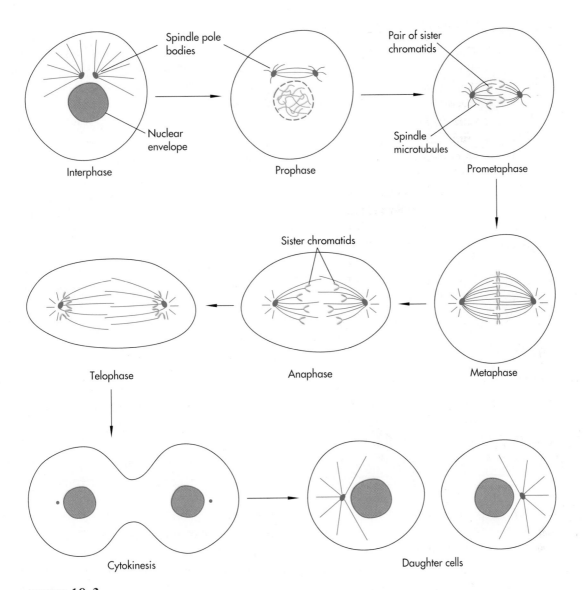

FIGURE **19-2**
Mitosis in animal cells. During G1, S, and G2, the nucleus is in interphase. Outside the
nucleus are two spindle pole bodies which will nucleate the microtubules forming the two
halves of the mitotic spindle. During interphase, these bodies nucleate cytoplasmic microtubules. At prophase the chromatin condenses into visible chromosome fibers; the nuclear envelope (membrane) starts to break down; and spindle microtubules nucleate around the pole
bodies, pushing the pole bodies to opposite sides of the nucleus. By prometaphase, the nuclear envelope has disappeared, and each pole body is organizing a set of microtubules. The
two sets of microtubules constitute the spindle. At this stage, each chromosome, which has
become completely condensed, consists of two duplicate DNA molecules called sister chromatids (before S, each chromosome is one DNA molecule and one chromatid; after S, each
chromosome is two molecules and two chromatids). Each chromatid has a centromere and an
associated structure called a *kinetochore* at which microtubules attach. Eventually each kinetochore is attached to many microtubules. At metaphase all the chromosomes are balanced
halfway between the poles in a plane called the *metaphase plate*. After all the chromosomes are
aligned, some signal causes the sister chromatids suddenly to separate. This is the beginning
of anaphase. The chromosomes (now single chromatids) move along the microtubule bundles
to the poles. At telophase, the chromosomes have moved to the poles, and the poles are
separated by their maximum distance. The nuclear envelope reforms. At cytokinesis, the cytoplasm pinches to give two separate cells. At some point, the spindle poles must be duplicated (not shown).

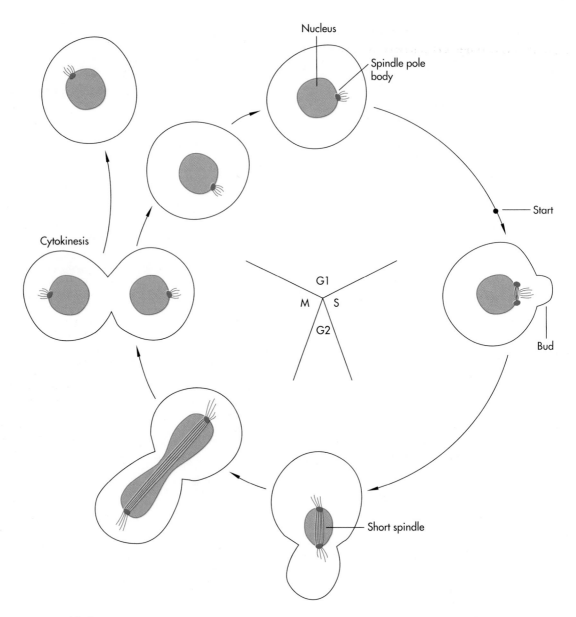

FIGURE **19-3**

The *S. cerevisiae* cell cycle. The shape of a cell shows its position in its division cycle. The bud emerges at the beginning of S phase and enlarges during G2 and M. Yeast mitosis differs from animal mitosis. Most strikingly, *S. cerevisiae* has a "closed" mitosis—the nuclear envelope never breaks down. The spindle microtubules are *inside* the nucleus, and the spindle pole bodies are embedded in nuclear envelope. Also, the mitotic spindle is present at an earlier stage than in most other organisms. Chromosomes do not become visible and apparently do not condense. Finally, cell division is by budding—the new daughter cell arises as an outgrowth from the mother. After Start, most cellular growth is directed into the bud.

growth. For most cells, this mechanism works at a control point in G1. When G1 phase cells have grown sufficiently, and have satisfied various other conditions, they become irreversibly committed to division. This "point of no return" is called the *commitment point* or, in the yeasts, *Start*. This is the main control point in the eukaryotic cell cycle. The mechanism of commitment, and the signals that cause commitment, are very active areas of research.

Yeasts Are Good Systems for Studying Cell Cycle Control

Since all eukaryotes appear to have similar cell cycles, it makes sense to study the cell cycle in the simplest possible experimental system that can be manipulated with the most powerful tools of analysis. The yeasts *Saccharomyces cerevisiae* (also known as *budding yeast, baker's yeast,* or *brewer's yeast*) and *Schizosaccharomyces pombe* (*fission yeast*) have proved excellent experimental systems for studying cell cycle control. The genetics of yeast is very well characterized, and there are highly developed systems for manipulating yeast genes with recombinant DNA (see Chapter 12). The power of yeast genetics allows easy selection for mutants with almost any phenotype, and since the corresponding wild-type genes can be readily cloned, altered in the test tube, and put back into yeast, the actual function of any particular gene can be rigorously tested in the living cell. Yeast go through a generation in just 2 hours and can be grown in large numbers for biochemical analysis.

Another advantage of yeast is that the position of an individual cell in the cell cycle can be deduced from the cell's morphology, that is, its shape and size. In fission yeast, the length of the cell is extremely well correlated with cell cycle position. In budding yeast (Figure 19-3), the appearance of a bud marks the beginning of S phase, and the size of the bud is correlated with progress through S, G2, and M. The morphology of the nucleus distinguishes G2 from M in both yeast types.

Much of the rest of this chapter deals with various genes and proteins discovered in yeasts. These genes and proteins are named according to the conventions shown in Table 1.

TABLE **19-1** Nomenclature for Yeast Genes and Proteins		
S. cerevisiae:	*CDC28*	Dominant wild-type gene
	cdc28-1	Recessive mutant gene (allele 1)
	Cdc28	The protein encoded by the *CDC28* gene
S. pombe:	*cdc2*$^+$	wild-type gene
	cdc2	Mutant gene
	Cdc2$^+$	The protein encoded by the *cdc2*$^+$ gene

Gene names are italicized. A yeast gene is named with three letters and a number. In addition, different alleles of the same gene may be named with a second number. For instance, *abc1-6* is the sixth known mutant allele of gene *ABC1.* For *Saccharomyces cerevisiae,* dominant alleles (usually the wild-type alleles) are named with uppercase letters. Recessive (usually mutant) alleles are named with lowercase letters. For *Schizosaccharomyces pombe,* all genes are named in lowercase letters. Wild-type alleles are distinguished with a superscript "+." Mutant alleles lack the "+" and may occasionally, for emphasis, have a superscript "−." There is no established method for naming the protein product of a yeast gene. In this chapter, the protein will be given the same name as the gene, but with only the first letter in uppercase, and without italicization.

Temperature-Sensitive Mutants Are a Valuable Tool in Studying the Cell Division Cycle

Mutant yeast completely lacking some cell cycle function such as DNA replication cannot divide and are inviable. Therefore, completely defective mutants cannot be isolated or studied. However, it is generally possible to isolate conditional mutants—that is, cells containing mutant genes that are functional under one condition and nonfunctional under another. The most common type of conditional mutant is the temperature-sensitive (ts) mutant. A temperature-sensitive mutant makes a protein that is functional at one temperature (for example, 23°C) but nonfunctional at another (for example, 37°C). The mutant cell dies at the higher temperature and so is said to be "temperature-sensitive."

(a)

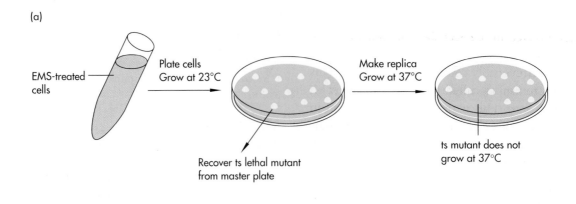

(b)

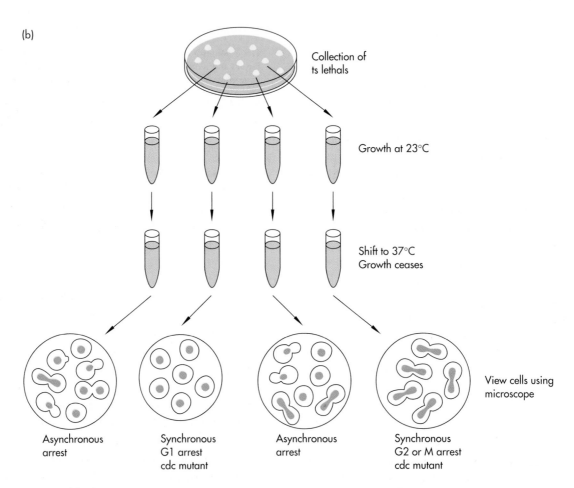

FIGURE **19-4**

Isolation of cdc mutants. (a) Cells were mutagenized with the DNA methylating agent ethyl methanesulfonate (EMS). Mutagenized cells were grown at 23°C and then spread on agar plates and grown into colonies at 23°C (the permissive temperature). These colonies were replica-plated onto a second set of plates, which were incubated at 37°C (the nonpermissive temperature). Colonies that failed to grow at 37°C were noted—such colonies were temperature-sensitive (ts) lethal mutants. Cells were recovered for study from the master plate kept at 23°C. (b) A large collection of ts lethal mutants was obtained and studied. Cdc mutants were defined as mutants for which 95 percent of the cells arrested with the same terminal phenotype—95 percent of the cells arrested unbudded, or 95 percent of the cells arrested with a single bud, etc. To find the cdc mutants, individual colonies of ts lethals were grown in liquid culture at 23°C, shifted to 37°C for a few hours, and examined by microscopy. For most mutants, individual cells arrested at random points in the cycle (asynchronous arrest), but for some mutants, all cells arrested at the same point of the cycle (synchronous arrest), the step mediated by the wild-type allele of their defective gene.

In the early 1970s, ts cell division cycle (cdc) mutants were isolated in *S. cerevisiae*. The mutants were found as depicted in Figure 19-4. Cells were grown at 23°C, then shifted to 37°C, and the morphology of the ts lethal mutants (dead cells) obtained in this screen was examined using a microscope. Most of the ts lethals gave a mixture of morphologies—the cells had died throughout the cell cycle. These were "ordinary" ts lethals—the mutant genes were probably not specifically involved in cell cycle control. However, some ts lethals died with a uniform morphology. For instance, *cdc35* mutants arrested as small, round, unbudded cells early in G1 phase; *cdc28* mutants arrested as large, unbudded cells late in G1 phase; and *cdc4* and *cdc7* cells arrested as budded cells at the beginning of S phase. The cells arrest uniformly because there is only one part of the cell cycle that they cannot pass through. The *cdc4* mutant, for instance, cannot initiate DNA synthesis, though it can replicate DNA perfectly well once initiation has occurred. When the temperature is shifted from 23°C to 37°C, a *cdc4* mutant cell will progress from whatever stage of the cell cycle it is in at the time, until it reaches the point at which it should initiate DNA synthesis. The cell will bud, and it will continue to grow in mass and size (since metabolism and enlargement are not specific to any cell cycle phase), but it will not be able to replicate its DNA, and it will not progress any farther in the cell cycle. Thus, all the cells in the culture will accumulate at this same cell cycle point.

Approximately 60 different cdc genes have been identified in *S. cerevisiae* (Figure 19-5), and a somewhat smaller number in *S. pombe*. Many of these genes have been cloned by complementation (see Chapter 12). For instance, to clone the *CDC28* gene from *S. cerevisiae*, random fragments of DNA from wild-type yeast were cloned into a plasmid vector capable of replicating in yeast. This library was used to transform a temperature-sensitive *cdc28* yeast strain at 23°C, the permissive temperature. Thousands of transformants were obtained; each contained a plasmid with a different piece of wild-type yeast DNA. These transformants were spread on plates and incubated at 37°C. Since the cells were *cdc28* mutants, most of them died. However, the transformants (about 1 in 5000) that carried the wild-type *CDC28* gene on a plasmid survived and grew into colonies. The plasmid carrying the *CDC28* gene was recovered from the surviving colonies. The DNA sequence of many *CDC* genes has been determined, and the functions of the genes have been examined.

The *CDC28* Gene Encodes a Ubiquitous Protein Kinase

A series of important discoveries arose from analysis of the *CDC28* gene. The original *cdc28* mutation caused arrest at Start, the G1 commitment point. Additional work showed that some alleles of *cdc28* could cause arrest at the boundary between G2 and M, as well as at Start. Thus, *CDC28* was unusual in that it was needed at two points in the cell cycle. One other such gene was known—the *cdc2*⁺ gene of *S. pombe*. Like *CDC28*, *cdc2*⁺ was required for Start and also for the G2-M transition. When the two genes were cloned and sequenced, it was found that they encoded extremely similar proteins. A comparison of the Cdc28 and cdc2⁺ protein sequences with other proteins in computer databases showed that the Cdc28 and Cdc2⁺ proteins were members of the protein kinase family. A protein kinase is an enzyme that can transfer a phosphate group from ATP onto another protein. The Cdc28 and Cdc2⁺ proteins were most closely related to protein kinases that transfer the phosphate to the hydroxyl group of a serine or threonine residue in a protein. As we learned in Chapters 17 and 18, phosphorylation often increases or decreases the activity of the phosphorylated protein; this is a very widespread method of regulating protein activity. Thus, protein kinases serve to regulate other proteins. An enzymatic assay for protein kinase activity (Figure 19-6) showed that the Cdc28 and the Cdc2⁺ proteins were indeed able to phosphorylate other proteins in vitro. Cloning of these genes therefore led to the first major insight into the underlying mechanisms of cell cycle control— protein phosphorylation was a critical event.

Although *S. cerevisiae* and *S. pombe* are both yeasts, they are only very distantly related. That a similar cell cycle gene was important in both yeasts suggested it might occur in other organisms as well. Was such a gene present in humans? In a spectacular application of recombinant DNA technology, the technique of complementation was used to search for the human gene. A human cDNA library was transformed into an *S. pombe* temperature-sensitive *cdc2* mutant. Transformants were found that were able to grow at 37°C, and the human cDNA insert in these cells was isolated and sequenced. The encoded human protein (Cdc2Hs; *Hs* for *Homo sapiens*) was almost identical with both Cdc2⁺ and Cdc28 and could function in place of either

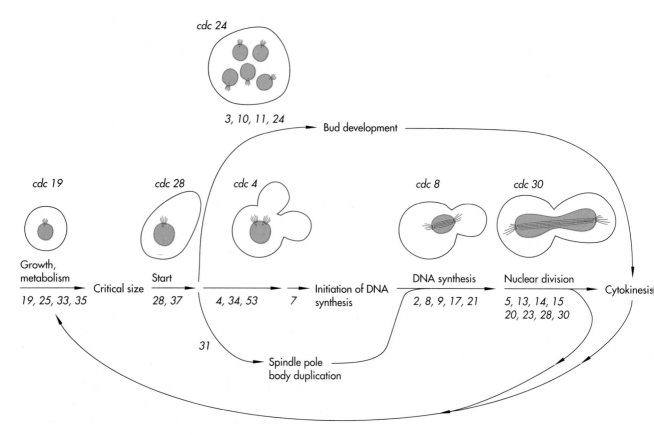

FIGURE **19-5**

A functional map of the *S. cerevisiae* cell cycle. The positions at which various cdc mutants—listed by number at the arrows—arrest are shown (for example, *cdc28* cells arrest at Start; *cdc17* mutants arrest in DNA synthesis), and the terminal arrest morphology of selected mutants is illustrated. A series of arrows that connect into one line shows a series of dependent events (for example, nuclear division depends on prior DNA synthesis and also on spindle pole body duplication). Arrows not connecting into a single line show independent events (for example, after Start has occurred, the pathways of bud development, DNA synthesis, and spindle pole body duplication are independent of each other). Only a small fraction of the known cdc mutations are shown here. Most of the corresponding cdc genes have been cloned and sequenced, and in some cases the function of the protein is known. For instance, *CDC19* encodes pyruvate kinase, *CDC35* encodes adenylate cyclase, *CDC28* and *CDC7* encode protein kinases, *CDC34* encodes a ubiquitin-conjugating enzyme involved in protein degradation, *CDC2* and *CDC17* encode different DNA polymerases, and *CDC9* encodes DNA ligase.

of them! This was the first time a human gene had been isolated by complementing a yeast mutation. We now know that every eukaryotic organism carries a gene closely related to *CDC28* and *cdc2*[+], and many organisms carry several such genes. Collectively, these genes are now referred to as the Cdc2 family, and in the rest of this chapter we will refer to the encoded proteins as Cdc2, regardless of their source.

The Protein Kinase Activity of the Cdc2 Protein Varies with the Cell Cycle

The availability of Cdc2 clones led not only to increased understanding of the structure and genetic function of the genes, but also to analysis of the Cdc2 proteins. Cloned genes were inserted into *E. coli*

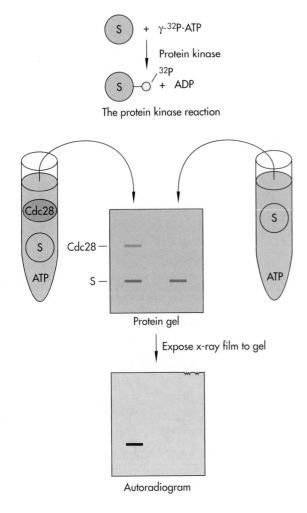

The protein kinase reaction

Protein gel

Expose x-ray film to gel

Autoradiogram

FIGURE **19-6**

Cdc28 and Cdc2$^+$ are protein kinases. Cdc28 protein was mixed with a substrate protein (S) and with γ-^{32}P-ATP. Typical substrate proteins are histone H1 and casein. In the control, the Cdc28 was omitted from the reaction. The reaction products were separated on an acrylamide gel. Since γ-^{32}P-ATP is a small molecule, it runs off the bottom of the gel. X-ray film was exposed to the gel. For the reaction mixture containing Cdc28, a band was observed on the x-ray film at the position of the substrate protein, showing that ^{32}P was transferred to the substrate by the Cdc28 protein kinase.

kinase activity changed. Antibodies were used to recover the Cdc2 protein from extracts made at different cell cycle stages, and the amount of kinase activity was measured by adding ^{32}P-ATP and histone H1 as substrates. Transfer of the isotopically labeled phosphate groups to H1 protein is an indication of protein kinase activity. In humans, *S. pombe*, *Xenopus* (a frog), and clams, a large amount of kinase activity was found in extracts from cells in mitosis, but little or no activity was found in cells in other portions of the cell cycle (Figure 19-7). Thus, it is the *activity* but not the *amount* of the protein that is regulated.

The Cdc2 protein has been analyzed by gel filtration chromatography, which separates proteins according to size. Gel filtration columns yielded at least two different fractions containing Cdc2. One fraction contained proteins of about 34 kDa, the size of the Cdc2 polypeptide. This fraction contained most of the Cdc2 protein, but had little or no protein kinase activity. The other fraction contained proteins of very high molecular weight and protein complexes. This fraction contained almost all the Cdc2 protein kinase activity, but only a small amount of the protein. This strongly suggested that the active form of Cdc2 is in a complex with some other protein or proteins.

MPF Contains Cdc2

In the frog *Xenopus laevis*, immature oocytes are arrested in the G2 phase of meiosis I, and have to go through a process of maturation to generate the mature egg (which is arrested in metaphase of meiosis II).

expression vectors, which provided a simple way to produce large amounts of the proteins. The proteins were injected into animals, and the animals mounted an immune response to them. The result was the production of highly specific antibodies that recognize and bind Cdc2 proteins, even in very crude cell lysates.

Using such antibodies to Cdc2, investigators could examine the behavior of the proteins in cells at different stages of the cell cycle. When extracted proteins were separated by gel electrophoresis and transferred to a nitrocellulose membrane, and the amount of Cdc2 protein was measured using an antibody (*Western blot* analysis), the same amount of this protein was found throughout the cell cycle (Figure 19-7). However, although the amount of protein was constant, its protein

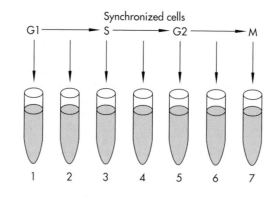

Synchronized cells

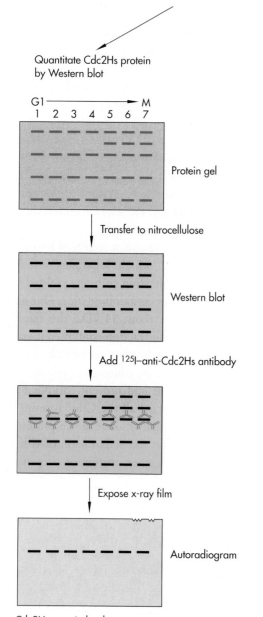

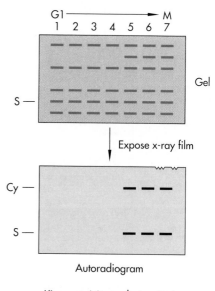

FIGURE 19-7

Cdc2Hs kinase activity oscillates; Cdc2Hs protein levels are constant. Human cells were synchronized at various points in the cell cycle, and extracts were made. The Cdc2Hs protein was precipitated with an anti-Cdc2Hs antibody. The immunoprecipitates were analyzed for the amount of Cdc2Hs (left panel) by Western blot analysis. The immunoprecipitated protein was run on an acrylamide gel and then transferred to a nitrocellulose membrane. The blot was probed with a radioactively labeled anti-Cdc2Hs antibody so that the Cdc2Hs band could be seen by exposing x-ray film to the Western blot. This showed that the amount of Cdc2Hs protein is constant all through the cycle. The immunoprecipitates were also analyzed for protein kinase activity (right panel) using the assay shown in Figure 19-6. Only the immunoprecipitates from mitotic cells were capable of phosphorylating the substrate (S), which in this case was histone H1. In addition, a second phosphorylated protein was seen (Cy). This protein was in the immunoprecipitate, and eventually proved to be a cyclin (Figure 19-11).

(a)

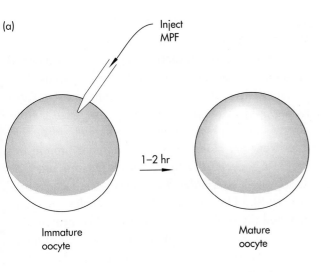

FIGURE **19-8**
(a) The maturation promoting factor (MPF) assay. When cytoplasm is taken from a mature oocyte and microinjected into an immature oocyte, it causes maturation. The activity responsible was called *maturation promoting factor,* or *MPF.* To assay MPF, test extracts are injected into immature oocytes; development of a white spot is a positive response. The white spot occurs because during maturation, the meiotic spindle apparatus migrates from a central cellular position to a position just under the surface of the oocyte, displacing the pigment granules normally found there. (b) Oocyte and embryonic development. The level of MPF activity [MPF] in cells from the time of the immature oocyte to the early embryonic cleavages is shown. MPF activity peaks with each meiotic or mitotic metaphase. In the mature oocyte (arrested in meiotic metaphase II), MPF activity is somehow stabilized in an active form by another activity called *cytostatic factor,* or *CSF.*

(b)

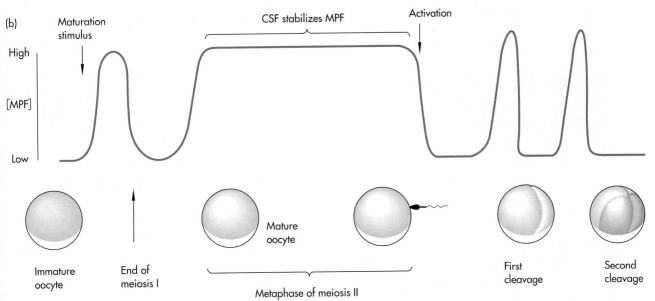

This maturation is normally induced by the hormone progesterone. However, in the early 1970s it was found that when cytoplasm from a mature egg was injected into immature oocytes, it caused meiosis and maturation (Figure 19-8). The activity responsible, which was clearly not progesterone, was called *maturation promoting factor* (*MPF*).

It was then found that mitotic cells from all tested eukaryotes contained an activity that would function as MPF in the frog oocyte assay. That is, extracts of mitotic cells from yeast, starfish, humans and other organisms all had MPF activity. Extracts of cells in G1 or S did not. Conversely, authentic MPF from mature frog eggs induced mitosis when injected into other cells. Thus, MPF appeared to be a ubiquitous eukaryotic inducer of meiosis and mitosis.

Learning how MPF works was impeded by the difficulty in obtaining for study large quantities of MPF, present in vanishingly small quantities in cells. After a long struggle, however, MPF was purified. It is a high-molecular-weight protein complex with protein kinase activity. Studies using anti-Cdc2 antibodies and Western blot analysis have shown that *Xenopus* Cdc2 is a major component of the MPF complex. Since this MPF component has been cloned and made available in essentially unlimited quantities, our understanding of MPF's role in cell cycle control has accelerated dramatically.

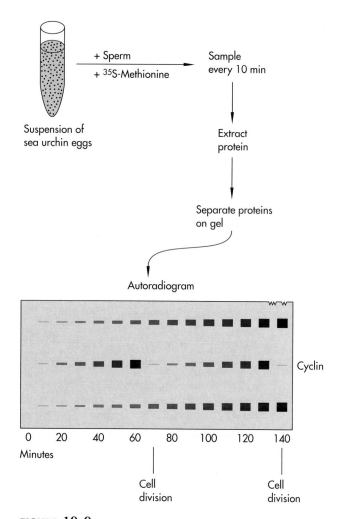

Suspension of
sea urchin eggs

+ Sperm
+ ^{35}S-Methionine

Sample
every 10 min

Extract
protein

Separate proteins
on gel

Autoradiogram

Cyclin

0 20 40 60 80 100 120 140
Minutes

Cell
division

Cell
division

FIGURE **19-9**
Identification of cyclin. Sea urchin eggs were mixed with
^{35}S-methionine and then fertilized synchronously. Samples
were taken at 10-minute intervals, extracts were run on an
SDS-polyacrylamide gel, and the gel was exposed to x-ray
film. The pattern of radioactively labeled protein reflects the
pattern of new protein synthesis. The vast majority of labeled proteins simply accumulate ^{35}S linearly with time as
shown in the top and bottom rows of bands. However, one
protein oscillated in abundance, with dramatic loss of
protein at every round of mitosis. This protein was named
cyclin.

Cyclins Are Cloned

While Cdc28 and Cdc2$^+$ were being studied in the
yeasts, and MPF was being studied in frogs, an apparently unrelated set of experiments was being done
in sea urchins. Large numbers of sea urchin eggs can
be isolated and fertilized simultaneously with sperm,
providing tens of thousands of cells undergoing synchronous cell cycles. These cells can be used to exam-

ine various biochemical properties of the cell cycle.

When the simultaneously dividing cells were grown
in the presence of ^{35}S-methionine, newly synthesized
proteins became radioactively labeled and could be
visualized by gel electrophoresis followed by autoradiography. Most proteins were synthesized continuously and simply accumulated more and more ^{35}S as
time went on. However, one prominent protein behaved in a cyclical fashion: it accumulated and then
disappeared, accumulated and disappeared again (Figure 19-9). The timing of these cycles precisely
matched the timing of cell division. The protein disappeared at each metaphase-anaphase transition in
mitosis. Because of this cyclical behavior, the protein
was named *cyclin*.

This unusual behavior and the abundance of cyclin
in fertilized eggs allowed cyclin cDNAs to be cloned.
With the availability of a cyclin cDNA, a decisive
experiment became possible. The cDNA was inserted
next to a bacteriophage promoter in a plasmid, and
from this plasmid large amounts of cyclin mRNA
were made in vitro (see Chapter 7). This mRNA was
injected directly into immature *Xenopus* oocytes (Figure 19-10). The injected cyclin mRNA forced the
oocytes through meiosis, precisely mimicking MPF.
Control injections of other mRNAs had no effect. So
cyclin alone appeared to be capable of driving the
oocytes through meiosis. It was already known that
injection of total clam or sea urchin oocyte mRNAs
(which included cyclin mRNAs) could trigger meiosis
in oocytes, but was cyclin mRNA also the active ingredient in these preparations? This question was answered by using antisense cyclin DNA and RNase H
to selectively deplete the naturally occurring cyclin
mRNA in these preparations (Figure 19-11). This procedure caused these preparations to completely lose
their ability to trigger cell division in the oocytes.
Thus, cyclin abundance in cells does not simply correlate with division, cyclin actually causes division.

Cyclin and Cdc2 Form
a Protein Complex

Cyclin genes from clam and sea urchin were sequenced, and the sequences were used to determine
the amino acid sequences of the encoded proteins.
These proteins were similar to each other, but not to

(a)

(b)   Mix total mRNA with antisense cyclin oligonucleotide

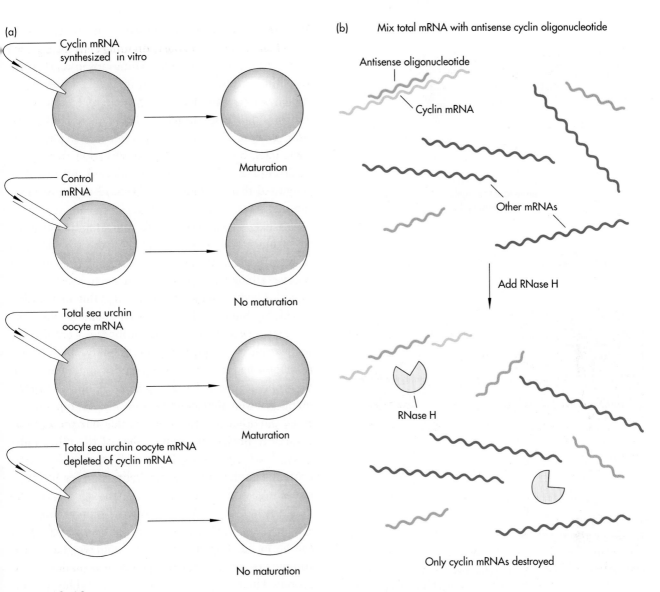

FIGURE **19-10**

Cyclin mRNA, like MPF, causes oocyte maturation. (a) Cyclin mRNA has MPF activity. A cyclin cDNA was cloned into a plasmid next to a bacteriophage promoter, and large quantities of the cyclin mRNA were made using the bacteriophage RNA polymerase. As a control, the cyclin cDNA was also cloned in backwards and an antisense transcript was made. Injection of oocytes with the sense transcript, but not with the antisense (control) transcript, caused oocyte maturation, exactly as MPF does. Injection of total sea urchin oocyte mRNA also caused maturation. When total sea urchin oocyte mRNA was specifically depleted of its cyclin mRNA (see part b), the remaining messages were not capable of inducing maturation. (b) Depletion of cyclin mRNA. Cyclin mRNA was specifically removed from total mRNA using an antisense cyclin DNA oligonucleotide and the enzyme RNase H. This enzyme can degrade an RNA strand that is part of an RNA–DNA hybrid duplex but not a single-stranded RNA. A synthetic DNA oligonucleotide complementary to part of the cyclin mRNA was mixed with and hybridized to total mRNA. When RNase H was added, it degraded the portion of the cyclin mRNA hybridized to the antisense oligonucleotide, destroying the function of the cyclin mRNA. Other mRNAs were not affected.

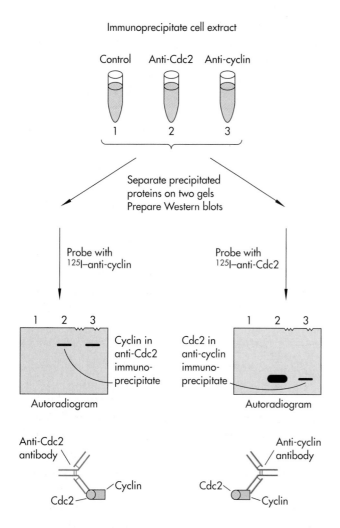

Immunoprecipitate cell extract

Control   Anti-Cdc2   Anti-cyclin

1   2   3

Separate precipitated
proteins on two gels
Prepare Western blots

Probe with
125I–anti-cyclin

Probe with
125I–anti-Cdc2

1 2 3

Cyclin in
anti-Cdc2
immuno-
precipitate

Autoradiogram

Cdc2 in
anti-cyclin
immuno-
precipitate

1 2 3

Autoradiogram

Anti-Cdc2
antibody

Cyclin

Cdc2

Anti-Cdc2 immunoprecipitate

Anti-cyclin
antibody

Cdc2

Cyclin

Anti-cyclin immunoprecipitate

FIGURE **19-11**
Cdc2 forms a complex with cyclin. Extracts were made
from developing clam embryos. Portions of the extract were
immunoprecipitated with an anti-Cdc2 antibody, or with an
anti-cyclin antibody, or with a control antibody. Half of
each immunoprecipitate was run on each of two gels, and
the resolved proteins on each gel were transferred to a nitrocellulose filter for Western blot analysis. One of the blots
was probed with an anti-cyclin antibody (left panel). This
detected cyclin in the lane where the anti-cyclin immunoprecipitate had been run, as expected; but in addition, cyclin
was seen in the lane containing the anti-Cdc2 precipitate.
The other Western blot was probed with an anti-Cdc2 antibody (right panel). In this case also, Cdc2 protein was seen
in both the anti-Cdc2 and the anti-cyclin lane. These experiments strongly suggest that Cdc2 and cyclin associate
with each other and form a reasonably stable complex, such
that when one of the proteins is directly recognized and
precipitated by specific antibody, the associated protein is
carried along in the precipitate as well. Since most cells
have an excess of Cdc2 protein (that is, they have
monomeric Cdc2 as well as Cdc2 in cyclin complexes),
the amount of Cdc2 protein immunoprecipitated by
anti-Cdc2 is relatively large (right panel).

any other proteins in the computer database. Shortly
afterward, however, the *S. pombe cdc13*+ gene was sequenced, and it was a cyclin. Mutations in *cdc13* cause
arrest in mitosis. Genetic experiments had previously
shown strong interactions between *cdc13* mutations
and *cdc2* mutations. Some alleles of *cdc13* could suppress certain alleles of *cdc2* and vice versa. These genetic data, together with the observation that cyclin
behaved like MPF when microinjected into oocytes,
suggested that the Cdc2+ and Cdc13 proteins might
interact physically.

Once again antibodies were central to demonstrating a physical association between the two proteins
(Figure 19-11). An anti-Cdc2 antibody was used to
recover all the Cdc2 protein from a clam extract. This
material had protein kinase activity. But in addition
to Cdc2 protein, it also contained cyclin. In the converse experiment, anti-cyclin antibodies were used to
recover all cyclin protein from the extract. This material also had protein kinase activity and, in addition
to cyclin, contained Cdc2 protein. Control experiments showed that neither antibody directly recognized the other protein. The fact that one protein was
invariably present in preparations of the other indicated that Cdc2 and cyclin existed together in a tight
protein complex. This Cdc2-cyclin complex had protein kinase activity.

Detailed biochemical studies now support this conclusion—a stable Cdc2-cyclin protein kinase complex
can be purified from a variety of sources. This large
complex is MPF. Free Cdc2 protein has no detectable
protein kinase activity without cyclin. This is why
injection of cyclin mRNA into oocytes mimics the
effect of MPF. The newly added cyclin complexes
with free Cdc2 in the oocytes, and active MPF is
created. A highly simplified model showing one way
in which Cdc2 and cyclin could lead to a cell cycle
oscillation is shown in Figure 19-12.

Cyclin Destruction Is Required to Inactivate the Kinase Activity

When an egg is fertilized, the first 10 to 15 rounds of
cell division are extraordinarily rapid. For instance,
embryonic *Drosophila* or *Xenopus* cells have a cell cycle
time of 10 to 20 minutes, compared with several hours
for adult cells. These rapid divisions are possible because the egg contains stockpiles of the ingredients

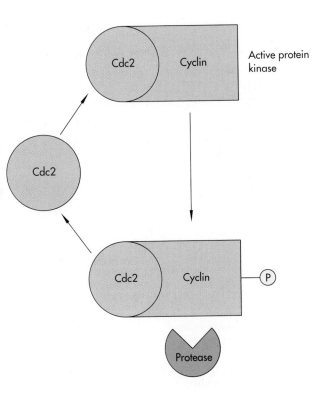

FIGURE **19-12**

A cyclin machine. A simple oscillator composed of Cdc2, cyclin, and a protease could be built as shown here. Initially, Cdc2 is present, but there is no cyclin, and no kinase activity. As the cell grows, cyclin is synthesized and accumulates. This cyclin binds to and activates Cdc2. The resulting kinase activity phosphorylates many substrates which push the cell into mitosis. The cyclin also gets phosphorylated, and this phosphorylation makes it a substrate for some protease. Finally, the protease destroys the phosphorylated cyclin, returning the machine to its initial state. This hypothetical model is presented only to show that simple oscillators capable of driving the cell cycle can, in theory, be constructed using types of molecules already known. This particular model may not be correct, and it certainly omits several known regulatory features. Such a model would require time delays in the transitions between states: otherwise, individual molecules would go through this cycle asynchronously, and there would be no net oscillation.

required for cell division. Eggs are therefore a very rich source of the proteins involved in cell division. In fact, extracts from *Xenopus* eggs can perform many cell cycle processes in vitro. When a demembranated nucleus is added to one of these "cycling" extracts, the DNA of the nucleus is replicated and chromosome condensation occurs. A mitotic spindle forms, and the two sets of chromosomes separate to opposite poles. Nuclear membranes form around the two sets of nuclei. The two nuclei will then go through DNA synthesis and divide again.

The ability of these extracts to produce cycling is blocked if the mRNA in them is destroyed by a very light treatment with ribonuclease, showing that new protein synthesis is required. However, if the mRNA is destroyed, and then cyclin mRNA is added back to the extracts along with RNase inhibitors, the extracts can cycle again. This suggests that cyclin is the only newly synthesized protein required. The cyclin in these extracts behaves much as it does in cells—it increases in abundance by continuous protein synthesis until the demembranated nuclei enter anaphase, and then it is destroyed. Also, the protein kinase activity of the Cdc2-cyclin complex increases to a peak at mitosis and then disappears when the cyclin is destroyed. Thus, these extraordinary extracts faithfully perform in a test tube the complex events of mitosis.

The availability of an extract that recapitulates mitosis provided an opportunity to use recombinant DNA techniques to analyze how cyclin works (Figure 19-13). Mutant cyclin mRNAs were synthesized in vitro using standard techniques. These synthetic cyclin mRNAs were added to extracts from which the endogenous cyclin mRNA was depleted. An mRNA encoding a cyclin mutant lacking the first 90 amino acids of the protein had a remarkable effect. Addition of the mRNA for this cyclin caused the extract to proceed normally to mitosis. However, at metaphase, the mutant cyclin was not destroyed, kinase activity did not decline, and the extracts stalled in metaphase. This result suggested that the amino-terminal region of the cyclin molecule contains a signal for destruction of the protein. If the signal is missing, the cyclin cannot be destroyed. And if the cyclin is not destroyed, then kinase activity does not decline, and exit from metaphase cannot take place. Thus, cyclins control two points in mitosis: their presence is needed for mitosis to begin, and their absence is needed for mitosis to end.

A Different Set of Cyclins Regulates Cdc28 in G1 Phase

Cdc28 of *S. cerevisiae* and Cdc2$^+$ of *S. pombe* are needed at two points in the cell cycle—Start and mitosis. Both yeasts have a set of cyclins that act at mitosis. However, these cyclins do not seem to be needed for Cdc28 or Cdc2$^+$ to act at Start, just as the cyclins identified in other organisms seem to be required only at mitosis.

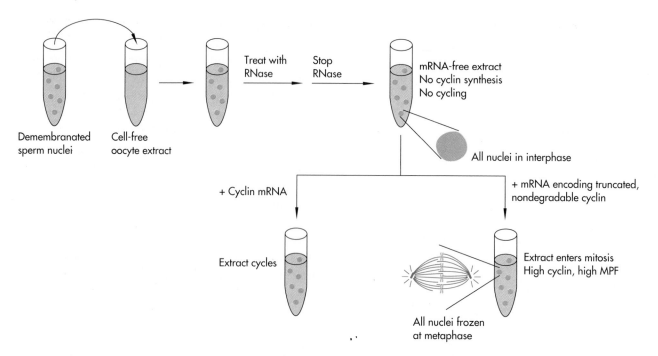

FIGURE **19-13**

Cyclin synthesis and destruction control mitosis. An oocyte extract that goes through multiple rounds of the cell cycle can be made. To see if cyclin was needed for cycling, the mRNA in the extracts was destroyed by a very light treatment with RNase. The RNase action was then blocked with an inhibitor, RNasin. (The extent of the RNase treatment is very important: too much RNase destroys tRNAs and ribosomes and thus prevents any future protein synthesis.) When all mRNAs were destroyed in this way, the extracts did not cycle. No cyclin was present; MPF activity was very low, and added nuclei stayed in interphase. Next, cyclin mRNA (made in vitro using the cloned gene and a bacteriophage RNA polymerase) was added to the extract. Cyclin protein was synthesized, and the extract began to cycle. This shows that cyclin is the *only* newly synthesized protein required for cell cycle progress. When, instead, a truncated mRNA encoding a cyclin missing the first 90 amino acids of the wild-type protein was added to the mRNA-depleted extracts, a remarkable result was obtained. The extracts synthesized the truncated cyclin and developed MPF and kinase activity; and nuclei went to a metaphase morphology, as with the normal cyclin mRNA. However, the cycle ceased at that point. It became clear that the truncated cyclin could not be proteolytically degraded and that in the continued presence of the indestructible cyclin, the extract remained frozen at metaphase.

What, then, are Cdc28 and Cdc2$^+$ doing in the G1 phase at Start?

In *S. cerevisiae*, genetic experiments have identified three genes called *CLN1, CLN2,* and *CLN3* that help regulate Start. When an extra copy of any of these genes is put into a yeast cell, the cell goes through Start at a smaller than normal size, suggesting that the normal relationship between growth and division has been changed. If all three genes are mutated, the cells are unable to go through Start at all and die in G1 phase. Sequencing of these three genes shows that they are distantly related to the mitotic cyclins, but they seem to act only at Start and not at mitosis. Since they work in G1 phase, they have been called G1 cyclins. It appears that the Cln1, Cln2, and Cln3 proteins associate with the Cdc28 protein and help regulate its G1 activity, just as mitotic cyclins regulate the G2-M activity.

The finding of G1 cyclins in yeast sparked a hunt for G1 cyclins in mammalian cells. Since the principal control point for mammalian cell growth is in G1, G1 cyclins might be a common target for disorders of cell

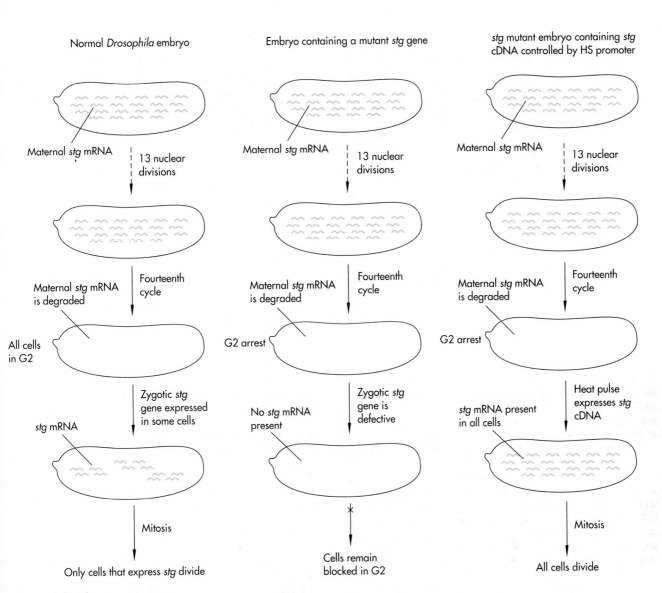

FIGURE **19-14**
Activation of the *string* (*stg*) gene triggers mitosis in *Drosophila* embryos. In a fly embryo, the first 13 rounds of nuclear division proceed synchronously and quickly, driven by products deposited in the egg by the mother. One of these maternally supplied products is *stg* mRNA. When an *stg* probe was hybridized to embryos fixed at different stages of development, *stg* mRNA was found in the embryos through the first 13 cycles but disappeared as the cells reached the G2 phase of cycle 14. In normal embryos, the embryos' own *stg* genes then switch on in discrete patches of cells, which divide soon thereafter (this occurs in a precise and identical pattern in all embryos). In embryos containing mutant *stg* genes, functional *stg* RNA and protein are not made, so no further cell divisions occur and the embryo dies. To test whether turning on the *stg* gene is sufficient to trigger mitosis in cycle 14, mutant embryos were engineered to carry an *stg* cDNA expressed under the control of a heat-shock (HS) promoter. When the engineered embryos reached cycle 14, they stopped in G2 as expected. At this point, the embryos were heated to 37°C for 25 minutes, which switches on the heat-shock promoter. The *stg* mRNA quickly appeared in all cells of the embryo, and over the next 15 minutes, all entered mitosis. This experiment established that *stg* gene expression is a mitotic trigger in fly embryos. This is consistent with the known function of the closely related *cdc25⁺* gene in *S. pombe*. Related genes have now been isolated from the human genome.

growth such as cancer. Very quickly, candidate G1 cyclin genes were isolated from mammalian cells. Several groups used yeast complementation to isolate human cDNAs that could substitute for the yeast G1 cyclins; sequencing of these clones showed that they were indeed new cyclins. But cyclins started turning up elsewhere, too. One group was studying cDNAs for genes activated several hours after quiescent cells are treated with growth factors; among these genes was a new cyclin. Yet another lab cloned a gene situated near a chromosomal translocation found in certain human tumors; it, too, was a cyclin. In fact, the new cyclin genes isolated by these three unrelated approaches were the same.

Whether or not these new genes will prove to be authentic G1 cyclins in mammalian cells, these experiments show that there is a sizeable family of cyclin genes in the mammalian genome. Together with the yeast work, the picture that is emerging is of a squadron of different Cdc2-cyclin complexes deployed to control critical events throughout the cell cycle.

Other Cdc Proteins Are Also Involved in Regulating Cdc2 Kinase Activity

Other cdc genes are involved in the regulation of the Cdc2-cyclin complex. For instance, the $cdc25^+$ gene of *S. pombe* is required for mitosis; mutants lacking $cdc25^+$ do not divide. The $cdc25^+$ gene behaves genetically as an activator of $cdc2^+$. Indeed, $cdc25$ mutants have a Cdc2$^+$-cyclin complex, but the complex has no protein kinase activity. A dramatic demonstration of the importance of $cdc25^+$ in the control of mitosis came unexpectedly from cloning experiments in *Drosophila*. In the fly mutant called *string*, mutant embryos underwent normally the 13 rapid nuclear divisions that begin embryogenesis (see Chapter 21), but they proceeded no further, stalling in the G2 phase of the fourteenth cycle. Cloning and sequencing of the *string* gene revealed that it was closely related in sequence to $cdc25^+$, suggesting that it might be involved in the control of cell division in the embryo.

The role of *string* as a mitotic trigger in *Drosophila* embryogenesis became clear with further experiments. Embryos moving through the first 13 cycles were found to contain abundant *string* mRNA uniformly distributed throughout the embryo. This mRNA was deposited into the egg by the mother. After the thir-

teenth cycle, the maternal *string* mRNA abruptly disappeared. In normal embryos, *string* mRNA soon reappeared, but instead of appearing uniformly in all cells of the embryo, *string* mRNA appeared in stages in an intricate and reproducible spatial pattern. This pattern precisely anticipated the pattern of mitosis in the embryo: 20 minutes after a group of cells turned on their *string* genes, they divided. In mutant embryos, *string* mRNA is never produced, and the cells never divide again, despite an abundance of cyclins. This strong correlation between *string* expression and subsequent mitosis suggested that *string* triggers mitosis. This hypothesis could be easily tested in flies (Figure 19-14). A synthetic gene was constructed in which the *string* cDNA was placed under the control of a *heat-shock* promoter, which is a promoter activated by an increase in temperature. Using gene transfer methods detailed in Chapter 20, this switchable *string* gene was introduced into living flies carrying a chromosomal *string* mutation. When mutant embryos were hatched, they proceeded normally to the fourteenth cycle and stalled. When the embryo was heated to activate the heat-shock promoter, *string* mRNA appeared in all cells and all cells divided. Thus, turning on the *string* gene is sufficient to drive these embryonic cells to mitosis. The prediction that this gene functions as a mitotic trigger was fulfilled.

Recombinant DNA Provides a Common Language for Different Experimental Systems

Before the application of recombinant DNA techniques to the study of cell cycle control, researchers in different systems—yeast, sea urchins, frogs, and mammals—had little reason to think their discoveries had anything in common. Cloning of the *CDC28* and $cdc2^+$ genes quickly revealed that at the molecular level cell cycle control was a very highly conserved mechanism in all eukaryotic cells. Facts learned from yeast genetics, for example, could be rapidly applied to biochemical studies of the mammalian cell cycle. From the different experiments possible in different organisms arose an explosion of knowledge of the precise molecular controls of the cell cycle. Studies of the cell cycle will continue to accelerate as new genes are rapidly cloned and their structures and functions determined.

Reading List

Original Research Papers

THE CELL CYCLE

Nurse, P. "Genetic control of cell size at cell division in yeast." *Nature*, 256: 547–551 (1975). [Review]

Pringle, J. R., and L. H. Hartwell. "The *Saccharomyces cerevisiae* Cell Cycle." In J. N. Strathern, E. W. Jones, and J. R. Broach, eds., *The Molecular Biology of the Yeast* Saccharomyces: *Life Style and Inheritance.* Cold Spring Harbor Laboratory, Cold Spring Harbor, N. Y., 1981, pp. 97–142. [Review]

Nurse, P. "Universal control mechanism regulating onset of M-Phase." *Nature*, 344: 503–508 (1990). [Review]

ISOLATION OF cdc MUTANTS

Hartwell, L. H., R. K. Mortimer, J. Culotti, and M. Culotti. "Genetic control of the cell cycle in yeast: V. Genetic analysis of cdc mutants." *Genetics*, 74: 267–286 (1973).

Nurse, P., and Y. Bisset. "Gene required for G1 for commitment to cell cycle and in G2 for control of mitosis in fission yeast." *Nature*, 292: 558–560 (1981).

CDC28 AND *cdc2*+ GENES

Lorincz, A. T., and S. I. Reed. "Primary structure homology between the product of yeast cell division control gene *CDC28* and vertebrate oncogenes." *Nature*, 307: 183–185 (1984).

Lee, M. G., and P. Nurse. "Complementation used to clone a human homologue of the fission yeast cell cycle control gene *cdc2*." *Nature*, 327: 31–35 (1987).

Dunphy, W. G., L. Brizuela, D. Beach, and J. Newport. "The *Xenopus* cdc2 protein is a component of MPF, a cytoplasmic regulator of mitosis." *Cell*, 54: 423–431 (1988).

Gautier, J., C. Norbury, M. Lohka, P. Nurse, and J. Maller. "Purified maturation-promoting factor contains the product of a *Xenopus* homolog of the fission yeast cell cycle control gene *cdc2*+." *Cell*, 54: 433–439 (1988).

ISOLATION OF GENES ENCODING CYCLIN

Evans, T., E. T. Rosenthal, J. Youngblom, D. Distel, and T. Hunt. "Cyclin: a protein specified by maternal mRNA in sea urchin eggs that is destroyed at each cleavage division." *Cell*, 33: 389–396 (1983).

Swenson, K. I., K. M. Farrell, and J. V. Ruderman. "The clam embryo protein Cyclin A induces entry into M Phase and the resumption of meiosis in *Xenopus* oocytes." *Cell*, 47: 861–870 (1986).

Pines, J., and Hunt T. "Molecular cloning and characterization of the mRNA for cyclin from sea urchin eggs." *EMBO J.*, 10: 2987–2995 (1987).

Nash, R., G. Tokiwa, S. Anand, K. Erickson, and A. B. Futcher. "The *WHI1*+ gene of *Saccharomyces cerevisiae* tethers cell division to cell size and is a cyclin homolog." *EMBO J.*, 7: 4335–4346 (1988).

Hadwiger, J. A., C. Wittenberg, H. E. Richardson, M. D. Lopes, and S. I. Reed. "A family of cyclin homologs that control the G1 phase in yeast." *Proc. Natl. Acad. Sci. USA*, 86: 6255–6259 (1989).

Richardson, H. E., C. Wittenberg, F. Cross, and S. I. Reed. "An essential G1 function for cyclin-like proteins in yeast." *Cell*, 59: 1127–1133 (1989).

Wittenberg, C., K. Sugimoto, and S. I. Reed. "G1-specific cyclins of S. cerevisiae: cell cycle periodicity, regulation by mating pheromone, and association with the p34CDC28 protein kinase." *Cell*, 62: 225–237 (1990).

Matsushime, H., M. F. Roussel, R. A. Ashmun, and C. J. Sherr. "Colony-stimulating factor 1 regulates novel cyclins during the G1 phase of the cell cycle." *Cell*, 65: 701–713 (1991).

Motokura, T., T. Bloom, H. G. Kim, H. Juppner, J. V. Ruderman, H. M. Kronenberg, and A. Arnold. "A novel cyclin encoded by a *bcl1*-linked candidate oncogene." *Nature*, 350: 512–515 (1991).

Xiong, Y., T. Connolly, B. Futcher, and D. Beach. "Human D-type cyclin." *Cell*, 65: 691–699 (1991).

THE Cdc2+-CYCLIN COMPLEX

Booher, R., and D. Beach. "Interaction between *cdc13*+ and *cdc2*+ in the control of mitosis in fission yeast: dissociation of the G1 and G2 roles of the *cdc2*+ protein kinase." *EMBO J.*, 6: 3441–3447 (1987).

Draetta, G., and D. Beach. "Activation of cdc2 protein kinase during mitosis in human cells: cell cycle-dependent phosphorylation and subunit rearrangement." *Cell*, 54: 17–26 (1988).

Booher, R. N., C. E. Alfa, J. S. Hyams, and D. H. Beach. "The fission yeast cdc2/cdc13/suc1 protein kinase: regulation of catalytic activity and nuclear localization." *Cell*, 58: 485–497 (1989).

Draetta, G., F. Luca, J. Westendorf, L. Brizuela, J. Ruderman, and D. Beach. "cdc2 protein kinase is complexed with both Cyclin A and B: evidence for proteolytic inactivation of MPF." *Cell*, 56: 829–838 (1989).

Murray, A. W., and M. W. Kirschner. "Cyclin synthesis drives the early embryonic cell cycle." *Nature,* 339: 275–280 (1989).

Murray, A. W., M. J. Solomon, and M. W. Kirschner. "The role of cyclin synthesis and degradation in the control of maturation promoting factor activity." *Nature,* 339: 280–286 (1989).

cdc25 AND *STRING*

Russell, P., and P. Nurse. "*cdc25*$^+$ functions as an inducer in the mitotic control of fission yeast." *Cell,* 45: 145–153 (1986).

Edgar, B. A., and P. H. O'Farrell. "Genetic control of cell division patterns in the *Drosophila* embryo." *Cell,* 57: 177–187 (1989).

Edgar, B. A., and P. H. O'Farrell. "The three postblastoderm cell cycles of *Drosophila* embryogenesis are regulated in G2 by *string.*" *Cell,* 62: 469–480 (1990).

Moreno, S., P. Nurse, and P. Russell. "Regulation of mitosis by cyclic accumulation of p80^{cdc25} mitotic inducer in fission yeast." *Nature,* 344: 549–552 (1990).

Sadhu, K., S. I. Reed, H. Richardson, and P. Russell. "Human homolog of fission yeast cdc25 mitotic inducer is predominantly expressed in G2." *Proc. Natl. Acad. Sci. USA,* 87: 5139–5143 (1990).

20

Genes That Control the Development of *Drosophila*

F ew phenomena fascinate more than embryonic development—the intricate and precise program by which a large and complex animal arises from a single fertilized egg cell. How does the embryo mark and measure space and time, so that organs and tissues develop on schedule and in the right locations? How are the three dimensions of space and one of time encoded in the one-dimensional DNA molecule? These are among the most profound questions in biology.

Studying development in mammals is difficult, because it occurs tucked out of sight in the uterus, it spans weeks or months, and genetic analysis is slow and complicated. Thus, many developmental biologists have focused their effort on model systems in which embryonic development is easier to study. The hope has been that the principles uncovered in the study of organisms like fish, flies, frogs, and worms would apply to mammalian development as well. And as the molecular basis of development has emerged from these systems, this hope has been realized—many of the same molecules and some of the same mechanisms underlie development in all animals.

In this chapter we will focus on embryonic development in *Drosophila*. We choose *Drosophila* because of the profound ways in which recombinant DNA experiments with *Drosophila* have fleshed out principles and ideas established by

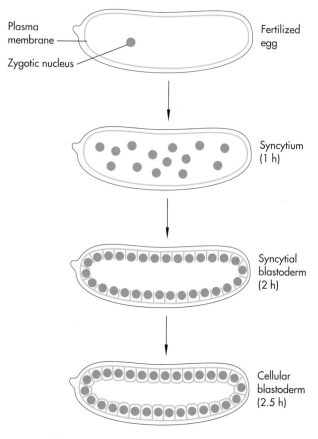

Plasma membrane

Zygotic nucleus

Fertilized egg

Syncytium (1 h)

Syncytial blastoderm (2 h)

Cellular blastoderm (2.5 h)

FIGURE **20-1**
Early *Drosophila* embryogenesis. Fertilization begins with the penetration of sperm into egg and the fusion of the male and female pronuclei to form the zygotic nucleus. The nucleus undergoes a series of rapid divisions in the shared cytoplasm of the embryo (the *syncytium*). After 10 divisions, the nuclei migrate to the periphery of the embryo, forming the *syncytial blastoderm*. After three more divisions, membranes form around each of the nuclei, forming the *cellular blastoderm*. Beginning with the next cycle, cell division becomes asynchronous.

decades of classical genetics and embryology. Moreover, *Drosophila* developmental genes have led us directly to genes involved in mammalian development. Although many aspects of *Drosophila* development have been elucidated through elegant combinations of classical and molecular experiments, we will confine our discussion to just one aspect of early embryonic development—how the early embryo is endowed with the patterning information that eventually directs the construction of body parts to their correct positions in the animal.

We will learn that very early in development, each embryonic cell achieves a unique identity that deter-

mines its subsequent developmental fate. This subdivision of the early embryo was not apparent under the light microscope; only when molecular probes became available was it discovered that the morphologically identical cells in these early embryos were already expressing distinct combinations of genes. The development of these unique gene expression patterns is due to the action of a hierarchy of regulatory genes, many of which encode transcription factors.

Serendipity and Systematic Genetic Screens Identify Genes That Control Development

The life cycle of *Drosophila* consists of embryogenesis, three larval stages, a pupal stage, and the adult stage. The total development period from egg to adult lasts about 10 days. After fertilization there are 13 rapid nuclear divisions in a single shared cytoplasm. The nuclei are divided into cells by membranes that grow in from the surface of the embryo, giving rise to the *blastoderm* stage embryo, which is composed of about 6000 identical-looking cells surrounding a large mass of yolk in the interior (Figure 20-1). After a series of cell movements to form the tissue layers and organs, a larva hatches at about 22 hours. The larva persists for about 4 days, before forming the pupa. Metamorphosis, the most magical step of insect development, occurs within the pupa. Most of the larval tissues, except the brain and a few others, are destroyed and used to fuel the construction of new adult organs. Inside the larva, cells that were set aside at an earlier stage have grown into small lumps of tissue called *imaginal discs*. There are discs for each leg and wing, for eyes and antennae, for mouthparts and genitalia. Thus, the larva is nearly a distinct organism from the adult fly, serving to store and nourish the cached cells for adult organs. After several days as a pupa, the completed adult fly emerges.

The study of *Drosophila* genetics dates back nearly a century. The earliest developmental mutants were isolated as flies with obvious derangements in the organization of their bodies. Some of these mutations, termed *homeotic,* led to dramatic transformations of one body part into another. For example, a dominant homeotic mutation in a gene called *Antennapedia* (*Antp*) caused legs to sprout from the head in place of an-

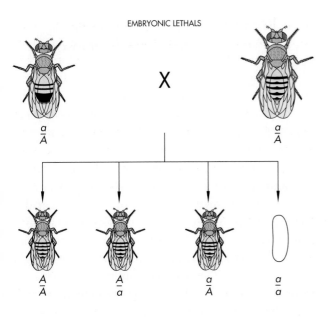

EMBRYONIC LETHALS

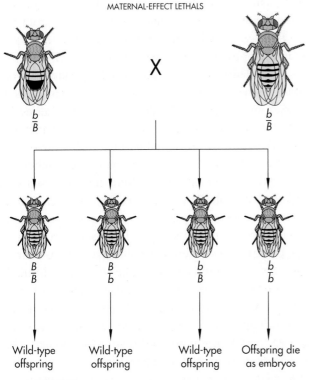

MATERNAL-EFFECT LETHALS

Wild-type offspring   Wild-type offspring   Wild-type offspring   Offspring die as embryos

FIGURE **20-2**
Genetic screen for embryonic lethals. Flies are treated with a mutagen, leading to the development of heterozygous mutants (*a/A*). Heterozygotes are bred; 3 of 4 offspring develop normally, but 1/4, which inherit both mutant alleles, die as embryos. Inspection of the terminal phenotype of the embryos allows classification of the mutation. In flies carrying maternal-effect mutations, all offspring of the first mating develop normally, but 1/4 of the females of this generation produce dead embryos. These females carry homozygous mutations in a maternal-effect gene.

tennae. Most mutations in developmental control genes, however, do not permit flies to survive to adulthood. Thus, a series of systematic genetic screens was performed to identify *embryonic lethal* mutations, which in their homozygous state lead to death of the embryo (Figure 20-2). Visual inspection of the terminal embryos and classical genetic experiments divided these embryonic lethal mutations into several classes (Figure 20-3).

The first class are the *maternal-effect* genes, so called because the phenotype of the embryo reflects the

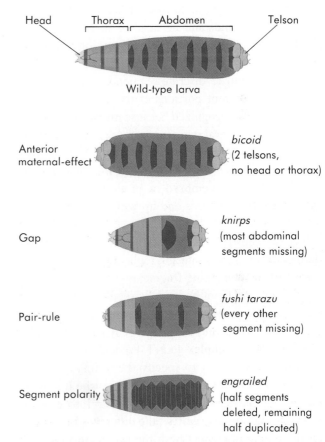

FIGURE **20-3**
Phenotypes of developmental mutants. Shown at the top is a normal *Drosophila* larva, with the wild-type segmentation pattern of head, thorax, abdomen, and telson. Shown below are the phenotypes of selected mutants. In *bicoïd* mutants, anterior-posterior polarity is lost, resulting in the formation of larvae with two telsons and no head or thorax. In gap mutants, several segments are lost; in *knirps* mutants, for example, most of the abdominal segments are missing. In pair-rule mutants, such as *fushi tarazu*, every other segment is missing. In segment polarity mutants, such as *engrailed*, half of every segment is missing and replaced with a mirror-image of the remaining half.

genotype of the mother rather than the embryo. Mothers carrying homozygous maternal-effect mutations are themselves normal but produce mutant embryos. These genes have this unusual property because they encode products that are deposited into the egg by the mother. If the mother is unable to provide these products because she is a mutant, the embryos display the mutant phenotype. As we will learn shortly, maternal-effect genes are responsible for setting up the initial *polarity* of the embryos, that is, which end is the front and which is the back, which the top and which the bottom. Mutations occurring in the maternal-effect gene *bicoid (bcd)*, for example, result in embryos with no front-to-back polarity: they have two rear ends and no head or thorax.

The second class of genes is responsible for establishing *segmentation* of the embryo, the subdivision of the embryo into 15 discrete segments destined to develop into different portions of the adult body. The *gap* genes were recognized because mutations resulted in the loss of several segments of the embryo, forming a large gap in the normal segmentation pattern. Mutations in the *pair-rule* genes led to loss of segment-sized pieces of the embryo in an alternating pattern, so that mutant embryos end up with half the number of segments. Mutations in the *segment polarity* genes affect the substructure of each segment, usually deleting part of the segment and replacing it with a mirror image of the remaining fragment.

The third class of embryonic regulatory genes are the homeotic genes we have already mentioned. These genes, which fall into two large gene complexes, the Antennapedia complex (ANT-C) and the bithorax complex (BX-C), act to specify the ultimate fate for each segment. This is why mutations in the homeotic genes sometimes transform one body part into another.

As we will discuss shortly, the different phenotypes of these mutants reflect both the timing and location of their expression during early embryonic development. They act together as a genetic cascade to generate progressively finer subdivisions in the developing embryo. The interactions among these genes are complex. Many were inferred from classical embryology and genetics, but our molecular view of how these genes act has sharpened considerably as these genes have yielded, reluctantly at first, to the efforts of molecular cloners.

The First *Drosophila* Development Genes Were Cloned by Chromosome Walking

Biologists are quick to exploit the peculiarities of their experimental organisms. One odd but invaluable feature of flies is the production of *polytene chromosomes.* Found in the salivary glands of the fly, the polytene chromosomes are actually about 1000 copies of each chromosome aligned in precise register, so that differences in chromosome condensation (the relative concentrations of DNA and protein in the chromatin) are easily visible under the light microscope as light and dark bands. The banding pattern is reproducible from fly to fly, and each band has been identified with a number. Mutations that result from gross chromosomal rearrangements, such as inversions, translocations, and large deletions, are simple to spot in the polytene chromosomes, and if a mutant allele under study is associated with a rearrangement of this type, then its approximate position on the chromosome can be easily deduced from examining the polytene chromosomes in mutant flies. In addition, when cloned probes are available, they can be hybridized directly to polytene chromosomes to determine the band position where the corresponding gene resides.

The polytene chromosomes gave researchers a foothold in their efforts to clone the genes of the bithorax complex by chromosome walking. They knew from examining the polytene chromosomes in many bithorax mutants that deletions and inversions that affected BX-C were located in band 89E on chromosome 3, but they had no probes that hybridized to 89E to begin their walk. Instead, they jumped into BX-C from a walk in search of the *rosy* eye-color gene at 87E (Figure 20-4). They knew of a mutant fly strain that carried a chromosome inversion that joined 89E to 87E, so they hunted for a clone from their *rosy* walk that spanned the inversion site. With this clone in hand, they constructed and screened a library from the strain carrying the inversion and isolated a clone that carried the 89E-87E joint fragment. Now they had a probe in 89E and the BX-C walk could begin. The *rosy*/BX-C chromosome walk was the first ever, and it established both the important principles and the methods of this widely used technique.

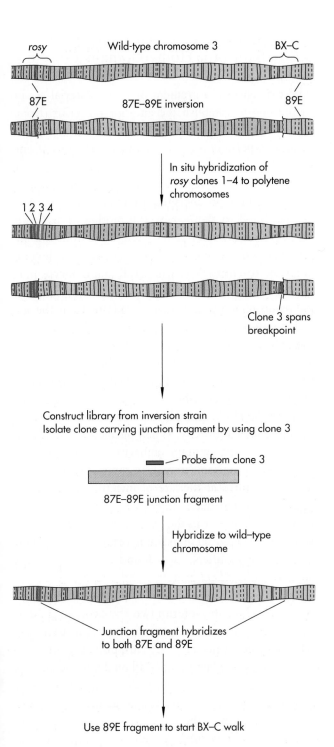

FIGURE **20-4**

Jumping from *rosy* to the bithorax (BX-C) complex. Shown at the top are sections of polytene chromosome 3 from a wild-type fly and a mutant in which the section between 87E and 89E has been inverted. To locate clones from 87E that are immediately adjacent to 89E in the inversion chromosome, investigators hybridized DNA from several clones, labeled 1–4, to the polytene chromosomes. On the normal chromosome, these clones hybridize in linear order. On the inversion chromosome, however, clones 1 and 2 hybridized to their normal positions, but clone 4 hybridized beyond the inversion breakpoint, and clone 3 hybridized to sequences on both sides of the break, suggesting that it physically spanned the breakpoint. A library was constructed with DNA taken from the inversion strain and screened with a clone 3 probe, allowing investigators to retrieve one of the junction fragments from the inverted chromosome. This fragment should contain sequences from both 87E and 89E, which was confirmed by hybridizing it to normal chromosome 3. The 89E fragment was used to initiate the bithorax walk.

known genes of BX-C span more than 300 kb, though only 10–20 kb are found in the various mRNAs produced from these genes (Figure 20-5). The remaining DNA appears to have a regulatory function, and many homeotic mutations affect not the coding sequences but rather the large introns and intergenic regions of the complex. Presumably this regulatory DNA is required to integrate the positional information that in turn dictates where in the embryo each of the homeotic genes is expressed. We discuss below how this information may be communicated to the homeotic gene complexes.

As genomic fragments were isolated from the homeotic gene complexes, they were tested for hybridization to embryonic mRNAs, and soon cDNAs for individual genes within the BX-C and ANT-C were isolated. When the first cDNAs were compared, it was found that they shared a 180-bp region of similar sequence, which was originally termed the *homeo box*. We now know that the homeo box encodes a conserved protein domain, the *homeodomain,* related to the helix-turn-helix DNA-binding domain of bacterial repressors. This observation suggested, and biochemical studies later confirmed, that the homeotic genes encode DNA-binding proteins that act as transcription factors.

When a homeo box probe was hybridized to a Southern blot of *Drosophila* genomic DNA, a large number of fragments hybridized, suggesting that the

The Homeodomain Is Shared among *Drosophila* Developmental Genes

In all, researchers cloned approximately 400 kb from BX-C and 300 kb from ANT-C. These complex loci each contain several protein-coding genes spread out over an unusually large amount of DNA. The three

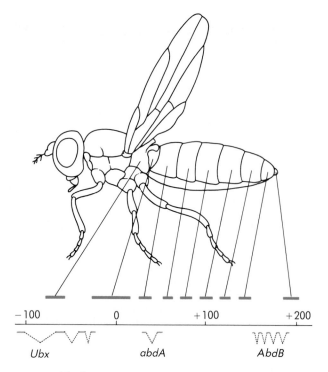

FIGURE 20-5

Map of the bithorax complex. The continuous line represents genomic DNA cloned from BX-C. The "0" indicates the position from which the bidirectional walk was initiated. The numbers indicate length in kilobase pairs. Marked in orange are the three protein-coding genes in the complex, with a somewhat simplified view of the location and splicing patterns of the exons (these genes give rise to several different mRNAs). The blue bars indicate genetically defined regulatory regions that lead to homeotic transformations of discrete fly segments, because of their effects on the expression of the protein-coding genes of the complex. Remarkably, there is a precise linear correlation between these regulatory regions and the segments of the fly they affect.

Drosophila genome carried other homeodomain genes. These genes were quickly isolated using homeo box probes, and in these collections of clones were found several genetically defined (but not yet cloned) genes. These included not just homeotic genes, but segmentation and maternal-effect genes as well. The discovery of this conserved motif probably eliminated hundreds of man-years of effort that would have been required to clone all these genes by chromosome walking.

Amazingly, the homeo box probe also hybridized to genomic DNA from other animals, including *Xenopus* (frogs), chickens, mice, and humans. And quickly these genes, too, were isolated. The genes were remarkably similar in sequence to the *Drosophila* genes. The homeodomain of one of the mouse genes differs in only one amino acid from the *Antp* gene of *Drosophila*. At the end of this chapter, we will return to the mammalian developmental genes that were isolated because of their similarity to the *Drosophila* genes.

Anterior-Posterior Polarity Arises from a Gradient of a Homeodomain Protein

The existence of a gradient of some material set up in the egg to establish the anterior-posterior (front-to-back) polarity of the embryo had been inferred from genetic experiments and direct manipulations of embryos. For example, several maternal-effect mutations result in embryos that have lost their polarity, suggesting that the products required to establish polarity are provided to the egg by the mother. This could also be shown by removing cytoplasm from the anterior pole of an embryo: the embryo did not develop anterior structures. You will recall that embryos born of a mother homozygous for a *bicoid* mutation have the same phenotype. When cytoplasm from the anterior pole of a wild-type embryo was injected into the anterior pole of a *bicoid* embryo, formation of the anterior structures was restored. These observations suggested that the *bicoid* gene product functions as a *morphogen* that establishes the polarity of the embryo.

With the cloning of the *bicoid* gene, this hypothesis could be tested directly. When purified *bicoid* mRNA was injected into mutant embryos, it acted just like wild-type cytoplasm, triggering the formation of anterior structures at the site of injection. This observation suggested that *bicoid* RNA or protein was the morphogen present in the cytoplasm. Indeed, when a *bicoid* probe was used in in situ hybridizations to whole embryos, *bicoid* mRNA was found exclusively at the anterior pole. *Bicoid* protein, synthesized at the pole, diffuses toward the other end of the embryo, forming a gradient along the anterior two-thirds of the embryo. Sequencing of the *bicoid* gene showed that it encodes a homeodomain protein and, thus, probably functions as a transcription factor. Since all nuclei in the embryo reside in a shared cytoplasm, the bicoid gradient results in a gradient of the transcription factor in the nuclei along the length of the embryo.

How are *bicoid* mRNAs (and similarly polarized mRNAs at the other end of the embryo) confined to limited locations in the free cytoplasm of the embryo? The mechanism remains unknown, but for *bicoid* deletion analysis indicates that correct localization requires a sequence in the 3′ untranslated region of the *bicoid* mRNA. Additional maternal-effect genes that have effects on embryonic development similar to those of *bicoid* may be involved in this localization, because in these mutants, in situ hybridization indicates that *bicoid* mRNA is not correctly localized. Intriguingly, the protein product of one of the genes required for correct development of posterior structures resembles in sequence RNA-binding proteins and may therefore act to bind and retain mRNAs in the posterior pole. Continued analysis of the genes responsible for localizing maternal factors to the anterior and posterior of the embryo will eventually reveal the mechanisms responsible.

Gap Genes Define the First Subdivisions of the Embryo

The embryo, born with one signal emanating from the anterior pole and another, which we have not discussed, from the posterior, undergoes 11 rounds of nuclear division fueled by maternal stores of mRNA and protein before the first zygotic genes are activated. Among these are the gap genes, whose role is to interpret the crude positional information provided by the maternal gradients and transmit that information to the next set of genes in the segmentation gene heirarchy. Three of the principal gap genes, *hunchback* (*hb*), *Krüppel* (*Kr*), and *knirps* (*kni*) have been extensively studied. Cloning of these genes showed that they also encode transcription factors (of the zinc finger family), suggesting that they transmit signaling information to the segmentation genes by directly modulating their transcription. *In situ* hybridization of the gap gene clones to embryos revealed that they were expressed in broad bands across the embryos, the first sign of segmentation in the embryo.

How do the gap genes resolve the two unidirectional gradients into these bands? The simplest idea is that their transcription is directly regulated by the maternal factors. Expression of *hunchback*, which is primarily seen in the anterior half of the embryo, depends genetically on *bicoid*. In *bicoid* mutants, there is no expression of *hunchback* in the anterior of the embryo. In these same mutants, the band of *Krüppel* expression, normally confined to the middle of the embryo, spreads toward the anterior (Figure 20-6). This suggests that *Krüppel* expression is repressed by *hunchback*, *bicoid*, or both. One view of how the boundaries between the domains of gap gene expression become established is by mutual repression among the gap genes. Thus, it is a combination of regulation by the

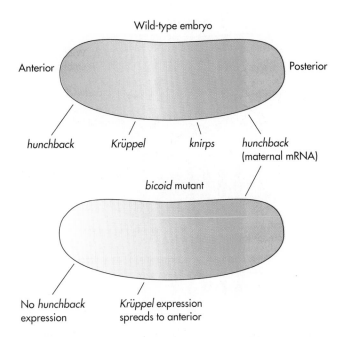

FIGURE **20-6**
Expression domains of the gap genes in the embryo. Expression of specific genes in the embryo can be visualized either by in situ hybridization, in which RNA probes are hybridized to mRNA in embryos fixed on microscope slides, or by immunocytochemistry, in which the protein products of the genes are detected with specific antibodies. These methods give similar but not identical results. The three principal gap genes are expressed in broad bands across the embryo. The *hunchback* protein is found in a gradient extending from the anterior pole into the middle of the embryo. *Krüppel* protein is found in a central band, and *knirps* immediately posterior to *Krüppel*. A domain of *hunchback* expression is also found at the posterior pole, but this is due to the deposit of maternal *hunchback* mRNA into the oocyte, whereas the others result from zygotic gene expression. Note that the domains of expression overlap. In *bicoid* mutants, anterior expression of *hunchback* is lost, because *bicoid* is required to activate *hunchback* expression, and the domain of *Krüppel* expression spreads toward the anterior, because it is normally repressed by *hunchback* and *bicoid*.

maternal factors and interactions among the gap genes themselves that resolves the anterior-posterior gradient into bands of zygotic transcription factors.

Segmentation Genes Divide the Embryo into Stripes

The next genes in the cascade of events that resolves the embryo into segments are the pair-rule genes. With the cloning of *fushi tarazu* (*ftz*), a homeodomain-con-

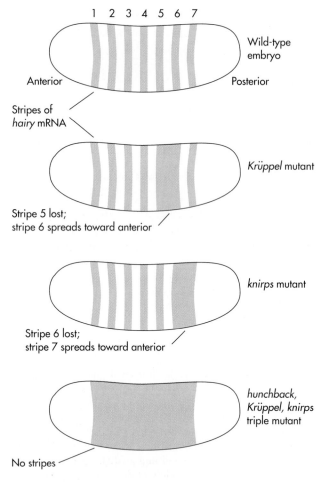

1 2 3 4 5 6 7

Wild-type embryo

Anterior — Posterior

Stripes of
hairy mRNA

Krüppel mutant

Stripe 5 lost;
stripe 6 spreads toward anterior

knirps mutant

Stripe 6 lost;
stripe 7 spreads toward anterior

hunchback,
Krüppel, knirps
triple mutant

No stripes

FIGURE 20-7

Stripes of expression of the *hairy* gene. In normal embryos, *hairy* mRNA and protein are found in seven stripes. This pattern is built up a stripe at a time by complex interactions among the gap gene proteins. This is seen by examining *hairy* stripes in different gap gene mutants. In *Krüppel* mutants, there is an apparent fusion of stripes 5 and 6; in *knirps* mutants, apparent fusion of 6 and 7. This is a simplified view: there are other stripe abnormalities in these mutants. In a triple mutant lacking the *hunchback*, *Krüppel*, and *knirps* genes, all periodicity of *hairy* expression is lost, and expression is uniform through the central portion of the embryo.

taining pair-rule gene located in the ANT-C, the first stunning in situ hybridization pictures showing embryos with zebralike stripes appeared. The *fushi tarazu* mRNA is found in seven sharp stripes, each four cells wide, encircling the embryo. This striped pattern of expression is shared by all the pair-rule genes, although their boundaries differ, and it is a dramatic demonstration that the apparently uniform-looking embryo has already developed precise positional information.

Just as the broad bands of gap gene expression grew out of a simple maternal gradient by interactions with the maternal factors and among those of the gap genes themselves, the stripes appear as a result of the combined actions of the gap gene proteins and interactions among the pair-rule genes themselves. This principle was established by examining the stripes of pair-rule gene expression in mutants lacking gap or pair-rule genes (Figure 20-7). In gap gene mutants, for example, the stripes of expression of the pair-rule genes are changed. Thus, the gap genes must be responsible for establishing pair-rule gene stripes. In addition, one pair-rule gene, *hairy*, is expressed in seven stripes that are precisely complementary to the stripes formed by another pair-rule gene, *runt*. This mutually exclusive interaction is apparently due to the mutual repression by the two genes. When one gene is eliminated by mutation, expression of the other spreads to the previously silent interstripe regions.

Soon after the pair-rule genes resolve into their unique stripe patterns, the embryo cellularizes—that is, cell walls form around the nuclei. This locks in each cell's particular complement of pair-rule, gap, and maternal-effect proteins and shuts out any further effects of cytoplasmic gradients of transcription factors. At this point, different cells in the embryo express different combinations of pair-rule genes. It is thought that this information is sufficient to specify each cell's developmental fate through the action of the homeotic genes. The sequential pathway through which this positional information is established is summarized in Figure 20-8.

The *bicoid* Protein Directly Regulates the Transcription of *hunchback*

Thus far, we have discussed primarily genetic experiments that have begun to unravel the regulatory cascade of genes that divides the *Drosophila* embryo into segments. But interpretation of these experiments would have been difficult without the molecular probes that let researchers see with their own eyes the effects mutations in one gene had on the expression pattern of others. Recall that before the availability of these probes, there was no way even to know that there were periodic patterns of gene expression in the developing embryo. As in other areas of research, how-

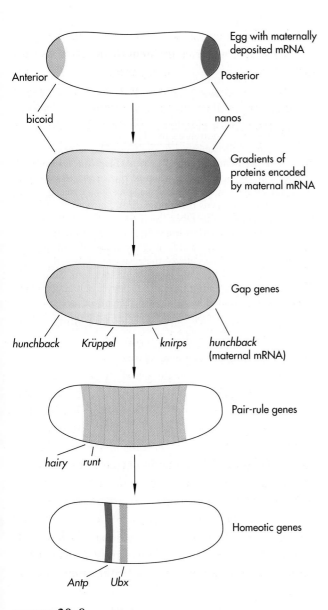

Anterior

Posterior

Egg with maternally deposited mRNA

bicoid

nanos

Gradients of proteins encoded by maternal mRNA

Gap genes

hunchback   Krüppel   knirps   hunchback (maternal mRNA)

Pair-rule genes

hairy   runt

Homeotic genes

Antp   Ubx

FIGURE **20-8**
A summary of the hierarchy of positional gene expression in early embryogenesis. The egg contains tightly localized maternal mRNAs for *bicoid* and *nanos*, at the anterior and posterior poles, respectively. Translation of these mRNAs, coupled with diffusion of the proteins, leads to stable gradients of the proteins along the length of the embryo. These proteins act on the gap genes, which also interact among themselves, to develop broad, overlapping bands of gap gene proteins. These proteins, in turn, regulate expression of the pair-rule genes, which are activated in stripes. Here we show the pattern of expression for two of these genes, *hairy* and *runt*, which form alternating stripes. Two other pair-rule genes, *even-skipped* and *fushi tarazu*, also form alternating stripes (not shown), but they are displaced relative to the *hairy-runt* stripes. Thus, every cell along the embryo is expressing a different combination of the pair-rule genes. All these gene products act to specify the expression of the homeotic genes in particular bands that eventually give rise to the visible segments of the adult fly.

ever, the real power granted to researchers by recombinant DNA is the ability to manipulate the structure of a gene and return the modified gene to the animal to see what happens.

Drosophila researchers have a remarkable tool at their disposal to get genes into flies—the P element. The P element, as we learned in Chapter 10, is a transposable element that jumps from one location to another in the *Drosophila* genome. P element–based vectors have been engineered to make it easy for researchers to jump the gene they choose into the fly genome and generate pure strains of flies that carry and express this gene as any other (Figure 20-9). P elements are used for examining the effects of genes on specific processes in the fly, they are used for studying the regulation and organization of promoters and enhancers, and they have been used as enhancer traps to locate new genes with interesting patterns of expression (we discuss how enhancer traps work in Figure 10-14 and present a specific example in Chapter 21).

Investigators studying the regulation of the gap gene *hunchback* by the maternal morphogen *bicoid* have made extensive use of P element–mediated transformation. We have mentioned that genetic evidence suggested that *hunchback* expression was controlled by *bicoid*. Researchers constructed a series of fusion genes in which the bacterial *lacZ* gene was placed under the control of various fragments of the regulatory DNA from the *hunchback* gene. They transformed these fusion genes into flies using P elements and examined the pattern of β-galactosidase expression in embryos produced by these flies. They found that some of the fusion genes directed β-galactosidase expression to the anterior portion of the embryo, whereas others did not (Figure 20-10a). These experiments identified a short element of about 100 bp in the *hunchback* promoter that was sufficient to confer *bicoid*-dependent anterior expression in the embryo.

Meanwhile, investigators had used the cloned *bicoid* gene to produce the *bicoid* protein in *E. coli*, and they used this protein in DNA-binding experiments with fragments of the *hunchback* promoter. They mapped several high-affinity and low-affinity binding sites for the *bicoid* protein in the *hunchback* promoter. When they compared the ability of the high-affinity and low-affinity sites to direct β-galactosidase expression in embryos, they found that the high-affinity sites caused β-galactosidase expression to occur from the anterior

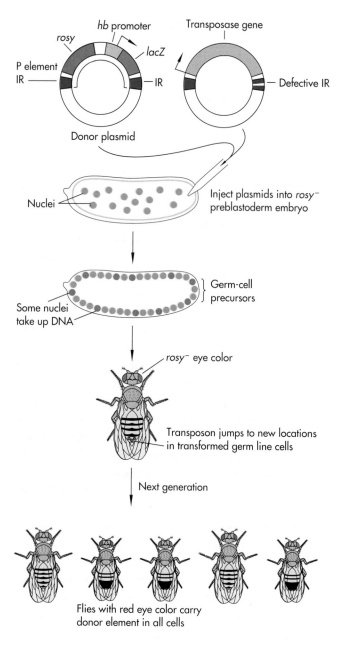

P element IR

rosy

hb promoter

lacZ

IR

Donor plasmid

Transposase gene

Defective IR

Inject plasmids into *rosy⁻* preblastoderm embryo

Nuclei

Some nuclei take up DNA

Germ-cell precursors

rosy⁻ eye color

Transposon jumps to new locations in transformed germ line cells

Next generation

Flies with red eye color carry donor element in all cells

FIGURE 20-9

The use of P element transformation to study gene expression in embryos. One plasmid, the donor, carries the reporter gene to be studied, here the *hunchback* (*hb*) promoter fused to the *E. coli lacZ* gene, flanked by the inverted terminal repeats (IR) of the P element transposon. This element also contains the *rosy* eye-color gene, which allows transformed flies to be spotted easily. The bracket indicates the sequence that will transpose in the flies. The second plasmid carries a P element derivative that produces the transposase enzyme but, owing to inactivation of one of the IRs, cannot itself transpose. The mixture of plasmids is injected into preblastoderm embryos from *rosy⁻* flies. Some of the embryonic nuclei take up the DNA, including cells at the posterior pole destined to form the germ line cells of the adult fly. P elements transpose only in germ line cells. Thus, in transformed germ line cells of the resulting adults, the donor element jumps to new chromosome locations. Some of the offspring of this adult, therefore, carry a stable genomic P element insertion. These flies are recognized because normal red eye color is restored by the *rosy* gene on the P element. The transformed flies in turn pass the P element on to their embryos, where the pattern of β-galactosidase expression during development can then be examined.

Expression in Stripes Is Encoded by Discrete Cis-Acting Regulatory Elements

We have already mentioned that the development of the correct stripes of expression of the pair-rule genes requires appropriate expression of the gap genes. In reciprocal experiments, investigators examined the structure of mutant alleles of the *hairy* gene that had unusual phenotypes. Most *hairy* mutations resulted in loss of alternate segments (the phenotype of pair-rule mutations) along the whole length of the embryo. These unusual mutants, however, lost only one or a few segments. These alleles contained deletions in putative regulatory sequences upstream of the gene, suggesting that individual regulatory elements direct expression to individual stripes. Are these elements where the products of the gap genes act?

To sort these issues out, investigators inserted individual DNA fragments from the *hairy* upstream region into a *lacZ* vector, as done with the *hunchback* promoter, and put these constructions into flies by P element–mediated transformation. They found that

end more deeply toward the posterior end of the embryo than did the low-affinity sites (Figure 20-10b). This observation shows how regulatory DNA sequences can direct the expression of a gene to a specific position along a gradient. Low-affinity sites are bound by the *bicoid* protein and transcription is activated only where the protein concentration is high, whereas high-affinity sites can function at the lower *bicoid* protein concentrations found farther down the gradient, towards the posterior end of the embryo.

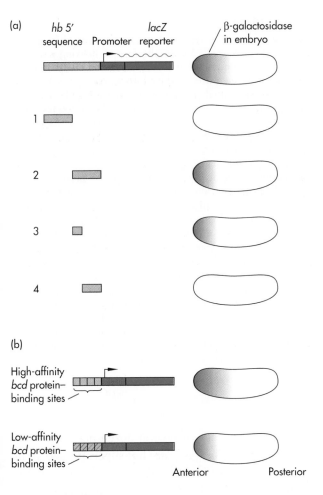

(a)

hb 5' sequence Promoter lacZ reporter   β-galactosidase in embryo

1

2

3

4

(b)

High-affinity bcd protein–binding sites

Low-affinity bcd protein–binding sites

Anterior   Posterior

FIGURE **20-10**

Showing that regulatory elements in the *hunchback* (*hb*) gene direct expression to anterior regions of the embryo. (a) A fusion gene was constructed, consisting of 5' regulatory sequences from the *hunchback* gene, a simple promoter lacking regulatory elements of its own, and the *E. coli lacZ* gene as a reporter. This fusion gene was transformed into embryos as described in Figure 20-9. Staining of the embryos with antibodies against β-galactosidase showed that the protein was expressed in the anterior portion of the embryo in a pattern similar to that seen for *hunchback* itself. By testing individual fragments of the *hunchback* sequence in the same *lacZ* reporter, a small fragment, fragment 3, was found to carry sequences sufficient to direct β-galactosidase expression to the anterior of the embryo. The role of *bicoid* binding sites in this regulation was investigated in the experiment shown in (b). The *lacZ* fusion gene was placed under the control of either four high-affinity binding sites or four low-affinity sites for the *bicoid* protein from fragment 3 of the *hunchback* gene. The gene with the high-affinity binding sites directed β-galactosidase expression over a wider range (that is, from anterior regions into more posterior regions of the embryo) than did the gene with the low-affinity sites. This is because the high-affinity sites are able to bind *bicoid* protein at the lower protein concentrations found farther down the gradient.

different fragments directed β-galactosidase expression to different stripes (Figure 20-11a). For example, one fragment led to β-galactosidase expression only in stripe 7 and another in stripe 6. To determine how the gap genes influence expression in stripe 6, they transferred the reporter gene with the fragment that produced only stripe 6 to embryos mutant for *Krüppel* and *knirps* (Figure 20-11b). When this reporter gene was inserted in a *knirps* mutant, stripe 6 disappeared, suggesting that *knirps* is normally an activator of stripe 6. In a *Krüppel* mutant, stripe 6 expanded, suggesting that *Krüppel* normally represses stripe 6. And in embryos engineered to express *Krüppel* protein throughout the embryo, no stripe 6 formed at all, again indicating that *Krüppel* represses this stripe.

Experiments with *Krüppel* and *knirps* proteins produced in *E. coli* show that both proteins bind to sequences within the stripe 6 element of the *hairy* gene. If the *Krüppel* protein is a repressor and *knirps* an activator, then stripe 6 expression should occur only when *Krüppel* protein levels are too low to repress it but *knirps* levels are sufficiently high to activate it. This is precisely the situation in the vicinity of stripe 6, where the concentration of the *knirps* protein is high and *Krüppel* is falling off sharply (Figure 20-11c). Just as different regulatory sites in *hunchback* can sense different concentrations of the *bicoid* protein, the affinities of the gap protein binding sites in the stripe elements of *hairy* determine the position along the anterior-posterior axis where the conditions are appropriate for expression.

The *fushi tarazu* pair-rule gene behaves differently from *hairy*, responding not to the direct action of the gap genes but instead to the periodic signal established earlier by *hairy* and two other pair-rule genes, *runt* and *eve*. When various *fushi tarazu* promoter fragments were fused to *lacZ* and transformed into flies, the entire seven-stripe pattern mapped to a single element, termed the *zebra* element. When this element was deleted, all seven stripes were lost. The zebra element is a complex sequence that binds many proteins, and its inner workings are not yet apparent. A further wrinkle in the *fushi tarazu* gene is an autoregulatory element that responds to levels of the *fushi tarazu* protein itself. This autoregulation establishes a positive feedback loop that sharpens and strengthens the periodic pattern of *fushi tarazu* expression.

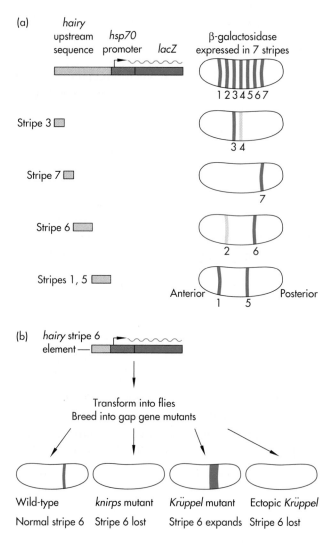

(a)

hairy upstream sequence | hsp70 promoter | lacZ

β-galactosidase expressed in 7 stripes

1 2 3 4 5 6 7

Stripe 3

3 4

Stripe 7

7

Stripe 6

2 6

Stripes 1, 5

Anterior 1 5 Posterior

(b) hairy stripe 6 element

Transform into flies
Breed into gap gene mutants

Wild-type | knirps mutant | Krüppel mutant | Ectopic Krüppel

Normal stripe 6 | Stripe 6 lost | Stripe 6 expands | Stripe 6 lost

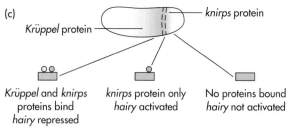

(c)

knirps protein

Krüppel protein

Krüppel and knirps proteins bind hairy repressed | knirps protein only hairy activated | No proteins bound hairy not activated

FIGURE **20-11**

Making a *hairy* (*h*) stripe. (a) A series of fusion genes that carry fragments of DNA from the *hairy* 5′ flanking sequence were tested as described in Figures 20-9 and 20-10. Individual regulatory elements produced discrete stripes. To test the role of individual gap genes in stripe 6 regulation, the fusion gene that produced only stripe 6 was introduced into embryos lacking the indicated gap gene, *knirps* or *Krüppel*. *knirps* mutants gave no stripe 6, showing that *knirps* protein activates this element, and *Krüppel* mutants gave an enlarged stripe 6, showing that *Krüppel* protein represses this element. In an embryo engineered to produce *Krüppel* protein throughout the embryo (*ectopic* expression), the stripe did not appear, again implicating *Krüppel* protein as a repressor of the stripe 6 element. (c) Stripe 6 occurs where the concentration of *knirps* protein is high and *Krüppel* protein is falling off sharply. Anterior to the stripe the *Krüppel* protein concentration is sufficiently high to bind to the *hairy* promoter and repress transcription, despite the presence of *knirps* protein. Posterior to the stripe, the concentration of *knirps* protein is too low to bind to the promoter and activate it. Only in a single narrow domain of the embryo is the *knirps* protein concentration high enough to activate *hairy* and the *Krüppel* protein concentration too low to repress it. This is where stripe 6 (dashed line) forms.

The Specificity of Homeotic Genes Remains Unexplained

By the time the homeotic genes switch on at the cellular blastoderm stage, things have gotten pretty complicated in the embryo. Although the cells still look the same under the light microscope, we now know that a cascade of several dozen maternal-effect, gap, and segmentation genes has superimposed an intricate pattern of differential gene expression onto the embryo, so that each cell is different from its neighbor. It is this rich lode of positional information that is mined by the homeotic genes, whose principal responsibility is to specify the subsequent development of each embryonic segment.

Mutations in homeotic genes can cause dramatic transformations of one body part into another. How is this specificity of action achieved? Part of the answer is in the complex regulation of homeotic gene expression. The homeotic genes contain enormous stretches (over 200 kbp) of regulatory DNA that integrate the positional information supplied by the segmentation genes. From the mapping of many homeotic mutations to the physical maps of the homeotic genes complexes, it was apparent that some of these mutations affected these regulatory regions and not the protein-coding sequences. The effect of such mutations was to shift homeotic protein expression to different locations in the embryo, resulting in homeotic transformations of the affected segments.

However, the expression pattern of these proteins is not the whole story. The proteins themselves have

FIGURE **20-12**

The homeodomain swap. Investigators constructed a *De-formed* (*Dfd*) gene expression vector by placing a *Dfd* cDNA under the control of a heat-shock promoter (hsp 70). The expression vector was placed in the genome by P element–mediated transformation. When embryos carrying the *Dfd* vector were subjected to heat shock, the heat-shock promoter switched on, causing *Dfd* protein to be produced throughout the embryo. This resulted in transformation of thoracic segments into head. Next the researchers constructed a *Dfd/Ubx* expression vector, identical to the *Dfd* vector except that the homeodomain of *Dfd* was replaced with that of *Ubx*, a homeotic gene that specifies thoracic development. Now heat-shock produced the chimeric *Dfd/Ubx* protein in the embryo. In these larvae, head segments were transformed into thorax. Thus, it is the homeodomain that determines the developmental specificity of homeotic genes.

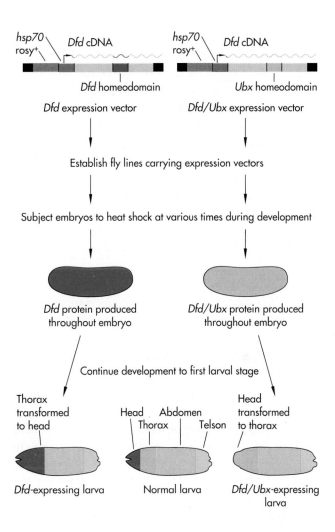

different activities. This was directly demonstrated by expressing homeotic genes throughout the embryo as fusion genes under the control of a heat-shock promoter (*ectopic* expression). Such fusion genes were placed in P element vectors and transformed into flies. When embryos carrying such a fusion gene are incubated at elevated temperature, the heat-shock promoter turns on, leading to a burst of production of the homeotic protein everywhere in the embryo (Figure 20-12). When this was done with the *Deformed* gene (*Dfd*), a member of the ANT-C that normally directs the development of the head, head parts developed in the thorax. When the homeodomain of *Deformed* was swapped with the homeodomain of a BX-C gene, *Ubx*, thorax formed where the head should be. Thus, this swap, which changed only 22 amino acids in the *Deformed* protein, completely reprogrammed its developmental specificity from head to thorax, the normal specificity of the *Ubx* protein.

Still unresolved is how different homeodomains achieve these distinct specificities of action. The sequences of the *Deformed* and *Ubx* homeodomains are very similar, and in the *recognition helix*, the protein domain that determines the DNA-binding specificity of the proteins, the sequences are identical. In fact, the DNA-binding properties of homeodomain proteins have been extensively studied, and many seem to bind to the same target sequences, despite their different biological activities. Resolving how these proteins act with such specificity and precision remains a major challenge in developmental biology.

Formation of the Embryo's Ends Requires a Protein Kinase–Dependent Signal Transduction Pathway

In addition to the two organizing systems that establish the anterior-posterior polarity of the embryo, a separate regulatory system organizes the terminal structures. Mutations that inactivate a gene called *torso* result in embryos lacking the structures normally found at either end. Certain alleles of *torso* are dominant gain-of-function alleles; that is, they produce excess *torso* activity. These embryos are all head and tail and have no middle. How *torso* works became evident when it was cloned and sequenced. The *torso* protein appears to be a transmembrane protein with protein-tyrosine kinase activity, like the mammalian growth factor receptors we discussed in Chapters 17 and 18. Thus, this protein probably acts by transmitting an external signal to cells in the embryo termini, and we can understand the gain-of-function *torso* mutants as alterations that activate kinase activity in the absence of its normal signal. This is precisely the type of activation that converts genes for mammalian growth factor receptors into oncogenes that drive uncontrolled cell growth. Interestingly, in situ hybridization of a *torso* probe to embryos showed that *torso* mRNA (and presumably the protein as well) is distributed uniformly throughout the embryo. Thus, it must be the signaling molecule that activates the *torso* protein that is restricted spatially to the ends of the embryo.

Mutations in a second gene, with the colorful name *lethal (1) polehole*, produce a phenotype similar to *torso* mutations, and in combination with gain-of-function *torso* alleles, *polehole* mutations prevent the development of the all-head-and-tail phenotype. Thus, the polehole protein must be a target for *torso* action (Figure 20-13). Remarkably, when *polehole* was cloned and sequenced, it turned out to be very similar to the mammalian c-*raf* protooncogene. This gene encodes a serine- and threonine-specific protein kinase. In mammals, there is biochemical evidence that the c-*raf* protein associates with and is activated by growth factor receptors. A popular model based on this biochemical data is that Raf is an intermediate in a signaling pathway triggered by growth factors. The genetics of *torso* and *polehole* suggests that this model is probably correct. This is a classic example of how

evolution conserves useful regulatory systems. And it illustrates again how recombinant DNA can quickly link disparate areas of biological research, as we saw in the case of the cell cycle (Chapter 19).

Dorsal-Ventral Polarity Is Achieved by Regulated Transport of a Transcription Factor to the Nucleus

Mother flies supply their embryos with one more piece of positional information, telling them which end is up (dorsal) and which down (ventral). Classical genetic analysis has identified a collection of mutants in which this information is lost, leading to the formation of *dorsalized* embryos, in which all cells adopt dorsal fates. Transplantation of cytoplasm from wild-type to mutant embryos, as was done in studies of the anterior-posterior axis, suggested that embryos are endowed with a morphogenetic gradient that determines dorsal-ventral polarity. High concentrations of the morphogen are required to specify development of ventral tissues. Loss of the morphogen leads to development of dorsal tissues throughout the embryo.

At least 12 different genes are required to set up this gradient, but the principal genes are *dorsal, cactus,* and *Toll*. Genetic studies suggest that *dorsal* encodes the morphogen. However, when *dorsal* was cloned, allowing its mRNA to be visualized in embryos, investigators were surprised to find this mRNA uniformly distributed. When antibodies to the *dorsal*

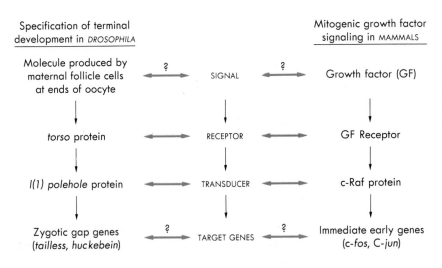

FIGURE **20-13**
A common signal transduction pathway for terminal polarity in *Drosophila* embryos and mitogenic signaling in mammalian cells. Red arrows indicate proteins known to be similar in structure in both systems. Blue arrows suggest possible relationships.

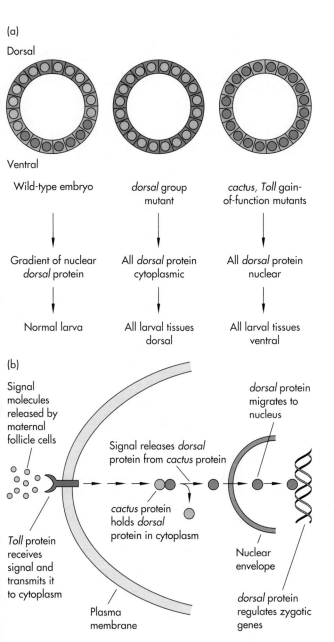

(a)

Dorsal

Ventral

Wild-type embryo	*dorsal* group mutant	*cactus, Toll* gain-of-function mutants
↓	↓	↓
Gradient of nuclear *dorsal* protein	All *dorsal* protein cytoplasmic	All *dorsal* protein nuclear
↓	↓	↓
Normal larva	All larval tissues dorsal	All larval tissues ventral

(b)

Signal molecules released by maternal follicle cells

Toll protein receives signal and transmits it to cytoplasm

Plasma membrane

Signal releases *dorsal* protein from *cactus* protein

cactus protein holds *dorsal* protein in cytoplasm

dorsal protein migrates to nucleus

Nuclear envelope

dorsal protein regulates zygotic genes

FIGURE **20-14**

Specification of dorsal-ventral polarity. Polarity is established by a gradient of localization of the *dorsal* protein. (a) In a wild-type embryo, seen here end-on, *dorsal* protein is in the nuclei of ventral cells and in the cytoplasm of dorsal cells. Nuclear *dorsal* protein specifies the development of ventral tissues (note that the gene is named for the phenotype of a loss-of-function mutation). Loss-of-function mutations in 10 different genes (*dorsal* group mutants) abolish this gradient, and *dorsal* protein accumulates in the cytoplasm of all cells (hence, all cells develop into dorsal tissues). Loss-of-function mutations in *cactus* and rare gain-of-function alleles of *Toll* result in *dorsal* protein migrating to the nuclei in all cells (hence, ventral development throughout). (b) The presumed signaling pathway controlling dorsal localization. *Toll* is thought to encode a transmembrane receptor for a signal released by follicle cells outside the embryo ends. Most of the other dorsal group genes act to relay the signal from the *Toll* protein to a cytoplasmic complex of *cactus* and *dorsal* proteins. The signal causes this complex to dissociate, freeing the *dorsal* protein to migrate to the nucleus, where it regulates the expression of zygotic genes involved in dorsal and ventral development.

protein were developed and used to locate the protein in embryos, researchers made an extraordinary discovery. Although the protein could be found throughout the embryo, on the dorsal side it was in the cytoplasm and on the ventral side it was in the nucleus (Figure 20-14a). So the gradient was not in absolute protein level, as was the case for *bicoid*, but in the localization of the protein to the nucleus. In dorsalized mutants, *dorsal* protein was found only in the cytoplasm, suggesting that the other dorsalizing genes set up this gradient. In certain dominant gain-of-function alleles of *Toll* (which lead to ventralized embryos) *dorsal* protein was found only in the nuclei, suggesting that a signal from the *Toll* protein directs the *dorsal* protein to the nucleus. And in *cactus* loss-of-function mutants (which also lead to ventralized embryos), the *dorsal* protein was also found primarily in nuclei, suggesting that the *cactus* protein might anchor the *dorsal* protein in the cytoplasm.

Sequencing of the *dorsal* and *Toll* genes explained some of these observations (Figure 20-14b). The *dorsal* gene is closely related to another vertebrate protooncogene, c-*rel*, and to the gene for a mammalian transcription factor, NF-κB. Thus, the *dorsal* protein must be a transcription factor that acts to regulate zygotic genes that specify dorsal and ventral development. But, more intriguingly, the c-*rel* protein and NF-κB both show the same cytoplasmic-nuclear shuffling seen with the *dorsal* protein, and the biochemistry of this process is relatively well understood for the mammalian proteins. NF-κB, for example, is retained in the cytoplasm by another protein, IκB. Phosphorylation of IκB prompts it to release NF-κB to go to the nucleus. It is conceivable that the *cactus* protein, which seems to retain the *dorsal* protein in the cytoplasm, is similar to IκB. And one of the dorsalizing genes may encode a protein kinase that phosphorylates the *cactus* protein.

The *Toll* gene encodes a transmembrane protein that is likely to be a receptor of some kind. Like the *torso* protein, therefore, the *Toll* protein may act by receiving a spatially localized signal and transmitting it down the line to the *dorsal* protein. Transplantation experiments suggest that such a signal may emanate from the follicle cells that surround the egg. Thus, the dorsal-ventral gradient is set up by a positional cue that comes from outside the egg.

Drosophila Developmental Genes Help Isolate New Vertebrate Developmental Genes

We began this chapter by saying that scientists choose to study model systems for complex processes like development in the hope that the principles elucidated from simpler systems will apply to more complex systems as well. This idea has been resoundingly vindicated by the exploration of *Drosophila* development, shown most clearly by the use of *Drosophila* gene probes to retrieve related genes from mammals. We mentioned earlier that homeo box probes from the first cloned homeotic genes hybridized with genomic fragments in other species. Many of the vertebrate homeodomain genes, termed *HOX* genes, have now been cloned. Analysis of the HOX genes suggests that many are true homologues of the corresponding *Drosophila* genes, with similar structures and functions.

The first similarity is in amino acid sequence. In the homeodomain, which we know encodes the DNA-binding activity and much of the biological specificity of the fly homeotic genes, the HOX genes are closely related to their fly counterparts. The homeodomain of the mouse HOX 1.1 gene differs from *Antennapedia* in only a single amino acid. HOX 2.2 differs in only five. The human HOX 4.2 gene differs from the *Deformed* gene in just six of its homeodomain positions and is about 40 percent identical overall.

The second and perhaps more astonishing similarity is in the organization and expression pattern of the gene clusters that harbor the HOX genes (Figure 20-15). In flies, the homeotic genes are found in two large clusters, ANT-C and BX-C, located near to one another on chromosome 3. The individual genes in the fly complexes are aligned in a physical order that

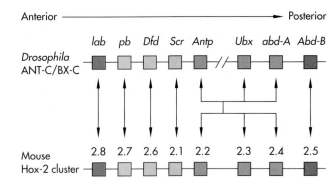

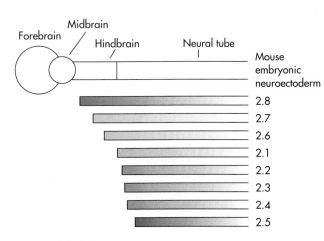

FIGURE **20-15**
The conserved arrangement of *Drosophila* homeotic and mouse HOX genes. The top line is a representation of the genes in the ANT-C and BX-C. Although not immediately adjacent to one another, these complexes are found in the same region of chromosome 3, and the insertion of DNA between them is thought to be a recent event in *Drosophila* evolution. The second line is the mouse HOX-2 cluster. Genes most closely related in sequence are shown in the same color. The genes marked in orange are all about equally related to one another. Note that the order is the same in both complexes and corresponds with the domains of gene expression along the anterior-posterior axis of the embryo. This is shown at the bottom for the primitive nervous system of the mouse embryo. The HOX genes are expressed in overlapping domains with sharp anterior boundaries that correspond to their positions in the gene cluster.

reflects their order of action along the embryo—the genes that specify head development at one end, abdominal development at the other. In mammalian HOX gene clusters, this order is precisely matched. And the expression patterns of the HOX genes, revealed by in situ hybridization to mouse embryos, is similarly ordered. The HOX genes are expressed in overlapping domains with sharp anterior boundaries along the long axis of the embryo. Remarkable as it seems, humans and flies have apparently inherited from a common ancestor some 600 million years ago an intact system for specifying positions along the anterior-posterior axis.

These striking similarities in the structure, organization, and expression of the HOX genes suggest that they perform similar homeotic functions in vertebrates. Current technology has allowed scientists to put this hypothesis to the test. They have produced transgenic mice that express HOX genes ectopically throughout the embryo, much as the "fly people" did with their genes, and these mice develop abnormalities that resemble homeotic transformations. Furthermore, in a dramatic demonstration of functional conservation across half a billion years of evolution, scientists have expressed the HOX 2.2 and HOX 4.2 genes in flies. These genes elicit the same transformations as their fly homologues, *Antennapedia* and *Deformed*.

Scientists studying *Drosophila* development stand on the shoulders of the giants of *Drosophila* genetics of decades past, but by adopting and pioneering new recombinant DNA methodologies, they have rapidly pushed forward our molecular view of how the positional and temporal patterns underlying embryo development are encoded in the genome. Recombinant DNA has been a particularly potent tool in studying development for several reasons. First, it has allowed us to isolate the genes involved, and from simply examining gene sequence we can often infer function. Second, it has allowed us to perform critical tests of the models developed by geneticists and embryologists, by providing new ways to see the inner workings of the embryo and enabling genes to be modified and returned to living animals for study. Third, the simple phenomenon of cross-hybridization has allowed us to leapfrog millions of years on the evolutionary tree to the genes that control development in mammals. As we have learned from studies of cell cycle control (Chapter 19) and will see again in our discussion of the brain (Chapter 21), evolution is stingy with its ideas. Once it finds an efficient way to do something, construct an embryo, for instance, it uses the same method over and over again.

Reading List

General

Nüsslein-Volhard, C., and E. Wieschaus. "Mutations affecting segment number and polarity in *Drosophila*. "Nature, 287: 795–801 (1980).

Ingham, P. W. "The molecular genetics of embryonic pattern formation in *Drosophila*." *Nature*, 335: 25–34 (1988). [Review]

Melton, D. A. "Pattern formation during animal development. "*Science*, 252: 234–241 (1991). [Review]

Original Research Papers

ISOLATION OF HOMEOTIC GENES

Peifer, M., F. Karch, and W. Bender. "The bithorax complex: control of segmental identity." *Genes and Devel.*, 1: 891–898 (1987). [Review]

Kaufman, T. C., M. A. Seeger, and G. Olsen. "Molecular and genetic organization of the *Antennapedia* gene complex of *Drosophila melanogaster*." *Adv. Gen.*, 27: 309–362 (1990). [Review]

Bender, W., M. Akam, F. Karch, P. A. Beachy, M. Peifer, P. Spierer, E. B. Lewis, and D. S. Hogness. "Molecular genetics of the bithorax complex in *Drosophila melanogaster*." *Science*, 221: 23–29 (1983).

Bender, W., P. Spierer, and D. S. Hogness. "Chromosomal walking and jumping to isolate DNA from the *Ace* and *rosy* loci and bithorax complex in *Drosophila melanogaster*." *J. Mol. Biol.*, 168: 17–33 (1983).

Garber, R. L., A. Kuroiwa, and W. J. Gehring. "Genomic and cDNA clones of the homeotic locus Antennapedia in *Drosophila*." *EMBO J.*, 2: 2027–2036 (1983).

Scott, M. P., A. J. Weiner, T. I. Hazelrigg, B. A. Polisky, V. Pirrotta, F. Scalenghe, and T. C. Kaufman. "The molecular organization of the *Antennapedia* locus of *Drosophila.*" *Cell,* 35: 763–776 (1983).

McGinnis, W., R. L. Garber, J. Wirz, A. Kuroiwa, and W. J. Gehring. "A homologous protein-coding sequence in *Drosophila* homeotic genes and its conservation in other metazoans." *Cell,* 37: 403–408 (1984).

McGinnis, W., M. Levine, E. Hafen, A. Kuroiwa, and W. J. Gehring. "A conserved DNA sequence found in homeotic genes of the *Drosophila* Antennapedia and bithorax complexes." *Nature,* 308: 428–433 (1984).

ANTERIOR-POSTERIOR POLARITY

Nüsslein-Volhard, C., H. G. Frohnhöfer, and R. Lehmann. "Determination of anteroposterior polarity in *Drosophila.*" *Science,* 238: 1675–1681 (1987). [Review]

Manseau, L. J., and T. Schüpbach. "The egg came first, of course." *Trends Gen.* 5: 400–405 (1989). [Review]

Driever, W., and C. Nüsslein-Volhard. "A gradient of bicoid protein in *Drosophila* embryos." *Cell,* 54: 83–93 (1988).

Driever, W., J. Ma, C. Nüsslein-Volhard, and M. Ptashne. "Rescue of bicoid mutant *Drosophila* embryos by bicoid fusion proteins containing heterologous activating sequences." *Nature,* 342: 149–154 (1989).

Lehmann, R., and C. Nüsslein-Volhard. "The maternal gene *nanos* has a central role in posterior pattern formation of the *Drosophila* embryo." *Development,* 112: 679–693 (1991).

GAP AND SEGMENTATION GENES

Scott, M. P., and S. B. Carroll. "The segmentation and homeotic gene network in early *Drosophila* development." *Cell,* 51: 689–698 (1987). [Review]

Gaul, U., and J. Jäckle. "Role of gap genes in early *Drosophila* development." *Adv. Gen.,* 27: 239–275 (1990). [Review]

Hülskamp, M., and D. Tautz. "Gap genes and gradients—the logic behind the gaps." *Bioessays,* 13: 261–268 (1991). [Review]

Hafen, E., A. Kuroiwa, and W. J. Gehring. "Spatial distribution of transcripts from the segmentation gene *fushi tarazu* during *Drosophila* embryonic development." *Cell,* 37: 833–841 (1984).

Ingham, P. W., D. R. Howard, and D. Ish-Horowicz. "Transcription pattern of the *Drosophila* segmentation gene *hairy.*" *Nature,* 318: 439–445 (1985).

Ingham, P. W., D. Ish-Horowicz, and K. R. Howard. "Correlative changes in homeotic and segmentation gene expression in *Krüppel* mutant embryos of *Drosophila.*" *EMBO J.,* 5: 1659–1665 (1986).

Gaul, U., and H. Jäckle. "Pole region-dependent repression of the *Drosophila* gap gene *Krüppel* by maternal gene products." *Cell,* 51: 549–555 (1987).

Tautz, D. "Regulation of the *Drosophila* segmentation gene hunchback by two maternal morphogenetic centres." *Nature,* 332: 281–284 (1988).

TRANSCRIPTIONAL CONTROL IN THE EMBRYO

Carroll, S. B. "Zebra patterns in fly embryos: activation of stripes or repression of interstripes." *Cell,* 60: 9–16 (1990). [Review]

Pankratz, M. J., and H. Jäckle. "Making stripes in the *Drosophila* embryo." *Trends Gen.,* 6: 287–292 (1990). [Review]

Hiromi, Y., and W. J. Gehring. "Regulation and function of the *Drosophila* segmentation gene *fushi tarazu.*" *Cell,* 50: 963–974 (1987).

Howard, K., P. Ingham, and C. Rushlow. "Region-specific alleles of the *Drosophila* segmentation gene *hairy.*" *Genes and Devel.,* 2: 1037–1046 (1988).

Dearolf, D. R., J. Topol, and C. S. Parker. "Transcriptional control of *Drosophila fushi tarazu* zebra stripe expression." *Genes and Devel.,* 3: 384–398 (1989).

Driever, W., and C. Nüsslein-Volhard. "The *bicoid* protein is a positive regulator of *hunchback* transcription in the early *Drosophila* embryo." *Nature,* 337: 138–143 (1989).

Driever W., G. Thoma, and C. Nüsslein-Volhard. "Determination of spatial domains of zygotic gene expression in the *Drosophila* embryo by the affinity of binding sites for the *bicoid* morphogen." *Nature,* 340: 363–367 (1989).

Goto, T., P. Macdonald, and T. Maniatis. "Early and late periodic patterns of *even skipped* expression are controlled by distinct regulatory elements that respond to different spatial cues." *Cell,* 57: 413–422 (1989).

Stanojevic, D., T. Hoey, and M. Levine. "Sequence-specific DNA-binding activities of the gap proteins encoded by *hunchback* and *Krüppel* in *Drosophila.*" *Nature,* 341: 331–335 (1989).

Struhl, G., K. Struhl, and P. M. Macdonald. "The gradient morphogen bicoid is a concentration-dependent transcriptional activator." *Cell,* 57: 1259–1273 (1989).

Pankratz, M. J., E. Seifert, N. Gerwin, B. Billi, U. Nauber, and H. Jäckle. "Gradients of *Krüppel* and *knirps* gene products direct pair-rule gene stripe patterning in the posterior region of the *Drosophila* embryo." *Cell,* 61: 309–317 (1990).

SPECIFICITY OF HOMEOTIC GENE ACTION

Hayashi, S., and M. P. Scott. "What determines the specificity of action of *Drosophila* homeodomain proteins." *Cell,* 63: 883–894 (1990). [Review]

Kuziora, M. A., and W. McGinnis. "Autoregulation of a *Drosophila* homeotic selector gene." *Cell,* 55: 477–485 (1988).

Kuziora, M. A., and W. McGinnis. "A homeodomain substitution changes the regulatory specificity of the *Deformed* protein in *Drosophila* embryos." *Cell*, 59: 563–571 (1989).

Gibson, G., A. Schier, P. LeMotte, and W. J. Gehring." The specificities of *Sex combs reduced* and *Antennapedia* are defined by a distinct portion of each protein that includes the homeodomain." *Cell*, 62: 1087–1103 (1990).

Mann, R. S., and D. S. Hogness. "Functional dissection of *Ultrabithorax* proteins in *D. melanogaster.*" *Cell*, 60: 597–610 (1990).

ORGANIZATION OF EMBRYONIC TERMINI

Pawson, T., and A. Bernstein. "Receptor tyrosine kinases: genetic evidence for their role in *Drosophila* and mouse development." *Trends Gen.*, 6: 350–356 (1990). [Review]

Siegfried, E., L. Ambrosio, and N. Perrimon. "Serine/threonine protein kinases in *Drosophila.*" *Trends Gen.*, 6: 357–362 (1990). [Review]

Ambrosio, L., A. P. Mahowald, and N. Perrimon. "Requirement of the *Drosophila raf* homologue for torso function." *Nature*, 342: 288–291 (1989).

Sprenger, F., M. L. Stevens, and C. Nüsslein-Volhard. "The *Drosophila* gene *torso* encodes a putative receptor tyrosine kinase." *Nature*, 338: 478–483 (1989).

DORSAL-VENTRAL POLARITY

Gilmore, T. D. "NF-κB, KBF1, *dorsal*, and related matters." *Cell*, 62: 841–843 (1990). [Review]

Rushlow, C., and M. Levine. "Role of the *zerknült* gene in dorsal-ventral pattern formation in *Drosophila.*" *Adv. Gen.*, 27: 277–307 (1990). [Review]

Anderson, K. V., G. Jürgens, and C. Nüsslein-Volhard. "Establishment of dorsal-ventral polarity in *Drosophila*: genetic studies on the role of the *Toll* gene product." *Cell*, 42: 779–789 (1985).

Steward, R. "*Dorsal*, an embryonic polarity gene in *Drosophila*, is homologous to the vertebrate proto-oncogene c-*rel.*" *Science*, 238: 692–694 (1987).

Roth, S., D. Stein, and C. Nüsslein-Volhard. "A gradient of nuclear localization of the *dorsal* protein determines dorsoventral pattern in the *Drosophila* embryo." *Cell*, 59: 1189–1202 (1989).

Rushlow, C. A., K. Han, J. L. Manley, and M. Levine. "The graded distribution of the dorsal morphogen is initiated by selective nuclear transport in *Drosophila.*" *Cell*, 59: 1165–1177 (1989).

Steward, R. "Relocalization of the *dorsal* protein from the cytoplasm to the nucleus correlates with its function." *Cell*, 59: 1179–1188 (1989).

Ip, Y. T., R. Kraut, M. Levine, and C. A. Rushlow. "The dorsal morphogen is a sequence-specific DNA-binding protein that interacts with a long-range repression element in *Drosophila.*" *Cell*, 64: 439–446 (1991).

VERTEBRATE HOX GENES

Akam, M. "Hox and HOM: homologous gene clusters in insects and vertebrates." *Cell*, 57: 347–349 (1989). [Review]

Kessel, M., and P. Gruss. "Murine developmental control genes." *Science*, 249: 374–379 (1990). [Review]

Balling, R., U. Deutsch, and P. Gruss. "Undulated, a mutation affecting development of the mouse skeleton, has a point mutation in the paired box of *Pax1.*" *Cell*, 55: 531–535 (1988).

Balling, R., G. Mutter, P. Gruss and M. Kessel. "Craniofacial abnormalities induced by ectopic expression of the homeobox gene *Hox-1.1* in transgenic mice." *Cell*, 58: 337–347 (1989).

Graham, A., N. Papalopulu, and R. Krumlauf. "The murine and *Drosophila* homeobox gene complexes have common features of organization and expression." *Cell*, 57: 367–378 (1989).

Wolgemuth, D. J., R. R. Behringer, M. P. Mostoller, R. L. Brinster, and R. D. Palmiter. "Transgenic mice over-expressing the mouse homeobox-containing gene *Hox-1.4* exhibit abnormal gut development." *Nature*, 337: 464–467 (1989).

Kessel, M., R. Balling, and P. Gruss. "Variations of cervical vertebrae after expression of a *Hox-1.1* transgene in mice." *Cell*, 61: 301–308 (1990).

Malicki, J., K. Schughart, and W. McGinnis. "Mouse *Hox-2.2* specifies thoracic segmental identity in *Drosophila* embryos and larvae." *Cell*, 63: 961–967 (1990).

McGinnis, N., M. A. Kuziora, and W. McGinnis. "Human *Hox-4.2* and *Drosophila Deformed* encode similar regulatory specificities in *Drosophila* embryos and larvae." *Cell*, 63: 969–976 (1990).

21

The Genes Behind the Functioning of the Brain

The brain is the final frontier of biology. No organ in the body is more complex in structure and function. As with other organs, we seek to understand how the brain develops and how its individual cell types specialize. But unlike most other organs, which are built from only a few different types of cells, the brain consists of billions of neurons, each connected to others in a unique pattern, and therefore each neuron is different. As with other organs, we also seek to understand how its cells communicate. But unlike other organs, in which cells exchange only a few signals, communication is the business of the brain, and brain cells employ a complex array of chemical and electrical signals. And we seek to understand how the activities of individual brain cells are integrated to perform the key functions of the organ. But, for the brain, these functions are of the most profound sort. The brain is the control center for every other organ in the body. And, of course, it is the brain that defines who we are, allowing us to learn, think, create, and dream.

Recombinant DNA technology has only begun its assault on the brain and is still in its gathering phase. In this chapter, we will learn how the tools and methods discussed so far in this book are being marshaled to identify and clone the genes encoding the constituents of the brain. We will learn how researchers are identifying the molecules that control the development, architecture, and wiring of the brain, and the components of the brain's rich signaling systems. And we will examine how molecular studies are beginning to unravel integrated brain functions such as learning and memory. Finally, we will see how recombinant DNA technology is illuminating the degenerative processes that degrade brain function in diseases like Alzheimer's. In many ways, this chapter is about possibilities rather than accomplishments. But future decades promise spectacular breakthroughs in our understanding of the brain.

Genes Specifying the Development of Neurons Are Cloned

Development of the nervous system begins early in embryonic growth. As we learned in the last chapter, many aspects of embryonic development are best understood in *Drosophila,* and this is true for the nervous system as well. The *Drosophila* genome has been systematically screened for mutations that disrupt the development of the sensory nervous system. Many genes controlling this process have been identified by noting mutations that prevent or disorganize the development of sensory neurons. Other genes have been identified using the enhancer trap strategy (Figure 10-14) to find chromosomal loci expressed exclusively in neuronal precursors. These genes fall into several

groups: *proneuronal genes* that give cells the potential to become neurons, *neurogenic genes* that select individual precursor cells for further development into neurons while inhibiting the development of neighboring precursors, and *selector genes* that determine the type of neuron into which the precursor develops.

Cloning and sequencing of these regulatory genes led to the remarkable observation that virtually all were related to genes previously isolated from other organisms (Table 21-1). For example, it was found that the proneural genes *daughterless* (*da*) and *achaete-scute* (*AS-C,* actually a complex of four closely related genes) encode proteins with a helix-loop-helix (HLH) motif found in several mammalian transcription factors (see Chapter 9). This discovery, of course, strongly predicts that the proneural genes are themselves transcription factors that act to regulate other genes specifying neural development. Studies with the mammalian HLH proteins have shown that these proteins act as dimers (that is, the functional form of the protein contains two subunits). These can be either homodimers of identical subunits or heterodimers of different subunits. Having a small number of different proteins that can freely change partners makes possible a large number of different protein complexes, which can have quite different properties (Figure 21-1).

The biochemistry of the HLH proteins is perhaps best understood for the mammalian gene *myoD,* a powerful regulatory gene that can convert many types of

TABLE **21-1**
Drosophila Neuronal Development Genes

	GENE	MOTIF	POSSIBLE FUNCTION*
Proneural genes	*achaete-scute* (*AS-C*)	Helix-loop-helix	Transcription factor
	daughterless (*da*)	Helix-loop-helix	Transcription factor
Neurogenic genes	*notch*	Transmembrane + EGF repeat	Adhesion molecule or receptor
	enhancer of split [*E(spl)*]	Helix-loop-helix	Transcription factor
	enhancer of split [*E(spl)*]	G protein	Signal transduction
	big brain	Glycerol facilitator	Small molecule channel
	mastermind	?	?
	neuralized	?	?
Selector genes	*cut*	Homeobox	Transcription factor

* The sequences of these genes suggest how they must work. The proneural genes are related to several mammalian transcription factors. Many of the neurogenic genes specify proteins involved in cell-cell signaling.

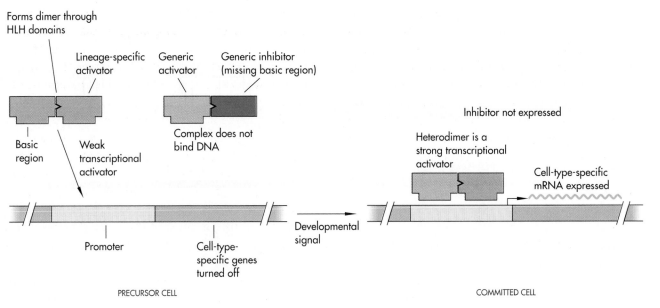

FIGURE **21-1**
Regulation of neuronal development by interacting helix-loop-helix (HLH) proteins. Differentiation of *Drosophila* sensory neurons and mammalian muscle cells follows a very similar scheme. Precursor cells contain three different HLH proteins. One is a generic activator found in all cells (in *Drosophila* this is the product of the *daughterless* gene; in mammals, the *E2A* gene). The second is lineage-specific activator protein expressed in a limited repertoire of cells (in *Drosophila* neuronal progenitors, the protein product of *achaete-scute;* in mammalian muscle precursors, *myoD* and related genes). The third is a generic inhibitor, also widely distributed (in *Drosophila* this is the product of the *emc* gene; in mammals, *Id*). This inhibitor protein differs from the other HLH proteins because it lacks a DNA-binding domain (basic region). Each of the proteins is able to interact with the others, but when all three are present, as occurs in the precursor cell, only two combinations form: a heterodimer of the generic activator and inhibitor and a homodimer of the lineage-specific activator. The heterodimer cannot bind DNA and is therefore inactive, and the homodimer is only weakly active, perhaps sufficient to maintain the potential to differentiate. After a signal to differentiate is received by the precursor cell, the inhibitor is no longer synthesized and disappears, allowing the two activator proteins to join, forming a potent transcription factor that turns on the genes required to form the new cell type. Among the genes turned on are the activator genes themselves, reinforcing the commitment to differentiate.

cells into muscle cells. The MyoD protein is expressed in muscle precursor cells. A dimer made up of two MyoD proteins is only weakly active as a transcription factor. MyoD is highly active, however, when partnered with a widely expressed HLH protein, E2A. But in precursor cells, E2A is tied up in a partnership with a third HLH protein, Id. Id is unique among the HLH family in that it lacks a DNA-binding domain, so that complexes that include Id cannot bind to DNA. Thus, precursor cells contain predominantly inactive E2A-Id heterodimers and weak MyoD homodimers. When the cells receive a signal to differentiate, transcription of Id stops and the Id protein disappears, freeing E2A

to form a potent transcription factor complex with MyoD. The MyoD-E2A complex then activates the genes required for muscle development.

The genetics of the *Drosophila* sensory nervous system suggests that the same types of interactions are controlling neuron development. The *achaete-scute* genes are expressed predominantly in neuronal precursors but can't work without the protein product of *da*, which is ubiquitously expressed and closely related in sequence to the protein E2A. And a genetically defined inhibitor of neuronal development, the product of the *emc* gene, has a structure like Id, lacking a DNA-binding domain. Expression of the *Drosophila*

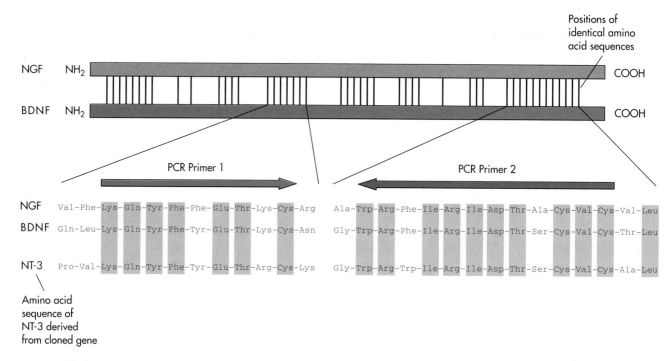

FIGURE **21-2**

Cloning of a novel neurotrophic factor by PCR. Two previously cloned neurotrophic factors, NGF and BDNF, have similar amino acid sequences in several regions (indicated by vertical lines). Sequences from two of these regions are displayed below. Investigators guessed that additional members of this gene family existed. Two degenerate oligonucleotides were synthesized to contain all the DNA sequences that encode the amino acids indicated by the arrows. These oligonucleotides were used as primers in a polymerase chain reaction. The amplified products were cloned and sequenced. Most carried fragments of the NGF and BDNF genes, but a few contained a new gene called *neurotrophin-3* (*NT-3*). This fragment was used as a probe to isolate the entire *NT-3* gene by hybridization. The authentic NT-3 sequence in the primer regions is shown in the bottom line. The shaded amino acids are identical in all three proteins.

proteins in vitro will likely confirm that these genetic interactions occur at the biochemical level.

Because HLH genes are involved in development of the *Drosophila* nervous system and of vertebrate muscle, a search was launched for similar genes that might control development of the mammalian brain. Investigators designed degenerate oligonucleotide primers based on amino acid sequences highly conserved in genes of the *achaete-scute* complex. They used these primers to amplify related genes from a rat neuronal precursor cell line (see below) by polymerase chain reaction (PCR, see Chapter 6). Two such genes were isolated, and analysis of the expression of their mRNAs showed that they were expressed in neuronal precursor cells in rat embryos. Thus, these genes may function as proneural genes in the development of the vertebrate nervous system. Here we see how the clon-

ing of genes from two quite different systems, fly sensory neurons and mammalian muscle, highlighted similarities that could be exploited to identify new genes in yet another developmental pathway.

Neurotrophic Factors Stimulate Neuron Growth and Differentiation

While researchers studying the development of the vertebrate nervous system have not had the genetic tools available to those studying flies, they have nevertheless learned much about the cell-cell interactions in the developing nervous system, through a combination of observation, manipulation, and biochemical analysis of developing embryos. One thing that became clear from these studies was that neuronal differentiation depended critically on secreted factors, like the polypeptide growth factors we discussed in Chapters 17 and 18. These factors are required for the differentiation of neuronal precursors into mature neurons, for the continuing survival of neurons, and for the programming of the neurotransmitter repertoire produced by each cell. The first such factor identified was *nerve growth factor,* or *NGF,* first isolated in the 1950s. NGF could be readily isolated because, for unexplained reasons, it is produced in large quantities in rat salivary gland. But despite abundant evidence that other neurotrophic factors existed, they remained elusive because they are produced in vanishingly small quantities.

However, cloning of rare neurotrophic factors has accelerated dramatically in recent years. Improved technology now allows us to determine an amino acid sequence from very small amounts of protein and to use this information to design DNA probes. Thus, 40 years after the discovery of NGF, a second neurotrophic factor, *brain-derived neurotrophic factor (BDNF),* was cloned using degenerate PCR primers based on amino acid sequences from BDNF protein painstakingly purified from pig brain. When the cloned BDNF cDNA was inserted into an expression vector and transfected into COS cells (see Chapter 12), the COS cells secreted into their culture medium an activity that could keep neurons alive, just as authentic BDNF does. Sequencing of the BDNF cDNA revealed that it was closely related to NGF. Based on this sequence comparison, several research groups designed degenerate PCR primers encoding the most highly conserved amino acids and cloned a third neurotrophic factor, *NT-3* (Figure 21-2). As with BDNF, recombinant NT-3 was shown to support the survival of neurons in culture. These neurotrophic factors, now available in essentially unlimited quantities, may prove valuable in the battle against neurodegenerative disorders like Parkinson's disease.

As we learned in Chapters 17 and 18, growth factors act on target cells through specific high-affinity cell surface receptor proteins. Isolation of these receptors is essential to understanding the biology of the factors. Thus, as neurotrophic factors are identified, attention turns to finding their receptors. The NGF receptor was the first to be cloned, thanks to the availability of monoclonal antibodies against the receptor. Transfection of genomic DNA from neuronal cells into nonneuronal cell lines was used to establish new cell lines

that expressed the receptor. The antibodies were used to identify or select rare transfected cells with receptor on their surfaces (Figure 21-3). A receptor cDNA was then cloned from the transfected cells. Expression of the cloned receptor cDNA in nonneuronal cells caused those cells to express NGF-binding sites on their surfaces, demonstrating that the cDNA in fact encoded a 75,000 dalton, low-affinity NGF receptor. Recently, the 140,000 dalton protein encoded by the *trk* protooncogene has been shown to be part of a high-affinity receptor complex for NGF. Other proteins related in structure to Trk bind BDNF and NT-3. Whether the low-affinity and high-affinity NGF receptors interact is currently under investigation.

Recombinant DNA can also help in the earliest stages of receptor identification. Investigators searching for cells that express the receptor for a newly cloned neurotrophic factor, CNTF, used recombinant DNA methods to produce a form of CNTF tagged with a short peptide sequence from an unrelated protein (Figure 21-4). They had at their disposal a high-affinity monoclonal antibody that recognized the peptide with which they had tagged CNTF. The tagged neurotrophic factor could be produced in high quantities and then incubated with cultures of candidate cell lines. Cells that bound the factor could be spotted using the monoclonal antibody that recognized the tag. This procedure could be adapted to select cell populations that express the receptor at high levels. Such cells, of course, provide an excellent starting point for receptor cloning.

Retroviruses Are Used to Mark and Immortalize Neurons

Another way recombinant DNA is expanding our view of how the brain develops is through the use of viral vectors that can infect developing neurons in living animals, marking them so that we can follow their development (we discussed this method in Chapters 12 and 14). Such marking experiments have led to fundamental observations about brain development. For example, they indicate that many different types of mature brain cells can originate from a single progenitor, and that this differentiation can occur quite late in brain development. Furthermore, they show that the migration patterns of neurons are fairly

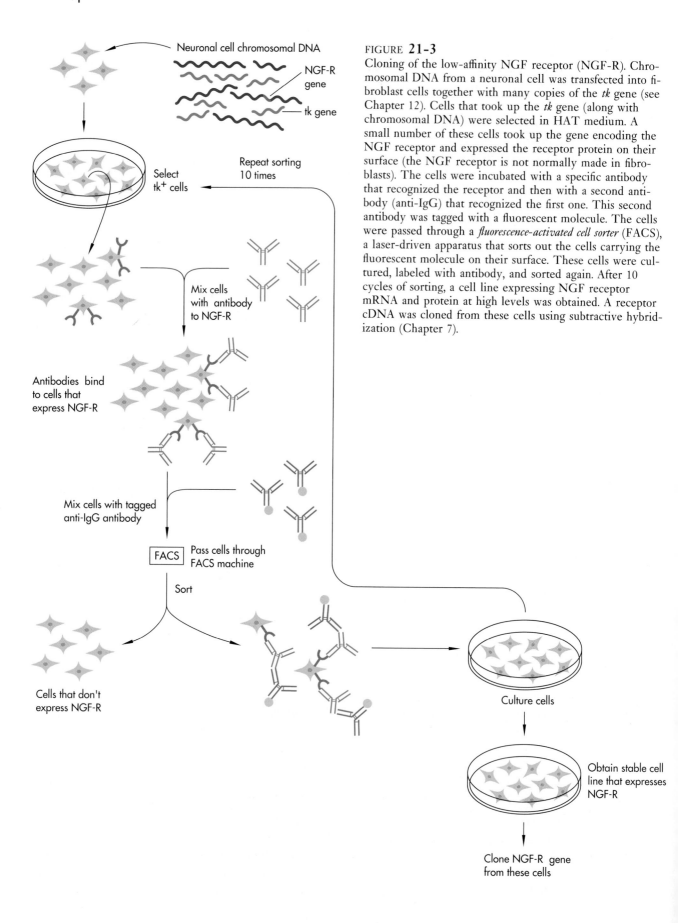

FIGURE **21-3**
Cloning of the low-affinity NGF receptor (NGF-R). Chromosomal DNA from a neuronal cell was transfected into fibroblast cells together with many copies of the *tk* gene (see Chapter 12). Cells that took up the *tk* gene (along with chromosomal DNA) were selected in HAT medium. A small number of these cells took up the gene encoding the NGF receptor and expressed the receptor protein on their surface (the NGF receptor is not normally made in fibroblasts). The cells were incubated with a specific antibody that recognized the receptor and then with a second antibody (anti-IgG) that recognized the first one. This second antibody was tagged with a fluorescent molecule. The cells were passed through a *fluorescence-activated cell sorter* (FACS), a laser-driven apparatus that sorts out the cells carrying the fluorescent molecule on their surface. These cells were cultured, labeled with antibody, and sorted again. After 10 cycles of sorting, a cell line expressing NGF receptor mRNA and protein at high levels was obtained. A receptor cDNA was cloned from these cells using subtractive hybridization (Chapter 7).

Neuronal cell chromosomal DNA

NGF-R gene

tk gene

Select tk+ cells

Repeat sorting 10 times

Mix cells with antibody to NGF-R

Antibodies bind to cells that express NGF-R

Mix cells with tagged anti-IgG antibody

FACS Pass cells through FACS machine

Sort

Cells that don't express NGF-R

Culture cells

Obtain stable cell line that expresses NGF-R

Clone NGF-R gene from these cells

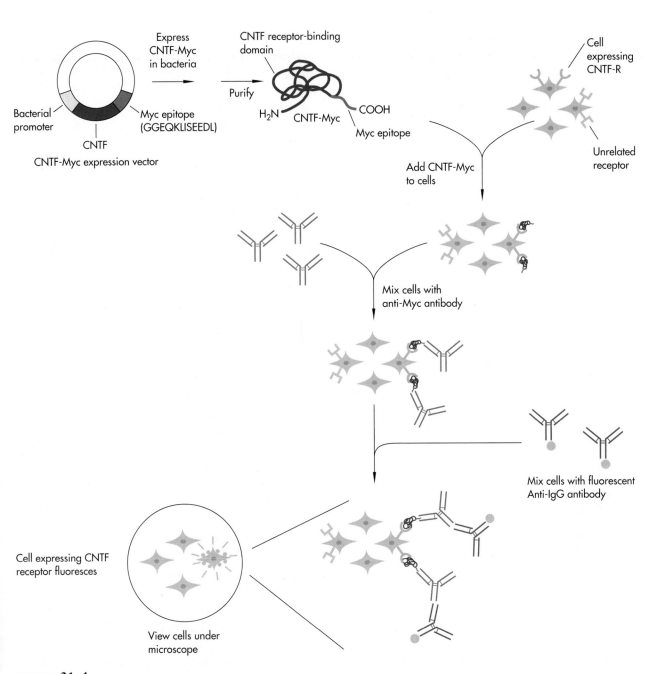

FIGURE **21-4**

Tagging the CNTF receptor. To find cells that expressed the receptor for the neurotrophic factor CNTF, investigators constructed a plasmid in which the CNTF sequence was fused to a short sequence from a different gene (expressing Myc). This allowed them to produce in bacteria a CNTF molecule carrying a short protein tag, or *epitope*, in this case a portion of the Myc protein, for which they had an antibody. The tagged CNTF (CNTF-Myc) was incubated with candidate cell cultures; it bound to cells only carrying the CNTF receptor. The presence of bound CNTF on cells was detected using the anti-Myc antibody and a fluorescently labeled second antibody (anti-IgG) that recognized the anti-Myc antibody. Cells expressing the CNTF receptor could be seen to fluoresce under a microscope. As with the procedure used to clone the NGF receptor (Figure 21-3), this procedure can be used to purify and enrich populations of cells expressing the CNTF receptor at a high level, facilitating characterization of the receptor protein and eventually cloning of the receptor cDNA.

constrained: marked cells are often found in a tight column of neuronal tissue. This is not an absolute constraint, however, for renegade cells that have moved a great distance away are also seen.

Retrovirus vectors can also be used to transfer activated oncogenes into neuronal cells. What can result is a cell line that retains many of the properties of the neuronal cell but now grows robustly in culture, thanks to the powerful influence of the oncogene. This is an extremely promising approach for studying the development and function of the brain, for *immortalization* by virally transduced oncogenes can allow the expansion of a population of rare neuronal precursor cells and of mature neurons, which normally never divide at all. Having a ready source of these cells makes them accessible to the palette of potent molecular methodologies now available to researchers in other fields. For example, the rat *achaete-scute*–like genes we discussed earlier in this chapter were cloned from a rat neuronal progenitor cell line generated by immortalization with an oncogenic retrovirus. Previously, cultures of these cells were prepared by dissecting out fetal adrenal glands, dissociating them into single-cell suspensions, marking them with a fluorescently tagged antibody to a specific cell-surface protein, and sorting out the marked cells using a fluorescence-activated cell sorter. Only a few cells could be obtained this way, and they could be maintained in culture for only a short time before they died. Once immortalized by an oncogene, however, these cells can be grown indefinitely and in essentially unlimited numbers, while retaining many of their normal properties.

Neuronal precursors have also been immortalized with a retrovirus carrying an SV40 T-antigen gene with a temperature-sensitive mutation. This resulted in a cell line that grew indefinitely with the properties of a neuronal progenitor cell at $33°C$, when T antigen is active. But when the cells were shifted to $39°C$, inactivating T antigen, they ceased dividing and differentiated into mature neurons (Figure 21-5). Now investigators can study the molecular events that occur during differentiation. An exciting possibility arises from the fact that the internal body temperature of animals is near the nonpermissive temperature for the T antigen mutant. Thus, cells expanded in culture at $32°C$ can be transplanted into the brains of living animals where they differentiate essentially normally,

making connections with neighboring brain cells. Researchers now have the opportunity to manipulate brain cells in culture using modern recombinant DNA methods and then to return them to a living brain for study. This has great therapeutic promise as well: implantation of NGF-producing cells into the brains of diseased animals prevents or reverses brain cell degeneration.

Cloning and Mutagenesis Establish Structural Models of the Voltage-Gated Ion Channels

Although communication between axon terminals and their target cells is usually via chemical neurotransmitters, electrical activity is the key to neuron function. Neurons send out long processes, as long as a meter for certain motor neurons, and getting a signal quickly from the cell body to the axon terminal cannot be accomplished by diffusion of a chemical messenger. Instead, it is accomplished by electrical impulses, termed *action potentials*. Action potentials are sharp changes in *membrane potential*, the voltage difference between the inside and outside of the cell. Action potentials propagate quickly in one direction along an axon, and the firing of an action potential is an all-or-nothing effect. When incoming signals exceed a threshold value, a neuron fires an action potential. Action potentials result from the concerted action of integral membrane proteins called *ion channels*. Ion channels allow ions to flow across the plasma membrane, which is normally impermeable to charged molecules. The firing of an action potential involves the sequential opening and closing of two types of channels, one selective for sodium ions, which move into the cell causing it to *depolarize* (become less negatively charged), and one for potassium ions, which flow out causing the cell to *repolarize*, or *hyperpolarize* (become more negatively charged). These channels are opened and shut (*gated*) by changes in the membrane potential or voltage. Determining the structure of these channels and how they are gated is one of the major challenges in neurobiology.

The first of the voltage-gated channels cloned was the sodium channel. It was cloned with oligonucleotides based on amino acid sequences from protein purified from the electric eel, a rich source of the

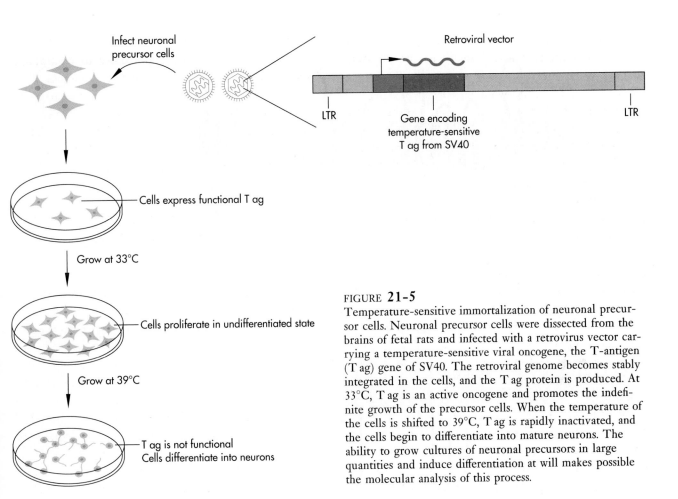

FIGURE **21-5**
Temperature-sensitive immortalization of neuronal precursor cells. Neuronal precursor cells were dissected from the brains of fetal rats and infected with a retrovirus vector carrying a temperature-sensitive viral oncogene, the T-antigen (T ag) gene of SV40. The retroviral genome becomes stably integrated in the cells, and the T ag protein is produced. At 33°C, T ag is an active oncogene and promotes the indefinite growth of the precursor cells. When the temperature of the cells is shifted to 39°C, T ag is rapidly inactivated, and the cells begin to differentiate into mature neurons. The ability to grow cultures of neuronal precursors in large quantities and induce differentiation at will makes possible the molecular analysis of this process.

sodium channel. Soon thereafter, the *Drosophila* voltage-gated potassium channel was cloned based on genetic studies of a fly mutant called *shaker*. These flies shake because of a potassium channel defect that causes their neurons to repolarize poorly after the firing of action potentials. The *shaker* locus was cloned by chromosome walking (see Chapter 7), and sequence analysis showed that it had a structure similar to that of the gene for the sodium channel. That the *shaker* gene indeed encoded a voltage-gated potassium channel was confirmed by injecting *shaker* mRNA into *Xenopus* oocytes (see Chapter 18) and showing that the injected oocytes acquired such a channel. Mammalian counterparts of these genes have now been cloned by low-stringency hybridization using *shaker* probes.

Dozens of channel cDNAs have now been cloned, and they encode complex proteins with related structures (Figure 21-6). The sodium channel protein is large, consisting of almost 2000 amino acids arranged into four repeated domains that each in turn encode six peptide helices that span the membrane to create a channel for ions to pass through. The potassium channel genes encode only a single six-helix domain. A functional potassium channel is thought to assemble from four of the six-helix subunits. Studies of channel function rely on transcribing cDNAs into mRNA with bacteriophage RNA polymerase and injecting the RNA into *Xenopus* oocytes. The channel proteins are produced and assembled into the oocyte membrane, and channel function can be measured by impaling the oocyte with a microelectrode. This simple assay allows investigators to put mutations into channel cDNAs and measure the functional properties of the mutant proteins. Such studies are getting to the heart of questions that have puzzled neurophysiologists for decades. For example, researchers have identified the portion of the channel protein that forms the ion pore

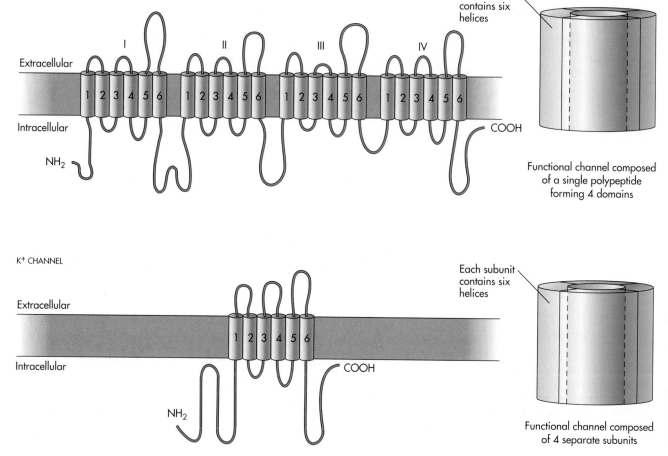

NA⁺ CHANNEL

Each domain contains six helices

Extracellular

Intracellular

NH₂

COOH

Functional channel composed of a single polypeptide forming 4 domains

K⁺ CHANNEL

Each subunit contains six helices

Extracellular

Intracellular

NH₂

COOH

Functional channel composed of 4 separate subunits

FIGURE **21-6**

The conserved structure of ion channels as revealed by gene cloning. Each sodium channel comprises a single long polypeptide chain that forms four domains (I–IV) each domain contains six helices (1–6) that span the cell membrane. The four linked domains fold to form the channel molecule. Calcium channels exhibit the same four-domain structure as sodium channels. In contrast, the potassium channel gene encodes only a single six-helix unit; four of these molecules assemble to form a functional channel. The similarities in the sequence and organization of the different channels suggest that they arose from a single common ancestor channel earlier in evolution.

that selectively passes one ion but not others—by substituting a few amino acids from one channel to another, they can transfer the ion conductance properties of the first channel to the second (Figure 21-7). Also under study is how the channel senses changes in voltage that signal it to open or shut the ion pore (Figure 21-8). Mutations in positively charged amino acids in the fourth transmembrane helix affected the voltage sensitivity of channel proteins, suggesting that changes in the position of this positively charged helix trigger channel opening.

Neurotransmitter Receptors Are Members of Large Gene Families

The neurotransmitters that convey signals from one cell to another fall into two general classes, small molecules (usually related to amino acids) and short peptides (such as the endorphins and enkephalins). We have already discussed the acetylcholine (ACh) receptors in Chapter 17. First purified and cloned from electric eel, the *nicotinic* ACh receptor (so called because it is activated by nicotine) is a *ligand-gated voltage*

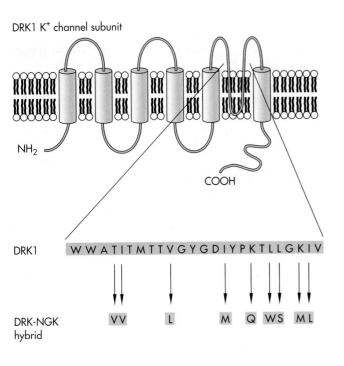

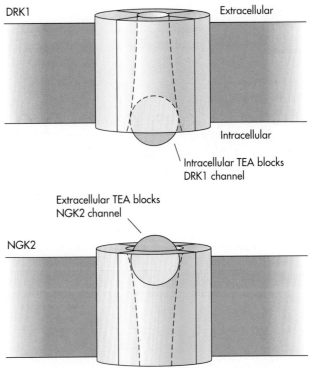

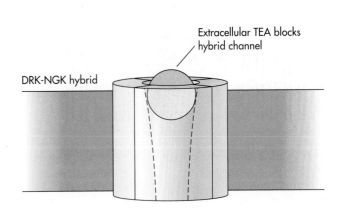

FIGURE **21-7**
Identification of the ion pore in the potassium channel. Two different potassium channels differ in their sensitivity to the drug TEA, which acts by blocking the pore of the channel. The DRK channel is more sensitive to TEA on its cytoplasmic side, whereas the NGK channel is blocked more readily by TEA on its extracellular side. The sequences of the two channel proteins differ by nine amino acids in the region thought to comprise the pore. By site-directed mutagenesis (Chapter 11), these nine amino acids in the DRK channel were replaced with the corresponding amino acids of the NGK channel. The hybrid protein, expressed in *Xenopus* oocytes, adopted the TEA sensitivity of the NGK channel, showing that this region indeed constitutes the pore.

channel: binding of ligand (ACh or nicotine) changes the conformation of the receptor, allowing sodium and potassium ions to pass through. The receptor is made up of multiple subunits encoded by distinct genes; each subunit consists of a protein that spans the membrane four times. By a combination of expression strategies and homology cloning, dozens of receptor genes of this family have now been cloned. These genes include ones encoding receptors for glutamate, glycine, and γ-aminobutyric acid (GABA); the last two receptors are channels for chloride ions. Cloning has un-

veiled the existence of distinct neurotransmitter receptors in numbers that had not been anticipated by neurophysiologists. Expression of each of these receptors in oocytes has shown that they can differ in ligand affinity and ion conductance properties. Distinct subunits can combine in any number of ways to produce channels with still different biochemical properties. The significance of this surprising receptor heterogeneity is not yet known, but individual receptor subtypes are expressed in unique patterns in the brain (this has been shown using in situ hybridization,

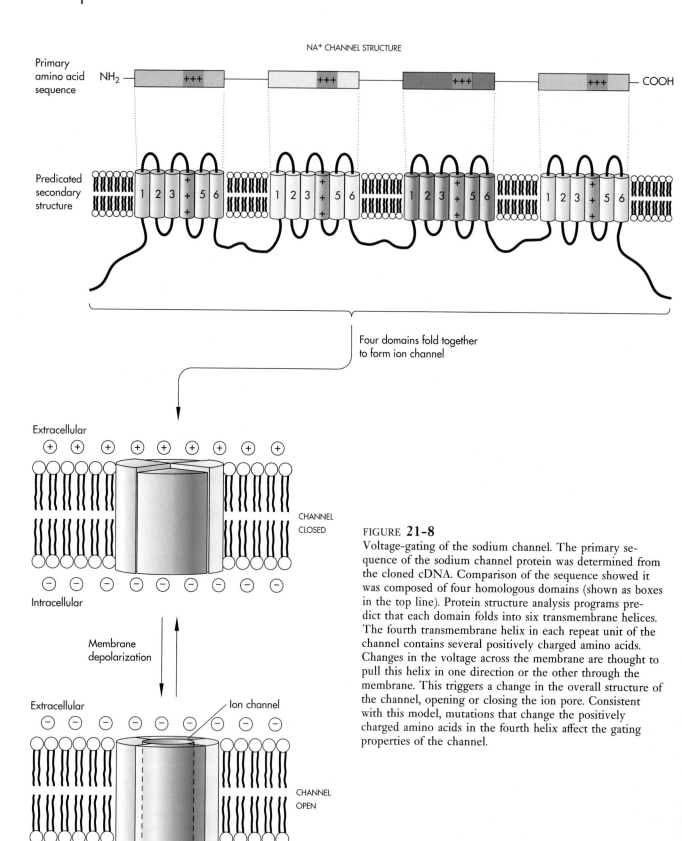

NA+ CHANNEL STRUCTURE

Primary amino acid sequence

Predicated secondary structure

Four domains fold together to form ion channel

Extracellular

CHANNEL CLOSED

Intracellular

Membrane depolarization

Extracellular

Ion channel

CHANNEL OPEN

Intracellular

FIGURE **21-8**

Voltage-gating of the sodium channel. The primary sequence of the sodium channel protein was determined from the cloned cDNA. Comparison of the sequence showed it was composed of four homologous domains (shown as boxes in the top line). Protein structure analysis programs predict that each domain folds into six transmembrane helices. The fourth transmembrane helix in each repeat unit of the channel contains several positively charged amino acids. Changes in the voltage across the membrane are thought to pull this helix in one direction or the other through the membrane. This triggers a change in the overall structure of the channel, opening or closing the ion pore. Consistent with this model, mutations that change the positively charged amino acids in the fourth helix affect the gating properties of the channel.

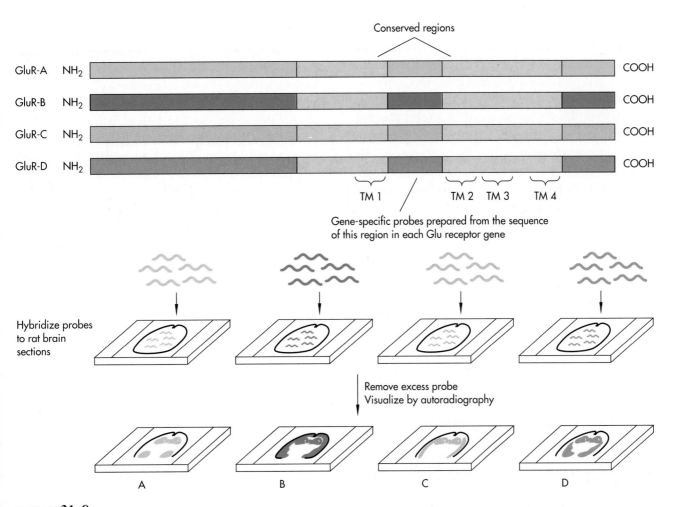

FIGURE **21-9**

Localization of glutamate receptor (GluR) expression in rat brain by in situ hybridization. A set of cDNAs was isolated that encoded different receptors (A–D) for the excitatory amino acid neurotransmitter glutamate. A comparison of the predicted amino acid sequences showed that they contained regions of sequence that were similar among all the receptors (conserved regions) and regions of unique sequence. The regions predicted to form the membrane-spanning helices are indicated as TM (transmembrane). To determine whether the different receptors were made in different parts of the brain, thin slices of rat brain were placed on microscope slides. Radioactively labeled RNA probes were prepared from the unique segments of each receptor cDNA and incubated individually with brain slices under conditions that allowed the probes to anneal to receptor mRNA anywhere it occurred in the tissue slice. Exposure of the slices to x-ray film generated a pattern of exposure that reflected the location of expression of each receptor. The patterns for the four receptors showed some regions in common, whereas other parts of the brain expressed only a single form of receptor. These patterns of expression can sometimes be helpful in deciphering the function of a cloned gene.

illustrated in Figure 21-9), and each receptor probably responds differently to drugs used to treat mental illness. The ability to produce each receptor in a uniform fashion by expression in oocytes provides a powerful way to examine which drugs affect which receptors.

Other neurotransmitter receptors belong to the large family of receptor proteins that span the membrane seven times, an evolutionarily ancient family that includes the bacterial rhodopsins and yeast pheromone receptors. These receptors are not themselves ion channels, but are, in many cases, coupled to ion channels via G proteins and soluble second messengers, as discussed in Chapter 17. Because the coupling of these receptors to ion channels is indirect, activation of these receptors generates slower and longer-lasting signals than for the ligand-gated channels. The flexibility in the type of signal triggered by a given neurotransmitter contributes to the complexity and potential information content of the signals it generates. The G protein–coupled receptors constitute an enormous gene family, the members of which are being isolated by a variety of homology cloning strategies. Indeed, these strategies have been so successful that there are now dozens of orphan receptors for which the ligand is not known. The brain receptor for the active ingredient of marijuana, Δ^9-tetrahydrocannabinol (THC), was found this way. A receptor cDNA cloned using an oligonucleotide probe based on the substance K receptor was expressed in oocytes and mammalian cells and tested for its ability to bind and respond to a collection of neurotransmitters and drugs. In situ hybridization studies using this receptor cDNA as a probe showed that the receptor was expressed in the same parts of the brain known to be affected by THC. When THC was tested on the receptor-expressing cells, it elicited a strong response. The sheer number of these receptors cloned in recent years makes plain the daunting complexity of the brain, but nevertheless gives us a firm starting point from which to launch a molecular attack on this complexity.

Homology Cloning Is Used to Isolate Components of Signal Transduction Pathways in Olfactory Neurons

Homology cloning has been especially successful at unraveling the molecular basis of smell. The mammalian olfactory system has the sensitivity to detect odorant molecules at concentrations of a few parts per trillion and the resolution to discriminate among 10,000 different odorants, some of which differ only in their chiral properties. How is this discrimination achieved, and how is the information processed and passed to the olfactory bulb in the brain? Odorant detection is accomplished by several million olfactory neurons located in the nose. Each cell extends processes to the surface of the nasal cavity and a single axon to the brain. Based on analogies to the visual system (Chapter 17), it was hypothesized that olfactory signal transduction occurs via a G protein–coupled second messenger cascade. Indeed, it was shown biochemically that some odorants induce a GTP-dependent stimulation of adenylate cyclase, the enzyme that produces cAMP. The stage was then set for homology cloning of the components of the odorant-stimulated signal transduction system.

The first component cloned was an olfactory-specific G_α protein. A rat olfactory cell cDNA library was screened with a degenerate oligonucleotide directed against a highly conserved sequence in G proteins. In addition to several known G proteins, a novel clone was identified. Using this clone as a probe for mRNA prepared from different rat tissues, investigators found that expression of this gene was confined to olfactory tissue. Moreover, expression was confined to the neurons. When the olfactory bulb of a rat was removed, resulting in the specific degeneration of the olfactory neurons, this G protein mRNA was no longer found in olfactory tissue. Expression of the cloned cDNA in a cell line lacking G proteins showed that it was capable of activating adenylate cyclase. Recently, an olfactory neuron–specific adenylate cyclase and a cAMP-gated ion channel have also been cloned, both by low-stringency hybridization with probes from related proteins. Together, these studies have rapidly contributed to the view of olfactory signal transduction shown in Figure 21-10.

The missing piece of the puzzle is the odorant receptor. On the basis of involvement of a G protein–coupled second messenger cascade in olfaction, it seemed likely that the odorant receptors would belong to the seven membrane–spanning domain family that includes the rhodopsins and neurotransmitter receptors. By the use of degenerate PCR primers based on conserved sequences in this family, a large collection of clones encoding receptors of this class has been isolated from olfactory cells. Hybridization studies suggest that there may be hundreds of these receptor genes in the genome. The abundance and structure of these genes, together with their exclusive expression in the olfactory epithelium, are consistent with a role as odorant receptors, and the availability of each piece

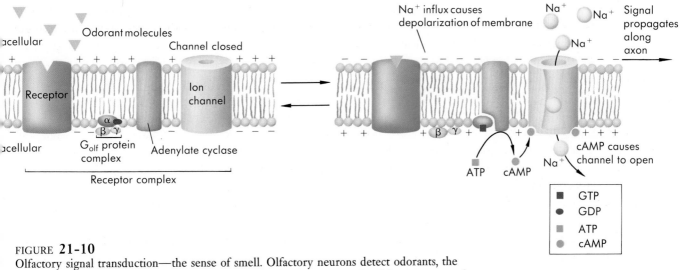

FIGURE **21-10**

Olfactory signal transduction—the sense of smell. Olfactory neurons detect odorants, the molecules we sense as odors, using a signal transduction apparatus related in structure and organization to other signaling systems (Chapter 17). But the individual components are specific and specialized for the olfactory system. Odorants bind to one or more receptor molecules (R) on the surface of olfactory cilia that extend into the mucous lining of the nasal cavity. There appear to be several hundred of these receptors, which fall into several subfamilies—all, however, of the seven-transmembrane class. The receptors are coupled to an olfactory-specific G protein, G_{olf}. Upon binding odorant, the receptor catalyzes the exchange of GDP for GTP by the α subunit of G_{olf}. This α subunit dissociates from $\beta\gamma$ and activates an olfactory-specific adenylate cyclase. Cyclic AMP, produced by adenylate cyclase, directly gates an ion channel, initiating an action potential that is propagated to the olfactory bulb in the brain for further processing. Other receptors activate a phospholipase (Chapter 17) instead of adenylate cyclase. Although there are hundreds of different olfactory receptor molecules, mammals can distinguish at least 10,000 different odors. Thus, each receptor must bind several odorant molecules, and discrimination among them must occur by information processing in the brain.

of the signaling cascade as a molecular clone should allow this role to be established unambiguously. With the isolation of odorant receptors, the analysis of how we can distinguish so many different odors can begin in earnest.

Learning and Memory Require Stable Changes in Neuron Function

Although all the signaling events we have discussed occur over time periods of milliseconds, our brains allow us to store information for decades. How are these signals converted to a stable form for storage in the brain? In other words, how do we learn and remember? This is perhaps one of the oldest problems in science. Now, some of the molecular mechanisms responsible for information storage in the brain are beginning to appear. One phenomenon under ag-

gressive study by scientists interested in learning and memory is *long-term potentiation*, or *LTP*. LTP is a long-lasting strengthening of synaptic signaling that may reflect the types of changes required for memory storage. LTP appears to involve a particular type of receptor for the excitatory amino acid neurotransmitter glutamate. Because of its importance in LTP, this receptor (called the *NMDA receptor*, for an agonist that specifically activates this class of receptors) is a popular target for gene cloners.

Another approach for studying the biochemistry of learning and memory uses the marine slug *Aplysia*, which is capable of crude forms of learning. Short-term learning in *Aplysia* appears to involve modifications of preexisting signal transduction proteins that change the firing pattern of the neurons involved. Components of the cAMP cascade are especially important in this response. Converting this short-term learning to longer-term storage, in contrast, requires

(a)

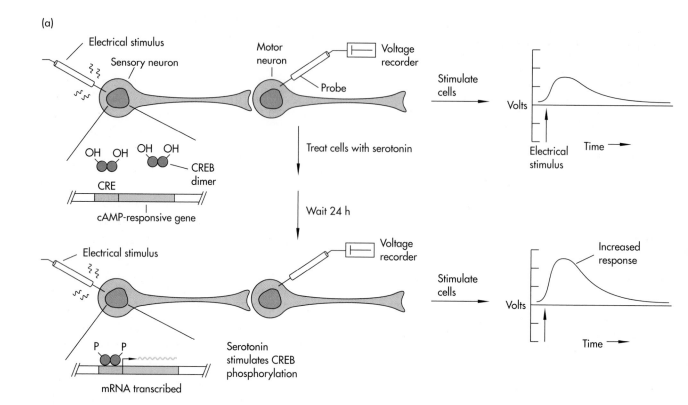

(b)

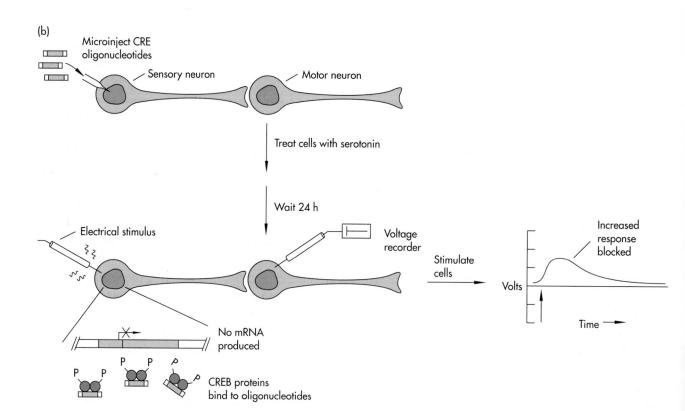

FIGURE **21-11** *(Facing page)*
Establishing a role for specific gene expression in long-term memory. Studies with defined neuron circuits in the sea slug *Aplysia* have established models for learning and memory. (a) When an *Aplysia* sensory neuron is electrically stimulated, it transmits a synaptic signal to a neighboring motor neuron, which can be monitored with a probe inserted into the motor neuron. If serotonin is applied five times in rapid succession, the sensory neuron undergoes a long-term *facilitation* that can persist for days: when the sensory neuron is subsequently stimulated, it elicits a much stronger response from the motor neuron. Evidence suggested a role for cAMP and for gene transcription in serotonin-mediated long-term facilitation. Control of transcription by cAMP is mediated by the CREB transcription factor, which binds to the cAMP responsive element (CRE) upon phosphorylation (see Chapter 17). (b) Thus, investigators injected the sensory neuron with oligonucleotides that bind the cAMP-responsive transcription factors that activate immediate early gene expression in response to serotonin. The oligonucleotides sequester the transcription factors, preventing them from reaching their target genes. When the sensory neuron was so injected prior to treatment with serotonin, it failed to establish long-term facilitation. Thus, activation of cAMP–regulated genes is required for long-term changes in neuronal signaling.

that neurons synthesize new mRNAs and proteins. What might these new products be? One hint comes from the discovery that a specific set of genes is rapidly turned on in neurons subject to signals. These genes, termed *immediate early genes,* have already been extensively studied, because they are also rapidly activated in cells treated with growth factors (see Chapters 17 and 18). Among these genes are protooncogenes such as c-*fos,* which are themselves transcription factors able to turn on new sets of genes. There are now many reported cases in which treatment of animals with stimuli known to elicit brain activity induces immediate early gene expression in precisely the same brain cells that had been implicated in the response by use of classical electrophysiological methods. This gene expression is visualized in brain slices using in situ hybridization with gene probes or using immunochemical methods that employ antibodies against the protein products of the genes.

That specific changes in gene expression may be required for the long-term changes associated with memory comes from an *Aplysia* experiment (Figure 21-11). Investigators injected an *Aplysia* sensory neuron with a double-stranded oligonucleotide that binds transcription factors required for the induction of immediate early gene expression by cAMP (see Chapter 9). Injection of this oligonucleotide blocked the long-term response of the neuron, showing that the cAMP signaling system functions both in the short-term and long-term modification of neuron function.

Perhaps the most promising avenue for studying learning and memory is a surprising one: *Drosophila.* Although it requires more ingenuity from the experimenters than the flies, it is possible to teach flies to avoid noxious stimuli (Figure 21-12). With this learning assay, investigators have methodically sorted through mutagenized flies to find ones that don't learn well. Several different mutants were identified, and the genes were named *dunce, rutabaga, cabbage, turnip,* and *radish.* A sixth mutant line of flies learned properly but promptly forgot what they learned; the gene for this defect was termed *amnesiac.* The chromosomal location of the *dunce* mutation was mapped genetically and the gene cloned by chromosome walking. The gene encodes a cAMP phosphodiesterase, an enzyme that destroys cAMP. The *rutabaga* mutants were deficient in adenylate cyclase, the enzyme that synthesizes cAMP. These discoveries dramatically affirm the importance of the cAMP signaling pathway in learning and memory.

Although the first fly learning and memory mutants were identified in the 1970s, molecular analysis of the genes has been slowed by their rather complex genetics and the difficult assays that must be performed to follow the mutant alleles in genetic experiments. Now, however, the powerful new tools for gene analysis in *Drosophila* (Chapter 20) are being used to study the learning genes and to isolate new ones. Antibodies to the Dunce protein have shown what portions of the fly's brain express this protein at highest levels. Several hundred fly lines carrying enhancer trap insertions have been screened to find flies with insertions expressed in these same brain cells. Several such lines have been found, and in some of them the flies are defective in learning, suggesting that the insertion ac-

(a)

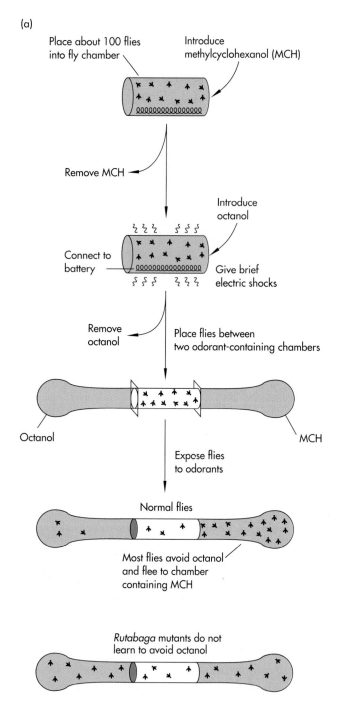

Place about 100 flies into fly chamber

Introduce methylcyclohexanol (MCH)

Remove MCH

Introduce octanol

Connect to battery

Give brief electric shocks

Remove octanol

Place flies between two odorant-containing chambers

Octanol

MCH

Expose flies to odorants

Normal flies

Most flies avoid octanol and flee to chamber containing MCH

Rutabaga mutants do not learn to avoid octanol

FIGURE **21-12**

Testing flies for learning and memory. (a) Flies are placed in a chamber and exposed first to one odorant, methyl-cyclohexanol (MCH), and then to a second, octanol. In the presence of octanol, they are given an electrical shock. To determine whether the flies have learned to associate octanol with the unpleasant shock, flies are placed in a chamber flanked by sources of both odorants. Normal flies flee the octanol for the MCH chamber. Files with learning mutations (such as *rutabaga*) distribute equally between the two chambers. (b) Most normal flies will "remember" the association of octanol with the shock if tested again in 30 minutes. Flies with memory mutations, like *amnesiac*, flee from octanol if tested immediately but have forgotten when retested 30 minutes later.

(b)

Train flies
Wait 30 minutes
Retest

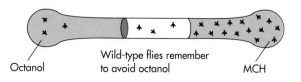

Octanol

Wild-type flies remember to avoid octanol

MCH

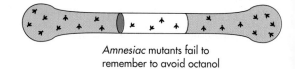

Amnesiac mutants fail to remember to avoid octanol

(a) Amyloid precursor protein cDNAs

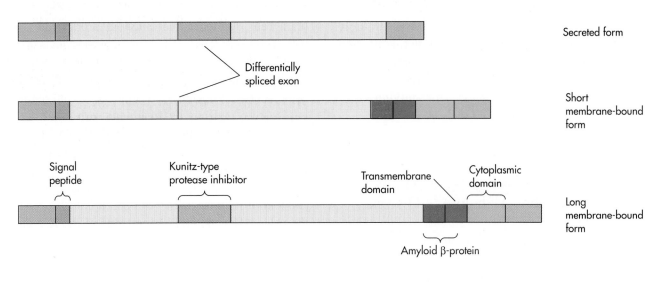

(b) Amyloid precursor protein

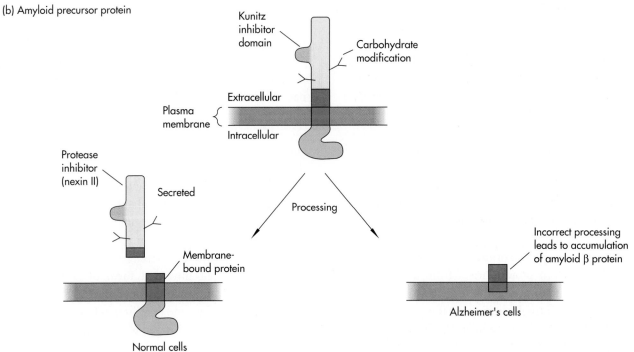

FIGURE 21-13

Organization and processing of the amyloid precursor protein. Amyloid protein accumulates in the plaques and tangles of Alzheimer's disease brains. (a) Analysis of cDNAs encoding the amyloid protein revealed that it arises from a much larger protein, the amyloid precursor protein (APP). Several forms of APP are produced owing to alternative splicing of the APP mRNA, as shown by cDNA cloning. (b) The largest mRNAs encode proteins that span the cell membrane. The large extracellular domain of one form functions as an inhibitor of proteases (enzymes that cleave proteins). In normal cells, this domain can be released from the cell surface by a proteolytic cleavage. In Alzheimer's disease, it is believed that the APP is cleaved improperly, generating the amyloid β protein.

tually landed within a learning gene. Of course, a probe from the enhancer trap vector can now be used to clone the gene. And the regulatory sequences from the gene, which apparently target expression to these particular neurons, can be used to direct expression of mutated versions of cloned learning genes back to these cells so their effects can be studied in living animals. The potential of *Drosophila* for studying learning and memory is just being realized.

Molecular Cloning and Gene Mapping Begin to Explore Alzheimer's Disease

Because we know so little about how the brain works, mental illness is particularly baffling to medicine. But as we learn more about brain development, the mechanisms of synaptic transmission, and the roles of the many neurotransmitters in brain function, rational treatments for mental illness will become possible. One of the major diseases of the brain under molecular scrutiny is Alzheimer's disease (AD), characterized by a progressive loss of memory, orientation, and judgment. Postmortem analysis of the brains of people stricken with AD revealed plaques of dead or dying neurons enmeshed in a tangle of fibers made up largely of a single short polypeptide, termed the *amyloid* protein. The amyloid protein was purified from diseased brain tissue, and its amino acid sequence was determined, allowing several research groups to design oligonucleotide probes to isolate amyloid protein cDNAs. The sequence of the cDNAs isolated showed that the amyloid protein arises from a much larger amyloid precursor protein (APP), which has the characteristics of a protein that spans the cell membrane. The protein fragment found in the amyloid plaques is apparently cleaved off the larger precursor by protease enzymes. Several different forms of the APP protein are produced by alternative splicing of APP mRNA (Figure 21-13). Some of these forms contain a sequence closely related to a family of known protease inhibitors, raising the possibility that problems with protein degradation contribute to the development of the disease.

Left unresolved is the fundamental question of whether the amyloid protein actually causes neuronal degeneration or is simply debris left behind by dying brain cells. Evidence that amyloid production contributes to neuronal cell death comes from a gene transfer experiment in which a neuronal precursor cell line was engineered to express a fragment of the APP gene. When these cells were induced to differentiate into mature neurons by treatment with NGF, they stopped differentiating after several days, degenerated, and died. Another indication that amyloid protein contributes directly to neuron degeneration was the observation that addition of synthetically produced amyloid protein fragments to neuron cultures caused the cells to degenerate. Examination of the sequence of the active fragment revealed that it was related to a family of known neuropeptides. Adding these neuropeptides to the cultures blocked the effect of the amyloid protein. These experiments suggest that amyloid protein may act by interfering with the action of normal neuropeptides.

The genetics of Alzheimer's disease is complex, and the results of genetic studies of inherited forms of AD have been conflicting, but recent discoveries begin to point to a causative role for amyloid protein. After the APP gene was cloned, it was mapped to human chromosome 21. An extra copy of chromosome 21 is the cause of Down syndrome (DS), a form of mental retardation, and DS patients develop symptoms similar to AD, including amyloid plaques, at an early age. Perhaps Alzheimer's and DS are both due simply to excessive production of the amyloid protein. This hypothesis was considerably strengthened when the gene responsible for an inherited form of AD was also mapped to chromosome 21 near the APP gene (we discuss methods for gene mapping in Chapter 26). But this neat picture deteriorated when more detailed genetic analyses showed evidence for recombination between the "Alzheimer's gene" and the APP gene, suggesting that the two genes were different. Recently, however, the gene responsible for inherited AD in two unrelated British families was once again mapped close to the APP gene. Using the polymerase chain reaction to amplify and sequence the APP gene from affected individuals in these families, researchers discovered that the gene carried a single nucleotide mutation that changed a valine to an isoleucine in the transmembrane domain of the amyloid protein. A mutation changing a nearby amino acid has also been found in individuals with a rare hereditary cerebral hemorrhage disorder. Although formal proof that APP mutations contribute to the development of Alzheimer's disease is still lacking, this is the strongest evidence to date for a causative role for the amyloid protein. Here again, we see how the tools of recombinant DNA have given us a foothold in our effort to understand and treat a mystifying and tragic disease.

Reading List

General

The Brain. Cold Spring Harbor Symp. Quant. Biol., Vol. 55. Cold Spring Harbor Laboratory, Cold Spring Harbor, N. Y., 1990.

Original Research Papers

GENES THAT REGULATE NEURONAL DEVELOPMENT

Weintraub, H., R. Davis, S. Tapscott, M. Thayer, M. Krause, Benezra, T. K. Blackwell, D. Turner, R. Rupp, S. Hollenberg, Y. Zhuang, and A. Lassar. "The *myoD* gene family: nodal point during specification of the muscle cell lineage." *Science*, 251: 761–766 (1991). [Review]

Benezra, R., R. L. Davis, D. Lockshon, D. L. Turner, and H. Weintraub. "The protein Id: a negative regulator of helix-loop-helix DNA binding proteins." *Cell*, 61: 49–59 (1990).

Ellis, H. M., D. R. Spann, and J. W. Posakony. "*extramacrochaetae*, a negative regulator of sensory organ development in *Drosophila*, defines a new class of helix-loop-helix proteins. " *Cell*, 61: 27–38 (1990).

Johnson, J. E., S. J. Birren, and D. J. Anderson. "Two rat homologues of *Drosophila achaete-scute* specifically expressed in neuronal precursors." *Nature*, 346: 858–861 (1990).

IMMORTALIZED NEURONAL CELLS

Cepko, C. L. "Immortalization of neural crest cells via retrovirus-mediated oncogene transduction." *Ann. Rev. Neurosci.*, 12: 47–65 (1989). [Review]

Frederiksen, K., P. S. Jat, N. Valtz, D. Levy, and R. McKay. "Immortalization of precursor cells from the mammalian CNS." *Neuron*, 1: 439–448 (1988).

Birren, S. J., and D. J. Anderson. "A v-*myc*-immortalized sympathoadrenal progenitor cell line in which neuronal differentiation is initiated by FGF but not NGF." *Neuron*, 4: 189–201 (1990).

Hammang, J. P., E. E. Baetge, R. R. Behringer, R. L. Brinster, R. D. Palmiter, and A. Messing. "Immortalized retinal neurons derived from SV40 T-antigen-induced tumors in transgenic mice." *Neuron*, 4: 775–782 (1990).

GENES ENCODING NEUROTROPHIC FACTORS AND RECEPTORS

Chao, M. V., M. A. Bothwell, A. H. Ross, H. Koprowski, A. A. Lanahan, C. R. Buck, and A. Sehgal. "Gene transfer and molecular cloning of the human NGF receptor." *Science*, 232: 518–521 (1986).

Radeke, M. J., T. P. Misko, C. Hsu, L. A. Herzenberg, and E. A. Shooter. "Gene transfer and molecular cloning of the rat nerve growth factor receptor." *Nature*, 325: 593–597 (1987).

Leibrock, J., F. Lottspeich, A. Hohn, M. Hofer, B. Hengerer, P. Masiakowski, H. Thoenen, and Y.-A. Barde. "Molecular cloning and expression of brain-derived neurotrophic factor." *Nature*, 341: 149–152 (1989).

Rosenthal, A., D. V. Goeddel, T. Nguyen, M. Lewis, A. Shih, G. R. Laramee, K. Nikolics, and J. W. Winslow. "Primary structure and biological activity of a novel human neurotrophic factor." *Neuron*, 4: 767–773 (1990).

Squinto, S. P., T. H. Aldrich, R. M. Lindsay, D. M. Morrissey, N. Panayotatos, S. M. Bianco, M. E. Furth, and G. D. Yancopoulos. "Identification of functional receptors for ciliary neurotrophic factor on neuronal cell lines and primary neurons." *Neuron*, 5: 757–766 (1990).

Kaplan, D. R., D. Martin-Zanca, and L. F. Parada. "Tyrosine phosphorylation and tyrosine kinase activity of the *trk* proto-oncogene product induced by NGF." *Nature*, 351: 158–160 (1991).

Klein, R., S. Jing, V. Nanduri, E. O'Rourke, and M. Barbacid. "The *trk* proto-oncogene encodes a receptor for nerve growth factor." *Cell*, 65: 189–197 (1991).

ION CHANNELS

Betz, H. "Ligand-gated ion channels in the brain: the amino acid receptor superfamily." *Neuron*, 5: 383–392 (1990). [Review]

Noda, M., et al. "Primary structure of *Electrophorus electricus* sodium channel deduced from cDNA sequence." *Nature*, 312: 121–127 (1984).

Goldin, A. L., T. Snutch, H. Lubert, A. Dowsett, J. Marshall, W. A. Auld, W. Downey, L. C. Fritz, H. A. Lester, R. Dunn, W. A. Catterall, and N. Davidson. "Messenger RNA coding for only the α subunit of the rat brain Na channel is sufficient for expression of functional channels in *Xenopus* oocytes." *Proc. Natl. Acad. Sci. USA*, 83: 7503–7507 (1986).

Papazian, D. M., T. L. Schwarz, B. L. Tempel, Y. N. Jan, and L. Y. Jan. "Cloning of genomic and complementary DNA from Shaker, a putative potassium channel gene from *Drosophila*." *Science*, 237: 749–753 (1987).

Leonard, R. J., C. G. Labarca, P. Charnet, N. Davidson, and H. A. Lester. "Evidence that the M2 membrane-spanning region lines the ion channel pore of the nicotinic receptor." *Science*, 242: 1578–1581 (1988).

Stuhmer, W., F. Conti, H. Suzuki, X. Wang, M. Noda, N. Yahagi, H. Kubo, and S. Numa. "Structural parts involved in activation and inactivation of the sodium channel." *Nature*, 239: 597–603 (1989).

Hartmann, H. A., G. E. Kirsch, J. A. Drewe, M. Tagilalatela, R. H. Joho, and A. M. Brown. "Exchange of conduction pathways between two related K$^+$ channels." *Science*, 251: 942–944 (1991).

Yellen, G., M. E. Jurman, T. Abramson, and R. MacKinnon. "Mutations affecting internal TEA blockade identify the probable pore-forming region of a K$^+$ channel." *Science*, 251: 939–942 (1991).

Yool, A. J., and T. L. Schwarz. "Alteration of ionic selectivity of a K$^+$ channel by mutation of the H5 region." *Nature*, 349: 700–704 (1991).

NEUROTRANSMITTER RECEPTORS

Hollman, M., A. O'Shea-Greenfield, S. Rogers, and S. Heinemann. "Cloning by functional expression of a member of the glutamate receptor family." *Nature*, 342: 643–648 (1989).

Keinanen, K., W. Wisden, B. Sommer, P. Werner, A. Herb, T. A. Verdoorn, B. Sakmann, and P. H. Seeburg. "A family of AMPA-selective glutamate receptors." *Science*, 249: 556–560 (1990).

Matsuda, L. A., S. J. Lolait, M. J. Brownstein, A. C. Young, and T. I. Bonner. "Structure of a cannabinoid receptor and functional expression of the cloned cDNA." *Nature*, 346: 561–564 (1990).

Nakanishi, N., N. A. Shneider, and R. Axel. "A family of glutamate receptor genes: evidence for the formation of heteromultimeric receptors with distinct channel properties." *Neuron*, 5: 569–581 (1990).

OLFACTION

Reed, R. R. "G protein diversity and the regulation of signalling pathways." *New Biol.*, 2: 957–960 (1990). [Review]

Jones, D. T., and R. R. Reed. "G$_{olf}$: an olfactory neuron specific-G protein involved in odorant signal transduction." *Science*, 244: 790–795 (1989).

Bakalyar, H. A., and R. R. Reed. "Identification of a specialized adenylyl cyclase that may mediate odorant detection." *Science*, 250: 1403–1406 (1990).

Dhallan, R. S., K.-W. Yau, K. A. Schrader, and R. R. Reed. "Primary structure and functional expression of a cyclic nucleotide-activated channel from olfactory neurons." *Nature*, 347: 184–187 (1990).

Buck, L., and R. Axel. "A novel multigene family may encode odorant receptors: a molecular basis for odor recognition." *Cell*, 65: 175–187 (1991).

LEARNING AND MEMORY

Sheng, M., and M. E. Greenberg. "The regulation and function of c-*fos* and other immediate early genes in the nervous system." *Neuron*, 4: 477–485 (1990). [Review]

Dash, P. K., B. Hochner, and E. R. Kandel. "Injection of the cAMP-responsive element into the nucleus of *Aplysia* sensory neurons blocks long-term facilitation." *Nature*, 345: 718–721 (1990).

ALZHEIMER'S DISEASE

Masters, C. L., G. Simms, N. A. Weinman, G. Multhaup, B. L. McDonald, and K. Beyreuther. "Amyloid plaque core protein in Alzheimer disease and Down syndrome." *Proc. Natl. Acad. Sci. USA*, 82: 4245–4249 (1985).

Goldgaber, D., M. I. Lerman, O. W. McBride, U. Saffiotti, and D. C. Gajdusek. "Characterization and chromosomal localization of a cDNA encoding brain amyloid of Alzheimer's disease." *Science*, 235: 877–880 (1987).

Kang, J., H.-G. Lemaire, A. Unterbeck, J. M. Salbaum, C. L. Masters, K.-H. Grzeschik, G. Multhaup, K. Beyreuther, and B. Muller-Hill. "The precursor of Alzheimer's disease amyloid A4 protein resembles a cell-surface receptor." *Nature*, 325: 733–736 (1987).

Tanzi, R. E., J. F. Gusella, P. C. Watkins, G. A. P. Bruns, P. S. George-Hyslop, M. L. Van Keuren, D. Patterson, S. Pagan, D. M. Kurnit, and R. L. Neve. "Amyloid β protein gene: cDNA, mRNA distribution and genetic linkage near the Alzheimer locus." *Science*, 235: 880–884 (1987).

Kitaguchi, N., Y. Takahashi, Y. Tokushima, S. Shiojiri, and H. Ito. "Novel precursor of Alzheimer's disease amyloid protein shows protease inhibitory activity." *Nature,* 331: 530–532 (1988).

Ponte, P., P. Gonsalez-DeWhitt, J. Schilling, J. Miller, D. Hsu, B. Greenberg, K. Davis, W. Wallace, I. Leiberburg, F. Fuller, and B. Cordell. "A new A4 amyloid mRNA contains a domain homologous to serine proteinase inhibitors." *Nature,* 331: 525–527 (1988).

de Sauvage, F., and J.-N. Octave. "A novel mRNA of the A4 amyloid precursor gene coding for a possibly secreted protein." *Science,* 245: 651–653 (1989).

Van Nostrand, W. E., S. L. Wagner, M. Suzuki, B. H. Choi, J. S. Farrow, J. W. Geddes, C. W. Cotman, and D. D. Cunningham. "Protease nexin-II, a potent antichymotrypsin, shows identity to amyloid β-protein precursor." *Nature,* 341: 546–549 (1989).

Esch, F. S., P. S. Keim, E. C. Beattie, R. W. Blacher, A. R. Culwell, T. Oltersdorf, D. McClure, and P. J. Ward.

"Cleavage of amyloid β peptide during constitutive processing of its precursor." *Science,* 248: 1122–1124 (1990).

Levy, E., M. D. Carman, I. J. Fernandez-Madrid, M. D. Power, I. Lieberburg, S. G. Van Duinen, G. TH. A. M. Bots, W. Luyendijk, and B. Frangione. "Mutation of the Alzheimer's disease amyloid gene in hereditary cerebral hemorrhage Dutch type." *Science,* 248: 1124–1126 (1990).

St. George-Hyslop, P. H., et al. "Genetic linkage studies suggest that Alzheimer's disease is not a single homogenous disorder." *Nature,* 347: 194–197 (1990).

Yankner, B. A., L. K. Duffy, and D. A. Kirschner. "Neurotrophic and neurotoxic effects of amyloid β protein: reversal by tachykinin neuropeptides." *Science,* 250: 279–282 (1990).

Goate, A., et al. "Segregation of a missense mutation in the amyloid precursor protein gene with familial Alzheimer's disease." *Nature,* 349: 704–706 (1991).

22

Recombinant DNA and Evolution

Questions about the origins of living organisms have played a central role in our intellectual history since the earliest times. But it was only in the nineteenth century, with the appearance of Darwin's theory of evolution by natural selection, that a scientific approach to these questions became possible. Evolutionary theory informs all of biology, and molecular genetics and biology are no exception. Genes contain a record of their own evolutionary history, a history that can be glimpsed in the structural and organizational similarities between genes and the preservation of functional motifs in different genes performing similar functions. We have repeatedly seen how this evolutionary history, written in DNA sequence, can be exploited for gene cloning and for deducing the functions of newly cloned genes. In this chapter we will look at what is known of the origins of living organisms and the ways in which genomes, genes, and their proteins have evolved. Recombinant DNA techniques have also provided biologists with new tools for examining controversial issues in evolutionary biology. This field, called *molecular systematics,* has revolutionized studies of species. It is a controversial field, because some interpretations of relationships based on molecular data conflict with traditional views. Finally, even the evolution of behavioral traits like altruism are being explored by using recombinant DNA techniques.

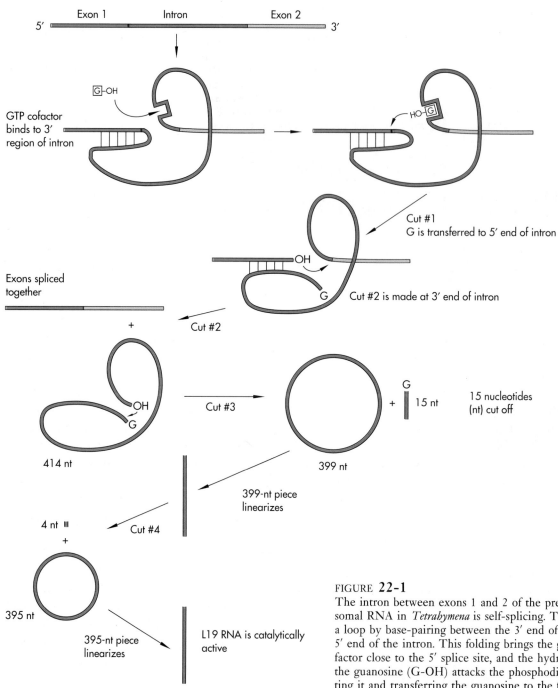

FIGURE **22-1**

The intron between exons 1 and 2 of the precursor of ribosomal RNA in *Tetrahymena* is self-splicing. The RNA forms a loop by base-pairing between the 3′ end of exon 1 and the 5′ end of the intron. This folding brings the guanosine cofactor close to the 5′ splice site, and the hydroxyl group of the guanosine (G-OH) attacks the phosphodiester link, cutting it and transferring the guanosine to the free 5′ end of the intron. The free hydroxyl group of the last nucleotide of exon 1 now attacks the 3′ splice site, cutting it and producing a free phosphate group on the first nucleotide of exon 2. Immediately, exons 1 and 2 join spontaneously. The excised intron then undergoes further reactions, in which first 15 and then 4 nucleotides (nt) are cut from the intron. The final product of these reactions, L19 RNA (so called because 19 nucleotides have been removed) is catalytically reactive. It can act as both a polymerase and a nuclease, depending on pH and the concentrations of the substrates.

Life May Have Originated in an RNA World

How did the first biologically important molecules arise? Forty years ago there was great excitement in the research world when it was shown that amino acids are made when electrical discharges occur in an atmosphere of ammonia and hydrogen, in the presence of water. These experiments were thought to reproduce the conditions present on the primeval earth, and it seemed reasonable to assume that more complex biological molecules would be formed in the resulting primordial soup. However, proteins cannot reproduce or evolve, and more recently this vision of life beginning with simple proteins has been supplanted by the speculation that nucleic acids are life's more likely starting materials.

This idea received a tremendous boost when RNA molecules were found to act as catalysts, a function previously thought to be an exclusive property of proteins. This extraordinary finding came from two sources. *RNase P* is an enzyme that processes the large precursors of transfer RNAs. The enzyme has both a protein and an RNA component, but it was shown that the RNA component alone was able to cut the precursor of the *E. coli* tyrosine tRNA. Another example of a catalytic function for RNA came from studies of the splicing process of the ribosomal RNAs of the ciliate *Tetrahymena* (Figure 22-1). The precursors of these molecules contain an intron; but this intron is cut out of the molecule by the action of the molecule itself, together with a guanosine molecule as a cofactor. This is called *group I* self-splicing and is characterized by the involvement of the guanosine cofactor. It occurs in fungal mitochondrial pre-mRNAs and pre-rRNAs and in some bacteriophage pre-mRNAs, as well as in *Tetrahymena*. Excision of the intron is not the end of the story, however, for the excised intron twice circularizes and linearizes and cuts another 19 nucleotides off itself. The resulting molecule (called L-19) is able to catalyze the formation of polynucleotides, act as a site-specific nuclease, and ligate short oligonucleotides! This observation was so unexpected that especially rigorous proof was required to demonstrate that the RNA was not contaminated by a protein (Figure 22-2). A second form of splicing occurs in the pre-mRNAs of mitochondrial genes of yeasts and fungi and in the chloroplasts of unicellular organisms like *Chlamydomonas*. This *group II* self-splicing does not require a cofactor, but instead the hydroxyl group of an adenine at the 3′ end of the intron cuts at the 5′ splice site. This mechanism involves the formation of a so-called *lariat structure* where the cut end of the intron loops back to join the RNA of the intron a short distance from the 3′ splice site (Figure 22-3). This is strikingly similar to the mechanism involved in the splicing of nuclear pre-mRNAs. However, nuclear pre-mRNAs require a complicated machinery for splicing, especially *spliceosomes*, complexes of ribonucleoprotein particles.

The interesting evolutionary question is why, if some RNAs can self-splice and mitochondrial RNA group II splicing is so similar to mRNA splicing, do mRNAs require a complicated cellular machinery for their splicing. One possible explanation is revealed by the different degrees to which splice site sequences are conserved in the different RNAs. The splice site recognition sequences are highly conserved in group I and II introns, and such conservation necessarily restricts sequence variations that may lead to new functions. If, however, sequences at splice sites are allowed to vary, then there must be other signposts that point to splice sites. The complex of proteins and RNA that constitutes the spliceosome may have arisen to do just this.

If RNA was the first molecule and had catalytic functions, then we must assume that it played a central role in the development of the biochemical processes essential to life. Traces of this central role might still be visible in modern cells, and this seems to be the case. Some of the modern functions of RNA are listed in Table 22-1. Furthermore, RNA, nucleotides, and ribose are involved in many steps in intermediary metabolism, for example, as cofactors. There are other data suggesting that RNA may be a very ancient molecule. Posttranscriptional *RNA editing* has been discovered in three species of protozoa. Comparisons of the genomic DNA and the mRNA sequence for some mitochondrial genes in these organisms showed that the mRNA molecules had nucleotides added or subtracted to produce the correct reading frame for

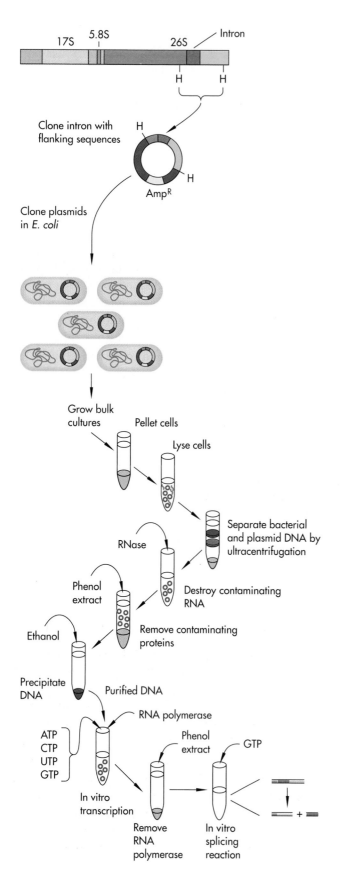

FIGURE **22-2**
Protein-free *Tetrahymena* RNA can self-splice. A portion of the ribosomal RNA genes, including the intron, was cloned into a plasmid. Bacteria containing the plasmid were grown in bulk culture, and plasmid DNA was separated from bacterial DNA by ultracentrifugation of the DNA through a cesium chloride gradient. The plasmid DNA was further purified. RNA was destroyed by treatment with RNase; phenol extractions were used to remove proteins; and pure DNA was precipitated by using ethanol. This DNA was used in an in vitro transcription system to produce RNA. The transcription reaction was treated with phenol to remove the *E. coli* RNA polymerase, and the purified RNA was then used in an in vitro splicing reaction. When the products of this reaction were analyzed by electrophoresis, a band was found that corresponded to the excised intron. Thus, self-splicing occurs in a system scrupulously clean of proteins.

TABLE **22-1**

Some Functions of RNA and Ribonucleotides

TYPE OF RNA*	FUNCTION
mRNA	Transfers information from genes to protein-synthesizing machinery
tRNA	Carries activated amino acids for protein synthesis
rRNA	Protein synthesis
U1, U2, U4/6, U5 snRNAs	mRNA splicing
M1 RNA	Catalytic unit of RNase P
Telomerase RNA	Template for telomere synthesis
Primer RNA	Initiation of DNA replication
7S RNA	Part of protein secretory complex
ATP	Carrier of energy-rich bonds
Coenzyme A	A key molecule in intermediate metabolism

* mRNA, messenger RNA; rRNA, ribosomal RNA; snRNA, small nuclear RNA; tRNA, transfer RNA.

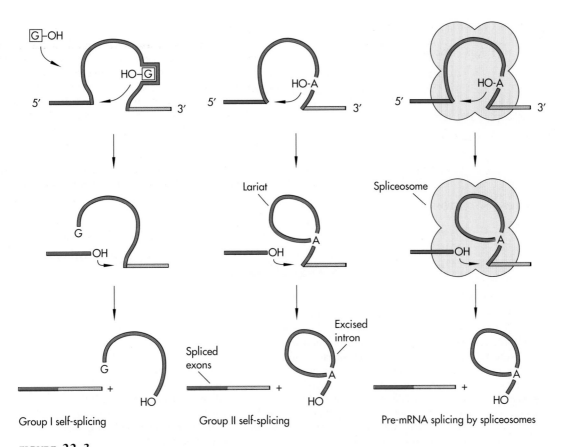

Group I self-splicing   Group II self-splicing   Pre-mRNA splicing by spliceosomes

FIGURE **22-3**
Comparison of the splicing of group I, group II, and pre-mRNA introns. As we have just
shown, group I self-splicing requires an added guanosine cofactor, and the 5′ end of the
intron remains free with the guanosine attached to it. The splicing reactions of group II and
pre-mRNA introns are similar to each other, in that the attack on the 5′ splice site is mediated by the hydroxyl group of an adenine nucleotide that is part of the intron sequence.
After cutting, the 5′ end of the intron forms a bond with this adenine, producing the lariat
structure. For pre-mRNA, this process requires the participation of a spliceosome, a complex
of several proteins.

translation (Figure 22-4). Remarkably a related phenomenon has been found in a mammalian gene. Apolipoprotein B is an essential component of the various
forms of plasma lipoproteins, and it is found in two
forms: apo-B100 is synthesized by the liver, and apo-
B48 is synthesized in the intestine. Apo-B48 is about
one-half the size of apo-B100; it comprises the amino
terminal half of apo-B100. This smaller molecule
could have been produced from a different gene, by
post-translational processing, or by differential splicing, but experiments had excluded these explanations.
Instead, the answer came from sequence comparisons

of apo-B cDNAs cloned from the liver and intestine.
These showed that there was a stop codon (TAA) in
the apo-B48 cDNA sequence where there was a glutamine codon (CAA) in apo-B100. Furthermore, when
the apo-B48 gene was analyzed, it was found that the
genomic DNA contained a glutamine codon and not
a stop codon. The conclusion is that the CAA codon
in the apo-B48 mRNA is edited to UAA. RNA editing
may be the vestige of an error-correcting mechanism
left over from a very early time when RNA replication
was an inaccurate process and careful editing was necessary to maintain correct genetic information.

ADDITION OF URIDINE

C. fasiculata COII gene

mtDNA ...AAGGTAGA   G A ACCTGGA...
mRNA   ...AAGGUAGAUUGUAUACCUGGA...

ADDITION OF CYTIDINE

P. polycephalum α subunit, ATP synthetase

mtDNA ...TGTC GTGCTTTAAATAC TTAGTCAAACCC TGTAGGTT...
cDNA   ...TGTCCGTGCTTTAAATACCTTAGTCAAACCCCTGTAGGTT...

ADDITION AND DELETION OF URIDINE

L. tarentolae COIII gene

mtDNA ...CG G A   G   G GTTTTGATTTTTGTTTGTTTTGTTG...
mRNA   ...CGUGUUAUUUUUGUUGGUG- - -UGA- - - - -G- -UG- - - -G-UG...

FIGURE **22-4**

RNA editing adds (or removes) nucleotides from transcribed RNA molecules. The examples show posttranscriptional modifications of RNA transcribed from the mitochondrial genes (mtDNA) of three organisms. The phenomenon was described first in the cytochrome oxidase subunit II gene (COII) of the protozoan *Crithidia fasiculata*. Comparisons of the RNA sequence with the genomic DNA sequence showed that uridines, for which there were no corresponding thymidines in the DNA sequence, had been inserted in the mRNA. This phenomenon is not restricted either to uridines or to protozoa. Cytidines are added to the ATP synthetase mRNA of the slime mold *Physarum polycephalum* (shown here as a cDNA). In some cases, both addition (green) and removal (orange) of uridines occurs in the same mRNA, as shown by the sequence from the COIII gene of the protozoan *Leishmania tarentolae*.

Data like these suggest that life may have begun in an RNA world; but there are difficulties with this view. Ribose can be synthesized by the polymerization of formaldehyde, and this reaction might have been possible in a prebiotic environment. However, over 40 different sugars are produced in this reaction, and ribose is not a major component of the mixture. Furthermore the nature of the world facing the first macromolecules is likely to have been very inhospitable. For example, RNA is very susceptible to hydrolysis, especially at the high temperatures and high concentrations of cations assumed for the prebiotic environment. It has been suggested instead that organic molecules did not arise in a primordial soup, but rather in special environments such as on the surface of clays or pyrites.

DNA—Why So Much?

Although we believe that DNA became the repository for genetic information because it is more stable than RNA, the steps in the transition from an RNA to a DNA world are lost in the mists of the prebiotic era. We can, however, learn something about the evolution of DNA genomes, once they became established, by studying present-day organisms.

An early unexpected finding of DNA studies of different species was that the amount of DNA did not seem to be correlated very strongly with what appeared to be the phenotypic complexity of the species. For example, taking the 4×10^6 bp of the *E. coli* genome as the unit of DNA measurement, *Drosophila* has about 40 units, and mouse and human beings about

1000 units. The newt, however, has 10,000 units, and some plants as much as 500,000 units! It is clear that in many cases the total amount of DNA does not reflect the numbers of different genes in a species. This excess of DNA over what is required for coding proteins has been called "junk" DNA, and some sequences, for example, those of pseudogenes (Chapter 8) do appear to be just that. However, we cannot assume that this DNA is "junk" just because we have not yet discovered its functions. In fact, we do know the functions of some of this DNA; for instance, the controlling elements of many genes are found in their noncoding 5′ regions. Some genes are present in multiple copies. These *amplified* genes may be genes whose product is required in very large quantities (histone genes are an example) or genes that became amplified in response to selection (drug-resistance genes are in this category). However, the functions of other excess DNA—for example, repetitive sequences and introns—remains speculative.

Genes Can Be Turned On (and Off) by Movable Elements

We learned in Chapter 10 how maize, yeast, and *Drosophila* have small genetic units called transposons that move around in the genome affecting gene expression. And while there is little genetic evidence for active transposons in human beings, genome sequencing revealed repetitive elements like the *Alu,* LINE, and SINE families that together make up as much as 25 percent of the genome. These elements have become distributed throughout the human genome by transposition (Chapter 7). The interesting question from an evolutionary perspective is why natural selection has not eliminated these elements. They are, after all, not functioning genes and are a burden on the cell that has to replicate them along with the "useful" DNA in genes. The answer appears to be that transposable elements are evolutionarily advantageous because they contribute to the generation of genetic diversity in eukaryotic genomes.

On a large scale, transposable elements can enhance chromosome rearrangements. This seems to be the case in yeast, where there is a correlation between the locations of Ty elements and the breakpoints of translocations, inversions, and other chromosomal abnormalities. These changes lead to a shuffling of large sections of chromosomes, thereby bringing together previously unlinked genes and modifying recombination frequencies. Transposable elements can also contribute to duplication and amplification of genes, because they themselves are targets for recombination events. Once again, as in other chapters, the globin gene cluster provides the example. The γ-globin gene is duplicated in the anthropoid lineage that includes *Homo sapiens,* gibbons, rhesus monkeys, and spider monkeys (Figure 22-5). Sequence analysis shows that the two γ gene loci are flanked by LINE elements (L1a and L1b) and are separated by an abnormal LINE sequence (L1ba). The latter is a hybrid made up of the 5′ region of L1b and the 3′ sequence of L1a. These LINE elements are absent from the galago, a prosimian. These are animals that diverged from the anthropoid lineage about 55 million years ago. The galago has only one γ gene, suggesting that the LINE elements may have been instrumental in the duplication of the γ gene in the anthropoids. It is probable that L1a inserted upstream of the single γ gene, and, independently, L1b inserted downstream of the gene (Figure 22-5). Recombination would then lead to duplication of the γ gene and to the creation of L1ba. What is the evolutionary significance of these changes? Both the ε and γ genes are expressed in the extra-embryonic yolk sac in the prosimian primates. These ε and γ genes have now diverged sufficiently that they may both be essential for different roles during the early embryonic life of these lower primates. In contrast, while the ε genes of the higher primates are still expressed in the yolk sac, the two γ genes are now expressed in the liver at a later stage of development. The duplication of the γ genes probably facilitated this change in the timing of γ-globin gene expression from the very early embryonic to later fetal development.

Another way in which transposable elements can generate genetic diversity is through inaccurate excision, a process resulting in the removal or addition of a small number of nucleotides from the integration site. This can happen surprisingly frequently. A sequence analysis of the insertion of an Spm/En element causing mutations in the *waxy* gene of maize showed that only one out of ten excisions was accurate. Six of the mutations still coded for a protein, and these differed from the wild type by one or two amino acids.

(a)

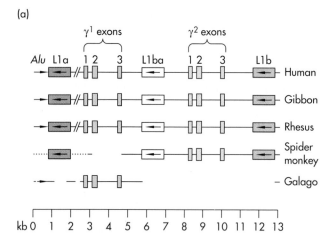

kb 0 1 2 3 4 5 6 7 8 9 10 11 12 13

(b)

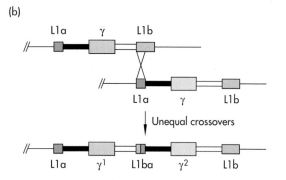

FIGURE 22-5
Repeat elements may be important in generating genetic diversity. (a) There are two γ-globin genes in primates, which have presumably arisen as a result of gene duplication. There are LINE elements, L1a and L1b, respectively upstream and downstream of the two genes (the orientation of the LINE elements is shown by the arrows). In between the two γ-globin genes is a peculiar sequence (L1ba) made up of the 5' end of L1b and the 3' end of L1a. The breaks in the sequences are gaps introduced to maximize the alignment between species. The galago has only one γ-globin gene, and this has arbitrarily been aligned with the γ¹ gene of the other species. The sequence for the 5' region of spider monkey is incomplete (dotted line). It appears that this duplication event was mediated by unequal crossing over, involving homologous sections of the L1a and L1b elements. The result was duplication of the ancestral γ-globin gene and formation of L1ba. (Fitch el al., 1991)

Such "new" proteins might be advantageous, and the mutation maintained. However, movements of transposable elements can result in deleterious mutations. Genetic evidence for active transposons in humans comes from two patients who have hemophilia A caused by the insertion of a LINE element in the Factor VIII gene (Chapter 27) and from a patient with neurofibromatosis who has an *Alu* element inserted in the NF gene. Nevertheless, provided this does not happen too often, transposable elements will contribute usefully to the generation of genetic diversity.

Introns Are Ancient Components of Genes

The extraordinary discovery that genes are made up of coding regions (*exons*) interspersed with noncoding regions (*introns*) cried out for an evolutionary explanation. The explanation must provide answers to two related questions: Are introns ancient partners of genes or newly acquired interlopers? And does this arrangement of genes into introns and exons have a functional significance? These questions are further complicated because different types of introns are present in different modern organisms and the evolutionary relationships among the intron types are not known. Nevertheless, data on the origins of introns have come from looking for introns in the present-day members of what are thought to be the most ancient groups of organisms. As we will describe later in this chapter, the living world has been divided into three great kingdoms: the archaebacteria, the eubacteria, and the eukaryotes. The archaebacteria have introns in their transfer and ribosomal RNAs, but these are very different from the group I and group II self-splicing introns of the eukaryotes. Until recently, the eubacteria were thought to have no introns at all. The "introns-are-ancient" hypothesis explains this by assuming that the bacteria have lost introns because of selective pressures to lose unnecessary DNA in developing the "streamlined" genomes characteristic of modern bacteria. According to the "introns-are-new" hypothesis, the modern bacteria lack introns because they never had them, introns entering the world after the separation of the eubacterial and eukaryotic kingdoms. This argument has been partially resolved in favor of the "introns-are-ancient" hypothesis by the finding of an intron in the leucine tRNA of eight species of cyanobacteria. This finding implies that introns were present in the genomes of organisms predating the development of eukaryotes. The fact that the same intron is found in eight different species suggests that it could be very ancient; indeed,

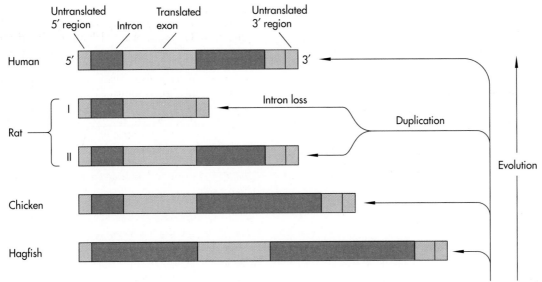

FIGURE **22-6**

An intron has been lost from one of two copies of the rat insulin gene. The basic structure of the insulin gene is clear from a comparison of hagfish, chicken, human, and rat insulin genes. There is a short untranslated 5′ sequence followed by an intron, then the coding sequence that is split by a second, long intron. The rat is unique in having two insulin genes, which must have arisen by duplication of the ancestral gene. Remarkably, one of these genes, by convention called gene I, has only one intron, whereas gene II resembles the insulin gene of other organisms. Rat gene I must have *lost* its second intron following duplication.

it must have been present early in the evolution of the cyanobacteria some 2.5 billion years ago!

There is also strong evidence that introns are not just inert genetic elements that have been carried around passively in the genomes since early times. For example, there is a single insulin gene in the hagfish, chicken and human, and this gene has two introns. The rat has two insulin genes, I and II; gene II has two introns, but gene I has only one (Figure 22-6). It seems that the single, original rat insulin gene duplicated to produce genes I and II, and that the former has *lost* one of its introns. Evidence for the *insertion* of introns comes from the α- and β-tubulin genes. These are very large families of genes present in all eukaryotes, and the two families arose by duplication of an ancestral gene. Comparisons of the distributions of introns in the tubulin gene of a variety of organisms show that they are found at 35 different places in the α and β families, with only one intron in common. This suggests that the introns were gained by the genes after the families separated. Furthermore, the distribution of introns within a family is best explained by the acquisition of a particular intron rather than by

the multiple losses of introns (Figure 22-7). So, interpretations of modern genome organization must take account of the fact that introns have been both gained and lost during evolution.

Exon Shuffling Contributes to Gene Evolution

It seems then that introns are ancient. But why have they persisted in eukaryotic genomes? Do they have a function, or are they "junk" DNA? Introns might have an important role in generating novel proteins if they facilitated recombination between exons that encode functional domains of proteins. This process, called *exon shuffling*, contrasts with gene duplication, a process in which an entire gene becomes available for evolutionary changes. There is some evidence that introns do mark important domains of proteins. For example, the different triosephosphate isomerases of chick and maize have six introns in common, and the introns occur at the boundaries of α helices and β-pleated sheets in the protein. The low-density-lipo-

(a) Distribution of introns at various codons in members of the α- and β-tubulin families

	2	4	5	9	13	16	17	19	20	21	33	35	41	56	58	59	62	76	90*	90*	95	126	134	177	208	211	257	319	327	351	353	407	412	437	448
α-tubulins																																			
Homo sapiens	+	–	–	–	–	–	–	–	–	–	–	–	–	–	–	–	–	–	+	–	–	+	–	–	–	–	–	–	–	–	–	–	–	–	–
Rattus norvegicus	+	–	–	–	–	–	–	–	–	–	–	–	–	–	–	–	–	–	+	–	–	+	–	–	–	–	–	–	–	–	–	–	–	–	–
Drosophila melanogaster	+	–	–	–	–	–	–	–	–	–	–	–	–	–	–	–	–	–	–	–	–	–	–	–	–	–	–	–	–	–	–	–	–	–	–
Physarum polycephalum	–	+	–	–	–	–	–	–	–	–	–	–	–	–	–	–	–	+	–	–	–	+	+	–	–	+	–	+	–	+	–	–	–	+	–
Chlamydomonas reinhardii	–	–	–	–	+	–	–	–	–	–	–	–	–	–	–	–	–	–	–	–	+	–	–	–	–	–	–	–	–	–	–	–	–	–	–
β-tubulins																																			
Homo sapiens	–	–	–	–	–	–	–	+	–	–	–	–	–	+	–	–	+	–	–	–	+	–	–	–	–	–	–	–	–	–	–	–	–	–	–
Neurospora crassa	–	–	+	–	+	–	–	+	–	+	–	+	–	+	–	–	–	–	–	–	–	–	–	–	–	–	–	+	–	–	–	–	–	–	–
Schizosaccharomyces pombe	–	–	+	–	–	–	–	+	–	+	–	+	–	+	–	–	–	–	–	–	–	–	–	–	–	–	–	–	–	+	–	–	–	–	–
Chlamydomonas reinhardii	–	–	–	+	–	–	–	–	–	–	–	–	–	–	–	+	–	–	–	–	+	–	–	–	–	+	–	–	–	–	–	–	–	–	–
Toxoplasma gondii	–	–	–	–	–	–	–	–	–	+	–	–	–	–	–	–	–	–	–	–	–	–	–	–	–	–	–	–	–	–	–	–	+	–	+	–

*An intron occurs at two different positions within codon 90.

(b) Ancestral gene lacks codon 126 intron

Intron at codon 126

- Human +
- Monkey +
- Rat +
- Chicken +
- Fruit fly –
- *Physarum* –
- *S. pombe* –
- *Chlamydomonas* –
- β-tubulins –

Intron acquired → +

α-tubulins

Ancestral tubulin gene

(c) Ancestral gene contains codon 126 intron

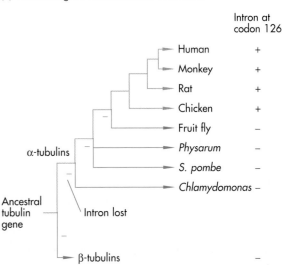

Intron at codon 126

- Human +
- Monkey +
- Rat +
- Chicken +
- Fruit fly –
- *Physarum* –
- *S. pombe* –
- *Chlamydomonas* –
- β-tubulins –

α-tubulins

Ancestral tubulin gene

Intron lost

FIGURE **22-7**

The tubulin family of genes provides evidence for insertion of introns. (a) Analysis of a large number of tubulin genes from a wide variety of organisms shows that introns are present in at least 35 different locations in these genes. Most tubulins have three introns, some have none, and a tubulin gene from *Aspergillus nidulans* has eight (not shown). The present-day distribution of these introns can be best explained on the hypothesis that the tubulins have *acquired* introns during evolution. (b) The distribution of the intron found at codon 126 in α-tubulin can be explained if the gene acquired the intron after the divergence of the vertebrates and invertebrates, as shown. (c) If it is assumed that this intron was present in the ancestral gene tubulin, then repeated loses (−) of the intron would have had to occur to generate the present distribution of this intron.

protein receptor (LDL-R) is made up of bits of other proteins stitched together to make a new protein. The LDL-R gene has 18 exons that can be grouped into six blocks that encode protein domains with different functions (Figure 22-8). One of these exon groups is homologous with part of the C9 complement factor, and another is homologous with a part of the EGF receptor and blood clotting Factors IX, X, and C. The advantage of constructing proteins in this modular fashion is that only a limited number of units are needed to generate the enormous diversity of proteins found in cells. "Useful" units are evolutionarily successful and will be kept and reused. The number of units has been estimated by first identifying exons and then determining which have been reused by comparing their sequences. The result of this calculation is that all the proteins so far sequenced are made up of only some 1000 to 7000 different exons! This estimate is highly controversial, but it indicates that the reassortment of exons to produce new proteins may be an important factor in protein evolution.

Gene Duplication Is a Driving Force in Evolution

The most important way in which genomes increase in size and complexity is through gene duplication and subsequent sequence divergence. The importance of gene duplication is that one copy of the duplicated gene can undergo mutation in the absence of selection, because the other copy can supply the protein needed to sustain cell function. This gradual accumulation of mutations in the absence of selective pressure is called *genetic drift*. The evolutionary significance of molecular genetic drift is that the mutations may lead to the acquisition of new functions. This is particularly clear in the case of the β-globin cluster, where five β-globins have evolved from an ancestral β-globin gene (Figure 22-9; see also Chapter 8). These different β-globin subunits have different physiological properties and are expressed at different times during development. For example, fetal hemoglobin F is composed of 2 α chains and 2 γ chains ($\alpha_2\gamma_2$). It has a higher affinity for

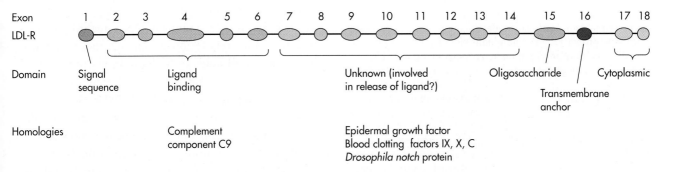

FIGURE **22-8**
Evidence of exon shuffling in the low-density lipoprotein receptor (LDL-R). The structure of the LDL-R exemplifies the modular construction of proteins, a feature that may have played an important role in evolution. The receptor contains six functionally distinct domains. Four of these are motifs common to many membrane proteins. There is an amino-terminal signal sequence that is needed for the protein to move to the cell surface; exon 16 codes for the hydrophobic transmembrane domain; exon 15 codes for a region that binds oligosaccharides and may keep the receptor away from the cell membrane; and exons 17 and 18 code for a cytoplasmic domain. LDL binds to the ligand-binding domain (exons 2 through 6). This region has striking homology to the complement component factor C9. Exons 7 through 14 code for a region of unknown function, but one that may be involved in mediating release of LDL once the LDL–LDL-R complex has been internalized by the cell. This domain has homology with a large number of other proteins, including epidermal growth factor.

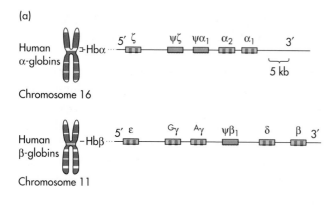

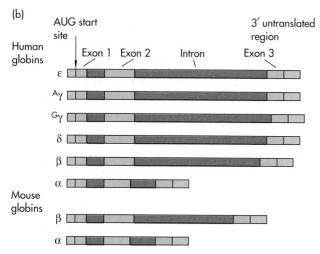

FIGURE 22-9
Gene duplication leads to the development of gene variants with new properties. (a) The globin gene family has provided a wealth of information about gene evolution. The present-day globin genes are in two clusters; the α-globin cluster covers about 30 kbp on chromosome 16, and the β-globin genes are spread over 50 kbp on chromosome 11. The order of the genes along the chromosomes (from left to right) is the order in which they are expressed during development. Pseudogenes (ψϱ, ψα₁, ψβ₁) are nonfunctional. (b) The overall morphology of the genes is remarkably constant, and comparisons of the exon-intron structures of the five β-globin genes, and of their sequences, show clearly that they are related.

oxygen than does hemoglobin A, the adult form ($\alpha_2\beta_2$). Thus transport of oxygen across the placenta is from the low-affinity hemoglobin A in the maternal circulation to the higher-affinity hemoglobin F of the fetus.

Genetic drift in duplicated genes may be desirable for generating genetic diversity, but in other situations it needs to be counteracted. For example, tandemly repeated genes like those for histones must be main-

tained free of mutations. This may come about in two ways. *Unequal crossing over* leads to the accumulation of extra copies of a tandemly repeated gene. If too many copies of a mutant form accumulate, the individual carrying those copies may be at a selective disadvantage, and those mutations would thus be eliminated from the population. Mutations may be corrected rather than eliminated by *gene conversion*. Here a break in a DNA strand of the normal allele leads to base pairing between that strand and the complementary strand in the mutant allele. Following repair and replication, the mutant sequence is replaced by the normal sequence. This outcome is more likely than the converse, in which the normal sequence is replaced by the mutant sequence, because there are more wild-type alleles than mutant alleles, so that the break that initiates the process is more likely to occur in the wild-type allele.

DNA Clocks Measure Rates of Evolution

Turning now to evolutionary studies—which endeavor to determine the ancestral relationships of living organisms—recombinant DNA techniques have had the same revolutionary impact there that they have had on all fields of biology. In evolutionary biology, amino acid and nucleotide sequence analyses have provided biologists with *molecular clocks* to supplement their analyses based on fossil and morphological data. The concept of a molecular clock is based on certain assumptions. One assumption is that sequence differences between the same genes and proteins in two species have accumulated since the species diverged from a common ancestor. Furthermore, the rates at which the differences accumulate are assumed to be equal. If this were not the case, the clocks would run at different speeds in the two species. The clock certainly runs at different speeds for different classes of nucleotides. For example, nucleotides located in a position where a change produces an alteration in an amino acid accumulate changes more slowly than do nucleotides located in a position where the changed codon codes for the same amino acid. The clock also runs much more slowly for molecules like histones, in which any amino acid change may have a significant effect on protein function, in this case the ability of

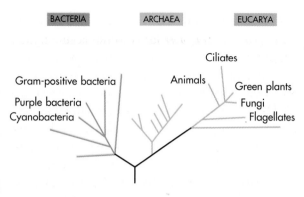

FIGURE **22-10**

The relationships of organisms revealed by rRNA sequence comparisons. The sequences of the ribosomal genes are known for several hundred organisms from throughout the living world. Analysis shows that there are distinct ribosomal sequences for eubacteria (common bacteria like *E. coli*), the archaebacteria (organisms that live in environments with extremes of temperature and salinity and utilize unlikely substances like sulfur), and eukaryotes (organisms with a nucleus). This analysis shows further that the three groups differ equally from one another. The new name proposed for the archaebacteria, and used in this figure, does not contain the suffix "bacteria" in recognition of this equality.

rRNA Sequencing Identifies the Three Great Kingdoms of Living Organisms

One of the questions that has occupied biologists since the time of the great systematist Linnaeus is how living organisms are related to one another. Many classification schemes have been proposed, but there was general agreement that the living world could be divided into two kingdoms: the Prokaryotes (bacteria; cells without nuclei) and Eukaryotes (cells with nuclei). It seemed like common sense to suppose that prokaryotes are more primitive than eukaryotes, and that the latter evolved from the former. However, this classification ignored the tremendous diversity of bacteria. Now data from sequence analysis of the ribosomal RNA (rRNA) genes from many organisms have shown that common sense is a fallible guide for understanding the relationships between organisms. 16S rRNA is popular with molecular systematists because this class of RNA is found in all organisms and the molecules are sufficiently long—about 1660 nucleotides—to allow accurate interspecies comparisons. These comparisons showed that those bacteria living in habitats with extremes of heat, pressure and salinity, and metabolizing sulfur and molecular hydrogen constituted one group—the *Archaebacteria*—while all others constituted the *Eubacteria*. Furthermore, the comparisons showed that the three kingdoms archaebacteria, eubacteria, and eukaryotes are coequal; that is, the differences between archaebacteria and eubacteria are as great as those between archaebacteria and eukaryotes (Figure 22-10). Recently, a controversial proposal has been made—to rename the three groups to take account of these relationships and to assign each group to a new taxonomic rank called a *domain*, with the term "kingdom" reserved for present kingdoms like animals and plants.

histones to bind to DNA. In addition, more than one change may have occurred at any given site. Corrections can be made for these effects, and molecular clocks have been used extensively to derive *phylogenetic trees* showing species relatedness and to estimate how many years have elapsed since species diverged. The trees are constructed by using the principle of *parsimony*, that is, the different species are placed on the tree so as to minimize the numbers of changes needed to move from one position on the tree to another.

The first measurements using molecular clocks based on nucleotide differences were made in the early 1970s, long before DNA sequencing had been developed. The only tool then available to measure these differences was DNA-DNA hybridization in solution, and many invaluable data were obtained using this technique. More recently comparison of gene sequences between different organisms has become possible with the development of rapid sequencing methods. There is controversy over which approach produces the most reliable results, but little disagreement that nucleic acid studies together with classical anatomical studies are producing exciting new results.

Recombinant DNA Techniques Have Sorted Out Relationships in the Primate Family

Some phylogenies based on molecular clock estimates have proved very controversial, and none is more so than that dealing with the evolutionary relationships between the primates, particularly the relationship of human beings to the great apes (orangutan, chimpan-

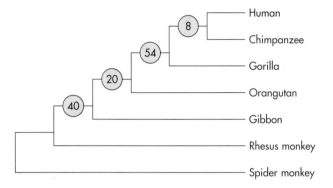

FIGURE **22-11**

Molecular analysis leads to a revised version of the evolutionary tree of human beings and the great apes. This interpretation of the relationships between the primates is based on sequencing 10.8 kbp of the β-globin genes in the animals shown (except for the gibbon, for which 6.5 kbp was sequenced). The sequences were aligned with one another, and the numbers of substitutions and deletions or insertions between pairs of species were determined. This tree was constructed by using maximum parsimony; that is, the species have been placed on the tree so as to minimize the number of changes necessary to move from one branch to another. The numbers on the branches show the difference between the most parsimonious value and the next most favorable value. Thus, grouping human beings and chimpanzees is favored by 8 over grouping human beings and gorillas, and the grouping of humans, chipanzees, and gorillas is favored by 54 over the grouping of humans and chimpanzees with orangutans. This arrangement is supported by other statistical analyses of these data and by measurements of relatedness based on DNA-DNA hybridization data. (Goodman et al., 1990)

zee, and gorilla). Traditionally, human beings have been placed in a family by themselves, and the chimpanzees, gorillas, and orangutans have been assigned to a second family. However, more recent morphological analyses place human beings, chimpanzees, and gorillas in a group, and in this group chimpanzees and gorillas are more closely related to each other than either is to human beings. These conclusions were thrown into disarray by findings based on measurements of DNA-DNA hybridization in solution. DNA isolated from one species was cut into small fragments that were radioactively labeled and then heated to separate the strands. These strands were then added to single-stranded DNA from a second species and cooled so that they anneal. The temperature of the mixture was again gradually raised so that the strands separated. The temperature required to dissociate the paired strands depends on the sequence similarities between the DNAs from the two species. Closely related species will form extensive duplexes and require high temperatures. DNA duplexes from more distantly related species will dissociate at lower temperatures. When used as a measure of primate relationships, DNA-DNA hybridization suggested that humans and chimpanzees are more closely related to each other than either is to gorillas. Humans and chimpanzees differ in only about 1.6 percent of their genomes.

Interpretation of DNA-DNA hybridization data is complicated by the complexity introduced by the use of whole genomes, (although this is also used as an argument in favor of this approach). As a consequence these results were severely criticized. An alternative approach is a detailed comparison of a smaller region by sequencing. In the case of the higher primates, the β-globin complex is an obvious choice, where over 10,000 bp of this region have been sequenced. Comparisons between chimpanzee, gorilla, and human beings confirm the interpretation of the DNA-DNA hybridization data; humans and chimpanzees are more closely related then either is to gorillas (Figure 22-11). These results of the analysis of the β-globin gene have recently been substantiated by analyses of the mitochondrial cytochrome oxidase II gene.

Mitochondrial DNA as a Molecular Clock

Mitochondrial DNA (mtDNA) is a favored molecule for molecular phylogenetic studies. Sequence analysis of nuclear DNA is complicated because of recombination that rearranges genes and because a molecular clock based on nuclear DNA changes runs slowly. Few sequence differences accumulate even over many generations, because nuclear DNA has a relatively low mutation rate. In contrast, mtDNA has a high mutation rate, about 10 times that of nuclear DNA, and nucleotide substitutions therefore accumulate quickly even between species that diverged quite recently. Furthermore, because mtDNA is maternally inherited and does not recombine, individuals sharing the same mtDNA sequences must have had a common female ancestor.

The most fascinating example of using mtDNA in phylogenetic studies comes from the data establishing a human Eve. Here 12 restriction enzymes were used to determine the patterns of fragments produced in seven regions of human mtDNA from 147 individuals from five different geographic regions. Comparisons established that there were 133 distinct types, and a phylogenetic tree was constructed on the basis of these data. The outcome of these calculations shows that these lineages could be linked to a single ancestral female who lived some 200,000 years ago in Africa. However, we have to remember the different contributions made by our female ancestors to our nuclear and mitochondrial genomes. Although all our mitochondrial DNA may have come from just one of our maternal great grandparents, she gave us just one-eighth of our nuclear genes! Furthermore, this conclusion rests on certain assumptions, for example the strict maternal inheritance of mtDNA. The polymerase chain reaction has now been used to determine that sperm *do* contribute mitochondria to the fertilized egg. Female mice of one strain were mated with male mice of a second strain, and the female hybrid offspring were than mated with the mice of the second strain. This back-crossing was repeated for 26 generations in the hope that this would lead to an increase in paternal mitochondria in these mice. Oligonucleotides specific for mitochondrial sequences in the two species were used as primers for a PCR, and paternal mitochondrial DNA was detected. The amount of paternal mtDNA was very small even after all the back-crossing, but this is not surprising, given that a sperm contains only about 50 mitochondria while the egg has over 100,000. Nevertheless, this means that mtDNA is not just maternally inherited and that the population from which mtDNA is drawn is larger than previously assumed. This means that the source of contemporary mtDNA could be more recent than the estimate of 200,000 years.

Some Intracellular Organelles Were Once Bacteria

Chloroplasts and mitochondria, both eukaryotic intracellular organelles, seem to have a life of their own in the cell. These organelles contain DNA that codes for organelle-specific proteins, and they replicate themselves. The idea that chloroplasts and mitochondria arose from bacterialike cells that were assimilated by eukaryotic cells is called the *endosymbiosis* theory.

Once again, rRNA gene sequences have been used to compare the genomes of these organelles with those of eukaryotes and modern prokaryotes. These comparisons show that the organelle genomes are different from the nuclear genomes of their eukaryotic host cells and closely resemble eubacterial genomes. Furthermore, the chloroplast and mitochondrial genomes differ and must be derived from different eubacteria! This conclusion means that at least *two* independent endosymbiotic events took place, one involving a cyanobacterium-like cell that gave rise to chloroplasts and the other involving a purple photosynthetic bacterium that gave rise to mitochondria.

The partners involved in endosymbiosis do not seem to be restricted to bacterial and eukaryotic cells. The phytoflagellate *Cryptomonas* contains a closed membrane system that encloses a chloroplast, cytoplasm, and the nucleomorph, which looks like a vestigial eukaryotic nucleus (Figure 22-12). On these morphological grounds, it was suggested that this complex structure is the remains of an endosymbiotic eukaryotic alga. Analysis of the 18S rRNA genes in these creatures shows that this supposition is probably correct. Two different fragments, Nu and Nm, are produced when the polymerase chain reaction is used to amplify 18S rRNA genes with total DNA extracted from cells as the starting material. One sequence is presumed to be derived from nuclear rRNA genes and the other fragment from nucleomorph rRNA genes. This analysis is supported by data from in situ hybridizations, using probes specific for rRNA genes of either prokaryotes or eukaryotes. The prokaryote-specific probe hybridized to ribosomes in the chloroplast and not to ribosomes in the nucleomorph. The eukaryote-specific probe labeled ribosomes in the nucleoluslike region of the nucleomorph and not those in the chloroplast. The crucial experiment will be to perform in situ hybridization with Nu- and Nm-specific sequences to see where they hybridize.

When the Nu and Nm sequences are incorporated into a phylogenetic tree, the Nu sequences are in a group including an amoeboid protozoan, whereas the Nm sequences are in group including red algae. These results are consistent with the hypothesis that *Cryptomonas* is the result of two symbiotic events; the first occurred between a eukaryotic cell and a photosynthetic prokaryote and was followed by a second union between the resulting eukaryotic alga and another eukaryotic host (Figure 22-12).

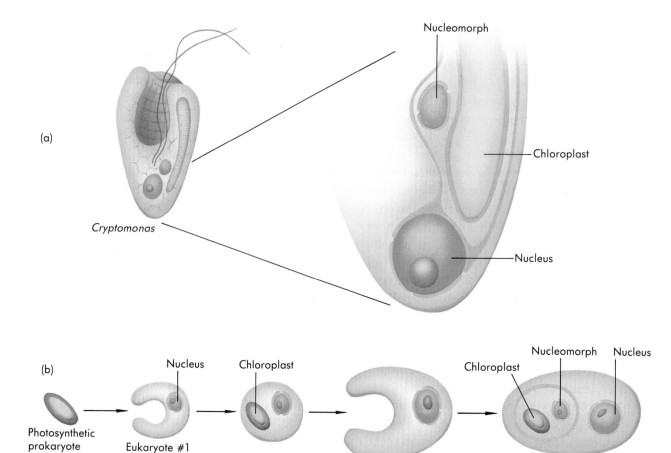

(a)

Nucleomorph

Chloroplast

Nucleus

Cryptomonas

(b)

Nucleus

Chloroplast

Chloroplast

Nucleomorph

Nucleus

Photosynthetic
prokaryote

Eukaryote #1

Eukaryote #2

FIGURE **22-12**

Hypothesized derivation of *Cryptomonas* from two endosymbiotic events. (a) *Cryptomonas* is a photosynthetic flagellate with two flagella and a nucleus. It contains a curious membrane-bound structure that encloses a nucleuslike nucleomorph and a chloroplast. Analysis of rRNA genes shows that there are two distinct forms of rRNA genes in these organisms, and in situ hybridization shows that prokaryote-like rRNA genes are present in the chloroplast but that both the nucleus and nucleomorph contain genes that hybridize to eukaryotic rRNA genes. These results are consistent with the scheme shown in (b): A photosynthetic prokaryote was assimilated by a eukaryotic cell, which in its turn underwent symbiosis with a second eukaryotic species. (McFadden, 1990)

DNA Fingerprinting Helps Us Understand the Genetic Basis of Altruism

Altruistic behavior, in which individuals put themselves at risk to help other individuals of their own species, has always been difficult to explain by natural selection. By their actions, altruistic individuals are apparently reducing their Darwinian fitness, that is, reducing the probability that they will leave offspring. (Strictly speaking, however, Darwinian fitness applies to populations and not to individuals. Genetically identical individuals in identical enviroments need not have the same Darwinian fitness because of chance variations in their lives). This paradox can be resolved by thinking in terms of genes rather than offspring. The theory of *kin selection* suggests that altruism arises because genetically related individuals will cooperate to ensure that genes held in common are passed on to the next generation. This theory is being tested directly by using DNA fingerprinting to determine the genetic relatedness of individuals behaving in an altruistic fashion.

Lions in Africa have been subjected to such an analysis. A pride of lions is a social structure comprising males and females, in which the males cooperate in maintaining the pride. Kin selection had not been invoked to explain pride structure because it was thought that all the males of a pride benefited reproductively, the males having equal access to females, no matter the size of the pride. However, DNA fingerprinting shows that the males' access to females is related to the size of the pride; males in prides with only two males fare equally well in breeding, whereas in larger prides at least one male fails to reproduce. Thus it is an "altruistic" act for males to be part of a large pride because they may be putting themselves at a reproductive disadvantage. As you might expect if kin selection is at work, DNA fingerprinting shows that males in larger prides are more closely related to one another than are males in small prides, who are usually unrelated. The close genetic relatedness of these nonbreeding males to the reproductively successful males suggests that the kin selection is helping maintain large coalitions. A similar finding comes from the naked mole rats of northeastern Africa, which have a social system unique among mammals. Each colony contains a single reproducing female, who has between one and three mates, whereas all the other members of the colony are concerned with the maintenance of the colony. DNA fingerprinting shows that the naked mole rats in a colony are the most inbred of free-living mammals known. This result is in accord with the kin selection theory. Ecological factors may have initiated colony life, but a very high degree of close genetic relatedness may be necessary for the reproductive sacrifices of most members of a colony.

So we can see how recombinant DNA techniques have illuminated the whole sweep of the evolutionary process. They have provided us with tantalizing clues on how life originated and the nature of the first living organisms; we have been able to analyze the ways in which genomes have changed and to use these data to clarify the evolutionary relationships between organisms. And now these early explorations of the genetic basis of altruism may bring a deeper understanding of an aspect of evolution that has intrigued biologists ever since the time of Darwin—the development of social behavior.

Reading List

Original Research Papers

THE PREBIOTIC ERA

Horgan, J. "In the beginning." *Sci. Am.,* 264: 116–125 (1991). [Review]

Pace, N. R. "Origin of life—facing up to the physical setting." *Cell,* 65: 531–533 (1991). [Review]

THE RNA WORLD

Cech, T. R. "RNA as an enzyme." *Sci. Am.,* 255: 64–75 (1986). [Review]

Benne, R. "RNA editing in trypanosomes: is there a message?" *Trends Genet,* 6: 177–181 (1990). [Review]

Lamond, A. I., and T. J. Gibson. "Catalytic RNA and the origin of genetic systems." *Trends Genet.,* 6: 145–149 (1990) [Review]

Kruger, K., P. J. Grabowshi, A. Zaug, A. J. Sands, D. E. Gottschling, and T. R. Cech. "Self-splicing RNA: autoexcision and autocyclization of the ribosomal RNA intervening sequence of Tetrahymena." *Cell,* 31: 147–157 (1982).

Guerrier-Takada, C., K. Gardiner, T. Marsh, N. Pace, and S. Altman. "The RNA moiety of ribonuclease P is the catalytic subunit of the enzyme." *Cell,* 35: 849–857 (1983).

Zaug, A. J., M. D. Been, and T. R. Cech. "The *Tetrahymena* ribozyme acts like an RNA restriction endonuclease." *Nature,* 324: 429–433 (1986).

Benne, R., J. van den Burg, J. P. J. Brakenhoff, P. Sloof, J. H. Van Boom, and M. C. Tromp. "Major transcript of the frame-shifted *coxII* gene from trypanosome mitochondria contains four nucleotides that are not encoded in the DNA." *Cell,* 46: 819–826 (1986).

Chen, S.-H., G. Habib, C.-Y. Yang, Z.-W. Gu, B. R. Lee, S.-A. Went, S. R. Siberman, S.-J. Cai, J. P. Deslypere, M. Rosseneu, A. M. Gotto, Jr., W.-H. Li, and L. Chan. "Apolipoprotein B-48 is the product of a messenger RNA with an organ-specific in-frame stop codon." *Science*, 238: 363–366 (1987).

Powell, L. M., S. C. Wallis, R. J. Pease, Y. H. Edwards, T. J. Knott, and J. Scott. "A novel form of tissue-specific RNA processing produces apolipoprotein B-48 in intestine." *Cell*, 50: 831–840 (1987).

Shaw, J. M., J. E. Feagin, K. Stuart, and L. Simpson. "Editing of kinetoplastid mitochondrial mRNAs by uridine addition and deletion generates conserved amino acid sequences and AUG initiation codons." *Cell*, 53: 401–411 (1988).

Mahendran, R., M. R. Spottswood, and D. L. Miller. "RNA editing by cytidine insertion in mitochondria of *Physarum polycephalum*." *Nature*, 349: 434–438 (1991).

Young, L. S., H. M. Dunstan, P. R. Witte, T. P. Smith, S. Ottonello, and K. U. Sprague. "A class III transcription factor composed of RNA." *Science*, 252: 542–546 (1991).

INTRONS—ORIGINS AND ROLE IN EVOLUTION

Belfort, M. "Self-splicing introns in prokaryotes." *Cell*, 64: 9–11 (1991). [Review]

Cavalier-Smith, T. "Intron phylogeny: a new hypothesis." *Trends Genet.*, 7: 145–148 (1991). [Review]

Bell, G. I., R. L. Pictet, W. J. Rutter, B. Cordell, E. Tischer, and H. M. Goodman. "Sequence of the human insulin gene." *Nature*, 284: 26–32 (1980).

Perler, F., A. Efstratiadis, P. Lomedico., W. Gilbert, R. Kolodner, and J. Dodgson. "The evolution of genes: the chicken proinsulin gene." *Cell*, 20: 555–556 (1980).

Dibb, J., and A. J. Newman. "Evidence that introns arose at proto-splice sites." *EMBO J.*, 8: 2015–2021 (1989).

Kuhsel, M. G., R. Strickland, and J. D. Palmer. "An ancient group I intron shared by eubacteria and chloroplasts." *Science*, 250: 1570–1573 (1990).

Xu, M.-Q., S. D. Kather, H. Goodrich-Blair, S. A. Nierzwicki-Bauer, and D. A. Shub. "Bacterial origin of a chloroplast intron: conserved self-splicing group I introns in cyanobacteria." *Science*, 250: 1566–1570 (1990).

MOVABLE ELEMENTS AND GENOME EVOLUTION

Finnegan, D. J. "Eukaryotic transposable elements and genome evolution." *Trends Genet.*, 5: 103–107 (1989). [Review]

Kazazian, H. H., C. Wong, H. Youssoufian, A. F. Scott, D. G. Phillips, and S. E. Antonarakis. "Hemophilia A resulting from the *de novo* insertion of L1 sequences represents a novel mechanism for mutation in man." *Nature*, 332: 164–166 (1988).

Fitch, D. H. A., W. J. Bailey, D. A. Tagel, M. Goodman, L. Sieu, and J. L. Slightom. "Duplication of the γ-globin gene mediated by repetitive L1 long interspersed repetitive elements sequences in an early ancestor of simian primates." *Proc. Natl. Acad. Sci. USA*, 88: 7396–7400 (1991).

Wallace, M. R., L. B. Andersen, A. Saulino, P. E. Gregory, T. W. Glover, and F. S. Collins. "A de novo *Alu* insertion results in neurofibromatosis type 1." *Nature*, 353: 864–866 (1991).

GENE DUPLICATION IN EVOLUTION

Efstratiadis, A., J. W. Posakony, T. Maniatis, R. M. Lawn, C. O'Connell, R. A. Spritz, J. K. DeRiel, B. G. Forget, S. M. Weissman, J. L. Slightom, A. E. Blechl, O. Smithies, F. E. Baralle, C. C. Shoulders, and N. J. Proudfoot. "The structure and evolution of the human β-globin family." *Cell*, 21: 653–668 (1980).

RIBOSOMAL RNA AND THE RELATIONSHIPS OF LIVING ORGANISMS

Woese, C. R. "Archaebacteria." *Sci. Am.*, 244: 98–125 (1980).

Lake, J. A. "Origin of the eukaryotic nucleus determined by rate-invariant analysis of rRNA sequences." *Nature*, 331: 184–186 (1988).

Woese, C. R., O. Kandler, and M. L. Wheelis. "Toward a natural system of organisms: proposal for the domains, Archaea, Bacteria, and Eucarya." *Proc. Natl. Acad. Sci. USA*, 87: 4576–4579 (1990).

THE MOLECULAR CLOCK AND PRIMATE EVOLUTION

Zuckerkandl, E. "On the molecular evolutionary clock." *J. Mol. Evol.*, 26: 34–46 (1987).[Review]

Diamond, J. M. "The future of DNA-DNA hybridization studies." *J. Mol. Evol.*, 30: 196–201 (1990). [Review]

Sibley, C. G., and J. E. Ahlquist. "The phylogeny of hominoid primates, as indicated by DNA-DNA hybridization." *J. Mol. Evol.*, 20: 2–15 (1984).

Sarich, V. M., and J. Marks. "DNA hybridization as a guide to phylogenies: a critical analysis." *Cladistics*, 5: 3–32 (1989).

Goodman, M., D. A. Tagle, D. H. A. Fitch, W. Bailey, J. Czelusniak, B. F. Koop, P. Benson, and J. L. Slighton. "Primate evolution at the DNA level and a classification of hominoids." *J. Mol. Evol.*, 30: 260–266 (1990).

Ruvolo, M., T. R. Disotell, M. W. Allard, W. M. Brown, and R. L. Honeycutt. "Resolution of the African hominoid trichotomy by use of a mitochondrial gene sequence." *Proc. Natl. Acad. Sci. USA*, 88: 1570–1574 (1991).

EXONS AND EXON SHUFFLING IN PROTEIN EVOLUTION

Gilbert, W. "Why genes in pieces?" *Nature*, 271: 501 (1978). [Review]

Doolittle, R. F. "The genealogy of some recently evolved vertebrate proteins." *Trends Biochem.* Sci. 10: 233–237 (1985). [Review]

Sudhof, T. C., D. W. Russell, J. L. Goldstein, M. S. Brown, R. Sanchez-Pescador, and G. I. Bell. "Cassette of eight exons shared by genes for LDL receptor and EGF precursor." *Science,* 228: 893–895 (1985).

Dorit, R. L., L. Schoenbach, and W. Gilbert. "How big is the universe of exons?" *Science,* 250: 1377–1382 (1990).

Pathy, L. "Exons—original building blocks of proteins?" *BioEssays,* 13: 187–192 (1991).

ORIGINS OF INTRACELLULAR ORGANELLES

Margulis, L. *Symbiosis in Cell Evolution.* W. H. Freeman and Co., New York, 1981.

Gray, M. W. "The evolutionary origins of organelles." *Trends Gen.,* 5: 294–299 (1989). [Review]

McFadden, G. I. "Evidence that cryptomonad chloroplasts evolved from photosynthetic eukaryotic endosymbionts." *J. Cell Sci.,* 95: 303–308 (1990).

Douglas, S. E., C. A. Murphy, D. F. Spencer, and M. W. Gray. "Cryptomonad algae are evolutionary chimaeras of two phylogenetically distinct unicellular eukaryotes." *Nature,* 350: 148–151 (1991).

MOLECULAR INSIGHTS ON HUMAN EVOLUTION

Cann, R. L., M. Stoneking, and A. C. Wilson. "Mitochondrial DNA and human evolution." *Nature,* 325: 31–36 (1987).

Kruger, J., and F. Vogel. "The problem of our common mitochondrial mother." *Hum. Genet.,* 82: 308–312 (1989).

Gyllensten, G. Wharton, A. Josefsson, and A. C. Wilson. "Paternal inheritance of mitochondrial DNA in mice." *Nature,* 352: 255–257 (1991).

Rienzo, A. D., and A. C. Wilson. "Branching pattern in the evolutionary tree for human mitochondrial DNA." *Proc. Natl. Acad. Sci. USA,* 88: 1597–1601 (1991).

GENETIC BASIS OF ALTRUISM

Reeve, H. K., D. F. Westneat, W. A. Noon, P. W. Sherman, and C. F. Aquadro. "DNA 'fingerprinting' reveals high levels of inbreeding in colonies of the eusocial naked mole-rat." *Proc. Natl. Acad. Sci. USA,* 87: 2496–2500 (1990).

Packer, C., D. A. Gilbert, A. E. Pusey, and S. J. O'Brien. "A molecular genetic analysis of kinship and cooperation in African lions." *Nature,* 351: 562–565 (1991).

23

Recombinant DNA
in Medicine
and Industry

As soon as the first successful cloning experiments were reported in 1973, applications for this powerful technology quickly followed. The significance of being able to produce large quantities of human proteins that were normally available in exceedingly small amounts, if at all, was not lost on scientists, physicians, and businessmen alike. In 1976 *biotechnology* became a reality as the methodologies for DNA cloning, oligonucleotide synthesis, and gene expression converged in a single experiment, in which a human protein was expressed from recombinant DNA for the first time. The protein was somatostatin, a 14 amino acid peptide neurotransmitter. The gene encoding somatostatin was not the natural gene but was synthesized chemically and cloned into a plasmid vector for expression in *E. coli*. Soon after followed the successful expression of human insulin for the treatment of diabetes, the first commercial product of the biotechnology industry. Instead of insulin extracted from the pancreases of pigs and cows, diabetics could now receive insulin identical to that normally produced by humans.

The ability to achieve such feats relied on the successes in all areas of molecular biology, including oligonucleotide synthesis, isolation of enzymes that cleave and join DNA, characterization of bacterial plasmids, and an understanding of gene expression. These methods have, of course, revolutionized research in biology and medicine, but what is equally important, they have spawned an entirely new industry, one devoted to the cloning and production of proteins of importance to both medicine and industry. Today, proteins are produced through recombinant DNA technology for treatment of numerous diseases—cancer, allergies, autoimmune disease, neurological disorders, heart attacks, blood disorders, infections, wounds, and genetic diseases—as well as for more prosaic tasks, such as use in laundry detergents and food production. In addition, entirely new approaches to drug design have emerged from recombinant DNA technology, as scientists have gained the ability to tinker with natural proteins to improve their function and to change them in subtle and useful ways.

Expression Systems Are Developed to Produce Recombinant Proteins

Cloning the gene or cDNA encoding a particular protein is only the first of many steps needed to produce a recombinant protein for medical or industrial use. The next step is to put the gene into a host cell for production. The development of expression systems has been an important research area in both industrial and academic laboratories. The most popular expression systems are the bacteria E. coli and Bacillus subtilis, yeast, and cultured insect and mammalian cells. We have learned in earlier chapters about the development of vectors and DNA transformation methods for these organisms. Here we will discuss the issues that are important for protein production. The choice of which cell is used depends on the project goals and on the properties of the protein to be produced.

Bacterial cells offer simplicity, short generation times, and large yields of product with low costs. And, especially with B. subtilis, the cells can be induced to secrete the product into the culture medium, thus greatly simplifying the task of purification. But expression in prokaryotic cells has several drawbacks. Although some proteins are expressed to high levels (greater than 10 percent of the mass of all bacterial proteins), they often fail to fold properly and hence form insoluble *inclusion bodies*. Protein extracted from these inclusion bodies is often biologically inactive.

Small proteins can sometimes be refolded into their active forms, but larger proteins usually cannot. A second problem is that foreign proteins are sometimes toxic to bacteria, so cell cultures producing the protein cannot be grown to high densities. This problem can often be circumvented by using an inducible promoter that is turned on to begin transcription of the gene for the foreign protein only after the culture has been grown. Third, bacterial cells lack enzymes that are present in eukaryotic cells and add posttranslational modifications, such as phosphates and sugars, to proteins. These modifications are often required for proper functioning of proteins. Researchers are addressing this problem by purifying the eukaryotic enzymes that carry out these modifications and using these enzymes to add the needed modifications to bacterially expressed proteins.

Yeast has been used for centuries by brewers and bakers, and now it toils for biotechnologists as well. As discussed in Chapter 13, yeast is a simple eukaryote that resembles mammalian cells in many ways but can be grown as quickly and cheaply as bacteria can. Yeast perform many of the posttranslational modifications found on human proteins and can be induced to secrete certain proteins into the growth medium for harvesting. A disadvantage of yeast is the presence of active proteases that degrade foreign proteins, thereby reducing the yield of product. Researchers are dealing with this problem, however, by constructing yeast strains in which the protease genes have been deleted.

Expression of heterologous proteins in insect cells by baculovirus vectors (as previously described in Figure 12-12) is a relatively new approach. The main advantages are high-level expression, correct folding, and posttranslational modifications similar to those in mammalian cells. A vaccine for the AIDS virus has been prepared by producing one of the HIV glycoproteins with this system. Although the cost of culturing insect cells is currently more than that for culturing bacteria and yeast, it is less than that for culturing mammalian cells.

Despite the significant advantages of producing human proteins in heterologous host cells, in some cases the best place to produce a mammalian protein is in mammalian cells. Great improvements have been made to promoters, vectors, transformation protocols, and host cell systems. Transient expression in mammalian cells (described in Figure 12-4) is often used for checking the function of a newly cloned gene and as a quick method for assessing the function of engineered proteins. The extracellular domains of cell-surface receptors (Chapter 17) have been engineered for secretion from cells by introducing a stop codon into the gene before the transmembrane domain sequence. These *soluble receptors* are valuable reagents for studying ligand binding in vitro and for screening for receptor agonists or antagonists, and they may eventually be used as therapeutics themselves. Although transient systems yield enough protein for laboratory experiments, stably integrated amplified genes in mammalian cells are used for the large-scale production of proteins such as tissue plasminogen activator, which we describe later.

Insulin Is the First Recombinant Drug Licensed for Human Use

The first licensed drug produced through genetic engineering was human insulin. An important hormone that regulates sugar metabolism, insulin is produced by a small number of cells in the pancreas and secreted into the bloodstream. An inability to produce insulin results in diabetes, but daily injections of insulin are sufficient to reverse or at least allay the debilitating effects of the disease. Prior to production of the recombinant molecule, insulin for treatment of diabetes was obtained from the pancreases of pigs and cows. Although this insulin is biologically active in humans, the amino acid sequences are not identical to that of the human molecule. Thus, some patients produced antibodies against injected insulin, occasionally resulting in serious immune reactions. Because recombinant human insulin is identical to the natural product, immunogenicity should not be a problem.

In mammals, insulin is expressed as a single-chain *prepro-hormone*, which is secreted through the plasma membrane. A prepro-hormone contains extra amino acids not present in the mature hormone. Amino-terminal amino acids form the *pre* sequence and target the expressed protein for secretion. The *pro* sequence is a stretch of amino acids in the middle of the hormone sequence that is important for folding the polypeptide

chain into the correct structure. During secretion, these extra amino acids are cleaved from the prepro-hormone by cellular proteases to release the mature insulin molecule, consisting of two short polypeptide chains, A and B, linked by two disulfide bonds. The principal challenge in the production of recombinant insulin was getting insulin assembled into this mature form. The initial approach was to construct synthetic genes from oligonucleotides that separately encoded the A and B chains. These were individually inserted into the *E. coli* gene encoding β-galactosidase, so the bacteria produced large fusion proteins that had the insulin sequences tacked onto the end of the β-galactosidase enzyme (Figure 23-1). These large proteins were purified from bacterial extracts, and the insulin chains were released by treatment with cyanogen bromide, a chemical that cleaves peptide bonds following methionine residues. Because a methionine codon had been inserted at the boundaries between β-galactosidase and the insulin chains in the fusion proteins, cyanogen bromide treatment clipped intact insulin chains off the fusion proteins. These were purified, mixed, and reconstituted into an active insulin molecule. This approach was refined by producing a single β-galactosidase–insulin fusion protein, which could be cleaved in a single step to release mature insulin. A similar method is now in use for the commercial production of recombinant insulin.

Recombinant Human Growth Hormone is Produced in Bacteria by Two Methods

Growth hormone is a 191 amino acid protein that is produced in the pituitary gland and regulates growth and development. Children born with growth hormone deficiency—hypopituitary dwarfs—never achieve normal stature. Regular injections of growth hormone stimulate the growth of these children so that they reach near-normal heights. Unlike the situation with insulin, animal-derived growth hormones are ineffective. Only the human protein works, and for many years it was painstakingly extracted from the pituitaries of human cadavers. One unforeseen and

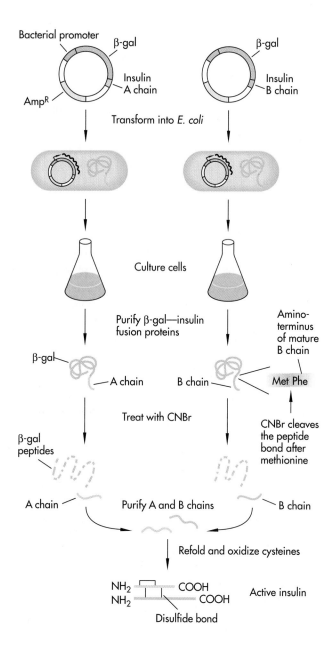

FIGURE **23-1**

Expression of human insulin in *E. coli*. Recombinant insulin was first made by expressing the A and B chains separately, then refolding them into a mature insulin molecule. A DNA fragment encoding each insulin chain was made by annealing two complementary oligonucleotides that had been chemically synthesized. Each fragment was ligated into a bacterial expression vector so that, when translated, the insulin chain would be fused to the carboxy terminus of the enzyme β-galactosidase (β-gal). The expression vectors were transformed into *E. coli*, and the β-gal–insulin fusion proteins accumulated inside the bacterial cells. The cells were harvested, and each β-gal–insulin fusion protein was purified. The insulin-coding DNA was synthesized so that it started with a methionine codon. This setup provided a way to cleave off the β-gal part from the insulin polypeptide. Treatment of the fusion protein with the chemical cyanogen bromide (CNBr) results in cleavage of peptide bonds after all methionines. In this way, the natural insulin peptides were obtained. Because β-gal contains other methionine residues, CNBr treatment cleaved it into many small peptides. The insulin chains were not cleaved further because they did not contain internal methionines. The A and B chains were purified and then mixed together to form active recombinant insulin.

a plasmid adjacent to a bacterial promoter. Like insulin, hGH is normally produced as a larger precursor protein containing an amino-terminal signal sequence. Because the human signal sequence would not be recognized by the bacterial secretion machinery, the 5′ end of the cDNA was reengineered with a synthetic DNA sequence enabling the bacteria to produce a nearly normal version of the mature human protein.

The first hGH expression vectors directed the production of the protein inside the cell. Purification required many steps to separate hGH from the thousands of intracellular bacterial proteins. Another way to produce the protein in bacteria is to engineer the protein so it is secreted. This can be done by linking the coding sequence for the desired protein to a signal sequence from a secreted bacterial protein, thus forming a *prehormone* (Figure 23-2b). Human growth hormone is produced by the bacteria and then secreted with the concomitant removal of the signal peptide by a bacterial protease. Secretion into the periplasm, where there are fewer proteins than inside the cell, makes purification simpler. The only difference between the secreted hGH and that produced intracellularly is the presence of an amino-terminal methionine on the intracellularly expressed molecule. Because the secreted

unfortunate consequence of growth hormone treatment, however, was the infection of a number of children with a fatal virus from one of the cadavers. Production of recombinant human growth hormone (hGH) would clearly provide a safe, reliable, and plentiful source of this drug.

The initial production of hGH was achieved by constructing a hybrid gene from the natural hGH cDNA and synthetic oligonucleotides that encoded the amino terminus of the mature form of the protein (Figure 23-2a). This coding sequence was ligated into

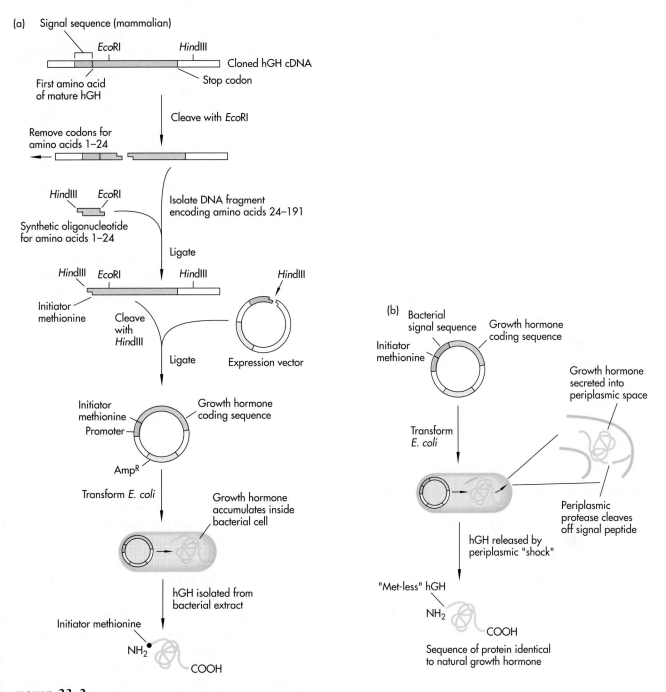

FIGURE **23-2**

Bacterial production of human growth hormone (hGH). (a) An expression vector was constructed for intracellular production of hGH. The coding sequence was constructed by isolating from the cDNA a DNA fragment that encoded amino acids 24–191 and ligating this to a synthetic oligonucleotide fragment that encoded amino acids 1–24. Following introduction of the expression vector into bacterial cells, recombinant hGH was produced inside the cells. The expressed protein behaved just like natural human growth hormone but contained the initiator methionine at the amino terminus. (b) A protein can be produced in bacteria without this extra methionine by targeting it for secretion. To do this, a DNA fragment encoding a bacterial *signal sequence*, which specifies secretion of a bacterial protein, was placed in front of the hGH coding sequence. Upon introduction of this vector into bacteria, hGH is produced, and the signal sequence targets the protein for secretion. The protein accumulates in the *periplasmic space* between the inner and outer bacterial membranes and can be released by hypotonic disruption of the outer membrane. In contrast to the intracellular form of hGH, the protein produced by this procedure does not contain an initiator methionine, since a periplasmic protease cleaved off the signal sequence.

form lacks this methionine, it is called *met-less* hGH. Bacterially expressed hGH has been administered to thousands of growth hormone–deficient children, who have benefited greatly from this recombinant drug.

A Hepatitis B Virus Vaccine Is Produced in Yeast by Expression of a Viral Surface Antigen

One of the successes of modern medicine is the development and implementation of vaccines against infectious diseases. Prior to the advent of recombinant DNA technology, two types of vaccines were used. *Inactivated* vaccines are chemically killed derivatives of the actual infectious agent. *Attenuated* vaccines are live viruses or bacteria altered so that they no longer multiply in the inoculated organism. Both types of vaccines work by presenting surface proteins (antigens) to B and T lymphocytes, which become primed to respond rapidly should the organism actually become infected, usually destroying the infectious agent before any damage is done (Chapter 16). However, these types of vaccines are potentially dangerous because they can be contaminated with infectious organisms. For example, a small number of children each year contract polio from their polio vaccinations. Thus, one of the most promising applications of recombinant DNA technology is the production of *subunit vaccines,* consisting solely of the surface protein to which the immune system responds. With a subunit vaccine, there is no risk of infection.

The first successful subunit vaccine was produced for hepatitis B virus (HBV), which infects the liver and causes liver damage and, in some cases, cancer. The virus particle is coated with a surface antigen, HBsAg, and infected patients carry large aggregates of this protein in their blood. Early experiments suggested that these aggregates would make a potent vaccine, but how could they be produced in quantities sufficient to vaccinate large populations against HBV? With the cloning of the HBV genome, the possibility of a subunit vaccine could be explored. Initial attempts to produce the HBsAg protein in *E. coli* failed, so researchers turned to yeast. The HBsAg gene was inserted into a high-copy yeast expression vector (Fig-

ure 13-3) and engineered, in this case, so that it would not be secreted (Figure 23-3). Yeast transformed with this plasmid produced large quantities of the viral protein (about 1–2 percent of the total yeast protein). By growing the yeast in large fermentors, it was possible to produce 50–100 mg of the protein per liter of culture. This recombinant protein closely resembled the natural viral protein; it even formed aggregates with properties similar to those of the immunogenic aggregates found in infected patients. The yeast protein is now used commercially to vaccinate people against HBV infection.

Vaccines against many human and animal pathogens are currently in various stages of development. Recombinant DNA technology has provided a safe means to work with and to inoculate children and adults with only noninfectious parts of infectious agents. In Chapter 25, we will discuss various strategies for the development of a vaccine against the AIDS virus.

Complex Human Proteins Are Produced by Large-Scale Mammalian Cell Culture

Most of the recombinant proteins we have discussed thus far in this chapter are relatively small and simple in both structure and function. Other proteins of medical interest are considerably more complicated in structure and function, and biologically active proteins have proved difficult to produce in bacteria and yeast. In these cases, biotechnology companies have resorted to using mammalian cells for protein production. Mammalian cells are finicky and expensive to grow, but they can be counted on to produce correctly modified, fully active proteins. Thus, much effort in the biotechnology industry has been devoted to setting up fermentor systems for large-scale culture of mammalian cells.

The first drug to be produced commercially by mammalian cell culture was *tissue plasminogen activator* or *tPA*, which is administered to heart attack victims. Tissue plasminogen activator is a protease, an enzyme that cleaves other proteins. It works by clipping *plasminogen,* an inactive precursor protein, to form *plasmin,* itself a potent protease that degrades *fibrin,* the protein

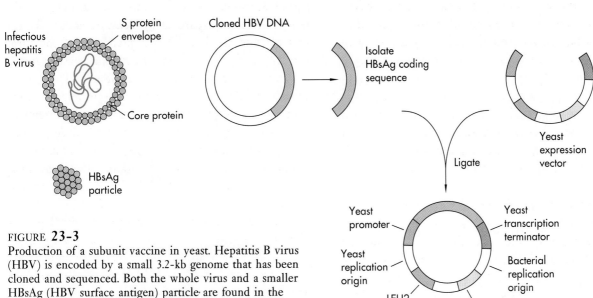

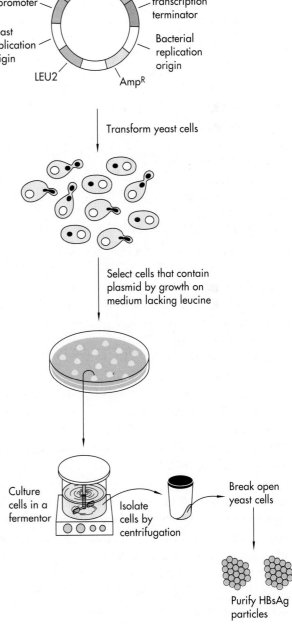

FIGURE **23-3**

Production of a subunit vaccine in yeast. Hepatitis B virus (HBV) is encoded by a small 3.2-kb genome that has been cloned and sequenced. Both the whole virus and a smaller HBsAg (HBV surface antigen) particle are found in the blood of infected patients. To prepare a vaccine against HBV, which has been difficult to propagate in culture, the HBsAg gene was cloned into a vector for expression in the yeast *Saccharomyces cerevisiae*. Transcription occurs from the strong promoter from the gene encoding alcohol dehydrogenase I. A transcription terminator was placed downstream. The vector contains replication origins and markers for both bacteria and yeast. Yeast transformed with this plasmid can be grown to high cell densities in fermentors. This process results in the accumulation of large amounts of HBsAg protein, which upon purification was found to aggregate into particles about 20 nanometers in diameter, resembling the particles found in HBV-infected patients.

that forms blood clots. Rapid administration of a plasminogen activator after a heart attack dissolves the life-threatening clots that lead to irreversible damage of heart muscle. Tissue plasminogen activator is commercially produced from a mammalian cell line carrying a stably integrated, highly amplified expression vector (Figure 23-4).

Another protein being produced by mammalian cell culture is Factor VIII, a protein required for normal clotting of the blood. Genetic defects in Factor VIII production are responsible for hemophilia. For many years, hemophiliacs have been treated with Factor VIII purified from human blood. With the contamination of the human blood supply by the AIDS virus, however, thousands of hemophiliacs became infected and hundreds died from AIDS. The Factor VIII cDNA had already been cloned before scientists found that the blood supply was contaminated with the AIDS virus. Recognition of the need for a safer source of Factor VIII accelerated efforts already under way to produce the protein using recombinant DNA methods. Like tPA, Factor VIII is a large and complex protein and can only be efficiently produced in mammalian cell culture. But the availability of recombinant protein will spare future generations of hemophiliacs from infectious agents that contaminate the blood supply.

Monoclonal Antibodies Function as "Magic Bullets"

We have discussed the use of biotechnology to produce novel vaccines that elicit antibody production by the body's immune system. As we learned in Chapter 16, antibodies are exquisitely selective proteins that can bind to a single target among millions of irrelevant sites. Researchers have long dreamed of harnessing the specificity of antibodies for a variety of uses that require the targeting of drugs and other treatments to particular sites in the body. It is this use of antibodies as targeting devices that led to the concept of the "magic bullet," a treatment that could effectively seek and destroy tumor cells and infectious agents wherever they resided.

The major limitation in the therapeutic use of antibodies is producing a useful antibody in large quantities. Initially, researchers screened *myelomas,* which are antibody-secreting tumors, for the production of

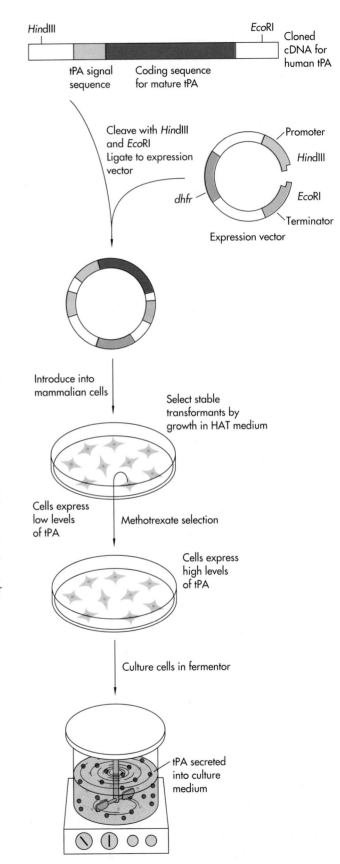

FIGURE **23-4**

Production of tissue plasminogen activator (tPA) by mammalian cell culture. The cloned cDNA for human tPA was ligated into an expression vector that contained a strong promoter and terminator. The vector was stably transfected into a mammalian cell line. The initial transformants secreted tPA into the culture medium, but the level of expression was very low. Cell lines that expressed tPA to high levels were obtained using methotrexate treatment, which selects for cells that have amplified the *dhfr* gene resident in the vector together with the linked tPA expression cassette (Chapter 12). High-expressing lines are grown in large fermentors and recombinant tPA is purified from the culture medium.

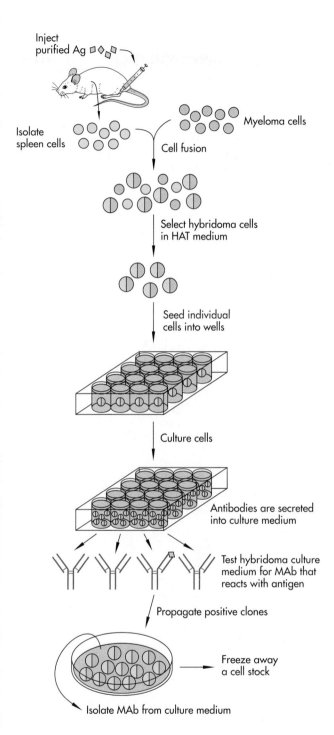

FIGURE **23-5**
Production of a monoclonal antibody (MAb). A mouse is inoculated with an antigen (Ag) of interest. This stimulates the proliferation of lymphocytes expressing antibodies against the antigen. Lymphocytes are taken from the spleen and fused to myeloma cells by treatment with polyethylene glycol. Hybrid cells are selected by growth in HAT medium (Chapter 12). The myeloma cells lack the enzyme HPRT and thus die in this medium unless they become fused with a lymphocyte, which expresses the missing enzyme. Unfused lymphocyte cells soon die off as well, because they do not grow for long in culture. Individual hybrid cells are transferred to the wells of a microtiter dish and cultured for several days. Aliquots of the culture fluids are removed and tested for the presence of antibody (Ab) that binds the antigen. Cells that test positive are cultured for monoclonal antibody production. Antibody-producing cell lines are stored frozen in liquid nitrogen (this process is called *cell banking*). Aliquots can be thawed out and cultured as needed.

is inoculated with the antigen to which an antibody is desired. After the animal mounts an immune response to the antigen, its spleen, which houses antibody-producing cells (lymphocytes), is removed, and the spleen cells are fused en masse to a specialized myeloma cell line that no longer produces an antibody of its own. The resulting fused cells, or *hybridomas,* retain properties of both parents. They grow continuously and rapidly in culture like the myeloma cell, yet they produce antibodies specified by the lymphocyte from the immunized animal. Hundreds of hybridomas can be produced from a single fusion experiment, and they are systematically screened to identify those producing large amounts of a desired antibody. Once identified, this antibody is available in limitless quantities. Monoclonal antibodies are already widely used for the diagnosis of infections and cancer and for the imaging of tumors for radiotherapy. And investigations into their use in the direct treatment of cancer, inflammation, and immune disorders are on the rise.

Human Antibodies That Recognize Specific Antigens Can Be Directly Cloned and Selected

One new application of monoclonal antibody technology is the generation of *abzymes,* antibodies that behave like enzymes to catalyze a chemical reaction.

useful antibodies. But they lacked a means to program a myeloma to produce an antibody to their specifications. This situation changed dramatically with the development of *monoclonal antibody* technology. The procedure for producing monoclonal antibodies, or MAbs, is shown in Figure 23-5. First, a mouse or rat

FIGURE **23-6**
Direct cloning of antibody cDNAs by PCR. To engineer an
antibody, the amino acid sequence of the variable domain
needs to be determined. This could be done by sequencing
a purified preparation of the heavy-(H) and light-(L) chain
proteins, but a simpler method is to deduce the sequence
from the cloned cDNA. In the past, a cDNA library was
prepared from hybridoma mRNA and screened with probes
from the constant regions of the H and L chain genes. A
simpler method has been developed that uses the PCR.
From a comparison of a large number of antibody
sequences, amino acids frequently found at the amino
termini of antibodies were identified. From this information,
a set of degenerate PCR primers was designed that correspond to all the possible sequences in this region. Because
the amino acids in the constant domains of different antibodies are nearly identical, only one PCR primer is needed
for the 3' end of each H and L chain sequence. To directly
clone the antibody cDNAs, cDNA is prepared by treating
hybridoma mRNA with reverse transcriptase, mixed with a
pair of PCR primers (in this case, for amplifying the
heavy chain sequences), and subjected to PCR. Without
knowledge of the amino terminus of the antibody chain, a
PCR had to be set up with each of the different 5' primers
until an amplified DNA fragment was obtained. The process can be simplified if the sequence of the first six or
seven amino acids of the antibody can be determined; this
is sufficient to design a single 5' PCR primer.

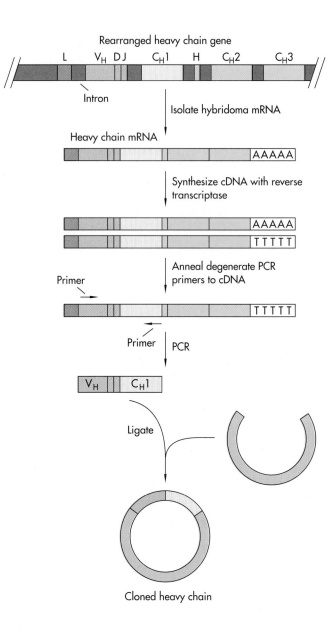

Enzymes catalyze reactions by stabilizing a chemical
structure intermediate between the substrate and
product, termed the *transition state.* Thus, if monoclonal
antibodies could be made to a transition state analogue—a molecule resembling the transition state of
a chemical reaction—then some of these antibodies
might have catalytic activity. The ability to produce
custom-designed catalysts would be very valuable, especially to the chemical and pharmaceutical industries.

Initial attempts to produce catalytic antibodies indicated that they were exceedingly rare and often not
found among the hybridomas produced by conventional monoclonal antibody technology. An excellent
fusion might produce several hundred different antibodies, but the entire repertoire of antibodies that
can be produced by the immune system is perhaps
100 million. How can the entire repertoire be tapped?

One strategy that shows promise is to bypass the
inefficient fusion step in hybridoma production and
directly clone antibody cDNAs from the lymphocytes
of immunized mice (Figures 23-6 and 23-7). Investigators inoculated a mouse with an antigen. They

recovered spleen cells from the mouse and used PCR
to amplify millions of cDNAs for antibody light and
heavy chains. The light- and heavy-chain cDNAs were
cloned separately into phage vectors and then recombined in vitro to generate a third, *combinatorial* library
of phage carrying random pairs of light and heavy
chains. The library was plated onto a bacterial lawn,
and the resulting phage plaques, each containing a
unique antibody, were screened with radioactively labeled antigen in a manner similar to that used for

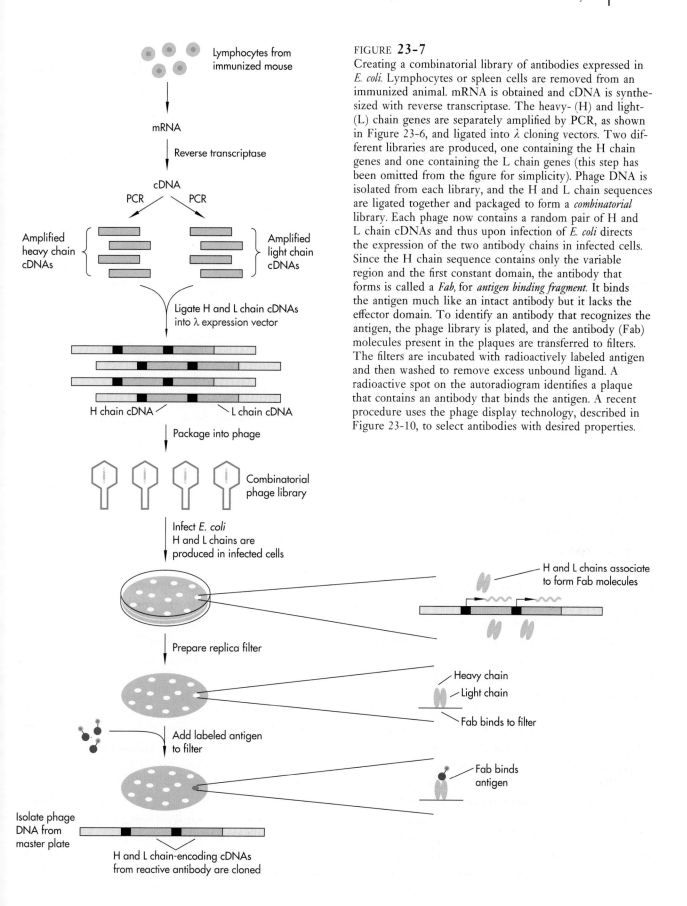

FIGURE **23-7**

Creating a combinatorial library of antibodies expressed in *E. coli.* Lymphocytes or spleen cells are removed from an immunized animal. mRNA is obtained and cDNA is synthesized with reverse transcriptase. The heavy- (H) and light- (L) chain genes are separately amplified by PCR, as shown in Figure 23-6, and ligated into λ cloning vectors. Two different libraries are produced, one containing the H chain genes and one containing the L chain genes (this step has been omitted from the figure for simplicity). Phage DNA is isolated from each library, and the H and L chain sequences are ligated together and packaged to form a *combinatorial* library. Each phage now contains a random pair of H and L chain cDNAs and thus upon infection of *E. coli* directs the expression of the two antibody chains in infected cells. Since the H chain sequence contains only the variable region and the first constant domain, the antibody that forms is called a *Fab,* for *antigen binding fragment.* It binds the antigen much like an intact antibody but it lacks the effector domain. To identify an antibody that recognizes the antigen, the phage library is plated, and the antibody (Fab) molecules present in the plaques are transferred to filters. The filters are incubated with radioactively labeled antigen and then washed to remove excess unbound ligand. A radioactive spot on the autoradiogram identifies a plaque that contains an antibody that binds the antigen. A recent procedure uses the phage display technology, described in Figure 23-10, to select antibodies with desired properties.

cloning cDNAs from an expression library (Figure 7-10). Out of a million phage plaques screened, 200 clones were identified that produced an antibody binding the antigen. Thus, with this approach, investigators were able to sample a million possible antibodies—at least a thousand times more than they could screen by conventional monoclonal antibody technology. Since phages in a particular plaque encode the antibody expressed in the plaque, it is a trivial matter to clone the heavy- and light-chain cDNAs from the phage DNA. These cDNAs can be placed into bacterial or mammalian expression vectors for production of large quantities of the selected antibody.

A recent modification of this method uses filamentous phages such as M13 instead of λ phage and allows display of the antibodies on the phage surface. This offers the advantage of being able to screen thousands more phage (because the screening can be done in solution) and to select phage that express tight-binding antibodies. We will discuss this method later and in Figure 23-10.

"Humanized" Monoclonal Antibodies Retain Activity But Lose Immunogenicity

Although swift progress is being made in the identification of monoclonal antibodies with potential therapeutic value, their use is limited by a problem we have already discussed in this chapter. Monoclonal antibodies are usually mouse proteins, and they are not identical to human antibodies. Thus, antibodies injected into a patient will eventually be recognized as foreign proteins and will be cleared from the circulation.

As we learned in Chapter 16, both chains of the antibody molecule can be divided into variable and constant regions. The variable regions differ in sequence from one antibody to another, and this is the region of the protein that binds the antigen. The constant region is the same among all antibodies of the same type. The first method used to reduce the immunogenicity of a mouse monoclonal antibody was simply to construct *chimeric* genes that encoded proteins in which the variable regions from the mouse

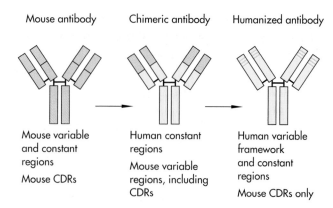

Mouse antibody   Chimeric antibody   Humanized antibody

Mouse variable and constant regions

Mouse CDRs

Human constant regions

Mouse variable regions, including CDRs

Human variable framework and constant regions

Mouse CDRs only

FIGURE **23-8**

Antibody engineering. The basic structure of a mouse monoclonal antibody (MAb) resembles that of a human antibody. However, there are numerous differences between amino acid sequences of the antibodies from the two species. These sequence differences account for the immunogenicity of mouse MAbs in humans. A *chimeric* MAb is constructed by ligating the cDNA fragment encoding the mouse V_L and V_H domains to fragments encoding the C domains from a human antibody. Because the C domains do not contribute to antigen binding, the chimeric antibody will retain the same antigen specificity as the original mouse MAb but will be closer to human antibodies in sequence. Chimeric MAbs still contains some mouse sequences, however, and may still be immunogenic. A *humanized* MAb contains only those mouse amino acids necessary to recognize the antigen. This product is constructed by building into a human antibody the amino acids from the mouse complementarity determining regions or CDRs.

antibody were fused to the constant regions from a human antibody. The chimeric antibody (Figure 23-8) retained its binding specificity but more closely resembled a natural human antibody.

This antibody, however, was not fully *humanized*, because it retained amino acid sequences from the mouse protein. Thus, scientists have set out to engineer fully humanized monoclonal antibodies that will be indistinguishable from natural molecules. Extensive studies of the three-dimensional structures of antibody molecules tell us that only a few of the one hundred amino acids in the variable region of an antibody actually contact the antigen; these regions of contact are referred to as *complementarity determining regions* (CDRs). Three CDRs each comprise the antigen-binding sites on the light and heavy chains. The rest of the variable region serves as a scaffold to anchor the CDRs in the correct positions. This breakdown of amino acids in the variable region into those serving recognition and those serving structural roles is also evident from simply comparing the sequences of many antibody molecules. Amino acid sequences in the CDRs are *hypervariable,* whereas the structural, or *framework,* amino acids differ little.

Thus, to make a fully humanized antibody, all that would be required in principle would be to use in vitro mutagenesis to transfer the CDR amino acid sequences from a mouse MAb to a natural human antibody (Figure 23-8). This method was used to humanize an antibody that recognizes an antigen on the surface of human lymphocytes. This humanized MAb is now in clinical trials as an immunosuppressant and for treatment of lymphoid tumors. Another potentially valuable MAb binds a growth factor receptor found in large numbers on the surface of many breast tumor cells. Laboratory experiments showed that this antibody could block the growth of these cells in culture and caused tumors seeded in mice to regress. Unfortunately, the first humanized versions of this antibody bound the receptor protein but failed to block the growth of breast carcinoma cells. Investigators suspected that the problem was with the framework amino acids, and they used computer modeling to design amino acid substitutions that would strengthen the antibody-antigen interaction. Several such variant antibodies were produced and tested; one bound the receptor 250 times more tightly than did the original antibody and successfully blocked tumor cell growth in culture. This antibody is now being produced in large quantities for clinical trials.

Protein Engineering Can Tailor Antibodies for Specific Applications

Humanizing monoclonal antibodies is an example of the emerging technology of *protein engineering,* that is, a process using recombinant DNA to modify the structure of natural proteins to improve or change their function. Antibodies are particularly attractive candidates for protein engineering, because their structure

is understood in great detail and because their potential for use in medicine is enormous. Another way in which antibodies are being engineered is by changing their *effector domains,* the regions of the heavy chain that specify antibody function—for example, killing of cells marked by the antibody. In this way, the mode of action of a monoclonal antibody can be reprogrammed. One promising strategy is to replace the effector domain entirely with a sequence encoding a toxin. An antibody-toxin fusion protein would deliver the toxin specifically to cells bearing the target antigen. This product could be an exceptionally potent treatment for cancer and for viral diseases such as AIDS. Antibody engineering is also being used to construct *bispecific antibodies.* In these antibodies, each of the two arms recognizes a different antigen, thus allowing an antibody to bridge the two antigens. For example, a bispecific antibody could recognize a tumor cell protein with one arm and a protein on the surface of a killer T cell with the other, thereby bringing the killer cells directly to the tumor (Figure 23-9).

Protein Engineering Is Used to Improve a Detergent Enzyme

Subtilisin is a serine protease produced by bacteria. Due to its broad specificity for proteins that commonly soil clothing, this enzyme was developed for commercial use in laundry detergents. (It is subtilisin that is prominently advertised as the enzyme additive in modern detergents.) But the first detergents containing subtilisin suffered from a serious drawback: they could not be used with bleach, because bleach inactivates the enzyme. Biochemical analysis determined that loss of activity was due to the oxidation of a methionine at position 222. Once this happened, the modified enzyme lost 90 percent of its activity. Because they knew which amino acid was bleach sensitive, however, scientists decided to see whether a variant of subtilisin could be produced that was no longer sensitive to bleach.

To do this, site-directed mutants were constructed in the gene encoding subtilisin. The strategy was simply to substitute, one at a time, each of the non–wild-type amino acids at residue 222. The mutant genes

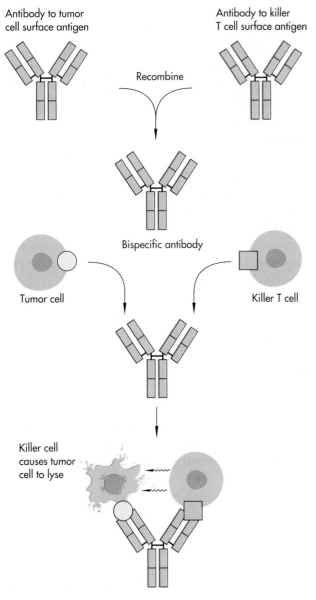

Antibody to tumor
cell surface antigen

Antibody to killer
T cell surface antigen

Recombine

Bispecific antibody

Tumor cell

Killer T cell

Killer cell
causes tumor
cell to lyse

FIGURE **23-9**
A bispecific antibody. By using recombinant DNA, the cDNAs for antibodies to two different antigens can be engineered to make an antibody in which each arm recognizes a different antigen. Thus it is possible to recombine antibodies to a surface antigen on tumor cells and to a protein on cytotoxic T cells to make a bispecific antibody that brings the two cells together to facilitate killing of the tumor cells.

were cloned into expression vectors and the 19 different subtilisin derivatives were expressed. Biochemical analysis showed that the cysteine-222 enzyme was

even more active than the wild-type protein, but it was also inactivated by bleach. The next most active variant was the alanine-substituted enzyme, which was 53 percent as active as wild-type subtilisin. This variant exhibited no detectable bleach sensitivity, so detergents containing this engineered subtilisin can now be used with bleach. This new variant of subtilisin is an example of a *second-generation* molecule, a molecule specifically engineered for a new desirable trait. Protein engineers are currently at work on a third-generation molecule that exhibits decreased temperature sensitivity so that it can be used in hot water.

This experiment points out the power of recombinant DNA as a tool for the engineering of natural products. Changing the properties of a protein was all but impossible prior to the development of recombinant DNA techniques. Now it is not only possible, but easy. It is a routine exercise for protein engineers to generate hundreds of variants of a natural protein for testing. These changes can be educated guesses based on detailed knowledge of the structure of a protein; alternatively, changes can easily be made on a purely random basis. And, as we will see in the next section, a combination of structural information with random mutagenesis and a powerful selection for improved protein function can have dramatic results.

Growth Hormone Variants with Improved Binding Are Selected by Phage Display

To engineer an improved subtilisin enzyme, researchers were aided by the knowledge that only one specific amino acid had to be changed. Thus, they could systematically vary that amino acid to find the one that worked the best. But more complex challenges face protein engineers. Is it possible, for example, to engineer antibodies with higher affinity for antigen; to design an inhibitor that tightly binds to and blocks a cell-surface protein or an enzyme inside a cell; to generate a growth factor or hormone with increased affinity for its receptor? Alterations of this sort require several amino acid changes, and with 20 possible amino acids at each position, the number of variants that

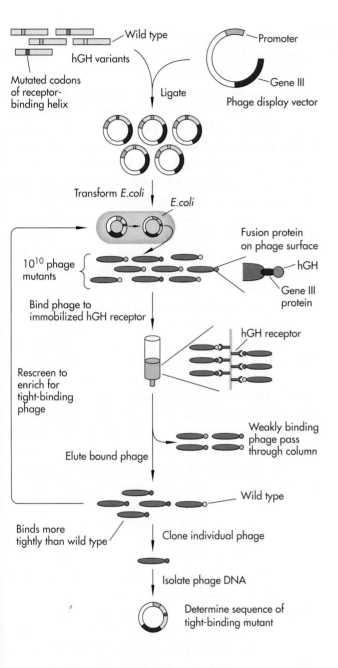

FIGURE **23-10**

Expression of proteins and peptides on the surface of filamentous phage. A library of randomly mutated hGH cDNAs was ligated into an M13-based phagemid vector so that hGH was fused to the carboxy-terminal domain of the M13 gene III protein. The carboxy terminus of the gene III protein associates with the phage particle, and the amino terminus, containing the hGH variants, is displayed on the outer surface of the phage. The library of phagemids is introduced into *E. coli*, and ampicillin-resistant colonies are obtained. These *E. coli* are then infected with a helper phage that induces the production of phagemid particles. Only 1–10 percent of the phage particles contain an hGH–gene III fusion protein, and these contain only one hGH fusion molecule per phage. This ensures that the phage retain sufficient wild-type gene III protein in their coats to remain infectious. hGH-phage were passed through a column containing the hGH receptor covalently linked to plastic beads. Only the phage expressing hGH were retained. The nonbinding phage lacking hGH passed through the column. The bound phage were isolated, cultured in *E. coli*, and passed again over the column. Repeated rounds of selection resulted in the identification of hGH variants that bound the receptor with exceptionally high affinity.

need to be screened is enormous (for changes at just 3 amino acids, there are 8000 different combinations; for 10 amino acids, 10^{13} different proteins are possible). Clearly, these variants cannot be made and tested one at a time, and a method for direct selection of improved proteins is needed.

Researchers have used a new approach to select variants of human growth hormone with increased affinity for growth hormone receptor (Figure 23-10).

From structural studies and extensive mutagenesis of hGH, they knew what portions of the amino acid sequence were important for receptor binding. They synthesized degenerate oligonucleotides that encoded all possible amino acids at these positions and ligated the pool of oligonucleotides in place of the natural hGH sequence. The resulting pool of variant hGH cDNAs was fused to the reading frame of gene III in the filamentous phage M13. Gene III encodes a minor phage coat protein expressed on the surface of the phage, and incorporation of the hGH cDNA into this gene results in the display of the hGH variants on the surface of the phage, one variant per phage. This technique is known as *phage display*.

Now it was a simple matter to pass this library of more than 10^{11} different phage over a column containing the hGH receptor. Phage displaying weakly binding hGH variants were washed off the column, and phage displaying tightly binding variants were recovered with a more stringent wash. This population of tight-binding phage was amplified by infection of *E. coli* and passed over the column a second time. The selection was repeated for a total of six rounds, each round enriching for the phage displaying hGH variants with highest affinity for the receptor bound to the column. At this point, individual phage were cloned, the affinities of their hGH variants were measured directly, and the sequences of the hGH cDNAs were examined. Among these variants was one that bound its receptor about 10 times more tightly than natural hGH did. When selected amino acids from another region of hGH that had been randomized were introduced into this variant, the resulting hGH molecule bound to the hGH receptor over 50 times more tightly than the wild-type hGH did. This process is being repeated in the hope to obtain even more tightly binding variants.

The ability afforded by techniques such as phage display to correlate protein structure and function in a systematic way makes possible new methods of finding novel drugs. If researchers have a good idea what combination of amino acids gives the best fit to the binding site on a receptor, the next step in *rational drug design* would be to design, or even select, a small peptide that binds as well as the larger protein. And then, using computer modeling to display the molecular contacts between ligand and receptor, researchers can attempt to design and synthesize small nonprotein molecules that make the same contacts. The end-product would be a small organic molecule that could be produced more cheaply than a recombinant protein, yet would retain the full biological activity of the protein hormone. And, more important, such molecules could be administered orally, thus eliminating the major disadvantage of most recombinant protein therapeutics—that they must be delivered directly into the bloodstream by injection. This type of rational drug design contrasts sharply with the conventional approach to drug discovery now in use in the pharmaceutical industry, in which an inventory of completely unrelated compounds is tested at random until an active compound is found.

New Technologies Promise New Approaches to Drug Design

The biotechnology industry is in its infancy, and its successes to date follow directly from developments in molecular biology that are already nearly two decades old. The recombinant drugs currently in clinical use arise from what is by now conventional technology—gene cloning, expression, and mutagenesis to improve protein function. These methods will continue to turn out new drugs such as erythropoietins to treat anemia caused by kidney disease, DNase to treat cystic fibrosis, or colony-stimulating factors (CSFs) to increase white blood cell production during chemotherapy.

But the true promise of biotechnology is in novel technologies that are only now being developed. We have mentioned efforts to design catalytic antibodies that can accelerate chemical reactions in both medical and industrial applications. This is but one example of a whole new approach to protein engineering in which novel activities can be placed on unrelated protein scaffolds, using random mutagenesis coupled with selection methods like phage display. Similar goals may be achieved by the engineering of *ribozymes*, RNA molecules with catalytic activity, and the use of the polymerase chain reaction to select nucleic acid molecules that bind tightly to targets of medical importance. Another strategy that may see widespread application is treatment with antisense DNA and RNA to inhibit the expression of oncogenes in tumors or of viral genes in infected patients. And a variety of new technologies based on viral vectors promise new approaches for vaccines and gene therapy.

Many of these techniques now work in the test tube, and the principal challenge facing biotechnology companies is to turn these laboratory techniques into commercially viable processes.

Reading List

General

Hall, S. S. *Invisible Frontiers: The Race to Synthesize a Human Gene.* Atlantic Monthly Press, New York, 1987.

Hood, L. "Biotechnology and medicine of the future." *J. Am. Med. Assoc.*, 259: 1837–1844 (1988).

Goeddel, D. V. (ed.) Systems for Heterologous Gene Expression. *Meth. Enzymol.,* Vol. 185, Academic Press, New York, 1990.

Original Research Papers

EXPRESSION OF HUMAN PROTEINS IN E. COLI

Itakura, K., T. Hirose, R. Crea, A. Riggs, H. L. Heyneker, F. Bolivar, and H. Boyer. "Expression in *E. coli* of a chemically synthesized gene for the hormone somatostatin." *Science,* 198: 1056–1063 (1977).

Goeddel, D. V., H. L. Heyneker, T. Hozumi, R. Arentzen, K. Itakura, D. G. Yansura, M. J. Ross, G. Miozzari, R. Crea, and P. Seeburg. "Direct expression in *Escherichia coli* of a DNA sequence coding for human growth hormone." *Nature,* 281: 544–548 (1979).

Goeddel, D. V., D. G. Kleid, F. Bolivar, H. L. Heyneker, D. G. Yansura, R. Crea, T. Hirose, A. Kraszewski, K. Itakura, and A. D. Riggs. "Expression of chemically synthesized genes for human insulin." *Proc. Natl. Acad. Sci. USA,* 76: 106–110 (1979).

EXPRESSION IN YEAST

Valenzuela, P., A. Medina, W. J. Rutter, G. Ammerer, and B. D. Hall. "Synthesis and assembly of hepatitis B virus surface antigen particles in yeast." *Nature,* 298: 347–350 (1982).

Hitzeman, R. A., D. W. Leung, L. J. Perry, W. J. Kohr, H. L. Levine, and D. V. Goeddel. "Secretion of human interferons by yeast." *Science,* 219: 620–625 (1983).

Sabin, E. A., C. T. Lee-Ng, J. R. Shuster, and P. J. Barr. "High-level expression and in vivo processing of chimeric ubiquitin fusion proteins in *Saccharomyces cerevisiae*." *Bio/Tech.,* 7: 705–709 (1989).

EXPRESSION IN INSECT CELLS

Luckow, V. A., and M. D. Summers. "Trends in the development of baculovirus expression vectors." *Bio/Tech.,* 6: 47–55 (1988). [Review]

Medin, J. A., L. Hunt, K. Gathy, R. K. Evans, and M. S. Coleman. "Efficient, low-cost protein factories: expression of human adenosine deaminase in baculovirus-infected insect larvae." *Proc. Natl. Acad. Sci. USA,* 87: 2760–2764 (1990).

EXPRESSION OF PROTEINS IN MAMMALIAN CELLS

Gorman, C. M. "Mammalian cell expression." *Curr. Opin. Biotech.,* 1: 36–43 (1990). [Review]

Pennica, D., W. E. Holmes, W. J. Kohr, R. N. Harkins, G. A. Vehar, C. A. Ward, W. F. Bennett, E. Yelverton, P. H. Seeburg, H. L. Heyneker, D. V. Goeddel, and D. Collen. "Cloning and expression of human tissue-type plasminogen activator cDNA in *E. coli*." *Nature,* 301: 214–221 (1983).

Paborsky, L. R., B. M. Fendly, K. L. Fisher, R. M. Lawn, B. J. Marks, G. McCray, K. M. Tate, G. A. Vehar, and C. M. Gorman. "Mammalian cell transient expression of tissue factor for the production of antigen." *Prot. Eng.,* 3: 547–553 (1990).

VACCINES

Brown, F. "From Jenner to genes—the new vaccines." *Lancet,* 335: 587–590 (1990). [Review]

Bolognesi, D. P. "Approaches to HIV vaccine design." *Trends Biotech.* 8: 40–45 (1990). [Review]

Berman, P. W., T. J. Gregory, L. Riddle, G. R. Nakamura, M. A. Champe, J. P. Porter, F. M. Wurm, R. D. Hershberg, E. K. Cobb, and J. W. Eichberg. "Protection of chimpanzees from infection by HIV-1 after vaccination with recombinant glycoprotein gp120 but not gp160." *Nature,* 345: 622–625 (1990).

PROTEIN ENGINEERING RECOMBINANT PRODUCTS

Estell, D. A., T. P. Graycar, and J. A. Wells, "Engineering an enzyme by site-directed mutagenesis to be resistant to chemical oxidation." *J. Biol. Chem.* 260: 6518–6521 (1985).

DesJarlais, R. L., G. L. Seibel, I. D. Kuntz, P. S. Furth, J. C. Alvarez, P. R. Ortiz de Montellano, D. L. DeCamp, L. M. Babe, and C. S. Craik. "Structure-based design of nonpeptide inhibitors specific for the human immunodeficiency virus 1 protease." *Proc. Natl. Acad. Sci. USA,* 87: 6644–6648 (1990).

Abrahmsen, L., J. Tom, J. Burnier, K. A. Butcher, A. Kossiakoff, and J. A. Wells. "Engineering subtilisin and its substrates for efficient ligation of peptide bonds in aqueous solution." *Biochemistry,* 30: 4151–4159 (1991).

Bennett, W. F., N. F. Paoni, B. A. Keyt, D. Botstein, A. J. S. Jones, L. Presta, F. M. Wurm, and M. J. Zoller. "High resolution analysis of functional determinants on human tissue-type plasminogen activator." *J. Biol. Chem.,* 266: 5191–5201 (1991).

Cunningham, B. C. and J. A. Wells. "Rational design of receptor-specific variants of human growth hormone." *Proc. Natl. Acad. Sci. USA,* 88: 3407–3411 (1991).

MONOCLONAL ANTIBODIES

Milstein, C. "Monoclonal antibodies." *Sci. Am.,* 243: 66–74 (1980). [Review]

CLONING AND EXPRESSION OF ANTIBODIES

Pluckthun, A. "Antibodies from *Escherichia coli*." *Nature,* 347: 497–498 (1990). [Review]

Huse, W. D., L. Sastry, S. A. Iverson, A. S. Kang, M. Alting-Mies, D. R. Burton, S. J. Benkovic, and R. A. Lerner. "Generation of a large combinatorial library of the immunoglobulin repertoire in phage lambda." *Science,* 246: 1275–1289 (1989).

Chaudhary, V. K., J. K. Batra, M. G. Gallo, M. C. Willingham, D. J. FitzGerald, and I. Pastan. "A rapid method of cloning functional variable-region antibody genes in *Escherichia coli* as single-chain immunotoxins." *Proc. Natl. Acad. Sci. USA,* 87: 1066–1070 (1990).

Mullinax, R. L. et al. "Identification of human antibody fragment clones specific for tetanus toxoid in a bacteriophage λ immunoexpression library." *Proc. Natl. Acad. Sci. USA,* 87: 8095–8099 (1990).

Berg, J., E. Lotscher, K. S. Steimer, D. J. Capon, J. Baenziger, H. M. Jack, and M. Wabl. "Bispecific antibodies that mediate killing of cells infected with human immunodeficiency virus of any strain." *Proc. Natl. Acad. Sci. USA,* 88: 4723–4727 (1991).

Wood, C.R., A. J. Dorner, G. E. Morris, E. M. Alderman, D. Wilson, R. M. J. O'Hara, and R. J. Kaufman. "High level synthesis of immunoglobulins in Chinese hamster ovary cells." *J. Immunol.,* 145: 3011–3016 (1990).

CATALYTIC ANTIBODIES

Jacobs, J. W. "New perspective on catalytic antibodies." *Bio/Tech.,* 9: 258–262 (1991). [Review]

Benkovic, S. J., J. A. Adams, C. L. Borders, K. D. Janda, and R. A. Lerner. "The enzymic nature of antibody catalysis: development of multistep kinetic processing." *Science,* 250: 1135–1139 (1990).

Bowdish, K., Y. Tang, J. B. Hicks, and D. Hilvert. "Yeast expression of a catalytic antibody with chorismate mutase activity." *J. Biol. Chem.,* 266: 11901–11908 (1991).

HUMANIZING ANTIBODIES

Winter, G., and C. Milstein. "Man-made antibodies." *Nature,* 349: 293–299 (1991). [Review]

Jones, P. T., P. H. Dear, J. Foote, M. S. Neuberger, and G. Winter. "Replacing the complementarity-determining regions in a human antibody with those from a mouse." *Nature,* 321: 522–525 (1986).

Riechmann, L., M. Clark, H. Waldmann, and G. Winter. "Reshaping human antibodies for therapy." *Nature,* 332: 323–327 (1988).

Carter, P., L. Presta, C. M. Gorman, J. B. Ridgway, D. Henner, W. L. T. Wong, A. M. Rowland, C. Kotts, M. E. Carver, and M. H. Sheppard. "Humanization of an anti-p185[HER2] antibody for human cancer therapy." *Proc. Natl. Acad. Sci. USA,* 89: 4285–4289 (1992).

PHAGE DISPLAY

Smith, G. P. "Filamentous fusion phage: novel expression vectors that display cloned antigens on the virion surface." *Science,* 228: 1315–1317 (1985).

Cwirla, S. E., E. A. Peters, R. W. Barrett, and W. J. Dower. "Peptides on phage: a vast library of peptides for identifying ligands." *Proc. Natl. Acad. Sci. USA,* 87: 6378–6382 (1990).

Clackson, T., H. R. Hoogenboom, A. D. Griffiths, and G. Winter. "Making antibody fragments using phage display libraries." *Nature,* 352: 624–628 (1991).

Lowman, H. B., S. Bass, N. Simpson, and J. A. Wells. "Selecting high affinity binding proteins by monovalent phage display." *Biochemistry,* 30, 10832–10838 (1991).

NEW RECOMBINANT DRUGS

Shak, S., D. J. Capon, R. Hellmiss, S. A. Marsters, and C. L. Baker. "Recombinant human DNase I reduces the viscosity of cystic fibrosis sputum." *Proc. Natl. Acad. Sci. USA,* 87: 9188–9192 (1990).

Takaue, Y., et al. "Effects of recombinant human G-CSF, GM-CSF, IL-3, and IL-1 alpha on the growth of purified human peripheral blood progenitors." *Blood,* 76: 330–335 (1990).

Shepard, H.M., G. D. Lewis, J. C. Sarup, B. M. Fendly, D. Maneval, J. Mordenti, I. Figari, C. E. Kotts, M. A. Palladino, A. Ullrich, and D. slamon. "Monoclonal antibody therapy of human cancer: taking the HER2 protooncogene to the clinic." *J. Clin. Immunol.,* 11: 117–127 (1991).

Watson, S. R., C. Fennie, and L. A. Lasky. "Neutrophil influx into an inflammatory site inhibited by a soluble homing receptor-IgG chimaera." *Nature,* 349: 164–167 (1991).

FUTURE TRENDS

Sarver, N., E. M. Cantin, P. S. Chang, J. A. Zaia, P. A. Ladne, D. A. Stephens, and J. J. Rossi. "Ribozymes as potential anti-HIV-1 therapeutic agents." *Science,* 247: 1222–1225 (1990).

Tuerk, C., and L. Gold. "Systematic evolution of ligands by exponential enrichment: RNA ligands to bacteriophage T4 DNA polymerase." *Science,* 249: 505–510 (1990).

Han, L., J. S. Yun, and T. E. Wagner. "Inhibition of Moloney murine leukemia virus-induced leukemia in transgenic mice expressing antisense RNA complementary to the retroviral packaging sequences." *Proc. Natl. Acad. Sci. USA,* 88: 4313–4317 (1991).

Miller, P. S. "Oligonucleoside methylphosphonates as antisense reagents." *Bio/Tech.,* 9: 358–362 (1991).

Schultz, J. S. "Biosensors." *Sci. Am.,* 265: 64–69 (1991).

24

Generation of Agriculturally Important Plants and Animals

Recombinant DNA technology not only has enhanced the health of humans, but also has contributed to exciting developments in agricultural biotechnology. Using the methods we described in the previous chapters, researchers have produced transgenic plants and animals with desirable properties such as resistance to disease. Flowers with exotic shapes and colors have been genetically engineered by transgenic expression of pigment genes. Recombinant growth hormones are now available for farm animals, resulting in leaner meat, improved milk yield, and more efficient feed utilization. In the future, transgenic plants and animals may serve as bioreactors for the production of medicinals or protein pharmaceuticals. The same gene cloning methods that produced the pharmaceutical biotechnology industry also spawned new companies trying to apply these procedures to make improved agricultural products. The success of these companies relies on continued research into the genes that control biological processes in plants and animals, and the development of recombinant agricultural products that are useful and safe. In contrast to the results of recombinant DNA experiments that are carried out in laboratories, recombinant agricultural products have to be released into the environment. This can be done only after extensive, rigorous testing to ensure their safety. This chapter highlights experiments that represent the first steps in applying recombinant DNA technology to agricultural problems.

Plants Expressing a Viral Coat Protein Resist Infection

Plant viruses are a serious problem for many of the major agricultural crops. Infections can result in reduced growth rate, crop yield, and quality. Through a standard genetic trick, termed *cross-protection,* infection of a plant with a strain of virus that produces only mild effects protects the plant against infection by more damaging strains. Although the mechanism of cross-protection is not entirely known, it is thought that a particular viral-encoded protein is responsible for the protective effect. The cloning of viral cDNAs for plant virus proteins presented plant molecular biologists with the opportunity to study this phenomenon in more detail.

The first experiment was performed on tobacco. The tobacco mosaic virus (TMV) is an RNA virus about 6.5 kb in size. By cloning viral cDNAs, it was established that the TMV genome encodes four polypeptides: two replicase subunits, a coat protein, and a protein important for cell-to-cell movement. Transgenic plants expressing the TMV coat protein (CP) were produced by *Agrobacterium*-mediated gene transfer (Figure 24-1). When the generated plants expressing the viral CP were challenged with TMV, they exhibited increased resistance to infection. Whereas control plants developed symptoms in 3 to 4 days, transgenic CP-expressing plants resisted infection for up to 30 days. However, the delay of symptoms was shorter when increased doses of virus were applied.

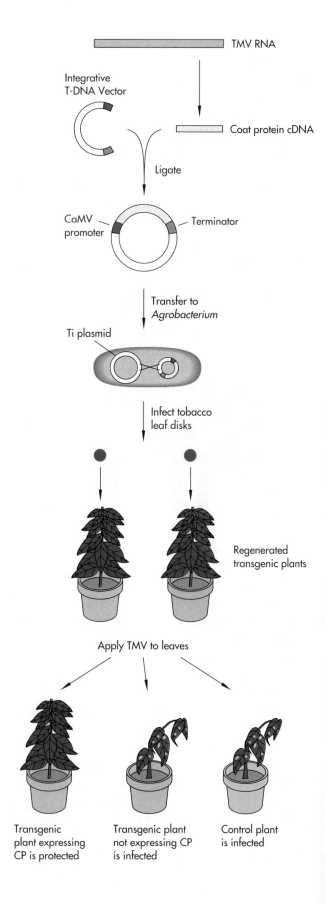

FIGURE **24-1**
Protection of plants from tobacco mosaic virus (TMV) infection by transgenic expression of the TMV coat protein (CP). A cloned cDNA encoding the coat protein from TMV was ligated into an integrative T-DNA vector with a cauliflower mosaic virus (CaMV) promoter. The plasmid was transferred into tobacco by *Agrobacterium*-mediated DNA transfer into leaf disks (see Chapter 15). Regenerated transgenic tobacco plants were obtained that expressed various levels of TMV coat protein in their leaves. Two transgenic plants, one expressing high levels of TMV coat protein and the other expressing undetectable levels, were infected with TMV, as was a nontransgenic control plant. After 2 weeks, the control plant and the nonexpressor exhibited symptoms of TMV infection. In contrast, the coat protein–expressing transgenic plant resisted infection.

Similar behavior is seen in natural cross-protection. The ability of a plant to resist infection appeared to be correlated with the level of CP expression. The protective agent was shown to be the viral coat protein itself, rather than the viral RNA. In some cases, it was found that the expression of a coat protein from one virus resulted in resistance to other related viruses. More recent experiments have demonstrated that effective protection against viral infection of crops such as potato, alfalfa, and tomato can also be achieved by CP expression. Transgenic tomato plants showed comparable levels of resistance in field studies and exhibited agronomic characteristics similar to normal plants.

Transgenic expression in tobacco of cDNA encoding a TMV replicase protein also produced resistant plants. Surprisingly, protection seemed to be due to the transcribed RNA molecules and not the encoded protein. Other approaches that are being investigated include expression of ribozymes to cleave viral RNAs, and transgenic expression of antisense RNAs.

Insects Fail to Prey on Plants Expressing a Bacterial Toxin

As any weekend gardener knows, plants are susceptible to damage by insects, and farmers lose billions of dollars of crops to insects. Currently, the major weapons against the attackers are chemical insecticides. However, it has been desirable to reduce the use of chemical pesticides. Natural microbial pesticides, such as certain species of *Bacillus thuringiensis* (Bt), have been used in a limited fashion for over 30 years. Upon sporulation, these bacteria produce a crystallized protein that is toxic to the larvae of a number of insects. The toxin protein does not harm nonsusceptible insects and has no effect on vertebrates. Unfortunately, production of the bacterial spores for commercial use is limited and the protective effect is short-lived.

Knowledge of this natural system led plant molecular biologists to develop plants that expressed the Bt toxin within their cells. The crystal protein is normally expressed as a large, inactive pro-toxin about 1200 amino acids in length. Upon digestion by susceptible larva, proteases in the insect's gut cleave the protein

into an active 68,000-dalton fragment. The toxin acts by binding to receptors on the surface of midgut cells and blocking the functioning of these cells. Initial studies demonstrated that tobacco and tomato plants expressing the Bt toxin gene killed larvae of tobacco hornworms. However, because the expression levels were so low, protection was not attained against less sensitive, but agronomically important, insect pests like the larvae of the cotton bollworm. Recently, cotton plants have beeen protected by using a strong promoter to increase expression of Bt toxin mRNA and by modifying part of the Bt toxin coding sequence so that it is efficiently translated in plants (Figure 24-2). These resistant cotton plants may have commercial use if they pass the rigors of field testing. Research is currently under way to alter the toxin protein so that it is effective against a wider range of insect species.

A second approach to the development of insect-resistant plants has been through the transgenic expression of serine protease inhibitors. These proteins are present in a number of plants such as tomato, potato and cowpea, and function to deter insects by inhibiting serine proteases in the insect digestive system. The effectiveness of this approach against insect larvae from a wide range of insect species has been demonstrated in tobacco.

Herbicide-Tolerant Transgenic Plants Allow More Effective Management of Weeds

The presence of weeds in a field of crop plants can reduce yield by over 10 percent. Since weeds are nothing more than an undesirable plant, it has been difficult to find ways to destroy them without also affecting the crop plant. Weed killers, also called *herbicides,* are not very selective, so their current use relies on differential uptake between the weed and the crop plant or on application of the herbicide before planting a field. With the ability to introduce DNA into plants, researchers are trying to create herbicide-tolerant crops by three strategies: increasing the level of the target enzyme for a particular herbicide, expressing a mutant enzyme that is not affected by the compound, or expressing an enzyme that detoxifies the herbicide.

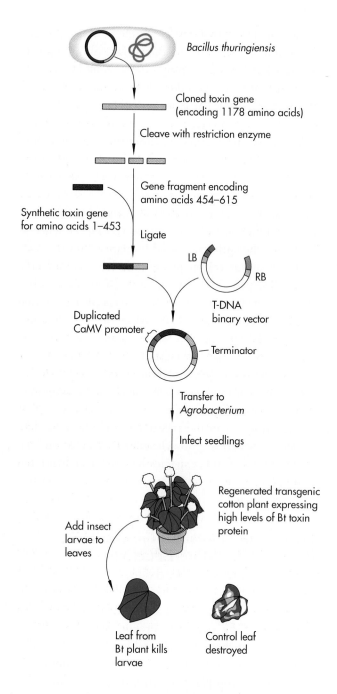

Leaf from
Bt plant kills
larvae

Control leaf
destroyed

FIGURE **24-2**

Killing insect larvae with transgenic cotton plants that express the *Bacillus thuringiensis* (Bt) toxin gene. The 1178 amino acid Bt crystal protein, an insect toxin, is encoded on a 75-kb plasmid and is produced upon bacterial sporulation. Previous studies had shown that the full-length crystal gene was poorly expressed in transgenic plants. In order to improve expression, a truncated gene encoding the functional amino-terminal portion (amino acids 1 to 615) was cloned into a T-DNA binary vector between LB and RB (see Chapter 15). To further increase expression in plants, a synthetic gene encoding amino acids 1 to 453, containing codons preferred in plants, was substituted for the natural sequence. (Although most amino acids can be encoded by more than one codon, analysis of coding sequences of cDNAs from different organisms showed strong preferences for particular codons. This is thought to reflect the concentration of the corresponding amino acyl–tRNAs in the cell. The transgenic production of some proteins often increases when the cDNA used contains the preferred codons.) The synthetic sequence was ligated to a natural DNA fragment encoding amino acids 454 to 615. In addition, the T-DNA vector contained a duplicated CaMV promoter, which has been shown to increase transcription about fivefold. The T-DNA vector was transferred into cotton seedlings (also called *hypocotyls*) by *Agrobacterium* infection, and transgenic cotton plants were regenerated. The efforts to improve expression and translation of the Bt toxin apparently paid off, as some of these plants produced 100 times more toxin than did transgenic plants containing the full-length toxin gene. The new transgenic plants exhibited protection against larvae from a number of species, including the relatively insensitive beet armyworm.

Of the large number of herbicides in use today, only a few of the cellular targets have been characterized. The herbicide glyphosate, the active ingredient in the commercial weed killer Roundup, kills plants by inhibiting 5-enolpyruvylshikimate 3-phosphate synthase (EPSPS), a chloroplast enzyme in the pathway for biosynthesis of essential aromatic amino acids. Roundup is currently the most widely used weed killer because it is active in very low doses against a broad spectrum of weeds. It is also much safer for the environment than earlier herbicides, since it is rapidly degraded by soil microorganisms. Transfer of a cloned cDNA encoding the petunia EPSPS into plants increased the level of enzymatic activity to about 20 times that found in nontransgenic plants. Overexpression of the enzyme allowed the transgenic petunias to grow in the presence of four times the level of herbicide that kills wild-type plants. Unfortunately, the growth rate of the treated transgenic plants was slower than that of untreated plants.

Another strategy for creating herbicide-tolerant plants has used mutant forms of bacterial EPSPS enzymes containing amino acid substitutions that make them less sensitive to inhibition by glyphosate (Figure 24-3). Genes encoding these mutant enzymes have been cloned from glyphosate-resistant bacteria and expressed in plants. In laboratory studies, these trans-

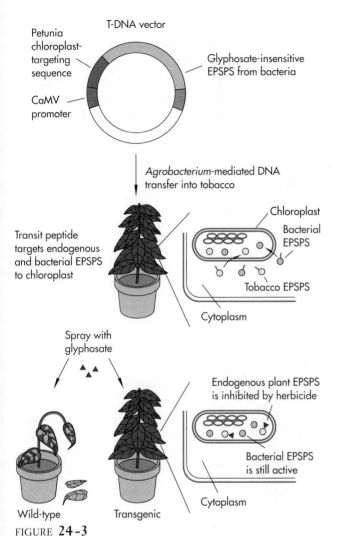

FIGURE 24-3

Engineering herbicide-resistant plants. The enzyme 5-enol-pyruvylshikimate 3-phosphate synthase (EPSPS), important for synthesis of aromatic amino acids in bacteria and plants, is inhibited by glyphosate, the active ingredient in the weed killer Roundup. The gene encoding a glyphosate-resistant EPSPS from bacteria was cloned into a T-DNA expression vector and introduced into tobacco by *Agrobacterium*-mediated gene transfer. Since EPSPS in plants is synthesized in the cytoplasm and then transferred into chloroplasts, a chimeric gene was constructed in which a segment encoding a 72-amino acid transit peptide from petunia EPSPS was fused to the amino terminus of the bacterial EPSPS coding sequence. Expression of the chimeric gene was controlled by the cauliflower mosaic virus (CaMV) 35S promoter. Transgenic plants expressed both the endogenous plant EPSPS and the bacterial glyphosate-insensitive enzyme. Biochemical studies demonstrated that the transit peptide had properly targeted the bacterial enzyme to chloroplasts. When sprayed with glyphosate, wild-type plants are killed because EPSPS is inhibited. The transgenic plants tolerated levels of glyphosate four times higher than levels that killed wild-type plants because the bacterial EPSPS still functions in the presence of the herbicide.

genic plants exhibited tolerance to levels of glyphosate that killed wild-type plants. The plant EPSPS enzyme is synthesized in the cytoplasm and is translocated to the chloroplasts by proteins that recognize and bind to the amino-terminal 72 amino acids, called the *transit sequence*. In order to target the bacterial enzyme, which lacks this sequence, to the chloroplasts of transgenic plants, a gene segment encoding the plant transit sequence was fused to the coding sequence for the bacterial enzyme. Biochemical analysis of the resistant bacterial enzymes showed that the mutations that impart tolerance to glyphosate reduce EPSPS activity. By protein engineering (Chapter 23), researchers hope to create improved forms of bacterial and plant EPSPS enzymes that exhibit a high level of glyphosate resistance yet retain normal enzymatic activity.

The third strategy for creating herbicide-tolerant plants is by transgenic expression of enzymes that convert the herbicide to a form that is not toxic to the plant. Some plants have developed their own detoxifying systems for certain herbicides; however, these activities in plants are encoded by a complex set of genes that has not yet been fully characterized. Certain bacteria have been found that naturally degrade herbicides, often by the action of a single enzyme. Several of these bacterial detoxifying enzymes have been shown to function in plants. Since glyphosate is currently the most widely used weed killer, researchers are searching for a bacterial enzyme that will degrade glyphosate. Some researchers believe that expression of a detoxifying enzyme in plants will not affect the agronomic properties of the plants as much as expression of increased levels or insensitive variants of enzymes necessary for metabolic processes will.

Flowers Exhibiting New Colors and Patterns Can Be Obtained by Genetic Engineering

The floral industry is a multibillion-dollar business that is beginning to experiment with recombinant DNA as a way to engineer flowers with new properties. Currently, most ornamental flowers are bred by vegetative propagation of cuttings, resulting in many individual plants with identical genetic background.

However, one of the problems with this approach is that desirable characteristics cannot be changed in a directed way. With the ability to introduce genes into plant cells and obtain regenerated plants, flowers with new colors, shapes, and growth properties can be engineered.

The first such experiment introduced into petunia a gene from maize that encoded an enzyme in the pathway for production of anthocyanin, a pigment that colors maize kernels purple. The cDNA was transferred by protoplast transformation into a petunia variant that was light pink in color because it contained

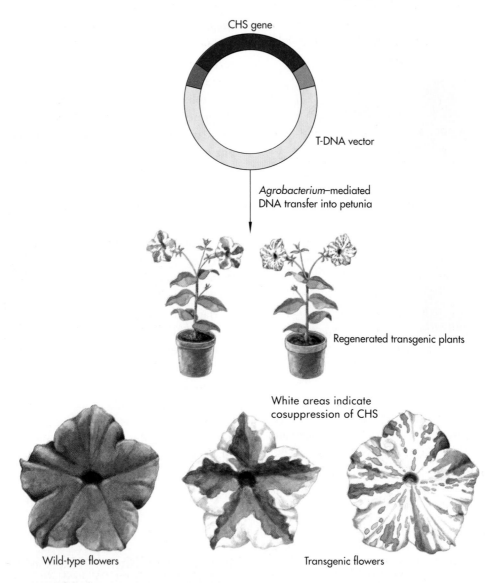

FIGURE 24-4

Introduction of an additional copy of petunia pigment gene inhibits expression of endogenous alleles and the transgene. Chalone synthetase (CHS) is an enzyme in the biosynthetic pathway for anthocyanin, a purple pigment found in flowers and maize kernels. An experiment was performed in which a cloned petunia CHS gene was introduced into a petunia plant. Scientists thought that by increasing the gene dosage the plants would produce a flower with a darker-purple color or perhaps a new color. To their surprise, some of the transgenic plants produced flowers with uncolored sectors, producing strikingly unusual patterns. In some cells, the presence of the additional CHS gene resulted in the complete inhibition of CHS mRNA transcription. Expression of other genes in the anthocyanin pathway were not affected. The mechanism of this phenomenon, termed cosuppression, is not yet understood but is not simply due to expression of the transgene.

a mutation in one of its pigment genes. Of 15 regenerated plants, two had uniformly brick-red flowers and four plants had flowers with brick-red sectors. Northern analysis of RNA from the transgenic plants demonstrated that the maize gene was indeed expressed. Another example involves roses. A blue rose has never been obtained because rose plants lack the enzymes that synthesize the pigment for blue flower coloration. A gene encoding a protein involved in biosynthesis of the blue pigment, delphinidum, has been cloned from a blue petunia. Experiments are under way to produce transgenic roses with blue flowers by introducing this gene into a rose plant. Future experiments aim to expand the spectrum of coloring of certain floral species by introducing the genes for entire pigment biosynthesis pathways.

The manipulation of flower coloring by DNA transfer into plants may not always produce the desired result. An experiment was performed in which a second copy of a petunia pigment gene was introduced into a petunia plant with colored flowers (Figure 24-4). It was expected that increased production of the encoded enzyme might produce flowers with a deeper purple color. A surprise came when the regenerated plants containing the additional gene produced flowers with either uncolored sectors or flowers lacking color entirely. Somehow the presence of the additional petunia gene had led to the inactivation of both it and the endogenous gene. The mechanism of this phenomenon, termed *cosuppression,* is not yet understood. Cosuppression has also been observed in potatoes and tobacco. This may provide an alternative strategy to antisense methods for controlling plant gene expression.

The Potential Use of Plants to Produce Proteins Is of Commercial Importance

Once gene transfer methods had been developed for plants, scientists were interested in using plants as a system for production of heterologous proteins. As we learned in Chapter 23, production of a protein drug in mammalian cell culture is an extremely expensive process requiring thousands of liters of cells to produce enough protein for commercial use. On the other hand, plants are cheap to grow, and huge quantities of protein

could be obtained from a single field. To date, only laboratory experiments have been performed that test the ability to express heterologous proteins in various plant species. For example, enkephalin, a human neuropeptide, and human serum albumin have been expressed in plants.

Another potential use is the expression of mouse monoclonal antibodies in plants (Figure 24-5). Antibodies against a plant toxin or herbicide could be

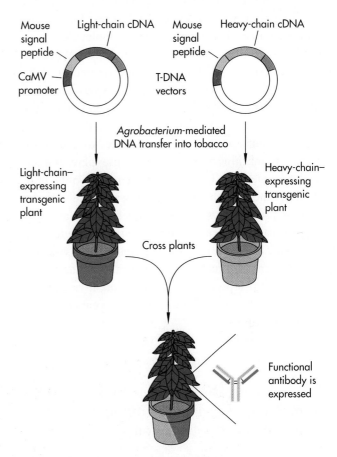

FIGURE **24-5**
Plants as bioreactors to produce antibodies. Cloned cDNAs encoding the light and heavy chains from a mouse monoclonal antibody were ligated into separate T-DNA vectors and placed under control of a constitutive CaMV promoter. The plasmids were transferred separately into tobacco plants by *Agrobacterium* infection. Transgenic plants containing the light- and heavy-chain genes were sexually crossed to produce progeny plants that contained both genes. Examination of protein extracted from leaves demonstrated the expression of functional antibody molecules in these progeny plants. Other experiments showed that the presence of a signal sequence was necessary for high-level expression. These results suggest that the plant secretion machinery can recognize the mouse signal peptide.

produced in plants to protect the plant from these agents. Alternatively, monoclonal antibodies for clinical therapy could be produced in plants. An experiment was performed in which the heavy (H) and light (L) chains that make up an antibody were expressed separately in tobacco plants. The transgenic plants were produced by *Agrobacterium*-mediated gene transfer. Two plants, one expressing the H chain and the other expressing the L chain, were sexually crossed to produce progeny that expressed both the H and the L chains. Leaves from these plants were tested for production of intact antibody molecules. The results indicated that functional antibody molecules were produced in these plants and represented about 1.5 percent of the total cellular protein extracted from the leaves. An interesting finding in these experiments was that functional antibodies were formed only when the H and L chain cDNAs contained signal sequences from the mouse antibody. This suggested that the plant secretion machinery recognizes these heterologous leader sequences. These experiments are only pilot studies, since much higher expression levels of the foreign protein must be achieved before plant bioreactors can compete with fermentation procedures in use today. In addition, since much of the cost for producing a drug, such as tissue plasminogen activator, stems from postproduction purification, it remains to be seen whether plant bioreactors will be economically competitive.

Recombinant Bovine Growth Hormone Stimulates Milk Production and Improves Feed Utilization

One of the first applications of recombinant DNA technology to animals was production of recombinant bovine growth hormone (rbGH) by expression in *E. coli*. Farmers had found over 60 years ago that milk production of dairy cows increased after the cows were injected with an extract made from cow pituitaries. The active molecule was later found to be growth hormone, also called *somatotropin*. The commercial use of growth hormone to stimulate milk production became cost-effective only after rbGH could be made in large quantities in bacteria. The rbGH is prepared

in essentially the same way as human GH (Figure 23-2). Experiments using pure rbGH clearly demonstrated what had been found with the impure pituitary extracts, namely, that milk production increased by at least 14 percent and that the amount of feed utilized per gallon of milk decreased.

As with all drugs that are used on animals, the Federal Drug Administration (FDA) of the United States requires the manufacturer to demonstrate that food derived from treated animals is safe for human consumption. In addition, the drug must be shown to be safe for the animals, effective for its intended purpose, and environmentally safe. Thus before milk or meat from a growth hormone–treated cow can be sold to the general public, experimental trials must be conducted, and the results of these studies must be assessed by the FDA. On the basis of analysis of data from over 10 years of studies on rbGH, the FDA has recently concluded that milk from rbGH-treated cows is safe for human consumption.

Other proteins of agricultural importance have been produced via recombinant DNA technology. Subunit vaccines against animal viruses, such as foot-and-mouth disease virus, have been produced by expression of viral coat proteins in bacteria and yeast, using a strategy similar to that for a hepatitis B vaccine for humans (Chapter 23).

Procedures for Generation of Transgenic Farm Animals Are Developed

Once the methods for production of transgenic mice were established, scientists were interested in developing similar procedures for obtaining transgenic livestock. The first experiments were done using rabbits, pigs, and sheep. At the time, the most reliable method for introduction of DNA into the germ line of mice was by microinjection into the nuclei of eggs (see Chapter 14), so this was tried with fertilized eggs from these farm animals. A special technique, called *differential interference contrast (DIC) microscopy*, was required to visualize the nuclei of these eggs. Although the nucleus in rabbit and sheep eggs can be seen directly by this technique, pig and cow eggs are opaque and the nuclei are difficult to see, even under DIC microscopy. This problem was solved by briefly spinning these eggs in a centrifuge. This concentrates pigments in the cytoplasm to one side of the egg so the nucleus was now visible for microinjection.

The first experiments introduced DNA containing a gene for human growth hormone (hGH) under transcriptional control of the metal-inducible metallothionein (MT) promoter. These studies paralleled the early experiments on transgenic mice. Human GH was chosen because sensitive assays could be used to detect expression of hGH mRNA and production of hGH protein in the transgenic animals. In addition, it was thought that the expression of hGH in these animals might be detected by an increased size of the animal as for transgenic mice. The results from the initial studies on rabbits, pigs, and sheep indicated that the injected DNA integrated into the genome of all three species and hGH was detectable in the serum from transgenic pigs and rabbits. However, the frequency with which an injected egg produced a transgenic animal was low. Only 1 of 200 injected eggs would produce a transgenic animal. This is about 10 to 15 times lower than typical results on mice. These early experiments used conditions that were optimized for mice, and now procedures are being developed to improve the frequency of obtaining a transgenic farm animal. Currently, pigs are used for most experiments because the efficiency of obtaining a transgenic pig is generally higher than for sheep or cows, the litter size is large, and the gestation period is shorter. Recently, a transgenic cow has been generated by microinjecting DNA into the pronucleus of a fertilized embryo. The embryo was cultured in vitro to the morula or blastocyst stage and then transferred to a recipient mother. About 1000 zygotes were injected with DNA, of which 129 developed into embryos and were transferred to cows. Of 19 calves born, two contained DNA integrated into their genome. Researchers are using PCR to identify which blastocysts contain cells that have integrated the injected DNA. This will further improve the commercial feasibility of obtaining a transgenic cow.

The ability to introduce genes into farm animals is now being put to practical use. A transgenic animal that naturally expresses large amounts of growth hormone need not take the drug by injection as is done now with dairy cows. Previous experiments showed that pigs treated with recombinant GH grew faster and produced meat that was leaner, a trait of commercial importance. Transgenic pigs expressing growth hormone exhibited these same features and were more efficient in utilizing feed than were nontransgenic siblings. However, these transgenic pigs developed debilitating physiological problems, probably because the expression of growth hormone was continual throughout the life of the animal. Normally, growth hormone is produced only during a 2-month period. Expression of the GH gene might be restricted to this time period by using zinc to turn the MT promoter on and off. However, expression of the GH gene in these transgenic pigs appears to be constitutive. Induction of the MT promoter by including zinc in the feed increased GH levels only twofold. Experiments are under way to identify other regulated promoters so that transgene expression can be controlled.

Pharmaceutical Proteins Can Be Produced in Transgenic Animals

As we learned in Chapter 23, the production of certain therapeutic proteins can be achieved only by mammalian cell culture, where the protein is folded or processed properly. Expression of these proteins in transgenic animals may be an alternative method to this costly technology. One idea being evaluated is the secretion of recombinant proteins into the milk of transgenic sheep or cows. Milk contains very high levels of proteins such as casein, β-lactoglobulin, and whey acidic protein. Production of these proteins is tightly regulated by promoters that limit gene expression only to cells of the mammary gland. The regulatory sequences from the genes of milk-specific proteins have been cloned and used to control the expression of heterologous genes in transgenic animals (Figure 24-6). Transgenic animals may someday serve as bioreactors that continually secrete high levels of a desired protein into their milk. The protein would be harvested simply by milking the animals, and then standard chromatographic procedures would be used to purify it.

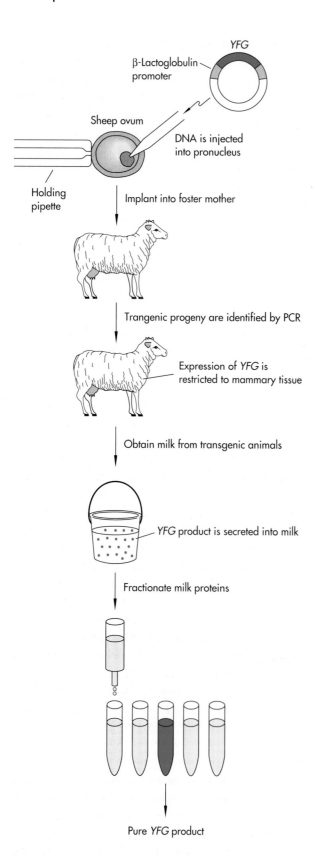

Examples of pharmacologically active proteins produced by this approach include factor IX, interleukin-2, tissue plasminogen activator (tPA), and urokinase. Although the levels of proteins in milk from transgenic mice can be extremely high, up to 25 g/L, the concentrations of proteins attained in the milk of transgenic farm animals have been too low to be commercially useful. However, a recent study reported the expression of human α_1-antitrypsin in transgenic sheep at about 35 grams per liter with use of the sheep β-lactoglobulin gene promoter. These levels may be improved further if additional genomic sequences can be found to enhance promoter activity. There may also be sequences in the human gene that decrease transcription from the sheep's milk-specific promoter. As with any transgenic experiment, the site of integration can affect transcriptional activity.

This technology might also be used to improve the protein content of milk from transgenic dairy cows. For example, an increase in the concentration of casein may result in higher yields of cheese per liter of milk. Another idea being tested is the secretion of proteins into the blood of animals. An exciting application of this approach has been the production of transgenic pigs expressing human hemoglobin for use as a blood substitute. The hemoglobin can be extracted and purified from the pig's red blood cells. A large pig can donate 20 pints of blood over the course of a year without detrimental health effects, thereby yielding 500 to 1000 g of purified human hemoglobin. Efforts are now under way to incorporate the recombinant hemoglobin into a synthetic blood.

FIGURE **24-6**

Production of a pharmaceutically important protein in the milk of transgenic sheep. *YFG* (your favorite gene) encodes a protein that is of therapeutic importance, such as tissue plasminogen activator used to dissolve blood clots in humans (Chapter 23). The gene is placed under control of the β-lactoglobulin promoter, active only in mammary tissue, and is introduced into sheep ova by microinjection of the expression vector into the nucleus. The injected ova are implanted into foster mothers, and progeny expressing the transgene are identified by PCR amplification of chromosomal DNA using primers from the sequence of *YFG*. Transgenic sheep express *YFG* only in mammary tissue and secrete high levels of the *YFG* protein into the milk, from which it can be purified.

Farm Animals May Be Protected from Viral Infection by Transgenic Expression of Viral Coat Protein

Infections of farm animals by bacteria and viruses are especially serious because the animals are usually in close contact, so infections spread rapidly. Experiments are being conducted to generate disease-resistant farm animals. One approach, similar to that described earlier in this chapter for plants, is the transgenic expression of a viral coat protein. Transgenic chickens have been produced that express the *env* gene, encoding the viral envelope glycoprotein from an avian leukosis virus (ALV). The nucleus of the chicken egg could not be injected with DNA, since it is very small and is obscured by the egg yolk. Instead, DNA was introduced by infection of fertilized eggs with a recombinant retrovirus. Once the chicks hatched, they were tested for integration of proviral DNA into the germ line by using radiolabeled ALV cDNA as a hybridization probe. Twenty-three transgenic chickens were obtained, each containing ALV provirus DNA integrated at a different genomic position. Most of the transgenic chickens expressed infectious viruses and developed ALV-induced pathologies. One line of transgenic chickens did not produce infectious virus because of a mutation in the proviral DNA. Immunological tests showed that this line expressed only the viral envelope protein, and these chickens were resistant to infection when challenged with ALV. The likely mechanism of protection is that the envelope protein binds to the cellular receptor for ALV, thereby blocking entry of the infectious virus into the cell. The use of a replication-competent virus, as in these experiments, will probably not be the method of choice for development of commercially important transgenic chickens. Nevertheless, these experiments suggest that viral resistance can be engineered into chickens.

Several other approaches to generate disease-resistant animals are currently being evaluated. Transgenic cows that express β-interferon, a cytokine that stimulates resistance to viral attack, are being tested for resistance to a virus that causes diarrhea. The transgenic expression of antibodies against a particular pathogen may be another way to prevent disease. Animals that have been bred to be more resistant to a particular disease have been shown to contain endogenous *disease-resistance genes*. For example, certain alleles of the mouse *Mx1* gene encode a protein that interferes with influenza virus replication. Introduction of the *Mx1* gene into susceptible mice renders them resistant to viral attack. Normally, when a virus infects an animal, it causes a local elevation in the level of interferon. By placing the *Mx1* gene under an interferon-inducible promoter, the gene was turned on only at the site of an infection, thereby stopping further propagation of the virus. These results have stimulated the search for other endogenous genes that offer protection to disease.

The Implementation of Agricultural Biotechnology Requires Continued Research and Social Discussion

An idea that spawned much of the early excitement in agricultural biotechnology was the creation of transgenic plants that expressed bacterial nitrogen-fixing genes so that the plants would produce their own fertilizer. Unfortunately, this idea is far from practical implementation, because so many genes are involved in this complicated process. In this chapter we have highlighted promising laboratory experiments that with further study may be practical. Several of the systems discussed in this chapter are ready for introduction into a commercial setting but, first, must pass tests that demonstrate environmental safety and, second, must be accepted by farmers and consumers.

The examples described so far represent only the beginning of what is possible. Basic research on plants and animals will continue to uncover the molecular details underlying biological pathways. The results of some of these experiments will find their way into practical application. For example, by cloning the genes encoding seed proteins, the nutritional value of plants may be improved by expressing engineered seed proteins. Seeds from plants such as soybean and corn are the major protein food source for animals. Introduction of genes conferring drought resistance into staple crops grown in Third World countries has the potential for alleviating famine. Transgenic plants expressing a defined set of enzymes may serve as bioreactors for the synthesis of medicinals. Genes involved in the production of desirable oils from undomesticated crops are being introduced into traditional oil-producing plants, where large quantities of these oils can be produced. Another potential product is frost-resistant plants, created by transgenic expression of proteins such as the antifreeze protein found in cold water fish.

The possibilities for advances in animal biotechnology are equally unbounded. Transgenic fish have been produced, opening the way for introduction of genes conferring increased growth or disease resistance. The biology of wool production in sheep has been studied intensively in countries such as Australia, New Zealand, South Africa, and the United Kingdom. The major factor affecting wool production is limitation of nutrients required for follicle cell proliferation and amino acids that compose the proteins of the wool fiber. Recombinant DNA experiments are under way to improve wool growth and quality by improving the amino acid content of feed plants, or by introducing amino acid biosynthetic genes into the genome of bacteria present in the sheep's rumen or into the genome of the sheep itself.

Genetic engineering of plants and animals by humans has been going on from the dawn of civilization. The many varieties of domesticated animals and the acres of corn covering the prairies testify to the skills of the breeder and farmer using the classical methods of genetic manipulation. Recombinant DNA methods provide a powerful set of tools to augment these genetic methods; however, the aims of agricultural genetic research are not achieved until the new strains of plants and animals have moved out of the laboratory. Slowing the practical application of such research is the fact that people are worried about the environmental hazards of releasing plants and animals genetically modified by recombinant DNA techniques, and there are also concerns about the safety of eating and drinking products of genetic engineering. These social and political considerations are unavoidable and place a special onus on those doing this research to explain what is being done, and to take exceptional precautions to ensure the safety and environmental acceptance of new strains of plants and animals. The agronomic applications of recombinant DNA techniques have the potential for bringing about a "green revolution" to benefit people throughout the world.

Reading List

General

Gasser, C. S., and R. T. Fraley. "Genetically engineering plants for crop improvement." *Science,* 244: 1293–1299 (1989).

Lindow, S. E., N. J. Panopoulos, and B. L. McFarland. "Genetic engineering of bacteria from managed and natural habitats." *Science,* 244: 1300–1307 (1989).

Mol, J., A. Stuitje, A. Gerats, A. van der Krol, and R. Jorgensen. "Saying it with genes: molecular flower breeding." *Trends Biotech.,* 7: 148–153 (1989).

Pursel, V. G., C. A. Pinkert, K. A. Miller, D. A. Bolt, R. G. Campbell, R. D. Palmiter, R. L. Brinster, and R. E. Hammer. "Genetic engineering of livestock." *Science,* 244: 1281–1288 (1989).

Altenback, S. B., and R. B. Simpson. "Manipulation of methionine-rich protein genes in plant seeds." *Trends Biotech.,* 8: 156–160 (1990).

Gadani, F., L. M. Mansky, R. Medici, W. A. Miller, and J. H. Hill. "Genetic engineering of plants for virus resistance." *Arch. Virol.,* 115: 1–21 (1990).

Hennighausen, L., L. Ruiz, and R. Wall. "Transgenic animals—production of foreign proteins in milk." *Curr. Opin. Biotech.,* 1: 74–78 (1990).

Oxtoby, E., and M. A. Hughes. "Engineering herbicide tolerance into crops." *Trends Biotech.,* 8: 61–65 (1990).

R. A. Pedersen, A. McLaren, and N. L. First, eds. *Animal Applications of Research in Mammalian Development. Cur-*

rent Communications in Cells and Molecular Biology, vol. 4. Cold Spring Harbor Laboratory, Cold Spring Harbor, N. Y., 1991.

Brunke, K. J., and R. L. Meeusen. "Insect control with genetically engineered crops." *Trends Biotech.*, 9: 197–200 (1991).

Fraley, R., and J. Schell. "Plant biotechnology." *Curr. Opin. Biotech.*, 2: 145–210 (1991).

Original Research Papers

INTRODUCTION OF DNA INTO CROP PLANTS

Hinchee, M. A. W., D. V. Connor-Ward, C. A. Newell, R. E. McDonnell, S. J. Sato, C. S. Gasser, D. A. Fischoff, D. B. Re, R. T. Fraley, and R. B. Horsch. "Production of transgenic soybean plants using *Agrobacterium*-mediated DNA transfer." *Bio/Tech.*, 6: 915–922 (1988).

McCabe, D. E., W. F. Swain, B. J. Martinelli, and P. Christou. "Stable transformation of soybean (*Glycine max*) by particle acceleration." *Bio/Tech.*, 6: 923–926 (1988).

Toriyama, K., Y. Arimoto, H. Uchimiya, and K. Hinata. "Transgenic rice plants after direct gene transfer into protoplasts." *Bio/Tech.*, 6: 1072–1074 (1988).

Fromm, M. E., F. Morrish, C. Armstrong, R. Williams, J. Thomas, and T. M. Klein. "Inheritance and expression of chimeric genes in the progeny of transgenic maize plants." *Bio/Tech.*, 8: 833–839 (1990).

Mullins, M. G., F. C. Tang, and D. Facciotti. "*Agrobacterium*-mediated genetic transformation of grapevines: transgenic plants of *Vitisrupestris scheele* and buds of *Vitis vinifera* L." *Bio/Tech.*, 8: 1041–1045 (1990).

Raineri, D. M., P. Bottino. M. P. Gordon, and E. W. Nester. "*Agrobacterium*-mediated transformation of rice (*Oryza sativa* L.)." *Bio/Tech.*, 8: 33–38 (1990).

VIRUS-RESISTANT PLANTS

Powell Abel, P., R. S. Nelson, B. De, N. Hoffmann, S. G. Rogers, R. T. Fraley, and R. N. Beachy. "Delay of disease development in transgenic plants that express the tobacco mosaic virus coat protein gene." *Science*, 232: 738–743 (1986).

Golemboski, D. B., G. P. Lomonossoff, and M. Zaitlin. "Plants transformed with tobacco mosaic virus non-structural gene sequence are resistant to the virus." *Proc. Natl. Acad. Sci. USA*, 87: 6311–6315 (1990).

Kaniewski, W., C. Lawson, B. Sammons, L. Haley, J. Hart, X. Delannay, and N. E. Tumer. "Field resistance of transgenic russet burbank potato to effects of infection by potato virus X and potato virus Y." *Bio/Tech.*, 8: 750–754 (1990).

Day, A. G., E. R. Bejarano, K. W. Buck, M. Burrell, and C. P. Lichtenstein. "Expression of an antisense viral gene in transgenic tobacco confers resistance to the DNA virus tomato golden mosaic virus." *Proc. Natl. Acad. Sci. USA*, 88: 6721–6725 (1991).

Hill, K. K., N. Jarvis-Eagan, E. L. Halk, K. J. Krahn, L. W. Liao, R. S. Matheson, D. J. Merlo, S. E. Nelson, K. E. Rashka, and L. S. Loesch-Fries. "The development of virus-resistant alfalfa, *Medicago sativa* L." *Bio/Tech.*, 9: 373–377 (1991).

INSECT-RESISTANT PLANTS

Hilder, V. A., A. M. R. Gatehouse, S. E. Sheerman, R. F. Barker, and D. Bouter. "A novel mechanism of insect resistance engineered into tobacco." *Nature*, 330: 160–163 (1987).

Johnson, R., J. Narvaez, G. An, and C. Ryan. "Expression of proteinase inhibitors I and II in transgenic tobacco plants: effects on natural defense against *Manduca sexta* larvae." *Proc. Natl. Acad. Sci. USA*, 86: 9871–9875 (1989).

Perlak, F. J., R. W. Deaton, T. A. Armstrong, R. L. Fuchs, S. R. Sims, J. Y. Greenplate, and D. A. Fischoff. "Insect resistant cotton plants." *Bio/Tech.*, 8: 939–943 (1990).

HERBICIDE-RESISTANT PLANTS

Comai, L., D. Facciotti, W. R. Hiatt, G. Thompson, R. E. Rose, and D. M. Stalker. "Expression in plants of a mutant *aroA* gene from *Salmonella tryphimurium* confers tolerance to glyphosate." *Nature*, 317: 741–744 (1985).

De Block, M., J. Botterman, M. Vandewiele, and J. Dockx, C. Thoen, V. Gossele, N. R. Movva, C. Thompson, M. Van Montagu, and J. Leemans. "Engineering herbicide resistance in plants by expression of a detoxifying enzyme." *EMBO J.*, 6: 2513–2518 (1987).

Cheung, A. Y., L. Borograd, M. Van Montagu, and J. Schell. "Relocating a gene for herbicide tolerance: a chloroplast gene is converted into a nuclear gene." *Proc. Natl. Acad. Sci. USA*, 85: 391–395 (1988).

Lee, K. Y., J. Townsend, J. Tepperman, M. Black, C. F. Chui, B. Mazur, P. Dunsmuir, and J. Bedbrook. "The molecular basis of sulfonylurea herbicide resistance in tobacco." *EMBO J.*, 7: 1241–1248 (1988).

Stalker, D. M., K. E. McBride, and L. D. Maljy. "Herbicide resistance in transgenic plants expressing a bacterial detoxification gene." *Science*, 242: 419–423 (1988).

GENETIC ENGINEERING OF FLOWERS

Meyer, P., I. Heidmann, G. Forkmann, and H. Saedler. "A new petunia flower color generated by transformation of a mutant with a maize gene." *Nature*, 330: 677–678 (1987).

Napoli, C., C. Lemieux, and R. Jorgensen. "Introduction of a chimeric chalone synthase gene into petunia results in reversible co-suppression of homologous genes *in trans.*" *Plant Cell.*, 2: 279–289 (1990).

van der Krol, A. R., L. A. Mur, M. Beld, J. N. M. Mol, and A. R. Stuitje. "Flavonoid genes in petunia: addition of a limited number of gene copies may lead to a suppression of gene expression." *Plant Cell.*, 2: 291–299 (1990).

PLANTS AS BIOREACTORS

Battey, J. F., K. M. Schmid, and J. B. Ohlrogge. "Genetic engineering for plant oils: potential and limitations." *Trends Biotech*, 7: 122–125 (1989). [Review]

Hiatt, A., R. Cafferkey, and K. Bosdish. "Production of antibodies in transgenic plants." *Nature*, 342: 76–78 (1989).

Vandekerckhove, J., J. van Damme, M. van Lijsebettens, J. Botterman, M. De Block, M. Vandewiele, A. De Clercq, J. Leemans, M. van Montagu, and E. Krebbers. "Enkephalins produced in transgenic plants using modified 2S seed storage proteins." *Bio/Tech.*, 7: 929–932 (1989).

Sijmons, P. C., B. M. M. Dekker, B. Schrammeijer, T. C. Verwoerd, P. J. M. van den Elzen, and A. Hoekema. "Production of correctly processed human serum albumin in transgenic plants." *Bio/Tech.*, 8: 217–221 (1990).

GENETIC ENGINEERING IN TRANSGENIC FARM ANIMALS

Chen, T. T., and D. A. Powers. "Transgenic fish." *Trends Biotech*, 8: 209–215 (1990). [Review]

Rodgers, G. E. "Improvement of wool production through genetic engineering." *Trends Biotech.*, 8: 6–11 (1990). [Review]

Hammer, R. E., V. G. Pursel, C. E. Rexroad, R. J. Wall, D. J. Bolt, K. M. Ebert, R. D. Palmiter, and R. L. Brinster. "Production of transgenic rabbits, sheep and pigs by microinjection." *Nature*, 315: 680–683 (1985).

Krimpenfort, P., A. Rademakers, W. Eyestone, A. van der Schans, S. van den Broek, P. Kooiman, E. Kootwijk, G. Platenburg, F. Pieper, R. Strijker, and H. de Boer. "Generation of transgenic dairy cattle using *'in vitro'* embryo production." *Bio/Tech.*, 9: 844–847 (1991).

GROWTH HORMONE AND MILK PRODUCTION

Bauman, D. E., P. J. Eppart, M. J. DeGeeter, and G. M. Lanza. "Responses of high-producing dairy cows to long-term treatment with pituitary somatotropin and recombinant somatotropin." *J. Dairy Sci.*, 68: 1352–1362 (1985).

Juskevich, J. C., and C. G. Guyer. "Bovine growth hormone: human food safety evaluation." *Science*, 249: 875–883 (1990).

SUBUNIT VACCINES FOR ANIMALS

Kleid, D. G., D. Yansura, B. Small, D. Dowbenko, D. M. Moore, M. J. Grubman, P. D. McKercher, D. O. Morgan, B. H. Robertson, and H. L. Bachrach. "Cloned viral protein vaccine for foot-and-mouth disease: responses in cattle and swine." *Science*, 214: 1125–1129 (1981).

ANIMALS AS BIOREACTORS

Pittius, C. W., L. Hennighausen, E. Lee, H. Westphal, E. Nicols, J. Vitale, and K. Gordon. "A milk protein gene promoter directs the expression of human tissue plasminogen activator cDNA to the mammary gland in transgenic mice." *Proc. Natl. Acad. Sci. USA*, 85: 5874–5878 (1988).

Clark, A. J., H. Bessos, J. O. Bishop, S. Harris, R. Lathe, M. McClenaghan, C. Prowse, J. P. Simons, C. B. A. Whitelaw, and X. Wilmut. "Expression of human antihemophilic Factor IX in the milk of transgenic sheep." *Bio/Tech.*, 7: 487–492 (1989).

Meade, J., L. Gates, E. Lacy, and N. Lonberg. "Bovine alpha$_{s1}$-casein gene sequences direct high level expression of active human urokinase in mouse milk." *Bio/Tech.*, 8: 443–446 (1990).

Wright, G., A. Carver, D. Cottom, D. Reeves, A. Scott, P. Simons, I. Wilmut, I. Garner, and A. Colman. "High level expression of active human alpha-1-antitrypsin in the milk of transgenic sheep." *Bio/Tech.*, 9: 830–834 (1991).

DISEASE-RESISTANT ANIMALS

Salter, D. W., and L. B. Crittenden. "Artificial insertion of a dominant gene for resistance to avian leukosis virus into the germ line of the chicken." *Theor. Appl. Gen.*, 77: 457–461 (1989).

Arnheiter, H., S. Skuntz, M. Noteborn, S. Chang, and E. Meier. "Transgenic mice with intracellular immunity to influenza virus." *Cell*, 62: 51–61 (1990).

25

Marshaling Recombinant DNA to Fight AIDS

I n 1981 the Centers for Disease Control (CDC) began to receive reports of an increasing incidence of two very rare conditions. One was a form of pneumonia caused by *Pneumocystis carinii* and the other a very unusual skin cancer named *Kaposi's sarcoma.* Some patients were afflicted by both conditions. It was determined that the immune system of these patients was impaired, and in 1982 the CDC recognized a new disease called *acquired immune deficiency syndrome (AIDS).* It became clear that an infectious agent was involved and that transmission of the disease required the transfer of bodily fluids from an infected to an uninfected individual. Thus while AIDS is primarily a sexually transmitted disease, some people who had received blood transfusions or blood products such as Factor VIII, used in the treatment of hemophilia were also affected. Blood supplies are now protected, but intravenous drug users who share needles remain a major risk group. It was in 1983, only 2 years after the appearance of the previously unknown disease, that the causative agent of AIDS was isolated. It was found to be a human retrovirus that was later named *human immunodeficiency virus (HIV-1).* Subsequently a second virus, HIV-2, was discovered in Africa. Thus began the research that has revealed the genetic complexity of HIV and dissected the molecular details of its life cycle.

AIDS is a catastrophic epidemic that no country, and no group of individuals, has escaped. The tools of recombinant DNA are playing an essential role in discovering how the AIDS virus acts, and in using this knowledge to develop therapies and vaccines. It is an ongoing battle and we have been able to select only a few topics from a vast amount of research to show how scientists are fighting AIDS. A cautionary note is necessary because this is a fast-moving field. Much of the research described in this chapter has yet to be confirmed, and the interpretations of many of the findings are controversial. Furthermore, the clinical significance of many of the studies has yet to be determined, and it may be many years before findings from laboratory research can be translated into practical therapies.

Human Immunodeficiency Virus Is the Cause of AIDS

What is the evidence that the human immunodeficiency virus is the cause of AIDS? The most straightforward evidence is that HIV can be isolated from almost all patients with AIDS, and that patients with more advanced disease have higher amounts of virus. Also, antibodies to HIV are found in asymptomatic individuals who later go on to develop AIDS. Furthermore, recipients of blood contaminated with HIV subsequently develop AIDS, even if they do not have any of the other risk factors associated with AIDS. Similarly, medical and laboratory staff with needlestick injuries involving HIV-infected blood have developed AIDS. About 30 percent of HIV-infected mothers transmit the virus to their unborn children, who subsequently develop AIDS. The uninfected siblings of these children do not develop AIDS. Thus, the common link between AIDS-affected individuals in very different social groups is the presence of HIV. Although the principal modes of transmission of HIV differ in different populations, epidemiological data currently show an absolute correlation between HIV infection and the subsequent development of AIDS; AIDS does not appear in a new location without evidence of prior HIV infection. Epidemiological studies,

and attempts to predict the course of the epidemic, are difficult because of the interval—up to 8 to 10 years—between infection with HIV and the development of AIDS. Indeed, it is not yet certain that all individuals infected with HIV will go on to develop AIDS. It is clear, however, that there is a very high mortality among individuals who develop AIDS.

HIV Is a Retrovirus

HIV has an RNA genome that is replicated via a DNA intermediate once the virus is inside the target cell. It is therefore a *retrovirus*, and has a life cycle similar to the RNA tumor viruses described in Chapter 12 (Figure 25-1). The first step is for the virus to penetrate the cell membrane. HIV has an outer lipid layer—acquired when a viral particle buds from the plasma membrane of a cell—surrounding a protein core that contains the RNA genome. The lipid envelope has a number of viral proteins embedded in it, and one of these binds to a cell surface receptor on the target cell for HIV infection. The virus lipid envelope then fuses with the cell membrane, and the core of the virus enters the cell. Once in the cell, reverse transcriptase (brought in by the virus) synthesizes a double-stranded DNA version of its RNA genome. It is this *provirus* that integrates into the host cell DNA and becomes a permanent part of the host cell's genome. The provirus is replicated along with the host DNA every time the cell divides.

Two identical structures, the *long terminal repeats* (*LTRs*), are generated at each end of the provirus in the course of reverse transcription of the HIV genomic RNA (Chapter 12). These contain enhancer and promoter sequences. Most retroviruses have a genome consisting of just three coding regions, *gag, env* and *pol*. As we will describe shortly, the group of viruses to which HIV belongs have complex genomes that include regulatory genes. At the 5' end of the genome is *gag*, which codes for a precursor polyprotein (Pr55) that is cut to make 4 smaller internal structural proteins. The protease that does this is coded for by *pol*, and it is part of a second long polyprotein (Pr160) produced from a *gag-pol* RNA. The enzymes reverse

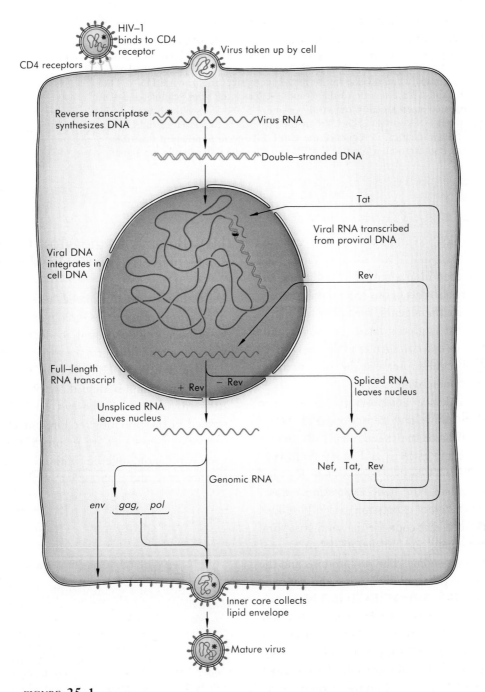

FIGURE **25-1**

Life cycle of HIV-1. The HIV virion is shown coated with glycoprotein (gp) 120 and bound to the cell surface CD4 receptor. The virus is taken up by the cell, and the viral RNA genome and reverse transcriptase (represented by the asterisk) are released. Reverse transcriptase synthesizes a double-stranded DNA copy that enters the nucleus and becomes integrated in the cell DNA. The HIV is now called a *provirus*. The regulation of RNA transcription depends on Tat and Rev, the products of the *tat* and *rev* genes. In the early phase, in the absence of Rev only spliced RNAs leave the nucleus, and these are translated to produce Tat, Rev, and Nef. Tat acts back on the provirus to stimulate production of RNA; when the level of Rev reaches a critical level, full-length, unspliced RNAs leave the nucleus. These code for the proteins of the viral core, including the structural proteins and reverse transcriptase, and for the proteins of the viral envelope. The former assemble around full-length RNA to form the viral core, while the envelope proteins move to the cell membrane. As the virus buds from the cell, it acquires a lipid coat, carrying the gp120 protein. The role of *nef* is not understood, except that it seems to be important in maintaining a high level of infection in vivo.

FIGURE **25-2**

The genes of the HIV provirus in detail. Most retroviruses contain only three genes, *gag, env*, and *pol*. HIV-1 has these genes plus at least six others. Their functions are listed in Table 25-1. The *tat* and *rev* genes are split, and their mRNAs are spliced. The long terminal repeats (LTR) are generated when reverse transcriptase synthesizes the proviral DNA. The sequences labeled U5 and U3 are unique to the 5′ and 3′ ends of the viral genome, and R is a short sequence repeated at each end of the viral genome. U5 and U3 are duplicated during synthesis of the proviral DNA so that each LTR in the provirus is composed of the U5, R and U3 regions. Although identical in sequence the two LTRs perform different functions; the 5′ LTR promotes transcription, while the 3′ LTR provides the polyadenylation signal.

transcriptase, which synthesizes double-stranded DNA using the genomic RNA as a template, and integrase, which then inserts the DNA copy into the host cell DNA, are also part of this second polyprotein. The third of the common genes, *env*, codes for a large protein, gp160, that is glycosylated (has sugar units added to it). It is cut into two smaller molecules (gp120 and gp41) by a host cell protease as the virus particles bud from the cell membrane.

In some cells, enhancer and promoter sequences in the LTR become active, and transcription of the integrated provirus begins. In the *early stage,* fully spliced viral RNAs leave the nucleus and move to the cytoplasm of the cell. These first RNAs code for viral proteins that regulate HIV gene expression. Subsequently, in the *late stage* of expression, partially spliced RNAs are produced that code for the viral structural proteins and enzymes, and full-length RNA is produced for the viral genome. The *gag* proteins and the enzymes coded by *pol* assemble around the genomic RNA to form the inner core of the virus. These cores bud from the cell surface, and as they do so, they acquire the *env* gp120 and gp41 glycoproteins and a lipid coat from the cell membrane. Most retroviruses are nonlytic; that is, they do not kill the host cell. However, HIV is a *lentivirus,* a group of retroviruses that can have cytopathic effects.

HIV infects at least three types of cells. The primary target is the helper T lymphocytes that carry the CD4 receptor. It is the loss of these cells that leads to the immune deficiency of AIDS patients. The other cells infected by HIV are macrophages and dendritic cells. These play an important role in carrying HIV to the central nervous system.

HIV Belongs to the Most Complex Class of Retroviruses Yet Discovered

Once the virus was isolated, research began immediately to investigate its biology, in the hope that the knowledge could be used to develop therapies and vaccines. One of the first surprises that came from the detailed analysis of HIV was that its 9.2-kbp genome is far more complex than those of other retroviruses. All retroviruses have the *gag, env,* and *pol* genes described in the previous section, but HIV has at least four other genes that are involved in regulating viral gene expression, together with other accessory genes (Figure 25-2 and Table 25-1). HIV-1 RNA is alter-

TABLE **25-1**

Functions of the Genes of HIV-1

GENE	FUNCTION
env	Encodes virus coat proteins
gag	Encodes proteins of the inner core
nef	Function not known but acts in vivo to maintain high level of infection
pol	Encodes viral enzymes, including reverse transcriptase
rev	Encodes Rev, which regulates transfer of unspliced RNA to cytoplasm
tat	Encodes Tat, which induces high-level expression of HIV genes
vif	Increases infectivity of virus particles
vpr	Encodes transcriptional activator
vpu	Participates in viral assembly and budding

natively spliced into a bewildering array of transcripts. The complexity of the life of HIV-1 inside the cell offers targets for therapies, especially the three genes *tat* (trans-activator), *rev* (regulator of virion protein expression), and *nef* (negative-regulatory factor—as we shall see later, this is probably a misnomer) that play a major role in HIV-1 infection. It is important to remember that research on HIV has been carried out with a limited number of isolates and cloned DNA. Most commonly used is the original LAV/HTLV-IIIB isolate, and the H9 cell line. It appears that the behavior of HIV freshly isolated from patients differs from that of viruses produced from cell lines, and it is not clear just how representative the properties of the latter are of HIV in vivo.

The *tat* Gene Regulates Synthesis of HIV RNA

The first of these genes to be analyzed in detail was *tat*, although even now its mechanism of activity is not precisely understood. By analogy with the other human retroviruses, HTLV-I and HTLV-II, it was suspected that HIV-1 produced factors that increased LTR-directed transcription of viral genes. This was found to be the case when a reporter gene under the control of an LTR was used to assay gene expression in uninfected and infected cells; expression was increased several hundredfold in the latter cells. The location of the gene involved in this trans-activation was pinpointed by making a series of mutations in a cloned HIV-1 genome and determining the minimum deletion that destroyed trans-activation (Figure 25-3). Small deletions of sequences just 5′ to the start of the *env* gene were all that was needed to destroy transactivation. Comparisons of the deletion data with the transcripts from that region showed that an exon coding for 72 amino acids was essential for transactivation.

The product of the *tat* gene is a protein called Tat. Its target was determined by deleting sequences in the HIV-1 LTR. However, the experiments were complicated by the presence of the HIV-1 enhancer in the same region. The effects of deleting the HIV-1 enhancer were distinguished from the effects of de-

leting the Tat target sequence by substituting an enhancer from another virus for the HIV-1 enhancer. When this was done, the Tat binding region was localized to a short sequence found at the 5′ end of all HIV-1 transcripts immediately downstream of the

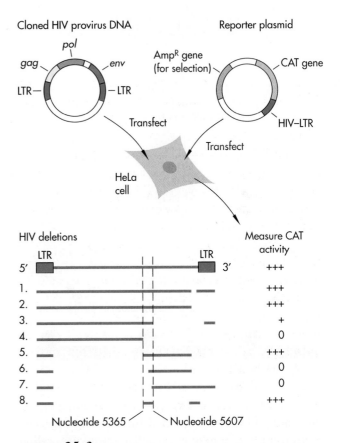

FIGURE **25-3**
Locating the *tat* gene. The assay for *tat* activity involves the cotransfection of a plasmid containing portions of the HIV genome (HIV plasmid) together with a reporter plasmid that contains the CAT gene. HIV plasmids containing different deletions of the HIV genome were used in the experiments (the lines in the lower part of the figure represent DNA sequences present in the plasmids). If the HIV plasmid contains a functional *tat* gene, the Tat protein produced by it will bind to the Tat-binding sequence in the HIV LTR and stimulate transcription of the CAT gene. The levels of CAT enzyme in the transfected cells can be measured accurately. Any deletion that removed wholly or partly the region between nucleotides 5365 and 5607 (deletions 4, 6, and 7; region indicated by dashed lines) had a drastic effect on CAT expression. On the other hand, a plasmid that contained just that small region between 5365 and 5607 was effective in stimulating production of CAT.

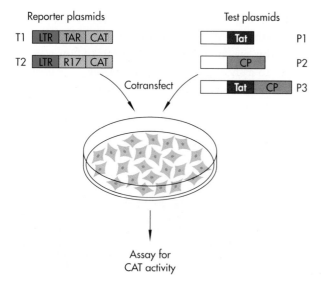

Reporter plasmids

Test plasmids

Cotransfect

Assay for
CAT activity

Target sequence	Protein	Activation
TAR (T1)	Tat (P1)	+++
R17 (T2)	Tat (P1)	0
R17 (T2)	CP (P2)	0
R17 (T2)	Tat/CP (P3)	+

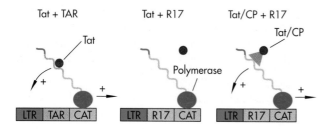

Tat + TAR   Tat + R17   Tat/CP + R17

FIGURE **25-4**
The only function of the TAR sequence that binds Tat is to position Tat at the right place for it to function. These experiments involved substituting another RNA-protein binding system in place of Tat-TAR, and targeting Tat to the new RNA sequence by making a fusion protein between Tat and the substitute protein. In this case the coat protein (CP) of bacteriophage R17 which binds to the R17 operator and so regulates expression of the bacteriophage after infection of *E. coli* was used. The assay measured Tat activation of a plasmid that contains a CAT reporter gene, either plasmid T1, which contains the HIV LTR including TAR or plasmid T2, in which TAR has been replaced in the LTR by the sequence (R17) that is the binding site for the R17 coat protein. Three test plasmids were used. Each contained the SV40 promoter. P1 contained the coding sequence for Tat, P2 the sequence for coat protein, and P3 the sequence coding for a fusion protein containing the first 67 amino acids of Tat linked to 127 amino acids of the coat protein. Each reporter plasmid was transfected into cells with each of the test plasmids. The results show that coat protein binding to the R17 sequence in the LTR cannot activate CAT, but that if Tat is fused with the coat protein, activation occurs; that is, the coat protein has bound to the R17 RNA sequence, and the Tat part of the fusion protein has activated transcription of CAT. Thus, TAR is not required for Tat trans-activation. (The arrows pointing left and right in the bottom panel indicate activation of polymerase binding and stabilization of transcription, respectively.)

start site. The RNA encoded by this sequence can fold to form a stem-loop secondary structure, called the *TAR element* (for trans-acting responsive or target element). Mutational analysis shows that the stem and a trinucleotide bulge close to the tip of the loop are essential for Tat binding in vitro.

The only function of TAR seems to be to bind Tat in the immediate vicinity of the promoter (Figure 25-4), and the arginine-rich central portion of Tat has been shown to bind specifically to the stem bulge. However, this is not the full story, because mutations in the loop that had a very strong in vivo effect had little effect on the binding of Tat to TAR. Other cellular proteins were invoked to account for this, and it now appears that at least one such protein may have been identified. The overall effect of Tat is to increase the levels of viral RNAs both by increasing transcription initiation and by stabilizing transcription. The first effect may be mediated by Tat's taking part in the formation of a complex of proteins, including other initiator factors, at the HIV promoter. The second effect must result from Tat's stopping premature termination of transcription by somehow stabilizing the polymerase complex as it moves along the proviral DNA. Whether Tat mediates both these effects simultaneously or independently is unknown. Experiments using *Xenopus* oocytes show that Tat also acts post-transcriptionally. A reporter gene consisting of the chloramphenicol acetyltransferase gene (CAT) under the control of an HIV LTR was injected into oocytes. When Tat protein was added, there was a very large stimulation of CAT activity but only a moderate increase in CAT mRNA levels over control levels without Tat. In some experiments, the reporter gene was injected first so that CAT mRNAs could be transcribed. Transcriptional inhibitors were then injected, followed shortly after by Tat. CAT activity was still stimulated by Tat, showing that it must

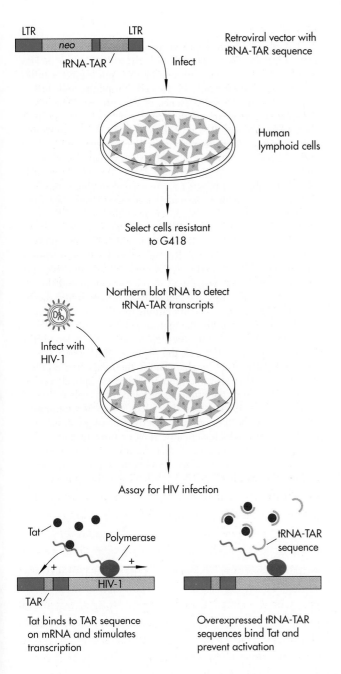

FIGURE **25-5**

Preventing in vitro HIV infection by "mopping up" Tat through the overexpression of the TAR sequence. Gene expression in HIV is stimulated by the binding of Tat to the TAR sequence. One way in which this relationship might be exploited therapeutically is to overexpress the TAR sequence in cells so that the Tat protein is prevented from interacting with its true target, the TAR sequence in viral transcripts. In this experiment, a retroviral vector expressing a tRNA-TAR chimeric RNA molecule was used to infect human lymphoid cells. G418-resistant cells were isolated and Northern blotting done to detect cells synthesizing the tRNA-TAR RNA. These cells together with control cells lacking the retroviral vector were infected with the HIV-1 virus. Unlike in the control cells, activation was prevented in the cells containing the vector because the highly expressed tRNA-TAR transcripts bound most of the TAT, leaving little for activation of transcription of the HIV genome.

the whole cascade of events leading to viral replication could be stopped. One way might be to use the TAR sequence to "mop up" Tat protein in the cell, and so prevent it from turning on transcription. This strategy seems to have worked, at least in a tissue culture model system (Figure 25-5). The TAR sequence was linked to a modified tRNA gene, and this recombinant gene was inserted into the LTR of a retroviral vector. Cell lines expressing high levels of the tRNA-TAR molecules were isolated and infected with HIV. Cells containing tRNA-TAR molecules were found to be resistant to infection. The intracellular location of the tRNA-TAR transcripts was not determined, but unprocessed tRNA transcripts have been shown to stay within the nucleus. The tRNA-TAR transcripts presumably become concentrated in the nucleus, thus maximizing their chance to interfere with Tat-TAR interaction.

be acting on pre-existing RNAs. This Tat activation is restricted to the nucleus and its mechanism is unknown.

The *tat* gene has a profound effect on HIV viral gene expression, increasing expression several thousandfold over expression from HIV genomes lacking *tat*. It is an obvious target for therapeutic intervention. If *tat* expression could be kept turned off in infected cells, or Tat interaction with TAR could be prevented,

Movement of HIV RNA from Nucleus to Cytoplasm Is Regulated by Rev

Another of the proteins encoded by HIV-1 that plays a crucial role in infection is Rev. It is involved in the transition from the production of early, highly spliced RNA to the later production of less extensively spliced RNA encoding structural proteins. Rev binds to a sequence called the Rev response element (RRE).

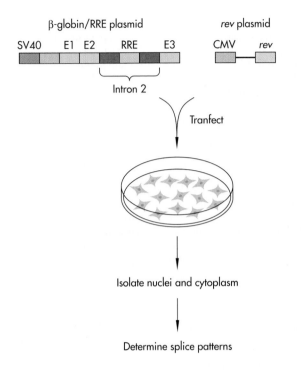

β-globin/RRE plasmid

| SV40 | E1 | E2 | RRE | E3 |

Intron 2

rev plasmid

| CMV | rev |

Tranfect

Isolate nuclei and cytoplasm

Determine splice patterns

	Cytoplasmic splice pattern	
β-globin/RRE plasmid	−Rev	+Rev
β-globin/RRE	S	S
β-globin/RRE 5′ and 3′ mutations	U	U
β-globin/RRE 5′ or 3′ mutations	None	U

FIGURE 25-6

Rev regulates whether spliced or unspliced mRNAs reach the cytoplasm. The change from early to late phases of HIV gene expression is characterized by the appearance of full-length, unspliced mRNAs in the cell cytoplasm, and this change depends on the presence of Rev protein. This was demonstrated experimentally by using a model system in which the effects of Rev on splicing of the transcript from the β-globin gene were examined. The test plasmid (β-globin/RRE) contained the SV40 promoter and exons 1, 2, and 3 of the rabbit β-globin gene. The *rev* response element (RRE) was inserted in what is the second intron of the β-globin gene. A second plasmid (*rev* plasmid) contained the *rev* gene driven by the human cytomegalovirus (CMV) promoter. When the β-globin/RRE plasmid was transfected into cells, spliced (S) β-globin was found in the cytoplasm of the cells whether or not Rev was present. If there were mutations in both 5′ and 3′ splice sites (5′ ss and 3′ ss), only unspliced (U) RNA was found in the cytoplasm and this transport was also Rev-independent. However, when the β-globin/RRE plasmids had mutations in either the 5′ or the 3′ sites, but not in both, the appearance of unspliced RNA in the cytoplasm became Rev-dependent. The interpretation of these experiments is that the β-globin/RRE mRNA is spliced rapidly in the nucleus and is then exported rapidly to the cytoplasm. However, mRNAs that are spliced inefficiently because of mutations in either 5′ or 3′ splice sites are retained in the nucleus and degraded. This retention may be caused by the formation of inactive splicing complexes on the mRNA, which somehow slow or prevent exit of the mRNA from the nucleus, so that the mRNA is exposed to degradation. Rev may act by dissociating these complexes, making the unspliced mRNA available for transport to the cytoplasm.

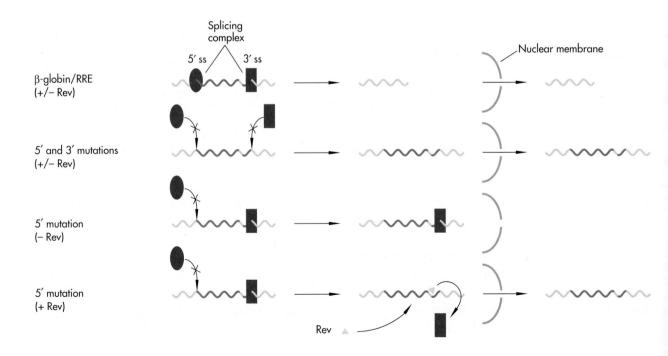

Rev does not seem to increase transcription of the RNAs for the structural proteins; the unspliced RNAs can be found in the nucleus even in the absence of Rev. It seems, however, that these unspliced RNAs are retained within the nucleus, and that Rev in some way releases this block in transport so that they can leave and travel to the cytoplasm. There is some evidence from a model experiment that it is the accumulation of splicing factors on these RNAs that keeps them in the nucleus (Figure 25-6). The β-globin premRNA is very efficiently spliced, so unspliced β-globin mRNA is not normally found in the cytoplasm. When a β-globin pre-mRNA with either the 5' or 3' splice site (but not both) mutated was joined to the RRE, this chimeric molecule appeared unspliced in the cytoplasm only when Rev was present. The hypothesis is that splicing factors attach at the normal splice site but are unable to carry out splicing because of the mutation at the other splice site. Instead, the splicing complexes remain attached to the pre-mRNA and block transport of the transcript from the nucleus. In the presence of Rev, however, this block is released. Splicing of HIV mRNA seems to be an inefficient process, so one can imagine that complexes of splicing factors bind to the HIV mRNA and prevent it from leaving the nucleus. Rev might act to displace these factors in the same way that it seems to do in the β-globin model. Here is another part of the HIV life cycle, unique to the virus, that could be a target for drug therapy.

The *nef* Gene May Be Essential for in Vivo Replication of Pathogenic AIDS

Tissue culture studies have failed to consistently demonstrate a function for the Nef protein, and viral replication in vitro is variably affected by mutations in *nef*. However, an experiment performed in vivo using the simian immunodeficiency virus (SIV) and rhesus monkeys has revealed a possible function for *nef*. An infectious clone of SIV containing a stop codon in *nef* (*nef*-stop) was modified in two ways (Figure 25-7). One clone, *nef*-open, was produced by altering the stop codon to produce a complete open reading frame. The second clone, *nef*-deletion, was produced

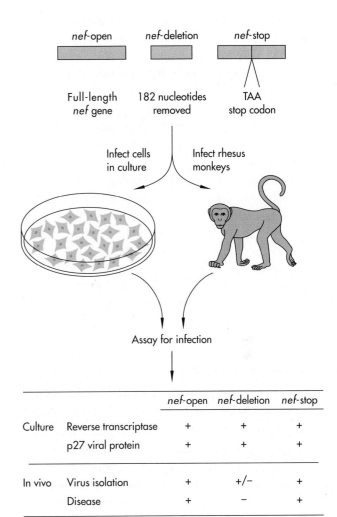

		nef-open	*nef*-deletion	*nef*-stop
Culture	Reverse transcriptase	+	+	+
	p27 viral protein	+	+	+
In vivo	Virus isolation	+	+/−	+
	Disease	+	−	+

FIGURE **25-7**
The function of the *nef* gene is shown by in vivo experiments. 182 nucleotides were deleted from the *nef* gene of an infectious SIV clone (*nef*-open) to make *nef*-deletion. A second version of the clone was made that had a stop codon (TAA) in the *nef* gene (*nef*-stop). All three clones were equally efficient at infecting cells growing in tissue culture. Infection was measured by the appearance of reverse transcriptase in the culture medium (the standard assay for retroviral replication) or by radioimmunoassay of the p27 protein in the medium. The same three clones were also used to infect rhesus monkeys. Here the results were very different. Recovery of virus from animals infected with the *nef*-deletion clone was very difficult, and none of these monkeys developed simian AIDS. In contrast, virus could be recovered much more easily from animals infected with *nef*-stop and *nef*-open, and these animals developed simian AIDS. Sequence analysis of the *nef* gene in the viruses recovered from the animals infected with *nef*-stop showed that the stop codon TAA had mutated to recreate the full *nef*-open reading frame. The *nef* gene seems to be involved in viral replication in vivo and in the development of disease.

by introducing a deletion into the *nef* gene. All three viruses replicated equally well in cultured cells, but the results were quite different when rhesus monkeys were infected with these viruses. Viruses were recovered from three animals receiving *nef*-stop, and all the clones analyzed were revertants in which the stop codon had been lost, permitting production of Nef. Recovery of viruses from animals receiving *nef*-deletion clones was much more difficult, and it was estimated that these animals contained at least a 100-fold lower virus concentration than those animals receiving the *nef*-open or *nef*-stop viruses. These differences in viral infection paralleled the clinical differences between these animals. All the *nef*-deletion animals remained healthy, while 5 out of 12 monkeys receiving *nef*-open or *nef*-stop died with AIDS-like symptoms within 7 months after infection. These results show that *nef* is involved in the development and maintenance of an active infection in vivo, and that it too should be considered a therapeutic target.

The three other HIV regulatory proteins—*vif, vpr,* and *vpu*—have been the subject of less intensive investigation. The *vif*-deficient viruses can replicate normally in the cell, but they are 1000-fold less infectious than wild-type virus for cells growing in tissue culture. *vif* may have a similar role in vivo, and interfering with *vif* expression could have some effect on the spread of HIV within an infected person. However, spread of the virus by fusion of infected and uninfected cells is not affected by *vif*, so the effect of anti-*vif* therapies might be small. The *vpr* gene has been shown to act like a weak transcriptional activator in cells in vitro, and *vpu* appears to be involved in release of mature virions from infected cells. Perhaps in vivo studies will be necessary to determine the precise role of these genes in the HIV life cycle.

AZT Works by Interfering with Viral DNA Synthesis

The knowledge that HIV-1 is a retrovirus immediately pinpointed reverse transcriptase synthesis of proviral DNA as a target for therapy. Reverse transcription is essential for viral infection, and reverse transcriptase is specific to retroviruses. Indeed, at present, inhibition of reverse transcriptase is the only therapy that has a demonstrable beneficial effect on AIDS patients. *Azidothymidine (AZT)* is a modified form of thymidine in which the hydroxyl group on the sugar ring has been replaced by an azido (N_3) group (Figure 25-8). Such compounds are called *dideoxynucleosides,* and, as suggested by the similarity of the names, they work in a manner similar to the dideoxynucleotide terminators used in the Sanger method for DNA sequencing. Reverse transcriptase incorporates AZT instead of thymidine into the growing DNA strand. However, because the missing hydroxyl group is essential for making the sugar-phosphate backbone of the DNA strand, reverse transcriptase cannot join the next nucleotide to the AZT and DNA synthesis is halted (Figure 25-8). Why does AZT not interfere with the host cell's DNA replication? It seems that when both AZT and thymidine triphosphate are available, the cell's DNA polymerases preferentially use thymidine triphosphate, while the HIV reverse transcriptase preferentially incorporates AZT. So the concentration of AZT needed to inhibit HIV replication in cells in culture is abut 10 to 20 times less than the concentration that is toxic to mammalian cells. However, some blood precursor cells in the bone marrow are killed by AZT, and this limits the dosage of AZT that can be used.

AZT treatment appears to be effective in prolonging the lives of people with AIDS by 1 or 2 years. The first randomized, placebo-controlled trial of AZT in the USA was halted 6 months into the trial when it was found that 19 of the placebo group, but only 1 of the treatment group, had died. Other studies have simply compared treatment and nontreatment groups. For example, one analysis found that the median survival following diagnosis of those receiving AZT was 770 days, compared with 190 days for those who did not receive the drug. This type of analysis is complicated by the fact that the group receiving AZT probably had more access to other forms of AIDS treatment and care. An accurate estimate of the value of AZT awaits the results of controlled trials going on in Europe. Nevertheless, the increase in survival time since 1987 reflects the antiviral therapy that became widely available that year.

However, AZT-resistant forms of reverse transcriptase can arise from mutations in the *pol* gene; AZT-resistant viruses have been cultured from patients who have been treated for 6 months or more with AZT.

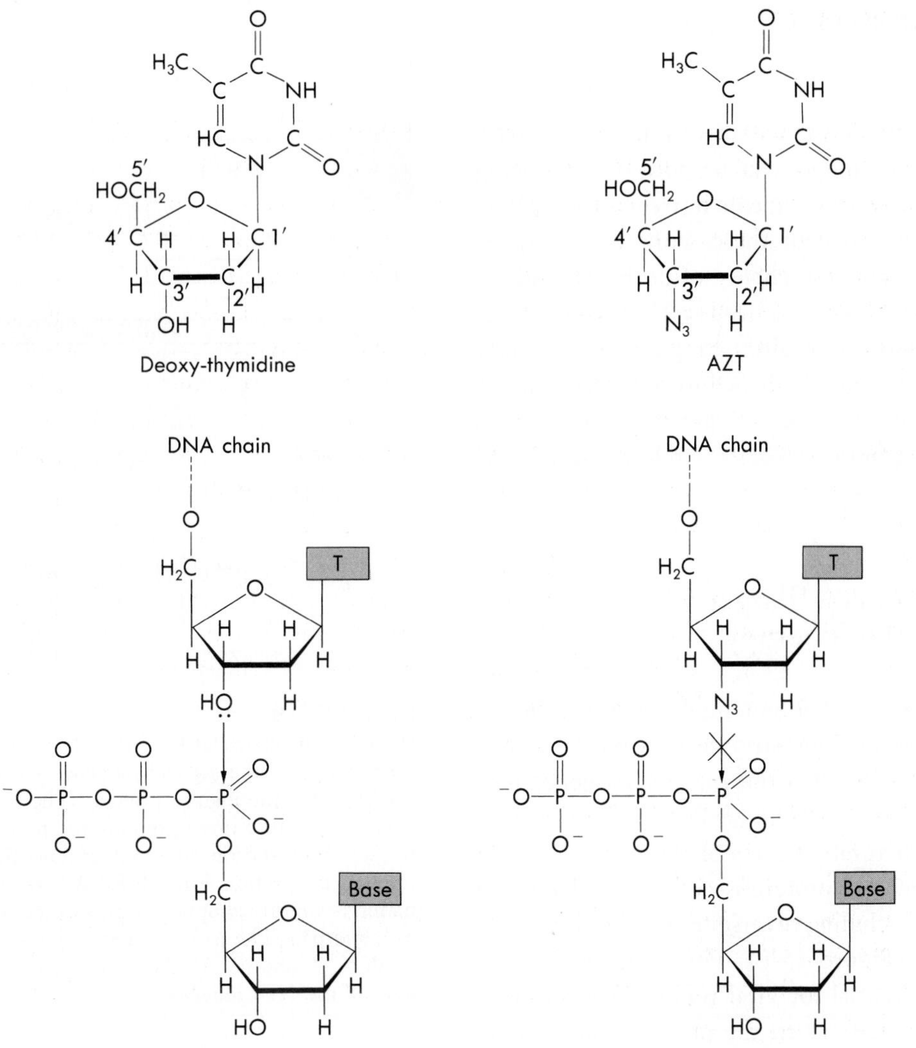

FIGURE 25-8

AZT works as a chain terminator during the synthesis of the proviral DNA by reverse transcriptase. AZT is an analogue of thymidine in which the hydroxyl group on carbon 3 of the deoxyribose ring has been replaced by an azido group (N_3). Nucleosides are phosphorylated before they are added to the growing DNA chain. A phosphodiester bond is made between the hydroxyl group on the last nucleotide incorporated, and the incoming deoxyribonucleoside triphosphate. If the last nucleotide incorporated was AZT, the hydroxyl group is missing, the next nucleotide cannot be added, and synthesis stops. AZT is a dideoxypyrimidine analog. There are also dideoxypurine analogues that have similiar activities. One of these, dideoxyinosine, has been useful in treating childhood AIDS.

The reverse transcriptase genes of HIV isolated from five patients before and after development of AZT resistance were compared. All five AZT-resistant isolates had three mutations in common, and a fourth mutation was common to three of them. These mutations were introduced into wild-type virus by site-directed mutagenesis, and this genetically engineered virus then exhibited AZT-resistance, albeit at a lower level than the naturally occurring resistant viruses. These same mutations have been detected in the *pol* genes of viruses freshly isolated from infected white blood cells, by using the polymerase chain reaction. This shows that the mutations are not simply an in vitro artifact of the cell culturing required to isolate virus. Fortunately, resistance to AZT does not confer resistance to other dideoxynucleotides that are being developed as therapeutic agents to treat AIDS. Alternating regimes of one of these with AZT may be effective in reducing the chance of developing drug-resistant forms of HIV, and because these toxic drugs have different side effects, alternating treatments might give the patient's body a chance to recover from the effects of one drug while still being treated with a second. Dideoxyinosine (ddI) is undergoing clinical trials.

HIV Protease Is a Target for AIDS Drug Therapy

HIV protease is another potential target for therapeutic intervention. This enzyme cuts the large polyproteins produced from the *gag* and *pol* genes into several smaller proteins. The *gag* Pr55 precursor gives rise to four structural proteins of the viral coat, and the *gag-pol* Pr160 polyprotein is cut to produce three viral enzymes, including reverse transcriptase and the protease itself (Figure 25-9). This posttranslational processing is essential for viral replication; defective virions are produced when the proteases of retroviruses are inactivated by mutations in *pol*.

The HIV protease is an aspartic protease, and synthetic inhibitors have been developed for other members of this family of enzymes. These inhibitors are short peptides that are substrates for the enzymes, except that the two amino acids on either side of the peptide bond cut by the enzyme are replaced and modified. The inhibitor binds to the active site of the enzyme but cannot be cut, and in so doing prevents the true substrate from gaining access to the enzyme. This is *competitive inhibition*. Protease inhibitors have been developed that are active against purified HIV protease in the test tube, and active in cells in tissue culture. A T-cell line that is permanently infected with HIV was treated with these protease inhibitors, and the synthesis of HIV-specific proteins was analyzed by Western blotting. Staining of the blots with antibodies to Pr55 proteins and to reverse transcriptase showed that processing of both polypeptides was inhibited. The inhibitors were also able to prevent HIV infection of cells in culture, presumably because of

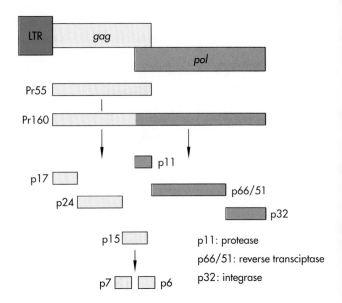

FIGURE **25-9**
Protease processing of Pr55 and Pr160, two polyproteins produced from the *gag* and *pol* genes and cut by the HIV protease into the mature proteins. The *gag-pol* Pr160 precursor is cut into three proteins: the protease itself, reverse-transcriptase, and the integrase enzyme that is involved in proviral integration into cell DNA. The *gag* Pr55 precursor undergoes more complicated processing, first producing p17, p24, and p15. The p15 segment undergoes another step to produce p7 and p6. All these *gag* proteins are structural proteins of the HIV inner core.

the lack of reverse transcriptase and because of the lack of the structural proteins derived from Pr55 which are necessary for proper assembly of viral particles. How these inhibitors will perform in vivo has yet to be established. There are likely to be further developments in the design of efficient inhibitors now that the precise three-dimensional structure of the HIV-1 protease has been determined by x-ray crystallography.

HIV Infects and Kills T Lymphocytes That Have the CD4 Receptor

The characteristic feature of AIDS is the profound immunodeficiency that leaves an infected person susceptible to *opportunistic infections*. These are infections by organisms that under normal circumstances are completely controlled by our immune system, such

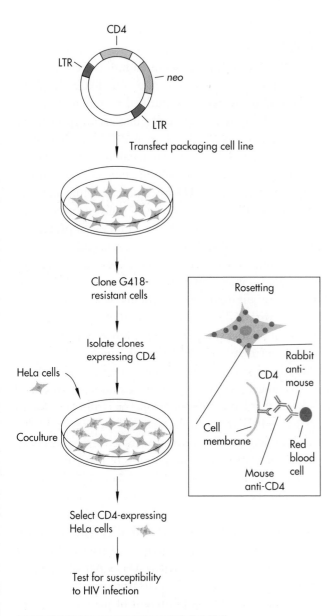

Rosetting

Cell membrane

CD4

Rabbit anti-mouse

Red blood cell

Mouse anti-CD4

Assay	HeLa	HeLa-CD4
Reverse transcriptase	−	+
Syncytia formation	−	+
Viral proteins in culture medium	−	+
Virus in cells	−	+

FIGURE **25-10**

Showing that the cell surface antigen CD4 is the HIV receptor on helper T cells. HeLa cells cannot be infected by HIV, and they lack the CD4 antigen. The question asked in this experiment was whether expression of CD4 in HeLa cells was sufficient to make them susceptible to HIV infection. A plasmid containing the elements of a retroviral vector and the cDNA for CD4 was transfected into a packaging cell line. G418-resistant cells were isolated and tested for CD4 expression by the following rosetting technique: A mouse antibody to CD4 was added to the cells, and colonies binding the antibody (and therefore expressing cell surface CD4) were identified by adding human red blood cells conjugated with rabbit anti-mouse immunoglobulin. The blood cells stick to colonies which have bound the mouse antibody, forming rosettes. These packaging cells produce a retrovirus carrying the CD4 gene that can infect human cells. The retroviral-producing cells were treated with mitomycin to prevent them from dividing, and cocultured with HeLa cells. G418-resistant cells were again selected, and colonies expressing CD4 were detected using the rosetting technique. These CD4-positive HeLa cells were then tested to see if they could now be infected by HIV. By four different assays it was clear that expression of CD4 alone was sufficient to render HeLa cells susceptible to HIV infection. So CD4 is a receptor for HIV.

as *Pneumocystis carinii*, which causes what is usually a very rare form of pneumonia. (The increasing occurrence of *P. carinii* infections was the first sign of the AIDS epidemic.) The very first clinical description of AIDS noted that patients had low numbers of lymphocytes bearing the CD4 cell surface antigen (Chapter 16). Healthy individuals have about 800 CD4 cells per milliliter, while AIDS patients typically have as few as 200 CD4 cells per milliliter when they begin to develop opportunistic infections.

The reason for the decline in CD4 cells became clear when laboratory studies later showed that the CD4 molecule is the cell surface receptor for HIV. There are both cytotoxic T cells and helper T cells, but in general only the latter carry CD4 antigen on their surface. The two types of T cell can be separated and grown in tissue culture, and it is only the helper T cells that can be infected by HIV. (A characteristic feature of HIV infection of susceptible cells in tissue culture using laboratory stains of virus is the development of *syncytia*—very large, abnormal, multinucleated cells formed by the fusion of infected cells with uninfected CD4 cells. This phenomenon is used as an assay for HIV). Infection of human peripheral blood lymphocytes by HIV-1 was blocked by monoclonal antibodies to CD4, indicating that this receptor plays a role in infection. Confirmation that CD4 is indeed the HIV receptor came from a recombinant DNA experiment (Figure 25-10). HeLa cells are a famous human cancer cell line grown in tissue culture. These cells do not normally express CD4, and they cannot be infected by HIV. In this experiment the cloned CD4 gene was introduced into HeLa cells so that they expressed it on their surfaces. These CD4-HeLa cells could now be infected by HIV and showed the same pathological response as infected T cells—the formation of cell syncytia. This is not the whole story, because mouse cells bearing the human CD4 receptor did not become infected or form syncytia, even though further investigation showed that these mouse cells bind HIV and that the cells can support viral DNA replication if HIV DNA is transfected into them. It is clear that CD4 alone is not sufficient for cells to take up HIV and that some component in addition to CD4 is involved in the internalization of the virus by human cells.

Of course, T cells do not carry the CD4 antigen for the purpose of binding HIV particles. Rather the virus has exploited the presence of a molecule that has a very important role in the immunological defense of the body. T cells bearing CD4 are key coordinators of the immune response, interacting with other immune cells, and CD4 is a key molecule involved. Thus the cells killed by HIV are precisely those whose loss promotes the spread of the infection within a patient, and renders the patient susceptible to other infections.

Soluble CD4 Molecules Can Be Used to Prevent HIV Infection

The interaction between HIV and the helper T cell is a lock-in-key interaction. The lock is the CD4 molecule on the surface of the T cell, and the key is the HIV virion coat glycoprotein gp120. It was hypothesized that HIV could be prevented from binding to and infecting T cells if its gp120 glycoprotein was already locked by free CD4 molecules that were supplied as a drug. Such experiments were carried out and showed that a recombinant soluble form of CD4 (sCD4) can block HIV infection of cells in vitro. (Figure 25-11). The soluble sCD4 was genetically engineered by removing the coding regions for the

transmembrane and cytoplasmic portions of the molecule and then expressing it in mammalian cells in culture. The sCD4 was secreted into the culture medium and was shown to form a complex with gp120. If HIV is incubated with purified recombinant sCD4 before adding it to CD4 cells, infection is prevented. sCD4 molecules have a short half-life in vivo, but a much longer half-life when combined with the Fc portion of an immunoglobulin molecule. In model experiments, this sCD4-Fc chimeric molecule was able to protect chimpanzees when given prior to injection of a purified preparation of HIV, but protection may be more difficult with cell-associated virus.

Although these preliminary experiments are encouraging, some early data from tests of sCD4 in patients are disappointing. Patients with AIDS were treated with sCD4, but this had no effect on the virus levels in plasma and peripheral white blood cells. Several experiments suggest that the difference between these in vivo experiments and those in tissue culture is that the latter make use of laboratory-maintained isolates that have different properties from HIV found in patients. For example, primary isolates from the patients were much more resistant to the effects of sCD4 than were laboratory strains of HIV, and in one experiment, a primary isolate was about 100-fold more resistant than the same isolate after 1 year in culture. This suggests that prolonged culture of HIV leads to the selection of viral strains with different properties from viruses in vivo. Furthermore, tissue culture experiments have shown that sCD4 is not effective in protecting brain or muscle cells from infection. These data imply that sCD4 treatment of patients may not be so efficacious as had been hoped.

CD4-Toxin Conjugates Specifically Kill HIV-Infected Cells

For a number of years, cancer researchers have been working to develop *immunotoxins* for cancer therapy. These are hybrid molecules made up of a toxin, like ricin, that kills mammalian cells, and a monoclonal antibody directed against the target cell. The antibody delivers the toxin specifically to the target cell, which

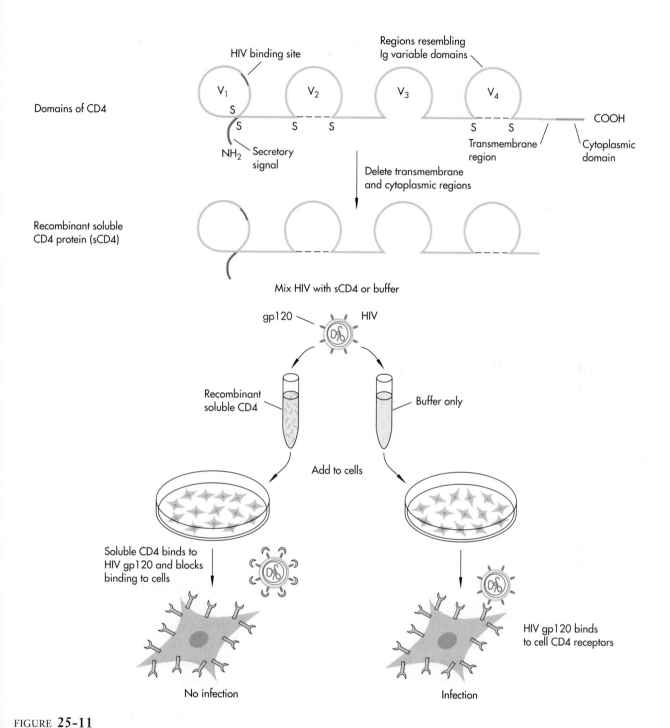

FIGURE **25-11**

Blocking HIV infection of cells in culture, using a soluble CD4. CD4 is a member of the immunoglobulin superfamily of proteins. A recombinant form of CD4, lacking the transmembrane and cytoplasmic domains, is soluble. The soluble CD4 was produced in tissue culture cells and isolated by affinity chromatography. Infectious HIV virions were incubated with soluble CD4 or buffer before being added to cultures of HIV-susceptible cells. Infection was assayed by the appearance of viral proteins in the culture medium and by the induction of syncytia formation. By both measures, soluble CD4 inhibited HIV infection in culture. At high concentrations, sCD4 may also inhibit HIV infection because it induces shedding of gp120 from HIV virions.

is killed. This is an attractive approach for HIV therapy, because the specificity of the HIV-CD4 interaction can be exploited to deliver cytotoxins to HIV-infected cells. In one set of experiments, the exotoxin A of *Pseudomonas* was used. This toxin has three domains. Domain I is responsible for the cell specificity of the toxin, domain II is involved in uptake of the bound toxin, and domain III interacts with elongation factor 2 to inhibit protein synthesis. An expression plasmid was made in which part of the CD4 molecule was expressed in *E. coli* as a fusion protein with exons II and III of the exotoxin gene. Purified fusion protein killed only cells expressing gp120, while the intact toxin killed both control and gp120 cells.

These experiments are not wholly successful because the cells must already be infected and expressing gp120 before they can be killed. Infected cells are already releasing virus, so the HIV infection eventually spreads through a culture. However, HIV can be eradicated in a culture if reverse transcriptase inhibitors are added together with the CD4-toxin. The former prevent viral replication in infected cells, while the latter kills those cells. Once again, however, there is a long way to go before the results of these experiments in vitro can be translated into practical therapies for patients.

Transport of HIV Proteins to the Cell Surface Can Be Inhibited

Other experiments are using recombinant CD4 molecules to interfere with viral replication. When they are synthesized, proteins like CD4 and gp120 have to be moved to the interior (lumen) of the endoplasmic reticulum for transport to the cell surface. It was observed that the amount of CD4 on the surfaces of infected cells was inversely related to the amount of virus produced, perhaps because CD4 and gp120 form an intracellular complex that cannot move easily to the cell membrane. Viral release might be inhibited if gp120 was prevented from reaching the cell surface at all by binding it with mutant CD4 molecules that cannot be transported out of the endoplasmic retic-

ulum. To do this, plasmids expressing two mutant forms of CD4 were made by in vitro mutagenesis (Figure 25-12). One form (sCD4) had both the transmembrane and the cytoplasmic domains of CD4 removed so that this molecule was fully soluble, as we have seen. The second form had a short peptide sequence—Lys-Asp-Glu-Leu (KDEL in the single-letter amino acid code)—added to it that retains proteins within the lumen of the rough endoplasmic reticulum (sCD4-KDEL). These plasmids were transfected into cells, and, as expected, sCD4 was released from the cells, while the sCD4-KDEL molecule was retained by cells. A plasmid vector expressing only gp120 was transfected into cells expressing one or the other of the two forms of CD4 and retention of gp120 was measured. The gp120 was trapped only in cells expressing sCD4-KDEL. Syncytium formation, a sensitive measure of gp120 surface expression, was blocked if the cells were expressing sCD4-KDEL. This is an interesting experiment using a model system, but much further research is needed. For example, what will happen when there is active replication of the whole virus in T cells, and how will the sCD4-KDEL gene be delivered to a patient's cells?

Simian Immunodeficiency Virus and Animal Models Are Useful in Studying AIDS

A major block to studying the pathogenesis of HIV and to finding effective therapies to counteract infections is the absence of a suitable animal model. The chimpanzee and gibbon apes are the only animals so far tested that can be infected with HIV. However, although infected animals become HIV-positive, they do not develop an AIDS-like disease. There are, however, simian immunodeficiency viruses (SIVs) that are closely related to the HIVs and attack the same target cells. These viruses do not usually cause disease in their natural hosts (African monkeys) but do produce AIDS-like diseases on injection into macaque monkeys from Asia. The isolation of molecular clones of SIVs that induce an acute illness similar to AIDS in rhesus

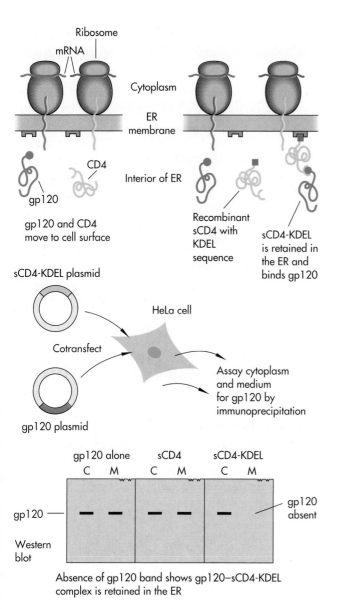

gp120 and CD4
move to cell surface

Recombinant
sCD4 with
KDEL
sequence

sCD4-KDEL
is retained in
the ER and
binds gp120

sCD4-KDEL plasmid

HeLa cell

Cotransfect

Assay cytoplasm
and medium
for gp120 by
immunoprecipitation

gp120 plasmid

Absence of gp120 band shows gp120–sCD4-KDEL
complex is retained in the ER

FIGURE **25-12**

Intracellular trapping of gp120 using a recombinant CD4 molecule. Following their synthesis on ribosomes attached to the endoplasmic reticulum (ER), proteins like gp120 and CD4 move into the ER and are transported to the cell surface. Other molecules that act within the ER are retained there by a special amino acid sequence at their carboxyl terminus. This sequence is Lys-Asp-Glu-Leu (KDEL, in the amino acid single-letter code). A soluble form of recombinant CD4 was made in which the transmembrane and cytoplasmic domains were replaced by the KDEL sequence (sCD4-KDEL). It was reasoned that this molecule would bind to the receptors in the ER and be retained there, as shown in the upper panel. A plasmid expressing sCD4-KDEL was transfected into HeLa cells, together with a plasmid expressing gp120 (lower panel). Cells were grown in medium containing radioactively labeled methionine to label newly synthesized proteins. Labeled gp120 in the cytoplasm of the cells and in the culture medium was immunoprecipitated and analyzed by electrophoresis. The labeled gp120 was found in both cells (C) and culture medium (M) when the gp120 plasmid was transfected alone or with soluble CD4. When cotransfected with sCD4-KDEL, gp120 was found only in the cells, suggesting that gp120-sCD4-KDEL complexes are indeed being retained in the ER.

and partially restore immune functions to these mice. Injection of such mice with HIV leads to an acute infection of the human tissues that is suppressed by treatment with AZT. These so-called human-*scid* mice may prove to be useful for testing drug therapies.

Recombinant HIV Proteins May Be Effective as Immunogens for AIDS Vaccines

Recombinant DNA analyses of HIV have opened many possibilities for AIDS therapy, but what of vaccines, the classical approach to controlling viral infections? Here, too, recombinant DNA techniques are being used extensively, to identify and to make HIV proteins that can be used as antigens. Two types of immunogen are used for vaccine production—whole virus, either killed or attenuated (alive but made innocuous), and purified viral proteins. In the case of

and pig-tailed macaques has opened the way to experimental analysis of the in vivo roles of the various genes specific to lentiviruses like HIV and SIV.

Another experimental model for AIDS studies is being developed using the *scid* (*severe combined immunodeficiency*) mouse. Mice homozygous for this mutation are immunologically deficient, and components of the human immune system, including thymus, spleen, and lymph nodes, can be surgically implanted in them. The human tissues acquire a blood supply

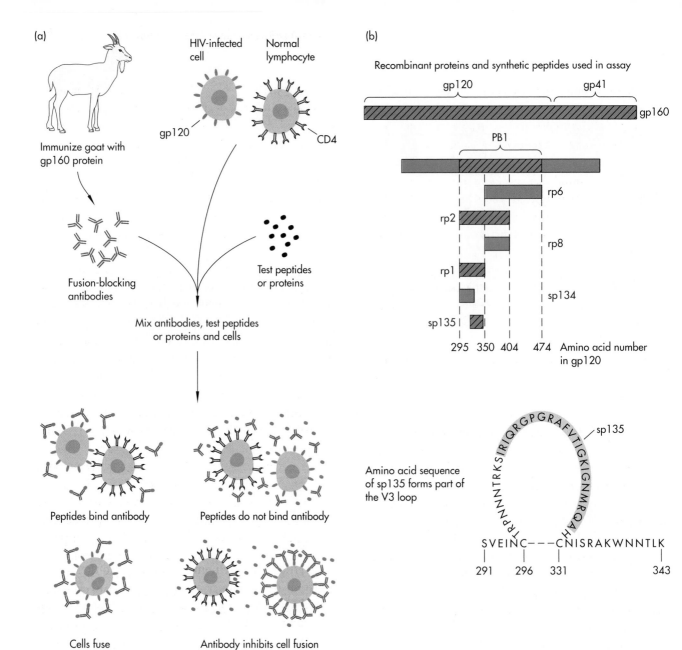

FIGURE **25-13**

Localizing the critical antigenic region of gp120. (a) The assay. One experiment to determine the location of the antigenic epitope of gp120 made use of recombinant proteins and synthetic peptides to block the cell fusion–inhibiting activity of sera raised against the envelope gp160 protein. Recombinant gp160 was made using a baculovirus expression vector and insect cells. This was injected into goats, and **antibodies inhibiting fusion of uninfected lymphocytes with cells chronically infected with HIV were isolated.** To determine which parts of the gp120 molecule are the antigens responsible for eliciting antibodies that can block the gp120-CD4 interaction involved in cell fusion, recombinant proteins and peptides were tested for their ability to block the activity of the antibodies. If the proteins contained the antigenic site, they were expected to bind to the antibodies, thereby blocking their activity and permitting the cells to fuse. If the proteins did not contain the antigenic site, the antibodies would be free to bind the gp120 on the HIV-infected cells, and so inhibit fusion. (b) Results of the assays. The entire gp160 protein bound the antibody, as did a protein (PB1) from the carboxy-terminal half of the molecule. (The cross-hatching indicates fragments that bound antibody.) Subregions of PB1 were tested, and finally a synthetic peptide (sp135), only 23 amino acids long, was found that was effective at blocking the fusion-inhibiting activity of the antibody. This sequence comes from the variable V3 loop.

HIV, there are concerns that a killed whole virus might not be an effective immunogen, and fears that a live, attenuated virus might be too dangerous. For these reasons, research has concentrated on developing vaccines based on viral protein subunits. However, we need to bear in mind that successful protection against SIV infection was only achieved using a whole killed virus and that there is only one example of a vaccine made using recombinant protein that has been approved for human use, the hepatitis B vaccine.

The HIV *env* glycoprotein gp120 has received the most attention because this molecule is located on the surface of the virus particle and is exposed to immune attack. In fact, AIDS patients mount an immune response to HIV infection and produce neutralizing antibodies to gp120. However, in the long term these antibodies are ineffective, perhaps because of changes in the virus itself. A key to developing a strategy for a vaccine based on gp120 was to determine what parts of the gp120 molecule are neutralization epitopes. This was done first by determining which monoclonal antibodies, made using peptides derived from the gp120 amino acid sequence, neutralized HIV. The immunogenic region was found to lie between two cysteines at positions 296 and 331. These make a disulfide bridge, so that a small loop is formed. This is known as the V3 loop, and it contains the *principal neutralizing domain* (PND) of gp120. Another approach is illustrated in Figure 25-13. Antibodies that inhibit HIV-induced cell fusion were made by immunizing goats with gp160 or recombinant proteins for portions of gp120. It was argued that these antibodies must be binding to critical regions of gp120 if they could inhibit cell fusion. These regions could be identified because synthetic peptides corresponding to those regions should block the actions of the antibodies. Again, the V3 loop was identified as the critical region of gp120. There are other neutralizing epitopes, for example the CD4-binding site and an epitope of gp41, but monoclonal antibodies specific for the V3 loop are much more potent.

Early studies based on a small number of analyses suggested that the PND of gp120 was highly variable. This was a serious problem because such variability suggests that a vaccine raised against one variant would be of limited use. However, recent sequence analysis of the loop has shown that the PND is more conserved than initially expected by some. DNA was extracted from the white blood cells of over 130 HIV-infected individuals, and the polymerase chain reaction was used to amplify the DNA sequence encoding the 36 amino acids from cysteine to cysteine of the PND. A consensus sequence was derived for over 200 PCR products, and each individual sequence was compared with the consensus. The Gly-Pro-Gly tripeptide at the tip of the loop was relatively conserved, and it was estimated that antibodies raised against the peptide Gly-Pro-Gly-Arg-Ala-Phe will weakly neutralize about 60 percent of randomly chosen virus isolates.

Recombinant proteins have been used as immunogens in chimpanzees, which were subsequently challenged with live HIV. In one study, gp120 and gp160 were produced from SV40 expression vectors in Chinese hamster ovary cells growing in tissue culture, and one or the other protein was injected into chimpanzees together with an adjuvant. (Adjuvants are nonspecific stimulators of the immune system, and are used to elicit strong responses to soluble antigens.) When neutralizing antibodies against the recombinant molecules were detected, the animals were challenged with live virus. The animals receiving the gp120 showed a decreased titer of gp120 antibodies, followed by an increase, suggesting that they were mounting an antibody response to the live HIV. This response protected the animals, because virus could not be recovered from the animals immunized with gp120. These results are encouraging, but at the same time confusing; gp160 did not protect immunized animals (HIV was recovered from them) even though gp160 includes gp120. It is not clear why the chimpanzees responded differently to the gp160 and gp120. The animals immunized with the latter protein had higher levels of antibodies directed against the gp120 principal neutralizing domain, and this difference might account for the difference in protection.

What of these recombinant proteins as immunogens in human beings? Clinical trials designed to test safety and immunogenicity of recombinant gp160 have been carried out with small numbers of healthy, non-HIV-infected volunteers. These trials have shown that the recombinant gp160 is safe, and that both cellular and humoral immune responses develop. In one trial, administration of gp160 is being used in an attempt to promote the immune response of already infected patients.

Kaposi's Sarcoma Is a Tumor Associated with AIDS

One of the primary diagnostic criteria for AIDS is the presence of a tumor of endothelial cells and mesenchymal spindle cells called *Kaposi's sarcoma* (*KS*). Before AIDS, this was a very rare lesion, occurring in fewer than 1000 individuals in the United States. The increased occurrence of Kaposi's sarcoma was one of the factors that alerted the Centers for Disease Control to the developing AIDS epidemic; of AIDS patients who are gay men about 25 percent develop Kaposi's sarcoma. AIDS-associated KS differs from the classical form in several ways. The tumors appear anywhere on the body instead of being restricted to the lower limbs, the disease becomes disseminated to the internal organs, and spread of the lesions seems to be more aggressive.

The pathogenesis of AIDS-associated KS is still poorly understood, but there are data suggesting that *tat* plays a part in initiating production of growth factors that drive the growth of KS cells. A line of transgenic mice was made by using a plasmid containing *tat* under the control of the LTR from HIV. The transgenic animals expressed *tat* in their skin and developed skin lesions similar to those of AIDS-associated KS; the cells of the mouse tumors resembled the spindle-shaped cells found in KS. However, cell lines derived from the tumors themselves did not express *tat*, suggesting that Tat secreted from expressing cells in vivo induces proliferation of neighboring skin cells that form the tumors. It has been shown that Tat is secreted by HIV-infected T cells in vitro, but its stimulatory effects on the growth of the spindle-shaped cells from KS tumors is small. Remarkably, development of KS-like lesions in these mice was restricted to males, paralleling the male predominance seen in both classical and AIDS-related KS.

However, interpretation of this experiment is confounded in light of what we know of Kaposi's sarcoma in AIDS patients. For example, AIDS patients who are hemophiliacs and became infected through receiving HIV-contaminated plasma are all male, but only very rarely do they develop KS. This suggests that infection by sexual transmission is an important factor in developing AIDS-related Kaposi's sarcoma

and implicates some other venereally transmitted organism in the development of KS.

The Origins and Evolution of the Human Immunodeficiency Viruses Are Revealed Through Recombinant Techniques

HIV and AIDS appeared with astonishing suddenness in 1981. It now appears that this abrupt pattern is somewhat deceptive. Retrospective analyses of sera collected for other purposes have identified antibodies to HIV in samples taken in 1959. The earliest clinical case seems to have been seen in 1959, in Manchester, England, where a man died of infections by cytomegalovirus and *Pneumocystis carinii*. Tissues had been prepared for histological examination, and the paraffin blocks were still in storage 30 years later. DNA was extracted from microtome sections cut from these blocks and tested for the presence of HIV by using the polymerase chain reaction with primers for *gag*. Careful negative and positive controls were used, and HIV was detected in the patient's tissues. The virus must of course predate these cases, and the story of the origin and historical epidemiology of HIV is likely to be fascinating.

The history of the molecular evolution of HIV is being approached using the classical methods of evolutionary biology, by drawing up a family tree of HIV and its relatives. Nucleotide sequences have been compared for viruses isolated from different species, for different isolates, and for different genes. The results of one study are shown in Figure 25-14. Here the amino acid sequences of reverse transcriptase were compared and the number of amino acid differences between the viruses determined. The family tree was constructed so as to minimize the numbers of changes that need to be made between members of the tree to reach the present distribution of sequences. This tree shows that HIV-1 and HIV-2 are in different groups. A lentivirus of chimpanzees has been isolated, cloned, and sequenced. Called SIV$_{cpz}$ (a misnomer, since chimpanzees are not simians but anthropoids), it contains the *vpu* gene found in HIV-1 and lacks the

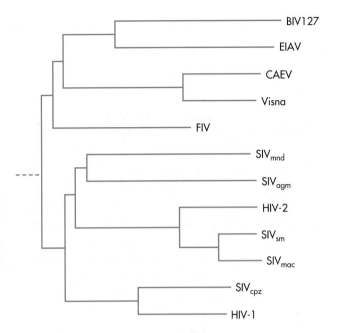

FIGURE 25-14

Family tree showing the relationships of the simian and human immunodeficiency viruses. Pairwise comparisons are made of the sequences of reverse transcriptase from the viruses shown. (In some cases, multiple isolates of the same virus type were analyzed; for example, seven independent isolates of HIV-1 were used.) The proportion of nucleotide differences between the pairs was calculated, and a correction applied to account for possible multiple substitutions at the same site. These values were used to construct the tree, where the lengths of the branches represent the number of nucleotide substitutions between the viruses (the greater the difference in end points, the greater the number of substitutions). The close relationship between the chimpanzee SIV and human HIV-1 is clear, as is the difference between HIV-1 and HIV-2. BIV127, bovine immunodeficiency virus; EIAV, equine infectious anemia virus; CAEV, caprine (goat) encephalitis-arthritis virus; Visna, visna lentivirus; FIV, feline immunodeficiency virus; SIV_mnd, mandrill; SIV_agm, African green monkey; HIV-2, human; SIV_sm, sooty mangabey; SIV_mac, rhesus macaque; SIV_cpz, chimpanzee; HIV-1, human.

vpx gene found in HIV-2 and SIVs. Clearly the relationships of the primate lentiviruses will become better established only when more sequence comparisons can be made on HIV-1–like viruses from other primates.

These types of analyses will be complicated by the extreme degree of variability between HIV-1 genomes even in the same individual. This variability is due to the rate of misincorporation of bases by reverse transcriptase. A consequence of this variability is that the population of viruses will evolve in response to selection pressures. For example, the variability among laboratory isolates, and between these and fresh isolates, may be due to different selective pressures during prolonged cell culture. A detailed analysis of this type of evolutionary change was made using four isolates obtained from one patient over a period of $2\frac{1}{2}$ years. Each isolate was cloned, and the *tat* genes of 20 clones from each isolate were sequenced. The data showed that *tat* genes of clones from the first two isolates were similar, but that there were changes between the second and third, and third and fourth isolates. The most striking finding came from comparisons of the *tat* sequences amplified directly from the patient's cells, with those after culture. It was clear that culturing had resulted in selection of genomes that were not representative of the spectrum of viruses in vivo. So once again we see that caution must be used in extrapolating from results using in vitro systems to what is going on in infected people.

Recombinant DNA Is at the Forefront of the Battle Against AIDS

Recombinant DNA is the key to attacking HIV and AIDS, first by analyzing the virus and understanding its life cycle, and now by engineering vaccines and designing drugs to block its actions in the cell. In the 8 years since HIV was isolated, we have gained knowledge of the virus and AIDS at an extraordinary rate, but the difficulties of translating this knowledge into clinical action must not be underestimated. Although a plethora of possible therapies is being developed, the progression from molecular and in vitro studies to studies in animals, to clinical trials, and finally to widespread use of therapeutic drugs and vaccines is of necessity a prolonged process. Nevertheless, the signs are encouraging. Several novel therapies are in development, some drugs are being tested in patients, AZT is proving partially effective, and recombinant proteins are in early clinical trials as vaccines. We can be optimistic that this unprecedented research effort will bear fruit.

Reading List

Original Research Papers

HUMAN IMMUNODEFICIENCY VIRUS

Various authors. "What science knows about AIDS." *Sci. Am.,* 259: 40–134 (1988). [Reviews]

Greene, W. C. "The molecular biology of human immunodeficiency virus type 1 infections." *N. E. J. Med.,* 324: 308–317 (1991). [Review]

Centers for Disease Control. "*Pneumocystis* pneumonia—Los Angeles." *Morb. Mortal. Wkly. Rpt.,* 30: 205–252 (1981).

Centers for Disease Control. "Kaposi's sarcoma and Pneumocystis pneumonia—New York City and California." *Morb. Mortal. Wkly Rpt.,* 30: 305–308 (1981).

Barre-Sinoussi, F., J. C. Cherman, F. Rey, M. T. Nugeyre, S. Chamaret, J. Gruest, C. Dauguet, C. Axler-Blin, F. Vezinet-Brun, C. Rouzioux, W. Rozenbaum, and L. Montagnier. "Isolation of a T-lymphotropic retrovirus from a patient at risk for Acquired Immune Deficiency Syndrome (AIDS)." *Science,* 220: 868–871 (1983).

Popovic, M., M. G. Sarngadharan, E. Read, and R. C. Gallo. "Detection, isolation, and continuous production of cytopathic retroviruses (HTLV-III) from patients with AIDS and pre-AIDS." *Science* 224: 497–500 (1984).

THE COMPLEXITY OF THE HIV GENOME

Cullen, B. R., and W. C. Greene. "Functions of the auxiliary gene products of the human immunodeficiency virus type 1." *Virol.,* 178: 1–5 (1990). [Review]

Rosen, C. A. "Regulation of HIV gene expression by RNA-protein interaction." *Trends Gen.,* 7: 9–14 (1991). [Review]

tat AND TAR

Rosen, C. A., J. G. Sodroski, and W. A. Haseltine. "The location of the *cis*-acting regulatory sequences in the human T-cell lymphotropic virus type III (HTLV-III/LAV) long terminal repeat." *Cell,* 41: 813–823 (1985).

Cullen, B. R. "*Trans*-activation of human immunodeficiency virus occurs via a bimodal mechanism." *Cell,* 46: 973–982 (1986).

Wright, C. M., B. K. Felber, H. Paskalis, and G. N. Pavlakis. "Expression and characterization of the *trans*-activator of HTLV-III/LAV virus." *Science,* 234: 988–992 (1986).

Braddock, M., A. Chambers, W. Wilson, M. P. Esnouf, S. E. Adams, A. J. Kingsman, and S. M. Kingsman. "HIV-1 TAT activates presynthesized RNA in the nucleus." *Cell,* 58: 269–279 (1989).

Laspia, M. F., A. P. Rice, and M. B. Mathews. "HIV-1 Tat protein increases transcriptional initiation and stabilizes elongation." *Cell,* 59: 283–292 (1989)

Selby, M. J., E. S. Bain, P. A. Luciw, and B. M. Peterlin. "Structure, sequence, and position of the stem-loop in *tar* determine transcriptional elongation by *tat* through the HIV-1 long terminal repeat." *Genes and Devel.,* 3: 547–558 (1989).

Sullenger, B. A., H. F. Gallardo, G. E. Ungers, and E. Gilboa. "Overexpression of TAR sequences renders cells resistant to human immunodeficiency virus replication." *Cell,* 63: 601–608 (1990).

rev AND REV

Malim, M. H., J. Hauber, R. Fenrick, and B. R. Cullen. "Immunodeficiency virus *rev trans*-activator modulates the expression of the viral regulatory genes." *Nature,* 335: 181–183 (1988).

Chang, D. D., and P. A. Sharp. "Regulation by HIV Rev depends upon recognition of splice sites." *Cell,* 59: 789–795 (1989).

Malim, M. H., J. Hauber, S.-Y. Le, J. V. Maizel, and B. R. Cullen. "The HIV-1 *rev trans*-activator acts through a structured target sequence to activate nuclear export of unspliced viral mRNA." *Nature,* 338: 254–257 (1989).

nef AND NEF

Ahmad, N., and Venkatesan, S. "*Nef* protein of HIV-1 is a transcriptional repressor of HIV-LTR." *Science,* 241: 1481–1485 (1988).

Hammes, S. R., E. P. Dixon, M. H. Malim, B. R. Cullen, and W. C. Greene. "Nef protein of human immunodeficiency virus type 1: evidence against its role as a transcriptional activator." *Proc. Natl. Acad. Sci. USA,* 86: 9549–9553 (1989).

Kim, S. Y., K. Ikeuchi, R. Byrn, J. Groopman, and D. Baltimore. "Lack of a negative influence on viral growth by the *nef* gene of human immunodeficiency virus type 1." *Proc. Natl. Acad. Sci. USA,* 86: 9544–9548 (1989).

Kestler, H. W., D. J. Ringler, K. Mori, D. L. Panicalli, P. K. Sehgai, M. D. Daniel, and R. C. Desrosiers. "Importance of the *nef* gene for maintenance of high virus loads and for development of AIDS." *Cell,* 65: 651–662 (1991).

DRUG THERAPIES INHIBITING REVERSE TRANSCRIPTION

Mitsuya, H., R. Yarchoan, and S. Broder. "Molecular targets for AIDS therapy." *Science,* 249: 1533–1544 (1990). [Review]

Furman, P. A., J. A. Fyfe, M. H. St. Clair, K. Wenhold, J. L. Rideout, G. A. Freeman, S. N. Lehrman, D. P. Bolognesi, S. Broder, H. Mitsuya, and D. W. Barry. "Phosphorylation of 3'-azido-3'-deoxythymidine and selective interaction of the 5'-triphosphate with human immunodeficiency virus reverse transcriptase." *Proc. Natl. Acad. Sci. USA,* 83: 8333–8337 (1985).

Fischl, M. A., D. D. Richman, M. H. Grieco, M. S. Gottlieb, P. A. Volberding, O. L. Laskin, J. M. Leedom, J. E. Groopman, D. Mildvan, R. T. Schooley, G. G. Jackson, D. T. Durack, D. King, and the AZT Collaborative Working Group. "The efficacy of azidothymidine (AZT) in the treatment of patients with AIDS and AIDS-related complex: a double-blind, placebo-controlled trial." *N. E. J. Med.,* 317: 185–191 (1987).

Larder, B. A., and S. D. Kemp. "Multiple mutations in HIV-1 reverse transcriptase confer high-level resistance to zidovudine (AZT)." *Science,* 246: 1155–1158 (1989).

Fitzgibbon, J. E., R. M. Howell, T. A. Schwartzer, D. J. Gocke, and D. T. Dubin. "In vivo prevalence of azidothymidine (AZT) resistance mutations in an AIDS patient before and after AZT therapy." *AIDS Res. and Hum. Retrovir.,* 7: 265–269 (1991).

Moore, R. D., J. Hidalgo, B. W. Sugland, and R. E. Chaisson. "Zidovudine and the natural history of the acquired immunodeficiency syndrome." *N. E. J. Med.,* 324: 1412–1416 (1991)

St. Clais, M. H., J. L. Martin, G. Tudor-Williams, M. C. Bach, C. L. Vauro, D. M. King, P. Kellain, S. D. Kemp, and B. A. Larder. "Resistance to ddI and sensitivity to AZT induced by a mutation in HIV-1 reverse transcriptase." *Science,* 253: 1557–1559 (1991).

THERAPIES DIRECTED AGAINST THE HIV PROTEINASE

Blundell, T. L., R. Lapatto, A. F. Wilderspin, A. M. Hemmings, P. M. Hobart, D. E. Danley, and P. J. Whittle. "The 3-D structure of HIV-1 proteinase and the design of antiviral agents for the treatment of AIDS." *Trends Gen.,* 15: 425–430 (1990). [Review]

Wlodawer, A., M. Miller, M. Jaskolski, B. K. Sathyanarayana, E. Baldwin, I. T. Weber, L. M. Selk, L. Clawson, J. Schneider, and S. B. H. Kent. "Conserved folding in retroviral proteases: crystal structure of a synthetic HIV-1 protease." *Science,* 245: 616–621 (1989).

Meek, T. D., D. M. Lambert, G. B. Dreyer, T. J. Carr, T. A. Tomaszek, Jr., M. L. Moore, J. E. Strickler, C. Debouck, L. J. Hyland, T. J. Matthews, B. W. Metcalf, and S. R. Petteway. "Inhibition of HIV-1 protease in infected T-lymphocytes by synthetic peptide analogues." *Nature,* 343: 90–91 (1990).

Roberts, N. A., J. A. Martin, D. Kinchington, A. V. Broadhurst, J. C. Craig, I. B. Duncan, S. A. Galpin, B. K. Handa, J. Kay, A. Krohn, R. W. Lambert, J. H. Merrett, J. S. Mills, K. E. B. Parkes, S. Redshaw, A. J. Ritchie, D. L. Taylor, G. J. Thomas, and P. J. Machin. "Rational design of peptide-based HIV proteinase inhibitors." *Science,* 248: 358–361 (1990).

CD4 AND THERAPEUTIC STRATEGIES EXPLOITING IT

Lifson, J. D., and E. G. Engleman. "Role of CD4 in normal immunity and HIV infection." *Immun. Rev.,* 109: 93–117 (1989). [Review]

Maddon, P. J., A. G. Dalgleish, J. S. McDougal, P. R. Clapham, R. A. Weiss, and R. Axel. "The T4 gene encodes the AIDS virus receptor and is expressed in the immune system and brain." *Cell,* 47: 333–348 (1986).

Smith, D. H., R. A. Byrn, S. A. Marsters, T. Gregory, J. E. Groopman, and D. J. Capon. "Blocking of HIV-1 infectivity by a soluble, secreted form of the CD4 antigen." *Science,* 238: 1704–1707 (1987).

Traunecker, A., W. Luke, and K. Karjalainen. "Soluble CD4 molecules neutralize human immunodeficiency virus type 1." *Nature,* 331: 84–86 (1988).

Clapham, P. R., J. N. Weber, D. Whitby, K. McIntosh, A. G. Dalgleish, P. J. Maddon, K. C. Deen, R. W. Sweet, and R. A. Weiss. "Soluble CD4 blocks the infectivity of diverse strains of HIV and SIV for T cells and monocytes but not for brain and muscle cells." *Nature,* 337: 368–370 (1989).

Buonocore, L., and J. K. Rose. "Prevention of HIV-1 glycoprotein transport by soluble CD4 retained in the endoplasmic reticulum." *Nature,* 345: 625–628 (1990).

Daar, E. S., X. L. Li, T. Moudgil, and D. D. Ho. "High concentrations of recombinant soluble CD4 are required to neutralize primary human immunodeficiency virus type 1 isolates." *Proc. Natl. Acad. Sci. USA,* 87: 6574–6578 (1990).

Ward, R. H. R., D. J. Capon, C. M. Jett, K. K. Murthy, J. Mordenti, C. Lucas, S. W. Frie, A. M. Prince, J. D. Green, and J. W. Eichberg. "Prevention of HIV-1 IIIB infection in chimpanzees by CD4 immunoadhesin." *Nature,* 352: 434–436 (1991).

CD4-TOXIN CONJUGATES KILL HIV-INFECTED CELLS

Chaudhary, V. K., T. Mizukami, T. F. Fuerst, D. J. FitzGerald, B. Moss, I. Pastan, and E. A. Berger. "Selective killing of HIV-infected cells by recombinant human CD4-*Pseudomonas* exotoxin hybrid protein." *Nature,* 335: 369–372 (1988).

Ashorn, P., B. Moss, J. N. Weinstein, V. K. Chaudhary, D. J. FitzGerlad, I. Pastan, and E. A. Berger. "Elimination of infectious human immunodeficiency virus from human T-cell cultures by synergistic action of CD4-*Pseudomonas* exotoxin and reverse transcriptase inhibitors." *Proc. Natl. Acad. Sci. USA,* 87: 8889–8893 (1990).

Till, M. A., V. Ghetie, R. D. May, P. C. Auerbach, S. Zolla-Pazner, M. K. Gorny, T. Gregory, J. W. Uhr, and E. S. Vitetta. "Immunoconjugates containing ricin A chain and either human anti-gp41 or CD4 kill H9 cells infected with different isolates of HIV, but do not inhibit normal T or B cell function." *AIDS,* 3: 609–614 (1990).

Tsubota, H., G. Winkler, H. M. Meade, A. Jakubowski, D. W. Thomas, and N. L. Letvin. "CD4-*Pseudomonas* exotoxin conjugates delay but do not fully inhibit human immunodeficiency virus replication in lymphocytes in vitro." *J. Clin. Invest.,* 86: 1684– 1689 (1990).

Zarling, J. M., P. A. Moran, O. Haffar, J. Sias, D. D. Richman, C. A. Spina, D. E. Myers, V. Kuebelbeck, J. A. Ledbetter, and F. M. Uckun. "Inhibition of HIV replication by pokeweed antiviral protein targeted to CD4$^+$ cells by monoclonal antibodies." *Nature* 347: 92–95 (1990).

SIMIAN IMMUNODEFICIENCY VIRUS AND ANIMAL MODELS

Letvin, N. L. "Animal models for AIDS." *Immun. Today,* 11: 322–326 (1990). [Review]

Dewhurst, S., J. E. Embretson, D. C. Anderson, J. I. Mullins, and P. N. Fultz. "Sequence analysis and acute pathogenicity of molecularly cloned SIV$_{SMM-PBj14}$." *Nature,* 345: 636–640 (1990).

Kestler, H., T. Kodama, D. Ringler, M. Marthas, N. Pedersen, A. Lackner, D. Regier, P. Sehgal, M. Daniel, N. King, and R. Desrosiers. "Induction of AIDS in rhesus monkeys by molecularly cloned simian immunodeficiency virus." *Science,* 248: 1109–1112 (1990).

McCune, J. M., R. Namikawa, C.-C. Shih, L. Rabin, and H. Kaneshima. "Suppression of HIV infection in AZT-treated SCID-hu mice." *Science,* 247: 564–566 (1990).

VACCINE DEVELOPMENT IN AIDS

Sonigo, P., M. Girard, and D. Dormont. "Design and trials of AIDS vaccines." *Immun. Today,* 11: 465–471 (1990). [Review]

Looney, D. J., A. G. Fisher, S. D. Putney, J. R. Rusche, R. R. Redfield, D. S. Burke, R. C. Gallo, and F. Wong-Staal. "Type-restricted neutralization of molecular clones of human immunodeficiency virus." *Science* 241: 357–359 (1988).

Richardson, N. E., N. R. Brown, R. E. Hussey, A. Vaid, T. J. Matthews, D. P. Bolognesi, and E. L. Reinherz. "Binding site for human immunodeficiency virus coat protein gp120 is located in the NH$_2$-terminal region of T4 (CD4) and requires the intact variable-region-like domain." *Proc. Natl. Acad. Sci. USA,* 85: 6102–6106 (1988).

Rusche, J. R., K. Javaherian, C. McDanal, J. Petro, D. L. Lynn, R. Grimaila, A. Langlois, R. C. Gallo, L. O. Arthur, P. J. Fischinger, D. P. Bolognesi, S. D. Putney, and T. J. Matthews. "Antibodies that inhibit fusion of human immunodeficiency virus–infected cells bind a 24-amino acid sequence of the viral envelope, gp120." *Proc. Natl. Acad. Sci. USA,* 85: 3198–3202 (1988).

Berman, P. W., T. J. Gregory, L. Riddle, G. R. Nakamura, M. A. Champe, J. P. Porter, F. M. Wurm, R. D. Hershberg, E. K. Cobb, and J. W. Eichberg. "Protection of chimpanzees from infection by HIV-1 after vaccination with recombinant glycoprotein gp120 but not gp160." *Nature,* 345: 622–625 (1990).

LaRosa, G. J., J. P. Davide, K. Weinhold, J. A. Waterbury, A. T. Profy, J. A. Lewis, A. J. Langlois, G. R. Dreesman, R. N. Boswell, P. Shadduck, L. H. Holley, M. Karplus, D. P. Bolognesi, T. J. Matthews, E. A. Emini, and S. D. Putney. "Conserved sequence and structural elements in the HIV-1 principal neutralizing determinant." *Science,* 249: 932–935 (1990).

Viscidi, R., E. Ellerbeck, L. Garrison, K. Midthun, M. L. Clements, B. Clayman, B. Fernie, and G. Smith. "Characterization of serum antibody response to recombinant HIV-1 gp160 vaccine by immunoassay." *AIDS Res. and Hum. Retrovir.,* 6: 1251–1256 (1990).

Cooney, E. L., A. C. Collier, P. D. Greenberg, R. W. Coombs, J. Zarling, D. E. Arditti, M. C. Hoffman, S.-L. Hu, and L. Corey. "Safety of and immunological response to a recombinant vaccinia virus vaccine expressing HIV envelope glycoprotein." *Lancet,* 337: 567–572 (1991).

Dolin, R., B. S. Graham, S. B. Greenberg, C. O. Tacket, R. B. Belshe, K. Midthun, M. L. Clements, G. J. Gorse, B. W. Horgan, and R. L. Atmar. "The safety and immunogenicity of a human immunodeficiency virus type 1 (HIV-1) recombinant gp160 candidate vaccine in humans." *Ann. Intern. Med.,* 114: 119–127 (1991).

Redfield, R. R., D. L. Birx, N. Ketter, E. Tramont, V. Polonis, C. Davis, J. F. Brundage, G. Smith, S. Johnson, A Fowler, T. Wierzba, A. Shafferman, F. Volvovitz, C. Oster, D. S. Burke, and the Military Medical Consortium for Applied Retroviral Research. "A phase I evaluation of the safety and immunogenicity of vaccination with recombinant gp160 in patients with early human immunodeficiency virus infection." *N. E. J. Med.,* 324: 1677–1684 (1991).

KAPOSI'S SARCOMA

Volberding, P. A. "Kaposi's sarcoma and the acquired immunodeficiency syndrome." *Med. Clin. N. Am.*, 70: 665–675 (1989). [Review]

Weiss, R. A. "The conundrum of Kaposi's sarcoma." *Eur. J. Cancer*, 26: 657–659 (1990). [Review]

Vogel, J., S. H. Hinrichs, R. K. Reynolds, P. A. Luciw, and G. Jay. "The HIV *tat* gene induces dermal lesions resembling Kaposi's sarcoma in transgenic mice." *Nature*, 335: 606–611 (1988).

Ensoli, B., G. Barillari, S. Z. Salahuddin, R. C. Gallo, and F. Wong-Staal. "Tat protein of HIV-1 stimulates growth of cells derived from Kaposi's sarcoma lesions of AIDS patients." *Nature*, 345: 84–86 (1990).

ORIGIN OF THE HUMAN LENTIVIRUSES

Meyerhans, A., R. Cheynier, J. Albert, M. Seth, S. Kwok, J. Sninsky, L. Morfeldt-Manson, B. Asjo, and S. Wain-Hobson. "Temporal fluctuations in HIV quasispecies in vivo are not reflected by sequential HIV isolations." *Cell*, 58: 901–910 (1989).

Gojobori, T., E. N. Moriyama, Y. Ins, K. Ikeo, T. Miura, H. Tsujimoto, M. Hayami, and S. Yokoyama. "Evolutionay origin of human and simian immunodeficiency viruses." *Proc. Natl. Acad. Sci. USA*, 87: 4108–4111 (1990).

Huet, T., R. Cheynier, A. Meyerhans, G. Roelants, and S. Wain-Hobson. "Genetic organization of a chimpanzee lentivirus related to HIV-1." *Nature*, 345: 356–359 (1990).

26

Mapping and Cloning Human Disease Genes

Human genetic diseases place a great burden, both financial and emotional, on affected families and on society. This is particularly true for those disorders affecting children. It has been estimated that about one-third of admissions to pediatric departments are associated with genetic disorders, and genetic diseases are a significant cause of death for children under the age of 15 years. Although most of the 3000 human inherited disorders that have been described are very rare, some genetic diseases are more common, and as many as 1 in 25 people of northern European origin are carriers of a cystic fibrosis mutation. On a global scale, genetic diseases have a tremendous impact on populations. A conservative estimate of the number of heterozygotes of the various anemias is put at 242 million, while there may be as many as 100 million affected by the glucose 6-phosphate dehydrogenase deficiencies.

Our understanding of human genetic diseases, and our ability to diagnose them, has been revolutionized by our ability to locate and clone human genes. Before the development of recombinant DNA and other techniques, the genetics of human beings was very difficult to study. Planned breeding experiments are not possible, and only by careful observation of special families could the inheritance of genes be analyzed. In 1911, for the first time, a human gene was mapped to a specific chromosome. Color blindness was recognized as having a sex-linked mode of inheritance (it affects only males), and it was assigned to the X chromosome because this was known to be associated with sex determination. Other genes, such as those for hemophilia, were also assigned to the X chromosome because of their sex-linked pattern of inheritance. However, it was not until 1967 that a gene (thymidine kinase) was assigned to an autosome (non-sex chromosome), by using the newly developed tools of somatic cell genetics.Other genes were mapped to chromosomes by using somatic cell hybrids, but it was not until cloned human genes became available in 1978 that extraordinarily rapid developments in human genetics began. As a consequence of these advances, human genes are being analyzed in a detail comparable with that in studies of viral and prokaryotic genes. In this chapter we will describe how some genes responsible for human diseases have been cloned, concentrating on approaches that have general applicability for human molecular genetics. Application of these cloned genes to the practical problems of diagnosing human inherited disorders will be covered in Chapter 27, and the progress being made in mapping and sequencing the entire human genome will be discussed in Chapter 30.

Human Genetic Diseases Have Simple and Complex Patterns of Inheritance

Cystic fibrosis (CF), sickle-cell anemia, β-thalassemia, and phenylketonuria are examples of *autosomal recessive* diseases (Figure 26-1). The genes for these disorders are carried on the autosomes, and individuals homozygous for the mutation are affected. Heterozygotes with one mutant gene are usually unaffected and are called *carriers*. There is a 1-in-4 chance that a child of parents who are both carriers will be affected, and a 1-in-2 chance that a child will be heterozygous for the gene. The mutant genes for autosomal recessive disorders may be quite common in the population.

Autosomal dominant disorders (Figure 26-1) are conditions like Huntington's disease (HD) and myotonic dystrophy. The genes for these diseases are carried on the autosomes, but only a single copy of the mutant gene is needed to cause the disease. Disorders in which there is late onset or variable expression (penetrance) of the mutant gene are an important group. In these cases, people who do not realize that they are affected

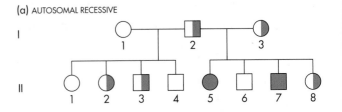

(a) AUTOSOMAL RECESSIVE

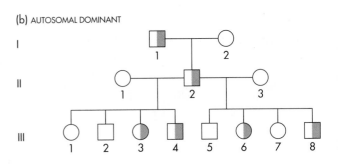

(b) AUTOSOMAL DOMINANT

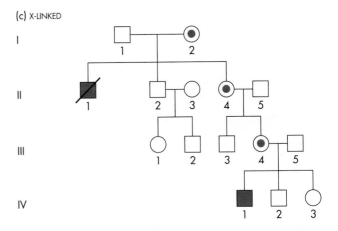

(c) X-LINKED

FIGURE **26-1**
Human inherited disorders. (a) *Autosomal recessive.* Only individuals with *two* copies of the mutant gene are affected (filled symbols), and they are the offspring of heterozygous carriers (I-2 and I-3). Heterozygotes with one copy of the gene are usually unaffected (half-filled symbols). (b) *Autosomal dominant.* All members of this family carrying a single mutant gene (half-filled symbols) are affected by the disease. The children of an affected individual have a 50 percent chance of inheriting the mutant gene, and males (□) and females (○) are affected equally. (c) *X-linked.* In disorders like Duchenne muscular dystrophy, where the boys die young and do not reproduce, the mutation is passed on only by their mothers, who have a single copy of the mutant gene (⊙). There is a 50% probability that the child of a carrier mother will inherit the mutant gene. Note how the disease can skip a generation (III) to reappear in an affected boy in the next generation (IV). Carrier mothers are often phenotypically normal. In some X-linked diseases, like hemophilia, it is possible for an affected male and a carrier female to have a female child homozygous for the mutation. The female homozygous for hemophilia has severe disease, just like an affected male.

may have children, and there is a 1-in-2 chance that any of their children will inherit the mutant gene and be affected by the disorder.

Diseases resulting from mutations in genes present on the X chromosome are known as *X-linked* disorders (Figure 26-1). The classic examples of X-linked diseases are hemophilia, resulting in abnormal blood clotting, and Duchenne muscular dystrophy (DMD), resulting in muscle degeneration and early death. Males have only one X chromosome, so that these disorders act like dominant mutations in males. In women, these mutations are like autosomal recessive mutations, and determining which women in a family are carriers is an essential part of genetic counseling for X-linked conditions. The characteristic feature of X-linked disorders is that male-to-male transmission of the mutation never occurs, as males pass their X chromosome only to their daughters.

These disorders are single-gene, or unifactorial, disorders in which the disease is presumed to be the result of the mutation of a single gene (although the same gene can be mutated in different ways in different patients). However, some of the commonest disorders are believed to have complex origins. *Polygenic* disorders result from mutations in any of a number of different genes, while *multifactorial* disorders arise from the interactions of environmental factors with multiple genes. Examples of these disorders include coronary heart disease, alcoholism, and schizophrenia. This group of disorders is difficult to analyze because of the complexity of the factors involved.

The Metabolic Basis Is Known for Some Human Inherited Diseases

The recessive mutations that cause simple Mendelian genetic diseases alter enzymes or other essential proteins such as hemoglobin, growth hormone, or blood-clotting factors. In about 600 of the more than 3500 single-gene diseases that have been identified by studies of family pedigrees, we know precisely which protein is defective. In sickle-cell anemia, it is the β-globin chain of hemoglobin; in phenylketonuria, the enzyme phenylalanine hydroxylase; in Tay-Sachs disease, the enzyme hexosaminidase A; in the Lesch-Nyhan syndrome, hypoxanthine phosphoribosyltransferase; in classic hemophilia, blood-clotting Factor

VIII; and so on. In some disorders—for example, Duchenne muscular dystrophy—the protein has been identified but little is yet known of its function. For the most part, however, we are still completely ignorant of the primary biochemical lesion in human genetic diseases, including Huntington's disease and those polygenic diseases, such as coronary artery disease, that are due to mutations in a number of genes.

A handful—but only a handful—of these genetic diseases can be successfully treated or managed because we understand their biochemistry. Therapeutic strategies include supplying deficient metabolites (for example, the vitamin biotin in certain diseases that inactivate carboxylase enzymes), supplying proteins (such as clotting Factor VIII in classic hemophilia), and designing strict diets to regulate the intake of nutrients that can no longer be metabolized properly (for instance, phenylalanine in the case of patients with phenylketonuria, who must also avoid artificial sweeteners that are metabolized to phenylalanine). However, the majority of these diseases—even those for which we know the primary biochemical lesion—remain untreatable. Either we cannot deliver the missing enzyme to the appropriate cells, or we still do not know exactly which of the many abnormal metabolites that accumulate causes the symptoms. In most cases the prognosis for afflicted children remains bleak. Thus, if patients and their families are to be helped, much needs to be learned, even about those inherited metabolic disorders for which the protein abnormality is known.

In cases in which the biochemical basis of an inherited disease is known, that information can be used to clone the gene involved (Table 26-1). For example, the Lesch-Nyhan syndrome, an extraordinary inherited disorder with mental retardation and bizarre behavioral changes including compulsive self-mutilation, is caused by abnormalities in the gene for hypoxanthine phosphoribosyltransferase (HPRT). This enzyme is used to reutilize, or "salvage," free guanine and hypoxanthine from the breakdown of nucleic acids. The mouse gene for HPRT was cloned using knowledge of the biochemical pathway involved to metabolically select cells that were overexpressing the enzyme (Figure 26-2). Once a mouse HPRT cDNA had been cloned, it was used at low stringency as a probe to identify homologous human cDNA in a human fetal liver cDNA library.

TABLE 26-1

Some Examples of Methods Used for Cloning Human Genes

METHOD OF CLONING	GENE PRODUCT	DISEASE
Transfection of a selectable gene	HPRT	Lesch-Nyhan
Antibodies to enrich for mRNA	β-hexosaminidase	Tay-Sachs
Protein sequence to predict oligonucleotide probes	Factor VIII	Hemophilia
Predicted oligonucleotides used in polymerase chain reaction	Urate oxidase	Gout
Probes derived from homologous animal gene	Phenylalanine hydroxylase HPRT	Phenylketonuria Lesch-Nyhan
Linkage analysis	Cystic fibrosis transmembrane conductance regulator (CFTR)	Cystic fibrosis
Structural abnormality in human chromosome and linkage analysis	Dystrophin	Duchenne muscular dystrophy
	NF1	Neurofibromatosis

"Positional Cloning" Uses the Location of a Gene on a Chromosome to Clone the Gene

Unfortunately, because the biochemical defects of most human inherited diseases are unknown, other methods have to be used to clone these genes. The cloning of a gene without any information about its protein product begins by determining the chromosomal location of the gene. This information is used to clone DNA sequences from that location, and these are used as probes to find the gene itself. Once the gene has been isolated, it can be sequenced and the sequence analyzed to determine the characteristics of the protein for which it codes. Originally called reverse genetics, this strategy is now better named *positional cloning*. Mapping human genes is carried out by physical methods that use cell hybrids for somatic cell genetics or that hybridize DNA sequences directly to chromosomes, or by analysis of chromosomal abnormalities that pinpoint the location of the mutation. Gene mapping is also carried out using linkage analysis, which compares within a family the inheritance of a mutant gene with the inheritance of DNA markers of known chromosomal location. Coinheritance of the disease gene and the marker suggests that they are physically close on the chromosome. This linked DNA sequence is used as the starting point to walk or jump to the gene by cloning DNA fragments that are even more tightly linked and, so, closer to the gene. The gene itself is identified by screening for coding sequences. This has become a very powerful tool in human genetics with the development of polymorphic DNA markers and new techniques for cloning and analyzing DNA.

Subchromosomal Mapping of Genes and Markers Can Be Accomplished with Somatic Cell Hybrids

Somatic cell hybrids were developed in the 1960s, when it was shown that cells growing in tissue culture could be made to fuse with each other by treating the culture with inactivated Sendai virus or with a chemical like polyethylene glycol. Hybrid cells can be made by fusion between human-mouse, human-hamster, and mouse-hamster cell combinations. When human-rodent cell fusions are made and the resulting hybrid cells are grown in culture, there is a progressive loss of human chromosomes until only one or a few human chromosomes are left. Hybrid cells containing only fragments of human chromosomes can be produced using human chromosomes with translocations and deletions, or the fragments can be produced experimentally by radiation (Chapter 30). Single hybrid cells

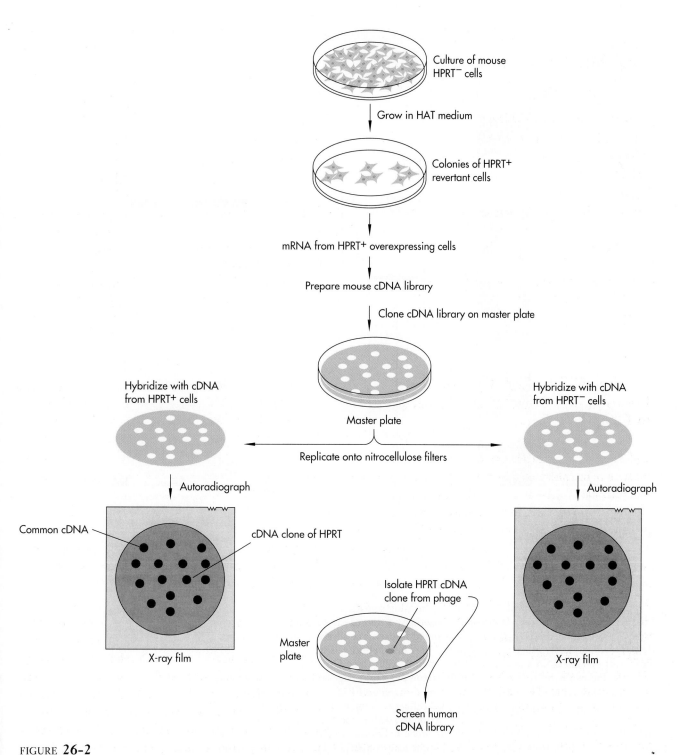

FIGURE **26-2**

Cloning the *HPRT* gene. Mouse cells with a homozygous mutation for HPRT were grown in a medium containing hypoxanthine, aminopterin, and thymidine (HAT). Only HPRT⁺ cells can grow in this medium, so the cells that survived and gave rise to colonies were HPRT⁺ revertants. In vitro translation of mRNA from such cells showed that the cells were overexpressing HPRT mRNA. This mRNA was used to prepare a cDNA library that was differentially screened with radioactively labeled cDNAs prepared from mRNA from the revertant and HPRT⁻ cells. A single clone that did not hybridize with HPRT⁻ cell cDNA and did hybridize with HPRT⁺ revertant cell cDNA was isolated and shown to contain HPRT cDNA by in vitro translation of hybrid-selected mRNA. This mouse cDNA was then used to screen a human cDNA library at low stringency to identify a human HPRT cDNA.

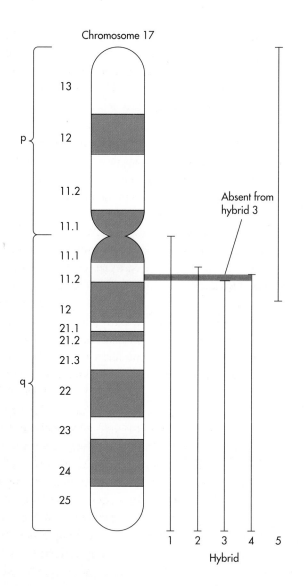

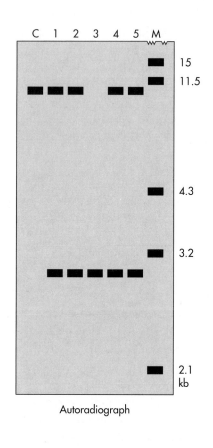

FIGURE 26-3

Mapping probes in the neurofibromatosis region using somatic human-rodent cell hybrids. Five cell hybrids (1–5), each containing a differing portion of the long arm (q) of chromosome 17, were used. The lengths of 17q included in each hybrid are shown by the lines (left). For example, hybrid 1 contains all of 17q, while hybrid 2 is missing segment 17q11.1 and a part of 11.2. DNA was extracted from human cells (lane C in the autoradiograph) and from the five hybrids (lanes 1–5 in the autoradiograph), digested with a restriction enzyme, and Southern blotted with a probe from the gene *EV12,* a candidate for the neurofibromatosis gene (right). The probe hybridizes to human DNA present in all hybrids except hybrid 3. (The second set of bands at about 3.0 kb is the rodent gene that cross-hybridizes with the human probe.) Since the probe hybridizes to DNA from hybrid 4, the sequence must come from that portion of 17q present in hybrid 4 but absent from hybrid 3, that is, a small section of 17q11.2. The bands in lane M are fragments of known size, marked in kilobases.

are isolated and grown to make lines of mouse cells containing just a small amount of human DNA. The human DNA in each cell line is characterized by using Southern blotting and probes of known chromosomal location. Panels of hybrid cell lines have been made in which each cell line carries the same human chromosome but with different deletions or translocations. These human chromosome fragments partially overlap, so that very small regions of the chromosome can be defined (Figure 26-3). It is then a routine matter to determine the chromosomal location of an unknown probe by hybridizing it to a Southern blot of DNA from such a panel of hybrid cells.

Cloned Genes and Markers Can Be Localized by in Situ Hybridization to Chromosomes

DNA sequences can be localized to subchromosomal regions by hybridizing radioactively labeled probes directly to chromosome spreads. Chromosome spreads are made using cells whose division has been blocked in metaphase by a chemical like colcemid that disrupts the mitotic spindle. The chromosomes are treated briefly with trypsin, and then stained with Giemsa. A pattern of light and dark bands develops on each chromosome, so that the chromosomes can be identified individually. Between 350 and 550 bands can be distinguished on the haploid human genome. Location of the radioactive probe is revealed by the distribution of silver grains in a photographic emulsion layered over the spread (the procedure is analogous to exposing an x-ray film), and the chromosomes are stained to reveal their banding patterns. In situ hybridization was first developed to map genes on *Drosophila* salivary gland chromosomes, where each gene is amplified some 1000-fold. The detection of single copies of human genes is more difficult but can be done by pooling the distribution of grains over as many as 30 metaphase spreads (Figure 26-4).

For metaphase and prometaphase chromosomes, sequences a few million base pairs apart can be resolved using radioactively labeled probes and autoradiography. Recently, sequences as close together as 50 kb have been resolved using fluorescently labeled probes and interphase chromosomes. One difficulty with this technique is that fluorescent labeling is incompatible with conventional staining methods used to identify chromosomes. This problem can be circumvented by using one or a series of previously mapped probes as landmarks on the chromosome. The probes may be biotinylated and detected by an avidin-fluorochrome complex, or they may be modified with a chemical such as dinitrophenol (DNP) and the modified molecule detected with an antibody to DNP. The positions at which the probes hybridize on the chromosome are determined relative to the tip of one arm of the chromosome, and the probes are ordered on the basis of these measurements. For example, the order of 50 clones containing DNA inserts from chromosome 11 determined by in situ fluorescent mapping was the same as that determined by other mapping techniques. Further developments have led to methods that produce banding patterns that can be used for directly identifying the subchromosomal location of fluorescently labeled probes. Cells can be treated with bromodeoxyuridine (BrdU), which is incorporated into DNA, before preparing chromosome spreads. When these chromosomes are treated with a fluorochrome-labeled antibody to BrdU, a pattern of bands similar to that seen with conventional staining is obtained. The locations and order of several probes can be determined simultaneously if they are labeled with different fluorochromes. This technique is becoming the method of choice for rapidly ordering newly isolated probes and as an independent check on maps produced by other means.

Chromosomal Abnormalities Provide Another Means of Locating Disease Genes

The location of a human disease gene can also be pinpointed if the disease arises as a consequence of the physical disruption of a gene. These disruptions may be large enough to be detected by light microscopy of chromosomes, and the detection of the same chromosomal abnormality in two or more patients with the same disease is strong evidence that the disease locus is at the point of disruption of the chromosome.

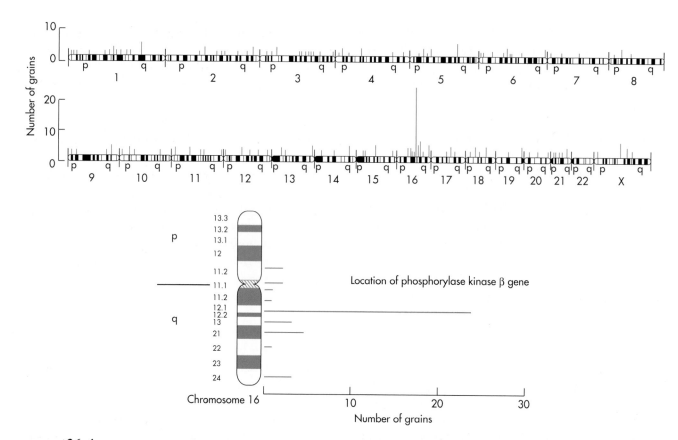

FIGURE **26-4**

Subchromosomal localization of the gene for phosphorylase kinase subunit β using in situ hybridization. The 23 human chromosomes (excluding the Y chromosome) are laid end to end and numbered 1 through 22, and X (upper panel). The short and long arms of each chromosome are marked p and q, respectively. In this example, the locations and numbers of silver grains on chromosomes in 61 cytogenetic preparations were determined and marked on the diagram. There is a clear peak on the long arm of chromosome 16, shown in detail on the enlarged diagram (lower panel). There is always a low background of silver grains that tends to mask the location of the probe in a single chromosome spread, but if approximately 60 percent of the counted silver grains are present at a single location, that is assumed to be the true location of the probe. (Francke et al., 1989)

Although the number of patients having an inherited disease as a result of a cytogenetic defect is small, the population of human beings being screened is sufficiently large that it has been possible to associate a number of diseases with specific chromosome defects (Table 26-2). However, chromosomal abnormalities may not be an infallible guide to the position of a human disease gene if mutations in more than one gene cause similar clinical phenotypes or if there are complex rearrangements at the site of the translocation.

Nevertheless, this information has been valuable

for analyzing some forms of cancer and cloning the genes involved. Chromosomal abnormalities led to the discoveries that in Burkitt's lymphoma the c-*MYC* oncogene is brought under the control of an immunoglobulin gene, and in chronic myelogenous leukemia the c-*ABL* gene is joined to the *BCR* gene (Chapter 18.) Some of these rearrangements are the basis of diagnostic tests (Chapter 27). Chromosomal rearrangements are much less common causes of other human inherited disease, but when found, they have greatly simplified the task of localizing a disease gene. The genes for Duchenne muscular dystrophy, chronic granulomatous disease, retinoblastoma, Wilms' tumor, and neurofibromatosis have all been cloned in this way, and intensive searches are being made for chromosomal abnormalities in other inherited disorders.

Restriction Fragment Length Polymorphisms Serve as Markers for Linkage Analysis

Clues to the location of a gene can come from comparing the inheritance of a mutant gene with the inheritance of markers of known chromosomal location within a family. Coinheritance, or genetic linkage, of disease gene and marker suggests that they are physically close together on the chromosome. Linkage is a familiar concept in genetics and dates back to the early studies on *Drosophila*, when it was shown that combinations of genes tended to be inherited as groups, linked together because they are close to each other on the same chromosome. Linkage is determined by analyzing the pattern of inheritance of a gene and a marker in suitable families. A marker must be polymorphic; that is, it must exist in different forms so that the chromosome carrying the mutant gene can be distinguished from the chromosome with the normal gene by the form of the marker it also carries. Until recently, polymorphisms could be detected only if

they were expressed by differences in the behavior of a protein, for example, by differences in enzyme activity or electrophoretic mobility. As a result, the use of linkage for genetic analysis in human beings has been very limited.

This situation changed dramatically with the realization that the sites recognized by restriction endonucleases could be polymorphic. Single base-pair changes are very frequent in the human genome, some estimates putting the frequency as high as 1 in 100 bp. The majority of single nucleotide changes are innocuous, although they may alter a restriction enzyme site so that an enzyme no longer cuts DNA at that site. For example, the restriction endonuclease *Bgl*II cuts DNA at a specific sequence:

```
...A GATCT...          ...A         GATCT...
...TCTAG A...    ⟶    ...TCTAG         A...
```

When one or more nucleotides in the endonuclease recognition sequence are altered, *Bgl*II will fail to cut the DNA strands at the altered site, and a longer DNA fragment is produced (Figure 26-5a). When a given restriction enzyme site is present in the DNA molecule of one chromosome but absent from the DNA mol-

TABLE **26-2**
Some Examples of Genetic Diseases Associated with Chromosomal Abnormalities

SYNDROME	DISEASE PHENOTYPE	CHROMOSOME ABNORMALITY*
Retinoblastoma	Tumors of the eye in children	Deletion of 13q14.11
Neurofibromatosis	Neural crest tumors	Constitutional translocation involving 17q11.2
Wilms' tumor	Kidney tumors in children	Deletion of 11p13
Prader-Willi	Neurological disorder	Deletions and duplications of 15q
DiGeorge syndrome	Thymus, parathyroid, and congenital heart disease	Abnormalities involving 22q11
Lowe syndrome	Congenital cataracts, mental retardation, kidney defects	Translocations with breakpoints at Xq25

* Some of these are very rare. For example, only two neurofibromatosis patients have been described with the 17q11.2 translocation, and only two females with Lowe syndrome have been found to have the Xq25 translocation.

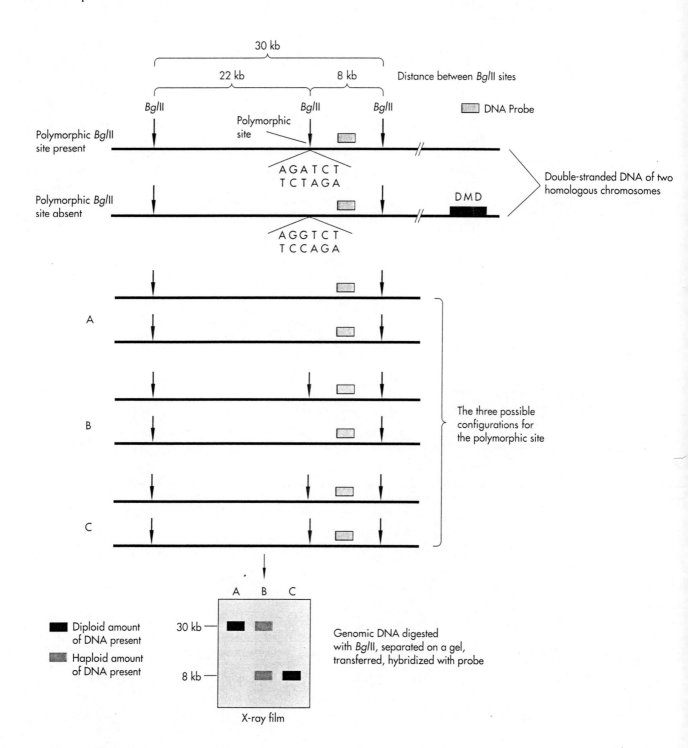

FIGURE **26-5a**
Restriction fragment length polymorphisms (RFLP). Each continuous line represents the double-stranded DNA molecule of one of the two homologous chromosomes. There can be up to three *Bgl*II sites (that is, sites that *Bgl*II recognizes and cuts) in this region of DNA, and one of these *Bgl*II sites is polymorphic. The sites are closely linked with the DMD (Duchenne muscular dystrophy) locus, which encodes dystrophin. The polymorphic site may be absent from the DNA of both chromosomes (A), present on one and absent from the other (B), or present on both (C). In its absence, a single fragment 30 kb long is produced by cleavage with *Bgl*II, and when the polymorphic site is present, fragments of 22 kb and 8 kb are produced. A DNA probe that anneals to the DNA strands at the position indicated hybridizes to the 30-kb and 8-kb fragments when used to probe a Southern blot.

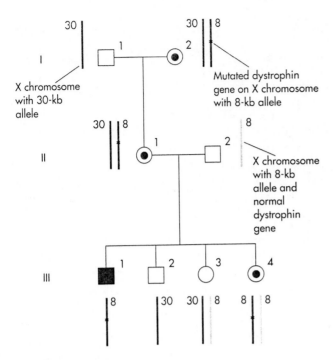

FIGURE 26-5b

Inheritance of an RFLP linked to an X-chromosome mutation (DMD, in this case). The probe–restriction enzyme combination detects a polymorphism giving fragments of sizes of 30 and 8 kb, as shown in Figure 26-5a. In this family, the mutation is carried on the chromosome with the 8-kb allele (that is, the allele with the restriction enzyme site that produces an 8-kb DNA fragment). The affected boy (III-1) carries the 8-kb allele on the chromosome that he inherited from his mother (II-1), who is a carrier (males inherit their X chromosome only from their mother). This chromosome had come from the grandmother (I-2). The unaffected boy (III-2) has inherited his mother's chromosome with the 30-kb polymorphism that had come from the grandfather (I-1), who was unaffected. Of the two daughters, III-3 has inherited her mother's normal X chromosome (carrying the 30-kb allele), while the other daughter (III-4) is also a carrier, having inherited the chromosome with the 8-kb allele and the mutant DMD allele. *Note that the RFLP is not the mutation itself.* This is shown clearly by II-2 and III-1, who both have the 8-kb fragment, but in II-2 this allele is linked with a normal dystrophin allele, while in III-1 it is linked with a mutated dystrophin gene.

ecule of the other, homologous chromosome, a shorter fragment is produced from the chromosome with the site, and a longer fragment from the other chromosome. We are now able to distinguish the two chromosomes in such an individual on the basis of this *restriction fragment length polymorphism (RFLP)*. The individual is heterozygous for this RFLP and, in the jargon of the geneticist, is said to be *informative* at that locus. Because the RFLP is inherited just like a gene, we can follow the individual chromosomes as they pass from generation to generation by tracing the inheritance of the marker fragments (Figure 26-5b).

A second form of DNA polymorphism results from variations in the number of tandemly repeated DNA sequences lying between two restriction sites. The sequences are 2 to 60 nucleotides long, and the number of alleles at these variable-number tandem repeat (VNTR) loci can vary from 2 to more than 20. These polymorphisms have become particularly important in the forensic applications of DNA techniques (*DNA fingerprinting*), and they will be discussed in more detail in Chapter 27.

The first restriction fragment length polymorphisms (RFLPs) to be found were associated with known genes (the human β- and γ-globin loci). The crucial step came with the realization that RFLPs were not necessarily associated with specific genes but were distributed throughout the genome, and in 1980 Botstein, White, Skolnick, and Davis made a detailed proposal for using RFLPs to map all the genes in the human genome. Linkage analysis using RFLPs propelled human genetics into the molecular age so that within just 7 years after the suggestion was made that RFLPs could be used for a human linkage map, a partial map of the human genome had been constructed.

One factor that is speeding the development of a linkage map for the human genome is the use of standard sets of families by gene mappers. One of the first efforts systematically to search out families suitable for linkage studies was made by Mark Skolnick, who established the Utah Genealogical Data Base using information from the Genealogical Society of Utah. This society is affiliated with the Mormon church and maintains detailed genealogical records of church members. The pedigrees are very extensive, data being available for several generations and each family containing many children. Families from Utah formed the core of the collection of families in Centre d'Etude du Polymorphisme Humaine (CEPH) in Paris. Founded in 1983 by Jean Dausset and Daniel Cohen, CEPH promotes the development of a human genetic map by making DNA samples available from a set of carefully selected families. The 40 families in CEPH have an average of over 8 children per family, and

DNA samples are available for all 4 grandparents in each of the 29 families in the reference panel. Lymphoblastoid cell lines are used for preparing DNA, which is then distributed to members of the CEPH Collaboration. The Collaboration is an international affair with over 60 participating laboratories throughout the world; the members are committed to contributing their linkage data to a CEPH database.

Linkage is Calculated from Frequency of Recombination

Remember that an RFLP is not a mutation that causes a genetic disease. Rather, an RFLP is used as a marker for a gene of interest; that is, it must be linked closely to the gene (Figure 26-5). When pairs of homologous chromosomes come together during meiosis, they exchange segments in a process called *recombination*. The farther an RFLP is from a gene, the more chance there is that there will be recombination between the gene and the RFLP. In a linkage analysis, the coinheritance of marker and gene are followed within a family. The probability that their observed inheritance pattern could occur by chance alone (that is, that they are completely unlinked) is calculated. The calculation is then repeated assuming a particular degree of linkage, and the ratio of the two probabilities (no linkage versus a specified degree of linkage) is determined. This ratio expresses the odds for (and against) that degree of linkage, and because the logarithm of the ratio is used, it is known as the *logarithm of the odds*, or *lod, score*. For practical purposes, a lod score equal to or greater than 3 is taken to confirm that gene and marker are linked. This represents 1000:1 odds that the two loci are linked. Calculations of linkage became much easier with the development of computer programs that were able to perform the many calculations necessary to take into account all the members of large pedigrees. These programs can perform multipoint analyses that calculate linkage between a large number of loci and produce a map of their order along the chromosome.

It is important to remember that markers are placed on a genetic map relative to each other, the order being determined by the recombination between them.

The unit of recombination is the *centimorgan* (c*M*), named for the great geneticist T. H. Morgan. Two markers are one centimorgan apart if they recombine in meiosis once in every 100 opportunities that they have to do so. The centimorgan is a genetic measure, not a physical one, but a useful role of thumb is that 1 cM is equivalent to approximately 10^6 bp. This relationship between centimorgans and base pairs is a dramatic illustration of the wide gap that exists between genetic and molecular studies. Geneticists consider a marker 5 cM from a gene to be useful for practical purposes such as genetic counseling, and yet the molecular biologist is still 5×10^6 bp away from the gene that is to be cloned! A great deal of research is devoted to developing techniques that are capable of manipulating very large DNA fragments in order to bridge this gap between genetic and physical distances (Chapter 29.)

Recombination has serious consequences when linkage analysis is used to trace the inheritance of a mutation in a family. The association between marker and gene will be lost if the crossover occurs in the region between the marker and the gene, and a misdiagnosis will result (Figure 26-6a). The accuracy of following a gene using RFLPs as markers can be increased substantially if informative RFLPs are available on each side of the gene. In this case a crossover can be detected because the pairing of flanking RFLPs on each chromosome is lost (Figure 26-6b). A double crossover between the flanking markers would not be detected, but if the RFLPs are less than 5 cM from the gene, the chance of a double crossover is 1 in 400 or less. Flanking markers are also important for cloning, because the investigator knows that the gene must lie between them.

Abnormal X Chromosomes Provide a Means of Cloning the Gene for Duchenne Muscular Dystrophy

Both linkage analysis and chromosomal abnormalities were used in the cloning of the gene for Duchenne muscular dystrophy (DMD), an X-linked disorder causing progressive muscle degeneration in young boys. There is no cure, and patients die in their late

(a) RECOMBINATION

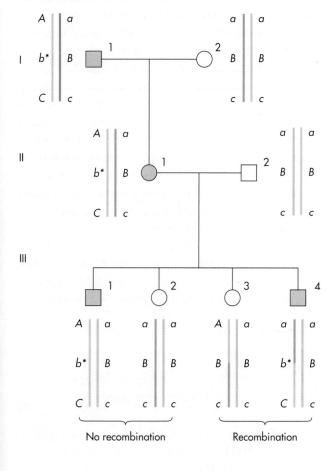

No recombination   Recombination

(b) RECOMBINATION WITH FLANKING MARKERS

No recombination   Recombination

FIGURE **26-6**

(a) Recombination and the importance of flanking markers as applied to an autosomal dominant disease. The marker known to be linked to the disease has two alleles, A and a. The disease allele is b^*, and the normal allele B. I-1 and II-1 are both affected, and b^* is linked to A in this family. In the absence of a crossover, III-1 (with A) is affected, and III-2 (with a) is unaffected. If there has been a crossover, the markers become unreliable. For III-3, the presence of A no longer indicates affected status, and conversely in III-4, a is now linked with the b^* mutation. Because we are examining only for A or a in this case, there is no way of determining whether a crossover has occurred. (b) Crossovers can be recognized if there are markers on each side of the disease locus. In this example, b^* is linked with both A and C in I-1 and II-1 (and the normal allele B with a and c). It can be seen that recombination has occurred in III-3 because although marker A is present, C is absent and has been replaced by c. In III-4, a and C are present, again showing that a crossover has occurred. It would not be possible to tell whether III-3 and III-4 were carrying b^* because the exact position of the crossover point between the two markers is unknown.

teens or early twenties. It is a relatively common disorder, occurring in about 1 in 3500 live male births. Becker muscular dystrophy is a milder form of muscular dystrophy that was thought to be caused by a mutation at a different locus from DMD. Despite a considerable amount of research over 30 years, very little progress had been made in understanding the pathogenesis of the disorder. Recombinant DNA techniques have now led to astounding progress in what seemed to be one of the most intractable of the inherited disorders.

Women who are heterozygous for an X-linked disorder have one X chromosome carrying a normal copy of the mutated gene and are not usually affected by the disease. However, women have been described who appear to have DMD, and it was found that some of these women had a chromosomal translocation, part of one of their X chromosomes having been exchanged with part of an autosome (Figure 26-7). The autosomes involved in these translocations differed, but remarkably the breakpoint of the X chromosome was always in band Xp21 of the short arm. This suggested that a gene involved in DMD was located at Xp21, and this was confirmed when RFLPs detected by probes mapping to Xp21 (Figure 26-8) showed linkage to DMD, albeit rather loosely at 15 cM.

(a)

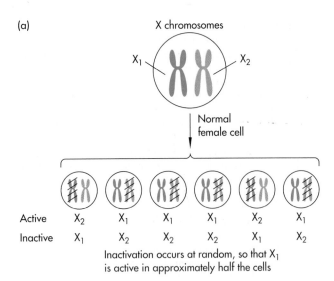

FIGURE **26-7**

X-inactivation and X-autosomal translocations. (a) In normal female cells, one of the X chromosomes is inactivated, so that there is only one functional X chromosome per cell, as in males. This normally occurs at random for any cell, so that a woman is a mosaic of cells as regards expression of her X chromosome (only the X chromosomes are shown). (b) The situation appears to be different for women with balanced X-autosomal translocations. In these women, the *normal* X chromosome is preferentially inactivated, while the so-called *derived* X chromosome, which is carrying trans-located autosomal DNA from the autosomal chromosome shown, remains active. In this way, each cell has the equivalent of two intact autosomes and one intact X chromosome. However, the break in the derived X chromosome mutates the gene located at the breakpoint; since the intact copy of the gene present on the normal chromosome has undergone inactivation, neither copy of this gene is functioning in these cells.

(b)

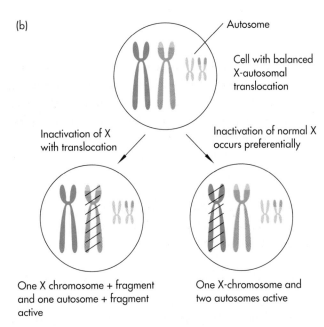

analysis showed that one clone, XJ1, was very closely linked with the DMD mutation, and it also detected deletions in boys with DMD, definitive evidence that a portion of the DMD gene had been cloned.

Another chromosomal abnormality was found in a young boy with a bizarre group of disorders including DMD. Cytogenetic analysis showed that there was a deletion covering a large portion of the Xp21 locus, a deletion so large that it affected the genes for several disorders. It was realized that this patient's DNA could be used in a subtractive hybridization experiment to enrich for genes deleted in the patient's DNA (Figure 26-9). Clones with DNA inserts that mapped to Xp21–Xp22 were found by using them as probes with a panel of DNA from hybrid cells containing human X chromosomes deleted for various regions; these clones were then screened against the patient's DNA and that of other unrelated DMD patients. One of these probes, pERT87, detected deletions in several DMD patients, thus showing that it came from the DMD gene.

As well as indicating the location of the DMD gene, one of these translocations provided a means of cloning the gene. The breakpoint on chromosome 21 in a patient with a t(X;21) translocation was in 21p12, a region of DNA coding for ribosomal RNA. As a result of the translocation, a block of ribosomal genes was now adjacent to the DMD gene, and these ribosomal gene sequences were used as a marker to clone X-chromosome sequences at the breakpoint. Linkage

cDNAs for the DMD Gene Were Cloned Using Two Strategies

Although the probes that had been isolated were from within the DMD gene, they showed recombination at 5 cM with the DMD mutation in some families, and

Portions of human X chromosome present in hybrid cells

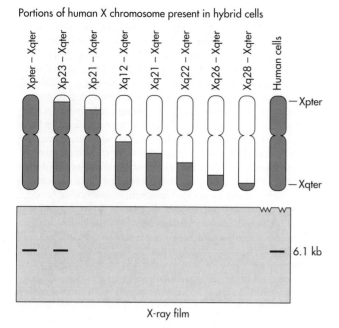

X-ray film

FIGURE **26-8**
Somatic cell mapping of a human Xp21 probe. A series of hybrids were made between mouse or Chinese hamster cells and human cells with different deletions of the X chromosome. The hybrid cells were selected for cells that had the X chromosome or portions of it as their only human DNA. Total DNA was prepared from each cell line and control parental cell lines, digested with *Eco*RI, and analyzed by Southern blotting using a probe from an X chromosome library. A band of 6.1 kb was detected in DNA from hybrid cells containing the whole X chromosome and in a control human fibroblast cell line. The band was missing from all DNA containing X chromosomes lacking the region between Xp21 and the tip of the chromosome (Xpter), showing that the probe came from this region.

so they could be as far as 5×10^6 bp from the mutation in these patients. This suggested that the gene was very large, and cloning a cDNA, which contains just the coding information of a gene, was an essential step for further analysis of the DMD gene. The cDNAs for the DMD gene were isolated from fetal and adult human skeletal muscle cDNA libraries using two strategies. The first strategy was to screen DNA from the mouse, hamster, chicken, cow, and cebus (a monkey) using probes from the pERT87 region. Some probes were found that hybridized to all these mammalian DNAs, suggesting that they contained conserved sequences coding for important exons. One of these probes detected a band on a Northern blot of mRNA from human fetal skeletal muscle and led to the cloning of the entire cDNA for the DMD gene. Searching for sequences conserved across species is an invaluable way to detect coding regions. In the second strategy, sequences from the XJ1.1 region were used to screen cDNA libraries prepared from human skeletal muscle. As was expected, it was found that the DMD gene is very large. The number of exons present in a gene can be determined by Southern blotting by using the cDNA for the gene as a probe (Figure 26-10). The DMD gene was found to contain at least 65 exons spread over about 2.5 million nucleotides. The average exon size is 200 bp, and the average intron size is 35,000 bp. One intron is 250 kb long! The cDNA probes detect genomic deletions in about 60 percent

of patients, and the use of these probes has revolutionized diagnosis of DMD (Chapter 27). Furthermore, the sequence of the cDNA has provided insights into the function of dystrophin, the DMD protein (discussed later in this chapter).

The Cystic Fibrosis Gene Was Cloned Using RFLP Analysis and Chromosome Jumping

Although linkage analysis helped confirm the location of the DMD gene, cloning the gene depended on using chromosomal rearrangements at Xp21. The gene for cystic fibrosis (CF) was the first human disease gene to be cloned solely on the basis of its position determined by linkage analysis. This tour de force required 4 years of work by a large number of laboratories. People with cystic fibrosis (CF) are diagnosed in childhood. The use of antibiotics and enzyme supplementation has extended the life span of some CF patients to as long as 30 years, but the disease is inevitably fatal. While most human single gene disorders are very rare, cystic fibrosis is remarkably common in Caucasians, affecting about one in 2500, with 1 person in 25 being a heterozygote. Cystic fibrosis is a disease for which population screening for heterozygotes might be justified if an appropriate test were available.

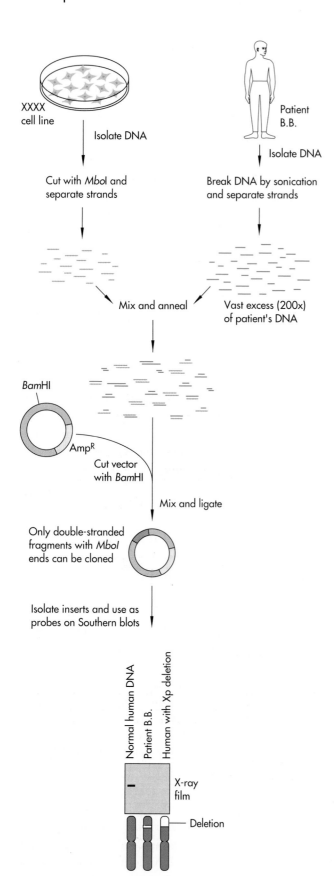

FIGURE **26-9**

Cloning DMD sequences by subtraction. DNA from a human cell line containing four X chromosomes was cut with the restriction enzyme *Mbo*I, while DNA from a DMD patient with a very large deletion of the DMD region was broken into pieces at random sites by sonication. A 200-fold excess of the patient's DNA fragments was added to the normal X-chromosome *Mbo*I fragments, and the DNA fragments were annealed for 37 hours in the presence of phenol to enhance reassociation. Under these conditions, three types of double-stranded molecules were produced. Most molecules were formed by reassociation of strands of the patient's DNA. Because these strands came from sonicated DNA, the ends of these double-stranded molecules were random sequences and thus could not be cloned by ligation into a vector. A smaller fraction were hybrid molecules between patient and cell line DNA. These double-stranded molecules could not be cloned because their ends were also incompatible with the cloning sites in the vector. The third class of double-stranded molecules were formed by reassociation of DNA strands from the cell line. These molecules were enriched for sequences present in the cell line and absent from the patient's DNA, because the vast excess of patient DNA had "mopped up" sequences common to both sources of DNA. Double-stranded molecules derived from the normal X chromosomes of the cell line were the only molecules in the annealing reaction that had ends that could be cloned into a vector, in this case a cloning site with *Bam*HI sticky ends that are compatible with the *Mbo*I sticky ends of the cell line-derived DNA. The origin of cloned inserts was determined by using them as probes in Southern blotting of DNA from a normal individual, from the patient, and from two rodent-human hybrid cell lines containing either a whole human X chromosome or an X chromosome with the tip deleted. A probe called pERT87 failed to hybridize to DNA from the patient or from the hybrid cell with the partially deleted X chromosome, showing that pERT87 comes from the deleted region.

The hunt for the CF gene began with screening a large number of CF families with a large number of RFLPs, looking for RFLPs that could be used to track the CF gene. The first clue to the location of the CF gene came when an RFLP, loosely linked to CF, was shown by in situ hybridization to be located on chromosome 7. Two further chromosome 7 RFLPs at 7q31–32 were found that were very tightly linked to CF and flanked the CF locus (Figure 26-11). It seemed a straightforward job to find the CF gene by simply cloning the DNA between the flanking markers and identifying the CF gene. In practice it was more dif-

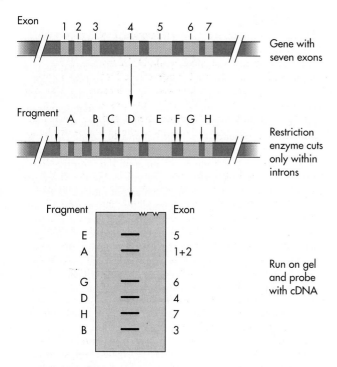

FIGURE 26-10

Exon mapping. The gene shown contains seven exons (1 to 7). A restriction enzyme is selected that cuts DNA fairly frequently but does not cut the gene's cDNA (thus, it does not cut the coding regions—exons—of the gene). This means that the gene is cut only within its introns, producing eight fragments (A to H). When this restriction enzyme–treated DNA is analyzed by Southern blotting using the labeled cDNA as a probe, only fragments containing exons are seen on the autoradiograph (so fragments C and F containing only intron DNA are not detected). Fragments may contain more than one exon if the enzyme does not cut within every intron (fragment A).

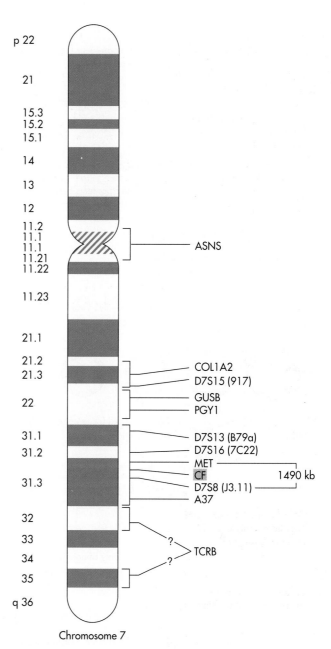

FIGURE 26-11

Probes in the CF region used for localizing and cloning the CF gene. Chromosome 7 is shown with its standard banding pattern. The CF gene had been mapped to the q31.2–q31.3 region and was known to lie between MET and D7S8, a distance of 1490 kb as determined by physical mapping. There are currently over 90 gene loci and over 400 polymorphic DNA fragments assigned to chromosome 7.

ficult, highlighting the problems involved in trying to detect a gene, even when its location has been narrowed down to a small region.

The successful strategy systematically cloned the region containing the CF gene and analyzed the expressed sequences in that region. The first step was to isolate a large number of probes from the region between the flanking markers, a distance of 1.5 million bp. Two probes detected polymorphisms linked with CF and narrowed down the region containing the CF gene to about 500 kbp. At this stage, chromosome "jumping" was used to move through this region. In

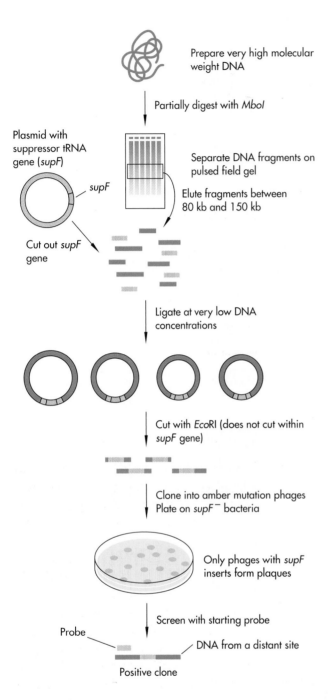

Plasmid with suppressor tRNA gene (*supF*)

supF

Cut out *supF* gene

Prepare very high molecular weight DNA

Partially digest with *Mbo*I

Separate DNA fragments on pulsed field gel

Elute fragments between 80 kb and 150 kb

Ligate at very low DNA concentrations

Cut with *Eco*RI (does not cut within *supF* gene)

Clone into amber mutation phages
Plate on *supF*⁻ bacteria

Only phages with *supF* inserts form plaques

Probe

Screen with starting probe

DNA from a distant site

Positive clone

FIGURE **26-12**
Making a jumping genomic DNA library. Large DNA fragments are generated by carrying out a partial digestion of very high molecular weight DNA using an enzyme such as *Mbo*I. The large DNA fragments are separated on pulsed field gradient gels (Chapter 29), and fragments between 80 and 150 kb are isolated from the gel. These fragments can be circularized by ligation at very low dilutions. If this is done in the presence of a tRNA suppressor gene (*supF*), a small percentage of the circularized molecules will incorporate the suppressor gene at the junction between the ends of the *Mbo*I fragment. *Eco*RI does not cut within the suppressor gene, so that digestion of the circularized molecules produces small DNA fragments containing the suppressor gene, flanked on either side by DNA sequences that were separated by a "jump" length of as much as 150 kb in the original *Mbo*I-cut DNA. These fragments are cloned into phage vectors with amber mutations (thus, the phage cannot replicate in the absence of a suppressor gene) and vectors with inserts can be selected for when plated on bacteria lacking a suppressor gene (*supF*⁻). The jumping library can be screened with a probe, and positive clones should contain a DNA fragment from a region the length of the jump from the probe.

clone was used to isolate further overlapping phage and cosmid clones by chromosome walking (Figure 26-13). One of these clones was the starting point for a second jump of about 75 kb, and further phage and cosmid clones were isolated using the clone at the end of this jump. In all, seven jumps were made in the direction of the CF locus and 49 recombinant phage and cosmid clones were isolated (Figure 26-14.) The end result of this tremendous effort was a contiguous stretch of 280 kb of DNA cloned into overlapping cosmid or phage clones. Eventually, over 500 kb of DNA encompassing the entire CF region was cloned and mapped.

contrast to chromosome walking, in which overlapping clones are isolated from a conventional phage library, a jumping library is constructed in such a way that each phage clone contains sequences that are widely separated on the chromosome (Figure 26-12). Two of the probes closely linked to CF were used as the starting points for jumping. One jump went almost 50 kb toward the CF locus, and a probe from this new

The Cystic Fibrosis Gene Was Identified by DNA Sequencing

The next problem was to determine whether any of these clones contained coding sequences. The strategy followed was that used in cloning the DMD cDNA, checking clones for cross-hybridization to sequences on Southern blots of bovine, mouse, hamster, and

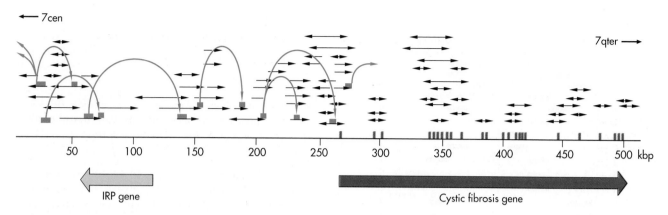

FIGURE **26-13**

Cloning the CF gene by jumping and walking. The long continuous line represents a DNA strand over 500 kb long (the length markers are at 50-kb intervals through the region of chromosome 7 containing the CF gene). The centromere (7cen) is to the left and the tip of chromosome 7 (7qter) to the right of the diagram. The curved arrows show the length and direction of each jump, and the solid boxes are the clones that were the starting and end points of the jumps. The horizontal arrows are overlapping phage and cosmid clones containing DNA isolated from the regions at the ends of each jump. The directions of the arrows show the direction of cloning; double-headed arrows indicate that cloning went in both directions from the starting clone. The location of the cystic fibrosis gene is indicated by the large arrow below the DNA strand, and the positions of its 24 exons are shown by the small vertical bars on the horizontal axis. The position of the *IRP* gene is shown by the large leftward-pointing arrow. [modified from J. L. Marx, "The cystic fibrosis gene is found," *Science*, 245: 923–925 (1989)]

chicken genomic DNA. Five probes from the CF region were found to cross-hybridize with sequences in other species, but four of these sequences were eliminated as candidates for the CF gene: Linkage data excluded two sequences, a third turned out to be a pseudogene, and a fourth conserved sequence failed to detect mRNA transcripts or clones in cDNA libraries. Analysis of the fifth sequence revealed the presence of a CpG island (Chapter 8), suggesting that this region contained the start of a gene, and a probe from this sequence was used to screen a cDNA library made from mRNA from cultured normal sweat gland cells. (An abnormality in sodium and chloride transport was thought to be the cause of CF, and as this is a primary function of sweat-gland duct cells, these were a favored tissue for CF studies). A single positive clone was found that detected an mRNA transcript on Northern blots, and further screening of cDNA libraries led to the cloning of a contiguous region of about 6.1 kbp that contained 24 exons.

It had to be shown that this cDNA was associated with the CF gene. This proved to be difficult because, unlike DMD, no deletions or other abnormalities

could be detected on Southern blots of genomic DNA, or on Northern blots of mRNA, from patients with CF. Instead, mutations had to be searched for by comparing the nucleotide sequence of the putative CF gene in a patient with that of a normal individual. The complete CF cDNA is over 6000 bp long, and just three base pairs were found to be deleted in the CF patient. This deletion results in the loss of a phenylalanine residue (Figure 26-14). The screening of many patients by using PCR to amplify the region containing the mutation and an allele-specific oligonucleotide probe to test for the presence of the deletion showed that this deletion accounted for 68 percent of the CF mutations. These results raised the expectation that population screening for CF would be possible if the remaining 30 percent of the mutations could be accounted for by a small number of additional mutations. Unfortunately, at least 60 different mutations have now been detected in patients who do not have the 3-bp deletion. Nevertheless, searching for the 3-bp deletion and a small number of other mutations, followed if necessary by RFLP linkage analysis, is already invaluable for diagnosis in affected families.

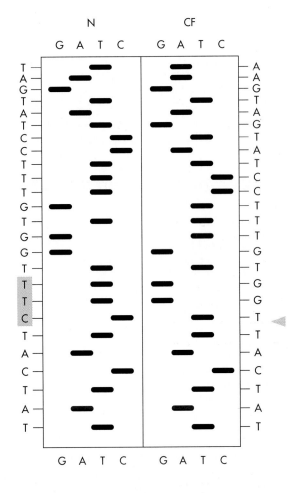

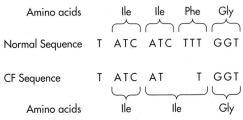

FIGURE **26-14**

The 3-bp deletion in CF. This mutation consists of the deletion of CTT in the tenth exon of the gene. The missing nucleotides are boxed in this autoradiograph showing this sequence in a normal individual (N) and a patient with cystic fibrosis (CF). Loss of three 3 bp maintains the reading frame so that in this case only a single phenylalanine is lost from the protein. The frequency of this mutation in CF patients ranges from as low as 30% in parts of southern Europe, to 95% in Denmark. In Northern Europe it is 75%, and in the U.S. population about 70%.

Clues about Protein Function Come from the Sequences of Cloned Disease Genes

Although a cloned gene can be invaluable for diagnostic purposes, human geneticists want to understand what is wrong with the protein involved and, armed with that knowledge, find a treatment. Fortunately, the cloned gene itself can lead to an understanding of the molecular pathology of the disease. The first thing done is to examine the nucleotide sequence for open reading frames, and to search by computer for similarities with other sequenced genes. There are now databases containing DNA and protein sequences that can be accessed directly by scientists using computers in their own laboratories.

The DMD cDNA was analyzed in this way, and the protein of the DMD locus, named *dystrophin,* was predicted to be a long, rodlike protein of 3685 amino acids. Database searches revealed homologies between domains of the dystrophin molecule and two cytoskeletal proteins, α-actinin and spectrin (Figure 26-15). It had long been suspected that the defect in DMD in some way involved the cell membrane, and the predicted structure suggests that dystrophin is a part of the muscle cytoskeleton, perhaps involved in anchoring the cytoskeleton to the cell membrane.

Cystic fibrosis was also believed to result from a defect in a membrane protein, probably one involved with the transport of chloride ions across the membrane. It was therefore very satisfying when the predicted structure of CF protein (called the *cystic fibrosis transmembrane conductance regulator, CFTR*) revealed it to be a transmembrane protein (Figure 26-16).

The power of sequence comparisons as a tool for finding clues about the functions of previously unknown proteins is shown particularly well in the case of neurofibromatosis 1 (NF1). This disease causes the development of tumors in the peripheral nerves, and there were no clues as to its molecular pathology. The NF1 gene was cloned, sequenced, and shown to be highly homologous to the mammalian and yeast proteins that interact with the *ras* oncogene. These proteins are involved in signal transduction, and the NF1 protein is probably involved in similar pathways.

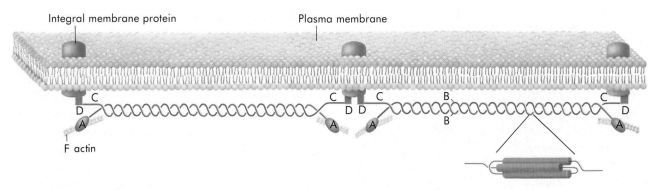

FIGURE **26-15**
A model of the dystrophin molecule, based on homologies with other proteins, experimental data, and speculation. Two pairs of dystrophin molecules are shown, the members of each pair forming an antiparallel helical structure. They lie on the cytoplasmic side of the plasma membrane, linking together three integral membrane proteins. Each dystrophin molecule has four domains (A to D). In the central portion (B), each dystrophin molecule is a triple helix, 2700 amino acids long, that has homologies with both α-actinin and spectrin. The amino-terminal domain (A) has 240 amino acids and has homology with that portion of the α-actinin molecule that binds actin filaments. Domain C resembles the carboxy terminus of α-actinin, and domain D is the carboxy terminus of the dystrophin molecule. Domain D has no homology with any previously described protein. (Koenig et al., 1988; Ervasti and Campbell, 1991)

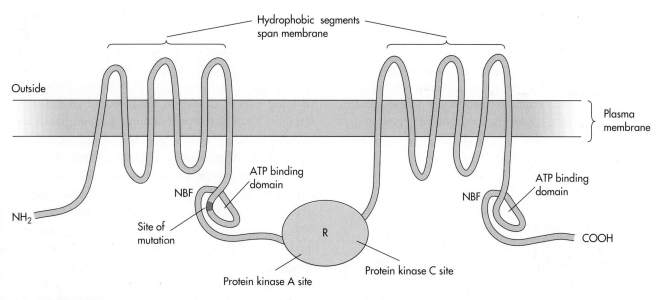

FIGURE **26-16**
A model of the cystic fibrosis transmembrane conductance regulator molecule, based on homologies with other proteins, experimental data, and speculation. The molecule has two membrane-spanning regions, each made of six hydrophobic segments. On the carboxy-terminal side of each membrane domain is a consensus sequence for nucleotide-binding folds (NBF) that bind ATP. The central portion of the molecule (R) has a 241 amino acid sequence characterized by the presence of a large number of charged residues. This R domain contains sites that can be phosphorylated by protein kinases A and C. The site of the three-base deletion in the first ATP binding domain is indicated. (Riordan et al., 1989)

Cloned Genes Are Used to Study Protein Expression and Function

Sequence analysis and finding homologies between sequences is very useful for suggesting functions for newly cloned genes, and the cloned gene itself can be used to glean further information about function. For example, the protein can be produced in *E. coli* by cloning the cDNA in an expression vector. The fusion protein can then be used to make antibodies with which to analyze the tissue expression of the protein and to determine its cellular location. In Duchenne muscular dystrophy, immunocytochemistry using such antibodies confirmed the prediction based on sequence comparison that dystrophin is located on the cytoplasmic surface of the muscle fiber plasma membrane. The best way to analyze function and to confirm that the cloned gene is indeed the disease gene, is by complementation, that is, showing that the cloned DNA contains sequences that can reverse the phenotype of mutant cells. Complementation tests may be technically difficult to carry out with large genes like the Duchenne muscular dystrophy gene, but even in this case a cDNA for mouse dystrophin has been used to express dystrophin in nonmuscle cells. The cystic fibrosis gene has been tested by complementation, and a wild-type gene has been shown to correct chloride transport in cells from CF patients in tissue culture (Chapter 12). This of course is the goal of gene therapy and will be discussed in more detail in Chapter 28

Cloning Genes for Polygenic and Multifactorial Disorders Is Difficult

Molecular genetics has been very successful in tackling many human inherited diseases, and although it is still not an easy task, it should be possible to clone any single-locus human gene by using the known, successful strategies and techniques for locating and cloning genes. This is not the case for the most common ailments that afflict us. Coronary heart disease, Alzheimer's disease, alcoholism, psychiatric illness, and autoimmune diseases are all complex disorders involving interactions among different genes, and between the genotype and the environment.

Genetic studies of these disorders are difficult when differential diagnosis fails to distinguish among the genetic subgroups involved. For example, narrow criteria for the diagnosis of schizophrenia may identify more genetically homogeneous families that can be studied by linkage analysis, but the numbers involved are small. And while larger and more numerous families can be found by including schizoid personality disorder and schizotypal phenotypes, some of the patients in these larger families may not have schizophrenia. Instead their behavior may be influenced by their environment, for example, simply by living in a family that includes schizophrenic individuals. The inclusion of "false affecteds," or phenocopies, in a pedigree has a disastrous effect on linkage analysis. Similar diagnostic problems apply to studies of alcoholism.

Studying inbred populations is one strategy that has been used to try to minimize genetic heterogeneity, but these small populations may have rare alleles that are not characteristic of the general population. This may partially account for the confused state of molecular linkage studies of mental disorders. There was great excitement when manic depression (bipolar affective disorder) in an Amish population was found to be tightly linked to chromosome 11 markers. However, other studies reported at the same time found no linkage of the same markers to manic depression in other North American families and in Icelandic families. These discrepant findings were originally explained away by assuming genetic heterogeneity in manic depression. Two years later, a reexamination of the families found that some individuals originally thought to be unaffected now had symptoms of manic depression. These revised diagnoses, together with new data on family members not previously studied, led to a significant downward revision of the lod score. The conclusion is that linkage between manic depression and markers on chromosome 11 is not proven.

Despite these complications, linkage analysis is a valuable tool for studying the genetics of complex disorders, provided that the difficulties and assumptions underlying the studies are recognized and that conservative measures of success are used. For example, a lod score of 3 is taken to show significant linkage in single-gene disorders, but it may be necessary to use a more stringent value for complex disorders. In the long term, it is expected that the human genome project will generate the RFLPs necessary for rapid, comprehensive mapping of these disorders. The practical outcome of this mapping may be the development of predictive tests for polygenic and multifactorial disorders, although variable environmental factors will confound the predictive power of tests for the latter group of disorders. The availability of such tests is likely to have a major impact on health care and insurance. Some of the societal consequences of human molecular genetics are discussed in Chapter 27.

Candidate Genes Can Be Used to Clone Human Disease Genes

An alternative strategy for cloning human disease genes is to choose *candidate* genes that may be involved. A candidate gene encodes a protein of known function, which, if mutated, could account in part for the pathogenesis of the disorder. The success of this approach depends, of course, on knowing enough about the possible molecular pathogenesis of the disorder to make an educated guess as to candidates. The best known example of the candidate gene approach is the work on familial hypercholesterolemia (FH) by Michael Brown and Joseph Goldstein. This is an autosomal dominant disorder, and patients who are homozyogous for mutations may begin to suffer heart attacks in childhood. Both FH homozygotes and heterozygotes have elevated levels of serum cholesterol. In a series of elegant experiments, Brown and Goldstein showed that the defect was in the binding of cholesterol by the low-density lipoprotein receptor on the cell surface. Subsequently, the receptor was cloned, and the mutations in the receptor gene have been analyzed in detail.

A recent successful example is cloning the gene for Marfan's syndrome, an inherited connective tissue disorder. Fibrillin, a connective tissue protein, was awarded candidate gene status because of its distribution and because of its abundance in tissues affected in Marfan's syndrome. Immunhistochemical studies showed that it was missing or at least greatly reduced in patients with Marfan's syndrome. A probe from the cloned fibrillin gene was mapped by in situ hybridization to 15q, the same region to which the Marfan locus had been mapped by linkage. The relationship between gene and the Marfan locus was confirmed when an RFLP in the fibrillin gene did not recombine with the disease in Marfan's syndrome families. Finally, the same mutation in the fibrillin gene was found in two unrelated patients with a sporadic form of Marfan's syndrome.

The candidate gene approach may have begun to unravel the complexities of Alzheimer's disease. There appears to be considerable genetic heterogeneity in Alzheimer's disease, with early onset in some families. There is a locus on chromosome 21 that segregates with Alzheimer's disease in some, but not all, families with early onset. This locus is close to the locus for amyloid precursor protein (APP), a tantalizingly finding because deposits of amyloid protein are found in the brains of patients with Alzheimer's disease. The Alzheimer's disease locus in a family with early onset was shown to be linked to a chromosome 21 locus, and PCR-based sequencing was used to analyze the APP gene in this family. A single mutation of a cytosine to a thymidine was found. A second unrelated family had the same mutation, and recently a third family has been reported. This is strong evidence that this mutation plays a part in the Alzheimer's disease of these families, but it is not the whole story. The majority of early-onset families tested, and all the late-onset families, did not have the mutation, showing that there are other genes involved in these other forms of Alzheimer's disease. The candidate gene approach has not been so successful in studies of schizophrenia and alcoholism.

Cloning Human Disease Genes Will Continue to Depend on Linkage Studies

Our current experience of cloning human disease genes by linkage highlights the present limitations and advantages of this approach. One difficulty of using linkage to locate genes is that genetic distances are much larger than the lengths of DNA molecules that can be analyzed readily by physical means. The work involved in cloning the cystic fibrosis gene (Figure 26-13) gives an idea of what is required to clone the

DNA represented by even short genetic distances. This gap between genetic and physical distances is being closed by technical developments. For example, pulsed field gel electrophoresis is becoming a routine technique for electrophoresis of DNA fragments hundreds or thousands of kilobases long. New vectors called *yeast artificial chromosomes* can be used to clone very large pieces of DNA, although it is still necessary to use phage and cosmid clones for fine-resolution cloning (Chapter 29)

RFLPs on either side of the disease locus are necessary to pinpoint the position of the disease gene. Searching for the gene responsible for Huntington's disease (HD) illustrates some of the problems that arise when flanking markers are difficult to find. Linkage analysis may be confounded by undetectable double recombination or gene conversion. And it is ultimately limited because recombination events become vanishingly rare as the physical distances between markers and gene become small.

When the region of DNA thought to contain the disease locus has been cloned, a number of candidate genes may be present, and it is not a trivial matter to recognize the disease gene itself. A great deal of excitement was generated early in the hunt for the CF gene by the finding of expressed sequences linked to CF. It was hoped that these were transcripts from the CF gene itself, but further studies showed that the transcripts were the same in normal individuals and CF patients. Cloning the gene for neurofibromatosis 1 was complicated by the presence of transcripts from three genes within a large intron of the NF1 gene. The most robust test to confirm that the correct gene has been isolated is complementation, but such assays do not exist for most human inherited diseases.

Despite these difficulties, the power of linkage analysis for locating genes is evident from the successes to date. It remains the only approach available for finding the genes of human diseases in the absence of either information on the protein involved or a chromosomal abnormality that pinpoints the gene. One of the most important consequences of this approach is that the probes and RFLPs discovered during the course of cloning can be used for diagnosis, long before the target gene is cloned. The gene for Huntington's disease is not yet cloned, but genetic counseling using RFLPs is already available. Many other examples will be discussed in the following chapter on the practical applications of human molecular genetics.

Our knowledge of the molecular genetics of human inherited diseases is going to increase rapidly as more RFLPs are found and more genes are cloned. As more is discovered, unexpected and highly informative connections will be made. The keys to this progress are collaboration and cooperation, as shown by the success of the CEPH Collaboration. The most ambitious of these coordinated efforts are the various nationally supported human genome projects, and the Human Genome Organization (HUGO) that is promoting international collaboration. The goals of the human genome projects and their probable impact on genetics and society will be discussed in Chapter 30.

Reading List

General

Emery, A. E. H. *Elements of Medical Genetics.* Churchill Livingstone, Edinburgh, 1983.

Scriver, C. R., A. L. Beaudet, W. S. Sly, and D. Valle, eds. *The Metabolic Basis of Inherited Disease.* McGraw-Hill Inc., New York, 1989.

Gelehrter, T. D., and F. S. Collins. *Principles of Medical Genetics.* Williams & Wilkins, Baltimore, 1990.

Weatherall, D. J. *The New Genetics and Clinical Practice,* 3rd ed. Oxford University Press, Oxford, Eng., 1991.

Original Research Papers

CLONING GENES FOR KNOWN BIOCHEMICAL DISORDERS

Brennand, J., A. C. Chinault, D. S. Konecki, D. W. Melton, and C. T. Caskey. "Cloned cDNA sequences of the hypoxanthine/guanine phosphoribosyltransferase gene from a mouse neuroblastoma cell line found to have amplified genomic sequences." *Proc. Nat. Acad. Sci. USA,* 79: 1950–1954 (1982).

Gitschier, J., W. I. Wood, T. M. Goralka, K. L. Wion, E. Y. Chen, D. H. Eaton, G. A. Vehar, D. J. Capon, and R. M. Lawn. "Characterization of the human factor VIII gene." *Nature*, 312: 326–330 (1984).

Myerowitz, R., and R. L. Proia. "cDNA clone for the α-chain of human β-hexosaminidase: deficiency of α-chain mRNA in Ashkenazi Tay-Sachs fibroblasts." *Proc. Natl. Acad. Sci. USA*, 81: 5394–5398 (1984).

SUBCHROMOSOMAL MAPPING OF GENES USING SOMATIC CELL HYBRIDS

Witkowski, J. A. "Somatic cell hybrids: a fusion of biochemistry, cell biology and genetics." *Trends Biochem. Sci.*, 11: 149–152 (1986). [Review]

Weiss, M. C., and H. Green. "Human-mouse hybrid cell lines containing partial complements of humans chromosomes and functioning human genes." *Proc. Natl. Acad. Sci. USA*, 58: 1104–1111 (1976).

O'Connell, P., D. Viskochil, A. M. Buchberg, J. Fountain, R. M. Cawthorn, M. Culver, J. Stevens, D. C. Rich, D. H. Ledbetter, M. Wallace, J. C. Carey, N. A. Jenkins, N. G. Copeland, F. S. Collins, and R. White. "The human homolog of murine *Evi-2* lies between two von Recklinghausen neurofibromatosis translocations." *Genomics*, 7: 547–554 (1990).

MAPPING CLONED GENES BY IN SITU HYBRIDIZATION TO CHROMOSOMES

Francke, U., B. T. Darras, N. F. Zander, and M. W. Kilimann. "Assignment of human genes for phosphorylase kinase subunits α (*PHKA*) to Xq12–q13 and β (*PHKB*) to 16q12–q13."*Am. J. Hum. Gen.*, 45: 276–282 (1989).

Fan, Y.-S., L. M. Davis, and T. B. Shows. "Mapping small DNA sequences by fluorescence *in situ* hybridization directly on banded metaphase chromosomes." *Proc. Natl. Acad. Sci. USA*, 87: 6223–6227 (1990).

Lichter, P., C.-J. C. Jang, K. Call, G. Hermanson, G. A. Evans, D. Housman, and D. C. Ward. "High-resolution mapping of human chromosome 11 by in situ hybridization with cosmid clones." *Science*, 247: 64–69 (1990).

Trask, B. J., H. Massa, S. Kenwrick, and J. Gitschier. "Mapping of human chromosome Xq28 by two-color fluorescence in situ hybridization of DNA sequences to interphase cell nuclei." *Am. J. Hum. Genet.*, 48: 1–15 (1991).

LOCATING DISEASE GENES USING CHROMOSOMAL ABNORMALITIES

Ledbetter, D. H., and Cavenee, W. K. "Molecular cytogenetics: interface of cytogenetics and monogenic disorders." In C. R. Scriver, A. L. Beaudet, W. S. Sly, and D. Valle, eds., *The Metabolic Basis of Inherited Disease.* McGraw-Hill, New York, 1989, pp. 343–371. [Review]

Friend, S. H., R. Bernards, S. Rogelj, R. A. Weinberg, J. M. Rapoport, D. M. Albert, and T. P. Dryja. "A human DNA segment with properties of the gene that predisposes to retinoblastoma and osteosarcoma." *Nature*, 323: 643–646 (1986).

Royer-Pokora, B., L. M. Kunkel, A. P. Monaco, S. C. Goff, P. E. Newburger, R. L. Baehner, F. S. Cole, J. T. Curnutte, and S. H. Orkin. "Cloning the gene for an inherited human disorder—chronic granulomatous disease—on the basis of its chromosomal location." *Nature*, 322: 32–38 (1986).

Call, K. M., T. Glaser, C. Y. Ito, A. J. Buckler, J. Pelletier, D. A. Haber, E. A. Rose, A. Kral , H. Yeger, W. H. Lewis, C. Jones, and D. E. Housman. "Isolation and characterization of a zinc finger polypeptide gene at the human chromosome 11 Wilms' tumor locus." *Cell*, 60: 509–520 (1990).

Cawthorn, R. M., R. Weiss, G. Xu, D. Viskochil, M. Culver, J. Stevens, M. Robertson, D. Dunn, R. Gesteland, P. O' Connell, and R. White. "A major segment of the neurofibromatosis type 1 gene: cDNA sequence, genomic structure, and point mutations." *Cell*, 62: 193–201 (1990).

Gessler, M., A. Poustka, W. Cavenee, R. L. Neve, S. H. Orkin, and G. A. P. Bruns. "Homozygous deletion in Wilms tumours of a zinc-finger gene identified by chromosome jumping." *Nature*, 343: 774–778 (1990).

Wallace, M. R., D. A. Marchuk, L. B. Andersen, R. Letcher, H. M. Odeh, A. M. Saulino, J. W. Fountain, A. Brereton, J. Nicholson, A. L. Mitchell, B. H. Brownstein, and F. S. Collins. "Type 1 neurofibromatosis gene: identification of a large transcript disrupted in three NF1 patients." *Science*, 249: 181–186 (1990).

LINKAGE ANALYSIS AND RESTRICTION FRAGMENT LENGTH POLYMORPHISMS

White, R., and J-M. Lalouel. "Chromosome mapping with DNA markers." *Sci. Am.*, 258: 40–48 (1988). [Review]

Ott, J. "Estimation of the recombination fraction in human pedigrees: efficient computation of the likehood for human linkage studies." *Am. J. Hum. Gen.*, 26: 588–597 (1974).

Kan, Y. W., M. S. Golbus, and A. M. Dozy. "Prenatal diagnosis of alpha-thalassemia: clinical application of molecular hybridization." *New Engl. J. Med.*, 295: 1165–1167 (1976).

Botstein, D., R. White, M. Skolnick, and R. Davis. "Construction of a genetic linkage map in man using restriction fragment length polymorphisms." *Am. J. Hum. Gen.* 32: 314–331 (1980).

Wyman, A. R., and R. White. "A highly polymorphic locus in human DNA." *Proc. Natl. Acad. Sci. USA*, 77: 6754–6758 (1980).

White, R., M. Leppert, T. Bishop, D. Barker, J. Berkowitz, C. Brown, P. Callahan, T. Holm, and L. Jerominski. "Construction of linkage maps with DNA markers for human chromosomes." *Nature, 313: 101–105 (1985).*

Nakamura, Y., M. Leppert, P. O'Connell, R. Wolff, T. Holm, M. Culver, C. Martin, E. Fujimoto, M. Hoff, E. Kumlin, and R. White. "Variable number of tandem repeat (VNTR) markers for human gene mapping." *Science,* 235: 1616–1622 (1987).

Dausset, J., H. Cann, D. Cohen, M. Lathrop, J.-M. Lalouel, and R. White. "Centre d'Etude du Polymorphisme Humain (CEPH): collaborative mapping of the human genome." *Genomics,* 6: 575–577 (1990).

CLONING THE GENE FOR DUCHENNE MUSCULAR DYSTROPHY

Emery, A. E. *Duchenne Muscular Dystrophy,* rev. ed. Oxford University Press, Oxford, Eng., 1988 [Review]

Witkowski, J. A. "Dystrophin-related muscular dystrophies." *J. Child Neurol.,* 4: 251–271 (1989). [Review]

Davies, K. E., B. D. Young, R. G. Elles, M. E. Hill, and R. Williamson. "Cloning of a representative genomic library of the human X-chromosome after sorting by flow cytometry." *Nature,* 293: 374–376 (1981).

Murray, J. M., K. E. Davies, P. S. Harper, L. Meredith, C. R. Mueller, and R. Williamson. "Linkage relationship of a cloned DNA sequence on the short arm of the X-chromosome to Duchenne muscular dystrophy." *Nature,* 300: 69–71 (1982).

Kunkel, L. M., A. P. Monaco, W. Middlesworth, H. D. Ochs, and S. A. Latt. "Specific cloning of DNA fragments absent from the DNA of a male patient with an X-chromosome deletion." *Proc. Natl. Acad. Sci. USA,* 82: 4778–4782 (1985).

Ray, P. N., B. Belfall, C. Duff, C. Logan, V. Kean, M. W. Thompson, J. E. Sylvester, J. L. Gorski, R. D. Schmickel, and R. G. Worton. "Cloning of the breakpoint of an X;21 translocation associated with Duchenne muscular dystrophy." *Nature,* 318: 672–675 (1985).

Koenig, M., E. P. Hoffman, C. J. Bertelson, A. P. Monaco, C. Feener, and L. M. Kunkel. "Complete cloning of the Duchenne muscular dystrophy (DMD) cDNA and preliminary genomic organization of the DMD gene in normal and affected individuals." *Cell,* 50: 509–517 (1987).

CLONING THE GENE FOR CYSTIC FIBROSIS

Knowlton, R. G., O. Cohen-Haguenauer, N. V. Cong, J. Frézal, V. A. Brown, D. Barker, J. C. Braman, J. W. Schumm, L-C. Tsui, M. Buchwald, and H. Donis-Keller. "A polymorphic DNA marker linked to cystic fibrosis is located on chromosome 7." *Nature,* 318: 380–382 (1985).

Wainwright, B. J., P. J. Scambler, J. Schmidtke, E. A. Watson, H-Y. Law, M. Farrall, H. J. Cooke, H. Eiberg, and R. Williamson. "Localization of cystic fibrosis locus to human chromosome 7 cen-q22." *Nature,* 318: 384–385 (1985).

White, R., S. Woodward, M. Leppert, P. O'Connell, M. Hoff, J. Herbst, J-M. Lalouel, M. Dean, and G. Vande Woude. "A closely linked genetic marker for cystic fibrosis." *Nature,* 318: 382–384 (1985).

Kerem, B.-S., J. M. Rommens, J. A. Buchanan, D. Markiewicz, T. K. Cox, A. Chakravarti, M. Buchwald, and L.-C. Tsui. "Identification of the cystic fibrosis gene: genetic analysis." *Science,* 245: 1073–1080 (1989).

Riordan, J. R., J. M. Rommens, B.-S. Kerem, N. Alon, R. Rozmahel, Z. Grzelczak, J. Zielenski, S. Lok, N. Plavsic, J.-L. Chou, M. L. Drumm, M. C. Iannuzzi, F. S. Collins, and L.-C. Tsui. "Identification of the cystic fibrosis gene: cloning and characterization of complementary DNA." *Science,* 245: 1066–1073 (1989).

Rommens J. M., and 14 coauthors. "Identification of the cystic fibrosis gene: chromosome walking and jumping." *Science,* 245: 1059–1065 (1989).

CLUES ABOUT PROTEIN FUNCTION FROM SEQUENCES OF CLONED DISEASE GENES

Smith, T. F. "The history of the genetic sequence databases." *Genomics,* 6: 701–707 (1990). [Review]

Koenig, M., A. P. Monaco, and L. M. Kunkel. "The complete sequence of dystrophin predicts a rod-shaped cytoskeletal protein." *Cell,* 53: 219–228 (1988).

Xu, G., P. O'Connell, D. Viskochil, R. Cawthorn, M. Robertson, M. Culver, D. Dunn, J. Stevens, R. Gesteland, R. White, and R. Weiss. "The neurofibromatosis type 1 gene encodes a protein related to GAP." *Cell,* 62: 599–608 (1990).

Ervasti, J. M., and K. P. Campbell. "Membrane organization of the dystrophin-glycoprotein complex." *Cell,* 66: 1121–1131 (1991).

USING CLONED GENES TO STUDY PROTEIN EXPRESSION AND FUNCTION

Hoffman, E. P., K. H. Fischbeck, R. H. Brown, M. Johnson, R. Medori, J. D. Loike, J. B. Harris, R. Waterston, M. Brooke, L. Specht, W. Kupsky, J. Chamberlain, C. T. Caskey, F. Shapiro, and L. M. Kunkel. "Characterization of dystrophin in muscle-biopsy specimens from patients with Duchenne's or Becker's muscular dystrophy." *N. E. J. Med.,* 318: 1363–1368 (1988).

Zubrzycka-Gaarn, E. E., D. E. Bulman, G. Karpati, A. H. M. Burghes, B. Belfall, H. J. Klamut, J. Talbott, R. S. Hodges, P. N. Ray, and R. G. Worton. "The Duchenne muscular dystrophy gene product is localized in sarcolemma of human skeletal muscle." *Nature,* 333: 466–469 (1988).

Drumm, M. L., H. A. Pope, W. H. Cliff, J. M. Rommens, S. A. Marvin, L.-C. Tsui, F. S. Collins, R. A. Frizzell, and J. M. Wilson. "Correction of the cystic fibrosis defect in vitro by retrovirus-mediated gene transfer." *Cell,* 62: 1227–1233 (1990).

Rich, D. P., M. P. Anderson, R. J. Gregory, S. H. Cheng, S. Paul, D. M. Jefferson, J. D. McCann, K. W. Klinger, A. E. Smith, and M. J. Welsh. "Expression of the cystic fibrosis transmembrane conductance regulator corrects defective chloride channel regulation in cystic fibrosis airway epithelial cells." *Nature,* 347: 358–363 (1990).

CLONING GENES FOR COMPLEX, POLYGENIC DISORDERS

Egeland, J. A., D. S. Gerhard, D. L. Pauls, J. N. Sussex, K. K. Kidd, C. R. Allen, A. M. Hostetter, and D. E. Housman. "Bipolar affective disorders linked to DNA markers on chromosome 11." *Nature,* 325: 783–787 (1987).

Lander, E. S. "Splitting schizophrenia." *Nature,* 336: 105–106 (1988).

Sherrington, R., J. Brynjolfsson, H. Petursson, M. Potter, K. Dudleston, B. Barraclough, J. Wasmuth, M. Dobbs, and H. Gurling. "Localization of a susceptibility locus for schizophrenia on chromosome 5." *Nature,* 336: 164–167 (1988).

Kelsoe, J. R., E. I. Ginns, J. A. Egeland, D. S. Gerhard, A. M. Goldstein, S. J. Bale, D. L. Pauls, R. T. Long, K. K. Kidd, G. Conte, D. E. Housman, and S. J. Paul. "Reevaluation of the linkage relationship between chromosome 11p loci and the gene for bipolar affective disorder in the Old Order Amish." *Nature,* 342: 238–243 (1989).

Kennedy, J. L., L. A. Giuffra, H. W. Moises, L. L. Cavalli-Sforza, A. J. Pakstis, J. R. Kidd, C. M. Castiglione, B. Sjogren, L. Wettenberg, and K. K. Kidd. "Evidence against linkage of schizophrenia to markers on chromosome 5 in a northern Swedish pedigree." *Nature,* 336: 167–170 (1989).

CANDIDATE GENES

Brown, M. S., and J. L. Goldstein. "A receptor-mediated pathway for cholesterol homeostasis." *Science,* 232: 34–47 (1986).

Blum, K. B., E. P. Noble, P. J. Sheridan, A. Montgomery, T. Ritchie, P. Jagadeeswaran, H. Nogami, A. H. Briggs, and J. B. Cohn. "Allelic association of human dopamine D_2 receptor gene in alcoholism." *J.A.M.A.,* 263: 2055–2060 (1990).

Bolos, A. M., M. Dean, S. Lucas-Derse, M. Ramsburg, G. L. Brown, and D. Goldman. "Population and pedigree studies reveal a lack of association between the dopamine D_2 receptor gene and alcoholism." *JAMA,* 264: 3156–3160 (1990).

Dietz, H. C., G. R. Cutting, R. E. Pyeritz, C. L. Maslen, L. Y. Sakai, G. M. Corson, E. G. Puffenberger, A. Hamosh, E. J. Nanthakumar, S. M. Curristin, G. Stetten, D. A. Meyers, and C. A. Francomano. "Marfan syndrome caused by a recurrent *de novo* missense mutation in the fibrillin gene." *Nature,* 352: 337–339 (1991).

Goate, A., M.-C. Chartier-Harlin, M. Mullan, J. Brown, F. Crawford, L. Fidani, L. Giuffra, A. Haynes, N. Irving, L. James, R. Mant, P. Newton, K. Rooke, P. Roques, C. Talbot, M. Pericak-Vance, A. Roses, R. Williamson, M. Rossor, M. Owen, and J. Hardy. "Segregation of a missense mutation in the amyloid precursor protein gene with familial Alzheimer's disease." *Nature,* 349: 704–706 (1991).

McKusick, V. A. "The defect in Marfan syndrome." *Nature,* 352: 279–281 (1991).

FUTURE PROGRESS IN CLONING HUMAN DISEASE GENES

U.S. Congress, Office of Technology Assessment. *Mapping Our Genes—The Genome Projects: How Big, How Fast?* Washington, D.C., U.S. Government Printing Office, 1988. [Review]

McKusick, V. A. "The Human Genome Organisation: history, purposes and membership." *Genomics,* 5: 385–387 (1989).

U.S. Department of Health and Human Services and U.S. Department of Energy. *Understanding Our Genetic Inheritance: The U.S. Human Genome Project.* Washington, D.C., U.S. Government Printing Office, 1990.

Pritchard, C., D. R. Cox, and R. M. Myers. "The end in sight for Huntington disease?" *Am. J. Hum. Genet.* 49: 1–6 (1991).

27

DNA-Based Diagnosis of Genetic Diseases

When the first edition of this book was published, applications of recombinant DNA techniques in human clinical genetics had only just begun, and DNA-based diagnosis was available for very few diseases. The situation has changed out of all recognition in the intervening years. The techniques described in the previous chapter have produced a wealth of cloned genes, RFLPs, and other markers that are being used now on a routine basis in an ever-increasing number of medical genetics laboratories. DNA-based diagnosis has many advantages. If the mutation causing a disease is known, the gene can be analyzed to determine if the mutation is present. Even if the gene has not been cloned, linkage analysis using RFLPs for closely linked probes provides invaluable data for genetic counseling, and DNA-based diagnosis can detect asymptomatic heterozygotes. Because of these factors, DNA-based tests have become the diagnostic methods of choice for inherited diseases, and over 200 genetic disorders can be diagnosed using recombinant DNA techniques. These techniques have also been used for other purposes, for example to identify individuals for forensic purposes. In this chapter, we will review the methods used to perform DNA-based diagnosis of inherited diseases, as well as the applications of these techniques in forensic science. We will also discuss some potential problems that may arise as a consequence of the very rapid applications of DNA technology in human biology.

Biochemical Markers Used for Early Diagnosis

Because inherited genetic disorders cause great suffering to the patients and their families, much effort has gone into developing methods for prenatal diagnosis of affected fetuses. When both parents are known to be heterozygous carriers—and all too often this discovery is made only after the birth of an affected child—further pregnancies can be monitored closely to determine if the fetus is affected. With this forewarning, therapies can be instituted soon after birth for disorders like phenylketonuria, where a phenylalanine-free diet is essential to promote normal development of the child. Alternatively the parents may choose the option of abortion if they feel that the burden of the disease on the child and the family will be too great to bear. Prior to the development of DNA-based diagnosis, prenatal diagnosis was limited to those genetic disorders for which tests for the enzymes or other relevant proteins were available (Table 27-1). More than 45 of these recessive genetic diseases can be diagnosed in this way.

The basic approach to prenatal diagnosis is to obtain samples of fetal tissues and to test them for biochemical defects or for chromosomal abnormalities. Some metabolites can be measured directly in the amniotic fluid, but for most diagnoses, fetal cells are required. For some disorders, like the hemoglobinopathies,

where the diagnosis can be made by using blood cells, samples of fetal blood can be collected from the umbilical cord. Another source of cells is the amniotic fluid. Cells that have been shed from the fetus into the amniotic fluid can be collected by centrifugation (Figure 27-1). Sometimes there are sufficient cells for direct analysis but it is usually necessary to culture the collected cells to get enough material for testing. Chorionic villus sampling, a relatively new technique, is performed between the eighth and twelfth weeks of gestation (Figure 27-1). The chorionic villi are part of the fetal component of the placenta, and about 30 mg of chorionic villi can be removed with little risk to the fetus. Cytogenetic analysis can be performed directly on this tissue, and as much as 3 μg of DNA can be isolated directly from each milligram of chorionic villus.

Direct diagnosis of metabolic disorders has had limited application. First, the enzyme or protein may be expressed only in specific tissues and therefore not detectable in amniotic fluid or in amniotic or chorionic villus cells. Second, culturing cells is a time-consuming business and fraught with difficulties; amniotic cells are not easy to grow, and cultures may become contaminated or die before there are sufficient cells for analysis. Third, testing has to be done for the abnormal gene product, so the opportunities for diagnosis are limited because relatively few of these products are known for human inherited disorders. DNA techniques circumvent some of these problems, and they have brought about a revolution in genetic diagnosis.

Mutations in Globin Genes Cause the Thalassemias

This revolution is exemplified by our detailed knowledge of the α and β-globin genes. The organization of the α-globin gene cluster on chromosome 16 and that of the β-globin cluster on chromosome 11 had been analyzed using hemoglobin from patients with mutations in these genes. But recombinant DNA techniques have provided a wealth of information about the molecular basis of these mutations, and in so doing have shown that human diseases can be genetically very heterogeneous. The thalassemias are diseases caused by abnormal synthesis of the globin chains.

TABLE **27-1**

Examples of Prenatal Diagnosis Based on Biochemical Markers

DISORDER	ENZYME ASSAY
Citrullinemia	Argininosuccinate synthetase
Gaucher's disease	Glucocerebrosidase
Hunter's syndrome	Iduronate sulfatase
Lesch-Nyhan syndrome	Hypoxanthine phosphoribosyltransferase
Metachromatic leukodystrophy	Arylsulfatase A
Phenylketonuria	Phenylalanine hydroxylase
Pompe's disease	α-1,4-Glucosidase
Sanfilippo B	α-N-Acetylglucosaminidase
Tay-Sachs disease	β-Hexosaminidase

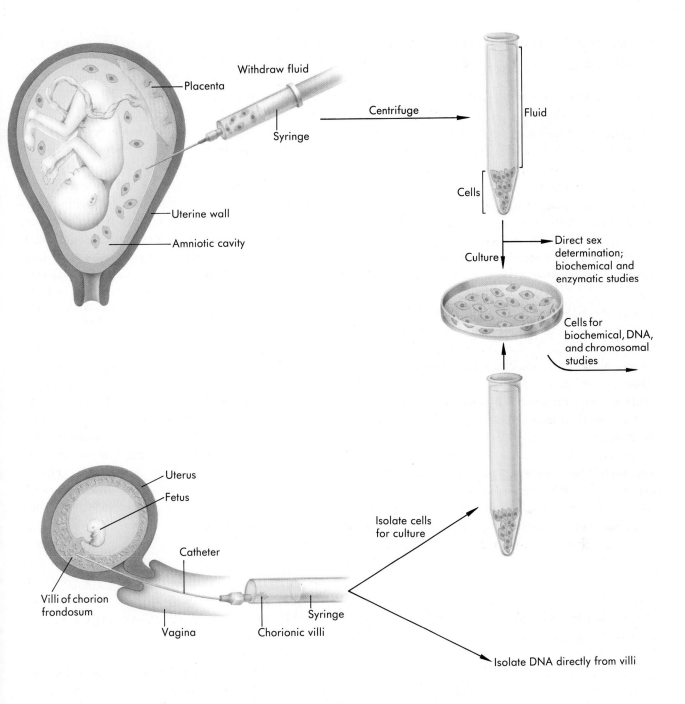

FIGURE **27-1**

Amniocentesis and chorionic villus sampling. (a) A sample of amniotic fluid (mostly fetal urine and other secretions) is taken by inserting a needle into the amniotic cavity during or around the sixteenth week of gestation. The fetal cells are separated from the fluid by centrifugation. The cells can be used immediately, or more usually they are cultured so that a number of biochemical, enzymatic, and chromosomal analyses can be made. The cultured cells can also be a source of DNA. (b) Chorionic villus sampling is performed between the eighth and twelfth weeks of gestation. A catheter is introduced through the vagina or transabdominally, and a small sample of chorionic villi is drawn into the syringe. DNA can be isolated directly from the tissue, or cell cultures can be established. Note that the various elements of this figure are not drawn to scale.

TABLE **27-2**

Examples of β-Thalassemias Mutations

MUTATION TYPE*	PHENOTYPE	ORIGIN
Nonsense mutations (▲)		
1 Codon 17 (A → T)	β^0	Chinese
2 Codon 39 (C → T)	β^0	Mediterranean
3 Codon 121 (G → T)	β^0	Polish
Frameshift mutations (△)		
4 −1 at codon 6	β^0	Mediterranean
5 −1 at codon 16	β^0	Indian
6 +1 at codon 71, 72	β^0	Chinese
Splice junction changes (↓)		
7 IVS-1, position 1 (G → A)	β^0	Mediterranean
8 IVS-1, 3′ end −25 bp	β^0	Indian
9 IVS-2, position 1 (G → A)	β^0	Mediterranean
10 IVS-2, 3′ end (A → G)	β^0	Black
Consensus changes (∗)		
11 IVS-1, position 5 (G → T)	β^+	Mediterranean
12 IVS-1, position 5 (G → C)	β^+	Mediterranean
13 IVS-1, position 6 (T → C)	β^+	Mediterranean
Internal change in IVS (#)		
14 IVS-1, position 110 (G → A)	β^+	Mediterranean
15 IVS-2, position 654 (C → T)	β^0	Chinese
16 IVS-2, position 745 (C → G)	β^+	Mediterranean
Internal change in exon (❙)		
17 Codon 24 (T → A)	β^+	Black
18 Codon 26 (G → A)	β^E	Southeast Asian
19 Codon 27 (G → T)	$\beta^{Knossos}$	Mediterranean
Promoter mutations (relative to start site) (●)		
20 Position −88 (C → T)	β^+	Black
21 Position −31 (A → G)	β^+	Japanese
22 Position −28 (A → C)	β^+	Kurdish
RNA-cleavage mutation/polyadenylation (○)		
23 AATAAA → AACAAA	β^+	Black
Cap site mutants (∨)		
24 Position +1 A → C	β^+	Asian Indian

* The numbers and symbols in this column refer to Figure 27-2.

These mutations cause abnormalities in transcription, RNA splicing, and mRNA stability and translation, resulting from point mutations and deletions. The β-thalassemias are characterized by a low level or total absence of β-globin, and over 50 different mutations have been found in the β-globin gene. Some of these are listed in Table 27-2, and their positions in the β-globin gene are shown in Figure 27-2.

β^0-Thalassemia is characterized by a complete absence of β-globin synthesis and is caused both by point mutations that give rise to translation termination codons and by small deletions or insertions in the β-globin coding sequence that cause a shift in the reading frame. The β^+-thalassemias are those conditions in

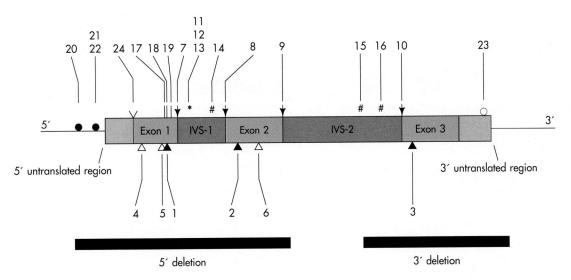

FIGURE **27-2**

Sites of some of the mutations found in the human β-globin gene. The different symbols and numbers correspond to the entries in Table 27-2. The exons, 5′ and 3′ untranslated regions, and introns (IVS-1 and IVS-2) are indicated. The location of the large deletions detected in two patients are indicated by the bars at the bottom. (After Orkin, 1987)

which some β-chains are made. Some $β^+$-thalassemias are due to mutations within the promoter region. These mutations affect transcription and result in levels of β-globin mRNA that are approximately 20 percent of normal levels.

Perhaps the most interesting β-thalassemias that have been discovered are those caused by abnormal RNA processing (Figure 27-3). Abnormal globin splicing has already been mentioned in Chapter 8. The human β-globin gene has two introns that must be spliced out. $β^0$-thalassemias have been described in which mutations at the 5′ donor splice junction at the beginning of either the first or the second intron in the mRNA prevent splicing at the correct site. Instead, other donorlike sequences (*cryptic sites*) in the RNA are used; this substitution leads to an incorrectly spliced (and untranslatable) mRNA. There are also mutations affecting the 3′ acceptor site for the second intron. Once again a cryptic site within intron 2 is used, so the mRNA contains part of the 3′ portion of intron 2. Where there are changes in the consensus sequences on either side of the conserved dinucleo-

tides of the splice sites, both the normal and cryptic sites are used and a $β^+$-thalassemia results. Mutations can occur *within* an intron to create a new 3′ (acceptor) splice site and within exons to create a new 5′ (donor) splice site. In both cases, the new splice sites can compete with the normal sites.

Detailed analyses have also been made of the Factor VIII (hemophilia) and hypoxanthine phosphoribosyltransferase (Lesch-Nyhan) genes, and a similar variety of mutations has been detected. In the case of hemophilia, a unique mutation was discovered. Two unrelated patients were found to have insertions in exon 14 of the hemophilia gene, presumably disrupting normal functioning of the gene (Figure 27-4). When the inserted DNA was cloned and sequenced, it was found to be derived from the so-called L1 elements. There are about 10^5 copies of these elements integrated throughout the human genome, and they are thought to be nonviral retrotransposons. The situation in these two patients appears to be analogous to the insertional mutagenesis caused by the integration of retroviruses (Chapter 10).

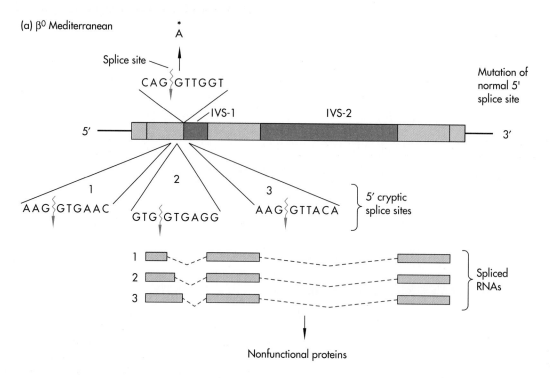

(a) β⁰ Mediterranean

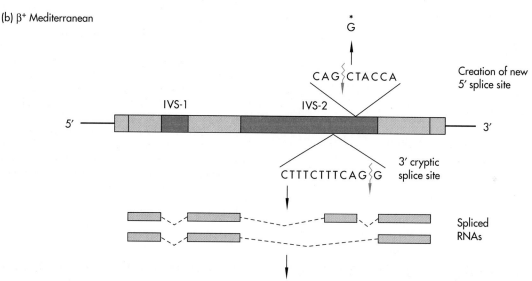

(b) β⁺ Mediterranean

FIGURE 27-3

Splicing mutations in the human β-globin gene. (a) β⁰ Mediterranean is caused by a G to A mutation at the first nucleotide of the first intron (IVS-1). (The exon-intron boundary is shown by the jagged arrow.) This change destroys the correct 5′ splicing site, but three other sequences (cryptic splice sites 1, 2, and 3) that do not normally function as splice sites are used instead. Splicing at these cryptic sites leads to the production of three β-globin mRNAs, none of which produces a functional protein. (b) β⁺ Mediterranean is caused by a C to G mutation at position 745 within the second intron (IVS-2). This change creates within the intron a new 5′ splice site, which splices to the correct 3′ splice acceptor at the beginning of the third exon. In addition, a cryptic 3′ splice acceptor within IVS-2 is used with the correct 5′ splice donor sequence at the end of the second exon. The consequence of using these new splice sites is that the mutant β-globin includes a new "exon" derived from sequences within IVS-2 that are normally removed by splicing. However, a small amount of correctly spliced β-globin mRNA is made, so the patient has the β⁺ rather than β⁰ phenotype. (Modified From Watson, J. D., N. H. Hopkins, J. W. Roberts, J. A. Steitz, and A. M. Weiner, *Molecular Biology of the Gene,* 4th ed. Benjamin-Cummings, Menlo Park, Calif., 1987)

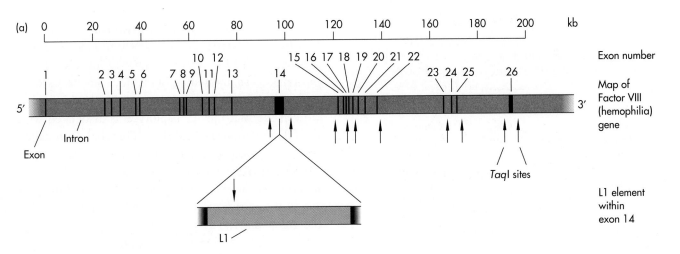

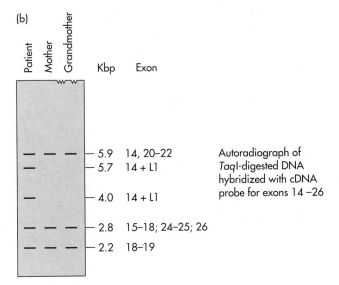

Autoradiograph of
*Taq*I-digested DNA
hybridized with cDNA
probe for exons 14 –26

FIGURE **27-4**
The mutation in the hemophilia (Factor VIII) gene caused by the insertion of an L1 element. (a) A map of the hemophilia gene, showing the locations of its exons (numbered 1–26) and *Taq*I sites (arrows). The L1 element lies within exon 14 and contains one *Taq*I site (arrow). (b) Digestion of normal DNAs (from mother and grandmother) with *Taq*I, followed by hybridization with a cDNA probe for exons 14 to 26 reveals three fragments from the hemophilia gene; their sizes are 5.9, 2.8, and 2.2 kbp. The band at 5.9 kbp actually contains two fragments of similar size, one including exon 14 and one covering exons 20–22. The patient's DNA has the latter fragment, but the fragment containing exon 14 is cut into two fragments because the L1 element inserted in his hemophilia gene contains a *Taq*I site. The L1 element increases the size of this exon 14 fragment from 5.9 to 9.7 kbp, and its two fragments are 5.7 and 4.0 kbp. (After Kazazian et al, 1988)

RFLPs and Linkage Analysis Are Used for Diagnosis

In the cases we have just discussed, the genes responsible for the diseases have been cloned, and we know now what mutations to look for in patients. But what of the thousands of human genetic disorders where the gene has not yet been cloned? Can recombinant DNA techniques still provide the medical geneticist with useful tools? Fortunately, the answer is yes. In the course of cloning a gene, affected families are analyzed by using RFLPs to determine how close a probe is to the disease locus (Chapter 26). If the RFLP is close to the locus, then the probe can be used to trace the inheritance of a mutation in an affected family and so determine which members of the family are carrying the mutation (Table 27-3). Of course, the reliability of the conclusions derived from linkage analysis depends on how tightly the RFLP and disease locus are linked. Five centimorgars (5 cM) is regarded as a minimum linkage for diagnostic purposes. A 5-cM linkage means that we can be 95 percent confident of our conclusions; that is, there is only a 1-in-20 possibility that a crossover between the RFLP and the disease locus has occurred. These probabilities based on linkage can be improved by incorporating information from other sources. For example, levels of serum creatine kinase (CK) are elevated in 95 percent of carriers of Duchenne muscular dystrophy (DMD) but in only 5 percent of noncarriers. Statistical calculations based on Bayesian logic can be used to reach an overall risk value by combining the probability of

TABLE **27-3**

Applications of RFLP Linkage Analysis That Use Gene Probes or Anonymous Probes

DISEASE	SOURCE OF PROBE
α_1-Antitrypsin*	Gene
Duchenne muscular dystrophy*	Gene
Factor X deficiency*	Gene
Friedreich's ataxia	Anonymous
Hemophilia*	Gene
Huntington's disease	Anonymous
Myotonic dystrophy	Anonymous
Phenylketonuria*	Gene

* Mutations in these genes are usually detected by direct analysis of the gene.

a woman being a carrier or a noncarrier based on her serum CK levels with the probability based on RFLP linkage analysis.

An example of an RFLP linkage analysis for Huntington's disease (HD) is shown in Figure 27-5. The DS410 locus is estimated to be 4 cM from the HD

locus, and the G8 probe from that locus detects four RFLPs with *Hin*dIII and one each with *Eco*RI and *Bgl*I. These six RFLPs define a *haplotype* for each chromosome of an individual, and they can be used to follow the inheritance of chromosome 4. In this family, we can see that the A2 haplotype is associated with the HD mutation (I-1 and II-2). This analysis is called *setting the phase*. The individual at risk (III-1) must have inherited the non-HD chromosome with the A1 haplotype from the affected parent (II-2), and the person is at low risk of being affected by HD. Note that the A2 haplotype is not always associated with the HD mutation. Individual II-3, who married into this HD family, has the A2 haplotype and the normal gene.

Diagnostic laboratories have used as many as ten polymorphic loci, detected by using six enzymes and eight probes, for the genetic diagnosis of DMD (Table 27-4). (One probe can be used to detect more than one RFLP.) The probes from within the gene (XJ1.1 and pERT series) show 5-cM recombination with DMD; the flanking markers (754 and C7) are at 10-cM recombination. The accuracy of diagnosis varies from 95 percent for a single RFLP within the gene, to better than 99 percent when both flanking markers are also informative. A typical analysis is shown in Figure 27-6. The haplotype of the affected boy shows which of his mother's X chromosomes is carrying the mutated gene. The male fetus has inherited an X chromosome like that of his affected brother, and in this case we can be more than 97 percent confident that the fetus is carrying the DMD mutation.

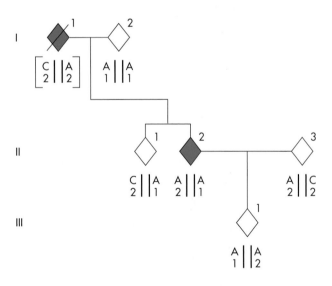

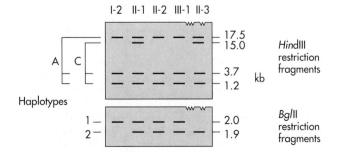

FIGURE **27-5**

RFLP linkage analysis for Huntington's disease by use of the G8 probe to analyze RFLPs for *Hin*dIII and *Bgl*II at the DS410 locus. (Symbols of affected individuals are blue.) The affected grandparent, I-1, is dead but the haplotypes of I-2, II-1, and II-2 can be used to infer the haplotype of I-1. I-2 is homozygous for A and 1 and must have contributed the A1 chromosome to II-1 and II-2. The latters' other chromosomes, C2 and A2, respectively, must have come from I-1. Because II-1 is not affected by HD, the chromosome with the C2 haplotype must be normal, so it is II-2's chromosome with the A2 haplotype that carries the HD mutation in this family. Because II-3 is homozygous for the *Bgl*I haplotype 2, III-1's A2 chromosome must have come from II-3. This means that III-1 inherited the normal chromosome with the A1 haplotype from the affected parent II-2. III-1 is at low risk (<5 percent) of carrying the HD mutation. (Note that because of the sensitivity of a HD diagnosis, it is customary to use nonstandard symbols to reduce the possibility of identifying families.) (After Meissen et al., 1988)

can be resolved by using cDNA probes to examine the DMD gene itself (see below).

Linkage Disequilibrium Can Be Used for Diagnosis

There is always a degree of uncertainty in a diagnosis based on linkage analysis because of the possibility of recombination. In some cases, however, the degree of

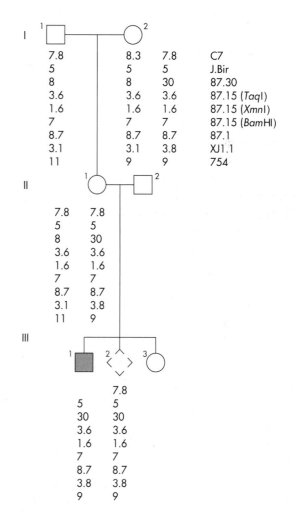

FIGURE **27-6**

Prenatal diagnosis of Duchenne muscular dystrophy. The mother (II-1) of the affected boy (III-1) is heterozygous for three loci (8/30; 3.1/3.8; 11/9), and these are used to identify her X chromosomes. Her affected son (III-1) has inherited her chromosome with the 30-, 3.8-, and 9-kbp alleles, and it is this chromosome that is carrying the DMD mutation. Her male fetus (III-2) has inherited the same maternal haplotype as his affected brother. There is a small chance that a recombination has occurred that has not been detected, but with these informative alleles, there is a greater than 97 percent probability that the fetus will be affected.

TABLE **27-4**

RFLPs Used Routinely in Diagnosing Duchenne Muscular Dystrophy

PROBE	RESTRICTION ENDONUCLEASE	ALLELES (kbp)
C7	*EcoRV*	8.3/7.8
J.Bir	*Bam*HI	21/5
pERT 87.30	*Bgl*II	30/8
pERT 87.15	*Taq*I	3.6/3.4
	*Xmn*I	2.8/1.6
	*Bam*HI	10/7
pERT 87.1	*Xmn*I	8.7/7.5
XJ1.1	*Taq*I	3.8/3.1
754	*Pst*I	11/9

Errors do occur in prenatal diagnosis based on RFLPs because the mutant gene and the marker haplotype can become uncoupled by undetected crossovers (Chapter 26). In the example shown in Figure 27-7, the mother was heterozygous for 5 RFLPs within the gene. Her male fetus inherited a chromosome with the haplotype of the maternal grandfather's X chromosome, so there was a better than 95 percent probability that the fetus would be unaffected. (This grandfather did not have DMD, so his X chromosome must be carrying a normal dystrophin gene.) However, laboratory tests shortly after the birth showed that the boy was affected by DMD. A crossover undetectable with these RFLPs must have occurred between the mother and son, so that the DMD mutation was transferred to what had been the normal chromosome.

Linkage analysis has to be performed within families, but not all families have the members needed for accurate analysis. Three-generation families including an affected grandparent and an affected parent are desirable for analysis in the case of Huntington's disease. However, in this disorder the affected grandparent will almost certainly have died before a definitive diagnosis is made in his or her children (the parents). It has been estimated that linkage analysis may not be appropriate for as many as 40 percent of those who are at risk of HD. For DMD, linkage analysis is unsatisfactory in so-called isolated cases, where there is a single affected boy in a family with no previous history of the disorder. A fetus inheriting the same haplotype as the single affected boy will be at intermediate risk (approximately 39 percent) of being affected. Fortunately, many of these ambiguous cases

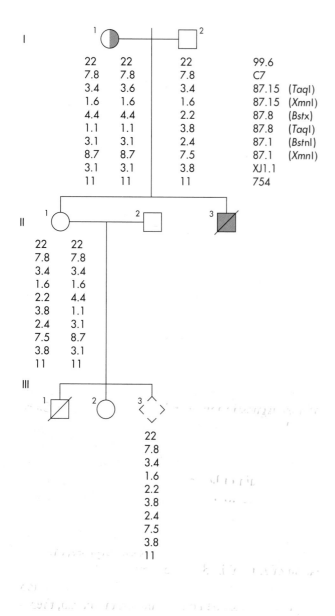

I				
	22	22	22	99.6
	7.8	7.8	7.8	C7
	3.4	3.6	3.4	87.15 (TaqI)
	1.6	1.6	1.6	87.15 (XmnI)
	4.4	4.4	2.2	87.8 (Bstx)
	1.1	1.1	3.8	87.8 (TaqI)
	3.1	3.1	2.4	87.1 (BstnI)
	8.7	8.7	7.5	87.1 (XmnI)
	3.1	3.1	3.8	XJ1.1
	11	11	11	754

II

22	22
7.8	7.8
3.4	3.4
1.6	1.6
2.2	4.4
3.8	1.1
2.4	3.1
7.5	8.7
3.8	3.1
11	11

III

22
7.8
3.4
1.6
2.2
3.8
2.4
7.5
3.8
11

FIGURE **27-7**
Recombination leading to a misdiagnosis. The mother II-1 was at high risk of being a carrier. Her affected brother (II-3) and her unaffected son (III-1) had both died, so the haplotypes of the X chromosomes carrying the normal and mutant dystrophin genes could not be determined directly. However, the grandfather's haplotype was determined, and it seemed that the X chromosome inherited by the fetus (III-3) from his mother carried the grandfather's haplotype. The grandfather of course was unaffected, and so the fetus was given a 95% probability of being unaffected. When the boy had been born, the level of creatine kinase in his serum was measured. Creatine kinase is an enzyme present in high concentrations in muscle and in the serum of patients with muscle diseases, presumably because it leaks out of damaged muscles. In this case the boy had a very high level of serum creatine kinase, indicating that he was affected by DMD. There must have been a crossover between the mother's X chromosomes that could not be detected because the mother's haplotype is uninformative for the region between the RFLPs for probe 87.15 and the flanking probes C7 and 99.6 that lie beyond the 3′ end of the gene. We now know that the distance between the 87.15 locus and the 3′ end of the dystrophin gene is about 1200 kbp.

uncertainty may be reduced if a specific haplotype (comprising a single allele of a polymorphic marker or a set of alleles of various markers) is almost always associated with a mutation in a gene. The allele and the mutation are then said to be in *linkage disequilibrium.* This situation is thought to arise because the mutation first occurred in an individual with a particular haplotype, and the mutation and the haplotype have remained associated in the population derived from that individual.

One of the best-known examples of the use of linkage disequilibrium for diagnosis is that reported by Kan and Dozy in their classic study of sickle-cell anemia published in 1978. There is an RFLP associated with the β-globin locus that can be detected

by using the enzyme *Hpa*I and a β-globin probe. Homozygous individuals have either a 7.6-kbp or 13.0-kbp fragment, and both fragments are detected in heterozygotes. The 13.0-kbp fragment was found in 70 percent of sickle-cell anemia patients but in only 3 percent of normal individuals. The association is not absolute, but within the defined population, it is very useful for diagnostic purposes. Subsequently, other examples of linkage disequilibrium have been discovered. For cystic fibrosis (CF) in a European population, 75 percent of chromosomes carrying the CF mutation are associated with one haplotype (Table 27-5). Twelve different haplotypes can be recognized in the phenylalanine hydroxylase gene in the Danish population, but 90 percent of mutations causing phenylketonuria are associated with just four of these RFLP haplotypes.

Exon Deletions Are Used for Direct Diagnosis of DMD

Genetic methods like linkage analysis using RFLPs are very useful, but as we saw in the case of DMD, crossovers can lead to diagnostic errors (Figure 27-7). The most certain way of performing DNA-based diagnosis is to determine physically that the gene has a

TABLE **27-5**

Linkage Disequilibrium in Cystic Fibrosis

DESIGNATION	HAPLOTYPE			CF CHROMOSOMES		NORMAL CHROMOSOMES	
	XV-2c	KM.19	MP6d-9	NO.	%	NO.	%
A	−	−	−	9	7	28	21
B	+	−	−	7	5	55	41
C	−	+	−	1	1	4	3
D	−	+	+	101	75	20	15
E	−	−	+	2	1.5	7	5
F	+	−	+	2	1.5	3	2
G	+	+	+	12	9	17	13

DNA samples were taken from the members of eighty European families with cystic fibrosis (CF). The samples were analyzed for three RFLPs: XV-2c (*Taq*I), KM.19 (*Pst*I), and MP6d-9 (*Msp*I). The + and − refer to the presence and absence of the restriction sites. Three RFLPs give rise to seven possible combinations (haplotypes). Haplotype D has a high probability of being associated with CF.

mutation. Direct analysis for mutations in at-risk individuals is very powerful, as became clear when cDNA probes became available for DMD diagnosis. The exons of the gene can be scanned for deletions or other abnormalities by cutting genomic DNA with an enzyme that does not cut within the exons and then probing with labeled cDNA (Figure 26-12). While the first genomic probes for DMD detected deletions in only 10 percent of patients, about 70 percent of DMD patients have deletions or duplications that can be detected using cDNA probes. (Figure 27-8). These structural changes can be easily recognized in the DNA of an at-risk fetus, and careful densitometry of autoradiographs can reveal the presence of only half-quantities of deleted exons in carrier mothers. These deletions are now detected by using the polymerase chain reaction (see later).

Gene Mutations Can Alter a Restriction Enzyme Site: Direct Diagnosis of Sickle-Cell Anemia

Another very useful but unfortunately rare way of detecting mutations directly is to use a restriction enzyme if the gene mutation alters the recognition site for that enzyme. One of the best known examples of this diagnostic approach is sickle-cell anemia. Sickle-cell anemia results from a mutation that changes a glutamic acid residue (coded by the triplet GAG), for a valine residue (coded by GTG) at position 6 in the β-globin chain of hemoglobin. (Note that by convention we refer to the sequence of the DNA strand corresponding to the mRNA, and not to the DNA strand that is used as the template to synthesize the mRNA). Fortunately, for diagnostic purposes, the mutation of A to T in the base sequence of the β-globin gene eliminates the site for a number of restriction enzymes. The sickle hemoglobin mutation can, therefore, be detected by digesting sickle-cell and normal DNA with a restriction enzyme that cuts the normal sequence, and performing Southern blot hybridization with a labeled β-globin probe (Figure 27-9). The restriction enzyme *Mst*II cuts at the sequence CCTNAGG (where N is any nucleotide) and generates a 1.1-kb β-globin gene fragment from normal DNA. This *Mst*II site is destroyed by the A to T change in sickle-cell DNA, and the 1.3-kb fragment produced instead is readily identifiable on a gel. Unlike a linkage study, this is a direct test for the mutation itself, and DNA samples from other family members are not needed. Unfortunately, mutations that both cause a disease and alter a restriction endonuclease recognition site are uncommon.

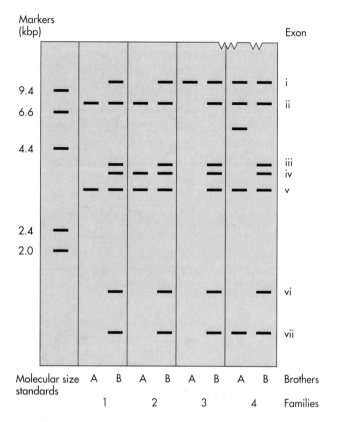

Markers
(kbp)

Exon

9.4
6.6

4.4

i
ii

iii
iv
v

2.4
2.0

vi

vii

Molecular size A B A B A B A B Brothers
standards
        1   2   3   4   Families

FIGURE **27-8**
Deletion mapping in DMD using cDNA probes. DNA from boys with DMD (lanes 1–4 A) and their unaffected brothers (lanes 1–4 B) were digested with the *Taq*I enzyme and hybridized with a cDNA probe that detects seven exons in normal DNA (i through vii). These affected boys have deletions for as many as six of these seven exons (lane 3A). Patient 4 A has an abnormal-sized fragment as well as missing exons. The first lane contains molecular weight standards (sizes are marked in kilobase pairs).

FIGURE **27-9** (*Right*)
Direct detection of the sickle-cell β-globin (βˢ) mutation by restriction enzyme digestion and Southern blotting. (a) The base change (A to T) that causes sickle-cell anemia destroys the recognition sequence for a number of restriction enzymes, including *Mst*II. (b) This enzyme cuts the normal globin gene and yields two fragments of 1.1 kbp and 0.2 kbp, which can be detected with a probe from the 5′ end of the β-globin gene. (c) A prenatal diagnosis of sickle-cell anemia using the *Mst*II polymorphism. The mother (M) and father (F) are carriers of sickle-cell anemia because they have 1.3- and 1.1-kbp fragments. Their unaffected child (C) is homozygous for the normal 1.1-kbp fragment. The fetus (Fe) has inherited a βˢ allele from each parent and will be affected by sickle-cell anemia. (After Chang and Kan, 1982).

Allele-Specific Oligonucleotide Probes Are Used to Detect Mutations

Another method for looking directly for a mutation uses probes that are designed to hybridize selectively to either the normal or the mutant allele. These *allele-specific oligonucleotide* probes (or ASOs) can be used for any disorder where the nucleotide sequence of the mutant and normal alleles are known.

α₁-Antitrypsin, produced in the liver, inhibits a protease called elastase, and the balance between the levels of α₁-antitrypsin and elastase is normally carefully controlled. In individuals with α₁-antitrypsin defi-

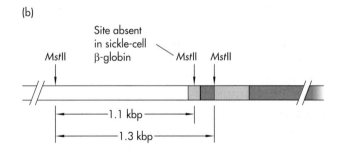

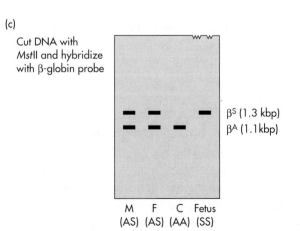

ciency, this balance is upset, the elastic fibers of the lung are slowly destroyed by elastase, and the patients suffer from pulmonary emphysema. The α_1-antitrypsin locus is called *Pi* (*p*rotease *i*nhibitor), and there is a complex allelic system controlling synthesis of α_1-antitrypsin. Homozygotes for the *M* allele (*MM*) have a normal phenotype, whereas homozygotes for the *Z* allele (*ZZ*) are severely affected.

The α_1-antitrypsin gene has been cloned, and the mutant *Z* gene has a single base change (G to A in the DNA strand corresponding to the mRNA) that leads to an amino acid substitution (glutamine to lysine) at residue 342 (Figure 27-10). The G to A substitution does not create or destroy a restriction enzyme site, so the detection strategy used with sickle-cell anemia is not possible. However, a 19-base-long synthetic oligonucleotide that is complementary to the normal α_1-antitrypsin gene sequence around the mutation site will have one mismatch with the mutant sequence. A similar oligonucleotide complementary to the mutant gene sequence is also synthesized. These oligonucleotides are labeled and used as probes to distinguish the normal and mutant sequences by changing the stringency of hybridization to a level at which each of the oligonucleotides will anneal stably only to the sequence to which it is perfectly complementary and not to the sequence with which it has the single mismatch. Heterozygotes can be recognized because both oligonucleotides hybridize to their DNA. However, this kind of ASO-based diagnosis is not an easy technique to use on a routine basis. Great care has to be taken to achieve the hybridization conditions in which the ASOs can distinguish a single base-pair mismatch.

Polymerase chain reaction (PCR) has now simplified the use of ASOs, because agarose electrophoresis and Southern blotting can be avoided. For an analysis of cystic fibrosis, two oligonucleotides complementary to sequences flanking the mutation site were used as primers to amplify a fragment across the mutation site. Two other oligonucleotides were synthesized, one complementary to the normal allele and the other complementary to the mutant sequence with the 3-bp deletion (Chapter 26 and Figure 27-11). These were used to determine whether the amplified DNA was normal or contained the mutation. Samples of the amplification mixture were transferred to an apparatus that applies the DNA in narrow bands to a nitrocel-

lulose filter ("slot blotting"). The filters were then hybridized with either the normal or the mutant radioactively labeled ASO. The resulting autoradiographs are very easy to interpret.

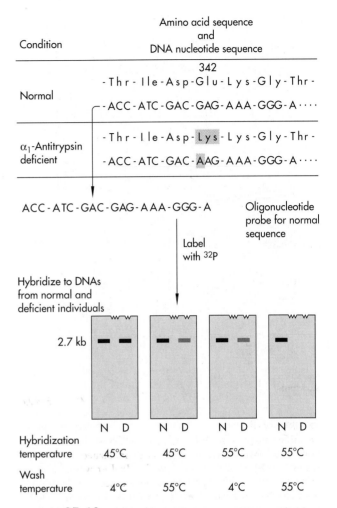

FIGURE **27-10**

Diagnosis of α_1-antitrypsin deficiency by using a synthetic oligonucleotide. DNA from a patient with α_1-antitrypsin deficiency was analyzed and found to have a single base change (G to A) in the protein-coding region. A synthetic 19-base oligonucleotide complementary to the normal gene (therefore having one mismatch with the mutant gene) was made. This oligonucleotide can distinguish between the normal (N) and mutant (deficient, D) gene when it is used as a probe in Southern blot analysis. Note that under conditions of lowered stringency (hybridization at 45°C, washing at 4°C) the probe hybridizes equally well with normal and mutant DNA. But as the stringency is increased (higher temperatures), the probe remains annealed only to the normal sequence.

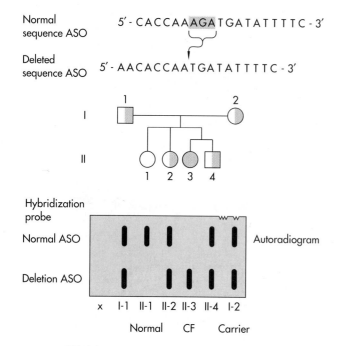

Normal sequence ASO
5'- CACCAA**AGA**TGATATTTTC- 3'

Deleted sequence ASO
5'- AACACCAATGATATTTTC- 3'

Hybridization probe

Normal ASO   Autoradiogram

Deletion ASO

x   I-1   II-1   II-2   II-3   II-4   I-2

Normal   CF   Carrier

FIGURE **27-11**

Diagnosis of cystic fibrosis by using PCR and allele-specific oligonucleotides as probes. DNA spanning the region of the 3-bp sequence deleted in many CF patients was amplified by using PCR. The PCR fragments were not separated by electrophoresis, instead a vacuum device was used to apply the DNA in narrow bands to a nitrocellulose filter. A sample from each patient was tested with two ASOs to determine whether the normal, deleted, or both sequences had been amplified. One ASO was complementary to the normal sequence, whereas the deletion ASO lacked the AGA nucleotides complementary to the deleted 3-bp sequence. The amplified sequence from the patient (II-3) hybridized only to the deletion ASO. Her parents (I-1 and I-2) must be heterozygotes. Indeed, both ASOs annealed to their DNA, a result showing that they each have one normal and one deleted allele. Of the patient's siblings, II-2 and II-4 are also heterozygotes, but II-1 has inherited the normal alleles from her parents, so her DNA anneals only to the normal ASO. The first lane (x) was a PCR with no DNA to check for contamination of the PCR reagents. (After Lemna et al., 1990)

The Ligase-Mediated Technique Detects Mutations

The ASO method exploits the mismatching between target sequence and probe to detect a mutation. The ligase-mediated method for detecting mutations exploits the fact that the ends of two single strands of DNA must be exactly aligned for DNA ligase to join

them (Figure 27-12). If the terminal nucleotides of either end are not properly base-paired to the complementary strand, then the ligase cannot join them. For the detection of the sickle-cell anemia mutation, the investigators synthesized oligonucleotides that annealed to the globin gene sequence 5' to the mutation site. The last (3') two bases of one oligonucleotide (N-oligo) were complementary to the normal (CT) sequence, and those of the other oligonucleotide (M-oligo) were complementary to the mutant (CA) gene sequence. A third oligonucleotide (C-oligo) was a perfect match for the sequence that is common to both β^A and β^S, immediately 3' to the mutation site. The N- and M-oligonucleotides were labeled with a biotin molecule, and the C-oligonucleotide was radioactively labeled. The N- and C-oligonucleotides anneal to single-stranded normal DNA, so they have flush ends that can be ligated. With single-stranded DNA from a patient with sickle-cell anemia as the template, however, the N- and C-oligonucleotides do not anneal with flush ends and cannot be ligated. The same reactions are performed with the M-oligo. Ligated molecules are isolated by passing the reaction mixture through a streptavidin column that binds only the biotinylated oligonucleotide; the radioactively labeled C-oligonucleotide remains on the column only if the two oligonucleotides have been ligated together. Fluorescent or enzymatic labels can be used rather than radioactive labels. These alternative labels would be very useful for the routine application of such a technique in diagnostic laboratories.

The Polymerase Chain Reaction Revolutionizes DNA-Based Diagnosis

As in so many other fields, the polymerase chain reaction has brought about a revolution in DNA-based diagnosis. PCR has many desirable features for diagnosis in a clinical setting (Table 27-6). First, sample preparation is minimal. As discussed previously, only nanogram amounts of DNA are needed to begin the reaction, so diagnosis can be performed on very small samples of blood or tissue (Chapter 6). PCR-based diagnoses have been performed with DNA from 200 μl of whole blood (hemophilia) and from one-half of a chorionic villus (cystic fibrosis). Another source of DNA is the blood sample taken from babies for neo-

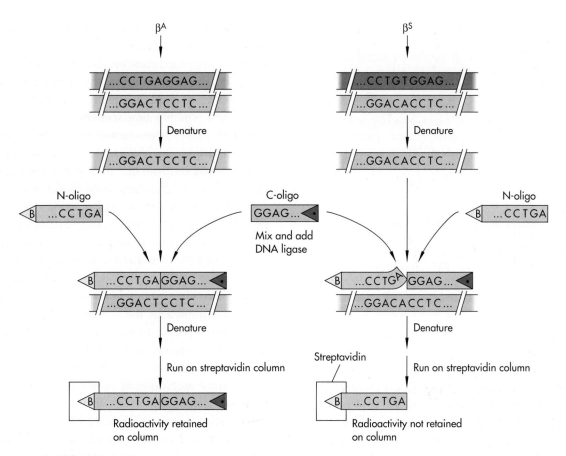

FIGURE **27-12**

Direct analysis of sickle-cell anemia by using DNA ligase. Oligonucleotides complementary to the normal sequence and mutant sequence, 5′ to and including the mutation site, were synthesized and labeled with biotin (B). A third oligonucleotide complementary to the common sequence (C) 3′ to the mutation site was synthesized and labeled with radioactivity (*). Here we show what happens with the oligonucleotide for the normal (N) sequence. The oligonucleotides were hybridized to separated strands of normal (β^A) or sickle-cell (β^S) DNA. The 5′ and 3′ oligonucleotides formed a flush junction that could be joined by DNA ligase only when they hybridized to the normal gene. The single base-pair mismatch (A · A) between the normal 5′ oligonucleotide and the mutation site is sufficient to prevent the ligase from acting. After denaturation, the reaction mixtures were run on a streptavidin column; radioactivity was retained on the column only if the 5′ biotinylated oligonucleotide had been joined to the 3′ radioactively labeled oligonucleotide. (Biotin and streptavidin have a very strong affinity for each other.) (After Landegren et al., 1988)

natal screening for phenylketonuria. These minute samples, taken by heel-prick, are stored as dried spots on Guthrie cards, named after R. Guthrie, who devised the test. These samples are an excellent source of DNA for diagnosis using PCR. Even a single saline mouth wash contains ample numbers of cells for DNA amplification. The cells are pelleted by centrifugation, lysed with detergent, and the lysate used immediately

for PCR without any further purification. The simplicity of this noninvasive procedure raises the possibility of using it to obtain DNA samples for a mass screening program of common genetic disorders like cystic fibrosis.

Second, the detection of PCR products is much easier than detecting single-copy sequences in genomic DNA because investigators can dispense en-

TABLE **27-6**

Examples of Human Inherited Disorders Diagnosed by Using the Polymerase Chain Reaction

Adenosine deaminase deficiency	Lesch-Nyhan syndrome
α_1-Antitrypsin deficiency	Maple syrup urine disease
	Ornithine transcarbamylase deficiency
Cystic fibrosis	Phenylketonuria
Fabry disease	Retinoblastoma
Familial hypercholesterolemia	Sandhoff disease
	Sickle-cell anemia
G-6-PD deficiency*	Tay-Sachs disease
Gaucher's disease	β-Thalassemia
Hemophilia A	δ-Thalassemia
Hemophilia B	von Willebrand disease

* G-6-PD, glucose-6-phosphate dehydrogenase.

tirely with radioactive probes. For example, DNA samples from families with hemophilia were amplified by using two sets of primers for the sequences involved in the *Xba*I and *Bcl*I polymorphisms in a single reaction mixture. Simple inspection of ethidium bromide–stained gels of the PCR products after digestion with either enzyme was sufficient to determine whether the restriction sites were present and to perform a linkage analysis (Figure 27-13). A diagnosis can now be made in a matter of hours rather than the days needed for the standard Southern procedures.

Another non-radioactive detection method uses fluorescently labeled ASOs as primers for PCR. Sickle-cell anemia was again the target mutation, and oligonucleotide primers that were exact matches for either the normal or the mutation site were synthesized. The normal oligonucleotide was labeled with a fluorescein fluorophore, and the other oligonucleotide was conjugated with rhodamine. A third oligonucleotide primer was a perfect match for sequences immediately 3′ to the mutation site. The amplification reaction was carried out with all three primers present and the products analyzed by electrophoresis on an agarose gel. Colored bands were visible on the gels; green (fluorescein) for the normal globin gene and red (rhodamine) for the sickle-cell mutation. DNA from a carrier of the sickle-cell trait contains both normal and mutant alleles and therefore produces a yellow band, the result of mixing the amplified products derived from each allele.

Third, the specificity of PCR means that several sequences can be amplified at the same time, provided care is taken to amplify sequences of different lengths so that the products can be distinguished easily after electrophoresis. A spectacular example of this so-called multiplex analysis is exon deletion analysis in DMD, where as many as nine sequences have been amplified in a single reaction mixture (Figure 27-14). This single PCR test detects deletions in up to 80 percent of all patients with dystrophin deletions. Multiplexing has also been used with the color detection system just described. Five sets of primers were used to amplify two mutations in β-globin and one mutation in α-globin. One primer in each set was specific for either the normal or the mutant allele, and each of these allele-specific primers was labeled with a different fluorochrome. All five amplifications were performed simultaneously on each patient sample, and the colors of the amplified products were determined by using a fluorometer. In all cases the appropriate products were detected (Table 27-7). These results are encouraging for DNA-based diagnosis of disorders for which it will be necessary to simultaneously test for several mutations in the same gene.

The most direct way to determine whether a mutation is present is to sequence the DNA, but this used to be impractical for diagnostic purposes because of the time and expense of cloning and the other manipulations required for dideoxy sequencing with M13 phage (Chapter 5). The development of methods for the direct sequencing of PCR products, using fluorescently labeled primers and automated sequencing machines, has reduced substantially the time and effort required for sequencing. Sequencing may yet become the standard tool for detecting mutations in human genetic diseases, and diagnoses will become molecular rather than phenotypic.

DNA Diagnostic Methods Are Used to Distinguish Tumor Types

As discussed in Chapter 18, cancer is a genetic disease, and so it is not surprising that the recombinant DNA tools that we have been describing in this chapter have been used in clinical oncology. At present, diagnosis

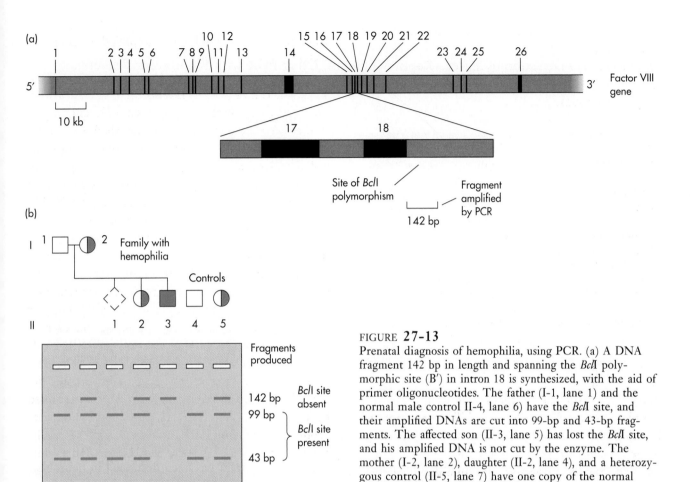

FIGURE **27-13**

Prenatal diagnosis of hemophilia, using PCR. (a) A DNA fragment 142 bp in length and spanning the *Bcl*I polymorphic site (B′) in intron 18 is synthesized, with the aid of primer oligonucleotides. The father (I-1, lane 1) and the normal male control II-4, lane 6) have the *Bcl*I site, and their amplified DNAs are cut into 99-bp and 43-bp fragments. The affected son (II-3, lane 5) has lost the *Bcl*I site, and his amplified DNA is not cut by the enzyme. The mother (I-2, lane 2), daughter (II-2, lane 4), and a heterozygous control (II-5, lane 7) have one copy of the normal gene (giving the 99-bp and 43-bp fragments) and one copy of the mutant gene (giving a single 142-bp fragment). The amplified fetal DNA (II-1, lane 3) is cut by the enzyme, a result showing that the male fetus is normal.

FIGURE **27-14**

Diagnosis of Duchenne muscular dystrophy using a multiple PCR to detect deleted exons. The 2000 kbp of the dystrophin gene are represented by the colored bar, and the positions of the XJ1.1 and some of the pERT (p87) series of probes are shown. Pairs of primers were synthesized to amplify the nine regions shown by the short black bars. These regions include those parts of the dystrophin gene that are most frequently deleted in DMD. DNA samples from patients are amplified using all nine sets of primers simultaneously in a single reaction, and the PCR products are separated on an agarose gel and stained with ethidium bromide. Patient 1 does not have a deletion for any of these regions, while patient 4 has a very large deletion covering over 1000 kbp. The other patients have deletions of varying size. The lane labeled "blank" is a control amplification performed without any added DNA sample to check whether the PCR reagents are contaminated with DNA. (After Chamberlain et al., 1990)

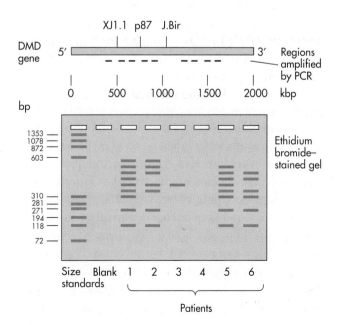

TABLE **27-7**

Simultaneous Detection of Three Globin Mutations, Using PCR with Fluorescent Primers

		β-Globin 4bp		β-Globin C-T		α-Globin
GENOMIC DNA	GENETIC STATUS	N	M	N	M	
α-globin deletion, hydrops fetalis		+	−	+	−	−
β-thalassemia, 4-bp deletion	Homozygous	−	+	+	−	+
	Heterozygous	+	+	+	−	+
β-thalassemia, C-T substitution	Homozygous	+	−	−	+	+
	Heterozygous	+	−	+	+	+

Fluorescently labeled, allele-specific oligonucleotides were used as primers to simultaneously perform five amplifications in the same reaction mixture. The reaction products were identified by using a fluorometer to measure the amount of fluorescence characteristic of each labeled primer. A "+" means that amplification took place with the primer and "−" means that no amplified products were detected. Hydrops fetalis is caused by a deletion of the α-globin gene. As expected, no amplification took place when the α-globin primer was used. The β-globin gene is normal in these patients, so the normal β-globin primers (N) did amplify this DNA, while the primers specific for the β-globin mutations (M) were negative. For the β-globin mutations, the M primers amplified DNA present in individuals homozygous and heterozygous for the mutations, but the N primers amplified DNA sequences only in the heterozygotes. (Normal homozygotes are not shown.)

of tumors relies on their histological appearance. While these histological criteria are useful for classifying tumor types, they have not been very successful at distinguishing between tumors of the same histological type that have very different clinical outcomes. This failure has very important practical consequences. Radiation and chemotherapy are used as postsurgical treatments for cancer patients, but these procedures invariably have severe side effects because they kill or impair the functioning of normal cells as well as any residual cancer cells that are the targets of the treatment. If the prognosis for patients could be assessed more accurately, then postoperative treatment could be tailored to the individual patient: patients with tumors of a type known to recur with high frequency would receive aggressive treatment with radiation and chemotherapy, whereas patients with good prognosis would receive minimal treatment.

With the cloning of oncogenes involved in human cancers, the hope has arisen that these will provide the basis for a more refined and accurate classification of tumors that will be directly related to the biological activities of the malignant cell. For example, amplification of N-*MYC* has been described in neuroblastoma types III and IV, which are invariably fatal; but apparently it is only rarely amplified in types I and II, which respond to treatment. In general, however, the number and variety of oncogenes involved in different tumors has confounded simplistic approaches to oncogene-based diagnosis. In breast cancer, amplification of the *NEU, MYC,* and *INT-2* oncogenes and over-expression of H-*RAS* appear to predict poor outcome, but the correlations between amplification and prognosis are not sufficiently consistent for clinical purposes. A similar situation is true for lung cancer, in which amplifications of *MYC,* L-*MYC,* and N-*MYC* and mutations in p53 have been detected. Recombinant DNA techniques are beginning to unravel the sequence of mutations that leads to colon cancer (Chapter 18). It may be possible to use this information to predict an individual's risk of developing colon cancer and to initiate treatment in good time.

Nevertheless, DNA-based diagnosis is proving to be very useful in oncology. For example, in inherited forms of cancer, cloned genes or linked RFLPs can be used to determine which members of a family are at risk for developing the cancer. This analysis was first achieved for a young child with retinoblastoma. The polymerase chain reaction and DNA sequencing have been used to detect point mutations in the retinoblastoma gene and to follow the course of patients who have received treatment for leukemias involving chromosomal translocations (Chapter 6). We can be confident that continued analysis of the oncogenes and their mutations involved in human cancers will lead to better diagnostic tests for cancer.

Novel Methods Are Developed for Screening for Mutations

All the methods we have just discussed test for specific mutations in a gene. For these, we have to have both the sequence of the gene and a detailed knowledge of what mutations are likely to be present. Fortunately, other methods have been developed to screen for unknown point mutations within a gene. Some of these methods, such as RNase A cleavage, chemical cleavage and modification, and density gradient gel electrophoresis, exploit characteristics of mismatched heteroduplexes formed between normal and mutant sequences. RNase cleavage, for example, uses the enzyme ribonuclease A (RNase A) to cut RNA–DNA hybrids wherever there is a mismatch between a nucleotide in the RNA strand and the corresponding nucleotide in the DNA strand (Figure 27-15). Typically, a radioactive RNA probe is made by using the normal sequence cloned in a vector with a phage RNA polymerase gene. The RNA strand anneals to the test genomic DNA and the mixture is treated with RNase A. If the DNA contains a mutation, then the enzyme cuts the RNA strand and two radioactive RNA fragments are detected on a denaturing gel. If the test DNA is normal, then a single RNA fragment corresponding to the intact RNA probe is detected. The chemical cleavage method is based on a similar principle but uses osmium tetroxide and hydroxylamine to cut the probe at mismatched C or T nucleotides.

A new technique called single-stranded conformation polymorphism (SSCP) analysis is proving to be especially useful. A single nucleotide difference between two short single-stranded DNA molecules in-

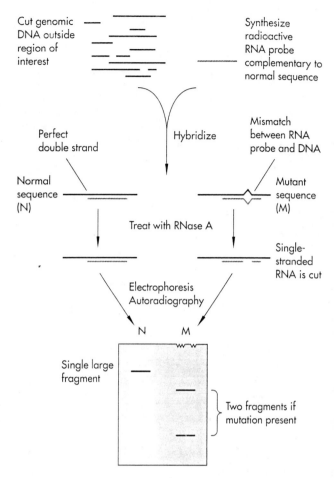

FIGURE 27-15
Detecting mutations by using RNase cleavage. Genomic DNA containing the sequence to be analyzed is cloned into a vector that encodes a phage RNA polymerase, and radioactive RNA is synthesized for use as a probe. Genomic DNA from a patient is digested by a restriction endonuclease that cuts at sites outside the region of interest, and this DNA is hybridized in solution with the RNA probe. If the patient's DNA contains the normal sequence, then a perfect RNA–DNA hybrid molecule is formed. If the patient's DNA contains a single nucleotide different from the normal sequence, then the RNA–DNA hybrid contains a mismatch. At the site of the mismatch, the RNA molecule is single stranded. The enzyme RNase A cuts RNA molecules only if they are single stranded, so when the hybrid molecules are treated with RNase A, the RNA strand of molecules with a mismatch is cut at the site of the mismatch. After denaturation, the samples are run on a gel, and autoradiography reveals the presence of the labeled RNA molecules. If the genomic DNA was normal, then an RNA molecule the same size as the original RNA probe is seen. If the genomic DNA contained a mismatch, then two RNA fragments are seen.

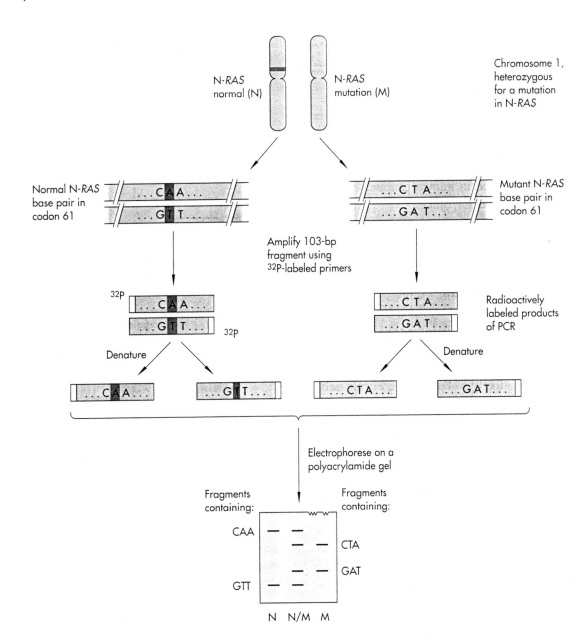

FIGURE **27-16**
Detecting mutations by using single-stranded conformation polymorphism (SSCP) analysis.
Genomic DNA was isolated from a cell line heterozygous for an A to T point mutation at
codon 61 in the N-*RAS* gene. A 103-bp fragment spanning the mutation was synthesized by
using the polymerase chain reaction. The 5′ ends of the primers used for the PCR were
radioactively labeled, so all the PCR fragments were labeled. After amplification, the double-
stranded fragments were denatured to separate the strands and electrophoresed on a polyacrylamide gel. Two bands appeared for the normal sample (N), corresponding to the two
strands of the amplified fragment. Similarly two bands appeared for the homozygous mutant
sample (M), but these were in different positions from those of the normal strands because
they differed by the single nucleotides T and A. The sample from the heterozygous cell line
(N/M) showed all four strands, two coming from the normal allele and two from the mutant
allele.

duces a difference in the conformations adopted by the two strands (Figure 27-16). Remarkably, the conformational difference is sufficient to produce changes in the molecules' electrophoretic mobilities on neutral polyacrylamide gels. The method was developed by using genomic DNA fragments produced by restriction digest, but it has now been simplified by using PCR to amplify those regions of a gene that might contain mutations. At present, most of these methods require a degree of technical sophistication that makes them unsuitable for routine applications, but this limitation is likely to change with continuing developments in instrumentation and robotics.

Genetic Testing May Bring Problems As Well As Benefits

All these technical advances make it more likely that genetic testing will become a routine diagnostic procedure in the future and that we may even have our own personal genetic profiles. This worries some who remind us that the eugenics movement earlier this century misused genetic information by trying to justify discrimination on the basis of genetic differences. The potential for such abuse is becoming greater as we learn more about the genes that influence our development and devise ways for controlling and manipulating them. We raise some of these issues here because it is important to be aware of them if they are to be resolved to the benefit of everyone in society.

One fear is that conditions we presently consider to be within the bounds of normal variation in humans will become regarded as unacceptable. Individuals may be stigmatized by their genetic constitution, irrespective of their actual health or behavior. For example, the discovery that criminal-mental institutions contained a higher than expected proportion of males with an XYY karyotype led to such individuals being stigmatized as "savage males." In fact the majority of XYY males are never in any conflict with the law. Will an increasing understanding of the genetic components of our mental and behavioral characteristics lead to *biological determinism*, that is, to the assumption that these characteristics cannot be shaped by our environment and culture because they have an inherited component?

DNA-based diagnosis itself is raising ethical issues. For example, in Huntington's disease (HD) how should we inform presently asymptomatic individuals that they will develop a devastating disease many years later? Screening for mutations in the general population raises a variety of problems. Some of these are practical—the question of how low the false-negative and false-positive rates must be for the test to be useful—and others are societal. There are worries about discrimination when a genetic disorder is prevalent in a minority group. The sickle-cell screening programs of the 1970s are an example. However, screening programs can be successful when they are implemented with the understanding and approval of the communities involved. Examples of well-accepted and beneficial programs include those for Tay-Sachs disease in the Ashkenazi Jewish population and for the thalassemias in Cyprus.

There is considerable concern that genetic information will be used unfairly in providing health care, life insurance, and employment. Might a couple at risk for having a child affected by a genetic disorder have difficulties finding health care coverage if they decline prenatal diagnosis? Individuals have every reason to be afraid that acknowledging that they carry a mutation will lead to increased premiums or loss of insurance coverage, with disastrous effects for their families; and genetic testing could be used to exclude at-risk individuals from group coverage. However, companies that choose not to make use of genetic testing will be at a disadvantage relative to companies that do and will lose money and attract at-risk individuals who must avoid the companies that are testing.

Genetic tests that can identify mutations predisposing individuals to illness may have serious consequences for employment. For example, the U.S. Air Force refused to accept individuals with sickle-cell trait, although there were no reliable data to indicate that these persons were at risk for suffering respiratory problems at high altitude. Suppose a person has inherited a mutation in one copy of a tumor suppressor gene. That person may not be allowed to work in environments with raised levels of mutagens, for fear that the remaining normal allele might become mutated. This may seem a wise and sensible precaution,

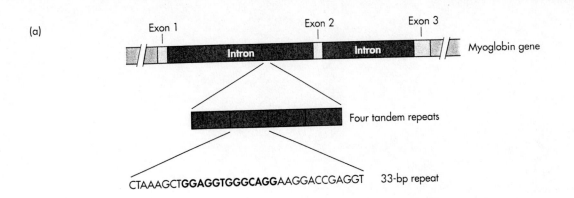

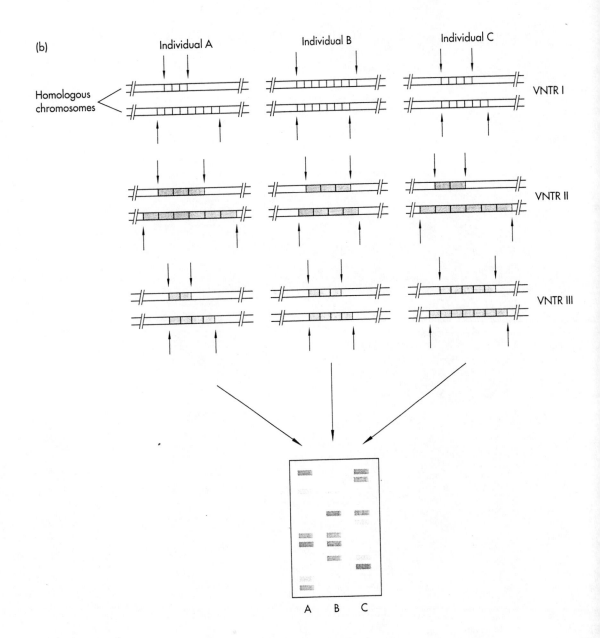

FIGURE 27-17

Variable number tandem repeat (VNTR) loci. (a) The VNTR locus in an intron of the myoglobin gene is composed of four repeats of a 33-bp sequence. The core sequence is shown in bold type. (b) The variability at a number of VNTR loci can generate diversity sufficient to identify individuals. Between two and eight tandem repeats of three different VNTR loci (I–III) have been detected by a hypervariable probe. Although different individuals may have some fragments in common, the chances that two individuals might have all fragments in common is low. Restriction endonuclease sites are indicated by arrows.

but there are worries that genetic testing to exclude at-risk workers may come to be a substitute for making the workplace safe.

Hypervariable or Variable Tandem Repeat Loci Can Be Used to Identify Individuals

The techniques of DNA diagnosis have found application in a quite different area, the identification of individuals. This is important in areas as diverse as identifying cell cultures, determining family relationships in studies of animal behavior, and, of course, in forensic science. Until recently, techniques for identifying individuals made use of protein polymorphisms. Isoenzymes were favored for cells in culture, and blood groups were analyzed in forensic samples. A whole series of such tests has to be used, and although it is possible to show definitively that two samples are different, the tests are rarely sensitive enough to establish identity of samples. The technique called DNA fingerprinting may circumvent these problems. This technique requires the analysis of a set of polymorphic loci, chosen so that the probability that two individual DNA samples with identical haplotypes could by chance have come from different individuals is very low. The most useful polymorphisms for forensic purposes are found at so-called hypervariable loci.

A hypervariable locus is made up of a variable number of identical sequences joined together in tandem (Figure 27-17). When DNA is digested with a restriction endonuclease that cuts the sequences flanking a hypervariable or variable number tandem repeat (VNTR) locus, the lengths of the DNA fragments produced in different individuals depend on the number of repeats at the locus; this number typically ranges from 3 to 20. There are many different VNTR loci in the genome, so the pattern of fragments from the VNTR loci in one individual is essentially unique for that individual.

The first highly variable probes to be used for individual identification were derived from a tandem repeat of a 33-bp sequence in an intron of the human myoglobin gene. When this repeat was used as a hybridization probe on Southern blots of human DNA, it detected the original sequence from the myoglobin gene and many other fragments. The same repeat identified clones in a human cosmid library, and se-

quence analysis of these showed that they contained a common core sequence of 10 to 15 bp. Two of the clones, each detecting a different pattern of fragments, revealed between 30 and 40 distinguishable bands when used together as probes on Southern blots. This variability is such that the pattern of bands is unique to an individual (except for monozygotic twins).

These first probes produce complicated patterns because they hybridize to fragments from many different VNTR loci in an individual. A single-locus probe is specific for just one VNTR locus and detects one or two fragments in an individual. However, there is a very wide range of fragment sizes for each locus; for example, one of these probes detected a minimum of 77 alleles at one locus, with repeats ranging from 14 to >500 in number. The fragments detected by one VNTR probe and their pattern of inheritance are shown in Figure 27-18. There are many other repetitive sequences in the human genome, including the dinucleotide repeat (CA) and its complement (GT). These repeats—usually written as $(C-A)_n \cdot (G-T)_n$—occur in some 100,000 blocks in the human genome (once every 50–100 kbp) and appear to be uniformly distributed throughout the genome. The value of n varies between 4 and 40, but most blocks have less than 25 repeats. The $(C-A)_n \cdot (G-T)_n$ repeats are very useful as markers because of their variability and because they can be analyzed quickly by using the polymerase chain reaction with primers for the DNA sequences flanking each block. The successful application of these $(C-A)_n \cdot (G-T)_n$ repeats has led to the use of a variety of other di-, tri-, and tetranucleotide sequences for mapping.

DNA probes were soon put to use in an immigration case where it was necessary to determine whether a boy was the son or the nephew of the woman who claimed him as her son. Conventional testing, using 17 protein polymorphisms, demonstrated that the boy and woman were related but could not determine their relationship more precisely. To do this, DNA samples were analyzed from the woman and two of her sisters and from the boy. Despite the lack of a sample from the father (there was some doubt about paternity), it was possible to show that the boy and the woman were related and that the relationship was that of son and mother (Table 27-8). The use of these probes in this way has led to the reuniting of many families who would otherwise have been kept apart.

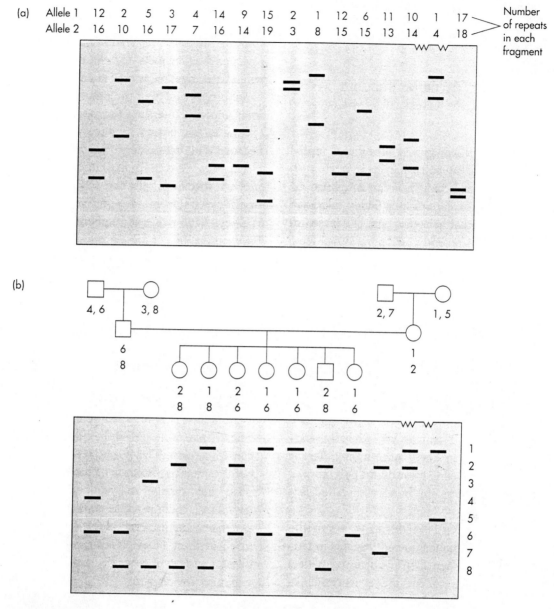

FIGURE 27-18

Fingerprinting, using variable number tandem repeat (VNTR) probes. (a) The probe pYNH24 detects 19 alleles in 16 unrelated individuals. No two individuals have the same pattern of fragments. (b) The fragments detected by VNTR probes are inherited in a normal Mendelian fashion and can be used in linkage analysis in the same way as a conventional RFLP (After Nakamura et al., 1987)

DNA Fingerprinting Is Used in the Courts

The potential of these probes for establishing identities in forensic science cases was obvious. The practicality of doing so was established when it was shown that DNA could be isolated from dried blood stains or from sperm in vaginal swabs that had been stored for as long as two years. Sufficient DNA for Southern blotting can be isolated from freshly pulled hair roots, but hair at the scene of a crime is not usually in this pristine condition. However PCR can be used to "rescue" DNA from shed hairs several months old containing less than 1 ng of DNA. Polymorphisms in mitochondrial DNA and the class II HLA gene $DQ\alpha$ have been analyzed this way. Obtaining sufficient

TABLE **27-8**
DNA Analysis to Determine Relationships

QUESTION	NUMBER OF SHARED FRAGMENTS	PROBABILITY*
Is the boy related to this family?	61	7×10^{-22}
Could an unrelated woman be the mother?	25	2×10^{-15}
Could the sister of the woman be the mother?	25	6×10^{-6}

* The probabilities are estimates of the likelihood that the relationships shown by the DNA typing patterns could have arisen by chance. For example, the probability that the boy could *by chance* have 61 fragments in common with the family is 7×10^{-22}.

quantities of intact DNA from forensic samples will always be a problem, and PCR may have a great impact in this area. The major concern over forensic applications of PCR is contamination of the evidentiary sample, both by other DNA at the crime scene and by other amplified DNA in the laboratory (Chapter 6).

The power of DNA fingerprinting was demonstrated in the very first murder case in which it was used. A man had been accused of two rape-murders committed three years apart and had made a confession. Samples of the rapist's semen were available from both attacks, and DNA fingerprinting by Alec Jeffreys showed, just as the police had surmised, that the same man was involved in both attacks. However, it was clear that the suspect was not the culprit. This result was, at first, unbelievable, and the DNA fingerprinting had to be repeated several times before he was released. Eventually the real murderer was caught, and DNA fingerprinting confirmed the identification.

DNA fingerprinting is now in use in many countries. In the United Kingdom, DNA fingerprinting is used to resolve family relationships in over 3000 immigration cases each year. DNA fingerprinting in the United Kingdom also has been fully accepted as a tool in forensic science, being used in several thousand forensic cases each year. In the United States, DNA fingerprinting has come under particularly close scrutiny in the courtroom. There have been objections to the quality of some of the DNA evidence that has been presented and also to studies of the population allele frequencies that underlie the calculations of the probability of identity. These are particularly important issues because DNA fingerprinting, and scientific evidence in general, carries great weight with juries. Nevertheless, DNA typing is a powerful piece of evi-

dence that can exonerate the innocent as well as convict the guilty. There is no doubt that it will achieve widespread acceptability when the practical and legal problems involved have been resolved.

Genetic Privacy Will Become an Important Issue

Forensic application of recombinant DNA techniques highlights another important issue, namely, privacy of genetic information. Several states are already collecting blood samples from all felons convicted of violent crimes. However, while a fingerprint inked on a card cannot be used for anything other than identification, a DNA sample contains the entire genetic information of an individual. As we learn more about human genes and their mutations, especially those involved in polygenic disorders, it will become possible to screen DNA samples for those mutations. This could lead to discrimination. It is argued that at present such testing is politically unacceptable, but political circumstances change, and what guarantees could be provided that this testing would never occur? And as more and more genetic information is stored in computer databanks, the likelihood that unauthorized individuals will gain access to these data increases. Our genetic make-up seems to be such an intimate and private part of ourselves that information about it should be accorded special status.

Nevertheless, we can expect that DNA testing will become standard diagnostic practice. Continuing developments in instrumentation and automation will hasten routine genetic testing, and new research techniques will be rapidly applied in DNA diagnosis. The polymerase chain reaction is an outstanding example

of the rapid transfer of a research tool from the specialist's laboratory into routine diagnostic practice. The list of cloned human genes continues to lengthen, and many of these are involved in inherited diseases. Recently, human cancer genes have been cloned, and studies of viral oncogenes are leading to a better understanding of the genetic pathways to cancer in human beings (Chapter 18). There are increasing efforts to use the tools of recombinant DNA and linkage analysis to dissect out the genetic and environmental components of complex polygenic disorders. There have been some successes. For example Brown and Goldstein discovered that mutations in the low density lipoprotein receptor contribute to familial hypercholesterolemia, which is one of the contributors to the overall incidence of heart attack. But at present there is also much confusion, especially in studies of alcoholism and mental disorders like schizophrenia and manic depression (Chapter 26). It may be many years before we have a useful understanding of the genetics of some of the commonest ills that afflict our society. It is to be hoped that the human genome projects will hasten the achievement of that understanding (Chapter 30).

Reading List

General

Emery, A. E. H. *An Introduction to Recombinant DNA.* John Wiley & Sons, Chichester, 1984.

Antonarakis, S. E. "Diagnosis of genetic disorders at the DNA level." *N. E. J. Med.,* 320: 153–163 (1989). [Review]

Weatherall, D. J. *The New Genetics and Clinical Practice,* 3rd ed. Oxford University Press, Oxford. 1991.

Original Research Papers

INBORN ERRORS OF METABOLISM

Goldberg, J. D., and M. S. Golbus. "Chorionic villus sampling." *Adv. Hum. Genet.,* 17: 1–25 (1988). [Review]

Scriver, C. R., A. L. Beaudet, W. S. Sly, and D. Valle, eds. *The Metabolic Basis of Inherited Disease.* McGraw-Hill, New York, 1989.

MRC Working Party on the Evaluation of Chorion Villus Sampling. "Medical Research Council European trial of chorion villus sampling." *Lancet,* 337: 1491–1499 (1991).

THE COMPLEXITY OF HUMAN DISEASE MUTATIONS

Orkin, S. H. "Disorders of hemoglobin synthesis." In *The Molecular Basis of Blood Diseases.* G. Stamatoyannopoulos, A. W. Nienhuis, P. Leder, and P. W. Majerus, eds., W. B. Saunders, Philadelphia, 1987, pp. 106–126. [Review]

Antonarakis, S. E. "The molecular genetics of hemophilia A and B in man." *Adv. Hum. Genet.,* 17: 27–59 (1988). [Review]

Kazazian, H. H., and C. D. Boehm. "Molecular basis and prenatal diagnosis of β-thalassemia." *Blood,* 72: 1107–1116 (1988). [Review]

Higgs, D. R., M. A. Vickers, A. O. M. Wilkie, I.-M. Pretorius, A. P. Jarman, and D. J. Weatherall. "A review of the molecular genetics of the human α-globin gene cluster." *Blood,* 73: 1081–1104 (1989). [Review]

Stout, J. T., and C. T. Caskey. "Hypoxanthine-guanine phosphoribosyltransferase deficiency: The Lesch-Nyhan syndrome and gouty arthritis." In C. R., Scriver, A. L. Beaudet, W. S. Sly, and D. Valle (eds.), *The Metabolic Basis of Inherited Disease,* McGraw-Hill, New York, (1989), pp. 1007–1028. [Review]

Kazazian, H. H., C. Wong, H. Youssoufian, A. F. Scott, D. G. Phillips, and S. E. Antonarakis. "Hemophilia A resulting from the *de novo* insertion of *L1* sequences represents a novel mechanism for mutation in man." *Nature,* 332: 164–166 (1988).

RFLPs AND LINKAGE ANALYSIS

Gusella, J. F., N. S. Wexler, P. M. Conneally, S. L. Naylor, M. A. Anderson, R. E. Tanzi, P. C. Watkins, K. Ottina, M. R. Wallace, A. Y. Sakaguchi, A. B. Young, I. Shoulson, E. Bonilla, and J. B. Martin. "A polymorphic DNA marker genetically linked to Huntington's disease." *Nature,* 306: 234–238 (1983).

Gitschier, J., D. Drayna, E. G. D. Tuddenham, R. L. White, and R. M. Lawn. "Genetic mapping and diagnosis of haemophilia A achieved through a *Bcl*I polymorphism in the factor VIII gene." *Nature,* 314: 738–740 (1985).

Farrall, M., X. Estivill, and R. Williamson. "Indirect cystic fibrosis carrier detection." *Lancet,* II: 156–157 (1987).

Meissen, G. J., R. H. Myers, C. A. Mastromauro, W. J. Koroshetz, K. W. Klinger, L. A. Farrer, P. A. Watkins, J. F. Gusella, E. D. Bird, and J. B. Martin. "Predictive testing for Huntington's disease with use of a linked DNA marker." *N. E. J. Med.,* 318: 535–542 (1988).

Ward. P. A., J. Hejtmancik, J. A. Witkowski, L. L. Baumbach, S. Gunnell, J. Speer, P. Hawley, S. Latt, U. Tantravahi, and C. T. Caskey. "Prenatal diagnosis of Duchenne muscular dystrophy: Prospective linkage analysis and retrospective dystrophin cDNA analysis." *Am. J. Hum. Genet.,* 44: 270–281 (1989).

GENETIC DIAGNOSIS USING LINKAGE DISEQUILIBRIUM

Kan, Y. W., and A. M. Dozy. "Antenatal diagnosis of sickle-cell anaemia by DNA analysis of amniotic-fluid cells." *Lancet,* II: 910–912 (1978).

Kan, Y. W., and A. M. Dozy. "Evolution of the hemoglobin S and C genes in world populations." *Science,* 209: 388–391 (1980).

Daiger, S. P., R. Chakraborty, F. Güttler, A. S. Lidsky, R. Koch, and S. L. C. Woo. "Polymorphic DNA haplotypes at the phenylalanine hydroxylase locus in prenatal diagnosis of phenylketonuria." *Lancet,* I: 229–232 (1986).

Cox, D. W., G. D. Billingsley, and T. Mansfield. "DNA restriction-site polymorphisms associated with the alpha 1-antitrypsin gene." *Am. J. Hum. Genet.,* 41: 891–906 (1987).

Estivill, X., C. McClean, V. Nunes, T. Casals, P. Gallano, P. J. Scambler, and R. Williamson. "Isolation of a new DNA marker in linkage disequilibrium with cystic fibrosis, situated between J3.11 (D7S8) and IRP." *Am. J. Hum. Genet.,* 44: 704–710 (1989).

DIAGNOSIS OF DUCHENNE MUSCULAR DYSTROPHY USING GENE DELETIONS

Kunkel, L. M., and 75 coauthors. "Analysis of deletions in DNA from patients with Becker and Duchenne muscular dystrophy." *Nature,* 322: 73–77 (1986).

Darras, B. T., M. Koenig, L. M. Kunkel, and U. Francke. "Direct method for prenatal diagnosis and carrier detection in Duchenne/Becker muscular dystrophy using the entire dystrophin cDNA." *Am. J. Med. Genet.,* 29: 713–726 (1988).

Baumbach, L. L., J. S. Chamberlain, P. A. Ward, N. J. Farwell, and C. T. Caskey. "Molecular and clinical correlations of deletions leading to Duchenne and Becker muscular dystrophies." *Neurology,* 39: 465–474 (1989).

MUTATIONS ALTERING RESTRICTION ENZYME SITES

Chang, J. C., and Y. W. Kan. "A sensitive new prenatal test for sickle-cell anemia." *N. E. J. Med.,* 307: 30–32 (1982).

Orkin, S. H., P. F. R. Little, H. H. Kazazian, and C. D. Boehm. "Improved detection of the sickle mutation by DNA analysis." *N. E. J. Med.,* 307: 32–36 (1982).

DIRECT DETECTION OF MUTATIONS: ALLELE-SPECIFIC OLIGONUCLEOTIDE PROBES

Kidd, V. J., M. S. Golbus, R. B. Wallace, K. Itakura, and S. L. C. Woo. "Prenatal diagnosis of α_1-antitrypsin defi-ciency by direct analysis of the mutation site in the gene." *N. E. J. Med.,* 310: 639–642 (1984).

Lemna, W. K., G. L. Feldman, B.-S. Kerem, S. D. Fernbach, E. P. Zevkovich, W. E. O'Brien, J. R. Riordan, F. S. Collins, L.-C. Tsui, and A. L. Beaudet. "Mutation analysis for heterozygote detection and the prenatal diagnosis of cystic fibrosis." *N. E. J. Med.,* 322: 291–296 (1990).

NOVEL DIAGNOSTIC METHODS

Landegren, U., R. Kaiser, J. Sanders, and L. Hood. "A ligase-mediated gene detection technique." *Science,* 241: 1077–1080 (1988).

Gibbs, R. A., P.-N. Nguyen, and C. T. Caskey. "Detection of single DNA base differences by competitive oligonucleotide priming." *Nucleic Acids Res.,* 17: 2437–2448 (1989).

Newton, C. R., A. Graham, L. E. Hiptinstall, S. J. Powell, C. Summers, N. Kalsheker, J. C. Smith, and A. F. Markham. "Analysis of any point mutation in DNA. The amplification refractory mutation system (ARMS)." *Nucleic Acids Res.,* 17: 2503–2516 (1989).

Wu, D. Y., L. Ugozzoli, B. K. Pal, and R. B. Wallace. "Allele-specific enzymatic amplification of β-globin genomic DNA for diagnosis of sickle cell anemia." *Proc. Natl. Acad. Sci. USA,* 86: 2757–2760 (1989).

DIAGNOSIS USING THE POLYMERASE CHAIN REACTION

Reiss, J., and D. N. Cooper. "Application of the polymerase chain reaction to the diagnosis of human genetic disease." *Hum. Genet.,* 85: 1–8 (1990). [Review]

Saiki, R. K., S. Scharf, F. Faloona, K. B. Mullis, G. T. Horn, H. A. Erlich, and N. Arnheim. "Enzymatic amplification of β-globin genomic sequences and restriction site analysis for diagnosis of sickle cell anemia." *Science,* 230: 1350–1354 (1985).

Kogan, S. C., M. Doherty, and J. Gitschier. "An improved method for prenatal diagnosis of genetic diseases by analysis of amplified DNA sequences." *N. E. J. Med.,* 317: 985–990 (1987).

Gibbs, R. A., P.-N. Nguyen, L. J. McBride, S. M. Koepf, and C. T. Caskey. "Identification of mutations leading to the Lesch-Nyhan syndrome by automated direct DNA sequencing of *in vitro* amplified cDNA." *Proc. Natl. Acad. Sci. USA,* 86: 1919–1923 (1989).

Chamberlain, J. S., R. A. Gibbs, J. E. Ranier, and C. T. Caskey. "Multiplex PCR for the diagnosis of Duchenne muscular dystrophy." In *PCR Protocols: A Guide to Methods and Applications.* M. A. Innis, D. H. Gelfand, J. J. Sninsky, and T. J. White, eds. Academic Press, New York, 1990, pp. 272–281.

Chehab, F. F., and Y. W. Kan. "Detection of sickle cell anaemia mutation by colour DNA amplification." *Lancet,* 335: 15–17 (1990).

DNA-BASED CANCER DIAGNOSIS

Johnson, B. E., D. C. Ihde, R. W. Makuch, A. F. Gazdar, D. N. Carney, H. Oie, E. Russell, M. M. Nau, and J. D. Minna. "*myc* family oncogene amplification in tumor cell lines established from small cell lung cancer patients and its relationship to clinical status and course." *J. Clin. Invest.,* 79: 1629–1634 (1987).

Varley, J. M., J. E. Swallow, W. J. Brammer, J. L. Whittaker, and R. A. Walker. "Alterations to either c-*erb*B-2 (*neu*) or c-*myc* proto-oncogenes in breast carcinomas correlate with poor short-term prognosis." *Oncogene,* 1: 423–431 (1987).

Yandell, D. W., T. A. Campbell, S. H. Dayton, R. Petersen, D. Walton, J. B. Little, A. McConkie-Rosell, E. G. Buckley, and T. P. Dryja. "Oncogenic point mutations in the human retinoblastoma gene: Their application to genetic counseling." *N. E. J. Med.,* 321: 1689–1695 (1989).

Leppert, M., R. Burt, J. P. Hughes, W. Samowitz, Y. Nakamura, S. Woodward, E. Gardner, J.-M. Lalouel, and R. White. "Genetic analysis of an inherited predisposition to colon cancer in a family with a variable number of adenomatous polyps." *N. E. J. Med.,* 322: 904–908 (1990).

SCREENING FOR MUTATIONS

Myers, R. M., Z. Larin, and T. Maniatis. "Detection of single base substitutions by ribonuclease cleavage at mismatches in RNA:DNA duplexes." *Science,* 230: 1242–1246 (1985).

Gibbs, R., and C. T. Caskey. "Identification and localization of mutations at the Lesch-Nyhan locus by ribonuclease A cleavage." *Science,* 236: 303–305 (1987).

Orita, M., Y. Suzuki, T. Sekiya, and K. Hayashi. "Rapid and sensitive detection of point mutations and DNA polymorphisms using the polymerase chain reaction." *Genomics,* 5: 874–879 (1989).

Abrams, E. S., S. E. Murdaugh, and L. S. Lerman. "Comprehensive detection of single base changes in human genomic DNA using denaturing gradient gel electrophoresis and a GC clamp." *Genomics,* 7: 463–475 (1990).

HYPERVARIABLE LOCI

Wyman, A. R., and R. White. "A highly polymorphic locus in human DNA." *Proc. Natl. Acad. Sci. USA,* 77: 6754–6758 (1980).

Jeffreys, A. J., J. F. Y. Brookfield, and R. Semeonoff. "Positive identification of an immigration test-case using human DNA fingerprints." *Nature,* 317: 818–819 (1985).

Jeffreys, A. J., V. Wilson, and S. L. Thein, "Hypervariable 'minisatellite' regions in human DNA." *Nature,* 314: 67–73 (1985).

Nakamura, Y., M. Leppert, P. O'Connell, R. Wolff, T. Holm, M. Culver, C. Martin, E. Fujimoto, M. Hoff, E. Kumlin and R. White. "Variable number tandem repeat (VNTR) markers for human gene mapping." *Science,* 235: 1616–1622 (1987).

Weber, J. L., and P. E. May. "Abundant class of human DNA polymorphisms which can be typed using the polymerase chain reaction." *Am. J. Hum. Genet.,* 44: 388–396 (1989).

DNA FINGERPRINTING AND FORENSIC SCIENCE

Ballantyne, J., G. Sensabaugh, and J. A. Witkowski (eds.). *DNA Technology and Forensic Science.,* Banbury Report # 32. Cold Spring Harbor Laboratory, Cold Spring Harbor, N. Y., 1989. [Review]

Gill, P., A. J. Jeffreys, and D. J. Werrett. "Forensic application of DNA 'fingerprints'." *Nature,* 318: 577–579 (1985).

Thompson, W. C., and S. Ford. "DNA typing: acceptance and weight of the new genetic identification tests." *U. Virginia Law Rev.* 75: 601–664 (1988).

Lander, E. S. "DNA fingerprinting on trial." *Nature,* 339: 501–505 (1989).

Wambaugh, J. *The Blooding.* William Morrow and Company, New York, 1989.

ETHICAL ISSUES IN HUMAN MOLECULAR GENETICS

Reilly, P. *Genetics, Law, and Social Policy.* Harvard University Press, Cambridge, Mass., 1977.

Kevles, D. *In the Name of Eugenics.* Alfred A. Knopf, New York, 1985.

Holtzman, N. A. *Proceed with Caution: Predicting Genetic Risks in the Recombinant DNA Era.* The Johns Hopkins University Press, Baltimore, 1989.

Beutler, E., D. R. Boggs, P. Heller, A. Maurer, A. G. Motulsky, and T. W. Sheehy. "Hazards of indiscriminate screening for sickle cell diseases." *N. E. J. Med.,* 285: 1485–1486 (1971).

Angastiniotis, M. A., S. Hyriakidou, and M. Hadjiminas. "How thalassemia was controlled in Cyprus." *World Health Forum,* 7: 291–297 (1986).

Brandt, J., K. A. Quaid, S. E. Folstein, P. Garber, N. E. Maestri, M. H. Abbott, P. R. Slavney, M. L. Franz, L. Kasch, and H. H. Kazazian. "Presymptomatic diagnosis of delayed-onset disease with linked DNA markers: the experience in Huntington's disease." *J. Amer. Med. Assoc.,* 261: 3108–3114 (1989).

Harper, P. S., and A. Clarke. "Should we test children for 'adult' genetic diseases?" *Lancet,* 335: 1206–1206 (1990).

Wilfond, B. S., and N. Fost. "The cystic fibrosis gene: medical and social implications for heterozygote detection." *J. Am. Med. Assoc.,* 263: 2777–2783 (1990).

28

Working Toward Human Gene Therapy

E ven though genetic counseling and prenatal diagnosis are invaluable for families who are known to have an inherited disorder, there exists a long-cherished hope that patients already afflicted might be treated by replacing their defective gene with the normal gene. For many years this prospect seemed to be very remote, more aptly belonging to the realms of science fiction. Only with the application of recombinant DNA techniques to human genetic diseases did research to develop practicable methods of gene therapy become possible. However, the obstacles to gene therapy are formidable. Much information is needed about the molecular pathology of a genetic disorder before researchers can decide whether the disorder is a suitable candidate for treatment, and as will be discussed later, this assessment cannot be made without consideration of various ethical and societal issues. The gene for the disorder must be cloned, and a safe, efficient way to introduce the gene into cells must be available. Research is needed to determine whether the gene has to be expressed in the affected tissue or whether release of the protein product at another site would be efficacious. In recent years, our increasing knowledge of the ways in which gene expression is controlled, together with our increasing technical skills in cloning and manipulating genes, have led to the solution of some of these problems. The ethical issues have been discussed at length and resolved to the satisfaction of those regulating this research. The first experiments in which genetically manipulated cells were introduced into patients began in 1989, and the first clinical trials of gene therapy, for adenosine deaminase deficiency, began in September 1990. In this chapter we will discuss some of the technical problems involved in gene therapy and how they are being overcome.

Gene Defects Have Been Corrected in Transgenic Animals

What evidence is there that normal function can be restored by introducing a normal gene into a cell carrying a mutated version of the gene? Experiments with transgenic mice have already shown that such therapy can be effective. An advantage of using mice for genetic studies is that lines of mice have been bred with inherited genetic defects that resemble or are homologous to human inherited disorders. For example, the *mdx* mouse has a mutation in the mouse homologue of the dystrophin gene, the same gene involved in human Duchenne muscular dystrophy. As described in Chapter 14, these mouse models may be produced by design by using homologous recombination, instead of searching for naturally occurring mutations.

When two mice of an inbred line carrying a recessive trait called *shiverer* are mated, the homozygous offspring develop tremors (hence the name *shiverer*) and die very young. The offspring are extremely deficient in myelin in the central nervous system as a consequence of a mutation in the gene for myelin basic protein (MBP). When a single copy of the normal MBP gene was introduced into *shiverer* embryos, the resulting transgenic mice had about 8 percent of normal levels of myelin (Figure 28-1). This level was insufficient to restore normal function, although the transgenic mice did live slightly longer than the mutant mice. Transgenic mice with two copies of the normal MBP gene produced 26 percent of normal levels of myelin, and these mice were phenotypically normal.

A single copy of the normal gene for a recessive trait caused by a mutated gene should be sufficient to correct the cell. (In the case of the experiment just

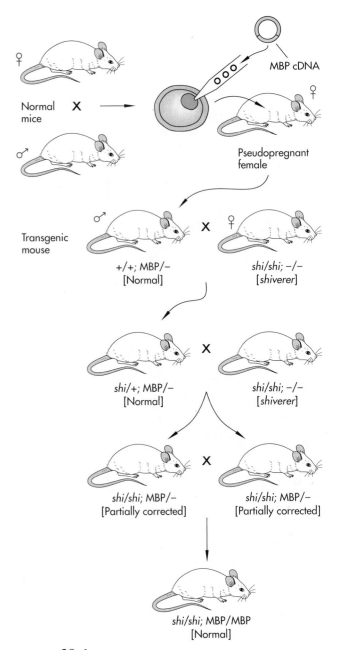

FIGURE **28-1**

Correcting the inherited disorder, *shiverer*, in mice. Transgenic mice were produced by injecting a plasmid containing the myelin basic protein (MBP) cDNA into fertilized eggs taken from normal females (+/+). A single male mouse with both copies of the wild-type gene (+/+) and carrying a single copy of the transgene (MBP/−) was obtained. This mouse was mated with homozygous female mice that completely lack the MBP gene (*shi/shi*). Some progeny (*shi/+*; MBP/−) of this mating were again mated with *shi/shi* homozygotes to produce mice in which the only copy of the MBP gene was the transgene (*shi/shi*; MBP/−). These mice were only partially corrected for the *shiverer* disorder, but those offspring of brother-sister matings that carried two copies of the transgene (*shi/shi*; MBP/MBP) were normal.

described, one normal copy was insufficient, probably because it was expressed at a low level.) For dominant traits created by a mutation, the mutant gene will have to be replaced or inactivated by the introduced normal gene. This replacement or inactivation can now be done experimentally by homologous recombination. The hypoxanthine phosphoribosyl transferase (HPRT) deficiency, although a recessive condition, has served as a model for this type of experiment. This deficiency results in the Lesch-Nyhan syndrome, and as we learned in earlier chapters there is a powerful selection available for HPRT$^+$ cells. Homologous recombination was used to replace the mutant HPRT gene in a line of HPRT$^-$ embryonic stem cells (Figure 28-2.) The cells were introduced into blastocysts, and a single male chimeric mouse carrying the corrected HPRT locus was obtained. Female offspring of the cross between this male mouse and females homozygous for the deletion inherited the corrected allele from their father (Figure 28-3.) Furthermore, the introduced gene had a low expression in liver and a high expression in brain, a pattern similar to that of the wild-type gene in a normal mouse.

Gene Therapy in Human Beings Raises Ethical Issues

These results in transgenic animals demonstrate that gene defects can be corrected by genetic manipulations of embryos. However, this approach is not feasible for gene therapy in human patients, for practical reasons; for example, the level of insertional mutagenesis caused by the integration of retroviral vectors into cellular DNA may be unacceptably high. Also, gene therapy in human beings is complicated by ethical considerations. A distinction has been drawn between genetic manipulations that involve only somatic cells and manipulations that might involve germ line cells. Gene therapy of somatic cells would be used to treat an already affected patient. The outcome is a genetic alteration that is restricted to the treated patient and is analogous to the treatment of genetic disorders by organ or tissue transplantation. For example, patients with hemophilia A show clinical improvement after receiving liver transplants for reasons unrelated to their hemophilia. Gene therapy involving germ line

cells is more controversial, because the modification is passed on to the children of the treated patient. This is considered by some to be ethically unacceptable, because, it is argued, we do not have the right to impose such a change on our descendants, no matter how well intentioned our reasons. It is not clear whether this opinion might hold under all circumstances. Some parents might be very eager to undergo a procedure that would remove a mutation forever from their family and all their descendants.

Another view considers whether gene therapy, by relieving human suffering, outweighs the risks involved. By this line of reasoning, gene therapy is justified if it is used to correct a genetic mutation that would otherwise result in a serious illness. On the other hand, the associated risks may outweigh the benefits if *gene enhancement* is used, not for the treatment of disease, but merely to supply a desired characteristic. In the latter case, it is not clear what criteria would be used to determine who would qualify for gene enhancement. It would almost certainly lead to further inequities in health care systems where access to care depends on financial status. However, there appears to be a general consensus that gene therapy for serious genetic disorders is justified, and considerable research has been performed to achieve this goal.

The Cystic Fibrosis Defect Can Be Corrected in Vitro

Many experiments have been undertaken, using cells in tissue culture, to find efficient ways to introduce human genes into cells so that they are stably incorporated without rearrangements or other alterations that might affect gene expression. There are now available a number of disabled vectors that cannot replicate after integration (Chapter 12), and many human genes have been successfully transferred to a variety of target cell types (Table 28-1). In some cases, genes have been transferred to cells taken from patients with inherited disorders. The introduced genes have been shown to produce the appropriate enzyme or protein.

The gene for cystic fibrosis (CF) was an obvious target for gene therapy. Abnormalities in chloride conductance channels had been described in vivo, and a

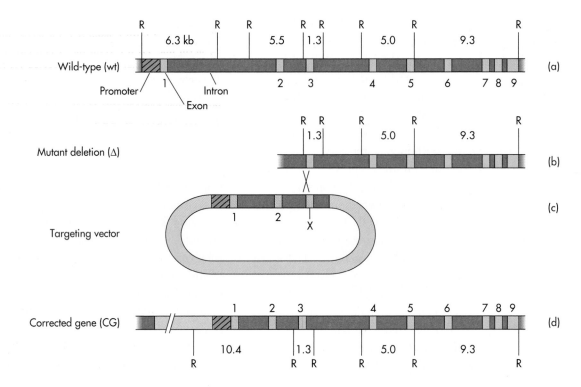

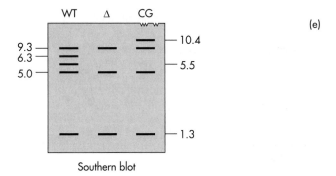

FIGURE **28-2**

Correcting an HPRT mutation in mice by homologous recombination. (a) The HPRT wild-type gene has nine exons separated by introns. The promoter region (hatched box) is immediately upstream of exon 1. There are eight sites for the restriction endonuclease *Eco*RI (R). When genomic DNA is cut with *Eco*RI, run on an electrophoresis gel, blotted, and probed with a cDNA for HPRT, five DNA fragments, each containing at least one exon, are revealed (WT lane in [e]). The sizes of these fragments in kilobases are shown above the gene. (b) The HPRT mutation used in these experiments is a deletion that removes the promoter and exons 1 and 2 (lane Δ in [e]). (c) The targeting vector used for homologous recombination includes the promoter, and exons 1, 2, and 3 of the HPRT gene. Exon 3 provides the homologous sequences through which recombination occurs. In addition, the HPRT gene fragment in the targeting vector has lost 6.7 kb from the first intron. (d) The deletion in the intron removes two *Eco*RI sites, so a unique *Eco*RI fragment of 10.4 kb is produced in targeted cells only if homologous recombination has occurred. Embryonic stem cells from a mouse with the deletion mutation are transfected with the targeting vector and selected for G418 resistance. (The vector includes a neomycin-resistance gene.) To promote recombination, the vector was linearized by cutting at an *Xho*I site (X) within exon 3. (e) Southern blotting is used to determine in which clones the vector DNA and endogenous deleted HPRT gene (Δ) have undergone recombination. Recombination results in the duplication of exon 3, but in the example shown, the corrected gene ([d] and lane CG in [e]) had undergone a deletion that removed this duplicated exon.

CF⁻ cell line expressing the same defects was available for assaying the activity of a CF gene introduced into the cells. In Chapter 12 we described how the defect in these cells has been corrected using a vaccinia virus vector (Figure 12-11), but retroviral vectors have been used as well. A complete cDNA for the cystic fibrosis transmembrane conductance regulator (CFTR) was constructed from three overlapping cDNA clones, and the full-length cDNA was cloned into a retroviral

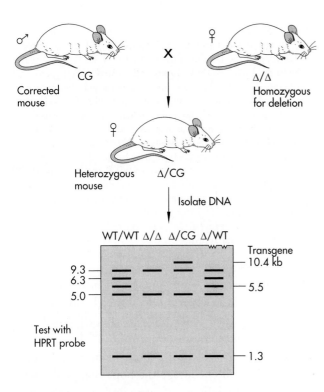

FIGURE **28-3**
The corrected HPRT gene (CG) is stably integrated into the germ line. HPRT is on the X chromosome, so males can be obtained that have the corrected HPRT gene with the characteristic 10.4-kb fragment as their only HPRT gene. The female offspring of matings between these males and females homozygous for the deleted HPRT gene (Δ/Δ) are heterozygous (Δ/CG).

vector (Figure 28-4.) The CF^- cells were infected with the retroviral vector carrying the CFTR gene, and cells with the integrated provirus were selected for G418 resistance. These cells expressed the CFTR gene. The most interesting data came from measurements of chloride channel function. Anion efflux from cells in response to adenylate cyclase stimulation provides a qualitative measure of chloride conductance pathways. In the absence of forskolin, a stimulator of adenylate cyclase, anionic efflux from CF^- cells and CF^- cells containing the introduced CFTR gene

TABLE **28-1**

Correction of Human Inherited Disorders in Cells in Tissue Culture

DISORDER	HUMAN GENE	TARGET CELL
Lesch-Nyhan syndrome	Hypoxanthine phosphoribosyl-transferase (HPRT)	Human HPRT$^-$ cells
Severe combined immunodeficiency	Adenosine deaminase (ADA)	Human ADA$^-$ skin fibroblasts; T cells; B cells
Severe combined immunodeficiency	Purine nucleoside phosphorylase (PNP)	Human PNP$^-$ skin fibroblasts
Gaucher disease	Glucocerebrosidase (GC)	Human GC$^-$ skin fibroblasts; bone marrow
Emphysema	α_1-Antitrypsin	Human liver cells
Short stature	Growth hormone	Human epidermal cells
Familial hypercholesterolemia (model)	Low density lipoprotein receptor	Hyperlipidemic rabbit fibroblasts; hepatocytes
Phenylketonuria	Phenylalanine hydroxylase	Mouse hepatoma cells
Citrullinemia	Argininosuccinate synthetase	Mouse fibroblasts
Thalassemia	β-Globin	Mouse fibroblasts; mouse erythroleukemia cells
Hemophilia (model)	Factor IX	Hemophilic dog skin fibroblasts

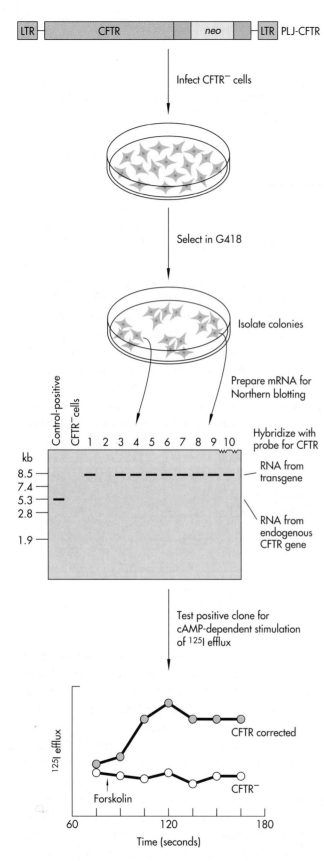

LTR — CFTR — neo — LTR PLJ-CFTR

Infect CFTR⁻ cells

Select in G418

Isolate colonies

Prepare mRNA for
Northern blotting

Hybridize with
probe for CFTR

RNA from
transgene

RNA from
endogenous
CFTR gene

Test positive clone for
cAMP-dependent stimulation
of ¹²⁵I efflux

FIGURE **28-4**
Correcting a cystic fibrosis mutation in vitro. A cystic fibrosis transmembrane conductance regulator (CFTR) gene was reconstructed from cDNA clones and cloned into a retroviral vector (PLJ-CFTR) containing a neomycin-resistance gene (*neo*). A cell line of airway epithelial cells (CFPAC-1) from a patient with CF was infected with the retrovirus, and cells with stably integrated provirus were selected for by their growth in G418. Messenger RNA was extracted from clones that grew in G418 and hybridized with a cDNA probe. One of the clones expressing CFTR mRNA was subjected to physiological studies of anion efflux, measured by using ^{125}I. Forskolin, a stimulator of adenylate cyclase, has no effect on efflux of ^{125}I from CFPAC-1 cells (○), but cells containing the introduced CF gene (●) respond normally. LTR, long terminal repeat.

conductance pathways were involved in the induced anion efflux. These experiments showed that chloride currents were detectable only in the CF⁻ +CFTR cells. Significantly, the responses of the CF⁻ +CFTR cells were comparable with those of normal human tracheal cells and encourage the hope that gene therapy for CF, with delivery of retroviral vectors directly to the lung epithelium by means of aerosols, will be possible.

Hematopoietic Cells Are Used for Expression of Human Genes in Animals

In the same way that a retrovirus acts as a vector to carry a gene into a cell (Chapter 10), so the cell can be regarded as a vector for the next stage of the process, carrying the gene into the patient's body. What are the requirements for these cells? They should be readily obtainable, grow well in culture, and be able to withstand the various manipulations involved in retrovirus infection. After infection, vector cells should be easy to return to the patient and should continue to live for many months, preferably for the life of the patient.

The cells of the bone marrow have many of these desirable features. The bone marrow contains stem cells that give rise to all cells of the hematopoietic series. Infection of these stem cells results in a continuous supply of cells containing the therapeutic gene. Investigators also have extensive experience in harvesting, manipulating, and transplanting bone marrow cells for treatment of hematopoietic diseases. In other words, the techniques for reconstituting the bone mar-

(CF⁻ +CFTR) is low. Upon addition of forskolin, efflux rapidly increases from CF⁻ +CFTR cells but remains low from CF⁻ cells. Electrophysiological techniques were used to determine whether chloride

row of patients and experimental animals are well worked out. One major drawback of using bone marrow cells as cellular vectors is that it has proved difficult to achieve high levels of expression of the introduced gene. In one set of experiments, the intact human β-globin gene with flanking 5′ and 3′ sequences was used in a retroviral construct. Bone marrow cells from female mice were infected with the virus, and these cells were then introduced into lethally irradiated male mice. When the mice were analyzed four to nine months later, all recipients had donor-derived cells in their spleen and bone marrow. Further analysis showed that while between 10 and 50 percent of the mouse erythrocytes synthesized human β-globin, the levels of human β-globin mRNA were only between 1.3 and 4.8 percent that of the transcript from the endogenous mouse gene. It is possible that these low levels of expression are a consequence of the absence from these constructs of some of the regions controlling β-globin expression. Other studies have located sequences 50 kb 5′ and 20 kb 3′ to the globin gene that increase expression of the β-globin in transgenic mice. It may be that expression of β-globin in hematopoietic cells will reach useful levels when these additional enhancer elements are included in the β-globin vectors.

Experiments achieving long-term expression of human adenosine deaminase in reconstituted mice have been more successful. The enzyme is expressed in many hematopoietic tissues, and the levels of expression are high. In one mouse, the human enzyme was 10 times the level of the endogenous mouse enzyme! This result is very encouraging for those designing clinical trials in patients. Calculations based on the amount of enzyme given to patients by conventional treatment indicate that genetically engineered hematopoietic cells will produce clinical improvement if plasma levels that are 25 percent of normal can be achieved.

Genetically Engineered Bone Marrow Cells Survive for Long Periods in Vivo

Convincing evidence that genes introduced in hematopoietic stem cells are expressed and that they can produce a therapeutic effect has come from experiments with a mutant dihydrofolate reductase gene (DHFR). Dihydrofolate reductase is an essential enzyme in the synthesis of deoxythymidylate (dTMP), which is required by dividing cells. Methotrexate (MTX) is a potent competitive inhibitor of DHFR and so has been used extensively in chemotherapy to kill dividing malignant cells. There is a cloned mutant DHFR gene (DHFR^r) that encodes a form of the enzyme resistant to MTX. This gene was introduced into murine bone marrow cells via a retroviral vector, and these cells were then transplanted to irradiated host mice (Figure 28-5). Two days after reconstitution, the primary transplant recipients were injected with methotrexate. Control mice rapidly became anemic as the MTX killed their bone marrow, and 60 percent of these mice were dead by 30 days. In contrast, although mice receiving the DHFR^r gene showed an initial fall in the number of white cells, these eventually reached normal levels. Two months later, bone marrow from these primary recipients was transplanted to a second set of irradiated mice that were then challenged with MTX. These mice were also resistant to MTX, an outcome showing clearly that a population of hematopoietic stem cells had been infected by the retroviral vector.

Skin Fibroblasts Are Target Cells for Gene Therapy

Another cell type that has been studied intensively as a vehicle for gene transfer is the skin fibroblast. These cells have been used in experimental treatments of patients with mucopolysaccharidoses, a group of disorders caused by inherited deficiencies of lysosomal enzymes. The treatments were unsuccessful, but they indicate that there are unlikely to be insurmountable obstacles to the introduction of genetically engineered fibroblasts into patients. Fibroblasts are easily obtainable, grow well in culture, and have been the subjects of many experiments; and skin fibroblasts can be efficiently infected with retroviral vectors. For example, in an early experiment, cells from a patient with Gaucher disease (glucocerebrosidase deficiency) were infected with a retroviral vector containing a glucocerebrosidase cDNA. Levels of enzyme in the infected

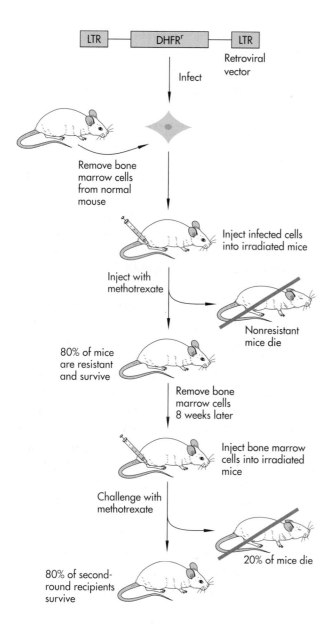

LTR — DHFRʳ — LTR
Retroviral vector

Infect

Remove bone marrow cells from normal mouse

Inject infected cells into irradiated mice

Inject with methotrexate

Nonresistant mice die

80% of mice are resistant and survive

Remove bone marrow cells 8 weeks later

Inject bone marrow cells into irradiated mice

Challenge with methotrexate

20% of mice die

80% of second-round recipients survive

FIGURE **28-5**

Genetically modified stem cells can produce an effect in animals. Bone marrow cells from mice were infected with a retroviral vector containing a mutated form of the gene for dihydrofolate reductase (DHFR). This mutant gene (DHFRʳ) encodes an enzyme resistant to methotrexate. Infected cells were injected into mice that had been irradiated to kill off their own bone marrow stem cells, and these primary recipients were challenged with methotrexate. Nonresistant control mice were killed by methotrexate, but 80 percent of mice receiving the genetically modified bone marrow cells survived. Bone marrow from these primary recipient mice was transferred to another set of mice that were again treated with methotrexate. About 80 percent of these secondary recipients also survived, a result showing that bone marrow stem cells in the primary recipients had been infected by the retrovirus and that these cells had maintained production of the DHFR enzyme through many cells divisions. LTR, long terminal repeat.

vector based on Moloney murine leukemia virus, containing a Factor IX cDNA driven by a cytomegalovirus enhancer-promoter. Infected cells produced large amounts of Factor IX, as determined by radioimmunoassay. This Factor IX, isolated by affinity chromatography, was fully functional in these clotting assays and was indistinguishable from authentic Factor IX prepared from plasma.

Hematopoietic cells are easy to return to the patient, and they have a natural site—the bone marrow—to populate. It is not clear however, how to implant skin fibroblasts in patients to achieve long-term survival of the implanted cells and to enable the protein to reach the affected tissues. Skin fibroblasts are thus clearly at a disadvantage relative to hematopoietic cells that are circulating in the blood. The most straightforward delivery routes for fibroblasts are injection into the peritoneal cavity or subcutaneous injection. In one experiment, NIH/3T3 cells with a human α_1-antitrypsin gene were injected intraperitoneally into mice. After four weeks, the mice were killed and the levels of human α_1-antitrypsin were measured in both the serum and in the lungs. The lung fluid contained approximately 29 percent of the level of human α_1-antitrypsin present in the serum of these mice, a result indicating that the enzyme had diffused from the peritoneal space to the bloodstream and thence to the target tissue, the lung epithelium. Mouse fibroblasts containing a human Factor IX gene have been embedded in three-dimensional collagen matrices, each matrix containing approximately

cells reached those of normal cells. In the case of another disorder, adenosine deaminase deficiency (ADA), skin fibroblasts from patients infected with a retrovirus carrying an ADA cDNA produced 12-fold more ADA than did normal cells. The investigators calculated that 4×10^8 of these fibroblasts would be needed to produce a therapeutic effect in patients. An important issue is whether skin fibroblasts can synthesize and secrete large amounts of a protein that is not a normal product of these cells. For example, can genetically engineered skin fibroblasts be used to produce Factor IX in hemophilia B patients? Human diploid fibroblasts were infected in vitro with a retroviral

4×10^6 cells. Two matrices were transplanted beneath the skin of mice. These gels were well vascularized by day 14 after implantation, and increasing amounts of human Factor IX were detectable in the plasma of recipient mice. However, it was discouraging that the levels of Factor IX declined and were undetectable by day 16.

This loss of activity in vivo has been analyzed in more detail by using a retroviral vector containing a neomycin-resistance gene and the human adenosine deaminase (hADA) gene. Skin fibroblasts were isolated from an inbred rat strain, infected with the vector, and infected cells selected with G418. The amount of hADA produced by each cell was such that as few as 5×10^7 cells could produce enough enzyme to treat a human patient. Cells were embedded in collagen matrices and transplanted either as a dermal graft or subcutaneously to the same strain of rats. Unfortunately, the loss of hADA activity was so rapid that it was undetectable by one month following transplantation. However, the polymerase chain reaction carried out with primers for the *neo* gene showed that cells containing the vector were still present in the grafts, and G418-resistant cells could be isolated by culturing cells from the grafts. The endogenous rat ADA gene was expressed at normal levels, an observation indicating that loss of hADA activity is the result of specific suppression of the human transgene. Determining whether this suppression of gene activity can be prevented by modification of the regulatory elements in the vector requires further research.

Hepatocytes May Be Used for Gene Therapy

A rather surprising target tissue for gene therapy is the liver. A large number of inherited metabolic diseases affect the liver, and liver transplantation has been tried in an effort to treat these and other conditions such as hypercholesterolemia and hemophilia. Techniques have been developed for isolating and culturing hepatocytes, so the appropriate target cells are available for retrovirus-mediated gene transfer. Examples of the use of hepatocytes for gene therapy are the experiments with the Watanabe strain of rabbits. These rabbits have clinical symptoms similar to those of fa-

milial hypercholesterolemia in humans. This inherited disorder is caused by mutations in the low density lipoprotein receptor (LDLR), and patients who are homozygous for mutations in the LDLR gene usually die of heart attacks before they are twenty years old. The Watanabe strain of rabbits also has an LDLR mutation, in this case a deletion of part of the gene. This is therefore an excellent animal model for gene therapy experiments in familial hypocholesterolemia. In these experiments, retroviral vectors carrying the human low density lipoprotein receptor gene were used to infect cultures of hepatocytes freshly isolated from Watanabe rabbits. Cell lines that were able to degrade low density lipoproteins at rates characteristic of normal cells were obtained. The main drawback to using hepatocytes is that methods for replacing them in vivo are not yet fully developed, even in animal models. One promising approach grows the hepatocytes on solid supports that are then transplanted to intraperitoneal or subcutaneous sites. Experiments have shown that genetically modified hepatocytes survived on these solid supports for up to two weeks following implantation. Other model experiments have used mouse hepatocytes from transgenic animals carrying either the human α_1-antitrypsin gene (hAAT) or the *E. coli* β-galactosidase gene. Hepatocytes from these transgenic animals were injected into the portal vein or directly into the spleen. Serum levels of hAAT showed that the cells survived at least one year, and staining to reveal β-galactosidase in liver sections of the recipient animals showed that a large fraction of the surviving hepatocytes reached the liver.

Gene Therapy Experiments Have Been Conducted in Large Animals

No matter how successful the experimental model systems of gene therapy are, the time will come when the methods have to be applied to human beings. This application presents several difficulties that are absent in experimental studies with mice or rats. The problems of scaling up from experiments in mice to large animals are not trivial. For example, even though a severely affected patient would benefit from Factor IX levels as low as 1 to 2 percent of normal, achieving this level would require that an 80-kg patient receive an implant containing as many as 2×10^9 cells (equivalent to about 10 mL of packed cells). As an intermediate step, gene therapy experiments have been performed in large animals like dogs and monkeys. These experiments in large animals further resemble those in human patients in that the cells have to come from the intended recipient of the modified cells. In one set of studies, bone marrow cells from macaque monkeys were infected with a vector carrying a human cDNA for adenosine deaminase. The donor animals received total body irradiation and then an infusion of the infected cells. Only low levels of expression of human ADA could be detected in these animals. Rather similar observations were provided by experiments with bone marrow cells that carried the human β-globin gene, suggesting that bone marrow cells may not be efficient cell vectors for gene therapy. Other cell types are being studied, including skin keratinocytes. In one study, canine keratinocytes were infected with a retroviral vector carrying a neomycin-resistance gene, and the infected cells were injected subcutaneously into the donor animal. Skin biopsies were taken, and the keratinocytes were isolated and grown in culture. Neomycin (G418)-resistant colonies could be detected for as long as 130 days after transplantation of the infected cells.

Myoblast Transfer Is Used to Treat Duchenne Muscular Dystrophy

Another kind of cell is available to those scientists working on gene therapy for muscle disorders like Duchenne muscular dystrophy (DMD). In the course of normal development, myoblasts—the stem cells of skeletal muscle—fuse to form multinucleate muscle fibers. A small number of myoblasts do not fuse, and these remain as single cells (called satellite cells) lying between the plasma membrane of the muscle fiber and the overlying extracellular matrix that covers the muscle fiber. Satellite cells retain the ability to fuse with each other and muscle fibers and are the stem cells that take part in muscle regeneration. Myoblasts might be the ideal cell vectors for delivering genes to muscle fibers. Experimental studies using the *mdx* mouse, which has a dystrophin mutation, have shown that a

limited amount of dystrophin synthesis occurs following injection of normal myoblasts into *mdx* muscles. However, it is unlikely that this will ever become an effective method of treatment for DMD patients. There are many obstacles to be overcome, not the least of which are scaling up from mouse muscles to those in boys and the possibility of immunologic rejection of the transplanted muscle cells. Furthermore, many different muscles, including cardiac muscle, are affected in boys with DMD, and the requirement to deliver sufficient numbers of myoblasts to all these different muscles is likely to be an insurmountable problem in myoblast therapy.

An obstacle on the road to gene therapy of DMD is that the gene and its cDNA are very large. An important step has now been taken with the construction of a full-length dystrophin cDNA that can be expressed in cells. Two cDNA libraries (A and B) were made in which reverse transcriptase was directed to begin synthesis of the first DNA strands at two points on the dystrophin mRNA. These points were specified by supplying two primers that hybridized to the middle (library A) or the 3' end (library B) of the mRNA. Clones encoding 7.8 kb of the 5' end of the gene and 8.0 kb of the 3' end were isolated from libraries A and B, respectively. These clones overlapped by 1.8 kb and were ligated together at a shared restriction site to give a full-length cDNA. This was cloned into an SV40 expression vector and transfected by electroporation into COS cells (Chapter 12). Western blotting using antibodies to dystrophin showed that a protein of the correct molecular weight for dystrophin was synthesized in the COS cells, and immunofluorescent studies showed that the protein was localized at the cell membrane. Yet, although these experiments demonstrate that even large genes like dystrophin can be manipulated in vitro, it will be a long time before such research will find clinical application.

Genes Can Be Delivered Directly to Target Sites in Vivo

One way to avoid the complications of developing cell-based systems for delivering genes to patients is

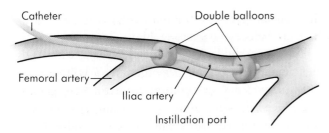

Catheter · Double balloons · Femoral artery · Iliac artery · Instillation port

FIGURE **28-6**
Direct delivery of retroviral vectors. A specially designed catheter has been used to deliver a retroviral vector carrying the gene for β-galactosidase to the endothelial cells of arteries. The catheter has two balloons that can be inflated. When the catheter has reached the correct site, the balloons are inflated so that a sealed compartment is created. A solution containing the retroviral vector carrying the β-galactosidase gene is introduced into this space and left there for 30 minutes. Between 10 days and 21 weeks later, segments of the blood vessels were examined for the presence of β-galactosidase activity by histochemical staining.

to deliver the retroviral vectors directly to the target cells. This technique has been shown to be an efficient way to infect the endothelial cells of blood vessel walls. These cells are an attractive target for gene therapy aimed at preventing atherosclerosis. They might be genetically modified to produce anticoagulants to prevent thrombus (blood clot) formation or angiogenic factors to promote repair of damaged vessels. A specially designed catheter was used to introduce a retroviral vector carrying the *E. coli lacZ* gene into blood vessels of the Yucatan pig, an experimental model for atherosclerosis. The catheter has two small balloons that, when inflated, create a sealed compartment (Figure 28-6). The retroviral vector was introduced into this compartment for 30 minutes. Sections of the blood vessel wall were then examined histochemically for β-galactosidase activity, and blue-staining cells, indicative of cells containing β-galactosidase expressed from the integrated provirus, were found up to 21 weeks after infection. It remains to be seen whether efficient ways can be found for direct delivery of retroviral vectors to other sites in vivo, for example, to the lungs in cystic fibrosis or to tumors.

Genetically Modified Lymphocytes Have Been Administered to Human Beings

A naturally occurring cell delivery system for malignant melanomas is already being used in clinical experiments. Tissue-infiltrating lymphocytes (TILs) are lymphocytes that are found infiltrating solid tumors and that can kill tumor cells when administered to the patient together with interleukin-2 (IL-2), a lymphocyte growth factor. When TILs were isolated from malignant melanomas, stimulated to grow in tissue culture by IL-2, and returned to the patient, they led to the regression of the melanoma in some patients. In trying to understand how TILs work in vivo, it is important to know how long the transplanted TILs survive in the patient and whether they reach the tumors. Retroviruses have been used successfully to mark cells for developmental studies (Chapter 14), and a similar strategy was proposed to follow TILs after infusion into patients. These experiments were the first in which human cells, genetically modified in vitro by using a retroviral vector carrying the neomycin-resistance gene, were to be introduced into patients, and the proposal was subjected to unparalleled scrutiny. Before the clinical experiments began, it had to be shown that TILs were infected efficiently and stably with the vector, that high levels of resistance to G418 were achieved, and that retroviral integration did not alter the growth characteristics of the cells. Five patients were treated with these TILs. Tumors were removed from each patient and TILs were grown in culture with interleukin-2. The cells were infected with a retroviral vector in which the *gag, pol,* and *env* genes were replaced with the neomycin-resistance gene (Chapter 12). The percentage of cells with a stably integrated *neo* gene was estimated to be between 1 and 11 percent. Patients received as many as 10^{11} TILs derived from their own tumors. Blood samples were taken on subsequent days, and PCR of the *neo* gene detected genetically modified TILs for up to 21 days in three patients, and for as long as 60 days in two patients. Moreover, cells containing the *neo* gene were found in samples of tumor tissue, but it is not clear whether these cells had homed in on the target tumors or had merely become trapped there by chance.

Nevertheless, the results show that it may be possible to use TILs as cell vectors to target tumors.

Human Adenosine Deaminase Deficiency is Treated by Gene Therapy

The experiments just described were intended to study the behavior of TILs in patients, and the genetic modification of these cells (introducing a retroviral vector with the neomycin gene) was done simply to provide a selectable marker to track the cells. However, data from these experiments demonstrated the feasibility and safety of introducing genetically engineered white blood cells into patients, and these experiments led to the first clinical trials of gene therapy. In September 1990, research workers at the National Institutes of Health began gene therapy treatment on patients with severe combined immunodeficiency caused by mutations in the adenosine deaminase gene. These children are unable to mount an immune response to infections and have to lead a very restricted life (the "boy in the bubble"). Without special care, they usually die before two years of age.

Now two children, aged 4 and 9 years, with ADA deficiency are receiving gene therapy. T lymphocytes are isolated from the children, a normal ADA gene is introduced by using a retroviral vector, and the cells are then returned to the patients. The children have been receiving gene-corrected cells every 1 to 2 months for up to one year and are showing significant clinical improvement. They are now able to mount an immune response and there has been a dramatic decrease in the number of infections the children suffer. The younger child, who has been treated for the longer time, is now leading a more normal life, attending school and participating in sports. These results are very encouraging, and if bone marrow stem-cell gene therapy becomes possible, repeated injections of modified cells may not be necessary. Whether this type or other forms of gene therapy will become standard care will depend on further research in the laboratory as well as in clinical trials. But the potential of gene therapy, discussed for so long, is now being realized.

Reading List

General

Friedmann, T., and R. Roblin. "Gene therapy for human genetic disease?" *Science,* 175: 949–955 (1972). [Review]

Ledley, F. D. "Clinical applications of somatic gene therapy in inborn errors of metabolism." *J. Inher. Metab. Dis.* 13: 587–616 (1990). [Review]

Miller, A. D. "Progress towards human gene therapy." *Blood,* 76: 271–278 (1990). [Review]

Tolstoshev, P., and W. F. Anderson. "Gene expression using retroviral vectors." *Curr. Opinions Biotech.,* 1: 55–61 (1990). [Review]

Verma, I. M. "Gene therapy." *Sci. Am.,* 262: 68–84 (1990). [Review]

Anderson, W. F. "Human gene therapy." *Science,* 256: 808–813 (1992). [Review]

Original Research Papers

CORRECTING GENE DEFECTS IN TRANSGENIC ANIMALS

Popko, B., C. Puckett, E. Lai, H. D. Shine, C. Readhead, N. Takahashi, S. W. Hunt, III, R. L. Sidman, and L. Hood. "Myelin deficient mice: expression of myelin basic protein and generation of mice with varying levels of myelin." *Cell,* 48: 713–721 (1987).

Readhead, C., B. Popko, N. Takahashi, H. D. Shine, R. A. Saavedra, R. L. Sidman, and L. Hood. "Expression of a myelin basic protein gene in transgenic shiverer mice: correction of the dysmyelinating phenotype." *Cell,* 48: 703–712 (1987).

Thompson, S., A. R. Clarke, A. M. Pow, M. L. Hooper, and D. W. Melton. "Germ line transmission and expression of a corrected HPRT gene produced by gene targeting in embryonic stem cells." *Cell,* 56: 313–321 (1989).

ETHICAL CONSIDERATIONS FOR GENE THERAPY IN HUMAN BEINGS

Walters, L. "The ethics of gene therapy." *Nature,* 320: 225–227 (1986).

Bontempo, F. A., J. H. Lewis, T. J. Gorene, J. A. Spero, M. V. Rgani, J. P. Scott, and T. E. Starzl. "Liver transplantation in hemophilia A." *Blood,* 69: 1721–1724 (1987).

Ledley, F. D. "Somatic gene therapy for human disease: background and prospects. Part II." *J. Pediatr.,* 110: 167–174 (1987).

"Gene therapy in man: recommendations of European Medical Research Councils." *Lancet,* I: 1271–1272 (1988).

Anderson, W. F. "Human gene therapy: why draw a line?." *J. Med. Philos.,* 14: 681–693 (1989).

Anderson, W. F. "Genetics and human malleability." *Hastings Center Report,* 20: 21–24 (1990).

CORRECTING GENETIC DEFECTS IN VITRO

Miller, A. D., D. J. Jolly, T. Friedmann, and I. M. Verma. "A transmissible retrovirus expressing human hypoxanthine phosphoribosyltransferase (HPRT): gene transfer into cells obtained from humans deficent in HPRT." *Proc. Natl. Acad. Sci, USA,* 80: 4709–4713 (1983).

Palmer, T. D., R. A. Hock, W. R. A. Osborne, and A. D. Miller. "Efficient retrovirus-mediated transfer and expression of a human adenosine deaminase gene in diploid skin fibroblasts from an adenosine deaminase-deficient human." *Proc. Natl. Acad. Sci. USA,* 84: 1055–1059 (1987).

Sorge, J., W. Kuhl, C. West, and E. Beutler. "Complete correction of the enzymatic defect of type I Gaucher disease fibroblasts by retroviral-mediated gene transfer." *Proc. Natl. Acad. Sci. USA,* 84: 906–909 (1987).

Miyanohara, A., M. F. Sharkey, J. L. Witztum, D. Steinberg, and T. Friedmann. "Efficient expression of retroviral vector-transduced human low density lipoprotein (LDL) receptor in LDL receptor-deficient rabbit fibroblasts in vitro." *Proc. Natl. Acad. Sci. USA,* 85: 6538–6542 (1988).

Osborne, W. R. A., and A. D. Miller. "Design of vectors for efficient expression of human purine nucleoside phosphorylase in skin fibroblasts from enzyme-deficient humans." *Proc. Natl. Acad. Sci. USA,* 85: 6851–6855 (1988).

Wilson, J. M., D. E. Johnston, D. M. Jefferson, and R. C. Mulligan. "Correction of the genetic defect in hepatocytes from the Watanabe heritable hyperlipidemic rabbit." *Proc. Natl. Acad. Sci. USA,* 85: 4421–4425 (1988).

Bordignon, C., S.-F. Yu, C. A. Smith, P. Hantzopoulos, G. E. Unger, C. A. Keever, R. J. O'Reilly, and E. Gilboa. "Retroviral vector-mediated high-efficiency expression of adenosine deaminase (ADA) in hematopoietic long-term cultures of ADA-deficient marrow cells." *Proc. Natl. Acad. Sci. USA,* 86: 6748–6752 (1989).

Axelrod, J. H., M. S. Read, K. M. Brinkhous, and I. M. Verma. "Phenotypic correction of factor IX deficiency in skin fibroblasts of hemophilic dogs." *Proc. Natl. Acad. Sci. USA,* 87: 5173–5177 (1990).

Fink, J. K., P. H. Correll, L. K. Perry, R. O. Brady, and S. Karlsson. "Correction of glucocerebrosidase deficiency after retroviral-mediated gene transfer into hematopoietic progenitor cells from patients with Gaucher disease." *Proc. Natl. Acad. Sci. USA,* 87: 2334–2338 (1990).

CORRECTION OF THE CYSTIC FIBROSIS MUTATION IN VITRO

Drumm, M. L., H. A. Pope, W. H. Cliff, J. M. Rommens, S. A. Marvin, L.-C. Tsui, F. S. Collins, R. A. Frizzell, and J. M. Wilson. "Correction of the cystic fibrosis defect in vitro by retrovirus-mediated gene transfer." *Cell,* 62: 1227–1233 (1990).

Gregory, R. J., S. H. Cheng, D. P. Rich, J. Marshall, S. Paul, K. Hehir, L. Ostedgaard, K. W. Klinger, M. J. Welsh, and A. E. Smith. "Expression and characterization of the cystic fibrosis transmembrane conductance regulator." *Nature,* 347: 382–386 (1990).

Rich, D. P., M. P. Anderson, R. J. Gregory, S. H. Cheng, S. Paul, D. M. Jefferson, J. D. McCann, K. W. Klinger, A. E. Smith, and M. J. Welsh. "Expression of cystic fibrosis transmembrane conductance regulator corrects defective chloride channel regulation in cystic fibrosis airway epithelial cells." *Nature,* 347: 358–363 (1990).

HEMATOPOIETIC STEM CELLS FOR GENE THERAPY

Dzierzak, E. A., T. Papayannopoulou, and R. C. Mulligan. "Lineage-expression of a human β-globin gene in murine bone marrow transplant recipients reconstituted with retrovirus-transduced stem cells." *Nature,* 331: 35–41 (1988).

Lim, B., J. F. Apperley, S. H. Orkin, and D. A. Williams. "Long-term expression of human adenosine deaminase in mice transplanted with retrovirus-infected hematopoietic stem cells." *Proc. Natl. Acad. Sci. USA,* 86: 8892–8896 (1989).

Corey, C. A., A. D. DeSilva, C. A. Holland, and D. A. Williams. "Serial transplantation of methotrexate-resistant bone marrow: protection of murine recipients from drug toxicity by progeny of transduced stem cells." *Blood,* 75: 337–343 (1990).

Wilson, J. M., O. Danos, M. Grossman, D. H. Raulet, and R. C. Mulligan. "Expression of human adenosine deaminase in mice reconstituted with retrovirus-transduced hematopoietic cells." *Proc. Natl. Acad. Sci. USA,* 87: 439–443 (1990).

SKIN FIBROBLASTS AS CELLULAR VECTORS FOR GENE THERAPY

Garver, R. I., A. Chytil, M. Courtney, and R. G. Crystal. "Clonal gene therapy: transplanted mouse fibroblast clones express human α_1-antitrypsin gene in vivo." *Science,* 237: 762–764 (1987).

Palmer, T. D., R. A. Hock, W. R. A. Osborne, and A. D. Miller. "Efficient retrovirus-mediated transfer and expression of a human adenosine deaminase gene in diploid skin fibroblasts from an adenosine deaminase-deficient human." *Proc. Natl. Acad. Sci. USA,* 84: 1055–1059 (1987).

St. Louis, D., and I. M. Verma. "An alternative approach to somatic gene therapy." *Proc. Natl. Acad. Sci. USA,* 85: 3150–3154 (1988).

Palmer, T. D., A. R. Thompson, and A. D. Miller. "Production of human factor IX in animals by genetically modified skin fibroblasts: potential for hemophilia B." *Blood,* 73: 438–445 (1989).

Palmer, T. D., G. J. Rosman, W. R. A. Osborne, and A. D. Miller. "Genetically modified skin fibroblasts persist long after transplantation but gradually inactivate introduced genes." *Proc. Natl. Acad. Sci. USA,* 88: 1330–1334 (1991).

HEPATOCYTES AS CELLULAR VECTORS FOR GENE THERAPY

Ledley, F. D., G. J. Darlington, T. Hahn, and S. L. C. Woo. "Retroviral gene transfer into primary hepatocytes: implications for genetic therapy of liver-specific functions." *Proc. Natl. Acad. Sci. USA,* 84: 5335–5339 (1987).

Wilson, J. M., D. M. Jefferson, J. R. Chowdhury, P. M. Novikoff, D. E. Johnston, and R. C. Mulligan. "Retrovirus-mediated transduction of adult hepatocytes." *Proc. Natl. Acad. Sci. USA,* 85: 3014–3018 (1988).

Wilson, J. M., D. E. Johnston, D. M. Jefferson, and R. C. Mulligan. "Correction of the genetic defect in hepatocytes from the Watanabe heritable hyperlipidemic rabbit." *Proc. Natl. Acad. Sci. USA,* 85: 4421–4425 (1988).

Anderson, K. D., J. A. Anderson, J. M. DiPietro, K. T. Montgomery, L. M. Reid, and W. F. Anderson. "Gene expression in implanted rat hepatocytes following retroviral-mediated gene transfer." *Somat. Cell Mol. Genet.,* 15: 215–217 (1989).

Armentano, D., A. R. Thompson, G. Darlington, and S. L. C. Woo. "Expression of human factor IX in rabbit hepatocytes by retrovirus-mediated gene transfer: potential for gene therapy of hemophilia B." *Proc. Natl. Acad. Sci. USA,* 87: 6141–6145 (1990).

K. P. Ponder, S. Gupta, F. Leland, G. Darlington, M. Finegold, J. DeMayo, F. D. Ledley, J. R. Chowdhury, and S. L. C. Woo. "Mouse hepatocytes migrate to liver parenchyma and function indefinitely after intrasplenic transplantation." *Proc. Natl. Acad. Sci. USA,* 88: 1217–1221 (1991).

GENE THERAPY EXPERIMENTS IN VIVO

Cornetta, K., R. Wieder, and W. F. Anderson. "Gene transfer into primates and prospects for gene therapy in humans." *Prog. Nucleic Acid Res.,* 36: 311–322 (1988). [Review]

Kantoff, P. W., A. P. Gillio, J. R. McLachlin, C. Bordignon, M. A. Eglitis, N. A. Kernan, R. C. Moen, D. B. Kohn, S.-F. Yu, E. Karson, S. Karlsson, J. A. Zwiebel, E. Gilboa, R. M. Blaese, A. Nienhuis, R. J. O'Reilly, and W. F. Anderson. "Expression of human adenosine deaminase in nonhuman primates after retrovirus-mediated gene transfer." *J. Exp. Med.,* 166: 219–234 (1987).

Flowers, M. E. D., M. A. R. Stockschlaeder, F. G. Schuening, D. Niederwieser, R. Hackman, A. D. Miller, and R. Storb. "Long-term transplantation of canine keratinocytes made resistant to G418 through retrovirus-mediated gene transfer." *Proc. Natl. Acad. Sci. USA,* 87: 2349–2353 (1990).

MYOBLAST TRANSFER FOR GENE THERAPY IN DUCHENNE MUSCULAR DYSTROPHY

Law, P. K., T. G. Goodwin, and M. Wang. "Normal myoblast injections provide genetic treatment for murine dystrophy." *Muscle & Nerve,* 11: 525–533 (1988).

Partridge, T. A., J. E. Morgan, G. R. Coulton, E. P. Hoffman, and L. M. Kunkel. "Conversion of mdx myofibres from dystrophin-negative to -positive by injection of normal myoblasts." *Nature,* 337: 176–179 (1989).

Law, P. K., T. E. Bertorini, T. G. Goodwin, M. Chen, Q. Fang, H.-J. Li, D. S. Kirby, J. A. Florendo, H. G. Herrod, and G. S. Golden. "Dystrophin production induced by myoblast transfer therapy in Duchenne muscular dystrophy." *Lancet,* 336: 114–115 (1990).

Morgan, J. E., E. P. Hoffman, and T. A. Partridge. "Normal myogenic cells from newborn mice restore normal histology to degenerating muscles of the mdx mouse." *J. Cell Biol.* 111: 2437–2449 (1990).

Lee, C. C., J. A. Pearlman, J. S. Chamberlain, and C. T. Caskey. "Expression of recombinant dystrophin and its localization to the cell membrane." *Nature,* 349: 334–336 (1991).

DIRECT DELIVERY OF GENES IN VIVO

Nabel, E. G., G. Plautz, F. M. Boyce, J. C. Stanley, and G. J. Nabel. "Recombinant gene expression *in vivo* within endothelial cells of the arterial wall." *Science,* 244: 1342–1344 (1989).

Wilson, J. M., L. K. Birinyi, R. N. Salomon, P. Libby, A. D. Callow, and R. C. Mulligan. "Implantation of vascular grafts lined with genetically modified endothelial cells." *Science,* 244: 1344–1346 (1989).

Nabel, E. G., G. Plautz, and G. J. Nabel. "Site-specific expression in vivo by direct gene transfer into the arterial wall." *Science,* 249: 1285–1288 (1990).

GENE THERAPY IN HUMAN BEINGS

Rosenberg, S. A., P. Spiess, and R. Lafreniere. "A new approach to the adoptive immunotherapy of cancer with tumor-infiltrating lymphocytes." *Science,* 233: 1318–1321 (1986).

Rosenberg, S. A., B. S. Packard, P. M. Aebersold, D. Solomon, S. L. Topalian, S. T. Toy, P. Simon, M. T. Lotze, J. C. Yang, C. A. Seipp, C. Simpson, C. Carter, S. Bock, D. Schwartzentruber, J. P. Wei, and D. E. White. "Use of tumor-infiltrating lymphocytes and interleukin-2 in the immunotherapy of patients with metastatic melanoma." *N. E. J. Med.,* 319: 1676–1680 (1988).

Kasid, A., S. Morecki, P. Aebersold, K. Cornetta, K. Culver, S. Freeman, E. Director, M. T. Lotze, R. M. Blaese, W. F. Anderson, and S. A. Rosenberg. "Human gene transfer: characterization of human tumor-infiltrating lymphocytes as vehicles for retroviral-mediated gene transfer in man." *Proc. Natl. Acad. Sci. USA,* 87: 473–477 (1990).

Rosenberg, S. A., P. Aebersold, K. Cornetta, A. Kasid, R. A. Morgan, R. Moen, E. M. Karson, M. T. Lotze, J. C. Yang, S. L. Topalian, M. J. Merino, K. Culver, A. D. Miller, R. M. Blaese, and W. F. Anderson. "Gene transfer into humans—immunotherapy of patients with advanced melanoma, using tumor-infiltrating lymphocytes modified by retroviral gene transduction." *N. E. J. Med.,* 323: 570–578 (1990).

29
Studying Whole Genomes

As every chapter in this book demonstrates, extraordinary insights into life's processes have come from the molecular analysis of genes, and there is every reason to believe that a complete knowledge of the genetics of an organism will be essential for a full understanding of complex biological phenomena like the organization of the cell, the development of organisms, and the functioning of the brain. There are now under way projects aimed at achieving just that, analysis of the complete genomes of a number of organisms. We know that many new genes of unknown function will be discovered, and that the experimental elucidation of the functions of those genes, and of how they interact in the cell, will be a great intellectual and experimental challenge for many years. The organisms being analyzed include *E. coli,* yeast, *Drosophila,* and the mouse because so much is already known of their genetics; *C. elegans* because of its structural simplicity and because it is amenable to detailed genetic analysis; *Arabidopsis* as a model for other plants; and, of course, human beings because of our need to learn more about genetic diseases and because of our innate curiosity to learn about ourselves.

However, it is one thing to map, clone, and sequence individual genes, and quite another to undertake a systematic effort to map, clone, and sequence an entire genome. The task is formidable even for small genomes like that of *E. coli,* and the difficulties increase rapidly with larger genomes (Table 29-1). Many technical innovations will be needed to complete these projects in a reasonable time, and at a reasonable cost. An immense amount of data will be produced, and the computational aspects of analyzing whole genomes require as great an effort as the work at the laboratory bench. There are also broader issues concerning the organization and funding of genome projects and, for the human genome projects, additional concerns about the societal impact of knowing the whole human genome. These issues will be discussed in the next chapter.

In this chapter, we describe the techniques that are being used to manipulate and analyze very large pieces of DNA. Some of these techniques are already well established, while a great deal of ingenuity and energy is being expended in devising new methods to overcome the technical difficulties inherent in tackling whole genomes. We will describe the two complementary approaches being followed: the bottom-up approach, by which small units of DNA are assembled to make larger pieces, and the top-down strategy, by which elucidation of gene organization begins by progressively breaking down large units.

Very Large Pieces of DNA Can Be Separated by Pulsed Field Gel Electrophoresis (PFGE)

Physical analysis of eukaryotic genomes required the development of ways to analyze very large fragments of DNA by electrophoresis. Gel electrophoresis followed by Southern blotting is an essential tool for analyzing DNA, but standard methods of agarose gel electrophoresis separate DNA molecules over only a limited size range. Molecules less than 10 kbp long are separated very well, those in the 10- to 30-kbp region are separated poorly, and those larger than 30 kbp are not separated at all. This limitation has been overcome with the development of *pulsed field gel electrophoresis (PFGE)*, in which the electrical field driving the long DNA molecules through the gel periodically changes its orientation. PFGE can separate molecules as large as yeast chromosomes (200 kbp to 3000 kbp).

TABLE 29-1		
The Subjects of Genome Projects and Their DNA Content (in base pairs)		
ORGANISM	CLASSIFICATION	BASE PAIRS
E. coli	Bacterium	4,000,000
Saccharomyces	Yeast	14,000,000
Arabidopsis thaliana	Plant	100,000,000
Caenorhabditis elegans	Nematode	100,000,000
Drosophila melanogaster	Insect	165,000,000
Mus musculus	Mammal	3,000,000,000
Homo sapiens	Mammal	3,500,000,000

The mechanism of PFGE is not understood. It is thought that separation of molecules of this size depends on the relative ease or difficulty that molecules experience when reorienting themselves in response to changing directions of the applied electrical field. Some remarkable experiments have been performed in which single, fluorescently labeled DNA molecules are directly observed moving through a gel matrix. In these experiments, electrophoresis takes place in a chamber on the stage of a fluorescence microscope, and DNA molecules can be seen to be extending and coiling, and reorienting, as the electrical field changes direction. Computer simulations that use mathematical models of DNA molecules duplicate these experimental results quite accurately (Figure 29-1). The size range of molecules that can be separated is determined by the length of time that the electrical field is applied before it is reoriented. The longer the field is applied in a given direction, the larger the size range of molecules that can be separated.

A critical step in PFGE is the isolation of very high molecular weight DNA molecules—essentially intact chromosomal DNA. Cells are first embedded in blocks of low-temperature gelling agarose and then treated with enzymes and other reagents to free the DNA from cell protein and RNA. The DNA molecules suffer less breakage within the gel matrix than they would if in free solution. While still in the agarose blocks, the DNA is digested with enzymes that cut very rarely in mammalian DNA. For example, *Not*I and *Sfi*I have octanucleotide recognition sequences, and these enzymes produce fragments hundreds of thousands of

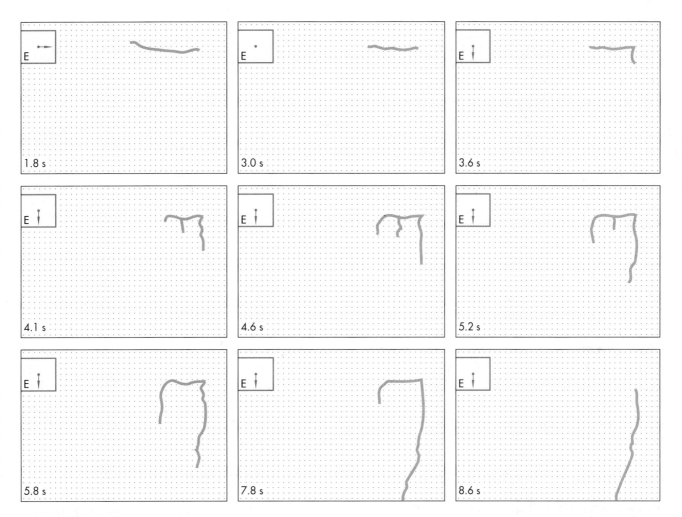

FIGURE **29-1**
Computer simulations of DNA molecules reorienting during pulsed field gel electrophoresis. These models closely resemble experimentally observed, fluorescently labeled molecules. The dots in the background are the gel matrix, and the direction of the electric field is shown by the pointer "E." At 3.0 s the electric field is switched off. At 3.6 s, when the electric field is switched through 90°, the leading end of the DNA molecule reorients first. A kink develops in the middle of the molecule shortly thereafter (4.1 s), and then the other end reorients (4.6 s). Eventually the leading end becomes dominant (7.8 s), and the molecule again migrates through the gel (8.6 s). Retardation of molecules of different size is related to kink development. More, and longer, kinks develop in longer molecules, and more time is needed for reorientation. (After Smith et al., 1990)

base pairs long when they cut mammalian DNA. The DNA fragments produced by such *rare cutters* range in size from 5×10^4 to over 9×10^6 bp. The agarose blocks containing the digested DNA are embedded directly in the agarose slab used for electrophoresis. Megabase DNA fragments may require several days of electrophoresis to be resolved.

PFGE Is Used to Make Large-Scale Physical Maps

Large scale genomic maps were originally genetic maps, in which the distances between markers are measured by recombination frequencies and the unit of distance is the centimorgan. These maps are an

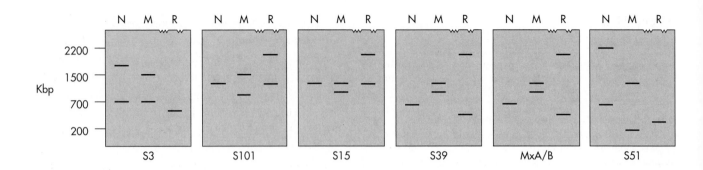

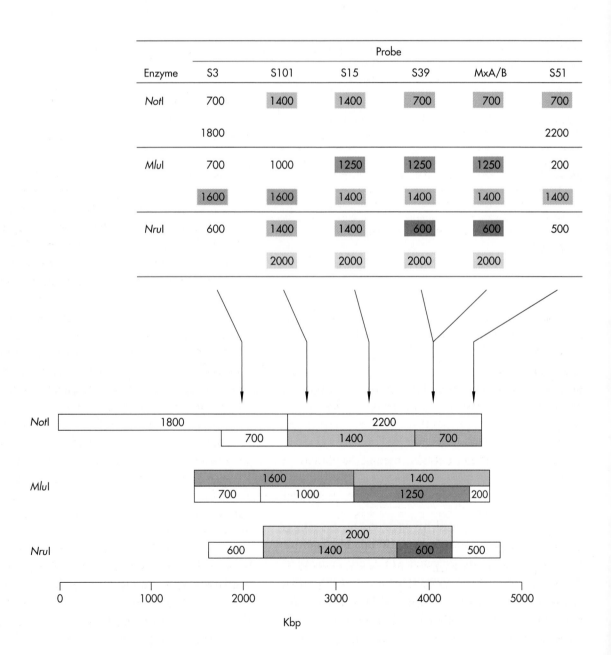

Mapping by pulsed field gel electrophoresis. This figure uses selected data from experiments mapping part of the long arm of chromosome 21. The original experiments used DNA both from a cell line and from white blood cells, and analyzed the DNA using a total of six enzymes and over 20 probes. We show simplified data for just three rare-cutting enzymes—*Not*I (N), *Mlu*I (M), and *Nru*I (R)—and six probes. The fragments detected on Southern blots are shown schematically in the first part of the figure, and the fragment sizes are listed below the blots. The arrows indicate the positions on the DNA strand where the probes bind. For each enzyme, fragments of the same color indicate that the same fragment is detected by more than one probe. For example, probes S101 and S15 both detect the same 1400-kb *Not*I fragment (green). In some cases, the enzymes cut only a fraction of the DNA, probably because of differing levels of methylation. In such cases, an enzyme will produce two bands in a given gel because the probe hybridizes to both the cut and the uncut sections of DNA. For example, for *Mlu*I probe S3 detects a 700-kb fragment, S101 detects a 1000-kb fragment, and both probes detect the same partially cut 1600-kb fragment that encompasses the two smaller fragments. In some cases the fragment is so difficult to detect on an x-ray film that it shows up with only one probe (for example the 2200-kb *Not*I fragment detected by probe S51). It is very difficult to measure the precise sizes of DNA fragments on pulsed field gels, which is one reason why the fragment sizes do not always add up correctly. These fragments can be put together to produce a restriction enzyme map covering 5000 kb, as shown at the bottom of the figure.

unreliable guide to *physical* distances on a chromosome because recombination rates vary in different regions of the genome. True physical maps, where the distance between markers is measured in nucleotides, can now be made using PFGE. The distance between two markers can be measured by determining the sizes of DNA restriction fragments that carry the markers; the smallest such fragment that carries both markers is an estimate of the distance between them. If combinations of probes and restriction enzymes are used, a physical map can be built up by determining which fragments have markers in common.

Because genetic markers will usually be megabases apart, rare cutting enzymes have to be used to produce large fragments, and PFGE has to be used to separate them. The sites for rare cutting enzymes are often subject to methylation, and some of the enzymes cannot cut sites that are methylated. This results in *partial cutting* if the sites are methylated, and the appearance

of large fragments that encompass smaller fragments that would be produced by complete cutting. This can be confusing, but the partial fragments are very useful because they link the smaller fragments. Since the patterns of methylation can vary between tissues, it is often helpful to analyze DNA from more than one tissue. The autoradiographs of pulsed field mapping gels are not easy to read, and our example, mapping five megabases on chromosome 21 (Figure 29-2), is a simplified version of the experimental data. Nevertheless, PFGE physical mapping has become an essential tool for finding one's way around the human and other genomes.

Putting Together the Cloned Genome of *E. coli* Requires Finding Overlapping Segments

At the other end of the scale from the large DNA fragments analyzed by PFGE are the cloned inserts of between 20 and 40 kbp of DNA in phage and cosmid clones. In the bottom-up approach, many thousands of such clones are made from genomic DNA. Indeed, a genomic DNA library must contain several genomes' worth of DNA to ensure that it contains all the genes of the organism. As genes are too long to be encompassed by a single cosmid clone, these clones must be ordered and linked together to reconstruct the original sequence. The linking is not done physically, but in computers, and reassembled contiguous stretches of DNA are called *contigs*. The ultimate contigs of cloned DNA are individual chromosomes.

How can this be done? One strategy being used to assemble cosmid clones identifies each clone by the pattern of restriction fragments that it generates and then orders the clones by searching for fragment sizes common to pairs of clones (Figure 29-3), thus indicating areas of overlap. The only complete genome assembled in this way is that of *E. coli*. Clones were digested with eight restriction enzymes, the fragments separated on agarose gels, and the Southern blots hybridized with a radiolabeled probe from one arm of the vector. The patterns of fragments on the autoradiographs were digitized, and computer algorithms were used to compare fragment sizes of pairs of clones.

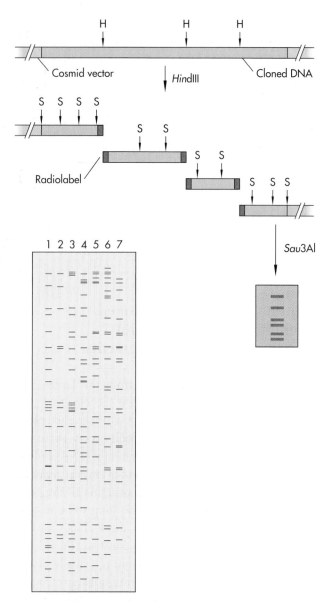

Pairs of clones with five restriction fragments in common were judged to be overlapping. This first step in the analysis led to 70 contigs, ranging in size from 20 to 180 kbp. Probes prepared from clones at the ends of contigs were used to screen arrays of clones. A clone that hybridizes to probes from the ends of two independent contigs is a candidate clone for linking the two contigs. This strategy led to the consolidation of the initial 70 contigs to just 7, ranging in size from 175 to 1570 kbp. This was a remarkable feat, for the total amount of DNA assembled in this restriction map is 4704 kbp. Reassuringly, the restriction map and the genetic map are in close agreement, and further analysis should lead to the integration of these maps.

A different strategy uses a large number of short oligonucleotides synthesized with random sequences as probes to characterize cosmid clones (Figure 29-4). The hybridization of a clone with a probe is recorded as "1" for a positive reaction, and "0" for a negative reaction. Clones containing DNA sequences in common will hybridize to the same subset of the total set of oligonucleotides used. By examining the patterns of "0" and "1" for all clones, the clones can be ordered. This type of analysis has been tested successfully in a model system, using the 153-kbp genome of herpes simplex virus (HSV) as the source of DNA. The application of this and other techniques to the human genome will depend on automation to speed up analysis and improve accuracy. Automated ma-

FIGURE **29-3**
Ordering cosmids by fingerprinting. In this example, the cloned DNA in a cosmid vector has three HindIII sites (H). The cut ends of the fragments are end-labeled with a radiolabeled nucleotide. A second digest is performed using Sau3A1 (S), which cuts more frequently than HindIII, so a set of smaller fragments is produced, only some of which are endlabeled. The fragments are sorted on a polyacrylamide sequencing gel, and the labeled fragments are detected by autoradiography. The sets of labeled fragments depend on the distributions of HindIII and Sau3A1 sites, and are characteristic for each cosmid clone (hence a "fingerprint"). The panel at the lower left shows the fingerprint patterns for seven different clones as they appeared on a computer screen during analysis of the patterns. Clones showing 5 or more restriction fragments were assumed to overlap. (After Coulson et al., 1986)

FIGURE **29-4** (*Right*)
Binary fingerprinting using oligonucleotide probes with random sequences. Twenty replicate filters carrying the same seven cosmid clones (A–G) are each probed with one of twenty different oligonucleotides. (1–20). In the binary pattern, positive hybridizations are listed as "1" and negatives as "0." For example, cosmid A hybridizes to four of the probes—2, 7, 14, and 19. This information can be used to determine which cosmids carry common sequences. Probe 2 hybridizes to cosmids A and C; probe 14 hybridizes to cosmids A, C, and D; and probes 7 and 19 hybridize to cosmids A and D. The cosmids can be ordered according to their overlapping patterns of hybridization. In this simplified "thought" experiment, there is insufficient information to order the probes unambiguously; for example, it cannot be determined whether the order is 6-13 or 13-6. In a model experiment, each filter carried 384 cosmids of herpes simplex virus type I and was probed with 1 of 22 oligonucleotides from known locations.

Hybridization of probes

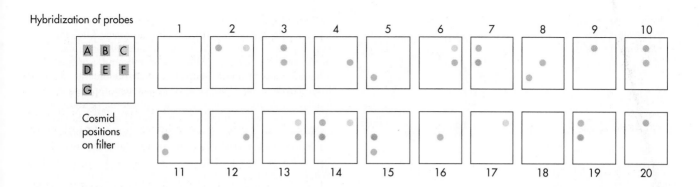

Cosmid positions on filter

Binary pattern of hybridization

	Oligonucleotide probe																			
Clone	1	2	3	4	5	6	7	8	9	10	11	12	13	14	15	16	17	18	19	20
A	0	1	0	0	0	0	1	0	0	0	0	0	0	1	0	0	0	0	1	0
B	0	0	1	0	0	0	0	0	1	1	0	0	0	0	0	0	0	0	0	1
C	0	1	0	0	0	1	0	0	0	0	0	0	1	1	0	0	1	0	0	0
D	0	0	0	0	0	0	1	0	0	0	1	0	0	1	1	0	0	0	1	0
E	0	0	1	0	0	0	0	1	0	1	0	0	0	0	0	1	0	0	0	0
F	0	0	0	1	0	1	0	0	0	0	0	1	1	0	0	0	0	0	0	0
G	0	0	0	0	1	0	0	1	0	0	1	0	0	0	1	0	0	0	0	0

Data analyzed to determine overlaps

	Oligonucleotide probe																			
Clone	1	4	12	6	13	17	2	14	7	19	11	15	5	8	16	3	10	9	20	18
A	0	0	0	0	0	0	1	1	1	1	0	0	0	0	0	0	0	0	0	0
B	0	0	0	0	0	0	0	0	0	0	0	0	0	0	0	1	1	1	1	0
C	0	0	0	1	1	1	1	1	0	0	0	0	0	0	0	0	0	0	0	0
D	0	0	0	0	0	0	0	1	1	1	1	1	0	0	0	0	0	0	0	0
E	0	0	0	0	0	0	0	0	0	0	0	0	0	1	1	1	1	0	0	0
F	0	1	1	1	1	0	0	0	0	0	0	0	0	0	0	0	0	0	0	0
G	0	0	0	0	0	0	0	0	0	0	1	1	1	1	0	0	0	0	0	0

Map of overlapping contigs

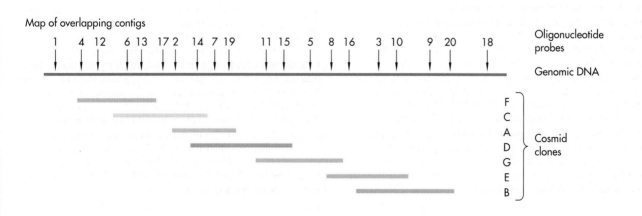

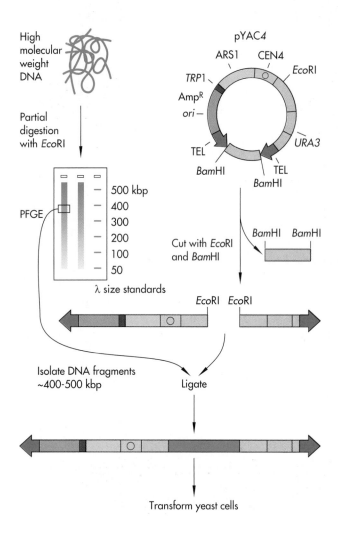

High
molecular
weight
DNA

Partial
digestion
with *Eco*RI

PFGE

500 kbp
400
300
200
100
50

λ size standards

pYAC4

ARS1   CEN4
TRP1   *Eco*RI
Amp^R
ori
URA3
TEL   TEL
*Bam*HI   *Bam*HI

Cut with *Eco*RI
and *Bam*HI

*Bam*HI   *Bam*HI

*Eco*RI   *Eco*RI

Isolate DNA fragments
~400-500 kbp   Ligate

Transform yeast cells

FIGURE **29-5**
Cloning DNA in yeast artificial chromosomes. The cloning
vector pYAC4 contains the elements needed for replication
of the vector as a linear artificial chromosome in yeast cells.
There are a centromere (CEN4) and two telomeres (TEL)
from *Tetrahymena*. ARS1 is an autonomously replicating sequence, equivalent to an origin of replication. In addition,
there are two genes for selection of transformed yeast cells:
URA3 and *TRP1*. The *ori* is a bacterial origin of replication
for growth in bacterial cells, and there is also a gene for
ampicillin resistance (Amp^R) for selection in bacterial cells.
When cut with *Eco*RI and *Bam*HI, the fragment between the
telomere sequences is lost, and two arms are produced with
*Eco*RI ends for cloning. Very high molecular weight DNA is
isolated and partially digested with *Eco*RI. The fragments
are run on a pulsed field gel, and that part of the gel containing fragments in the desired size range is cut out. (The
size markers for pulsed field gels are concatamers of λ
phage, so the marker fragments are multiples of 50 kbp.)
The DNA fragments are extracted from the gel and ligated
with the YAC arms. YACs containing cloned inserts are
separated from ligated vector lacking inserts by a second
size fractionation using pulsed field gel electrophoresis.
Spheroplasts—yeast cells stripped of their cell walls—are
transformed with the YACs.

chines for producing high-density arrays of cosmids
on filters have been devised, and densitometry of gels
is replacing manual analysis.

Yeast Artifical Chromosomes (YACs) Are Used for Cloning Huge DNA Fragments

Gene mapping and analysis based on cosmids is labor
intensive because so many cosmid clones are needed
to encompass the whole human genome. Now *yeast artificial chromosomes (YACs)* have been developed that can
clone hundreds of kilobases of DNA. A YAC vector
contains the following elements—a yeast centromere,
a sequence where the replication of yeast DNA is
initiated, two sets of yeast telomere sequences, se-

lectable markers, and a cloning site. These sequences
are assembled in a plasmid so that when the vector is
cut with the appropriate enzymes, two separate "arms"
are produced, each arm carrying a telomere sequence
and a selectable marker. Very high molecular weight
genomic DNA is partially digested with a rare-cutting
enzyme, fragments smaller than 200 kbp are removed
by electrophoresis or density gradient centrifugation,
and the large remaining fragments are then ligated
with the YAC arms (Figure 29-5). Following transfection into yeast cells, yeast clones containing both
arms of the YAC are selected using the selectable genes
present on each arm. The YACs behave as normal
yeast chromosomes, and replicate each time the cell
goes through mitosis. YAC clones can be screened on
a large scale by pooling many clones and using the
polymerase chain reaction to identify specific genes
(Figure 29-6). The clones of a pool that is positive
are then tested individually. Alternatively, clones can
be plated at high densities on filters and screened by
hybridization. YACs can be easily visualized as an
extra band on a pulsed field gel that separates the
chromosomes of a yeast clone, and their identity can
be confirmed by Southern blotting using total human
DNA or sequences for a specific gene as probes.

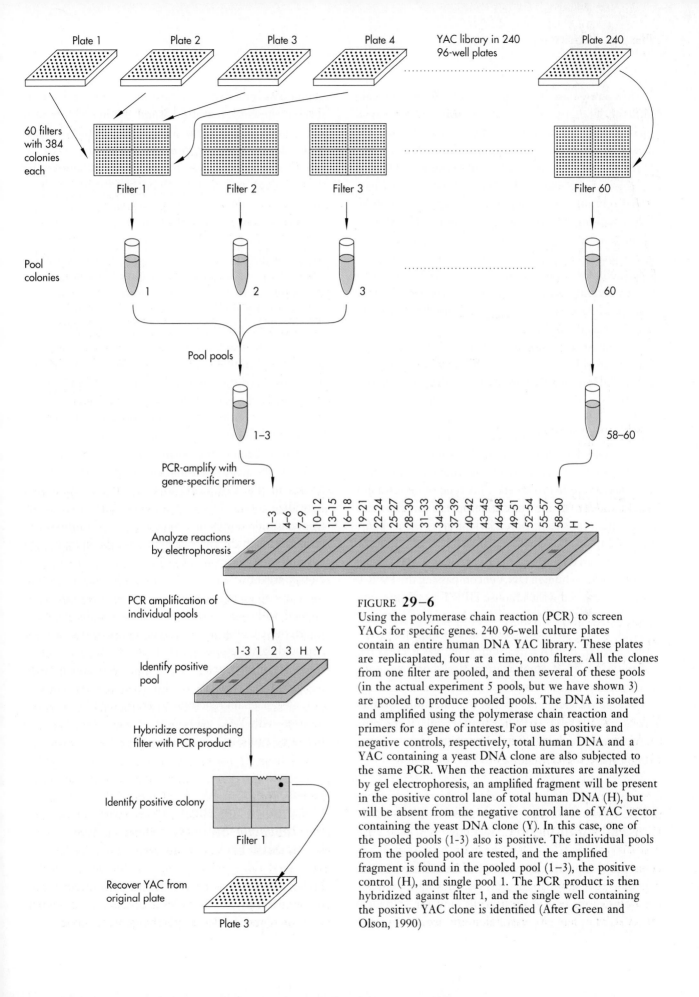

FIGURE **29–6**

Using the polymerase chain reaction (PCR) to screen YACs for specific genes. 240 96-well culture plates contain an entire human DNA YAC library. These plates are replicaplated, four at a time, onto filters. All the clones from one filter are pooled, and then several of these pools (in the actual experiment 5 pools, but we have shown 3) are pooled to produce pooled pools. The DNA is isolated and amplified using the polymerase chain reaction and primers for a gene of interest. For use as positive and negative controls, respectively, total human DNA and a YAC containing a yeast DNA clone are also subjected to the same PCR. When the reaction mixtures are analyzed by gel electrophoresis, an amplified fragment will be present in the positive control lane of total human DNA (H), but will be absent from the negative control lane of YAC vector containing the yeast DNA clone (Y). In this case, one of the pooled pools (1-3) also is positive. The individual pools from the pooled pool are tested, and the amplified fragment is found in the pooled pool (1–3), the positive control (H), and single pool 1. The PCR product is then hybridized against filter 1, and the single well containing the positive YAC clone is identified (After Green and Olson, 1990)

How useful are YACs going to be for large-scale cloning? The first results are promising. Statistical analysis suggests that YACs do not have a bias for accepting or excluding particular human DNA sequences, and this contrasts with cosmids that are known to clone selectively. About 10 percent of yeast clones contain more than one YAC, but the individual YACs in a single yeast clone can be isolated by PFGE. A much more serious problem is *cocloning*, in which a YAC contains DNA from two different locations. It is estimated that as many as 60 percent of YACs in some libraries could be like this, and this problem must be taken into account when YACs are used to generate long-range physical maps.

The integrity of human DNA in YAC clones has been verified by showing that the restriction maps of the cloned human DNA are similar to the same regions in genomic DNA. Furthermore, it can be shown that human genes cloned in YACs still code for functional proteins. One YAC clone contains the human glucose 6-phosphate dehydrogenase (G6PD) gene. When this human DNA is transfected into hamster cells, an active human enzyme is produced. This type of experiment can be done by fusing yeast spheroplasts directly with mammalian cells using polyethylene glycol, the chemical used to produce mammalian cell hybrids. Spheroplasts, prepared from yeast cells carrying a YAC clone with 680 kbp of human DNA encompassing the HPRT locus, were fused with a mouse HPRT$^-$ cell line, and HPRT$^+$ cells were selected for using HAT medium (Chapter 12). cDNA, synthesized using poly(A)$^+$ mRNA from the HAT-resistant cells, was used as the substrate for PCR amplification of a short sequence known to be different between the mouse and human HPRT genes. Sequencing the amplification product showed that these HPRT$^-$ mouse cells were now expressing human HPRT enzyme that had to be functional, since the cells grew in HAT medium.

YACs Are Used to Link the Cosmid Contigs of *C. elegans*

One problem with the maps produced using cosmid clones is that there are many gaps. Closing these gaps becomes more and more difficult, and for statistical reasons, complete closure of a cosmid map for a large genome is practically impossible. This problem is made even worse because cosmid genome libraries do not contain all the sequences present in the genome. YAC clones, however, provide a way of joining up cosmid contigs, because a single YAC clone may be large enough to contain sequences from the ends of adjacent but nonoverlapping cosmid contigs. Two strategies were employed in the *C. elegans* mapping project. In the first, large YAC clones were hybridized against a grid of cosmid clones. A grid is a regular array of clones spotted onto a filter. The clones can be spotted at very high densities, so large numbers of clones can be hybridized simultaneously to a probe. In this case, cosmids were grouped together if they hybridized to a single YAC clone. The second strategy was a more directed approach. A grid of YAC clones was hybridized successively with probes from cosmids that were at the ends of the already assembled contigs; YAC clones positive for these probes must contain sequences from the end of at least one cosmid contig. DNA from each of the positive YAC clones was used to screen the array of cosmid clones. Because the DNA inserts in the YACs are so large, the cosmids that hybridize to the DNA of a single YAC may come from cosmid contigs that have not yet been linked by cosmid clones. In this way, previously independent cosmid contigs can be assembled. Almost 1800 YAC clones are now on the *C. elegans* physical map, and there are 50 contigs with some 90 megabases of DNA aligned with the genetic map. How are the gaps between these contigs to be closed? These contigs usually end with YAC clones. The sequences at the ends of the YAC inserts are being determined and used to design primers for PCR. The primers are used in polymerase chain reactions to prepare probes that can be used with YAC grids. YAC clones hybridizing to the same probe are candidates for joining contigs.

Cosmid and YAC Clones Are Ordered along the Chromosomes of *Drosophila* Salivary Gland Cells by in Situ Hybridization

Different organisms have different advantages for genetic analysis. One factor contributing to the popularity of *Drosophila* as a subject for geneticists is that

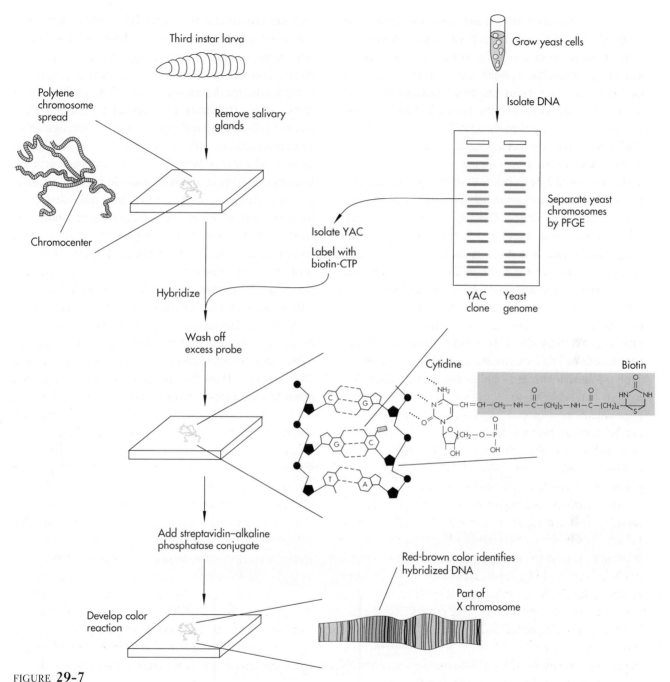

FIGURE **29-7**

Mapping *Drosophila* YACs on polytene chromosomes. Salivary glands are dissected from third instar larvae, and squash preparations are made of the cells so as to release and spread out the chromosomes. There are four chromosomes in *Drosophila*, but chromosome 4 is very small. The chromosomes form a characteristic pattern radiating from the chromocenter formed where the proximal heterochromatin of the 4 chromosomes comes together. The YAC DNA is separated from the other yeast cell chromosomes by pulsed field gel electrophoresis, and a DNA probe is synthesized using all four nucleotides, one of which carries a molecule of biotin. Following hybridization of the probe to the chromosome preparation, a streptavidin–alkaline phosphatase conjugate is added to the preparation. The streptavidin binds to the biotin, and a color reaction is developed by the addition of an appropriate substrate for the alkaline phosphatase. The location on the chromosome of the YAC insert is shown by the red-brown color of the reaction product.

the cells of the salivary glands of *Drosophila* larvae contain huge chromosomes. These *polytene* chromosomes arise because the DNA strands fail to separate when a chromosome replicates, so that each polytene chromosome consists of several hundred identical DNA molecules arranged in parallel. The banding pattern on these chromosomes is particularly clear, and for 60 years, *Drosophila* geneticists have used these bands as landmarks for mapping genes and chromosomal abnormalities. Now molecular geneticists are using these polytene chromosomes to map and order cosmid and YAC clones. The big advantage of these chromosomes is that unlike human chromosomes, for which bands may contain many millions of base pairs, the average length of a band in *Drosophila* is 22 kbp. This means that an average YAC clone will span several of these bands, and that contigs of overlapping YAC clones can be assembled by in situ hybridization. This is a "top-down" strategy that begins with an ordered set of large clones that can then be progressively broken down into smaller and smaller units for analysis and sequencing.

YAC libraries of *Drosophila* DNA are made just as other genomic libraries are, and here again *Drosophila* has advantages for molecular geneticists. The *Drosophila* genome is only some 165 million base pairs (for comparison, the human genome is 3000 million base pairs), and a *Drosophila* YAC library containing 3 genome equivalents requires only some 1500 clones. In situ hybridization of YACs is carried out by using pulsed field gel electrophoresis to isolate the YAC from the other yeast chromosomes in the cell, labeling it as a probe using biotin linked to a nucleotide, hybridizing this probe to cytological preparation of salivary glands, and detecting the probe with streptavidin–alkaline phosphatase (Figure 29-7). Almost 500 YAC clones have already been localized to polytene chromosomes (Figure 29-8), and soon there will be an ordered set of YAC clones covering the entire *Drosophila* genome. It is estimated that over 90 percent of the *Drosophila* genome is cloned in cosmids, YACs and chromosome walks. As unique sequences are found for these YACs, they can be used as oligonucleotide probes to link the YACs to the cosmid

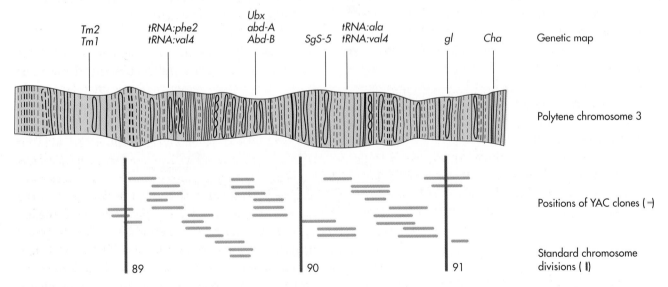

FIGURE **29-8**
Position of *Drosophila* YAC clones mapped on a polytene chromosome. This figure shows the portion of chromosome 3 that includes the *bithorax* complex (*Ubx*, *abd-A* and *Abd-B*) and illustrates the frequency and intricacy of the banding patterns on these chromosomes. This region is covered by overlapping YAC clones, each of which spans several bands. The numbers along the bottom of the figure refer to the system devised by Calvin Bridges in 1935 to identify regions of the chromosomes. Along the top of the figure are listed some of the genes that have been mapped to this area. [Modified from B. R. Jasny et al., "Genome maps 1991," *Science*, 254: 247–262 (1991)]

contigs being assembled in the bottom-up approach also being used to analyze the *Drosophila* genome.

An Entire Yeast Chromosome Has Been Sequenced

All the projects we have been discussing are concerned with cloning, but what of sequencing itself? After all, it is not until we get down to nucleotides that we can read the genetic code script. The most exciting example of large-scale sequencing so far accomplished is that of the whole of yeast chromosome III. This feat was accomplished by the European Yeast Genome Sequencing Network, a consortium of 36 laboratories in Europe. A total of 390 kbp was sequenced, which included 80 kbp of overlaps between the clones, giving a final figure of 315 kbp. Each clone was sequenced 4 times on both strands, and the level of error was low, at about 1 in 1000 bp. But what was hidden in this sequence? Computer analysis found 182 open reading frames (ORFs) that were more than 100 amino acids long. When these sequences were compared with amino acid sequences already in the databases, as many as 65 percent of the computer-detected chromosome III ORFs were found to be for unknown proteins. Experiments using, for example, mutation analysis are under way to determine the functions of these genes. This work shows clearly that large-scale sequencing is going to generate an extraordinary amount of new genetic data. All this will have to be analyzed experimentally to determine which sequences are genes, what proteins they encode, and what those proteins do. This new knowledge will lead to an unprecedented exploration of how cells work.

A Multiplex Method Speeds Up DNA Sequencing

While the sequencing of yeast chromosome III and of several hundred kilobases of *E. coli* DNA are undoubtedly steps in the right direction, they also point to the difficulties that arise from the magnitude of sequencing whole genomes. Sequencing on the genome scale—even that of *E. coli*—is going to require improvements in sequencing methods to increase speed and to reduce costs. The costs of present sequencing techniques are high because they are labor-intensive—each clone to be sequenced has to be carried individually through all the steps of the process. Multiplex DNA sequencing is an ingenious technique that reduces the numbers of samples that have to be handled by uniquely identifying each of 20 samples, and then mixing them for processing (Figure 29-9). Subsequent steps—Maxam-Gilbert chemical cleavage reactions, electrophoresis, transfer of DNA fragments to filters—are carried out on the pool of 20 samples. The pattern of fragments for one sample is revealed by hybridizing the filter with the specific probe that identifies that sample. The filter is washed and rehybridized with probes for each sample in turn. Experimental observations show that the DNA fragments are retained on the filters after as many as 50 hybridization and washing cycles.

Automated DNA Sequencing Greatly Speeds Up the Process

Multiplex sequencing increases the numbers of clones that can be analyzed simultaneously, and the yeast chromosome III experience shows that large-scale projects can be carried out by manual sequencing in many laboratories, provided there is strong organization and coordination. But this will not be sufficient by itself, and it is generally agreed that automation is essential for large-scale sequencing. Machines have already been developed that can perform two of the rate-limiting steps in DNA sequencing: the detection of DNA fragments and the translation of the fragment pattern into sequence information. These automated sequencing machines use nucleotides tagged with fluorochromes rather than radioactive nucleotides (Figure 29-10). Four different dyes are used, and when these are excited by a laser, they emit light at different wavelengths. The dyes can be used to label the M13 universal sequencing primer or each of the four dideoxy chain terminators. Because each dideoxy reaction mixture is identified by a different label, the mixtures can be pooled and run on a single lane, rather

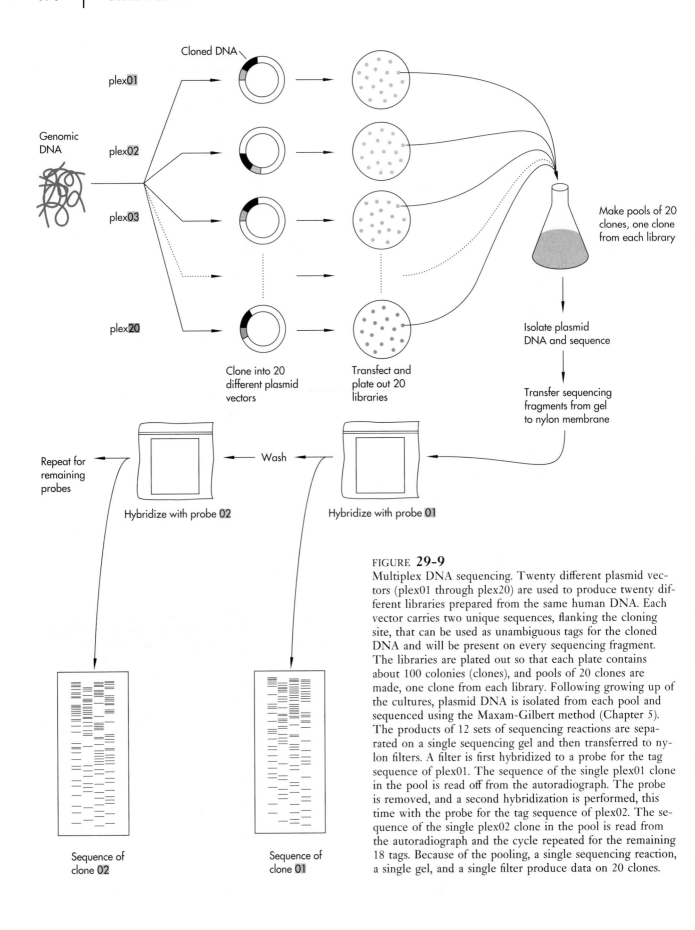

FIGURE **29-9**

Multiplex DNA sequencing. Twenty different plasmid vectors (plex01 through plex20) are used to produce twenty different libraries prepared from the same human DNA. Each vector carries two unique sequences, flanking the cloning site, that can be used as unambiguous tags for the cloned DNA and will be present on every sequencing fragment. The libraries are plated out so that each plate contains about 100 colonies (clones), and pools of 20 clones are made, one clone from each library. Following growing up of the cultures, plasmid DNA is isolated from each pool and sequenced using the Maxam-Gilbert method (Chapter 5). The products of 12 sets of sequencing reactions are separated on a single sequencing gel and then transferred to nylon filters. A filter is first hybridized to a probe for the tag sequence of plex01. The sequence of the single plex01 clone in the pool is read off from the autoradiograph. The probe is removed, and a second hybridization is performed, this time with the probe for the tag sequence of plex02. The sequence of the single plex02 clone in the pool is read from the autoradiograph and the cycle repeated for the remaining 18 tags. Because of the pooling, a single sequencing reaction, a single gel, and a single filter produce data on 20 clones.

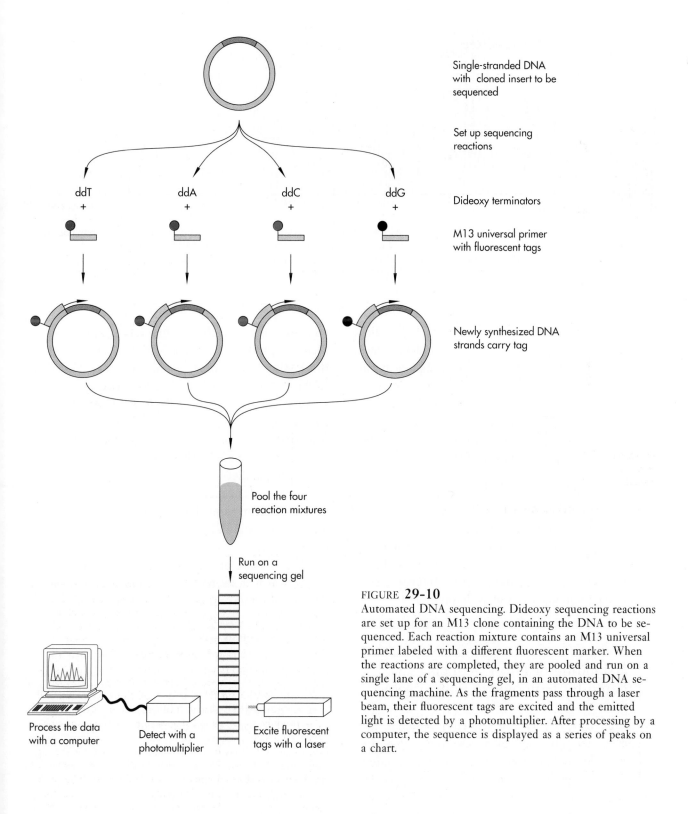

Single-stranded DNA with cloned insert to be sequenced

Set up sequencing reactions

Dideoxy terminators

M13 universal primer with fluorescent tags

Newly synthesized DNA strands carry tag

Pool the four reaction mixtures

Run on a sequencing gel

Process the data with a computer

Detect with a photomultiplier

Excite fluorescent tags with a laser

FIGURE **29-10**
Automated DNA sequencing. Dideoxy sequencing reactions are set up for an M13 clone containing the DNA to be sequenced. Each reaction mixture contains an M13 universal primer labeled with a different fluorescent marker. When the reactions are completed, they are pooled and run on a single lane of a sequencing gel, in an automated DNA sequencing machine. As the fragments pass through a laser beam, their fluorescent tags are excited and the emitted light is detected by a photomultiplier. After processing by a computer, the sequence is displayed as a series of peaks on a chart.

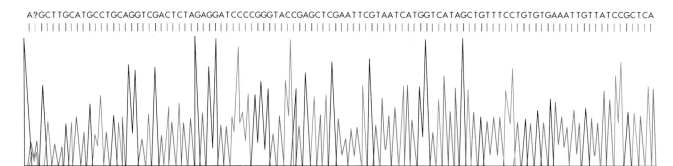

A?GCTTGCATGCCTGCAGGTCGACTCTAGAGGATCCCCGGGTACCGAGCTCGAATTCGTAATCATGGTCATAGCTGTTTCCTGTGTGAAATTGTTATCCGCTCA

FIGURE **29-11**
Output from an automated DNA sequencing machine. Each of the four colored traces represents one of the four nucleotides: red = thymidine; green = adenine; black = guanine; blue = cytosine. The order of bases in the DNA is shown by the succession of peaks and is listed at the top of the printout. When the computer of the sequencer is unable to interpret the output from the photomultipliers, the unread base is flagged with a yellow question mark (?).

than on the four separate lanes necessary with radioactive labeling. The fluorescently tagged dideoxy fragments migrate down the gel and pass through the beam of a laser. As they do so, the fluorochromes are excited by the laser and the light they emit is detected by a photomultiplier. Thus these machines read the gels in real time during electrophoresis. The emission peaks overlap, and computer analysis is necessary to resolve them. The typical processed output consists of a chart with four tracings, the order of the peaks corresponding to the ladder of fragments on conventional sequencing gels (Figure 29-11). Scientists cloning and sequencing the *C. elegans* genome are making intensive use of automated DNA sequencers. They have completed 121 kbp, and a further 160 kbp are at various stages of analysis and assembly.

Understanding of DNA Sequences Is Furthered by Homology Comparisons

No matter how fast we can sequence genomes, the sequences themselves are of no use unless they are quickly made available to other scientists, and unless we can make sense of what the sequences contain. *Informatics*—collecting, organizing, and analyzing sequences—will become a critical part of the genome projects. The first thing any molecular geneticist does with some newly derived sequence is to look for similarities to sequences already in the databanks. At present, the main databanks for DNA sequence are GenBank, at the Los Alamos National Laboratory, and the EMBL Data Library at the European Molecular Biology Laboratory, Heidelberg. These databases share sequences, and they have increased rapidly in size during the 1980s. By late 1991, GenBank contained 65,000,000 bases, of which about 20 percent are human sequences. The use of these databases has increased rapidly in parallel with the development of ever more powerful, and cheaper, personal computers and workstations. Scientists no longer require access to large mainframe computers for searching and analyzing sequence data.

Making sense of sequence data is not so simple as it might at first seem. Take, for example, the identification of coding regions within a sequence of genomic DNA. One strategy is to look for so-called *CpG islands*. These are regions where the incidence of the CpG dinucleotide is much higher than elsewhere in the genome, and such islands are frequently associated with genes. In the mouse, for example, it is estimated that some 30,000 genes have CpG islands in their upstream regions. As we discussed in Chapter 8, these CpG islands appear to be involved in regulation of gene expression. The cytosine in such CpG pairs can be methylated to form 5-methylcytosine, and genes whose associated CpG islands are methylated are usually inactive. A second strategy uses the sequences for the 5' and 3' splice sites as markers for intron and exon boundaries. Unfortunately, there are not unique sequences for either site. However, a careful search of the sequences present at several thousand 5′ and 3′ splice sites reveals so-called *consensus sequences:*

```
                         A
5′ splice site   A G│G U   A G U
                         G

                 U U      U U U U U U
3′ splice site       U U              X C A G│G
                 C C      C C C C C
```

The vertical line marks the junction between exon and intron. "X" can be any nucleotide, and where there are two nucleotides at the same place, the one on the upper line is the more common. However, since about 10 percent of sites do not conform to the consensus sequences, the consensus sequences do not provide an infallible guide to coding sequences. Furthermore, genes cannot be recognized by an invariant arrangement of introns and exons, presumably because splicing is able to cope with wide variations in intron/exon patterns.

Protein-coding regions can also be recognized by their nucleotide composition, which differs from that of noncoding regions. For example, protein-coding regions are not interrupted by stop codons. A nucleotide sequence can be read in three different ways (*reading frames*) because three nucleotides code for an amino acid:

are parts of genes. The situation is further complicated because important sequences, for example, those controlling a gene, may not lie in ORFs.

A nucleotide sequence can be converted to an amino acid sequence for comparison against a database. The similarities between proteins have often arisen as a result of gene duplication, leading to families of closely related proteins (Chapter 22). In these cases, searching for homologies between a new sequence and those already in the databases may lead to the recognition that the new protein is a member of an existing family of proteins. Some of these homologies have been very unexpected and influential in furthering research. One example is the relationship between an oncogene, *v-sis,* and platelet-derived growth factor (Chapter 18). Amino acid sequences are constrained by the need for different regions of a protein to perform specific functions. α-Helices; β-pleated sheets; and the zinc fingers, helix-turn-helix, and helix-loop-helix of DNA-binding proteins are examples. These *motifs* can be searched for and, when found, indicate a function for the protein. Such homology searching is limited because the existing databases contain a nonrandom selection of proteins. The failure to find homology between sequences in the databases and a newly discovered ORF does not mean that the latter is not a coding sequence. It may simply be the first protein to have been found with that particular sequence.

mRNA sequence	C U U A G C G U A G C U A C U A G A C U A G
Frame 1	C U U/A G C/G U A/G C U/A C U/A G A/C U A/G
Frame 2	C U/**U A G**/C G U/A G C/U A C/**U A G**/A C U/A G
Frame 3	C/U U A/G C G/**U A G**/C U A/C U A/G A C/**U A G**/

Translation of frames 2 and 3 is interrupted by two stop codons (**UAG**) in each reading frame. Frame 1 does not contain stop codons and is called an *open reading frame (ORF).* However, as there are three stop codons in the genetic code of sixty-four codons, a stop codon is likely to occur once every 21 codons, just by chance. ORFs average 200–250 bp in the mammalian genome, but as mammalian exons average 180 bp, the difficulty is to determine which open reading frames

The use of databases to interpret the functional significance of new DNA sequences will continue to increase and become ever more informative as more sequences become available for analysis and comparison. Analysis will also improve as new algorithms are devised. Parallel processing computers are being used for sequence analysis, and new computers are being designed that are dedicated to performing these analyses very efficiently.

Novel Methods Will Be Required for Large-Scale Sequencing of DNA

Automated machines are able to sequence faster than manual methods, but systems that sequence DNA much faster and cheaper than the present generation of automated machines must be developed for genome-scale sequencing. One promising technique carries out electrophoresis in a gel in a long capillary tube rather than in the conventional slab. Very high electrical field strengths can be used, and rapid, high-resolution separation of fragments is now being achieved. The principles of capillary gel electrophoresis are being applied to slab gels. The HUGE (for horizontal ultrathin gel electrophoresis) technique uses acrylamide gel slabs as thin as 10 μm. This technique combines the high resolution and speed of the capillary technique with the high sample capacity achieved by running multiple samples on slab gels.

The problem of sequencing whole genomes is such that some quite extraordinary techniques are being considered. For example, it may be possible to use the scanning tunneling microscope (STM) to read the order of nucleotides in a DNA strand directly. The STM works by maintaining a constant potential difference between a very fine probe and the surface it is scanning, in this case a DNA molecule. The potential difference is kept constant by moving the probe closer or farther away from the surface, and a three-dimensional picture of the DNA molecule is built up using the movements of the probe. When strands of synthetic DNA made up of only adenine are scanned, the imidazole rings of the deoxyribose nucleosides in the backbone can be distinguished from the purine rings of the adenine bases.

Probably the most remarkable sequencing technique under development is aimed at sequencing a single DNA molecule, one nucleotide at a time, but at a rate of some 100 to 1000 bases per second! A fluorescently labeled DNA molecule, about 40 kbp long, will be fed out into a laminar flow of buffer containing an exonuclease. The enzyme cuts off a single base at a time, and these are identified by a laser-photomultiplier system capable of detecting single molecules. Whether such a technique will ever reach the stage at which it can be used routinely remains to be seen, but such research illustrates the ingenuity being brought to bear on the problem of sequencing 3,000,000,000 bp in a reasonable time and at a reasonable cost.

Reading List

General

U.S. Congress, Office of Technology Assessment. *Mapping Our Genes—The Genome Projects: How Big? How Fast?* OTA-BA-373. U.S. Government Printing Office, Washington, DC., 1988.

Original Research Papers

ELECTROPHORESIS OF VERY LARGE DNA MOLECULES

Lai, E., B. W. Birren, S. M. Clark, M. I. Simon, and L. Hood. "Pulsed field gel electrophoresis." *BioTechniques,* 7: 34–42 (1989). [Review]

Schwartz, D. C., and C. R. Cantor. "Separation of yeast chromosome-sized DNAs by pulsed field gradient gel electrophoresis." *Cell,* 37: 67–75 (1984).

Vollrath, D., J. Nathans, and R. W. Davis. "Tandem array of human visual pigment genes at Xq28." *Science,* 240: 1669–1672 (1988).

Smith, S. B., S. Gurrieri, and C. Bustamente. "Fluorescence microscopy and computer simulations of DNA molecules in conventional and pulsed-field gel electrophoresis." In E. Lai and B. W. Birren, eds., *Electrophoresis of Large DNA Molecules: Theory and Applications.* Cold Spring Harbor Laboratory, Cold Spring Harbor, N. Y., 1990, pp. 55–79.

Burmeister, M., S. Kim, E. R. Price, T. De Lange, U. Tantravahi, R. M. Myers, and D. R. Cox. "A map of the distal region of the long arm of human chromosome 21 constructed by radiation hybrid mapping and pulsed-field gel electrophoresis." *Genomics,* 9: 19–30 (1991).

ORDERING COSMID CLONES

Lehrach, H., Drmanac, J. Hoheisel, Z. Larin, G. Lennon, A. P. Monaco, D. Nizetic, G. Zehetner, and A. Poustka. "Hybridization fingerprinting in genome mapping and sequencing". In K. E. Davies and S. M. Tilghman, eds., *Genetic and Physical Mapping I: Genome Analysis.* Cold Spring Harbor Laboratory, Cold Spring Harbor, N. Y., 1990, pp. 39–81. [Review]

Coulson, A., J. Sulston, S. Brenner, and J. Karn. "Toward a physical map of the genome of the nematode *Caenorhabditis elegans.*" *Proc. Natl. Acad. Sci. USA*, 83: 7821–7825 (1986).

Kohara, Y., K. Akiyama, and K. Isono. "The physical map of the whole *E. coli* chromosome: application of a new strategy for rapid analysis and sorting of a large genomic library." *Cell*, 50: 495–508 (1987).

Craig, A. G., D. Nizetic, J. D. Hoheisel, G. Zehetner, and H. Lehrach. "Ordering of cosmid clones covering the Herpes simplex virus type I (HSV-1) genome: a test case for fingerprinting by hybridisation." *Nuc. Acids Res.*, 18: 2653–2660 (1990).

CLONING DNA USING YEAST ARTIFICIAL CHROMOSOMES

Hieter, P., C. Connelly, J. Shero, M. K. McCormick, S. Antonarakis, W. Pavan, and R. Reeves. "Yeast artificial chromosomes: promises kept and pending." In K. E. Davies, and S. M. Tilghman, eds., *Genetic and Physical Mapping I: Genome Analysis.* Cold Spring Harbor Laboratory, Cold Spring Harbor, N. Y., 1990, pp. 83–120. [Review]

Schlessinger, D. "Yeast artificial chromosomes: tools for mapping and analysis of complex genomes." *Trends Gen.,* 6: 248–258 (1990). [Review]

Burke, D. T., G. F. Carle, and M. V. Olson. "Cloning of large segments of exogenous DNA into yeast by means of artificial chromosome vectors." *Science,* 236: 806–812 (1987).

Brownstein, B. H., G. A. Silverman, R. D. Little, D. T. Burke, S. J. Korsmeyer, D. Schlessinger, and M. V. Olson. "Isolation of single-copy human genes from a library of yeast artificial chromosome clones." *Science,* 244: 1348–1351 (1989).

Abidi, F. E., M. Wada, R. D. Little, and D. Schlessinger. "Yeast artificial chromosomes containing human Xq24-Xq28 DNA: library construction and represention of probe sequences." *Genomics,* 7: 363–376 (1990).

D'Urso, M., I. Zucchi, A. Ciccodicola, G. Palmieri, F. E. Abidi, and D. Schlessinger. "Human glucose-6-phosphate dehydrogenase gene carried on a yeast artificial chromosome encodes active enzyme in monkey cells." *Genomics,* 7: 531–534 (1990).

Green, E. D., and M. V. Olson. "Systematic screening of yeast artificial-chromosome libraries by use of the polymerase chain reaction." *Proc. Natl. Acad. Sci. USA,* 87: 1213–1217 (1990).

Huxley, C., Y. Hagino, D. Schlessinger, and M. V. Olson. "The human HPRT gene on a yeast artificial chromosome is functional when transfered to mouse cells by cell fusion." *Genomics,* 9: 742–750 (1991).

USING YACs FOR LONG-RANGE ORDERING OF CLONES

Coulson, A., Y. Kozono, B. Lutterbach, R. Shownkeen, J. Sulston, and R. Waterston. "YACs and the *C. elegans* genome." *BioEssays,* 13: 413–417 (1991). [Review]

Coulson, A., R. Waterston, J. Kiff, J. Sulston, and Y. Kohara. "Genome linking with yeast artificial chromosomes." *Nature,* 335: 184–186 (1988).

Green, E. D., and M. V. Olson. "Chromosomal region of the cystic fibrosis gene in yeast artificial chromosomes: a model for human genome mapping." *Science,* 250: 94–98 (1990).

MAPPING AND CLONING THE *DROSOPHILA* GENOME

Merriam, J., M. Ashburner, D. L. Hartl, and F. C. Kafatos. "Toward cloning and mapping the genome of *Drosophila.*" *Science,* 254: 221–225 (1991). [Review]

Lefevre, G. "A photographic representation and interpretation of the polytene chromosomes of *Drosophila melanogaster* salivary glands." In M. Ashburner and E. Novitski, eds., *The Genetics and Biology of Drosophila.* Academic, New York, 1976, pp. 31–66.

Siden-Kiamos, I., R. D. C. Saunders, L. Spanos, T. Majerus, J. Ttreanear, C. Savakis, C. Louis, D. M. Glover, M. Ashburner, F. C. Kafatos. "Towards a physical map of the *Drosophila melanogaster* genome: mapping of cosmid clones within defined genomic divisions." *Nuc. Acids Res.* 18: 6261–6270 (1990).

Ajioka, J. W., D. A. Smoller, R. W. Jones, J. P. Carulli, A. E. C. Vellek, D. Garza, A. K. Link, I. W. Duncan, and D. L. Hartl. "*Drosophila* genome project: one-hit coverage in yeast artificial chromosomes." *Chromosoma,* 100: 495–509 (1991).

SEQUENCING A YEAST CHROMOSOME

Goffeau, A., and A. Vassarotti. "The European project for sequencing the yeast genome." *Res. Microbiol.,* 142 (1991), in press.

Oliver, S. G., et al. "The complete DNA sequence of yeast chromosome III." *Nature,* 357: 38–46 (1992).

SEQUENCING *C. ELEGANS*

Sulston, J., et al. "The *C. elegans* genome sequencing project: a beginning." *Nature*, 356: 37–41 (1992).

MULTIPLEX DNA SEQUENCING

Church, G. M., and W. Gilbert. "Genomic sequencing." *Proc. Natl. Acad. Sci. USA*, 81: 1991–1995 (1984).

Church, G. M., and S. Kieffer-Higgins. "Multiplex DNA sequencing." *Science*, 240: 185–188 (1988).

AUTOMATED DNA SEQUENCING

Trainor, G. L. "DNA sequencing, automation, and the human genome," *Anal. Chem.*, 62: 418–426 (1990). [Review]

Smith, L. M., J. Z. Sanders, R. J. Kaiser, P. Hughes, C. Dodd, C. R. Connell, C. Heiner, S. B. H. Kent, and L. E. Hood. "Fluorescence detection in automated DNA sequence analysis." *Nature*, 321: 674–679 (1986).

Prober, J. M., G. L. Trainor, R. J. Dam, F. W. Hobbs, C. W. Robertson, R. J. Zagursky, A. J. Cocuzza, M. A. Jensen, and K. Baumeister. "A system for rapid DNA sequencing with fluorescent chain-terminating dideoxynucleotides." *Science*, 238: 336–341 (1987).

Brumbaugh, J. A., L. R. Middendorf, D. L. Grone, and J. L. Ruth. "Continuous, on-line DNA sequencing using oligodeoxynucleotide primers with multiple fluorophores." *Proc. Natl. Acad. Sci. USA*, 85: 5610–5614 (1988).

Wilson, R. K., C. Chen, N. Avdalovic, J. Burns, and L. Hood. "Development of an automated procedure for fluorescent DNA sequencing." *Genomics*, 6: 626–634 (1990).

ANALYZING DNA SEQUENCES

Bird, A. P. "CpG-rich islands and the function of DNA methylation." *Nature*, 321: 209–213 (1986). [Review]

Doolittle, R. F. *Of URFs and ORFs*. University Science Books, Mill Valley, Calif., 1987.

Antequera, F., J. Boyes, and A. Bird. "High levels of de novo methylation and altered chromatin structure at CpG islands in cell lines." *Cell*, 62: 503–514 (1990).

Doolittle, R. F., ed. "Molecular evolution: computer analysis of protein and nucleic acid sequences." *Meth. Enzymol.* 183: 1–736 (1990).

Senapathy, P., M. B. Shapiro, and N. L. Harris. "Splice junctions, branch point sites, and exons: sequence statistics, identification, and application to genome project." *Meth. Enzymol.*, 183: 252–278 (1990).

NOVEL METHODS FOR ANALYZING DNA

Edstrom, R. D., X. R. Yang, G. Lee, D. F. Evans. "Viewing molecules with scanning tunneling microscopy and atomic force microscopy." *FASEB J.*, 4: 3144–3151 (1990). [Review]

Beebe, T. P., T. E. Wilson, D. F. Ogletree, J. E. Katz, R. Balhorn, M. B. Salmeron, and W. J. Siekhaus. "Direct observation of native DNA structures with the scanning tunneling microscope." *Science*, 243: 370–372 (1989).

Swerdlow, H., and R. Gesteland. "Capillary gel electrophoresis for rapid, high resolution DNA sequencing." *Nuc. Acids Res.* 18: 1415–1419 (1990).

Davis, L. M., F. R. Fairfield, C. A. Harger, J. H. Jett, R. A. Keller, J. H. Hahn, L. A. Krakowski, B. L. Marrone, J. C. Martin, H. L. Nutter, R. L. Ratliff, E. B. Shera, D. J. Simpson, and S. A. Soper. "Rapid DNA sequencing based upon single molecule detection." *GATA*, 8: 1–7 (1991).

Jacobson, K. B., H. F. Arlinghaus, H. W. Schmitt, R. A. Sachleben, G. M. Brown, N. Thonnard, F. V. Sloop, R. S. Foote, F. W. Larimer, R. P. Woychik, M. W. England, K. L. Burchett, and D. A. Jacobson. "An approach to the use of stable isotopes for DNA sequencing." *Genomics*, 9: 51–59 (1991).

30

The Human Genome Initiative—Finding All the Human Genes

The various projects to map and sequence the entire genomes of human beings and other organisms open up the most exciting research prospects ever in the biological sciences. Map and sequence data will generate a wealth of new knowledge about biological processes. We have seen that sequence data are invaluable in revealing functions for newly cloned genes and for generating insights into how organisms work. This approach is especially rewarding when sequence comparisons between different species point to evolutionarily conserved sequences that are likely to encode functionally important proteins. This is one of reasons that the genomes of species other than *Homo sapiens* are also the subject of intensive study. In addition to helping us to learn more about the basic mechanisms of life, we hope to learn more about the genetic basis of human diseases, especially about the genes that contribute to the development of polygenic disorders. It is these diseases—such as coronary heart disease and mental disorders (Chapter 26)—that are very common, and an understanding of the interactions between genes and the environmental factors involved could lead to significant improvements in preventative medicine. Detailed knowledge of disease genes will help us unravel their pathologies and may suggest new rationales for treatment.

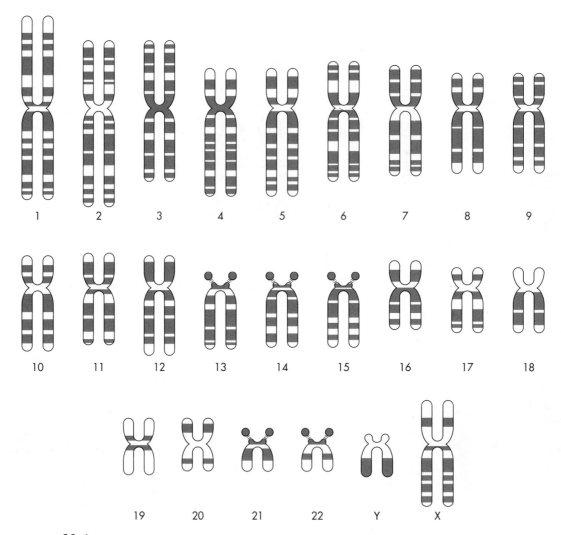

1 2 3 4 5 6 7 8 9

10 11 12 13 14 15 16 17 18

19 20 21 22 Y X

FIGURE **30-1**

The haploid human genome. This is a schematic drawing of 1 of each of the 23 human chromosomes, showing the pattern of staining seen with the Giemsa banding method. Chromosomes are first treated with trypsin and then stained with Giemsa. The patterns of light and dark bands are characteristic for each chromosome; and translocations, deletions, and other structural abnormalities can be identified. Typically 400 bands can be seen per haploid genome, and each band represents on average 7.5×10^6 bp, or twice as many base pairs as in the entire *E. coli* genome! Chromosome 1 constitutes 8.4 percent, and the Y chromosome about 2.0 percent, of the human genome. Taking the *E. coli* genome as a unit of genome size, a cytogenetic band is 2 genome units, and the Y chromosome is 15 genome units.

Until recently, human gene mapping and sequencing were done by individual investigators working on particular parts of the genome, sometimes in collaboration, but more often in competition, with other scientists. But the human genome is immense. The 23 human chromosomes (Figure 30-1) have a total genetic size of some 3300 cM, and the total number of base pairs is about 3 billion. To put this in context, even

the whole sequence of *E. coli*, a mere 4 million bp, is not yet known. The formidable size of the human genome has persuaded many scientists that a coordinated, directed approach is essential if the information in the human genome is to be acquired in a reasonable time; this was the impetus for the organized genome projects under way in various countries. In this chapter we will describe what progress has been

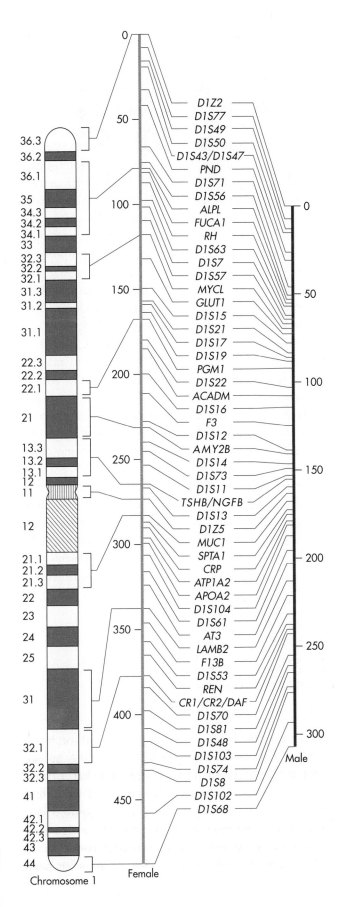

Chromosome 1

Female

Male

FIGURE **30-2**
High-resolution genetic linkage map of human chromosome 1, prepared by the CEPH Consortium. A drawing of the standard cytogenetic map is shown at the left, and the positions of 58 markers are shown. There are two linkage maps, one for females and one for males. Recombination is higher in human females than in males, so the *genetic map* of chromosome 1 is longer in females (about 500 cM) than in males (about 305 cM). This figure shows only the reference markers. The map is based on an analysis of 126 polymorphisms at 101 loci.

made with large-scale studies of the human genome, concentrating on some of the latest results in mapping, cloning, and sequencing.

Making a High-Resolution Genetic Map of Humans Uses Reference Markers

There has been remarkable progress in mapping genes and genetic markers since 1980, when mapping the human genome was first proposed. The Human Gene Mapping (HGM) workshops keep track of polymorphic loci and other genetic markers for the human genome. At the first workshop, HGM 1, in 1973, there were just 10; by HGM 10.5, held in 1990, 1867 genes and 4859 anonymous DNA fragments had been mapped! (An anonymous DNA fragment is a cloned DNA sequence of unknown function that has been mapped to a chromosome and that can be used as a marker). It was originally suggested that 150 to 200 linked markers spaced at 20 cM intervals would be sufficient for mapping genes in the human genome, but it has become clear that markers need to be closer together than this; the current aim is to produce a map with markers on average 2 cM apart by 1996. Intermediate maps with *reference markers* unambiguously assigned to a locus and ordered on the chromosome are being prepared with a 10-cM resolution. The markers are being chosen for their high degree of polymorphism, so that they should be useful for linkage analyses in many families. Chromosome 1 provides a good example of the progress in producing these reference marker maps (Figure 30-2.)

Chromosome 1 is the largest human chromosome and makes up approximately 8.4 percent of the autosomal DNA. At HGM 10.5, 193 genes and 120 anonymous DNA polymorphisms were assigned to this chromosome. In a collaborative effort involving 11 laboratories, and 40 families from the Center du Polymorphisme Humain collection (Chapter 26), 58 loci could be accurately mapped on chromosome 1 so that there is less than a 1 in 1000 chance that the loci could be elsewhere on the chromosome. These loci are a mean of 6.7 cM apart, and there are only four gaps where the separation between loci is greater than 15 cM. As other chromosome 1 markers are cloned, they will be mapped on the chromosome using these reference markers. Because the reference markers are available to all investigators, they will provide a framework for integrating data from different laboratories to make a linkage map for the whole chromosome. This is good progress, but it does show the magnitude of the task ahead over the next 4 years if the whole genome is to be mapped with a resolution of 2 cM.

Human Chromosomes Are Separated from Each Other Using Cell Sorting Machines

There have also been significant technical advances in cloning human genes. Because the human genome is so large, much effort has gone into isolating DNA from smaller portions of the genome. For example, it is much more efficient to search for an X-linked gene in a library prepared from just the X chromosome, rather than in a library containing clones from all 23 chromosomes. *Chromosome-enriched libraries* are made from chromosomes that are sorted by using fluorescence-activated cell sorters, or FACS (Figure 30-3). These machines were originally designed as flow cytometers to measure the DNA content of single cells. They were adapted to sort and collect different cell types, usually different classes of white blood cells. For chromosome sorting, metaphase chromosomes are isolated from tissue culture cells that have been blocked in mitosis by treating the cultures with the drug colcemid. The mitotic cells attach loosely to the culture dish surface and can be collected by shaking the dishes. The cells are treated with a chromosome isolation buffer that helps maintain the structure of the chromosomes through the mechanical trauma of passing through the FACS and that inactivates the cell nucleases that will destroy the DNA. The isolated chromosomes are then stained with DNA-specific dyes that stain AT-rich regions (for example, Hoechst 33258), and ones that stain GC-rich DNA (for example, chromomycin A). The stained chromosomes fluoresce when illuminated with light of the correct wavelength from lasers. The Hoescht dye fluoresces when exposed to UV light (wavelengths of 351 and 363 nm), while the chromomycin fluoresces with light of wavelength 458 nm. The amounts and proportions of the dyes taken up vary for each chromosome, and a computer recognizes the fluorescence pattern characteristic for each chromosome. In practice some chromosomes, particularly chromosomes 9 through 12, are indistinguishable and cannot be separated. It is usual to sort as many as 1×10^6 chromosomes to make a genomic library. This requires several days with a standard FACS, although there are machines custom-built specifically for chromosome sorting that can produce this number of chromosomes in a few hours. An early example of the usefulness of chromosome-enriched libraries is the X-chromosome library that was the source of the probes that detected the first RFLPs linked to Duchenne muscular dystrophy (Chapter 26). Now chromosome-enriched libraries are available for many chromosomes, and continuing technical developments will lead to genomic libraries prepared from increasingly pure chromosome isolates.

DNA for Cloning Can Be Microdissected from Human Chromosomes

While a FACS isolates single chromosomes, microdissection can be used to cut out DNA from specific regions of chromosomes. This technique was first used with those chromosomes that are easy to identify visually, without staining, using phase contrast light microscopy, for example, the giant polytene chromosomes of *Drosophila* and those of a mouse strain in which the X chromosome can be recognized because most of the chromosomes are fused in pairs. Similarly, clones could be prepared using DNA dissected from

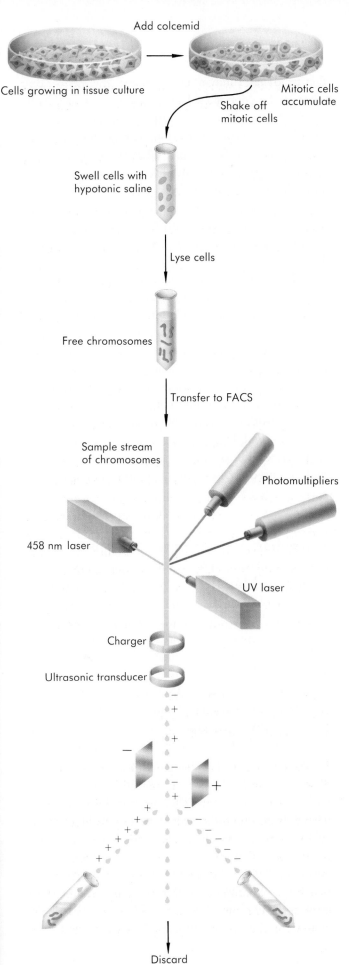

Add colcemid

Cells growing in tissue culture

Mitotic cells accumulate

Shake off mitotic cells

Swell cells with hypotonic saline

Lyse cells

Free chromosomes

Transfer to FACS

Sample stream of chromosomes

Photomultipliers

458 nm laser

UV laser

Charger

Ultrasonic transducer

Discard

FIGURE **30-3**
Principles of using a fluorescence-activated cell sorter (FACS) for sorting chromosomes. Cultures of cells are treated with colcemid, a mitotic inhibitor that blocks cell division in metaphase. Mitotic cells are rounded and only loosely attached to the culture dish surface, and pure preparations (better than 90 percent) of mitotic cells are collected by shaking the dish to dislodge them. The cells are gently broken open and the chromosomes stained with DNA-specific dyes like Hoechst 33258 and chromomycin A. The chromosome preparation is transferred to a FACS, where a narrow stream of solution containing the chromosomes passes through two laser beams that excite the fluorescent dyes. The photomultipliers monitor the emissions from each dye, and a computer determines the fluorescence of each individual chromosome as it passes through the laser beam. Human chromosomes stained with these dyes have characteristic fluorescence emissions so that the computer can distinguish between them. The computer directs the stream to be electrically charged depending on which chromosomes are to be collected. The stream is then broken into droplets by an ultrasonic transducer so that each droplet contains a single chromosome. If the drop is not charged, it passes straight through a magnetic field and is discarded. If the drop is electrically charged, it will be deflected by a magnetic field and collected in one of two collection tubes, depending on the charge it carries.

the short arm of human chromosome 2 because this also can be recognized in unstained preparations. Unstained preparations were used because it was thought that DNA subjected to the procedures required to reveal chromosome bands would be damaged or otherwise made unclonable. However, further research showed that cytogenetically stained chromosomes were a good source of DNA for cloning if PCR was used to amplify DNA from individual bands cut from the chromosomes. This meant that the technique could be applied to *all* human chromosomes. Spreads of human chromosomes are made and stained using the standard trypsin-Giemsa technique, and then chromosome fragments are cut from the appropriate bands by using ultrafine glass needles. DNA extraction (protease digestion, phenol extraction) and cloning reactions (restriction enzyme digestion, phenol extraction, ligation) are carried out in minute volumes of solution put within oil drops to reduce evaporation. These restriction fragments are ligated into a plasmid, or synthetic oligonucleotides ["linkers" (see Chapter 7)] are added to their ends. In either case, the fragments of unknown sequence are thus flanked by known sequences that are the targets for the PCR primers. The amplified DNA is cloned into a suitable vector, and the chromosomal location of the DNA in each clone is subsequently determined by using in situ hybridization or panels of somatic cell hybrids. The amount of DNA dissected from 20 chromosomes is estimated to be as little as 300 femtograms (300×10^{-15}g), but this is sufficient DNA for PCR amplification. Once again, the PCR turned a technique that had limited applications into a very useful method.

Somatic Cell Hybrids Serve as Sources for Purified Human Chromosome DNA

The importance of somatic cell hybrids in gene mapping and as sources of DNA for cloning was described in Chapter 26, and cell lines made by fusing mouse or hamster cells with cells from patients with chromosomal translocations and deletions have proved especially valuable. Recently a new method has been devised that circumvents the need to find patients with chromosomal abnormalities; rearranged and deleted chromosomes can now be produced at will by using x-irradiation to break up the chromosomes in human cells. The irradiated cells are then fused with rodent cells, and a selectable marker is used to identify hybrid clones containing human DNA from the region of the marker. Hybrid cells containing DNA from the region of the genes implicated in a complex of disorders— the Wilms tumor, aniridia, genitourinary anomalies, and mental retardation (WAGR) syndrome—were isolated in this way. The cell surface antigen gene MIC1 is on the short arm of chromosome 11 at 11p13, the same location as the WAGR syndrome. Hamster-human hybrid cells containing the short arm of chromosome 11 were irradiated and fused with Chinese hamster cells, and so-called radiation hybrids expressing the MIC1 cell surface antigen were selected by "panning" (see Chapter 7 and Figure 7-13). The amount of human DNA in these hybrids was narrowed down even further by discarding clones that expressed a second cell surface antigen, MER2, located at 11p15. One cell line contained about 3 mb of DNA from the 11p13 region, representing about a 900-fold purification from the rest of the genome.

This method relies on a selectable marker in the region of the human chromosome that the investigator can use to select the radiation hybrid cells containing the irradiated human chromosome fragments. But what can be done if a selectable marker is unavailable for the human DNA? A strategy that has been used to isolate DNA fragments from the region of the Huntington's disease (HD) gene offers a way out of this impasse (Figure 30-4). The chromosomes of a human-hamster hybrid cell line containing human chromosome 4 were fragmented with a lethal dose of radiation. These cells were fused with hamster cells that were HPRT⁻, and the cultures grown in HAT medium. The only cells that grow in this medium are radiation hybrid cells that contain an HPRT gene from the irradiated human-hamster hybrid cell line. (The lethal dose of radiation kills the latter unless they fuse with the hamster cells, and the hamster cells die because they are HPRT⁻). The surviving cells were screened with a probe specific for human LINE sequences, and the positive cells were tested with probes from the HD region. One cell line contained a 10-megabase fragment from the region of chromosome 4 believed to contain the HD gene. This technique does not rely on the retention of a selectable human marker and could be used for isolating DNA from any region.

X-Irradiated Fragments of Human Chromosomes Are Used for Gene Mapping

The x-irradiation hybrid method was originally devised for mapping genes, but it was of limited application because only a few markers were available at that time. This has changed dramatically with the cloning of many human DNA markers and with the availability of restriction fragment length and other polymorphisms. The power of this technique, named *radiation hybrid (RH) mapping*, has been demonstrated by the creation of a map spanning 40 megabases of human chromosome 21. Chinese hamster–human hybrid cells containing human chromosome 21 were x-irradiated. In situ hybridization with human genomic DNA showed that chromosome 21 was broken into an average of five pieces, each piece about 8 mb long. The irradiated cells were rescued by fusion with HPRT⁻ cells, and the surviving hybrid cells were screened with 18 chromosome 21 DNA probes (Figure

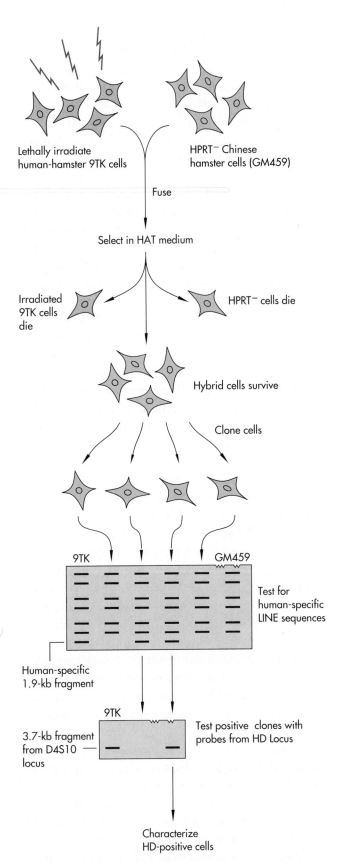

Lethally irradiate
human-hamster 9TK cells

HPRT⁻ Chinese
hamster cells (GM459)

Fuse

Select in HAT medium

Irradiated
9TK cells
die

HPRT⁻ cells die

Hybrid cells survive

Clone cells

9TK   GM459

Test for
human-specific
LINE sequences

Human-specific
1.9-kb fragment

9TK

3.7-kb fragment
from D4S10
locus

Test positive clones with
probes from HD Locus

Characterize
HD-positive cells

FIGURE **30-4**

Using x-irradiated hybrids for cloning. 9TK cells are a human-hamster hybrid cell line; the only human DNA present
is chromosome 4. These cells are x-irradiated to fragment
chromosome 4, and the dosage of radiation is sufficient to
kill the cells. The GM459 Chinese hamster cell line is deficient in hypoxanthine phosphoribosyl transferase (HPRT),
and these cells are killed when they are grown in medium
containing hypoxanthine, aminopterin, and thymidine [HAT
(Chapter 12)]. When the parental cells are fused and grown
in HAT medium, the only cells that can survive are
Chinese hamster cells that have acquired a hamster HPRT
gene from the irradiated hamster-human cell line. Clones of
surviving cells are first screened for the presence of human
DNA using probes for *Alu* or LINE repeat sequences.
Clones containing human DNA have a characteristic 1.9-kb
fragment. Positive clones were screened with probes mapping close to the Huntington's disease (HD) locus at the tip
of chromosome 4. For example, probe pHD2 from the
D4S10 locus detects a 3.7-kb band on Southern blots. In this
way other DNA sequences mapping in the HD region were
cloned.

30-5). The frequencies with which pairs of markers
were retained in each clone were calculated. This is
complicated because the cells may contain more than
one DNA fragment, and statistical methods are used
to determine the *breakage frequency* (Θ). This in turn
is used to calculate the distance between pairs of markers, the distance being measured in *centirays*, analogous
to the centimorgans of conventional mapping (Chapter
26). The best order of the markers was determined by
calculating the most probable of the 12 possible orders
for sets of four markers taken together. This produces
a probability score similar to the *lod* score of linkage
maps. The accuracy of this RH map was determined
by comparing it with a physical map of the region
produced by using pulsed field gel electrophoresis.
The two maps were similar, confirming the usefulness
of RH mapping.

Cloned Human DNA Fragments
Must Be Assembled into
Megabase-Sized Contigs

Cloning the entire human genome is not technically
difficult, and gene libraries, large enough to contain
the human genome several times over, are available

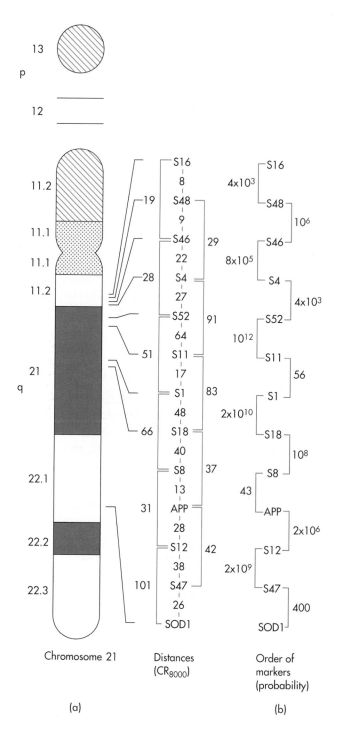

Chromosome 21

Distances
(CR$_{8000}$)

Order of
markers
(probability)

(a)

(b)

FIGURE **30-5**
Mapping genetic markers on human chromosome 21 using x-irradiated hybrids. Over 100 x-irradiation hybrid clones were made by the technique illustrated in Figure 30-4, except that the human-hamster hybrid cell line contained chromosome 21 as its only human genetic material. Each cell clone was tested with DNA markers from the long arm of chromosome 21. An analysis of the patterns of 13 retained markers leads to the map shown in (a), where the *distances* between pairs of markers (measured in centirays for the radiation dosage used) is determined by the frequency of breaks between markers, in a fashion similar to the way crossing over is used to establish distances (in centimorgans) in conventional mapping. In this case 8000 rads was used, so the unit is cR$_{8000}$. These can be used in turn to calculate the most likely *order* of the markers (b). For example, the probability that the order S8-APP-S12-S47 is correct compared with S8-S12-APP-S47 is 2 million to 1. The chromosome at p12 is composed of ribosomal genes that do not stain with conventional Giemsa staining, so that the DNA at the end of the chromosome appears separated from the main body of the chromosome.

in phage or yeast artificial chromosomes. However, these libraries contain DNA cloned at random from the genome. To make sense, the DNA in these clones must be arranged in the correct order so that the genome, or pieces sufficiently large to contain biologically interesting stretches of sequence, can be reconstructed. These reconstructions result in a series of overlapping clones in a computer database. Using some of the strategies outlined in Chapter 29, researchers have made progress in assembling some large sections of human chromosomes. Overlapping cosmid clones of chromosome 16 are being put together by a fingerprinting method which uses probes for the $(dC\text{-}dA)_n \cdot (dG\text{-}dT)_n$ repetitive sequence to identify restriction fragments common to two or more clones. The frequency of this repetitive sequence on chromosome 16 is such that there is a very high probability that clones with the same patterns of fragments hybridizing to a probe for $(dC\text{-}dA)_n \cdot (dG\text{-}dT)_n$ will have overlapping regions. Clones without $(dC\text{-}dA)_n \cdot (dG\text{-}dT)_n$ repeats have been further fingerprinted using other repetitive sequences like *Alu* and L1. 4000 clones have been fingerprinted in this way and assembled into 550 contigs, comprising about 60 percent of chromosome 16.

Sequence-Tagged Sites Identify Cloned DNA

Large-scale mapping and sequencing involve many laboratories, and the information generated must be collated. At present, this is difficult because different

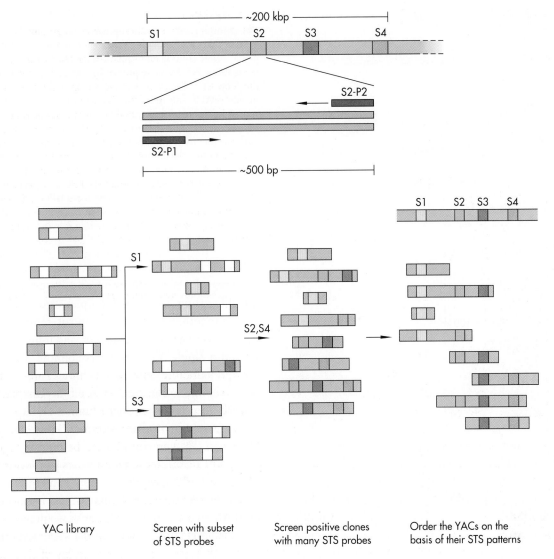

FIGURE **30-6**

Ordering YAC clones using sequence-tagged sites (STSs). A stretch of DNA some 200 kbp long has four STSs (S1 through S4). Each STS is about 500 bp long, and flanking sequences are known that can be used for making oligonucleotide primers for a polymerase chain reaction. S2 is shown in detail below (with primers designated P1 and P2). A YAC library is first screened by PCR using a subset of the STSs available (S1 and S3). Those clones that are positive are then screened with a large set of STSs, and the clones are ordered by the STSs that they contain.

laboratories are using different mapping and sequencing strategies. So-called *sequence-tagged sites* (STSs) may provide a common language to overcome these difficulties. An STS is a unique sequence, from a known location, that can be amplified using the polymerase chain reaction. Each STS is 200 to 500 bp long, and the sequences for the unique primers that will amplify each STS will be listed in databases. A laboratory will be able to obtain information about the sequence for

an STS from a database, synthesize the appropriate oligonucleotide primers, and "recover" the STS by PCR amplification of genomic DNA. One disadvantage is that unique sequences suitable for use as STSs have to be sought out. This requires knowledge, patience, and a degree of intuition; but interactive computer programs are being designed that may make the selection process faster and more efficient.

STSs can be used for ordering clones by searching

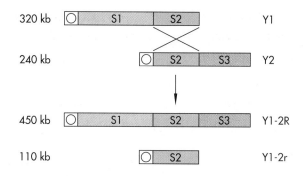

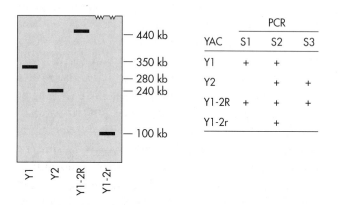

	PCR		
YAC	S1	S2	S3
Y1	+	+	
Y2		+	+
Y1-2R	+	+	+
Y1-2r		+	

FIGURE **30-7**

Reassembling genes by homologous recombination between YACs. Two YAC clones (Y1 and Y2) from the cystic fibrosis gene region contain the same sequence-tagged site (S2) and have about 110 kb of sequence in common. Y1 also contains the STS S1, while Y2 has STS S3. The yeast cells containing the YACs are mated to produce diploid cells, and spores in the resulting tetrads (Chapter 13) are analyzed for recombinants. The parental YACs are 320 and 240 kb long, and the recombinants should be 450 kb (Y1-2R) and 110 kb (Y1-2r) long. These molecules can be easily distinguished following pulsed field gel electrophoresis and Southern blotting with labeled total human DNA. Recombination adds STS S3 to Y1-2R and eliminates it from Y1-2r. As expected, STS analysis using the polymerase chain reaction shows that Y1-2R has S1, S2, and S3, while Y1-2r has only S2.

for clones common to two or more STSs. The strategy used in model experiments using the cystic fibrosis (CF) gene region is outlined in Figure 30-6. A YAC library was screened with STSs for sequences flanking and within the CF region. This screening identified clones likely to be carrying DNA of interest. Higher-resolution STS mapping was performed on these positive clones, and the clones were then ordered on the basis of the presence or absence of the STSs. The resulting map covered 1.5 mb of DNA and was very similar to that produced using conventional means. It has yet to be determined how straightforward this approach will be when exploring unknown regions of the genome, and it has been argued that approaches based on random oligonucleotide probes may be more efficient (Chapter 29).

Complete Human Disease Genes Can Be Reassembled in YACs

Intact genes, cloned in their entirety, are necessary for detailed studies of gene function. For example,

analyses of gene expression and regulation are impossible using less than complete genes. However, genes, together with their control regions, may be spread over very large distances, and it may be difficult to find a single YAC clone containing a large gene. Now, remarkable experiments have exploited the very high recombination rates in diploid yeast cells to assemble very large pieces of human DNA by homologous recombination between YACs that have a region of human DNA in common. For example, as many as three YAC clones, containing overlapping sequences from the CF region, have been sequentially recombined (Figure 30-7). A similar strategy was used successfully to construct a full-length version of the *BCL2* gene. The record so far for the reconstruction of a gene in a YAC is 2.4 mb of the Duchenne muscular dystrophy gene. However, this reconstructed gene is lacking a single exon, and recombinants will have to be checked carefully before being used for function studies. Nevertheless, this technique, combined with expression of YACs in mammalian cells, raises very exciting possibilities for analyzing the functions of even the largest human genes.

YACs Are Used to Clone the Telomeres of Human Chromosomes

One nice application of YACs is their use in cloning human telomere sequences. Telomeres are the structures at the ends of chromosomes, and they are essential for the proper replication of chromosomes. It

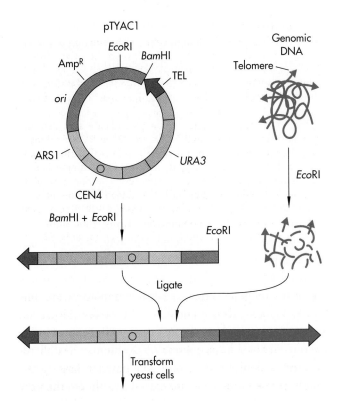

FIGURE **30-8**

Cloning human telomere sequences using a YAC. The plasmid pTYAC1, containing the various components needed for a yeast artificial chromosome, has one *Tetrahymena* telomere (TEL) and centromere (CEN4). When the plasmid is cut with both *Eco*RI and *Bam*HI, a single "arm," containing the telomere and centromere, is produced. After ligation with *Eco*RI-cut human genomic DNA, the only recombinant molecules that can be stably maintained in yeast are those that have acquired a human telomere sequence on one end. Transformed yeast colonies were screened by using pulsed field gel electrophoresis to separate their chromosomes, and Southern blotting was used to screen for clones containing human DNA. Positive clones were then screened for human telomere sequences by hybridization with a (TTAGGG)$_4$ oligonucleotide.

was known that telomeres from *Tetrahymena* (a freshwater protozoan) are made up of tandem arrays of the sequence TTAGGG, and that synthetic oligonucleotide probes complementary to this sequence hybridized to the tips of human chromosomes. This degree of conservation between species suggested that there might also be conservation of function. It was reasoned that if human telomeres could function in a yeast cell, it might be possible to use complementation to isolate human telomeres, and this proved to be the case (Figure 30-8). High-resolution in situ hybridization using

human DNA from these YACs showed that the cloned DNA hybridized to the telomeric regions of human chromosomes.

YACs Helped Unravel the Mysteries of the Fragile X Region

In Chapter 29, we discussed how YACs are being used to clone large pieces of DNA. Their effectiveness is evident from the recent cloning of a candidate gene for the fragile X syndrome, first described as recently as 1969. This is a form of mental retardation characterized by a structural abnormality of the X chromosome at Xq27, and it is a common disorder with a frequency of about 1 per 2000 males. (The frequency of Down's syndrome is about 1 per 1000 newborns.) Fragile X has a very puzzling pattern of inheritance. About 20 percent of males with the fragile site are phenotypically normal and are called "transmitting males." The degree to which children are affected depends on whether the fragile X is inherited from the transmitting male or from a female carrier. For example, all the daughters of a transmitting male receive a fragile X from their fathers, but they are phenotypically normal. However, their children—male and female—may be affected. The children of a mother who is mentally retarded are themselves at a greater risk of mental retardation than if they are born to a carrier who is not mentally retarded. These differences occur within the same family where the same mutation is apparently involved. Now recombinant DNA analysis is beginning to tell us what is going on in this unusual syndrome.

Cytogenetics, linkage studies, and somatic cell mapping (Chapter 26) mapped the fragile X site to the long arm of the X chromosome at Xq27.3. YACs encompassing the fragile X site were isolated, and probes from these YACs were found to hybridize to a DNA fragment that was altered in all families carrying the fragile X genotype. The sizes of the fragment *within* a family varied, suggesting that there is amplification or insertion of DNA at this site. Sequencing has shown that there is a trinucleotide sequence (CGG) that is repeated between 15 and 65 times in normal DNA but amplified in fragile X DNA. Presumably the instability of the fragile X site is due to the presence of this sequence, and the variable expression of the disorder is due to variability of this sequence even within a family.

The question remained as to which gene was disrupted in this syndrome. cDNAs for a gene called *FMR-1* (for fragile X mental retardation-1) have been cloned using probes from one of the fragile X YACs. One of the exons in the *FMR-1* gene is associated with the variably sized (CGG) fragment, and since the gene is expressed in the brain, it is a strong candidate as the source of the fragile X syndrome. *FMR-1* has a remarkable sequence suggesting that the protein has a continuous run of 30 arginine residues. This would be a highly basic protein, and, like histones, it may be involved in DNA binding. Variable methylation at the fragile X site and imprinting (Chapter 14) may be contributing factors in determining expression of the disease. Once again, detailed knowledge of a disease gene has led to the rapid development of a reliable diagnostic test that has already been used for prenatal diagnosis.

Large-Scale Sequencing of the Human HPRT Gene Was Done in Four Stages

Mapping and cloning the human genome are moving along quickly, but what of sequencing? Some indication of what is involved in systematic sequencing of a large genome comes from a simple calculation. Manual methods of sequencing generate a maximum of 100,000 base pairs per person per year, and at this rate, 30,000 person-years will be required to sequence the human genome. This is not satisfactory, and it is clear that very major improvements are going to be needed before a concerted effort to sequence the whole human genome can begin. Some of the techniques being developed were referred to in Chapter 29.

Some large-scale sequencing of human DNA is already under way and producing valuable data. For example, 57 kbp of the human HPRT locus has been sequenced using automated sequencing. The sequencing project was undertaken in four steps. In the first stage, sequencing was performed using standard procedures with randomly selected M13 clones (Chapter 7). This accounted for approximately 96 percent of the total sequence, and the clones could be ordered into 23 contigs. The inserts in the M13 clones were longer than the sequence derived from them, so selected clones at the ends of contigs were resequenced to try to obtain more of the sequence. The extra 1.3 percent of sequence reduced the contigs to 16. In the third step, selected clones at the ends of the contigs were sequenced using an M13 reverse primer so that sequence from the opposite end of the M13 inserts was obtained. Finally, the remaining gaps between the contigs were closed by using specific oligonucleotides as primers for the sequencing reactions. The end result was a sequence 56,736 nucleotides long, of which some 56,282 nucleotides were sequenced more than once, and on average each base was sequenced four times. This redundancy is unavoidable, and indeed desirable, as a way to check for errors. Almost finished is the complete sequence of the human T-cell receptor locus, about 90 kb. The sequence of the mouse T-cell receptor is complete (95 kb), and comparisons of the mouse and human sequences are already showing how much interesting biology will come from this approach.

A complementary strategy is to clone and sequence cDNAs as a shortcut to finding genes. This has the advantage that only expressed sequences are analyzed, and so avoids the difficulties of recognizing coding regions in genomic DNA. However, cloning cDNAs does not give the control regions for genes, and as the expression of many genes is developmentally regulated, it is difficult to estimate how many genes may be missed by cDNA cloning using cells from a tissue expressing only a subset of the developmentally regulated genes. Nevertheless, this approach is being followed in a project to clone the expressed genes of the brain. Here about 40 percent of the 600 cDNAs analyzed so far are not represented in the databases and, so, may be new genes.

Storing and Analyzing Genome Data Require Large Databases

Sequencing and mapping data are already accumulating at a rapid rate, and the problems of storing, retrieving, and analyzing these data are formidable.

Data relevant to genome analysis include ordered clone maps, gene linkage data, chromosomal location maps, protein sequences, and, for human diseases, a genetic disease database. The databases must be able to interact in an intelligent way so that a researcher can explore all relevant aspects of a topic. The data must be transferred from the laboratory to the databases as quickly as possible, and some scientific journals are helping this process by making submission of sequence and mapping data to public databases a prerequisite for publication of research articles.

The accuracy of sequence determinations will become more and more important as more and more research makes use of sequence data. Some of the "errors" in sequences result from the inherent genetic variability of individuals, so that there is not a *single* human genome. (The failure to appreciate this resulted in much popular discussion of *whose* genome was to be sequenced.) The best estimate of true experimental errors resulting from misreading of gels and mistakes in assembling the overall sequence is that between 1 and 3 nucleotides in 1000 are wrong. The consequences of these errors vary, depending on the uses being made of the sequences. For example, translation of nucleotide sequence into amino acid sequence is very sensitive to errors, while programs searching for homologies between sequences can tolerate errors better.

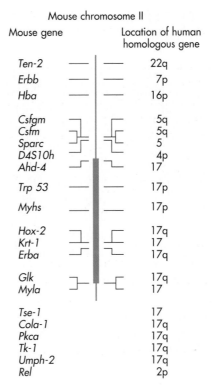

FIGURE **30-9**

Comparative gene mapping in mouse and human. Mouse chromosome 11 is shown with genes mapped to this chromosome listed in their correct locations. The locations of the human genes homologous to mouse chromosome 11 genes are listed opposite their mouse homologues. An extensive region of the mouse chromosome shows synteny with human chromosome 17. The genes listed below with their human homologues are known to be on mouse chromosome 11, but their positions on the chromosome map have not yet been determined.

Gene Mapping Can Be Facilitated by Comparing Species

We have given many examples of how molecular geneticists have exploited homologous relationships between genes for cloning and for guessing the probable functions of newly cloned genes. Genetic similarities between different species are not restricted to nucleotide sequences in homologous genes. The same set of genes grouped together on a chromosome in one species may be found grouped together on a chromosome in another species (Figure 30-9). For example, there are 13 genes on mouse chromosome 11 that have homologues on human chromosome 17. This is called *conserved synteny*. If the order of the genes (or markers) is the same in the two species, the genes demonstrate *conserved linkage*. Synteny and linkage conservation presumably reflect the preservation of ancient chromosome linkage groups in the mammalian ancestors, despite the rearrangements that occur over many generations. These preserved linkage groups can be used as guides to the locations of genes. This may be especially useful in speeding the mapping of genes of agriculturally important animals like cattle, sheep, and goats, animals in which little genetic analysis has been done at the molecular level.

Understanding Our Genome Will Benefit Humanity

We believe that the benefits of the new genetic knowledge are undeniable—in research, where the richness of sequence data will keep scientists at work for generations to come, and in human genetics, where there are many families who have healthy children as a consequence of prenatal diagnosis. But these benefits are tempered by other concerns; for example, DNA-based prenatal diagnosis may be unacceptable to those who believe abortion is wrong, and DNA storage may alarm those who see in it a threat to individual privacy. It is encouraging that both the National Institutes of Health and the Department of Energy human genome programs have set aside substantial funding to ensure that discussions of these issues take place, and that these debates involve the general public, not just experts. Many of the "big" projects of science—the Superconducting Super Collider, the space station, "Star Wars"—are seen to be at best irrelevant and at worst potentially catastrophic for humanity. The various plans to analyze the human genome have as their ultimate goal improving the lives of many thousands of people. Is it too much to hope that we as a society will be able to use the new knowledge of our molecular genetics in such a way as to enhance our duties and respect for others?

Reading List

General

U.S. Congress, Office of Technology Assessment. *Mapping Our Genes—The Genome Projects: How Big, How Fast?* OTA-BA-373. U.S. Government Printing Office, Washington, D. C., 1988.

Bishop, J. E., and M. Waldholz. *Genome.* Simon and Schuster, New York, 1990.

U.S. Department of Health and Human Services and U.S. Department of Energy. *Understanding Our Genetic Inheritance—The U.S. Human Genome Project: The First Five Years FY 1991–1995.* National Technical Information Service, U.S. Department of Commerce, Springfield, Va., 1990.

Original Research Papers

THE HUMAN GENOME PROJECTS

Cantor, C. R. "Orchestrating the Human Genome Project." *Science,* 248: 49–51 (1990). [Review]

Watson, J. D. "The Human Genome Project: past, present, and future." *Science,* 248: 44–49 (1990). [Review]

Dulbecco, R. "A turning point in cancer research: sequencing the human genome." *Science,* 231: 1055–1056 (1986).

Leder, P. "Can the human genome project be saved from its critics . . . and itself?" *Cell,* 63: 1–3 (1990).

Green, E. D., and R. H. Waterston. "The human genome project: prospects and implications for clinical medicine." *J. Am. Med. Assoc.,* 266: 1966–1975 (1991).

PRESENT STATE OF HUMAN GENOME MAPPING

Stephens, J. C., M. L. Cavanaugh, M. I. Gradie, M. L. Mador, and K. K. Kidd. "Mapping the human genome: current status." *Science,* 250: 237–244 (1990). [Review]

Dracopoli, N. C., P. O'Connell, T. I. Elsner, J. -M. Lalouel, R. L. White, K. H. Buetow, D. Y. Nishimura, J. C. Murray, C. Helms, S. K. Mishra, H. Donis-Keller, J. M. Hall, M. K. Lee, M. -C. King, J. Attwood, N. E. Morton, E. B. Robson, M. Mahtani, H. F. Willard, N. J. Royle, I. Patel, A. J. Jeffreys, V. Verga, T. Jenkins, J. L. Weber, A. L. Mitchell, and A. E. Bale. "The CEPH consortium linkage map of human chromosome 1." *Genomics,* 9: 686–700 (1991).

SEPARATING HUMAN CHROMOSOMES USING SORTING MACHINES

Gray, J. W., P. N. Dean, J. C. Fuscoe, D. C. Peters, B. J. Trask, G. J. van den Engh, and M. A. van Dilla. "High-speed chromosome sorting." *Science,* 238: 323–329 (1987). [Review]

Green, D. K. "Analysing and sorting human chromosomes." *J. Micros.,* 159: 237–244 (1990). [Review]

Deaven, L. L. "Chromosome-specific human gene libraries." In R. Dulbecco, ed., *Encyclopedia of Human Biology*, vol. 2: 455–464. Academic, New York, 1991. [Review]

Davies, K. E., B. D. Young, R. G. Elles, M. E. Hill, and R. Williamson. "Cloning of a representative genomic library of the human X-chromosome after sorting by flow cytometry." *Nature*, 293: 374–376 (1981).

CLONING BY CHROMOSOME DISSECTION

Ludecke, H. -J., G. Senger, U. Claussen, and B. Horsthemke. "Construction and characterization of band-specific DNA libraries." *Hum. Gen.*, 84: 512–516 (1990). [Review]

Scalenghe, F., E. Turco, J. -E. Edstrom, V. Pirotta, and M. Melli. "Microdissection and cloning of DNA from specific region of *Drosophila melanogaster* polytene chromosomes." *Chromosoma*, 82: 205–216 (1981)

Bates, G. P., B. J. Wainwright, R. Williamson, and S. D. M. Brown. "Microdissection of and microcloning from the short arm of human chromosome 2." *Mol. Cell. Biol.*, 6: 3826–3830 (1986).

Ludecke, H. -J., G. Senger, U. Claussen, and B. Horsthemke. "Cloning defined regions of the human genome by microdissection of banded chromosomes and enzymatic amplification." *Nature*, 338: 348–350 (1989).

Fiedler, W., U. Claussen, H. J. Ludecke, G. Senger, B. Horsthemke, A. Geurts-Van-Kessel, W. Goertzen, and R. Fahsold. "New markers for the neurofibromatosis-2 region generated by microdissection of chromosome 22." *Genomics*, 10: 786–791 (1991).

X-IRRADIATED SOMATIC CELL HYBRIDS FOR CLONING AND MAPPING

Goss, S. J., and H. Harris. "Gene transfer by means of cell fusion: I. Statistical mapping of the human X-chromosome by analysis of radiation-induced gene segregation." *J. Cell Sci.*, 25: 17–37 (1977).

Cox, D. R., M. Burmeister, E. R. Price, S. Kim, and R. M. Myers. "Radiation hybrid mapping: a somatic cell genetics method for constructing high-resolution maps of mammalian chromosomes." *Science*, 250: 245–250 (1990).

Cox, D. R., C. A. Pritchard, E. Uglum, D. Casher, J. Kobori, and R. M. Myers. "Segregation of the Huntington disease region of human chromosome 4 in a somatic cell hybrid." *Genomics*, 4: 397–407 (1990).

Glaser, T., E. Rose, H. Morse, D. Housman, and C. Jones. "A panel of irradiation-reduced hybrids selectively retaining human chromosome 11p13: their structure and use to purify the *WAGR* gene complex." *Genomics*, 6: 48–64 (1990).

Burmeister, M., S. Kim, E. R. Price, T. de Lange, U. Tantravahi, R. M. Myers, and D. R. Cox. "A map of the distal region of the long arm of human chromosome 21 constructed by radiation hybrid mapping and pulsed-field gel electrophoresis." *Genomics*, 9: 19–30 (1991).

ASSEMBLING HUMAN COSMID CLONES

Carrano, A. V., J. Lamerdin, L. K. Ashworth, B. Watkins, E. Branscomb, T. Slezak, M. Raff, P. J. de Jong, D. Keith, L. McBride, S. Meister, and M. Kronick. "A high-resolution, fluorescence-based, semiautomated method for DNA fingerprinting." *Genomics*, 4: 129–136 (1989).

Evans, G. A., and K. A. Lewis. "Physical mapping of complex genomes by cosmid multiplex analysis." *Proc. Natl. Acad. Sci. USA*, 86: 5030–5034 (1989).

Harrison-Lavoie, K. J., R. M. John, D. J. Porteous, and P. F. R. Little. "A cosmid clone map derived from a small region of human chromosome 11." *Genomics*, 5: 501–509 (1989).

Stallings, R. L., D. C. Torney, C. E. Hildebrand, J. L. Longmire, L. L. Deaven, J. H. Jett, N. A. Doggett, and R. K. Moyzis. "Physical mapping of human chromosomes by repetitive sequence fingerprinting." *Proc. Natl. Acad. Sci. USA*, 87: 6218–6222 (1990).

SEQUENCE-TAGGED SITES

Olson, M., L. Hood, C. Cantor, and D. Botstein. "A common language for physical mapping of the human genome." *Science*, 245: 1434–1435 (1989).

YACs AND THE CLONING OF HUMAN DISEASE GENES

Green, E. D., and M. V. Olson. "Chromosomal region of the cystic fibrosis gene in yeast artificial chromosomes: a model for human genome mapping." *Science*, 250: 94–98 (1990).

Silverman, G. A., E. D. Green, R. L. Young, J. I. Jockel, P. H. Domer, and S. J. Korsmeyer. "Meiotic recombination between yeast artificial chromosomes yields a single clone containing the entire BCL2 protooncogene." *Proc. Natl. Acad. Sci. USA*, 87: 9913–9917 (1990).

CLONING HUMAN TELOMERES USING YACs

Blackburn, E. H. "Structure and function of telomeres." *Nature*, 350: 569–573 (1991). [Review]

Moyzis, R. K., J. M. Buckingham, L. S. Cram, M. Dani, L. L. Deaven, M. D. Jones, J. Meyne, R. L. Ratliff, and J.-R. Wu. "A highly conserved repetitive DNA sequence, $(TTAGGG)_n$, present at the telomeres of human chromosomes." *Proc. Natl. Acad. Sci. USA*, 85: 6622–6626 (1988).

Riethamn, H. C., R. K. Moyzis, J. Meyne, D. T. Burke, and M. V. Olson. "Cloning human telomeric DNA fragments into *Saccharomyces cerevisiae* using a yeast-artificial-chromosome vector." *Proc. Natl. Acad. Sci. USA,* 86: 6240–6244 (1989).

CLONING AND ANALYZING THE FRAGILE X SITE

Kremer, E., M. Pritchard, M. Lynch, S. Yu, K. Holman, E. Baker, S. T. Warren, D. Schlessinger, G. R. Sutherland, and R. I. Richards. "Mapping of DNA instability at the Fragile X to a trinucleotide repeat sequence p(CCG)$_n$." *Science,* 252: 1711–1714 (1991).

Oberle, I., F. Rousseau, D. Heitz, C. Kretz, D. Devys, A. Hanauer, J. Boue, M. F. Bertheas, and J. L. Mandel. "Instability of a 550-base pair DNA segment and abnormal methylation in fragile X syndrome." *Science,* 252: 1097–1102 (1991).

Verkerk, A. J. M. H., M. Piertti, J. S. Sutcliff, Y.-H. Fu, D. P. A. Kuhl, A. Pizzuti, O. Reiner, S. Richards, M. F. Victoria, F. Zhang, B. E. Eussen, G.-J. B. van Ommen, L. A. J. Blonden, G. J. Riggins, J. L. Chastain, C. B. Kunst, H. Galjaard, C. T. Caskey, D. L. Nelson, B. A. Oostra, and S. T. Warren. "Identification of a gene (*FMR-1*) containing a CGG repeat coincident with a breakpoint cluster region exhibiting length variation in fragile X syndrome." *Cell,* 65: 905–914 (1991).

Rousseau, F., D. Heitz, V. Biancalana, S. Blumenfeld, C. Kretz, J. Boue, N. Tommerup, C. van der Hagen, C. DeLozier-Blanchet, M.-F. Croquette, S. Gilgenkrantz, P. Jalbert, M.-A. Voelckel, I. Oberle, and J.-L. Mandel. "Direct diagnosis by DNA analysis of the fragile X syndrome of mental retardation." *N. E. J. Med.,* 325: 1673–1681 (1991).

Sutherland, G. R., A. Gedeon, L. Kornman, A. Donnelly, R. W. Byard, J. C. Mulley, E. Kremer, M. Lynch, M. Pritchard, S. Yu, and R. I. Richards. "Prenatal diagnosis of fragile X syndrome by direct detection of the unstable DNA sequence." *N. E. J. Med.,* 325: 1720–1722 (1991).

LARGE-SCALE SEQUENCING OF HUMAN DNA

Edwards, A., H. Voss, P. Rice, A. Civitello, J. Stegemann, C. Schwager, J. Zimmermann, H. Erfle, C. T. Caskey, and W. Ansorge. "Automated sequencing of the human HPRT locus." *Genomics,* 6: 593–608 (1990).

Adams, M. D., J. M. Kelley, J. D. Gocayne, M. Dubnick, M. H. Polymeropoulos, H. Xiao, C. R. Merril, A. Wu, B. Olde, R. F. Moreno, A. R. Kerlavage, W. R. McCombie, and J. C. Venter. "Complementary DNA sequencing: expressed sequence tags and human genome project." *Science,* 252: 1651–1656 (1991).

STORING AND ANALYZING GENOME DATA

Doolittle, R. F. "Searching through sequence databases." *Meth. Enzymol.,* 183: 99–110 (1990). [Review]

Keil, B. "Cooperation between databases and scientific community." *Meth. Enzymol.,* 183: 50–60 (1990).

Waterman, M. S. "Genomic sequence databases." *Genomics,* 6: 700–701 (1990).

States, D. J., and D. Botstein. "Molecular sequence accuracy and the analysis of protein coding regions." *Proc. Natl. Acad. Sci. USA,* 88: 5518–5522 (1991).

Posfai, J., and R. J. Roberts. "Finding errors in DNA sequences." *Proc. Natl. Acad. Sci. USA,* 89: 4698–4702 (1992).

GENE FINDING BY COMPARING SPECIES

Nadeau, J. H. "Maps of linkage and synteny homologies between mouse and man." *Trends Gen.,* 5: 82–86 (1990). [Review]

Searle, A. G., J. Peters, M. F. Lyon, J. G. Hall, E. P. Evans, J. H. Edwards, and V. J. Buckle. "Chromosome maps of man and mouse IV." *Ann. Hum. Gen.,* 53: 89–140 (1989).

Erickson, R. P. "Mapping dysmorphic syndromes with the aid of the human/mouse homology map." *Am. J. Hum. Gen.,* 46: 1013–1016 (1990).

Threadgill, D. W., and J. E. Womack. "Syntenic conservation between humans and cattle. I. Human chromosome 9." *Genomics,* 8: 22–28 (1990).

IN THE FUTURE

Wexler, N. S. "The oracle of DNA." In L. P. Rowland, D. S. Wood, E. A. Schon, and S. DiMauro, ed., *Molecular Genetics in Diseases of Brain, Nerve, and Muscle.* Oxford University Press, New York, 1989, pp. 429–442.

Gilbert, W. "Towards a paradigm shift in biology." *Nature,* 349: 99 (1991).

Index

The letter "f" after a page number refers to a figure, "t" to a table.